Dermal
and
Ocular
Toxicology

Fundamentals and Methods

Editor
David W. Hobson
Battelle Memorial Institute
Columbus, Ohio

CRC Press
Boca Raton   Ann Arbor   Boston   London

Library of Congress Cataloging-in-Publication Data

Dermal and ocular toxicology: fundamentals and methods/editor, David W. Hobson
   p.   cm.
   Includes bibliographies and index.
   ISBN 0-8493-8811-2
   1. Dermatotoxicology—Technique. 2. Ocular toxicology—Technique.
I. Hobson, David W.
   [DNLM: 1. Dermatitis, Contact. 2. Drug Screening—methods. 3. Eye—drug
effects. 4. Skin—drug effects. 5. Skin Absorption. 6. Toxicology—methods.
WR 102 D435]
RL803.D45 1991
616.5—dc20
DNLM/DLC
for Library of Congress   91-16002
                                                   CIP

Developed by: Telford Press

Direct all inquires to CRC Press, Inc., 2000 Corporate Blvd., N.W., Boca Raton, Florida, 33431.

International Standard Book Number 0-8493-8811-2

Library of Congress Card Number 91-16002
Printed in the United States

PREFACE

In recent years, there has been increasing interest in the use of the skin and the eye in toxicologic studies as routes of exposure to various chemicals and chemical formulations. This interest has been the result of several factors:

- A desire to use the potential route of human exposure in the design and conduct of laboratory evaluations of toxicity

- Increasing interest in the environmental and occupational exposures

- A continuing desire to improve and refine methods used for assessing dermal and ocular toxicity

- Interest in the development of methods of evaluation for the transdermal delivery and transocular distribution of various chemicals and chemical formulations

Several excellent books have been written on various aspects of dermal and ocular toxicology. Most provide discussions of organ-specific dermal or ocular toxicology resulting from exposures to different types of toxicants and describe various methodologies used in dermal or ocular research; however, a need remains for a text created specifically to serve as a concise, current, and practical guide to the fundamentals and methods of dermal and ocular toxicology for individuals working in government, education, and private industry as an aid in evaluating or developing dermal or ocular studies. This book represents a first attempt at providing such a text.

In this book, scientists actively involved in different aspects of dermal and ocular toxicologic research have provided a coordinated set of practical reviews of each of their respective areas of expertise. Overall, the book represents approximately 3 years of effort and provides a chapter-coordinated introduction into the fundamentals of dermal and ocular toxicology. It is hoped that this text will serve as an aid to understanding the background and principles behind procedures currently used to assess dermal and ocular toxicity and might also serve as a useful reference for the design and conduct of future studies and procedures.

The book is divided into two general sections, dermal and ocular toxicology, each of which provide structured discussions of each subject with an objective to provide information necessary to understand or contemplate research studies with each target organ. For example, each section begins with a discussion of the toxicologically relevant anatomical and pathologic characteristics of each organ system. This is followed by discussions of various *in vivo* and *in vitro* techniques used to evaluate toxicity and chapters containing supporting information (i.e., statistical considerations, toxicokinetics, current perspectives on utilization, and research needs).

Because, in most instances, dermal and ocular toxicologic studies must be performed using laboratory animal models or tissues derived from laboratory animals, often with the intent that human risk be extrapolated from the results of these studies, major portions of the book are written from a comparative animal perspective. Thus, it should not be too surprising that many of the chapter authors have either considerable comparative animal experience or professional veterinary training in each of their respective areas of expertise.

Since this book is intended to serve as a practical instructional and methodologic guide, considerable emphasis has been placed on the introduction and usage of correct terminology to describe the anatomic structure and toxicologic pathology of each organ system. Nearly all methods mentioned are described along with pertinent references and, whenever deemed appropriate, helpful suggestions are provided to assist the reader in their successful implementation. An abbreviated reference format was selected in order to allow unlimited listing of such references without incurring significant additional cost to the volume.

In order to help ensure that the methodologic information presented remains current, plans for regular updates and future editions of the volume are in place and constructive suggestions from readers are welcomed. Please direct all suggestions directly to my attention. It is sincerely hoped that this book will be found to be a timely and useful introduction and guide to the fundamentals and methods of dermal and ocular toxicology.

David W. Hobson
Battelle Memorial Institute

CONTRIBUTORS

James A. Blank, Ph.D.
Battelle Memorial Institute
Columbus, Ohio

Robert D. Bruce
The Procter & Gamble Company
Miami Valley Laboratories
Cincinnati, Ohio

Leon H. Bruner, Ph.D.
The Procter & Gamble Company
Miami Valley Laboratories
Cincinnati, Ohio

**Richard H. Bruner,
D.V.M., D.A.C.V.P.**
Pathology Associates, Inc.
West Chester, Ohio

Michael P. Carver, Ph.D.
Colgate-Palmolive Company
Piscataway, New Jersey

**Stan W. Casteel, D.V.M.,
Ph.D., D.A.B.V.T.**
Veterinary Medical Diagnostic
  Laboratory
College of Veterinary Medicine
University of Missouri
Columbia, Missouri

George P. Daston, Ph.D.
The Procter & Gamble Company
Miami Valley Laboratories
Cincinnati, Ohio

William C. Eastin, Jr., Ph.D.
National Institute of
  Environmental Health Sciences
Research Triangle Park,
  North Carolina

Diane Essex-Sorlie, Ph.D.
College of Veterinary Medicine
University of Illinois
Urbana, Illinois

F. E. Freeberg, Ph.D.
The Procter & Gamble Company
Ivorydale Technical Center
Cincinnati, Ohio

Shayne Cox Gad, Ph.D., D.A.B.T.
G. D. Searle & Company
Skokie, Illinois

Keith Green, Ph.D., D.Sc.
Department of Ophthalmology
Medical College of Georgia
Augusta, Georgia

Richard H. Guy, Ph.D.
Departments of Pharmacy and
  Pharmaceutical Chemistry
University of California,
  San Francisco
San Francisco, California

Robert Hackett, Ph.D., D.A.B.T.
Alcon Laboratories, Inc.
Fort Worth, Texas

Jonathan Hadgraft
The Welsh School of Pharmacy
University of Wales
College of Cardiff
Cardiff, Wales

**H. Hugh Harroff, Jr., D.V.M.,
D.A.C.L.A.M.**
Administrative Director
The Children's Hospital
  Research Foundation
Columbus, Ohio

Gerry M. Henningsen, D.V.M., Ph.D., D.A.B.T., D.A.B.V.T.
Division of Biomedical and
 Behavioral Sciences
National Institute for Occupational
 Safety and Health
Robert A. Taft Laboratories
Cincinnati, Ohio

David W. Hobson, Ph.D., D.A.B.T.
Battelle Memorial Institute
Columbus, Ohio

John Kao, Ph.D., D.A.B.T.
Department of Drug Metabolism and
 Pharmacokinetics
SmithKline Beecham Pharmaceuticals
King of Prussia, Pennsylvania

George J. Klain, Ph.D.
Letterman Army Institute of Research
Division of Cutaneous Hazards
Presidio of San Francisco
San Francisco, California

Nancy Ann Monteiro-Riviere, M.S., Ph.D.
Cutaneous Pharmacology
 and Toxicology Center
College of Veterinary Medicine
 and Toxicology Program
North Carolina State University
Raleigh, North Carolina

Carl T. Olson, D.V.M., Ph.D., D.A.B.T., D.A.B.V.T.
Battelle Memorial Institute
Columbus, Ohio

William G. Reifenrath, Ph.D.
Letterman Army Institute of Research
Division of Cutaneous Hazards
Presidio of San Francisco
San Francisco, California

J. Edmond Riviere, D.V.M., Ph.D.
Cutaneous Pharmacology and
 Toxicology Center
College of Veterinary Medicine
 and Toxicology Program
North Carolina State University
Raleigh, North Carolina

John Shadduck, D.V.M., Ph.D.
Office of the Dean
College of Veterinary Medicine
Texas A&M University
College Station, Texas

Michael Stern, Ph.D.
Allergan Pharmaceuticals, Inc.
Irvine, California

John D. Taulbee, Ph.D.
The Procter & Gamble Company
Miami Valley Laboratories
Cincinnati, Ohio

David A. Wilkie, D.V.M., M.S., D.A.C.V.O.
Assistant Professor
College of Veterinary Medicine
The Ohio State University
Columbus, Ohio

Milton Wyman, D.V.M., M.S., D.A.C.V.O.
Professor
College of Veterinary Medicine
The Ohio State University
Columbus, Ohio

CONTENTS

DERMAL TOXICOLOGY

OCULAR TOXICOLOGY

Dermal Toxicology

1

Comparative Anatomy, Physiology, and Biochemistry of Mammalian Skin

NANCY ANN MONTEIRO-RIVIERE
Cutaneous Pharmacology and Toxicology Center
College of Veterinary Medicine and
Toxicology Program
North Carolina State University
Raleigh, North Carolina

I. INTRODUCTION

The primary function of skin is often considered to solely be the barrier between the well-regulated "milieu interieur" and the outside environment. This has led to the impression that the structure of the skin is likewise simple and solely focused on its barrier properties. The orientation of much of the historical research in percutaneous absorption and dermal toxicology has also reinforced this view, a fact supported by examining the model systems which have been used. However, the skin is a complex, integrated, dynamic organ which has functions that go far beyond its role as a barrier to the environment. A number of these are listed in Table 1. The paradigm of structure to function relationships is central to modern biology and infers, based on our current knowledge, that the structural and functional relationships of the skin are likely to be complex. Skin, the largest organ of the body, is anatomically divided into two principal components; the outer epidermis and the underlying dermis (Figure 1). Within these two primary divisions, numerous cell types and specialized adnexial structures can be identified. It is the purpose of this chapter to overview the structure and function of skin from this multifaceted perspective and to review the techniques used to sample and prepare skin for morphological study.

II. TECHNIQUES FOR SAMPLING AND PRESERVATION

For successful interpretation of skin morphology a standard and good procedure for biopsy must be followed. A local anesthetic (0.5% lidocaine) is injected subcutaneously. An excision biopsy involves making an elliptic section of skin with a scalpel or single-edge razor blade containing the area of interest, and removing it carefully with forceps. Most biopsies should contain normal and abnormal tissue. The wound

Table 1
Function of Skin

Environmental barrier
    —Diffusion barrier
    —Metabolic barrier
Temperature regulation
    —Regulation of blood flow
    —Hair and fur
    —Sweating
Immunological affector and effector axis
Mechanical support
Neurosensory reception
Endocrine
Apocrine/eccrine/sebaceous glandular secretion
Metabolism
    —Keratin
    —Collagen
    —Melanin
    —Lipid
    —Carbohydrate
    —Respiration
    —Biotransformation of xenobiotics
    —Vitamin D

area may or may not be sutured, depending on the intended disposition of the animal in the study protocol. An excision biopsy is usually larger than a punch biopsy and produces a greater sample for diagnostic purpose.

A punch biopsy is performed after the administration of a local anesthetic and a sharp stainless steel or disposable punch is used to create, loosen, and remove an area of skin (2 to 6 mm usually, depending on the diameter of the punch). The biopsy should be handled carefully. A pair of forceps should be used to lift up one end of the biopsy and, simultaneously, small scissors should be used to cut through the subcutaneous fat as close to the base as possible. Usually after a 2 to 4 mm biopsy, suturing is not required; however, placing two to three sutures is aesthetically pleasing. The use of an antibacterial ointment is recommended to prevent bacterial infections.

A shave biopsy is done when lesions or damage is suspected to be very superficial. After the administration of a local anesthetic, a fold of skin is lifted and a scalpel is used to cut through the skin surface.[1-4]

Following collection, the biopsy of choice can then be divided into sections and subsequently processed for either light microscopy, transmission (TEM) or scanning electron microscopy (SEM), enzyme histochemistry, or all four. These techniques are commonly used to study the integument for histology, pathology, diagnostic dermatology, or toxicologic pathology. Light microscopy examines the surface and structure of the cells in the skin at a low magnification. TEM is used to study the

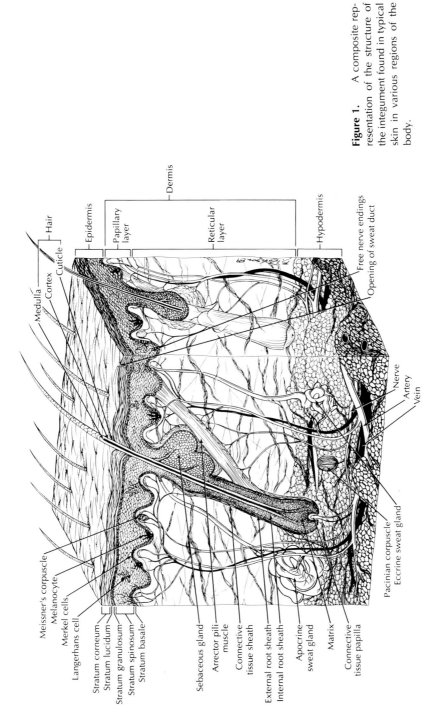

Figure 1. A composite representation of the structure of the integument found in typical skin in various regions of the body.

intrinsic nature of the cells and organelles in the skin. SEM allows one to examine the surface of skin in a three-dimensional context, and provides a larger viewing area than TEM. Histochemistry identifies and localizes chemical substances and enzymes at the cellular level.

A. Light Microscopy

Most mammalian skin biopsies are optimally fixed in 10% neutral buffered formalin for routine histologic evaluation. This all-purpose fixative is the most common because of its compatibility with most stains. Skin biopsies should be placed in fixative immediately after removal to prevent autolysis. Skin fixed in buffered formalin may be stored for at least several hours, or overnight, at room temperature, although storage for several months or even years is acceptable. Tissue samples may be rinsed in water to remove excess fixative and then placed in a tissue processor where they are dehydrated through increasing strengths of alcohols, cleared (xylene, toluene, Clear-Rite™, chloroform, or benzene) and then infiltrated and embedded in paraffin with the surface of the skin perpendicular to the intended plane of section. Sections are usually cut at 6 to 10 μm on a rotary microtome, floated in a water bath in which gelatin is added to aid section adhesion, transferred to a glass slide, and usually stained with hematoxylin and eosin for general evaluation.[5,6]

B. Transmission Electron Microscopy

The skin biopsy should be minced into 1 mm³ pieces and immediately placed in the appropriate fixative with the use of applicator sticks, not forceps, to minimize damage to the tissue. Tissue samples may be stored overnight or up to 1 week in this fixative at 4°C. Half strength Karnovsky's[7] fixative which consists of 2.5% glutaraldehyde and 2.0% paraformaldehyde solution in 0.2 M cacodylate buffer adjusted to a pH of 7.2 to 7.4 has been used by this author for porcine integument.[8,9] This mixture, or variations of it, consisting of both glutaraldehyde and paraformaldehyde has also been used in rats,[10] humans,[11-13] rabbits, and guinea pigs.[13,14] Tissue samples should be rinsed in buffer, postfixed in 1% osmium tetroxide in buffer for 1 h, then rinsed in buffer, dehydrated in a graded ethanol series, infiltrated, and embedded in Spurr's resin[15] or any other type of embedding media (i.e., epon, araldite). Thick sections approximately 1 μm thick should be cut on an ultramicrotome and stained with toluidine blue for general viewing by the light microscope. Ultra-thin samples are then cut at approximately 700 to 800 Å thick, mounted on copper grids, and stained with uranyl acetate and lead citrate for viewing with the transmission electron microscope at a determined accelerating voltage.

C. Scanning Electron Microscopy

Tissue samples should be fixed with a mixture of paraformaldehyde and glutaraldehyde[7,16] in buffer overnight at 4°C. Specimens must be postfixed in 1% osmium tetroxide in 0.1 M buffer, dehydrated through graded ethanol solutions, critical-point dried in 100% ethanol, mounted on stubs and sputter coated with gold palladium, and then examined with a scanning electron microscope.

D. Frozen Tissue

Frozen biopsies are best for rapid results and for enzyme histochemistry[17] techniques, immunocytochemistry, radioisotope localization, and for obtaining information on lipids which would normally be lost with normal light microscopic processing techniques.

Small skin biopsies are placed in an aluminum foil boat in which a few drops of commercial resin (i.e., OCT compound) has been added; the tissue is immediately placed for a few seconds in an isopentane well, cooled by liquid nitrogen in a Dewar flask until the biopsy has turned white. After the tissue has been frozen solid, it is wrapped in paraffin, placed in a tightly sealed bag, and stored in a −70°C freezer until further use. After quenching, the frozen biopsy may be mounted with a few drops of resin on an object disc and sections cut at 5 to 10 μm on a cryostat.[18] This procedure is widely used for enzyme histochemistry, routine studies in histology, cytology, and toxicologic pathology. One of its advantages is that it is rapid and is probably one of the best methods for preservation of tissue. However, if the tissue has been thawed, there may be loss of soluble enzymes or cofactors. Also, freezing and thawing can produce ice crystal formation within the skin tissue which can lead to ruptured cells, loss of enzyme activity, and morphologic artifacts.

E. Epidermal-Dermal Separation

Several methods have been described for separating the epidermis (epithelium) and dermis (connective tissue) so that either may be collected in a suitable manner for analysis. Table 2 lists several published techniques for complete epidermal-dermal separation. One must use discretion in employing some of these techniques depending on whether or not the skin is needed for morphological or biochemical analysis. Separation techniques can vary greatly and may be specific for species, temperature and concentration, and pH of reagents.

Skin prepared for *in vitro* use appears to offer similar barriers to topically applied compounds. The use of diffusion cells is a common method in which *in vitro* dermal penetration of topically applied xenobiotics can be estimated. Methods for preparing mouse skin for these experiments have utilized the dermatome, heat treatment, or trypsin and collagenase digestion.[19]

Separation of human skin (Figure 2) can be rapidly accomplished by immersion in heated water at 60°C for 1 min, followed by blunt dissection of the epidermis (Figure 3) and dermis (Figure 4). This technique, utilized by the author, is rapid and reproducible.

III. EMBRYOLOGY OF THE SKIN

In man, the integument is derived from two morphologically distinct layers: the epidermis from surface ectoderm and the dermis from mesoderm. The surface ectodermal cells proliferate in the second month to form the periderm (epitrichium). This cell layer consists of simple squamous epithelium with microvilli, which increases the epidermal surface area exposed to the amniotic fluid. The presence of

Table 2

Comparisons of Epidermal-Dermal Separation Techniques

Separation fluid	Species	Time (min)	Temp.	Ref.
EDTA	Human, guinea pig	120	37°C	20
EDTA	Mouse	90-150	37°C	21
Alkali (1% borate-carbonate)	Human, guinea pig	30	37°C	20
Ammonium hydroxide-ammonium chloride	Human, guinea pig	20-30	25°C	22
Sodium hydroxide/sodium chloride	Human	60-120	20°C	23
Acetic acid/NaCl	Human	120	4°C	23
Sodium thiocyanate	Human	5	37°C	23
Sodium iodide	Human	14	37°C	23
Ammonium thiocyanate	Human, guinea pig	20	37°C	20
Sodium bromide	Human, guinea pig	60	37°C	20
Sodium bromide	Human	25	37°	23
Sodium chloride	Human	6 h	37°C	23
Dithiothreitol	Human	120	4°C	24
Urea	Human	30	60°C	25
Dispase	Human	24 h	4°C	26
Lysozyme	Human, guinea pig	90	37°C	20
Trypsin	Human	4 h	37°C	27
Trypsin	Human, guinea pig	45	37°C	20
Keratinase	Human	30	37°C	27
Pancreatic elastase	Human	4 h	37°C	27
Fungal elastase	Human	4 h	37°C	27
Collagenase	Human	30	37°C	27
Pronase	Mouse, human	60	37°C	27
Heat	Human, guinea pig	2	49.2°C	20
Microwave oven	Mouse	10 s	600 W	28

microvilli on the amniotic surface suggests that the periderm cells are involved in the exchange of material between the fetus and amniotic fluid. The peridermal cells contain filaments and glycogen and the underside is attached by junctional complexes to the adjacent epithelial cells. The basal ectodermal cells continue to proliferate and undergo keratinization and desquamation. These exfoliative cells, along with sebum from the sebaceous glands and other cellular debris, form the vernix caseosa, a whitish protective substance that covers the fetal skin. At birth, all cell layers are present.[29-32] Cells of neural crest origin migrate into the epidermis during

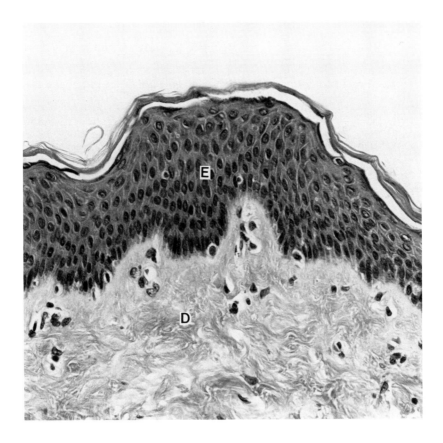

Figure 2.   Light micrograph of human breast skin. Note epidermis (E) and dermis (D). H & E, magnification × 330.

the early fetal period. Breathnach and Wyllie[33] noticed melanocytes and Langerhans' cells at 14 weeks in human fetal epidermis.

IV. EPIDERMAL KERATINOCYTES

The epidermis is composed of two primary cell types, the keratinocytes and the nonkeratinocytes, which include melanocytes, Merkel cells, Langerhans' cells, and indeterminate cells. The epidermis is separated from the underlying dermis by the basement membrane. Epidermal keratinocytes can be classified in layers above the basement membrane; starting with the stratum basale and continuing upward to the stratum spinosum, stratum granulosum, stratum lucidum, and ending with the outermost stratum corneum. The differences in cell structure seen among these layers primarily relate to the process of keratinization and formation of the complex stratum corneum barrier. When investigators generically refer to epidermal cells, they are

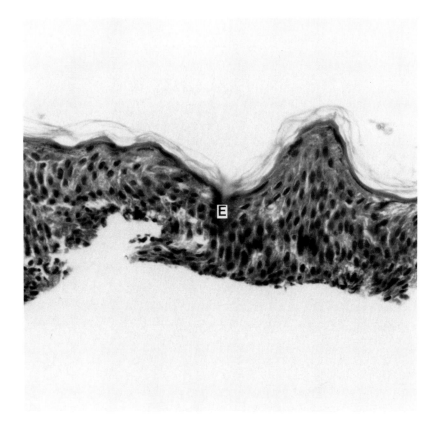

Figure 3.   Human breast epidermis (E) after heating in water for 1 min. H & E, magnification × 330.

generally speaking of the stratum basale cells (basal keratinocytes). Biochemically, these are the most active cells in the epidermis and form the precursors which ultimately form the intracellular protein and extracellular lipid components of the stratum corneum barrier.

A. Stratum Basale

The stratum basale consist of a single layer of columnar or cuboidal shaped cells that rest on the basement membrane (Figure 5). These cells are attached laterally to each other and to the overlying stratum spinosum cells by desmosomes and by hemidesmosomes to the irregular basement membrane (Figure 5). The nucleus of the stratum basale cells is large, oval, and occupies most of the cell. Its structure is similar to that of other nuclei, containing nuclear pores, one or two nucleoli, and chromatin. Ribosomes, rough endoplasmic reticulum, and other organelles are usually present. Mitochondria are numerous and are frequently seen just beneath the nuclei of stratum basale cells in some species.[8,34]

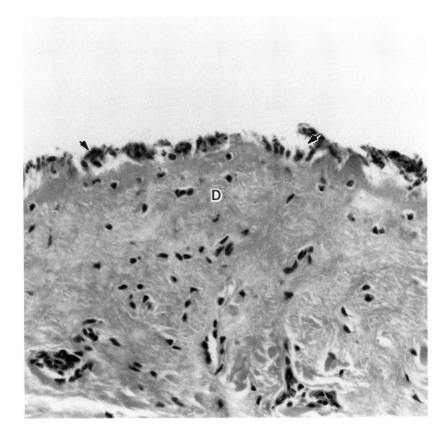

Figure 4.   Human breast dermis (D) after heating in water for 1 min. Note remnant of stratum basale cells (arrows) on surface. H & E, magnification × 330.

Tonofilaments and melanosomes are usually present in the cytoplasm of the stratum basale cells. The basal keratinocytes were assumed to be a homogeneous cell population by early morphologists, despite cell turnover kinetic data suggesting a heterogeneous population. It has been shown that there are two morphologically distinct types of basal keratinocytes found in monkey and human palm epidermis: the nonserrated cells and the serrated cells. The serrated cells are relatively large, columnar in shape, have a small (0.3 to 0.4) nuclear-cytoplasmic ratio and are found in the shallow rete ridges.[35,36] They are characterized ultrastructurally by numerous bundles of tonofilaments that extend all the way to the tips of the cytoplasmic projections and contain less melanin than the nonserrated cells. The other cell population, the nonserrated cells are located at the tips of the deep rete ridges and are smaller, cuboidal, and have a larger nuclear-cytoplasmic volume ratio (0.5 to 0.6), free ribosomes, melanosomes, and a paucity of tonofilaments.[35,36] Tritiated thymidine incorporation studies suggest that the nonserrated basal keratinocytes are slow

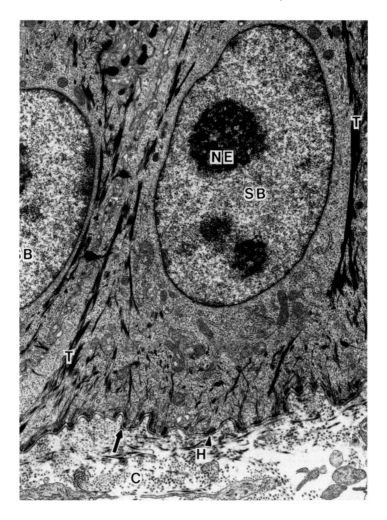

Figure 5.   Transmission electron micrograph of two stratum basale cells (SB). Tonofilaments (T), nucleolus (NE), basement membrane (arrow), cross and longitudinal sections of collagen (C), and hemidesmosomes (H). Magnification × 9000.

cycling stem cells which give rise to transient amplifying cells which can incorporate thymidine and then give rise to postmitotic cells. The serrated cells are thought to anchor the epidermis.[36] The rate of epidermal cell turnover in hairless mice, pigs, and humans have been compared using tritiated thymidine incorporation into nucleic acids and glycine incorporation into cellular proteins.[37] These results indicated that the epidermal cell turnover time was 30 days in humans, 26 to 27 days in pigs, and only 3 to 4 days in hairless mice.

Dark cells are also found among the basal keratinocytes. A dark cell is elongated and lies perpendicular to the basement membrane, has numerous tonofilaments, an indented nucleus, and has cytoplasm which is more electron dense than the adjacent keratinocytes. Dark cells have an increase in the number of melanosomes, filaments, and membrane-bound organelles, a higher surface density of nuclear membrane, and a larger number of desmosomes. The significance of dark cells is not known and has been the object of controversy. Some workers have interpreted them as being differentiating cells prior to migrating up through the epidermal cell layers.[38,39] Dark cells have been reported in both normal and abnormal tissues and their relevance to tumor promotion is not clear.[40,41] However, there is a good correlation to the induction of dark cells with some tumor promoters.[42,43] Dark basal keratinocytes have been observed in the mouse epidermis treated with one tumor promoter TPA (12-*o*-tetradecanoylphorbol-13-acetate).[42,44] Dark cells are not limited solely to the epidermis, but have been noted in other squamous epithelia such as the respiratory epithelium.[45,46]

B. Stratum Spinosum

The next succeeding outer layer is the stratum spinosum or "prickle cell layer", which consists of several layers of irregular polyhedral shaped cells (Figure 6). Desmosomes connect these cells to adjacent stratum spinosum cells and to the stratum basale cells below. Tonofilaments are more prominent in this layer than in the stratum basale. In addition to desmosomes, tight junctions (zona occludens) may also connect the spinosum cells to one another. In the upper stratum spinosum layers, membrane-coating granules are present.[38,47,48]

C. Stratum Granulosum

The third layer is the stratum granulosum, which consists of several layers (3 to 5) of flattened cells (Figure 7) lying parallel to the epidermal-dermal junction. This layer contains irregularly shaped, nonmembrane bound, electron-dense keratohyalin substance or granules which will ultimately form part of the intracellular matrix of the stratum corneum barrier. The number of organelles is decreased in comparison to the basal and spinous layers. Lysosomes are prominent in this layer. Desmosomes interconnect adjacent cells and tight junctions have been reported at the stratum granulosum and stratum corneum interface. Other intercellular sites of adhesions such as gap junctions have been reported in all layers of the epidermis, although they are rarely found on the lateral plasma membrane of the basal cells.[13,48]

A characteristic feature of the stratum granulosum and uppermost layers of the stratum spinosum cells is the presence of small granules known by a variety of names: Odland bodies, lamellated bodies, or membrane-coating granules. These granules are much smaller than mitochondria being 0.1 to 0.5 μm or 0.3 to 0.7 μm × 0.15 to 0.5 μm.[13,38] This difference in reported size could be attributed to variations in species. With high magnification, one can see a lamellar structure within a double-layered membrane. The lamellar subunits are 25 Å in width and are organized in parallel stacks. These granules are usually close to the Golgi complex and smooth endoplasmic reticulum. Higher in the epidermis, these granules increase in number and size, move toward the distal cell membrane, fuse with the plasmalemma, and

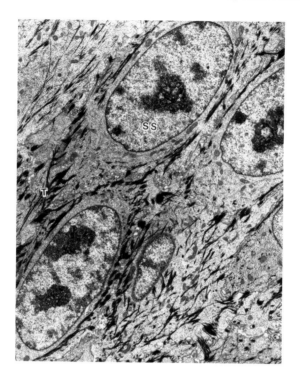

Figure 6.   Stratum spinosum (SS) cells, numerous tonofilaments (T) and desmosomes (arrows) in the epidermis. Magnification × 3960.

release their contents into the intercellular space. The biochemical composition and function of these membrane-coating granules will be reviewed later in a discussion concerning epidermal biochemistry. It has recently been postulated that these granules are responsible for the thickening of the stratum corneum cells and formation of the intercellular lipid matrix of the stratum corneum barrier. The granules release their substance, thereby forming an outer covering over the cell membrane of the protein-rich stratum corneum cells. As one can now appreciate, the term membrane-coating granule is appropriate. The deposition of their contents forms the lipid component of a complex barrier which prevents both the penetration of substances from the environment and the loss of body fluids.[13,38]

D. Stratum Lucidum

The human stratum lucidum is a constituent only of thick skin and hairless regions (i.e., plantar and palmar surfaces) and lies between the stratum granulosum and stratum corneum. A thin, clear, homogeneous line exists which consists of closely compacted cells whose nuclei and cytoplasmic organelles are no longer present. This clear or translucent layer consists of a semifluid substance known as eleidin.[49]

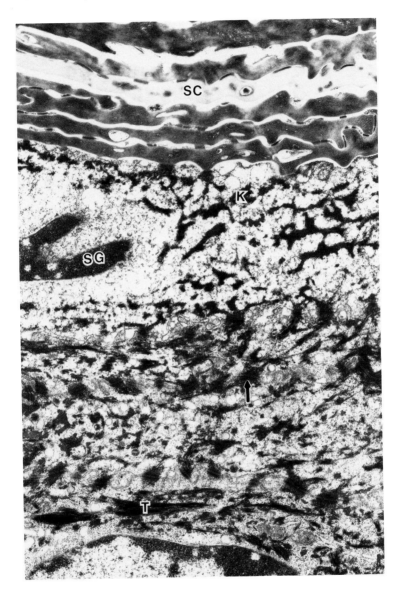

Figure 7. Stratum granulosum cell (SG), containing numerous keratohyalin (K) granules. Note tonofilaments (T) and stratum corneum (SC) cells. Magnification × 15,640.

E. Stratum Corneum

The stratum corneum consists of layers of cornified flattened cells (Figure 8). Nuclei and cytoplasmic organelles are not present. Each flattened, keratinized cell is about 30 μm in diameter and 0.5 to 0.8 μm in thickness. The number of keratinized

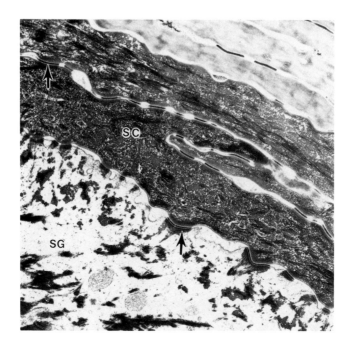

Figure 8.   Several layers of cornified flattened cells of the stratum corneum layer (SC). Note filamentous appearance of cells, desmosomes (arrows) and upper stratum granulosum (SG). Magnification × 15,000.

cells varies in different body areas and is specific for each individual.[50] Stratum corneum cells can assume two different appearances. Ultrastructurally, some may look homogeneous but at greater magnification, low density filaments embedded in a denser matrix can be seen. Others can assume a more filamentous appearance and contain dark-staining, loosely packed fibrils. Desmosomes occur throughout the stratum corneum but do not contain the dense plaques which are usually associated with the plasmalemma of the other layers. However, on the contact region between the stratum granulosum and stratum corneum (Figure 8), the dense plaque is associated with the stratum granulosum and not with the plasmalemma of the stratum corneum. Intercellular substance is present between the stratum corneum cells. This granular or reticular material is thought to be derived from the membrane-coating granules[38,47,51] and thus is a major component of the skin's barrier properties.

The stratum corneum has been divided into three sublayers; the basal, intermediate, and superficial layers. The basal layer is less dense than the other two. The intermediate is more opaque, while the superficial is intermediate in density. These differences may be artifactual due to tissue processing or secondary to changes of the opacity of the fibrils. In certain sites of the body, the stratum corneum sublayer densities can vary from the above. The shape of the corneal sublayers can differ.

Cells are usually elongated in the basal and intermediate sublayers while they are rounded and wrinkled in the superficial layers. The number of cell layers within the stratum corneum can vary between 15 to 37 in the abdominal region, 14 to 45 on the thigh, and 12 to 52 over the sacrum. The width of intercellular spaces may also vary from 20 nm to 2 μm in the abdominal and sacral regions. This difference could be due to artifact arising from fixation and processing.[52] The width of intercellular space that is entirely filled with membrane-coating granular substance is estimated to be 300 to 370 nm for the human plantar stratum corneum.[52] If this substance were not taken into account, the width of the intercellular spaces may be less and "artifact" space could not be differentiated from "true" space.

It has been postulated that the cells of the stratum corneum and stratum granulosum are organized into vertical columns. Light microscopic observations of both human and mouse skin show a highly organized spatial arrangement similar to a "stack of poker chips". This arrangement has also been described in the lumbar region of the guinea pig.[53] However, palmar and plantar surfaces lack this organization. Each cell interdigitates and overlaps laterally with the adjacent cells. TEM shows that desmosomes are abundant at these overlapped junctions. The mechanisms for this organization remain unknown.

Observations on the structure of these keratinized cells show them to be tetrakaidecahedral in shape. This 14-sided polygonal structure provides a minimum surface-volume ratio which allows for space to be filled by packing without interstices. Experiments performed on packed soap bubbles confirm this hypothesis because the bubbles immediately become tetrakaidecahedrons and form an ordered structure closely resembling mammalian epidermis. These cells spontaneously assemble into columns, apparently without direction from cellular layers below. It is important to know this spatial arrangement of the stratum corneum cells in order to understand transepidermal water loss since water loss is a function of the integrity and permeability of this layer.[54-62] The organization of the mammalian epidermis may also be explained by the epidermal proliferative unit. Each unit contains approximately 20 cells: 10 to 11 basal cells, 1 spinous cell, 2 granular cells, an identifiable basal cornified cell, and 5 to 7 cornified cells above it. In the central region of this unit in the basal cell layer is a Langerhans' cell with its dendrites extending toward the edges of the unit. Migration of the stratum basale cells occurs at the edges of the unit.[63,64]

V. KERATINIZATION AND FORMATION OF THE STRATUM CORNEUM BARRIER

Keratinization is the process by which epidermal cells differentiate. Keratin filaments, keratohyalin, and membrane-coating granules are formed in large numbers during keratinization. As the basal epithelial cells migrate upward, the volume of the cytoplasm increases, and organelles and tonofilaments are prominent. The tonofilaments are opaque and measure 5 nm in diameter. They aggregate into dense tonofibrils and filamentous substance between them may, at times, be visible. Loose tonofibrils may also be present. In the spinosum layers, membrane-coating granules are found, along with bundles of tonofibrils which usually have a compact appearance. Some tonofibrils may resemble those of the stratum basale layer.

The stratum granulosum (stratum intermedium) cells possess compact tonofibrils with dense staining regions which are irregular in shape (keratohyalin). Most tonofibrils are not discernible until they are transformed into keratohyalin. These keratohyalin granules may not be present in all of the stratum granulosum cells. If they are not present, then there exists fibrous substance in the basal and spinous layers. The keratohyalin granules exhibit differences in their appearance. They may be homogeneous or may exhibit a distinctive pattern of light and dark areas. They are usually associated with filaments and ribosomes. These keratohyalin granules become enlarged as the cells approach the surface. At this time, membrane-coating granules are seen to migrate toward the cell plasmalemma, fuse with it, and release their contents into the intercellular space. Organelles disintegrate and filaments arrange themselves into bundles.

The stratum corneum layer contains a fibrous substance similar to that found in basal cells, along with dark-staining tonofilaments that are 5 nm in diameter. Modification of the plasmalemma takes place and the intercellular space disappears in the stratum corneum. The plasmalemma is thicker (150 Å) than in the basal layer (70 Å) and is coated with the substance released from the membrane-coating granules.[65-67] The final product of the epidermal differentiation and keratinization process is a stratum corneum which consists of thick, plasma membrane-limited, protein-rich cells containing fibrous keratin and keratohyalin, surrounded by the extracellular lipid matrix. This forms the so-called "brick and mortar" structure which is the morphological basis for the heterogeneous, two-compartment stratum corneum model proposed by Elias.[68]

A. Biochemistry of the Epidermis

The epidermis is the primary component of the skin responsible for carbohydrate, lipid, and protein metabolism. Those areas of skin biochemistry which are integral for an understanding of specific cutaneous functions are discussed below in conjunction with a review of the structure of the cells associated with each function. This section will focus only on those generic areas of metabolism pertinent to gaining an appreciation of the metabolic capability and diversity of the epidermis.

The major end product of protein metabolism in the epidermis is keratin, the primary intracellular component of the terminally differentiated stratum corneum cells, and the major structural macromolecule of hair. Keratin is a fibrous structural protein largely composed of sulfur-rich cystine residues whose intrachain disulfide bridges stabilize its structure. Some 68% of total keratin composition in most species is made up of the following six amino acids; cystine, serine, glutamic acid, arginine, aspartic acid, and glycine. The relatively uncommon citrulline is also incorporated into keratin. The essential amino acids required for keratin biosynthesis are methionine and arginine. The mechanism of protein synthesis (mRNA translation on polyribosomes) which forms keratin and the basic histidine-rich keratohyalin is very similar to that for other proteins and will not be elaborated here. The secondary structure of mature keratin is an alpha helix in the resting state and a beta pleated sheet structure in a stretched state (hair). These helical protein chains are intertwined with each other to form hexagonally packed, seven-stranded cables which have a characteristic X-ray diffraction pattern. This water-insoluble structure is thermodynamically stable and resistant to the environment, making it a major component of the epidermal barrier.[69,70]

Proteolytic enzymes have been described in both human and animal skin and function in the normal catabolism of intracellular proteins and of extracellular collagen, elastin, fibronectin, and proteoglycans.[71] These enzymes also play a regulatory role, much as they do in other tissue types, in the metabolism of bioactive peptides (chemotactic factors, vasoactive peptides, hormones) and in regulating cell growth. Disease and inflammation modulate these activities. These proteases and their inhibitors, e.g., a kallikrien-like, kinin-releasing protease found in human skin, are especially important in inflammation since increases in vascular permeability with resultant tissue destruction are primary manifestations of many cutaneous toxicants.[72,73] Release of proteases and collagenases may be involved in the pathogenesis of blister formation. These reactions and their identification are complicated by the presence of migrating inflammatory cells (mast cells, leukocytes) present in skin.

Cutaneous metabolism is biochemically similar to that seen in other organs. Glucose may diffuse freely across epidermal cell walls and its utilization at physiological concentrations is proportional to substrate concentrations. The primary site of epidermal glucose utilization is the basal epidermal cell, with decreased consumption occurring at more peripheral layers. A difficulty inherent to quantitatively assessing glucose utilization is the inability to clearly isolate basal epidermal cells from less active superficial cell layers and the dermal connective tissue matrix. The presence of hair follicles in various stages of activity, and the variable metabolic activity of epidermal cells themselves (growth, wound healing, etc.), further confound this assessment. Epidermal glucose utilization is responsive to insulin; however, a precise quantitation of this effect is confounded by the above factors. Both the glycolytic pathway and the tricarboxylic acid cycle has been demonstrated to be present in skin. In addition, the pentose phosphate pathway, which may provide intermediates for nucleic acid synthesis, is also present and has increased activity in periods of epidermal growth. When glucose is present, lactic acid production via the Embden-Myerhoff pathway may be the primary metabolic pathway for glucose utilization.[74-76]

The anaerobic production of significant quantities of lactic acid, although inefficient from a bioenergetic viewpoint, is a characteristic of this tissue which results in a large cutaneous lactate pool. It has been estimated that when adequate glucose is present, 80 to 90% of the glucose utilized is converted to lactic acid. The accumulated lactic acid may then be recycled to the liver for subsequent utilization. The effects of this lactic acidosis on the cutaneous microenvironment cannot be ignored. Depending on the availability of other substrates or various physiological stimuli, other pathways of energy metabolism (fatty acids and amino acids) in the presence of glucose may also occur.[74-76] For example, in the presence of exogenous fatty acids, epidermal glucose oxidation of endogenous lipids also has been shown to occur.[78] A recent study using *in vitro* tissue slices concluded that glucose utilization, fatty acid synthesis, and insulin responsiveness are very similar at equivalent depths between man and the weanling pig.[79]

These findings would suggest a scenario whereby glucose utilization in the epidermis is modulated by the presence of fatty acids in the skin, with fatty acid oxidation by the tricarboxylic acid cycle predominating when glucose is absent, or fatty acid availability is high. Altered physiological states, including wound healing and some disease conditions, might tend to shift energy production to these more efficient pathways. In intermediate conditions, anaerobic glucose utilization with high lactate production may predominate. Epidermal energy requirements alone do not dictate the bioenergetic pathway employed, since in hyperthyroidism, increased glucose utilization results in a proportionate increase in lactic acid production.[80] An intriguing area for further investigation is the relationship of this predominantely glycolytic bioenergetic scenario to cytochrome-mediated xenobiotic biotransformation.

A byproduct of epidermal glucose metabolism, which is not present in normal adult skin, is glycogen. However, in proliferating states, in fetal tissue, after irradiation and insulin, and in numerous diseased conditions, abundant glycogen can be detected by histochemical staining (periodic acid-Schiff, or PAS) or direct biochemical analysis. The regulation of glycogen metabolism appears similar to other tissues studied.[74,75,81] Species comparisons of epidermal metabolism are difficult to assess.

As alluded to in the discussion on epidermal carbohydrate metabolism, fatty acids and lipids may be important substrates for energy production by the epidermis. The extracellular lipid matrix of the epidermal barrier is also a complicated structure which requires epidermal cells to possess the capacity to synthesize these lipids (see earlier discussion on Odland bodies or membrane-coating granules). In fact, skin has been considered one of the most active lipid-synthesizing tissues in mammals. The primary physiological substrate for lipid synthesis is glucose, although acetate is utilized in many *in vitro* tissue slice studies. Lipids identified in the epidermis include various neutral and phospholipids, ceramides, glycosylceramides, gangliosides, sterol esters and gangliosides, fatty acids, and alkanes. Phospholipids and sterols predominate in the basal epidermal layers, while sterols and other neutral lipids predominate in the stratum corneum. The remainder of the lipid is primarily sphingolipid. Glycosphingolipids are present in the basal layers, while nonpolar ceramides predominate in the outer stratum corneum. This change in lipid composition has a major role in determining permeability to topically applied compounds. Species differences probably exist in that wax esters were identified in rat epidermis but not in pig or human samples. Disease and nutritional deprivation also modulates the composition of epidermal lipids and would expect to also alter topical compound penetration and epidermal structure.[75,82-84]

As can be appreciated from this brief review of epidermal metabolism, the biochemistry of skin differs from other tissues in two principal ways: (1) glycolysis is the primary pathway for glucose utilization; and (2) synthesis of the protein and lipid barrier of the stratum corneum is relatively complex. It is important to acknowledge these differences when assessing the percutaneous absorption of xenobiotics or studying the mechanisms of cutaneous toxicity.

VI. MELANOCYTES

The melanocytes, which are derivatives of the neural crest, enter the epidermis about week 11 of fetal life. After 6 months of fetal age, they become situated on the basement membrane, along with the basal keratinocytes.[85] Melanocytes have several dendritic processes which either extend between adjacent keratinocytes or run parallel to the dermal surface.[38,86,87] Melanocytes can best be seen with the light microscope using dopa-treated sections. The ratio of melanocytes to keratinocytes in the skin of humans can vary from 1:11 in the arm and thigh to 1:4 in the cheek.[38]

A unique characteristic of the melanocyte is the presence of oval cytoplasmic granules referred to as melanosomes. Unlike the neighboring keratinocytes, the melanocyte does not have hemidesmosomes, desmosomes, or tonofilaments. A few cytofilaments may be present. The melanocyte nucleus is round. The cytoplasm is clear and contains most of the common organelles such as ribosomes, smooth and rough endoplasmic reticulum, mitochondria, and a well-developed Golgi complex.[33,38,88,89]

The melanocytes and associated keratinocytes form a structural and functional unit, the "epidermal melanin unit". It is this concept which aids in explaining the variations in skin color based on four fundamental biological processes: the synthesis and melanization of melanosomes in the melanocyte, the attachment of melanocytes to the keratinocytes, the transfer of melanosomes, and the degradation of melanosomes within the keratinocyte.[90,91]

The Golgi apparatus is so well developed that many investigators believe it plays a role in melanization. The enzyme tyrosinase has been shown to be involved in melanin biosynthesis. Tyrosinase is synthesized on the ribosomes, transferred to rough endoplasmic reticulum, and then to the Golgi apparatus where it is packaged into small vesicles. In these vesicles tyrosinase begins to form an ordered pattern and melanin biosynthesis takes place, thus forming a melanosome.[92-94] A spherical or ovoid cytoplasmic granule varying in size from one to several microns in diameter is called a melanin macroglobule or macromelanosome. These are frequently located in the cell body and often indent the nuclei of a melanocyte. This specific type of large melanosome has been observed in human and animal species with pigmentary disorders. There has been some disagreement about the nature and origin of these granules and studies by Nakagawa et al. have demonstrated that they are formed through autophagic processes and that their morphologic and enzymatic properties are consistent with lysosomes.[95]

Melanosomes go through four different stages before becoming fully melanized. Stage I consists of a large, spherical membrane-limited vesicle which contains tyrosinase and filaments possessing periodicity. Stage II is marked by elongation of the melanosome to an ellipsoidal shape with membranous filaments within, possessing periodicity. Stage III is characterized by an increase in electron-dense material, which obscures the internal structure of the particle. In the final stage, stage IV, the melanosome is totally electron opaque and no internal structure is evident.[93,96]

Melanosomes are oval and measure approximately 0.4×0.15 µm. Melanosomes of hair bulbs are larger, measuring 0.7×0.3 µm. If one were to cut a longitudinal section of a partially melanized melanosome (stage II or stage III), parallel lines with a 6 to 8 µm periodicity could be seen. The spacing of these lines become obscured as melanization progresses to stage IV.[33,38]

In addition to these four different stages of melanosomes, ten classes can be differentiated based on their chemical composition. Melanin forms 18 to 72% of the total weight of melanosomes. The melanin content of the melanosomes is directly correlated with their degree of melanization. The proteins isolated from the ten different classes of melanosomes contain most of the amino acids commonly present in proteins. Two unusual amino acids were found in some species; 3,4-dihydroxy-phenylalanine (dopa) and 2-aminoethane-1-sulfonic acid (taurine).[97]

Tyrosine is the precursor of melanin and the synthetic reaction sequence is as follows: dopa, dopaquinone, leukodopachrome, dopachrome, 5,6-dihydroxyindole, indole-5,6-quinone, and melanin. In the Raper-Mason scheme of melanin synthesis, the enzyme tyrosinase oxidizes tyrosine to dopa and also catalyzes the formation of dopaquinone.[98] This reaction occurs within the melanocyte. In albinism, there is a genetic defect (absence of tyrosinase) which prevents the oxidation of dopa.[99,100]

Little is known about melanosome transfer from the melanocyte to neighboring keratinocytes. The melanosomes migrate to the dendritic processes of the melanocyte, which then become pinched off and transferred to the keratinocyte. The area of contact between the keratinocyte and dendritic processes of the melanocyte undulates *in vitro*. This undulation of the membrane has been attributed to intracellular microfilaments.[101] Studies *in vitro* have demonstrated pigment transfer. When the fungal metabolite cytochalasin B is added to the culture medium, melanocytes shrink, undulations of the keratinocyte membrane cease, and microfilaments disappear. However, the attachment site between melanocytes and keratinocytes remains intact, while desmosomal attachments between keratinocytes split apart. This indicates that the melanocyte-keratinocyte attachment is stable and can withstand the tension, which the keratinocyte-keratinocyte desmosome attachment cannot. Other organelles are not affected by cytochalasin B treatment. The effects of cytochalasin B on microfilaments are reversible 1 h after removal of the drug.[90]

Jimbow et al. report that the cytoplasmic filaments present in the melanocytes of human skin are of the intermediate type (100 Å).[102] These filaments are located around the nucleus but are absent from the dendrites of normal (unirradiated) skin. After exposure to UV light the filaments were found distributed, along with the melanosomes, to the dendritic processes. These melanocytic filaments are similar in dimension and in distribution to neurofilaments. They have also been found in fibroblasts, muscle cells, and glial cells. Besides acting as a cytoskeleton, neurofilaments can function in axoplasmic transport, as well as elongation and contraction of neuronal dendrites. The cytoplasmic filaments in melanocytes could also be involved in the elongation and contraction of dendrites during melanosome transfer.[102]

Time lapse cinematography, light microscopy, SEM and TEM, and cell culture were used by Seiji et al. to study the mechanism of melanosome transfer between the melanocyte and keratinocyte.[103] The melanosome-packed dendritic tip of the melanocyte penetrates into, and is "pinched off" by the keratinocyte. The villus-like cytoplasmic projections of the keratinocyte enfold this "pinched off" portion of the melanocyte. The melanosomes are then surrounded by two membranes; the inner one belonging to the melanocyte and the outer one belonging to the keratinocyte. Few cytoplasmic constituents are present within the double membrane bound melanosomes. This pouch of melanosomes moves toward the nucleus of the keratinocyte where digestion of the inner (melanocyte-derived) membrane by lysosomes takes place. The reason for digestion of only the inner membrane was not explained. It is believed that melanosome complexes are similar to lysosomes because of their acid phosphatase activity and presence of degradation products. The final result is a single membrane pouch with melanosomes within. Melanosomes are released from this pouch in groups if they are small, and individually if they are large. They then acquire an outer membrane by an unknown mechanism. Melanosomes are usually randomly distributed within the cytoplasm of the keratinocytes, though sometimes they may localize over the nucleus, forming a cap-like structure. This localization presumably protects the nucleus from UV radiation.[96,103]

Toda et al. state that there are seven factors determining human skin color, which is one of the most obvious characteristics distinguishing race.[91] They are (1) the reflection coefficient of the skin surface, (2) the absorption coefficient of the epidermal and dermal cell constituents, (3) the scattering coefficient of the various cell layers, (4) the thickness of the stratum corneum, epidermis, and dermis, (5) the biochemical structure of the UV and visible light-absorbing components such as proteins (keratins, elastin, collagen, and lipoproteins), melanin in melanosomes, nucleic acids, urocanic acid, carotenoids, hemoglobin (reduced and oxidized), and lipids, (6) the number and spatial arrangement of melanosomes, melanocytes, and blood vessels, and (7) the quantity of blood and concentration of erythrocytes flowing through skin vessels.[91]

Epidermal melanocyte density, i.e., melanocyte: keratinocyte ratio, is characteristic for each body region and does not differ significantly among races.[104] The difference in skin color among various races is due to the number and distribution of melanosomes and their degree of melanization.

In newborns there are more melanosomes in black compared to white skin.[105] In addition, the melanosomes are fully melanized (stage IV) in Negroids and Mongoloids, while in Caucasoids the melanosomes are usually in stage II and stage III. Newborn Negroid skin possesses single large melanosomes even in the stratum corneum, while in Caucasoids they are in a melanosome complex and are not usually found in the stratum corneum.[105,106] In the newborn, the melanosomes seem to be randomly distributed throughout the keratinocyte and not localized around the perinuclear region.[105] In adult, nonirradiated skin, they are present in the basal region of the cell.

The presence of increased melanization in tropical-dwelling humans is considered an evolutionary adaptation which protects against UV radiation and prevents harmful levels of vitamin D synthesis.[99] In addition to the environmental factors such as sunlight which can influence skin color, genetic factors such as differences in number and size of melanosomes, and variations in the density of the epidermal melanin unit are probably important.[91,105,107]

Melanocyte counts can be in error due to the method used. Histochemical techniques, such as the measurement of tyrosinase activity, only determine the number of active melanocytes, while ultrastructural methods will identify active and inactive melanocytes based on morphological characteristics. For example, a melanocyte possessing only premelanosomes (no tyrosinase) would not be counted using cytochemical techniques but would be counted with electron microscopy.[85]

Melanocytes are not only found in the base of the epidermis, but also occur in the dermis. They have been identified in the upper part of the hair bulb, in the outer root sheath, and in the dermis adjacent to the hair follicle of the guinea pig.[108] In tropical Nigerian cattle, the White Fulami and N'Dama breeds, melanin pigments were found in all layers of the epidermis, and in the hair cortex, follicle sheaths and papillae.[109] In humans, melanocytes are found in the dermis in mongolian spots, blue nevi, nevi of Ota, and nevi of Ito.

Melanosomes can undergo qualitative and quantitative changes as the animal ages. Studies performed on *Macaca mulatta* choroidal melanocytes of 7, 13, and 15 years of age showed extensive alterations in melanosomes.[110] Ultrastructural observations showed a tendency of melanosomes to fuse into complexes and form vacuoles and the electron density of melanocytes was decreased. Approximately 10 to 15% of melanocytes had changes at 7 years, 50% had changes at 13 years, and more than 60% at 15 years of age. The differentiation and aging of melanocytes may be biochemically related to lysosomal enzymes and two additional enzymes; tyrosinase and gamma-glutamyl transpeptidase.[110]

Because of the central role melanocytes play in skin and hair pigmentation, some manifestations of cutaneous toxicology may be related to melanocyte dysfunction. Their unique enzymatic pathway for melanin synthesis makes them susceptible to toxicants which would not interact with keratinocyte metabolism. Compounds which cause hyper- and depigmentation are examples. In the field of carcinogenesis, the precursor cell for one of the primary skin tumors, the melanoma, is the melanocyte.

VII. MERKEL CELLS

In 1875, Merkel described a unique epidermal cell associated with an axonal ending near the base of the rete peg in the snout skin of a mole. He named these "tastzellen" (touch cells). He demonstrated this by the nonspecific osmic acid technique, which showed that this epithelial cell was associated with osmophilic nerve terminations.[111] He believed that these cells were cellular transducers of physical stimuli to nerve endings and analogous to the cellular transducers found in organs of special sensation, such as hair cells in the organ of Corti, and rod and cone cells in the retina.[112] Tretjakoff, in 1902, suggested that these specialized cells be termed Merkel cells, rather than touch cells.[113] When these cells are associated with neurites, they are called Merkel cell-neurite complexes.[114] Specialized spots of the epidermis containing many Merkel cell-neurite complexes are known as haarscheiben, according to Pinkus in 1902. Most anatomists refer to such a spot as a hair disc, tactile hair disc, or tylotrich pad, but physiologists use such terms as touch dome, touch corpuscle, touchspot, tactile pad, Iggo dome, Iggo-Pinkus dome,[115] hederiform endings, Pinkus corpuscles,[116] or Eimer's organ.[117]

Light microscopic observations of the Merkel cell in paraffin sections revealed them to be large cells located in the basal region of the epidermis with a characteristic vacuolated cytoplasm predominantly on their dermal side. Their associated neurite seems to conform to the shape of the Merkel cell. These cells are positive to PAS.[112]

This report of PAS activity could be a reflection of the glycogen in the neurite rather than a reaction of the Merkel cell granules.[118] Merkel cells can easily be recognized because their long axis is usually parallel to the skin's surface, making them perpendicular to the columnar epithelial cells above.

The haarscheibe is an elevated touch receptor found in mammalian skin. Among various species, they are structurally similar although they may differ in size, spacing, and the degree of elevation above the surrounding epithelium. A typical haarscheibe contains 25 Merkel cells.[119] In the cat and rat the haarscheibe can easily be seen with the naked eye, especially after depilation. In man a haarscheibe is difficult to identify with the unaided eye. When found, they seem to be abundant on the neck and abdomen.[115] When viewed under a dissecting microscope, they look like shiny domes and are more translucent than the surrounding epithelium.[119] Within the haarscheibe, tufts of blood vessels can occasionally be seen and these give it an erectile quality. The haarscheibe is more resistant to pressure than the surrounding dermis because it has a "denser dermal core and two layers of fibril containing pseudostratified columnar epithelium".[115] In rats they are 100 to 300 μm in diameter and 3 to 5 mm apart. In mice, rats, guinea pigs, and chinchillas, they appear to be associated with tylotrich hairs.[120] Their epithelium contains large basal keratinocytes which have a columnar orientation. Merkel cells are located in the basal layer of their epidermis.

Winkelmann and Breathnach[111] report that Merkel cells were identified by Perez and Perez, as well as Jalowy, through a silver technique in human fetal epidermis at the 7th month of intrauterine life. Breathnach and Robins,[121] using electron microscopy, found these cells as early as the 16th week in the finger tip and in the outer root sheath of the hair follicle. They also reported the presence of Merkel cells in the dermis, associated with Schwann cell axonal complexes, and also noted their passage across the epidermal-dermal junction. This evidence suggests that the cell migrates from dermis to epidermis during fetal life and that it may be of Schwann cell or of neural-crest origin.[111,122] Because of this migration, this cell cannot be a modified keratinocyte as once thought.

Recent studies by Winkelmann have shown esterase, hydrolase, and peptidase activity in Merkel cell complexes of the cat, guinea pig, and rabbit.[123] Using the Gomori technique, alkaline phosphatase was demonstrated in the hair disc of the oronasal area of the cat and in the tactile hair disc of the guinea pig. In the rabbit the hair disc Merkel cell complex showed cholinesterase activity. The Merkel-Ranvier hederiform ending of the rabbit palate showed 4-methoxyleucine aminopeptidase activity. Acetylcholinesterase and ATPase activity, normally found in nerve tissue, can also be selectively demonstrated in Merkel cell endings.[123] Histochemical studies indicate that Merkel complexes do not appear to contain any unique enzymes not found in neurons. The finding of acetylcholinesterase suggests that synaptic transmission between the Merkel cell and its neurite could be mediated by acetylcholine.

The Merkel cell is distinguished from the keratinocytes of the basal layer by its clear cytoplasm and lack of tonofilaments.[123] The cell possesses cytoplasmic protrusions onto the adjacent epithelial cells. These spine-like processes contain no organelles other than filaments. Desmosomes have not been seen on the surface of the processes.[124] Rough endoplasmic reticulum and microtubules are seen.[125] Microfilaments, free ribosomes, mitochondria, glycogen, lysosomes, and vacuoles are also found in the cytoplasm.[123] The cell is the same size, or slightly larger than the adjacent keratinocytes, and is less electron opaque. Unlike the surrounding keratinocytes, its nucleus is lobulated and irregular in outline.[112] The Merkel cell is connected to neighboring cells by desmosomes. The presence of desmosomes may lead one to classify them as keratinocytes. However, the absence of tonofilaments, coupled with the lack of evidence for its migration through the upper levels of the epidermis to become keratinized, strongly suggests that the Merkel cell is not a keratinocyte. Desmosome-like structures sometimes associated with the Schwann cell plasma membrane suggest that it could be an immigrant of Schwann cell lineage.[38] Hemidesmosomes have not been associated with the Merkel cell in adult epidermis, but Hashimoto[126] has reported them in the cells of human fetuses. The terminal axon associated with a Merkel cell is derived from a myelinated nerve. This nerve approaches the epidermis, loses its myelin sheath, but is still surrounded by the Schwann cell cytoplasm and basement membrane. It terminates as a flat meniscus on the basal aspect of the Merkel cell. The most prominent feature of the Merkel cell is the presence of round electron-dense granules. These are located on the opposite side of the nucleus to the Golgi apparatus and are approximately 80 to 100 nm in diameter, and are membrane bound.[112] The dense core is separated from the membrane by a clear zone of 8 to 10 nm. Another characteristic feature of the Merkel cell is the presence of an intranuclear rodlet. This rodlet was found in the mucous membrane of the human mouth, but was not found in human skin. Straile et al. observed this rodlet in the Merkel cells of rabbit skin and suggested it had a lattice-like arrangement of fine particles.[127] This rodlet has also been described in normal nerve tissue and nerve tumors. Tumors of the APUD cell system (*a*mine containing, amine *p*recursor *u*ptake and amino acid *d*ecarboxylase activity) often contain this peculiar intranuclear rodlet.[123]

Merkel cells were first thought to occur primarily in mammalian skin. However, several investigators have found them in other vertebrates including fish, amphibian larvae, tadpoles, lizards, and chickens. Merkel cells are usually found in the epidermis, except in the white leghorn chicken where they have been described in the dermis.[128]

There is an abundance of evidence to suggest that a Merkel cell functions as a sensory receptor of the nervous system. As previously mentioned, Merkel cells are associated with nerve endings and possess the intranuclear rodlet which can be found in nerve tissue. Functionally, Merkel cells are connected to neurons and the complex shows a slowly adapting response to physical force.[123] Other investigators studied this type of mechanical receptor in the tactile disc of cat skin.[111] They classified these structures into two types; a type I slowly adapting receptor in tactile discs of hairy skin, and a type II slowly adapting structure not associated with hair follicles or discs. The type I receptor only responds to a highly localized mechanical stimulus. Its response is then separated into two parts; a velocity-dependent and a force-dependent component.[111] Although it appears that Merkel cells are mechanical transducers, Munger feels that data to prove that this cell is the actual cellular transducer of tactile stimuli are still lacking.[112]

English demonstrated that denervating the haarscheibe (type I mechanical receptor) in cats caused degeneration of these structures. This was manifested by (1) a reduced number of Merkel cells in the haarscheibe, (2) cellular degeneration of persisting Merkel cells, (3) a decrease in the number of cytoplasmic dense core granules, (4) an increase in the number of agranular dendritic cells and Langerhans' cells, (5) a decrease in the number of keratinocytes covering the area, and finally, (6) a decrease in the number of transitional cells.[129] In a similar study, when the cutaneous nerve regenerates to the epidermis, Merkel cells are then found in the haarscheibe and the normal structure of the organ is restored.[119] This investigation indicates that Merkel cell physiology is intimately related to its neurite. Benkenstein, however, showed that Merkel cells in the sinus hairs of the upper lip of the rat did not change within 169 days of crushing the nerve.[130] The nerve terminal of the Merkel nerve ending degenerated 24 h after denervation and was phagocytized by adjacent keratinocytes.

Winkelmann and Breathnach[111] report that Mustakallio and Kiistala hypothesized that a Merkel cell's function in the human epidermis is that of monoamine storage. This was based on the Merkel cell dense core granule's structural similarity to monoamine granules found in other areas. Proof is lacking because conflicting results have been obtained when reserpine has been used to deplete these granules.[111] Lauweryns demonstrated serotonin in dense cored granules from neuroepithelial cells found in the rabbit respiratory mucosa.[131] These cells were comparable to Merkel cells because they possessed these granules, had an indented nucleus, and were associated with a neurite. If these observations are correct, the Merkel cells would be classified as sensory receptors employing serotonin as a neurotransmitter. Chen et al. showed that the granules found within a Merkel cell fused with a synapse-like structure that occurred between the Merkel cell and its neurite.[125] Digallic acid staining and goniometric tilting facilitated the resolution of the synaptic contact.[125] Structural analysis of the synaptic junction of Merkel cell axon complexes revealed a typical chemical synapse except for the presence of presynaptic clear vesicles.[132,133] Andres described these synapse-like structures in terms of thickening of the opposite areas of the membranes of the Merkel cell and the neurite.[134]

However, no one has yet determined the exact cytochemical properties of the substance in Merkel cell granules, nor has anyone been able to consistently demonstrate (using pharmacologic techniques) release of a transmitter substance from these granules. In addition, it has not been shown that a mechanical stimulus applied to a Merkel cell causes a generator potential to be formed which would release transmitter chemical from the granules into a synapse. It should be recalled that histochemical evidence implicates acetylcholine as the neurotransmitter.

Regardless of the identity of this substance, Merkel cells are intimately related to the nervous system, possibly as the mechanical receptors originally proposed by Merkel.[119,125,129,131] In line with this theory, many workers have attempted to demonstrate central neural connections to the haarscheibe, based on the assumption that mechanical receptors should have identifiable nerve tracts in the spinal cord and the brain. Some investigators have postulated that haarscheiben project to cerebral cortex by way of dorsal columns in cats and rabbits. However, others have only found type II receptors projecting to the dorsal columns. Tapper[135] abolished type I response by sectioning the spinocervical tract, while Brown and Franz[136] elicited no response in this tract by stimulating a haarscheibe. This confusion may be due to the investigator's failure to clearly distinguish between type I and type II slowly adapting receptors. There may also be a difference in hind-and forelimb projections.[115] It is, therefore, apparent that the precise neural circuitry involved in the processing of information from the haarscheibe has not been accurately determined.

A second hypothesis as to a Merkel cell's function is that they are support cells which exhibit a relationship to their neurites much like that which exists between a Schwann cell and its neurons.[137] This could indicate that the unmyelinated axon terminal is actually the primary sensory receptor. Embryological data does not refute this theory and the presence of desmosome-like structures supports it. Further, this theory could explain the discrepancies in the histochemical and ultrastructural studies.

A final hypothesis of Merkel cell function is that they are neurosecretory cells which secrete a polypeptide hormone as a member of the APUD system. This system includes the adrenal medulla cells, chromaffin cells, intestinal enterochromaffin cells, pituitary corticotrophes, melanotrophes, pancreatic beta islet cells, and the parafollicular cells of the thyroid. All of these are ultrastructurally similar to Merkel cells, contain serotonin, have a high esterase content, and are capable of taking up amine precursors.[123,138,139] In addition, Winkelmann believes that both APUD and Merkel cells are embryonically neural crest migrants to the epithelium.[123] It should be noted that histochemical studies of Merkel cells have demonstrated esterase activity and peptidase activity which could possibly be related to peptide production. What has yet to be demonstrated to classify these cells as members of the APUD system is the presence of a hormone product. Studies using immunohistochemistry on Merkel cells from sinus hair follicles of rats, demonstrated met-enkephalin.[140] This immunoreactivity was restricted to Merkel cells and was not present in its associated nerve axons or terminals. The demonstration of this neuroactive peptide reinforces the postulate that Merkel cells are members of the paraneuronal APUD system.[140] As far as this author is aware, Merkel cells have not been seriously considered as target sites for cutaneous toxicants.

VIII. LANGERHANS' CELLS

This intraepidermal nonkeratinocyte was discovered by Paul Langerhans in 1868.[141] Langerhans' cells are most commonly found in the upper spinous layer of the epidermis (Figure 9).[38,142,143] They have been identified as early as 15 weeks in the fetal epidermis[33] and they have also been identified in the oral mucosa, the human female genital tract, the sheep rumen, and in the upper alimentary tract of the mouse.[144] Not only are they found in stratified squamous epithelium, but Langerhans' cells have also been seen in lymphatic vessels, lymph nodes,[145] and in the dermis.[146]

Langerhans' cells were observed stained with a solution of gold chloride in the suprabasal layers of the epidermis. This technique, which is also used to identify nervous elements, led some investigators to think that Langerhans' cells were of neural origin.[38,143] Langerhans' cells have long dendritic processes which can traverse the intercellular space up to the granular cell layer.[89] Langerhans' cells exhibit ATPase activity, allowing them to be identified in the wall of the pilosebaceous canal. Quantitative studies on the guinea pig have shown that the mean Langerhans' cell population was not statistically different (i.e., 930 cells per mm² in red guinea pigs and 915 cells per mm² in albinos). In addition, the number of Langerhans' cells was similar in different regions of the guinea pig body.[147,148]

One of the best methods available for distinguishing Langerhans' cells from other epidermal cells is TEM. These cells have been classified as nonkeratinocytes because they lack the characteristic features, namely tonofilaments and desmosomes. They possess slender processes which penetrate the intercellular spaces among the prickle cells. These processes contain all the organelles which are normally found in the perikaryon.[149] Their nucleus (Figure 9) is indented and the cytoplasm contains Golgi complexes, smooth and rough endoplasmic reticulum, and lysosomes. The characteristic feature which distinguishes Langerhans' cells from other cells is the presence of a rod- or racket-shaped structure in the cytoplasm (Figures 10 and 11).[38,89,142,143,150,151] This structure, first described by Birbeck, is commonly known as the Birbeck or Langerhans' cell granule.[152] This specific granule (Figure 11) is shaped like a racket, having a "handle" portion and an expanded end. The "handle" consists of an outer limiting membrane with a central lamella which appears as a row

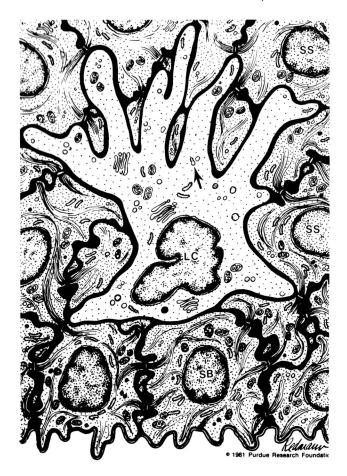

© 1981 Purdue Research Foundatic

Figure 9.   Schematic of Langerhans' cell (LC). Intercellular space is represented by black areas. Stratum basale cells (SB) are below and stratum spinosum (SS) cells are adjacent and above LC. Note dendritic processes, Langerhans' granule (arrow), and electron lucency of the cytoplasm. (From N. A. Monteiro-Riviere, *Swine in Biomedical Research,* Plenum Press, New York, 641 (1986). With permission.)

of particles exhibiting a 50 to 70 Å periodicity. The expanded end of the limiting membrane is usually clear, and on occasion may have a cross-striated pattern within.[38,89,142,152]

Rodriquez and Caorsi believe that the Langerhans' cell granule consists of two components; the Langerhans' cell disk (handle) and the Langerhans' cell vacuole (expanded end).[149] The disk has an array of particles centrally located with a 50 to 70 Å periodicity similar to what Zelickson and Breathnach et al. had previously described.[142,143,152] Further studies using microdensitometer tracings in conjunction with a zinc iodide osmium procedure showed the size and spacing of these par-

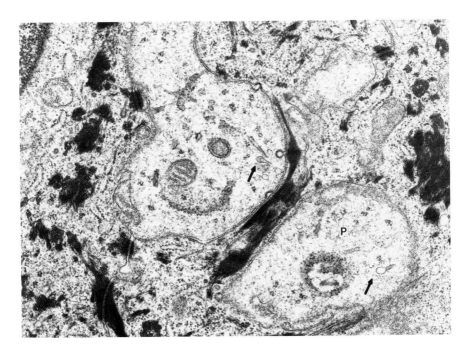

Figure 10. Langerhans' cell processes (P) traversing through stratum spinosum layer of pig skin. Small Langerhans' cell granules are present (arrow). Magnification × 25,500. (From N. A. Monteiro-Riviere, *Swine in Biomedical Research*, Plenum Press, New York, 641 (1986). With permission.)

ticles.[149] The Langerhans' cell disk limiting membrane has an external and internal leaflet. Attached to the inner surface of the internal leaflet are rows of "p" particles 7 nm in diameter and equally spaced at 7 nm. Extending between the "p" particles are filaments 3 nm in thickness. This arrangement results in a square with "p" particles in the corners and filaments for sides. The central region consists of smaller "s" particles which are 4 nm in diameter and are connected to the "p" particles by 4 nm filaments.[149]

The high concentration of Langerhans' cell granules near the Golgi complexes may suggest their origin.[89] On the other hand, Langerhans' cell granules were noted to be continuous with the cell membrane and also were seen to be connected to tubular structures, probably smooth endoplasmic reticulum.[149] The differences in attachment between the granules can be explained in two ways. First is that the Langerhans' cell granules are produced in the cell and migrate to the periphery where they attach to the cell membrane and liberate their contents into the surrounding intercellular space.[38,89] The second possibility is that they are infoldings of the plasma membrane which eventually get "pinched off" and become free floating in the cytoplasm. The first would suggest that the granule plays a role in excretion or secretion, while the second would infer that its function was related to absorption.[38,142]

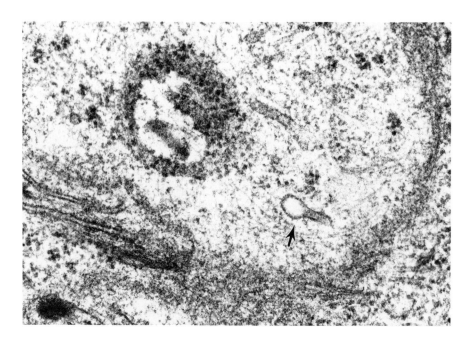

Figure 11.   Higher magnification of the same Langerhans' cell granule (arrow) seen in Figure 10. Magnification × 60,480. (From N. A. Monteiro-Riviere, *Swine in Biomedical Research,* Plenum Press, New York, 641 (1986). With permission.)

Analysis of the granule's limiting membrane indicated that it is a three-layered membrane measuring 9 nm in thickness.[149] In addition it is 17% thicker than the cell membrane and 30% thicker than the Golgi membranes. Histochemical studies have shown that the internal and external leaflets of the limiting membrane of the Langerhans' cell granule stain differently.[149] Therefore, one could assume that the Langerhans' cell membrane is unique.[149]

The morphological features of the Langerhans' cell and its granule are characteristic in the fetus as well as the adult except for the absence of the vacuole portion of the Langerhans' cell granule in the fetus. Premelanosomes were not identified in the fetus, but melanosomes have been found in Langerhans' cells in the adult.[33,38] The presence of a few melanosomes within Langerhans' cells enhanced the belief of early investigators that they were worn-out melanocytes. However, there is no evidence of Langerhans' cells synthesizing melanin. Instead, they probably acquire the melanosomes by phagocytosis.

Two types of Langerhans' cells have been observed in normal human skin. One is dendritic with a moderately electron dense cytoplasm and numerous Langerhans' granules but relatively few lysosomes and mitochondria. The second one appears less dendritic, occurs more frequently in the basal layers, has an electron-dense cytoplasm, fewer granules, and numerous mitochondria, some of which appear swollen, vacuolated, and without cristae.[146,153]

Another scheme classifies Langerhans' cells into four different types based on certain characteristic features: cell size, number of dendritic processes, and shape of the nucleus.[154] Shukla has reported a method of differentiating Langerhans' cells from melanocytes based upon number and thickness of dendritic processes.[155,156] The melanocytes have four to six relatively thick dendritic processes while the Langerhans' cells have two, three, or four thinner processes.

Several functions have been attributed to the Langerhans' cell over the past 100 years. It was once thought to be related to the nervous system because of staining characteristics by light microscopy.[141,157] However, newer methods have not demonstrated any type of neural connection. Langerhans' cells were thought to be derived from the neural crest; however, studies on mouse skin, experimentally deprived of neural crest components, demonstrated Langerhans' cells but no melanocytes, ruling out the neural crest as their origin.[158] Langerhans' cells were once considered to be effete melanocytes.[86] Since melanocytes are constantly being renewed, it was thought that as melanocytes increased with age and moved up through the epidermis, they lost their melanin. It was the terminal stage in this melanocyte life cycle which was felt to be the Langerhans' cell.[38,89] However, Langerhans' cells are capable of mitotic division because they possess numerous centrioles and therefore, would not be considered "effete or worn-out".[89] One theory holds that Langerhans' cells regulate epidermal cell proliferation.[159,160] Recently, it has been shown that they are derived from precursor cells originating in the bone marrow.[158,161,162] Katz and Tamaki believe Langerhans' cells probably originate from the monocyte-macrophage series because of similar surface marker characteristics such as the plasma membrane ATPase activity demonstrated by Wolff and Winkelmann.[163-165] Langerhans' cells are similar to macrophages in their ability to migrate.[166] In addition, they both have nonspecific esterase activity.[167] They have Ia histocompatibility antigens and Fc and C3b receptors.[160,167-169]

Langerhans' cells and macrophages also have been reported to transfer antigen *in vitro* upon lymphocyte stimulation. In the past it has been difficult to establish a cell line of Langerhans' cells, but now a cell line which is karyotypically, histochemically, ultrastructurally, and immunologically characteristic of Langerhans' cells exists.[167]

Since Langerhans' cell processes are usually directed toward the surface of the skin, they could play a role related to events occurring on the surface of the skin. It has been shown that Langerhans' cells pick up antigen in the skin and circulate it to draining lymph nodes.[166] Some authors believe Langerhans' cells are the body's first line of immunologic defense.[170] Functional properties of Langerhans' cells, in relation to the immune response, are under considerable investigation (see Chapter 5). Upon exposure to UV light, the epidermis becomes depleted of Langerhans' cells and the skin becomes immunologically unresponsive.[171,172] Langerhans' cells were the only epidermal cell type to exhibit ultrastructural changes in exposure to low levels of UV light. They lost their dendrites, became oval, possessed numerous vesicles and vacuoles, as well as had a condensed filamentous matrix. Higher doses abolished Langerhans' cell surface markers.[173] Studies also have shown a reduction in Langerhans' cell numbers in diseases such as psoriasis and eczema.[174] Langerhans' cells have also been reported in fibrotic lung disorders,[175] mycoses and fungoides,[176] atopic dermatitis,[177] and in a nondermatological disorder, eosinophilic granulomatosis.[178]

Reduction in Langerhans' cell surface density has been reported in two cutaneous sites; the hamster cheek pouch and the murine tail skin. Langerhans' cells are also absent from the cornea of hamsters, guinea pigs, and mice. The absence of Langerhans' cells from the cornea could explain persistent corneal infections caused by herpes simplex.[172,179] The hamster cheek pouch is considered an immunologically privileged site becuase skin grafts and tumors survive longer at this site than at any other. Murine tail skin allografts are rejected by recipient mice less often than the counterpart grafts prepared from the body wall.[172,179] These data would suggest that Langerhans' cells play an important role in modulating the immune response against foreign substances and malignant cells. In their absence, immunologic tolerance may develop.[166,171,180-182] It is now accepted that Langerhans' cells are functionally and immunologically related to the monocyte-macrophage series. They are capable of presenting antigen to lymphocytes. They are considered to be the initial receptors for cutaneous responses. Langerhans' cells are not thymus dependent because they are found in nude mice.[183]

Another type of cell, known as "veiled cells", are present in afferent lymph and has similar properties resembling Langerhans' cells.[184] An understanding of Langerhans' cell structure and function is important to cutaneous toxicology because of their role in modulating the immune function of the skin.

IX. INDETERMINATE CELLS

Indeterminate cells are different from the surrounding keratinocytes in that they possess dendrites and lack desmosomes such as the melanocytes and Langerhans' cells. They are usually located in the lower levels of the epidermis and occupy approximately 1% of it.[151] Indeterminate cells have also been called Type 3 cells[185] and alpha dendritic cells.[186,187]

Ultrastructurally, indeterminate cells are similar in many aspects to the Langerhans' cells in that they possess well-developed Golgi complexes and rough endoplasmic reticulum.[188] However, they do not have the Langerhans' (Birbeck) cell granule. The indeterminate cells have been shown along with similar cells in both fetal and postnatal dermis and are thought to be precursors of, or immature, Langerhans' cells.[158] Zelickson and Mottaz thought the indeterminate cell was a form of a melanocyte or an undifferentiated cell which could give rise to other epidermal dendritic cells or a new cell type.[151] Some investigators believe they are immature melanocytes.[176,188] Immunoelectron microscopic studies show that indeterminate cells have Ia antigens similar to Langerhans' cells.[188-190] These cells are also similar to Langerhans' cells in that they are derived from bone marrow.[191]

X. BASEMENT MEMBRANE

The epidermal-dermal junction is characterized by the presence of the basement membrane. Several functions have been attributed to the basement membrane. One is that it can act as a selective barrier between the epidermis and dermis, restricting some molecules and permitting the passage of others. It also provides support for the epidermis.

In light microscopy the basement membrane appears as a thin, homogeneous layer. Special stains, i.e., PAS, clearly show the epidermal-dermal interface. However, TEM shows a more structured organization of this membrane (Figure 12). It consists of four components: (1) the cell membrane of the basal epithelial cell which sometimes includes hemidesmosomes; (2) the lamina lucida; (3) the basal lamina; and (4) the sub-basal lamina with different types of fibrils.[192] The basement membrane of late fetal and postnatal life is not different compared to that of earlier stages except for the increase in number and increase in diameter of anchoring fibrils.[32,38]

The basal cell membrane of the epidermal-dermal junction is not always smooth. It may be irregular, forming finger-like projections into the dermis. The cell membrane consists of three layers, a thick internal leaflet, an electron lucent layer, and a thin external leaflet. Hemidesmosomes are sometimes found along the internal leaflet of the cell membrane. Occasionally, small vesicles (pinocytotic) are seen near and on the plasma membrane. The lamina lucida portion of the basement membrane is electron-lucent except in the hemidesmosome area, where one can see a sub-basal dense plaque and some fine filaments. The next layer is the basal lamina which is electron dense and may contain fibrillar components in an amorphous substance. The basal lamina is thicker and denser in regions beneath hemidesmosomes.

The last component of the epidermal-dermal junction is the sub-basal lamina containing three types of fibrous structures; anchoring fibrils, microfibrils in bundles, and single collagen fibers. Anchoring fibrils (Figure 12) extend from the basal lamina to the dermis, where they can end in a reticular network. The midportion of these fibrils have a banded appearance with an irregular periodicity. Anchoring fibrils are usually more numerous beneath hemidesmosomes. Fibrils arranged in bundles usually extend from the finger-like projections of the basal lamina into the dermis. Each fiber is approximately 8 to 11 nm in diameter, with a beaded periodicity. In cross-section, it resembles a hollow tubular structure which is similar to the microfibrils found near the elastic fibers. No proper term has been designated for these fibrils. Briggaman and Wheeler suggest that they should be called microfibril bundles. The collagen fibers lie near the basal lamina, occur singly, and are randomly oriented. They are not arranged into bundles like those of the deeper dermis.[192]

It has been postulated that the hollow tubular structure of the microfibrils is similar to the microfibrils of elastic tissue. Therefore, they could be an extension of the elastic tissue system. It was also thought that they were related to the oxytalan fibers.[192] Bundles of these tubular microfibrils have been related to the epidermal-dermal junction and are further identified as oxytalan fibers.[193]

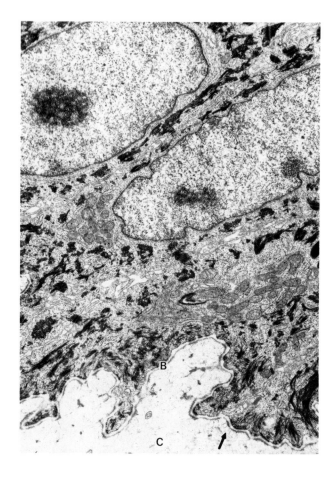

Figure 12.   Basement membrane (B) located between epidermis and dermis. Note the four basic components of the basement membrane, anchoring fibrils (arrow) beneath hemidesmosomes and collagen (C). Magnification × 10,500. (From N. A. Monteiro-Riviere, *Swine in Biomedical Research,* Plenum Press, New York, 641 (1986). With permission.)

The basement membrane of skin is synthesized by the basal cells of the epidermis. In addition, these basal cells depend on the semi-permeable nature of the basement membrane for allowing nutritive substances and oxygen to reach these from the dermis. In turn, the dermis depends on the epidermis for protection.[192,194-196] This mucopolysaccharide structure, similar to that of basement membranes in other tissue beds, may be the target of autoimmune reactions. Remarkable advances have been made in identifying and characterizing the basement membrane components. The basement membrane macromolecules that are ubiquitous components to all basement membranes include: type IV collagen, laminin, entactin/nidogen, and heparan sulfate proteoglycans.[300] Of additional relevance to cutaneous toxicology is its role in percutaneous absorption of large or possibly charged molecules. This phenomenon has not been seriously addressed.

XI. THE DERMIS

The dermis or corium lies beneath the basal lamina and extends to the hypodermis. The dermis is of mesodermal origin and consists largely of dense, irregular connective tissue (Figure 13). In this matrix, fibrous proteins such as collagen, elastin, and reticulin are embedded in an amorphous ground substance. The predominant cell types of the dermis are fibroblasts, mast cells, and macrophages. Plasma cells, fat cells, and extravasated leukocytes are often found. The dermis is traversed by blood vessels, lymphatics, and nerves. Glandular structures such as sebaceous and sweat glands are also found in this connective tissue matrix, along with hair follicles.

The dermis can be divided into two anatomical layers; an outer papillary and an inner reticular layer. The papillary layer is the thinnest and consists of loose connective tissue, while the reticular layer is thicker and consists of dense connective tissue.[197-199]

In man 15 to 20% of total body weight is dermis. The skin contains 18 to 40% of total body water, with most of this water found primarily in the dermis. The water content of the epidermis is about 12%, the papillary layer is 71%, the reticular layers 61%, and the subcutaneous layer 30%.[197] Several functions have been attributed to the dermis. It participates in mechanical support, in exchange of metabolites between blood and tissues, in fat storage, in protection against infection, and in tissue repair.[197,199]

A. Fibers

Collagen is formed in fibroblasts by protein biosynthesis and the molecules aggregate to form fibers. Similar to other proteins, collagen is synthesized on polyribosomes attached to endoplasmic reticulum. After a normal translation process, the incorporated residues of proline and lysine are hydroxylated and then glycosylated. These completed chains (pro-alpha-collagen) are released from the ribosomes into the cisternae of rough endoplasmic reticulum. Hydroxylation and glycosylation may continue to occur at this point. These chains assemble into a triple helix within the cisternae and are transported to the Golgi apparatus, where secretory vesicles are released. These fuse with the plasmalemma and their contents are released as a soluble precursor (procollagen) into the extracellular environment. Procollagen is activated in a series of steps which results in cleavage of a "tail-piece" to form the tropocollagen molecule. Tropocollagen aggregates and then is enzymatically cross-linked to form mature collagen fibers. Specific amino acid interactions insure that the tropocollagen molecules assemble in the proper orientation or register. This results in the banded appearance of the collagen fibril.[200-202] Collagen contains several amino acids such as 20% proline plus 4-hydroxyproline, 3-hydroxyproline, and hydroxylysine which are necessary for intermolecular cross-linking.

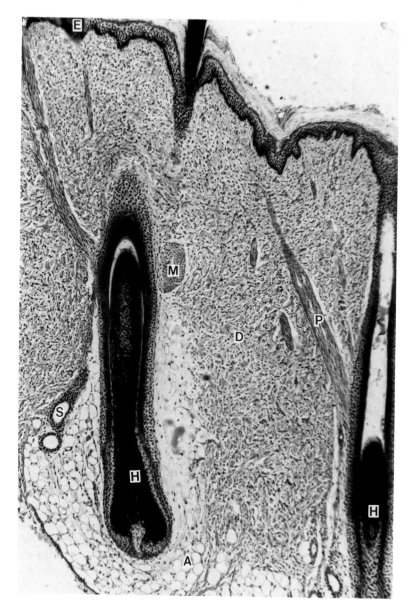

Figure 13. Low magnification for orientation of dermal components. Note epidermis (E), dermis (D), longitudinal section of two hair follicles (H), arrector pili (P), apocrine sweat gland (S), adipose cells (A), cross-section of the interfollicularmuscle (M), and hair (H). Pig skin, Bodian's Protargol; magnification × 100.

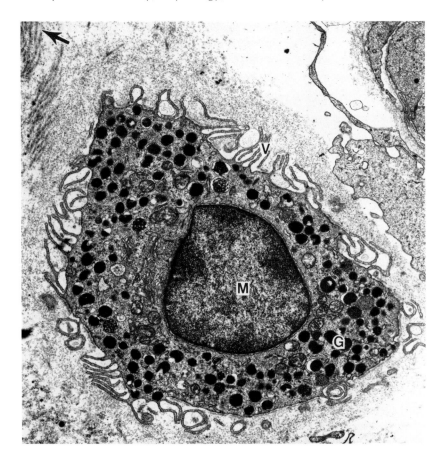

Figure 14. Mast cell (M) from an adult pig. Note numerous granules (G), microvilli (V), and collagen fibers (arrows). Magnification × 9200. (From N. A. Monteiro-Riviere, *Swine in Biomedical Research,* Plenum Press, New York, 641 (1986). With permission.)

Ultrastructurally, collagen fibers stained with heavy metal ions show a cross-banding of light and dark zones (Figures 5 and 14). Each collagen fibril exhibits an axial periodicity every 64 nm. This can be explained by the position of each collagen fibril, which is 280 nm in length, arranged side by side in a parallel fashion but staggered by one quarter of their length. With conventional staining, the areas of light and dark zones of the collagen fibril represent the gap region and overlap region, respectively. The reverse pattern is seen with negative staining.[38,197,201,202]

Five types of collagen fibers have been described in the human. The difference between the types is based on amino acid composition and sequence. Type I collagen is found in the dermis, bone, tendons, ligament, and dentin. Type II collagen is found predominantly in hyaline cartilage. Type III is found in the fetal dermis, uterus, and cardiovascular system, and immunofluorescent studies show it to be localized in the papillary dermis. Type IV is found in the basal lamina.[202-204] Type V collagen, also known as AB, is found in the cornea, dermis, placental membranes, and lung. This type of collagen is thought to be important in cell movement.[205]

In the fetus collagen fibers are fine and fibrillar, while in the adult they are coarser and aggregated into bundles. Collagen fibers continue to thicken until 25 years of age.[199]

Elastic fibers are found in tissues that are capable of expansion. Elastic fibers consist of two protein components; elastin which is the main component and an amorphous substance, and a microfibrillar protein which acts as a mold for the elastin. Elastin is synthesized in fibroblasts similarly to collagen. Microfibrils are first formed by the fibroblasts and act as a mold for the newly formed elastin which then become a fiber. Ultrastructurally, elastic fibers do not exhibit a periodicity like collagen fibers. Elastic fibers stained for electron microscopy appear as an amorphous mass, with a few discrete microfibrils appearing at the periphery. This mass is usually light staining, while the microfibrils are more electron dense.[38,199,202,204,206]

The number of elastic fibers varies from area to area but they are found in abundance in adults and newborns in the dermis of the scalp and face. They are found around hair follicles and help to attach the arrector pili muscles to the bulge of the hair follicles. Elastic fibers are located around eccrine and apocrine sweat glands and help to attach blood vessels to surrounding tissues.[199]

Elastic fibers found in the reticular dermis of man are coarser and thicker than those of the papillary layer, which branch repeatedly. Fine branches of the elastic fibers of the papillary layer extend to the basal lamina of the epidermis where they terminate as knob-like (globose) endings. Elastic fibers in the papillary layer of cheeks, eyebrows, and chins of humans are tightly woven around blood vessels. In the fingers, toes, palms, and soles, elastic fibers are very dense in the papillary layer and may surround individual Meissner corpuscles. Elastic fibers are found to run in various directions in the reticular dermis. In some areas, i.e., chest, thighs, axilla, breast, mons pubis, gluteal region, and abdomen, coarse elastic fibers run parallel to the surface in the reticular layer. In the skin of the chest, axilla, and mons pubis, some elastic fibers run up into the papillary layer from the reticular layer and form arcades in which straight branches radiate toward the epidermis.[207]

Elastic fibers show major changes in structure when exposed to UV radiation. In sun-exposed areas of the face, extensive alterations in elastic fibers of 20- and 30-year-old individuals were seen. Similar changes occurred in older individuals. The elastic fibers showed a thickening and tended to curl into amorphous masses. Areas directly beneath the epidermis (Grenz zone) sometimes lacked elastic fibers and appeared clear or contained damaged elastic segments.[207]

Elastic changes can occur in skin that has not been exposed to the sun. Studies on the axilla, breast, and genitalia showed changes in the elastic fiber architecture. Although there were differences attributed to individuals, profound changes occurred in people 50 years of age or more. Elastic fibers in the papillary layer were thicker and some of the distal branches were broken from their main structure, while some remained attached to the epidermis. Structural changes in elastic fibers do occur with aging in sun-exposed and nonexposed areas. UV radiation seems to expedite these alterations.[207]

Reticular fibers have been seen as thin, branching fibers in the dermis, usually more numerous beneath the basal lamina. The reticular fiber was once thought to be another fiber type present in the dermis. However, recent evidence using electron microscopy has shown the characteristic axial periodicity of 64 nm that is found in collagen fibrils. Therefore, it has been proposed that reticular fibrils may be collagen fibril templates.[199]

B. Ground Substance

The connective tissue fibers and cellular components of the dermis are embedded in an amorphous matrix, the ground substance. Proteoglycans are a major component of the ground substance. Various types of proteoglycans are found in the ground substance of skin; hyaluronate, dermatan sulfate (chondroitin sulfate B), and chondroitin sulfate A and C.[208] The ground substance also contains substances from the blood such as water, sugars, proteins, urea and inorganic ions, metabolic products of the parenchymal cells, and metabolic products from the connective tissue cells. One study has shown that dermal fibroblasts grown *in vitro* synthesized some components of the ground substance.[199]

The dermal ground substance shows little change with aging. Changes seem to occur in the young and then very little change is seen throughout life.[199] Jarrett reports that Loewi and Meyer found an increase in dermatan sulfate, hyaluronate and chondroitin sulfate in the adult pig in comparison with the embryonic pig.[201] The function of the ground substance is to maintain the homeostatic environment for cells and fibers. Also, it is thought to play a role in limiting the spread of bacteria because of its viscosity.[199,202]

C. Cells of the Dermis

One of the most important and most numerous cells in the dermis is the fibroblast. They are usually found among the collagen fibers. Young fibroblasts have an abundance of cytoplasm while old fibroblasts (fibrocytes) have very little. In light microscopy, fibroblasts are spindle shaped with long cytoplasmic processes at each end. Their nucleus is oval and contains two to four nucleoli. Fibroblasts are motile and capable of mitosis. They are derived from mesenchymal cells. The fine structure of fibroblasts include elongated processes and the usual complement of organelles. The most characteristic feature in the cytoplasm is the abundance of rough endoplasmic reticulum. This is expected of cells that synthesize the collagen protein for export. Fine filaments, sometimes occurring in bundles, are usually seen in the cytoplasm of the fibroblasts. An active or young fibroblast showing extensive rough endoplasmic reticulum and several nucleoli indicates that it is synthesizing protein for secretion. Fibroblasts secrete procollagen, glycosaminoglycans, and proelastin. Secretory vesicles can be seen all along the cell membrane of the fibroblast. These vesicles fuse with the cell membrane and release their contents (procollagen) into the extracellular space where it will be enzymatically converted to collagen.[38,199,204]

Mast cell population density varies between species and also among different regions in the same species. In the superficial dermis of humans, mast cells are found in high density contrasted to the high mast cell density seen in the deep dermis of rats.[209] Rang has reported the sensitivity to stress of the mass cell population density in the skin of pigs and also noted that the population is somewhat greater in the adult than in the young pig.[210] A quantitative study of mast cells from 1 through 14 weeks in pigs showed a peak at 7 weeks, followed by a moderate decrease.[8,34] This variation in density with age may be important when selecting pigs for comparative studies designed to assess mast cell function, such as allergy, histamine release, and drug hypersensitivity reactions. Characteristic mast cells can be seen in 19- and 28-week-old human fetuses. Mast cells tend to occur along blood vessels. Special stains are used to identify these cells and their granules for light microscopy. Mast cells are quite large (20 to 30 μm) and possess large granules containing histamine, heparin, or serotonin depending on the species.[211]

The fine structure of mast cells has been adequately described by Breathnach.[38] The nucleus is round and usually lies centrally within the cell (Figure 14). The mast cell has an irregular outline possessing numerous villous folds projecting from the plasmalemma. Common organelles present include mitochondria, a well-developed Golgi apparatus, and smooth and rough endoplasmic reticulum. In addition to the villous processes, another characteristic feature of the mast cell is the presence of a large number of cytoplasmic granules (Figure 14). Each granule is approximately 0.2 to 0.8 μm in diameter and is surrounded by a unit membrane. In humans the granules present a variety of appearances. Granules may have an internal structure which possess a crystalline arrangement, may have a central dense core in a homogeneous background, or may appear translucent. Some granules may contain lamellar bodies. The internal structure of the lamella may also assume different appearances. Some vacuoles may be noted within the mast cell and are thought to be present after a discharge or release of granules.[38,212-214] It is thought that the granules are derived from smooth endoplasmic reticulum.[215,216]

Another type of cell found in connective tissue is the macrophage. It is known by a variety of names such as clasmatocyte and histiocyte. The macrophage has been postulated to be derived from the monocyte. The life span may be 2 to 3 months, depending on the tissue. Macrophages can be identified by vital staining, an excellent method demonstrating macrophages "in action". Ultrastructurally, macrophages have an irregular shape. Their cell membrane has been modified into finger-like projections or pseudopods. The nucleus is indented and common organelles such as mitochondria and a small amount of rough endoplasmic reticulum are present. Macrophages contain numerous types of lysosomes with pinocytotic and phagocytotic vacuoles. The function of macrophages is to engulf and destroy bacteria or foreign substances, and to process antigens for presentation to the immune system. The functional activity of a macrophage can be appreciated by the complexity of its membrane modifications.[38,202,204]

The plasma cell is usually found in loose connective tissue and in great numbers in lymphatic tissue. It is derived from B-lymphocytes which differentiate in the connective tissue after antigenic stimulation. Antibody production is the function of the plasma cell (see Chapter 5). In routine staining for light microscopy, plasma cells can be seen to contain a large amount of cytoplasm relative to the size of the nucleus. The nuclear chromatin is distributed in clumps around the periphery of the cell. The nucleolus is prominent. The plasma cell can sometimes contain acidophilic granules termed Russel bodies. Ultrastructurally, plasma cells have an abundance of rough endoplasmic reticulum and prominent organelles such as a large Golgi region, mitochondria, polyribosomes, and finger-like extensions of the plasma membrane. The abundance of rough endoplasmic reticulum suggests protein secretion, a finding consist with their role of producing antibodies for release into the blood.[204]

Adipose cells (Figure 13) may accumulate along small vessels but are primarily located in the deeper dermis where they merge with the subcutaneous fatty tissue. Adipose or fat is defined as an area with a large number of fat cells. At the light microscopic level, they appear polygonal and range in size from 25 to 200 μm. The dehydration process usually extracts their lipid content and thus in light microscopic sections, they appear as empty spaces or holes. Ultrastructurally, the cytoplasm is only a thin layer surrounding the large central lipid inclusion. The presence of this inclusion displaces the nucleus to one side. Few organelles are present in the area of the nucleus, with mitochondria being the most obvious. Collagen or reticular fibers may be found along the cytoplasmic rim of the fat cell. Several fat cells may coalesce to give rise to a single large cell. Fat cells may also aggregate into lobules and be divided by septa within which blood vessels and nerves can travel. Mitotic activity in predetermined fat cells is complete by 2 to 3 weeks after birth. In the adult fat cells do not divide. It has been postulated that undifferentiated, spindle-shaped, mesenchymal cells resembling fibroblasts give rise to fat cells.[38,202]

XII. CUTANEOUS APPENDAGES

The components of the pilosebaceous system, hair follicles, and associated sebaceous glands are also found in the dermis (Figure 13). In man, the development of hair follicles occurs in the third fetal month. The epidermal cells become columnar and then invade the dermis. This downgrowth of epidermal cells (hair germ) becomes thicker and mesenchymal cells form around its base. The hair germ continues to penetrate deeper into the dermis and becomes the hair follicle, which will give rise to hair. Hairs develop first on the eyebrows, chin, and upper lip and within a few months, hairs cover the entire fetus. These first hairs are called lanugo hairs and are shed just before birth except in regions where they first developed. After birth, most of the hairs are of the vellus type. Terminal hairs, like those of the scalp, are coarse and develop during puberty in specific regions of the body. During development of the hair follicle, two epithelial bulges occur on the side of the follicle. The first will develop into the sebaceous gland and the second will serve as the point of attachment for the arrector pili muscle.[217] Another bulge may appear just above the sebaceous gland in some follicles and give rise to the apocrine gland.[218]

The hair follicle is composed of three primary layers: inner root sheath, outer root sheath, and connective tissue sheath. The first is composed of scale-like keratinized cells which interlock with cuticle cells of the hair. The second is continuous with the epidermis. Structurally, it resembles epidermis but also has glycogen in it. The third layer, the connective tissue sheath layer, is continuous with the papillary layer of the dermis and with the dermal papilla of the hair follicle. At the base of the hair follicle, termed the bulb, the dermal papilla and hair matrix are located. The matrix is composed of the germinative epithelial cells which give rise to the hair proper, and is the region responsible for biochemical regulation of hair growth. They are also the target cells for toxicants directly affecting hair growth.

Terminal hairs consist of a central portion, the medulla, which is poorly developed in humans. Surrounding the medulla is the cortex. This is the main constituent of the shaft. Cells in this layer are keratinized and carry the pigment of the hair. The outermost layer is the cuticle, which is also keratinized.[217,219]

The hair follicle undergoes regular changes with growth, atrophy, and inactivity. The growth phase is known as anagen. This involves the active division of matrix cells to produce new hair. The catagen phase is a short period where the hair bulb atrophies and mitotic activity of the matrix cell ceases. In the telogen phase, also known as the resting phase, the hair which was produced in anagen is anchored in place. The medulla, cuticle, and inner root sheath are absent during the telogen phase.[217,219-222]

Numerous species differences exist in regard to the temporal pattern of hair growth. It should be realized that hair in the active anagen phase would be more susceptible to cell cycle-specific toxins, such as cancer chemotherapeutic agents, than would hair follicles in a resting phase. The hormonal regulation of hair growth is a complex process involving interactions with gonadal sex steroids, thyroxine, and adrenal corticosteroids. These phenomenon are species, site, and hair-type dependent. Nutritional status and cutaneous blood flow also interact with this control. The extreme case is seen in those species which undergo annual molts (ferrets, mink, seals, etc.) that are synchronized to the season of the year or length of day. In these wild species all hair follicles abruptly shift to anagen, apparently under the direct control of thyroxine and corticosteroids, probably modulated through pituitary and pineal hormonal output. In laboratory and domestic species which do not undergo such a synchronized molt, hair growth patterns are more random but still under hormonal influence. Many species have continuous hair growth. Strain and breed differences are also prevalent, a fact well represented by the differences in canine hair growth between poodles and other breeds. Profound differences in hair growth may also occur at times where sex steroid levels abruptly change (puberty, pregnancy, lactation). In man, hair growth also has some seasonal activity and is under similar endocrine control as other species. Cessation of hair growth is a common manifestation of hypothyroidism. The so-called sexual hair of the pubis, axilla, and beard is responsive to androgens.[223-225]

The importance of the hair growth cycle to cutaneous toxicology and percutaneous absorption is obvious, in that response to specific agents may be dependent upon the state of hair growth present. If possible, hair growth should be synchronized within animals of a study. In laboratory animals this can be accomplished by clipping all animals a day before treatment, a process which induces a new hair growth cycle in all animals. This is especially important in tumor initiation and promotion protocols.

For some contemporary interest, a common cosmetic problem in our society today is male-pattern hair loss. It has been noted that patients taking minoxidil (Rogaine®, UpJohn Co.) for hypertension developed a thickening of hair in select individuals.[225a] A 2 to 3% solution of minoxidil topically applied to male subjects for 1 year had cosmetically acceptable hair growth.[225b] Minoxidil can promote hair growth in specific conditions such as alopecia areata and androgenetic alopecia.[225c] Although the mechanism of this drug is unclear, it is known that for minoxidil to be effective living hair follicles capable of stimulation and hypertrophy must be present.[225c]

Each hair follicle (Figure 13) is usually associated with a capillary plexus, a collar of nerve fibers, an arrector pili muscle, and a sebaceous gland. All hairs, except for the sexual hairs found in the beard and pubic regions, have arrector pili muscles. This smooth muscle tends to be a fan-shaped structure and is forked in hooved animals. The broad end of the arrector pili (Figure 13) muscle originates in the elastic network of the superficial papillary dermis and inserts on the outer root sheath of the follicle immediately below the sebaceous gland. Since hair sits at an obtuse angle to the skin surface and the arrector pili is situated at the lower portion of the hair follicle, one can seen how its contraction will cause the hair to erect into a position vertical to the skin surface. This muscle is supplied by postganglionic adrenergic sympathetic nerve fibers.[222]

Bell has reported the ultrastructure of the arrector pili muscle in the rhesus monkey, *Macaca mulatta*.[226] The smooth muscle cells are packed with myofilaments which are grouped in fusiform densities. Organelles form a cap over the nucleus. Glycogen and microtubules are present. The plasmalemma of the arrector pili smooth muscle cells exhibits zones of electron opacity and lucency. The electron lucent zones contain superficial vesicles or pinocytotic vesicles and coated vesicles. Tight junctions are found between these cells and small unmyelinated axons may be observed. Elastin fibrils attach the arrector pili muscles to the superficial papillary layer and to the hair follicle sheath. The ultrastructure of the arrector pili muscle is similar to that of other smooth muscle cells.[226-228]

The sebaceous glands develop from the outer bulge of the developing hair follicle. In man these glands are found over the entire body except for the palms and soles. They can be found at mucocutaneous junctions not associated with hair follicles. They vary in number and in size between individuals of different sex, age, and ethnic origin. In man sebaceous glands are richly supplied with blood vessels and may contain pigment cells is some species (primates). Nerves have not been associated with sebaceous glands.[217,219]

Sebaceous glands contain cells in various stages of differentiation. Morphologically, the gland consists of a group of lobes or acini. Three types of acinar cells have been described ultrastructurally. The structure of these cells changes progressively from the periphery to the center of each gland. The peripheral cell lies on the basement membrane that surrounds the acinus of the sebaceous gland. The second cell type is the partially differentiated cell which is actively synthesizing lipid. The third cell type is the fully differentiated cell which is large and full of lipid droplets. It is at this stage that the cell will rupture and release its entire secretion (lipid) into the sebaceous duct and then onto the surface of the skin. This process is termed holocrine secretion. The function of sebum, a complex mixture of lipids, is that of acting as an antibacterial agent. Also, in hairy mammals it may act as a waterproofing agent and may prevent moisture loss from the human epidermis.[217,219,229,230] As noted by Strauss et al.,[230a] differences can occur in sebum composition among mammals.

There are two types of sweat glands in man, the eccrine and apocrine glands. The eccrine sweat glands are distributed over the entire body surface except for a few specific regions. They are numerous in the palms and soles. These glands develop independently of hair follicles and are situated in the deep dermis. They traverse the epidermis via a duct to open directly onto the surface of the skin. These glands are simple coiled tubular structures with four major components: the terminal secretory portion, the coiled duct, the straight duct, and the intraepidermal duct. Three types of cells have been described in the eccrine sweat gland secretory coil region. At the periphery of the secretory portion are the myoepithelial cells, which sit on the basement membrane. Clear cells have been observed with lipid inclusions. Intracellular canaliculi pass in between adjacent clear cells. These cells are responsible for producing the aqueous sweat. The last cell type is the dark cell which produces mucin. On the luminal surface of the dark cell, microvilli are located.[229-232] Eccrine sweat glands are innervated by postganglionic sympathetic nerve fibers. Blood vessels surround the secretory portion and coil around the ducts of the sweat glands. Eccrine sweat glands are found only in a few mammals with hair. When present, they are usually restricted to the digital pads or the snout.[230]

The apocrine sweat glands are found in specific areas of the body in man: the axilla, the areola, the pubis, the perianal region, eyelids, and external auditory meatus. In hairy mammals such as Bovidae, Ovidae, Equidae, Suidae, and most carnivores, they are usually found over the entire body surface (Figure 13). Apocrine glands develop just above the sebaceous glands in the developing hair follicle. All apocrine glands run parallel to the hair follicle and empty in the piliary canal. In some primates (Lemuridae) and sometimes in man, they may open directly onto the epidermal surface.[217,219]

Apocrine glands are large, simple coiled tubular structures. The name apocrine is inappropriate because it was once thought that the apical portion of the cell was lost during secretion; however, electron microscopy has shown that the secretion is actually a product of the cell and that no cytoplasm is lost during the secretory process. The secretory portion of this gland consists of cuboidal to columnar cells which rest on the myoepithelial cells or on the basement membrane. In areas of the basement membrane which are devoid of myoepithelial cells, the plasmalemma of the secretory cell is highly convoluted. This folded area could provide for an exchange of metabolites. The apical portion of this cell has microvilli on its surface. Organelles are prominent, especially the Golgi apparatus which is usually located on the apical side of the nucleus. Mitochondria are located both on the basal side and apical side of the nucleus. They are greatly enlarged and have a homogeneous granular matrix. Large, dense granules are located on the apical side of the Golgi region and vary in size, density, and homogeneity. Rough endoplasmic reticulum and tonofilaments are abundant. Adjacent secretory cells are connected by junctional complexes.[229]

The ultrastructure of the human apocrine duct has been described by Hashimoto et al.[233] The apocrine duct is composed of two to three concentric layers of cells. The inner layer faces the lumen and contains microvilli on its surface. A clear zone of cytoplasm is located just beneath the microvilli. Just apical to the clear zone occurs a filamentous zone containing bundles of tonofilaments, vesicles, mitochondria, and ribosomes. Between the periluminal filamentous zone and the nucleus, another zone exists. The cell membrane on the lateral side of these inner cells is convoluted and sometimes desmosomes are present. The middle layer, if present, has the characteristics of the inner layer except for the presence of microvilli. The cells of the basal layer are less dense, have fewer tonofilaments, but have more glycogen. They are connected to adjacent basal cells by a convoluted plasmalemma with desmosomes. Some investigators suggest that the apocrine sweat duct consists of two epithelial cell layers: the luminal cell and the peripheral cell or the basal cell and the inner cell.[234]

The secretory product of the apocrine glands is a viscous substance. Studies have demonstrated proteins, sugars, pheromones, and ammonia in the apocrine secretion. The secretion in the human is odorless, but surface bacterial degradation produces odoriferous products. The significance of the glands in animals is related to communications between species probably as a sex attractant or as a territorial marker.[217,219]

XIII. VASCULAR SUPPLY

The function of small vessels in the dermis is that of thermal regulation and nutrition.[217,219,231,232,235] The blood vessels of the skin are limited to the dermis. The largest arteries (rete cutaneum) form a network in the subcutaneous layer. Branches arise from this and pass to deeper areas to supply the adipose tissue and to more superficial areas to supply the hair follicles. The superficial arteries traverse the reticular layer of the dermis and give off smaller branches to hair follicles, sebaceous glands, and sweat glands. Between the reticular and papillary layers they form another network composed of smaller arteries, the rete subpapillare (horizontal plexus). Arterioles from this rete subpapillare will supply capillary loops in the papillae. Capillary beds are commonly found immediately under the epidermis (Figure 15), in the matrix of hair follicles, and around sweat and sebaceous glands. Capillary loops in the dermal papillae can be referred to as a subepidermal plexus, papillary plexus, or papillary loops. In newborns these capillary loops are not found at the first day after birth but can be observed by the second. This gives an indication of their rapid rate of development. At 3 months, newborns have established a normal adult pattern of capillary anastomoses. In specific regions of the body alternative channels which allow blood to be passed from the arteriole to the venule exist. This

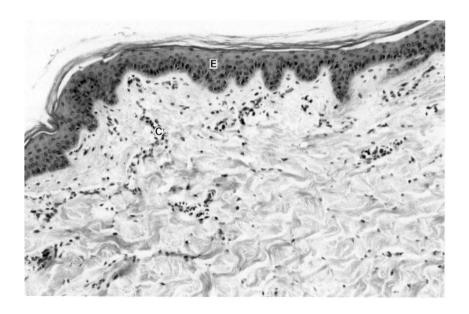

Figure 15. Skin from the back of a pig. Number of nucleated epidermal cells (E) closely resembles human skin compared to other species. Note capillaries (C) in the superficial papillary dermis. H & E, magnification × 155.

arrangement of vessels is commonly known as an arteriovenous anastomoses or shunt.[236] Large arteriovenous anastomoses are found in areas of skin such as the fingertips, toes, nose, and lips. These vascular structures, which are highly developed and surrounded by organized connective tissue, are called a glomus. The glomus functions in regulation of body temperature and peripheral blood circulation. The arterial segment of the glomus may be associated with nonmyelinated nerve fibers.[237] Extensive arteriovenous anastomoses have also been described in different species of seals.[238,239]

Several ultrastructural studies have been performed on the human dermal micro-circulation. Papillary dermal vessels in the human forearm which arise from the subpapillary plexus are composed of terminal arterioles, arterial and venous capillaries, and postcapillary venules.[240] Ultrastructural studies of the capillary loop in man showed it to be composed of an ascending venous limb. Each papilla is supplied by only one capillary loop.[241] Controversy exists over the presence of fenestrated capillaries. Bravermann and Yen[241] reported an absence of fenestrations while Takada and Hattori[235] reported that fenestrated capillaries did exist in the dermal papilla. Imayama used scanning and transmission electron microscopy to study rat skin and found that capillaries extremely close to the epidermis were fenestrated but those that were more distant from the epidermis were not fenestrated.[242] The presence of fenestrated capillaries has implications to the transport of substances into and out of the cutaneous circulation; the former in relationship to the absorption of large molecular weight substances, and the latter in respect to cutaneous localization of inflammatory mediators.

Age changes in capillary loops have been reported in the human scalp. A reduction in capillaries occurs with the aging and balding of scalp.[217,219] Similar studies on unexposed skin in older women showed that there was collapsing, disorganization, and total disappearance of some capillaries.[207]

The average blood flow to the skin has been reported to vary from 0.5 to 1.0 up to 100 ml/min/100 g, depending upon species, body site, technique of measurement, and temperature. Common techniques used to measure cutaneous blood flow include thermal dilution, radiolabeled isotope washout studies, photopulse plethysmography, ultrasonic Doppler angiography, and laser Doppler velocimetry. The local and systemic regulation of cutaneous blood flow is far beyond the scope of this review. Since cutaneous blood flow serves primarily as a thermoregulatory function, it would be expected that skin temperature would dramatically increase blood flow. An increase of 10- to 15-fold has been repeatedly demonstrated with temperature increases up to approximately 43 to 44°C, above which prolonged heating will decrease flow and result in vascular damage. Normal resting blood flow in humans is approximately 3 to 10 ml/min/100 g. Arteriolar capillary blood pressure ranges from 30 to 45 mmHg. Regional, species, and environmental variations in cutaneous blood flow should be taken into consideration when assessing the percutaneous absorption of certain compounds and the response of skin to cutaneous toxicants.[243-246]

XIV. NERVE SUPPLY

The nerve supply to the skin (Figure 16) is of extreme importance in order to sense certain stimuli from the environment. There are two major dermal nerve plexuses: one is superficial and lies just beneath the epidermis and the other is deeper with larger and coarser fibers. However, it is not always possible to demonstrate these two separate plexuses. Cutaneous nerves usually run with vessels. The cutaneous neural networks are so complex that it is almost impossible to trace a fiber with all of its ramifications. Nerve fibers may terminate in the dermal papilla and loop around themselves and then return to the superficial plexus. This is a common occurrence in hairy skin, but in glabrous skin special cutaneous receptors are usually present.[116,247]

Several attempts have been made to establish a method of classification of nerve receptors. Physiologists assign a classification based on their function.[248] Morphologists classify receptors based on their structure. Morphologically, cutaneous receptors can be classified into corpuscular endings and free nerve endings.[137]

The corpuscular receptors in skin are as follows: Merkel corpuscles, Pacinian corpuscles, Ruffini corpuscles, Meissner corpuscles, and Krause end-bulbs.[249] Free nerve endings may be found in the epidermis and dermis.[116,117,137,247] Free nerve endings can encircle hair follicles,[116,222,250] and in tactile hairs can enter the external root sheath.[204] Merkel corpuscles are found in the epidermis (see Section VII).

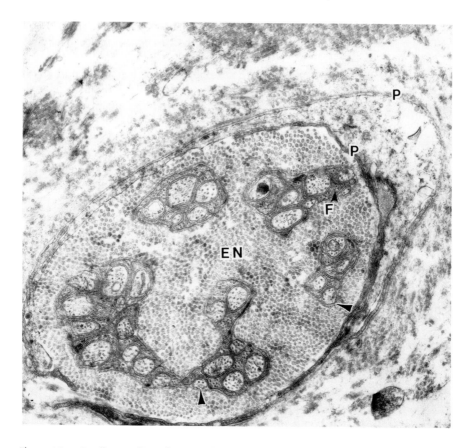

Figure 16. Small unmyelinated nerve with double perineurial sheath (P). Several unmyelinated axons and bare axons (arrows) are present. Endoneurial collagen (EN) occupies a large area. Cross-section of filaments (F) appearing as dark granules are present within the Schwann cell cytoplasm of a 7-week-old pig. Magnification × 14,750.

Pacinian corpuscles are found in the deeper layers of the dermis primarily in glabrous skin, i.e., fingers. These corpuscles are quite large, ranging from 0.5 to 2 mm in length.[249] Pacinian corpuscles may also be present in the mesentery of the cat, in the pancreas, lymph nodes, posterior abdominal wall, and in other places. These corpuscles are usually associated with blood vessels. The Pacinian corpuscle consists of a nerve fiber, an inner core, a subcapsular space, and a capsule. The afferent myelinated segments lose their myelin sheaths just inside the corpuscle. The middle segment of the axon is inside the inner core and contains mitochondria and neurofil- aments. The terminal portion of the axon ends in an expanded process containing vesicles and mitochondria. The inner core of the Pacinian corpuscle is made up of modified Schwann cells forming lamellae. The lamellae are connected by desmo- some-like structures. The lamellae may consist of 60 to 80 concentric rings and surrounding these is the subcapsular space. This space can vary in size, depending on the size of the corpuscle itself, and contains fibrocytes and collagen fibers. The capsule, which is the continuation of the perineurium, surrounds the subcapsular space. The capsule can possess several layers and depends on the location of the Pacinian corpuscle. The closer the corpuscle is to the surface of the skin, the fewer the number of layers there are in the capsule. The Pacinian corpuscle is a mechanoreceptor that is sensitive to mechanical displacement, i.e., deep pressure and vibrations.[116,117,217,219,249] Santini believes that the desmosome-like lamellar-axonal junction of the innermost lamella of the Pacinian corpuscle is the mechano-electrical transducer termed the "receptripse".[251]

Another type of encapsulated nerve receptor is the Ruffini corpuscle. These corpuscles lie in the subpapillary dermis and deep dermis and deep dermis of hairy and glabrous skin. They are fusiform structures and are thought to be mechanical receptors responding to movement in the surrounding collagen. A large myelinated fiber enters the corpuscle, loses its myelin sheath, and branches into smaller fibers which ramify among the collagen bundles. These collagen bundles make up the core of the receptor. The Ruffini corpuscle is surrounded by a thin capsule which is continuous with the endoneurial sheath.[116,249,252] There is much debate whether or not the Ruffini corpuscle actually exists. Some say it is an artifact due to metallic impregnation or that the structure seen is just an entanglement of nerve fibers.[116,252]

Meissner's corpuscles are located in the papillary layer of the dermis of glabrous skin. These corpuscles were first described in 1852 and are also known as Wagner-Meissner corpuscles, Dogiel's end-bulb, or as Ruffini's end-bulbs in older literature.[117] Each corpuscle is oval and approximately 100 μm long and 50 μm in diameter. Large myelinated nerve fibers (two to nine) enter the corpuscle at its base and along the sides, lose their myelin sheath, and then branch. The axoplasm is loaded with mitochondria and vesicles. These nonmyelinated axons spiral and form a bulbous ending among the flattened Schwann cells and collagen fibers. Surrounding this arrangement of expanded nerve endings and flattened cells is a connective tissue capsule which is continuous with the endoneurial sheaths of the afferent nerve fibers. The capsule is bound by elastic fibrils to the basal projections of the epidermis. The Meissner's corpuscle is a mechanoreceptor which responds to touch.[116,117,137,219,249]

The Krause end-bulb (mucocutaneous corpuscles), first described in 1859, is located in the transitional zone between the skin and mucous membrane.[116] The corpuscle consists of two to six unmyelinated fibers which arise from a myelinated afferent fiber. These unmyelinated fibers repeatedly branch within the corpuscle. A poorly developed capsule may surround the nonmyelinated endings. The precise structure and function of the Krause bulb is not known. Many investigators are doubtful as to the existence of this structure.

Pacinian corpuscles and Meissner corpuscles decrease in number with age.[116,253] The corpuscles that remain are usually larger. The Krause end-bulbs are thought by some investigators to be a stage in the growth or degeneration cycle of peripheral nerves.[116] Some investigators believe that Meissner's corpuscles, Pacinian corpuscles, mucocutaneous end organs, the hair follicle network, and free nerve endings are the only type of sensory receptors that are present in the skin of mammals. All other structures that have been described are considered artifacts due to shrinking, compression, section thickness, or plane of sectioning.[217,219,254]

Free nerve endings in the skin have been described by Langerhans, Ranvier, and Weddel.[141,255,256] They have been described in the snout skin of the opposum, the snout of the pig and other mammals, and in the human digital skin.[112,257-261] Intraepithelial axons have been identified in man and cat in the lower respiratory tract.[262]

Free nerve endings that penetrate the epidermis of the snout occur singly or in bundles in nocturnal animals. This may aid the animal in orientation as it moves.[259] Intraepidermal nerve fibers of snout epidermis of the pig course irregularly through the stratum basale and stratum spinosum. These fibers may divide and terminate in the upper stratum corneum layers.[263] The interfollicular epidermis of hairy skin consists of nerve fibers penetrating the epidermis in most laboratory animals except for the rat and mouse.[264] The myelinated nerve approaches the epidermis and loses its Schwann cell and continues to the surface surrounded by the basal lamina, thereby not directly exposing the axon to the intercellular space. Intraepidermal nerve endings continue to grow within the epidermis and then degenerate at the stratum corneum level.[112,137,264]

Intraepidermal nerve fibers are present in human skin and have been demonstrated in the forearm.[38,264,265] Bourlond reports that it is uncommon to find intraepidermal nerve fibers in human adult skin.[266] Chouchkov reports that free nerve endings that are present in human skin lie between epidermal cells that are associated with Merkel cell-neurite complexes.[267] Breathnach has observed intraepidermal nerves in the basal layer from the digital skin of 21-week-old fetuses.[265] He believes that their presence in fetal skin could reflect an "overspill" from the dermis during development.

Cauna reported that there are at least two morphologically distinct kinds of free nerve endings, one which is associated with hair follicles and the other which is in the subepidermal nerve network of hairy skin.[268] Cauna gives them the term "penicillate endings" because they are arranged in a tuft-like manner. These penicillate endings are derived from nonmyelinated nerve fibers. They are associated with the Schwann cell and may be surrounded by multiple layers of basal lamina, which is then surrounded by collagen fibers. The axons of these endings may vary in diameter and the axoplasm contains few organelles. The termination of these endings seems to be in the subepidermal regions and intraepidermal axons are wrapped by the keratinocyte membrane. These epidermal endings are very rare in the adult. The exact function of the penicillate nerve endings still remains to be determined. Their extensive neural surface area in the subepidermis suggests that they may be a type of receptor. A more recent study supports this hypothesis and suggests that it could be a receptor organ for itch. Patients suffering from urticaria with intensive itching, and normal skin where itching was induced, were investigated by electron microscopy.[269] Ultrastructural studies of these intensely itchy areas showed morphological changes and an abundance of glycogen in some of the penicillate endings. These findings suggest that one of the functions of penicillate nerve endings is that of mediating the sensory modality of itching.[269]

Three types of free nerve endings have been described in human digital skin based on their morphology. The open or coarse endings are incompletely enfolded by the Schwann sheath. One to four axons are present in the open ending and mitochondria, glycogen, and microvesicles are present in the axoplasm. The second type of ending is the beaded or varicose ending which is spindle shaped and also contains microvesicles, mitochondria, and glycogen particles. The third type is the plain or fine ending which has an extensive Schwann sheath with numerous cytoplasmic processes and few organelles present. These plain endings are sometimes found in the epidermis but usually follow capillary loops. This study showed that there are many types of free nerve endings in digital skin. The significance of the three morphologically distinct nerve endings is unknown.[261]

Aging causes the density of intraepidermal endings in the rat to decrease, probably due to the expansion of the skin.[253,264]

The nerve network consisting of unmyelinated nerve fibers around hair follicles is very complex and usually forms a circular plexus.[117,247,249,270] The hair follicle is supplied by several different stem fibers from the dermal plexus. Some fibers may extend up to the dermal papilla around the hair orifice.[271] Some large follicles may have up to 20 to 30 stem fibers and smaller follicles may have from 6 to 10 stem fibers. These fibers are myelinated until they reach the opening of the sebaceous gland where they lose their myelinated sheath and separate to form a collar or basketwork around the follicle. This collar is a double structure consisting of an outer layer that runs in a circular pattern and an inner layer that runs in a longitudinal pattern. Some investigators report that there are two types of nerve plexuses around the hair follicle; a circular plexus forming "palisades" with its branches and one that has only a palisade arrangement of nerve fibers and no circular plexus. The latter type is common in large hairs.[116,254] In the growth stage of hair (anagen) the nerve plexus becomes distended but during the resting stage (telogen), the hair follicle becomes involuted and the nerve network just collapses around the base.[250]

The autonomic nervous system supplies the skin and its appendages. The parasympathetic part of the autonomic nervous system does not venture into the skin. The sympathetic part is responsible for the innervation of the sweat glands, the smooth muscle of blood vessels, and arrector pili muscles.[116,247]

XV. SPECIES COMPARISONS

In general, the basic architecture of the integument is similar in all mammals. However, differences exist in the thickness of the epidermis (Table 3)[301] and dermis between species and within the same species in various regions of the body. Many differences exists in domestic animals in the arrangement of hair follicles and density of hair. Pig skin hair density, like man, is sparse (Figure 17) compared to rodent skin (Figure 18). The horse (Figure 19) and cow have a single hair follicle evenly

Table 3
Comparative Thickness of the Epidermis from the Ventral Abdomen in Nine Species[301]

Species	Epidermis (μm)	Stratum corneum (μm)	Epidermal cell layers (numbers)
Mouse	9.7 ± 2.3	3.0 ± 0.3	1.8 ± 0.3
Rat	11.6 ± 1.0	4.6 ± 0.6	1.4 ± 0.2
Rabbit	15.1 ± 1.4	4.9 ± 0.8	1.5 ± 0.1
Monkey	17.1 ± 2.2	5.3 ± 0.4	2.1 ± 0.1
Dog	22.5 ± 2.4	8.6 ± 1.9	2.3 ± 0.1
Cat	23.4 ± 10.2	4.3 ± 1.0	2.1 ± 0.8
Cow	27.4 ± 2.6	8.1 ± 0.6	2.4 ± 0.1
Horse	29.1 ± 5.0	7.0 ± 1.1	2.9 ± 0.4
Pig	46.8 ± 2.0	14.9 ± 1.9	4.5 ± 0.4

Note: Paraffin sections stained with H & E; n = 6, mean ± S.E. Comparable human data are not readily available.

distributed. In horses the average thickness of skin is 3.8 mm.[272] The superficial layer of the dermis is thick and penetrates deep into the skin. Usually each hair follicle in the horse is associated with two sebaceous glands and the branched sebaceous glands are present near body openings. Also, apocrine sweat glands are present in the horse.[272]

Comparative studies have been done between different breeds of cattle in which there are differences in thickness between certain breeds and sexes. Also, differences exists in the kind of pigment granules.[273] Cattle primary hair follicles (Figure 20) appear to be homologous to other mammals. Follicle density from animals in the same environment and age varied with body size.[274] Studies using frozen sections were performed on cattle skin and the epidermis and dermis were both found to be approximately 30 μm thick. The corneum contained 30 cell layers, while the nucleated epidermis was composed of only four.[275] Similar studies were done in sheep and the stratum corneum thickness was similar to cattle; however, the epidermis was only half as thick.[276]

The average epidermal thickness in dogs in hairy skin areas is 30 to 40 μm, while the thinnest in mongrel dogs is 26.5 μm.[277] The stratum corneum of canine foot pads is greatly thickened (Figure 21). This is similar to the thickened stratum corneum of foot pads of other species, and of human palms and soles. Dogs have compound follicles consisting of a single primary hair and a group of smaller secondary hairs. Up to 15 hairs can emerge from a single opening in the skin.[278] Follicles are in groups of three but can also occur in groups of two or four in the upper dermis.[274,277] In the deeper dermis follicle groups were massed together, appearing as if the hairs were in a single group.[277]

The domestic cat epidermis is thin, ranging from 12 to 45 μm with 25 μm being the average epidermal thickness in hairy areas. The cat follicle arrangement consists of a single large primary or guard hair surrounded by clusters of either two, three,

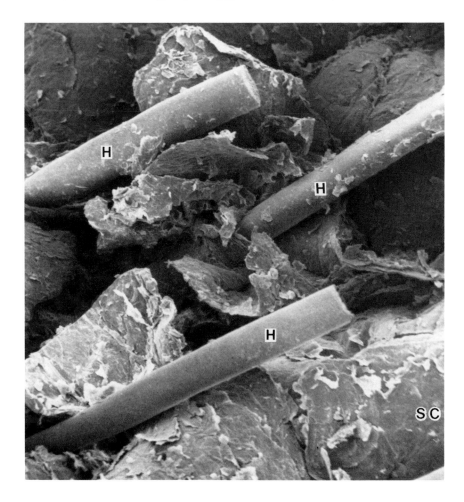

Figure 17. Scanning electron micrograph of pig skin showing low density of hair. Note the typical porcine hair follicle arrangement. Three hairs (H) are seen emerging from the stratum corneum (SC) layer of the epidermis. Note flaky arrangement of stratum corneum. Magnification × 75.

four, and five compound hair follicles (Figure 22). Within each compound follicle there are 3 primary hairs and 6 to 12 secondary hairs.[279]

Mouse epidermis is very thin compared to the domestic species (Figure 23). The number of viable nucleated cells in the epidermis varies from one to two cells in thickness. The stratum granulosum is usually devoid of nuclei and the stratum corneum consists of five to eight layers and is prominent. A true stratum spinosum as seen in other domestic animals and humans is absent.[280,281] Rat skin is similar to that of the mouse in that the epidermis consists of one or two nucleated layers (Figure 24). Ten to 15 hairs may be found in a hair follicle group.

Figure 18.   Scanning electron micrograph of mouse skin. Note the increased density of hairs (H) compared to Figure 11. Magnification × 430.

Guinea pig skin is commonly used to study immunology, inflammatory reactions, and wound healing because the physiological characteristics of its skin are felt to be similar to that of man. Hair follicles occur in clusters, with three to seven small follicles surrounding a large follicle.[282] The albino rabbit is considered to be the animal of choice to be used in Draize-type testing.[283] Rabbit epidermis is very thin and consists of one to two nucleated cell layers (Figure 25).

Clipping and shaving may affect assessment of skin thickness. Studies done on cattle skin demonstrated that clipping and shaving resulted in the removal of cell layers and loss of intercellular lipid from the stratum corneum.[284] A comparative study on pigs and rats was performed for hair removal both by clipping and chemical depilation.[285] Chemical depilation in rats showed no changes, while in pigs it caused partial or complete removal of the stratum corneum. Clipping proved to be the best

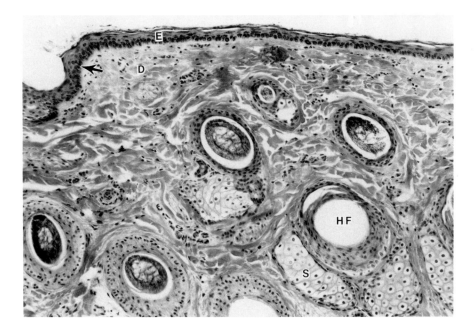

Figure 19. Light micrograph of horse back skin. Note epidermis (E), dermis (D), epidermal-dermal junction (arrow), cross-section of single hair follicle (HF), and sebaceous gland (S). H & E, magnification × 120.

method and least damaging, while chemical depilation in pigs caused partial or complete removal of the stratum corneum 2 h after treatment.[285]

The molecular structure of the fibrous proteins of hair and nail are thought to be similar.[286] Some investigators report that the bovine hoof is similar to the human nail, and that the bovine hoof may be an excellent model for studying keratinization.[286,287] However, disparate results have been reported in the amino acid analyses of the fibrous proteins in the bovine hoof which makes a comparison to other species difficult. The differences have been attributed to the different techniques used to isolate keratokyalin.[288] What is known is that the claws of dogs and cats, horns of ruminant species, hoofs of horses, ruminants and pigs, and nails of humans, all consist of hard keratin.[278] For more information on the biochemical composition of some of these keratinized structures, see Bernstein[288] and Baden and Kubilus.[286,287]

Numerous authors have alluded to the similarity of pig and human skin based on light microscopic observations (compare Figures 2 and 15). The morphology of fetal, newborn, and adult porcine integument (Figure 15) has been fully described by light microscopy.[289-292] Also, ultrastructural investigations on the postnatal development of porcine skin has shown it to be similar to that of man.[8,34] As previously presented, swine skin resembles human skin in having a sparse hair coat, a relatively thick epidermis, similar epidermal turnover kinetics, lipid composition, and carbohydrate

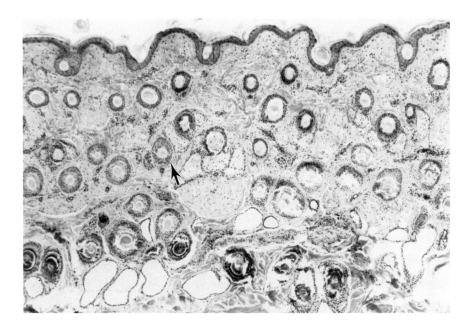

Figure 20.   Cow skin from the back of the animal. Note numerous primary hair follicles (arrow). H & E, magnification × 48.

biochemistry. The dermal microcirculation is reported to be similar to man, as is the arrangement of dermal collagen and elastic fibers.[293-296] Enzyme histochemistry is also similar.[297] Reported differences in the pig include a unique interfollicular muscle that spans the triad of hair follicles (Figure 26), the presence of only apocrine sweat glands (body surface), and a slightly thicker stratum corneum.[297-299]

XVI. SUMMARY

Skin is a complex and dynamic organ from both a morphological and a functional perspective. Toxicologists have generally focused on its barrier properties, which primarily reside in the stratum corneum. However, as can now be appreciated, the formation of this heterogeneous barrier is, in itself, a complex process whose final structure is the result of the successful process of epidermal differentiation and keratinization. Any toxicological insult or disease process which interferes with any stage of this process can result in altered barrier function. If percutaneous absorption is an investigator's focus, the stratum corneum is often assumed to be the rate-limiting barrier to topical compound penetration. Relatively simple experimental protocols (diffusion chambers) may be used since the corneum barrier can be isolated intact. However, if an *in vitro* system is to be used to investigate other aspects of cutaneous function, then viable skin must be utilized and all cell types should be accounted for. This is difficult and thus *in vivo* or more complicated animal models must be used.

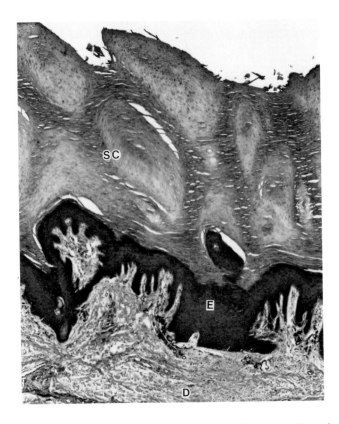

Figure 21.   Dog foot pad. Note the epidermis (E), dermis (D), and extremely thick stratum corneum (SC). H & E, magnification × 50.

Most of cutaneous toxicology has focused on basal epidermal cell function and structure. It is obvious that most cutaneous toxicants have their major site of action and effects on epidermal cells and their products. However, for specific compounds, other cell types may be involved. Attention might be focused on melanocytes for pigmentation dysfunction, on Langerhans' cells for cutaneous immunotoxicity, and possibly even on Merkel cells for cutaneous neurotoxicity. The structural integration of epidermis and dermis must also be considered.

Species differences in cutaneous structure and function must be taken into consideration before toxicological studies are designed. The most apparent difference is in thickness and complexity of the epidermis and stratum corneum for percutaneous absorption studies. Differences in biochemistry, microcirculation, hair density and structure, adnexal appendages, and distribution of epidermal cell types may also be important considerations. Finally, one must know the normal morphology of skin in the species selected for study before assessment of cutaneous toxicity can be properly interpreted.

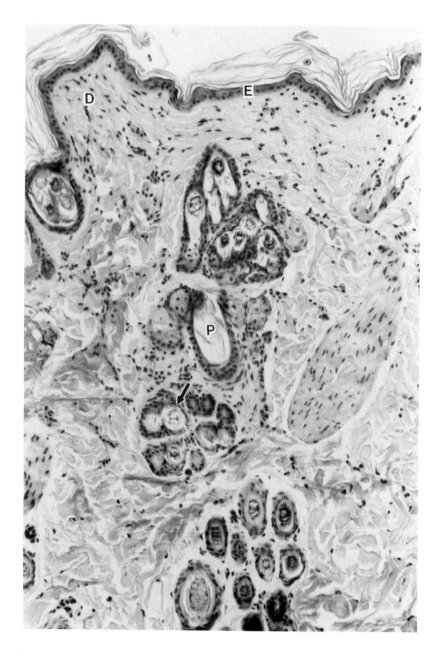

Figure 22.   Cat back skin. Epidermis (E), dermis (D), primary hair follicle (P) surrounded by compound hair follicles (arrow). H & E, magnification × 165.

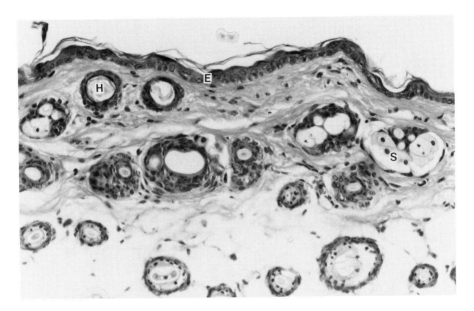

Figure 23. Mouse epidermis (E) showing a relatively small number of nucleated cells. Hair follicles (H) and sebaceous nuclei (S) can be seen in the dermis. H & E, magnification × 250.

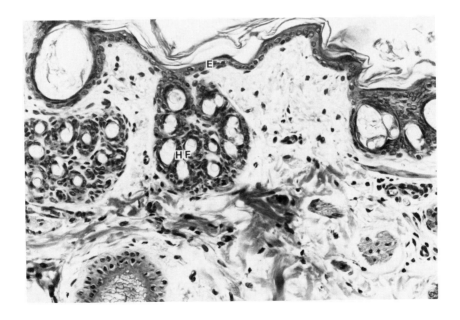

Figure 24. Skin from the back of a rat. Epidermis (E) is very thin and numerous hair follicles (HF) are seen. H & E, magnification × 114.

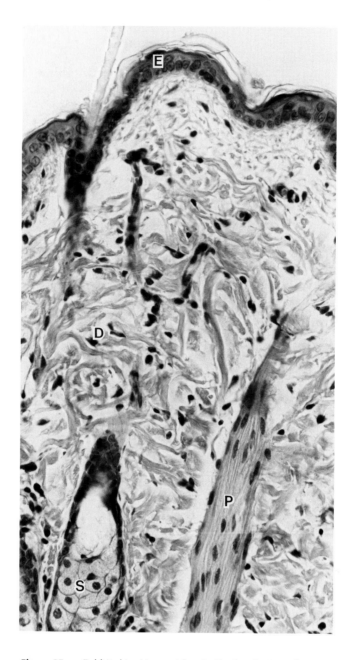

Figure 25. Rabbit skin. Note epidermis (E), dermis (D), sebaceous gland (S), arrector pili muscle (P), and hair (H). H & E, magnification × 330.

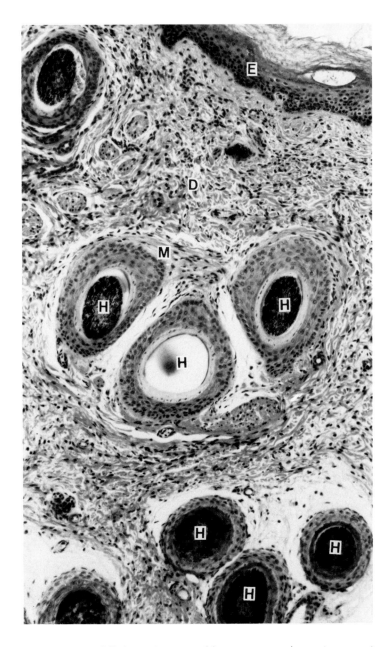

Figure 26. Hair follicles (H) in groups of three are commonly seen in young pigs. Note epidermis (E), interfollicular muscle (M), and dermis (D). H & E, magnification × 165.

REFERENCES

1. W.F. Lever and G. Schaumburg-Lever, *Histopathology of the Skin*. J.B. Lippincott, Philadelphia, 1 (1983).
2. G.H. Muller and R.W. Kirk, *Small Animal Dermatology*, 2nd ed. W.B. Saunders, Philadelphia, 163 (1976).
3. A. McQueen and E. Heyderman, *Histochemistry in Pathology*. M. Filipe and B.D. Lake, Eds., Churchill Livingstone, New York, 114 (1983).
4. J.K. Robinson, *Fundamentals of Skin Biopsy*. Year Book Medical Publishers, Chicago, 19 (1986).
5. L.G. Luna, *Manual of Histologic Staining Methods of the Armed Forces Institute of Pathology*, 3rd ed. McGraw-Hill, New York (1968).
6. G.L. Humason, *Animal Tissue Techniques*, 4th ed. W.H. Freeman, San Francisco (1979).
7. M.J. Karnovsky, *J. Cell Biol.* 27, 137A (1965).
8. N.A. Monteiro-Riviere, *Swine in Biomedical Research*. Plenum Press, New York, 641 (1986).
9. N.A. Monteiro-Riviere and T.O. Manning, *Proc. 45th Annu. Meet. Electron Microscopy Soc. Am.* G.W. Bailey, Ed. San Francisco Press, San Francisco, 948 (1987).
10. A.F. Hayward and A.P. Kent, *J. Ultrastruct. Res.* 84, 182 (1983).
11. K. Holbrook and G. Odland, *J. Invest. Dermatol.* 65, 16 (1975).
12. G.F. Murphy, R.S. Shepard, B.S. Paul, A. Menkes, R.R. Anderson, and J.A. Parrish, *Lab. Invest.* 49, 680 (1983).
13. E. Wolff-Schreiner, *Int. J. Dermatol.* 16, 77 (1977).
14. R.F. Vogt, Jr., A.M. Dannenberg, Jr., B.H. Schofield, N.A. Hynes, and B. Papirmeister, *Fund. Appl. Toxicol.* 4, S71 (1984).
15. A.R. Spurr, *J. Ultrastruct. Res.* 26, 31 (1969).
16. E.M. McDowell and B.F. Trump, *Arch. Pathol. Lab. Med.* 100, 405 (1976).
17. H. Troyer, *Principles and Techniques of Histochemistry*. Little, Brown, Boston, 25 (1980).
18. S.W. Thompson and R.D. Hunt, *Selected Histochemical and Histopathological Methods*. Charles C Thomas, Springfield, IL, 51 (1966).
19. R.E. Grissom, N.A. Monteiro-Riviere, and F.E. Guthrie, *Toxicol. Lett.* 36, 251 (1987).
20. L. Juhlin and W.B. Shelley, *Acta Dermatovener* 57, 289 (1977).
21. K.W. Baker and J.E. Habowsky, *J. Invest. Dermatol.* 80, 104 (1983).
22. J.P. Baumberger, *J. Natl. Cancer Inst.* 2, 413 (1942).
23. Z. Felsher, *J. Invest. Dermatol.* 8, 35 (1947).
24. E. Epstein, N.H. Munderloh, and K. Fukuyama, *J. Invest. Dermatol.* 73, 207 (1979).
25. R.B. Stoughton, 17th Annu. Meet. Soc. Investigative Dermatol. Chicago, IL, 395 (1956).
26. Y. Kitano and N. Okada, *Br. J. Dermatol.* 108, 55, (1983).
27. J.M. Einbinder, R.A. Walzer, and I. Mandl, *J. Invest. Dermatol.* 46, 492 (1966).
28. R.A. Mufson, L.M. DeYoung, and R.K. Boutwell, *J. Invest. Dermatol* 69, 547 (1977).
29. K.A. Holbrook and L.T. Smith, in *Morphogenesis and Malformations of the Skin. Birth Defects: Original Articles Series*, Vol. 17. R.J. Blandau, Ed. Alan R. Liss, New York, 9 (1981).
30. J. Langman, *Medical Embryology*, 3rd ed. 401 (1975).
31. K. Moore, *The Developing Human*, 2nd ed. W.B. Saunders, Philadelphia, 376 (1977).
32. A.S. Breathnach, *J. Invest. Dermatol.* 57, 133 (1971).
33. A.S. Breathnach and L. Wyllie, *J. Invest. Dermatol.* 44, 51 (1965).
34. N.A. Monteiro-Riviere and M.W. Stromberg, *Anat. Histol. Embryol.* 14, 97 (1985).
35. R.M. Lavker and T.T. Sun, *Science* 215, 1239 (1982).
36. R.M. Lavker and T.T. Sun, *J. Invest. Dermatol.* 81, 1215, (1983).
37. G.D. Weinstein, *Swine in Biomedical Research*. L.K. Bustad, R.O. McClellan, and M.P. Burns, Eds. Pacific Northwest Laboratory, Richland, WA, (1966).
38. A.S. Breathnach, *An Atlas of the Ultrastructure of Human Skin: Development, Differentiation and Post-Natal Features*. J.A. Churchill, London (1971).
39. A. Klein-Szanto, *J. Cutaneous Pathol.* 4, 275 (1977).
40. A.N. Raick, *Cancer Res.* 33, 269 (1973).

41. T.S. Argyris, *CRC Crit. Rev. Toxicol.* 14, 211 (1985).
42. A.J.P. Klein-Szanto, Morphological Evaluation of Tumor Promoter Effects on Mammalian Skin, in *Mechanisms of Tumor Promotion, Vol. II. Tumor Promotion and Skin Carcinogenesis.* T.J. Slaga, Ed. CRC Press, Boca Raton, FL, (1984).
43. T.J. Slaga, *Environ. Health Perspect.* 50, 3 (1983).
44. A.J.P. Klein-Szanto, S.K. Major, and T.J. Slaga, *Carcinogenesis* 1, 399 (1980).
45. A.J.P. Klein-Szanto, P. Nettesheim, A. Pine, and D. Martin, *Am. J. Pathol.* 103, 263 (1981).
46. A.J.P. Klein-Szanto, P. Nettesheim, and G. Saccomanno, *Cancer* 50, 107 (1982).
47. C. Selby, *J. Invest. Dermatol.* 29, 131 (1957).
48. K. Wolff and E. Wolff-Schreiner, *J. Invest. Dermatol.* 67, 39 (1976).
49. C.R. Leeson and T.S. Leeson, *Histology,* 3rd ed. W.B. Saunders, Philadelphia (1976).
50. K.A. Holbrook and G.F. Odland, *J. Invest. Dermatol.* 62, 415 (1974).
51. P. Parakkal and N. Alexander, *Keratinization — A Survey of Vertebrate Epithelia.* Academic Press, New York (1972).
52. I. Brody, *Int. J. Dermatol.* 16, 245 (1977).
53. N.A. Monteiro-Riviere, Y.C. Hwang, and M.W. Stromberg, *Am. J. Chin. Med.* 9, 155 (1981).
54. I. Blank, *J. Invest. Dermatol.* 18, 433 (1952).
55. A. Kligman, The Biology of the Stratum Corneum, in *The Epidermis.* W. Montagna and W. Lobitz, Eds. Academic Press, New York (1964).
56. F. Malkinson, in *The Epidermis.* W. Montagna and W. Lobitz, Eds. Academic Press, New York (1964).
57. D.N. Menton and A. Z. Eisen, *J. Ultrastruct. Res.* 35, 247 (1971).
58. T.C. Mackenzie, *J. Invest. Dermatol.* 65, 45 (1975).
59. D. Menton, *J. Invest. Dermatol.* 66, 283 (1976).
60. D. Menton, *Am. J. Anat.* 145, 1 (1976).
61. N. Hadley, *Am. Sci.* 68, 546 (1980).
62. A. Jarrett, *The Physiology and Pathophysiology of the Skin.* Vol. VI. Academic Press, New York, 2111 (1980).
63. C.S. Potten, *Cell Tissue Kinet.* 7, 77 (1974).
64. C.S. Potten, *Int. Rev. Cytol.* 69, 271 (1981).
65. I. Brody, *J. Ultrastruc. Res.* 2, 482 (1959).
66. I. Brody, *J. Cutaneous Pathol.* 6, 333 (1979).
67. G. Matoltsy and P. Parakkal, Keratinization, in *Ultrastructure of Normal and Abnormal Skin.* A.S. Zelickson, Ed. Lea & Febiger, Philadelphia (1967).
68. P.M. Elias, *J. Invest. Dermatol.* 80, 44s (1983).
69. S. Bresler, *Introduction to Molecular Biology.* Academic Press, New York (1971).
70. G. Rogers, *Ann. NY Acad. Sci.* 83, 408 (1959).
71. J.E. Fraki, G.S. Lazarus, and V.K. Hopsu-Havu, in *Biochemistry and Physiology of the Skin,* Vol. I., L.G. Goldsmith, Ed. Oxford University Press, New York, 338 (1983).
72. N. Toki and T. Yamura, *J. Invest. Dermatol.* 73, 297 (1979).
73. N. Toki and T. Yamura, *Arch. Dermatol. Res.* 267, 303 (1980).
74. R.K. Freinkel, in *Biochemistry and Physiology of the Skin.* Vol. 1. L.G. Goldsmith, Ed. Oxford University Press, New York, 328 (1983).
75. T.A. Johnson and R. Fusaro, *Adv. Metabol. Disord.* 6, 1 (1972).
76. R.K. Freinkel, *J. Invest. Dermatol.* 34, 37 (1960).
77. J.H. Herndon and J.S. Maguire, *Arch. Biochem. Biophys.* 119, 583 (1967).
78. H.J. Yardley and G. Godfrey, *J. Invest. Dermatol.* 43, 51 (1964).
79. G.J. Klain, *Swine in Biomedical Research,* Vol. 1. M.E. Tumbleson, Ed. Plenum Press, New York, 667 (1986).
80. R.K. Freinkel, *J. Invest. Dermatol.* 38, 31 (1962).
81. K. Adachi, *J. Invest. Dermatol.* 37, 381 (1961).
82. V.R. Wheatley, L.T. Hodgins, W.M. Coon, M. Kumarasiri, H. Berenzweig, and J.M. Feinstein, *J. Lipid Res.* 12, 347 (1971).

83. H.J. Yardley, *Biochemistry and Physiology of the Skin,* Vol. I, L.G. Goldsmith, Ed. Oxford University Press, New York, 363 (1983).
84. M.L. Williams and P.M. Elias, *CRC Crit. Rev. Ther. Drug Carrier Systems.* CRC Press, Boca Raton, FL, 3, 95 (1987).
85. Y. Mishima and S. Widlan, *J. Invest. Dermatol.* 46, 263 (1966).
86. R.E. Billingham and P.B. Medawar, *Philos. Trans. R. Soc. London* 237, 151 (1953).
87. R. Staricco and H. Pinkus, *J. Invest. Dermatol.* 28, 33 (1957).
88. A. Charles and J.T. Ingram, *J. Biophys. Biochem. Cytol.* 6, 41 (1959).
89. A.S. Zelickson, *Ultrastructure of Normal and Abnormal Skin.* Lea & Febiger, Philadelphia (1967).
90. M. Wikswo and G. Szabo, *Proc. 8th Int. Pigment Cell Conf.* 1, 27, (1973).
91. K. Toda et al., *Proc. 8th Int. Pigment Cell Conf.* 1, 66 (1973).
92. M. Seiji, H. Itakura, and K. Miyazaki, *Biology of Normal and Abnormal Melanocytes.* T. Kawamura, T. Fitzpatrick, and M. Seiji, Eds. University Park Press, Baltimore, 221 (1971).
93. K. Toda and T. Fitzpatrick, *Biology of Normal and Abnormal Melanocytes.* T. Kawamura, T. Fitzpatrick and M. Seiji, Eds. University Park Press, Baltimore, 265 (1971).
94. J. Brumbaugh, R. Bowers, and G. Chatterjee, *Proc. 8th Int. Pigment Cell Conf.* 1, 47 (1973).
95. H. Nakagawa, Y. Hori, S. Sato, T. Fitzpatrick, and R.L. Martuza, *J. Invest. Dermatol.* 83, 134 (1984).
96. T. Fitzpatrick, *Biology of Normal and Abnormal Melanocytes.* T. Kawamura, T. Fitzpatrick, and M. Seiji, Eds. University Park Press, Baltimore, 369 (1971).
97. J. Duchon, J. Borovansky, and P. Hach, *Proc. 8th Int. Pigment Cell Conf.* 1, 165 (1973).
98. M. Blois, *Biology of Normal and Abnormal Melanocytes.* T. Kawamura, T. Fitzpatrick, and M. Seiji, Eds. University Park Press, Baltimore, 125 (1971).
99. W. Quevedo, *Biology of Normal and Abnormal Melanocytes.* T. Kawamura, T. Fitzpatrick, and M. Seiji, Eds. University Park Press, Baltimore, 99 (1971).
100. P.D. Mier and D.W.K. Cotton, *The Molecular Biology of Skin.* Blackwell Scientific, London (1976).
101. M. Prunieras, *J. Invest. Dermatol.* 52, 1 (1969).
102. K. Jimbow, P. Davidson, M. Pathak, and T. Fitzpatrick, *Proc. 9th Int. Pigment Cell Conf.* 3, 13 (1976).
103. M. Seiji, K. Toda, K. Okazaki, M. Uzuka, F. Morikawa, and M. Sugiyama, *Proc. 9th Int. Pigment Cell Conf.* 3, 393 (1976).
104. M. Glimcher, R. Kostick, and G. Szabo, *J. Invest. Dermatol.* 61, 344 (1973).
105. I. Rosdahl and G. Szabo, *Proc. 9th Int. Pigment Cell Conf.* 3, 1 (1976).
106. K. Toda, M.A. Pathak, J.A. Parrish, T.B. Fitzpatrick, and W.C. Quevedo, *Nature* 236, 143 (1972).
107. M. Pathak, Y. Hori, G. Szabo, and T. Fitzpatrick, *Biology of Normal and Abnormal Melanocytes.* T. Kawamura, T. Fitzpatrick, and M. Seiji, Eds. University Park Press, Baltimore, 149 (1971).
108. R. Snell, *J. Invest. Dermatol.* 58, 218 (1972).
109. S.F. Amakiri, *Acta Anat.* 103, 434 (1979).
110. F. Hu, *J. Invest. Dermatol.* 73, 70 (1979).
111. R.K. Winkelmann and A.S. Breathnach, *J. Invest. Dermatol.* 60, 2 (1973).
112. B. Munger, *J. Cell Biol.* 26, 79 (1965).
113. D. Tretjakoff, *Z. Wiss. Zool.* 71, 625 (1902).
114. B.L. Munger, *J. Invest. Dermatol.* 69, 27 (1977).
115. K. Smith, *J. Invest. Dermatol.* 69, 68 (1977).
116. D. Sinclair, *The Physiology and Pathophysiology of the Skin.* A. Jarret, Ed. Academic Press, New York, 348 (1973).
117. Z. Halata, *Adv. Anat. Embryol. Cell Biol.* 50, 5 (1975).
118. K. Smith, *J. Comp. Neurol.* 131, 459 (1968).
119. K.B. English, *J. Invest. Dermatol.* 69, 58 (1977).
120. W. Straile, *Am. J. Anat.* 106, 133 (1960).
121. A.S. Breathnach and J. Robins, *J. Anat.* 106, 411 (1970).
122. A.S. Breathnach, *J. Invest. Dermatol.* 65, 2 (1975).

123. R.K. Winkelmann, *J. Invest. Dermatol.* 69, 41 (1977).
124. K. Kurosumi, U. Kurosumi, and K. Inove, *Arch. Histol. Jpn.* 42, 243 (1979).
125. S.Y. Chen, S. Gerson, and J. Meyer, *J. Invest. Dermatol.* 61, 290 (1973).
126. K. Hashimoto, *J. Anat.* 111, 99 (1972).
127. W.E. Straile, V.R. Tipnis, S.J. Mann, and W.H. Clark, *J. Invest. Dermatol.* 64, 178 (1975).
128. R. Saxod, *Am. J. Anat.* 151, 453 (1978).
129. K.B. English, *J. Comp. Neurol.* 172, 137 (1977).
130. M. Benkenstein, *Acta Anat.* 105, 409 (1979).
131. J.M. Lauweryns, M. Cokelaere, and P. Theunynch, *Science* 180, 410 (1973).
132. W. Hartschuh and E. Weihe, *J. Cutaneous Pathol.* 5, 313 (1978).
133. W. Hartschuh and E. Weihe, *J. Invest. Dermatol.* 75, 159 (1980).
134. K.H. Andres, *Naturwissenschaften* 54, 706 (1966).
135. D.N. Tapper, *Exp. Neurol.* 26, 447 (1970).
136. A.G. Brown and D.N. Franz, *Exp. Brain Res.* 7, 231 (1969).
137. N. Cauna, *The Skin Senses.* D. Kenshalo, Ed. Charles C Thomas, Springfield, IL, 15 (1968).
138. S. Bennett, *Arch. Histol. Jpn.* 40 (Suppl.) 317 (1977).
139. T. Fujita, *Arch. Histol. Jpn.* 40 (Suppl.) 1 (1977).
140. W. Hartschuh, E. Weihe, M. Buchler, V. Helmstaedter, G. Feurle, and W. Forssman, *Cell Tissue Res.* 201, 343 (1979).
141. P. Langerhans, *Virchows Arch. Pathol. Anat.* 44, 325 (1868).
142. A.S. Zelickson, *J. Invest. Dermatol.* 44, 201 (1965).
143. A.S. Breathnach, M. Birbeck, and J. Everall, *Ann. NY Acad. Sci.* 100, 223 (1963).
144. T.M. Alyassin and P. Toner, *J. Anat.* 122, 435 (1976).
145. I. Silberberg-Sinakin, *Int. J. Dermatol.* 16, 581 (1977).
146. A.S. Breathnach, *Clin. Exp. Dermatol.* 2, 1 (1977).
147. K. Wolff, *Zellen. Arch. Klin. Exp. Dermatol.* 218, 446 (1964).
148. K. Wolff and R.K. Winkelmann, *J. Invest. Dermatol.* 48, 504 (1967).
149. E. Rodriguez and I. Caorsi, *J. Ultrastruct. Res.* 65, 279 (1978).
150. A.S. Breathnach, W. Silvers, J. Smith, and S. Hegner, *J. Invest. Dermatol.* 50, 147 (1968).
151. A.S. Zelickson and J.H. Mottaz, *Arch. Dermatol.* 98, 652 (1968).
152. M. Birbeck, A. Breathnach, and J. Everall, *J. Invest. Dermatol.* 37, 51 (1963).
153. A.S. Breathnach, *Br. J. Dermatol.* 97, 14 (1977).
154. R.C. Shukla, *Nature* 211, 885 (1966).
155. R.C. Shukla, *Curr. Sci.* 34, 406 (1965).
156. R.C. Shukla, *Acta Morphol. Neerl-Scand.* 7, 214 (1970).
157. F. Ebling, *J. Invest. Dermatol.* 75, 3 (1980).
158. A.S. Breathnach, *J. Invest. Dermatol.* 75, 6 (1980).
159. T. Allen and C. Potten, *J. Cell Sci.,* 15, 291 (1974).
160. G. Stingl, S. Katz, I. Green, and E. Shevach, *J. Invest. Dermatol.* 74, 315 (1980).
161. K. Tamaki, G. Stingl, and S. Katz, *J. Invest. Dermatol.* 74, 309 (1980).
162. J. Frelinger and J. Frelinger, *J. Invest. Dermatol.* 75, 68 (1980).
163. S.I. Katz, K. Tamaki, and D.H. Sachs, *Nature* 282, 324 (1979).
164. K. Tamaki and S. Katz, *J. Invest. Dermatol.* 75, 12 (1980).
165. K. Wolff and R.K. Winkelmann, *J. Invest. Dermatol.* 48, 50 (1967).
166. I. Silberberg-Sinakin, *Int. J. Dermatol.* 16, 581 (1977).
167. B. Berman and D. France, *J. Invest. Dermatol.* 74, 323 (1980).
168. G. Rowden, *Br. J. Dermatol.* 97, 593 (1977).
169. G. Stingl, E.C. Wolff-Schreiner, W.J. Pichler, F. Gschnait, and W. Knapp, *Nature* 268, 245 (1977).
170. R.L. Baer, *J. Invest. Dermatol.* 74, 307 (1980).
171. J. Streilen, G. Toews, J. Gilliam, and P. Bergstresser, *J. Invest. Dermatol.* 74, 319 (1980).
172. P. Bergstresser, G. Toews, and J. Streilein, *J. Invest. Dermatol.* 75, 73 (1980).
173. W. Aberer, G. Schuler, and G. Stingl, *J. Invest. Dermatol.* 76, 202 (1981).
174. J. Schweizer and F. Marks, *J. Invest. Dermatol.* 69, 198 (1977).

175. O. Kawanami, F. Basset, V. Ferrans, P. Saler, and R. Crystal, *Lab. Invest.* 44, 227 (1981).
176. G. Rowden, T.M. Phillips, M.G. Lewis, and R.D. Wilkinson, *J. Cutaneous Pathol.* 6, 364 (1979).
177. H. Uno and J.M. Hanifin, *J. Invest. Dermatol.* 75, 52 (1980).
178. P. Lieberman, C. Jones, and D. Filippa, *J. Invest. Dermatol.* 75, 71 (1980).
179. P. Bergstresser, G. Toews, J. Gilliam, and J. Streilien, *J. Invest. Dermatol.* 74, 312 (1980).
180. J. Streilien, G. Toews, and P. Bergstresser, *J. Invest. Dermatol.* 75, 17 (1980).
181. R.L. Baer, *Year Book of Dermatology.* R. Dobson and B. Thiers, Eds. Year Book Medical Publishers, Chicago, 377 (1980).
182. G. Toews, *J. Invest. Dermatol.* 75, 78 (1980).
183. J.A.A. Hunter, D.J. Fairley, G.C. Priestly, and H.A. Cubie, *Br. J. Dermatol.* 94, 119 (1976).
184. B.M. Balfour, H.A. Drexhage, E.W.A. Kamperdijk, and E. Hoefsmit, *Ciba Found. Symp.* 84, 281, Pittman Medical, London (1981).
185. R. Snell, *Z. Zellforch. Mikrosk. Anat.* 66, 457 (1965).
186. Y. Mishima, H. Kawasaki, and H. Pinkus, *Arch. Dermatol. Forsch.* 243, 67 (1972).
187. Y. Mishima and M. Matsunaka, *Recent Progress in Electron Microscopy of Cells and Tissues.* E. Yamada, V. Mizuhira, K. Kurosumi, and T. Nagano, Eds. 290 (1976).
188. G. Rowden et al., *Br. J. Dermatol.* 100, 531 (1979).
189. G. Rowden, *J. Invest. Dermatol.* 75, 22 (1980).
190. J. Thorbecke, I. Sinakin-Silberberg, and T. Flotte, *J. Invest. Dermatol.* 75, 32 (1980).
191. R. Steinman, M. Witner, M. Nussenzweig, and L. Chen, *J. Invest. Dermatol.* 75, 14 (1980).
192. R. Briggaman and C. Wheeler, *J. Invest. Dermatol.* 65, 71 (1975).
193. F.G. Rodrigo and G. Cotta-Pereira, *Dermatologica* 158, 13 (1979).
194. A. Jarrett, *The Physiology and Pathophysiology of the Skin, Vol. 1.* A. Jarrett, Ed. Academic Press, New York, 1 (1973).
195. N. Kefalides, *J. Invest. Dermatol.* 65, 85 (1975).
196. N. Kefalides, R. Alper, and C. Clark, *Int. Rev. Cytol.* 61, 167 (1979).
197. T. Gillman, *An Introduction to the Biology of the Skin.* R.H. Champion, T. Gillman, A.J. Rook, and R.T. Sims, F.A. Davis, Philadelphia, 76 (1970).
198. M. Hussein, *Acta Anat.* 82, 549 (1972).
199. J. Pinto, *The Structure and Function of the Skin.* W. Montagna and P. Parakkal, Eds. Academic Press, New York, 96 (1974).
200. D. Jackson, *Advances in Biology of Skin, Vol. 10.* W. Montagna, J.P. Bentley, and R. Dobson, Eds. Appleton-Century-Crofts, New York, 39 (1968).
201. A. Jarrett, *The Physiology and Pathobiology of the Skin, Vol. 3.* A. Jarrett, Ed. Academic Press, New York, 810 (1974).
202. B. Goldberg and M. Rabinovitch, *Histology,* 4th ed., L. Weiss and R. Greep, Eds. McGraw-Hill, New York, 145 (1977).
203. W. Meigel, S. Gay, and L. Weber, *Arch. Dermatol. Res.* 259, 1 (1977).
204. A.W. Ham and D.H. Cormack, *Histology,* 8th ed. J.B. Lippincott, Philadelphia (1979).
205. K.S. Stenn, J.A. Madri, and F.J. Roll, *Nature* 277, 229 (1979).
206. A. Jarrett, *The Physiology and Pathophysiology of the Skin, Vol. 3.* A. Jarrett, Ed. Academic Press, New York, 847 (1974).
207. W. Montagna and K. Carlisle, *J. Invest. Dermatol.* 73, 47 (1979).
208. P.D. Mier and D.W.K. Cotton, *The Molecular Biology of Skin.* Blackwell Scientific, London (1976).
209. T. Cowen, P. Trigg, and R.A.J. Eady, *Br. J. Dermatol.* 100, 635 (1979).
210. H. Rang, *Zbl. Vet. Med. A.* 20, 546 (1973).
211. J.W. Combs, *J. Cell Biol.* 31, 563 (1966).
212. A. Weinstock and J. Albright, *J. Ultrastruct. Res.* 17, 245 (1967).
213. T. Kobayasi, K. Midtgard, and G. Asboe-Hansen, *J. Ultrastruct. Res.* 23, 153 (1968).
214. D. Lagunoff, *J. Invest. Dermatol.* 58, 296 (1972).
215. H. Fujita, C. Asagani, S. Murozumi, K. Yamamoto, and K. Kinoshita, *J. Ultrastruct. Res.* 28, 353 (1969).

216. A.S. Breathnach, *J. Invest. Dermatol.* 71, 2 (1978).
217. W. Montagna and P. Parakkal, Eds., The Piliary Apparatus, in *The Structure and Function of Skin*, 3rd ed. Academic Press, New York, 172 (1974).
218. A. Rook, *An Introduction to the Biology of the Skin*. R. Champion, T. Gillman, A. Rook, and R. Sims, Eds. F.A. Davis, Philadelphia, 164 (1970).
219. W. Montagna and P. Parakkal, *The Structure and Function of Skin,* 3rd ed. Academic Press, New York, (1974).
220. P. Parakkal and N. Alexander, *Keratinization — A Survey of Vertebrate Epithelia.* Academic Press, New York (1972).
221. R.I.C. Spearman, *The Integument: A Textbook of Skin Biology.* Cambridge University Press, London (1973).
222. R.I.C. Spearman, *The Physiology and Pathophysiology of the Skin,* Vol. 4. A. Jarrett, Ed. Academic Press, New York, 1255 (1977).
223. F.J. Ebling and P.A. Hale, *Biochemistry and Physiology of the Skin,* Vol. 1. L.A. Goldsmith, Ed. Oxford University Press, New York, 522 (1983).
224. J. Ling, *Q. Rev. Biol.* 45, 16 (1970).
225. J.E. Riviere, F.R. Engelhardt, and J. Solomon, *Gen. Comp. Endocrinol.* 31, 398 (1977).
225a. J. Civatte, B. Laux, N.B. Simpson, and C.F. Vickers, *Dermatologica* 175 (Suppl. 2), 42 (1987).
225b. R.L. DeVillez, *Dermatologica* 175 (Suppl. 2), 50 (1987).
225c. V.H. Price, *Dermatologica* 175 (Suppl. 2), 36 (1987).
226. M. Bell, *Advances in Biology of Skin,* Vol. 9. W. Montagna and R. Dobson, Eds. Pergamon Press, New York, 491 (1967).
227. A. Charles, *J. Invest. Dermatol.* 35, 27 (1960).
228. V.S. Orfanos, *Dermatology* 132, 445 (1966).
229. R.A. Ellis, *The Ultrastructure of Normal and Abnormal Skin.* A.S. Zelickson, Ed. Lea & Febiger, Philadelphia, 132 (1967).
230. R. Dobson, *The Structure and Function of Skin,* 3rd ed. W. Montagna and P. Parakkal, Eds. Academic Press, New York, 366 (1974).
230a. J.S. Strauss, D.T. Downing, and F.S. Ebling, *Biochemistry and Physiology of the Skin,* Vol. 1. L.A. Goldsmith, Ed. Oxford University Press, New York, 569 (1983).
231. R.H. Champion, *An Introduction to the Biology of the Skin.* R. Champion, T. Gillman, A. Rook, and R. Sims, Eds. F.A. Davis, Philadelphia, 114 (1970).
232. R.H. Champion, *An Introduction to the Biology of the Skin.* R. Champion, T. Gillman, A. Rook, and R. Sims, Eds. F.A. Davis, Philadelphia, 175 (1970).
233. K. Hashimoto, B. Gross, and W. Lever, *J. Invest. Dermatol.* 46, 6 (1966).
234. K. Kurosumi, *Arch. Histol. Jpn.* 40, 203 (1977).
235. M. Takada and S. Hattori, *Anat. Rec.* 173, 213 (1972).
236. T.J. Ryan, *The Physiology and Pathophysiology of the Skin,* Vol. 2. A. Jarrett, Ed. Academic Press, New York, 577 (1973).
237. T. Morishima and S. Hanawa, *Acta Dermatovener (Stockholm)* 58, 487 (1978).
238. G.S. Molyneux and M.M. Bryden, *Anat. Rec.* 191, 239 (1978).
239. M.M. Bryden and G.S. Molyneux, *Anat. Rec.* 191, 253 (1978).
240. A. Yen and I.M. Braverman, *J. Invest. Dermatol.* 66, 131 (1976).
241. I.M. Braverman and A. Yen, *J. Invest. Dermatol.* 68, 44 (1977).
242. S. Imayama, *J. Invest. Dermatol.* 76, 151 (1981).
243. J.M. Johnson, G. L. Brengelman, J.R.S. Hales, P.M. Vanhoutte, and C.B. Wenger, *Fed. Proc.* 45, 2841 (1986).
244. J.K. Kristensen and S. Wadskov, *J. Invest. Dermatol.* 68, 196 (1977).
245. D.M. Pence and C.W. Song, *Hyperthermia and Cancer Treatment,* Vol. 2. L.J. Anghileri and J. Robert, Eds. CRC Press, Boca Raton, FL, 1 (1986).
246. T.J. Ryan, *Biochemistry and Physiology of the Skin,* Vol. 2. L.A. Goldsmith, Ed. Oxford University Press, New York, 817 (1983).

247. R.K. Winkelmann, *Nerve Endings in Normal and Pathologic Skin*. Charles C Thomas, Springfield, IL (1960).

248. K.W. Horch, R. Tuckett, and P. Burgers, *J. Invest. Dermatol.* 69, 75 (1977).

249. A. Iggo, *Br. Med. Bull.* 33, 97 (1977).

250. L. Giacometti and W. Montagna, *Advances in Biology of the Skin*, Vol. 9. W.M. Montagna and R.L. Dobson, Eds. Pergamon Press, New York, 393 (1967).

251. M. Santini, *Sensory Functions of the Skin in Primates*. Y. Zotterman, Ed. Pergamon Press, New York, 37 (1976).

252. E.G. Jones and W.M. Cowan, *Histology*, 4th ed. L. Weiss and R. Greep, Eds. McGraw-Hill, New York, 283 (1977).

253. S.R. Macintosh and D.C. Sinclair, *J. Anat.* 125, 149 (1978).

254. W. Montagna, *J. Invest. Dermatol.* 69, 4, (1977).

255. L. Ranvier, *Q. J. Micr. Sci.* 20, 456 (1880).

256. G. Weddel, E. Palmer, and W. Pallie, *Biol. Rev.* 30, 159 (1955).

257. H.H. Wollard, *J. Anat.* 71, 54 (1936).

258. M.J. Fitzgerald, *J. Anat.* 95, 495 (1961).

259. W. Montagna, N. Roman, and E. MacPherson, *J. Invest. Dermatol.* 65, 458 (1975).

260. S.F. Amakiri, S.E. Ozoya, and P.O. Ogunnaike, *Acta Anat.* 100, 391 (1978).

261. N. Cauna, *Anat. Rec.* 198, 643 (1980).

262. R.M. Das, P.K. Jeffery, and J.G. Widdicombe, *J. Anat.* 126, 123 (1978).

263. N. Cauna, *J. Comp. Neurol.* 113, 169 (1959).

264. M.J. Fitzgerald, *The Skin Senses*. D.R. Kenshalo, Ed. Charles C Thomas, Springfield, IL, 61 (1968).

265. A.S. Breathnach, *J. Invest. Dermatol.* 69, 8 (1977).

266. A. Bourlond, *J. Invest. Dermatol.* 67, 106 (1976).

267. C.H. Chouchkov, *Acta Anat.* 88, 84 (1974).

268. N. Cauna, *J. Anat.* 115, 277 (1973).

269. N. Cauna, *Anat. Rec.* 188, 1 (1977).

270. R.K. Winkelmann, *The Skin Senses*. D. Kenshalo, Ed. Charles C Thomas, Springfield, IL, 38 (1968).

271. N. Cauna, *Prog. Brain Res.* 43, 35 (1976).

272. A.H. Talukdar, M.L. Calhoun, and A.W. Stinson, *Am. J. Vet. Res.* 33, 2365 (1972).

273. S. Goldsberry and M. L. Calhoun, *Am. J. Vet. Res.* 20, 61 (1959).

274. P.S. Blackburn, *Comparative Physiology and Pathophysiology of the Skin*. A.J. Rook and G.S. Walton, Eds. F.A. Davis, Philadelphia (1965).

275. D.H. Lloyd, W.B. Dick, and D. Jenkinson, *Res. Vet. Sci.* 26, 172 (1979).

276. D.H. Lloyd, S.F. Amakiri, and D. Jenkinson, *Res. Vet. Sci.* 26, 180 (1979).

277. A.J. Webb and M.L. Calhoun, *Am. J. Vet. Res.* 15, 274 (1954).

278. M.L. Calhoun and A.W. Stinson, Integument, in *Textbook of Veterinary Histology*. H.D. Dellmann and E.M. Brown, Eds. Lea & Febiger, Philadelphia (1981).

279. J.H. Strickland and M.L. Calhoun, *Am. J. Vet. Res.* 24, 1018 (1963).

280. K. Setala, L. Merenmies, L. Stjernvall, and M. Nyholm, *J. Natl. Cancer Inst.* 24, 329 (1960).

281. D. Tarin, *Int. J. Cancer* 2, 195 (1967).

282. J.E. Breazile and E.M. Brown, Anatomy, in *Biology of the Guinea Pig*. J.E. Wagner and P.J. Manning, Eds. Academic Press, New York (1976).

283. A.H. McCreesh and M. Steinberg, *Skin Irritation Testing in Animals in Dermatotoxicology*. F.N. Marzulli and H.I. Maibach, Eds. Hemisphere Publishers, New York, 6, 147 (1977).

284. D.H. Lloyd, W.B. Dick, and D. Jenkinson, *Res. Vet. Sci.* 26, 250 (1979).

285. T. Kadar and G.A. Simon, *Swine in Biomedical Research*. M.E. Tumbleson, Ed. Plenum Press, New York (1986).

286. H.P. Baden and J. Kubilus, *J. Invest. Dermatol.* 83, 327 (1984).

287. H.P. Baden and J. Kubilus, *J. Invest. Dermatol.* 81, 220 (1983).

288. I.A. Bernstein, *Biochemistry and Physiology of the Skin,* Vol. 1. L.A. Goldsmith, Ed. Oxford University Press, New York, 170 (1983).
289. E.H. Fowler and M.L. Calhoun, *Am. J. Vet. Res.* 25, 156 (1964).
290. J. Smith and M.L. Calhoun, *Am. J. Vet. Res.* 25, 165 (1964).
291. H. Marcarian and M.L. Calhoun, *Am. J. Vet. Res.* 118, 765 (1966).
292. M. Meyer, R. Schwartz, and K. Neurand, *Curr. Probl. Dermatol.* 7, 39 (1978).
293. P. Forbes, *Advances in Biology of the Skin.* Pergamon Press, New York, 419 (1969).
294. D.L. Ingram and M.E. Weaver, *Anat. Rec.* 163, 517 (1969).
295. W. Meyer, K. Neurand, and B. Radke, *Arch. Dermatol. Res. 270, 391 (1981).*
296. W. Meyer, K. Neurand, and B. Radke, *J. Anat.* 134, 139 (1982).
297. W. Meyer, R. Schwarz, and K. Neurand, *Curr. Probl. Dermatol.* 7, 39 (1978).
298. M.W. Stromberg, Y.C. Hwang, and N.A. Monteiro-Riviere, *Anat. Rec.* 201, 455 (1981).
299. R.L. Bronaugh, R.F. Stewart, and E.R. Congdon, *Toxicol. Appl. Pharmacol.* 62, 481 (1982).
300. R.A. Briggaman, *Epidermal-Dermal Junction: Structure, Composition, Function, and Disease Relationships.* A.N. Moschell, Ed. Progress in Dermatology, Evanston, IL, 24, 1 (1990).
301. N.A. Monteiro-Riviere, D.G. Bristol, T.O. Manning, R.A. Rogers, and J.E. Riviere, *J. Invest. Dermatol.* 95, 582 (1990).

2

Pathological Processes of Skin Damage Related to Toxicant Exposure

RICHARD H. BRUNER
Pathology Associates, Inc.
West Chester, Ohio

I. INTRODUCTION

This chapter discusses and illustrates pathologic changes in the integument of laboratory animals following exposure to toxic chemicals or materials. Fundamental objectives are to provide toxicologists and allied biomedical scientists with a working knowledge of the terminology, histotechnical applications, and histomorphic changes associated with dermal toxicants in research species. The discussion includes: (1) a survey of routine and special methods for collecting and processing skin specimens; (2) a dictionary of terminology relating to dermatopathology; (3) an illustrated outline of pathologic changes which typify dermal responses to various classes of toxicants; (4) a description of cutaneous neoplasms induced by chemical or physical agents; (5) a listing of special anatomic structures associated with the skin of laboratory species; and (6) a review of systemic diseases or spontaneous changes which might compromise *in vivo* dermatotoxicologic studies.

II. METHODS FOR SELECTING AND PROCESSING SKIN SPECIMENS

A broad assortment of special procedures is available to assist the dermatopathologist or toxicologist in the evaluation of skin specimens. Despite the proliferation of useful *in vitro* methods for evaluating skin damage, the biopsy continues to provide the most compelling and comprehensive test system for establishing dermatotoxic effects. In this regard, the biopsy often displays gross or microscopic changes which can be subjected to an array of special procedures for characterizing toxicologic phenomena or explaining pathogenic mechanisms. Most of these procedures begin with the collection of appropriate skin specimens.

A. Biopsy Procedures

Although punch biopsies are routinely used in diagnostic pathology, the surgically excised specimen usually provides greater tissue volume to examine or manipulate, and may provide a continuum of pathologic change from treated (abnormal) to juxtaposed untreated (normal) skin. Several general guidelines should be considered in harvesting skin biopsies:

1. When possible, avoid local anesthetics or aseptic surgical scrubs which might alter or obscure key diagnostic features.
2. In the experimental setting, include normal as well as abnormal skin (two sites may be required).
3. Generally, biopsies should be elliptical in shape with the long axis oriented parallel to the direction of hair flow. This provides a longitudinal microscopic view of hair follicle structures rather than undesirable cross sectional perspectives.
4. Skin specimens should be flattened on a piece of paperboard or photographic paper and gently stretched before immersion in fixative.
5. Appropriate (often multiple) fixatives for preserving skin specimens may include standard 10% buffered neutral formalin (BNF) for most routine histologic stains, or Bouin's, B5 or Zenker's along with frozen sections if immunohistochemical procedures are anticipated.
6. The worst biopsy specimen is no specimen at all. Usually, tissues that are inadvertently missed or improperly collected cannot be resurrected.

B. Routine and Special Stains

Although routine hematoxylin and eosin (H&E) stains are adequate for many histopathologic determinations, special stains may be required to characterize some dermatotoxic changes. For the **epidermal compartment**, the Ayoub-Shklar or Dane's methods may be helpful in identifying cytokeratins in poorly differentiated or neoplastic keratinocytes.[1] Gold chloride procedures have been used to distinguish epidermal Langerhans' cells and Fontana's technique has proved valuable in highlighting melanin granules.[1,2]

Periodic acid-Schiff (PAS) procedures can be used to identify substances in both the dermal and epidermal compartments which contain 1:2-glycol groupings or equivalent amino or alkylamino derivatives. Substances include glycogen, starches, cellulose, neutral mucosaccharides, glycolipids, unsaturated lipids, and phospholipids.[1-4] With the exception of the basal lamina and certain adnexal structures (e.g., pilosebaceous follicles and eccrine sweat glands), PAS positive substances are usually absent from the normal epidermis. Radiation or mechanical injury along with other pathologic stimuli, however, may result in increased PAS-positive materials within the epidermis.[4]

In the **dermal compartment**, a variety of special stains may be used to advantage. Some of the more common procedures include alcian blue and Mowry's colloidal iron for mucopolysaccharides, Gordon and Sweet's silver stain for reticulin fibers and Luna's aldehyde fuchin for elastic fibers. Luna's elastic stain has proved to be useful in demonstrating delicate elastic fibers in the mouse dermis.[5] In addition to connective tissue fiber stains, many special procedures have been developed for dermal inflammatory cells such as the Alpha-Naphthol AS-D chloracetate esterase reaction (Alpha-N AS-D CA ER) for differentiating tissue mast cells.[6]

C. Immunohistochemical Procedures

The rapidly expanding use of tagged polyclonal and, especially, monoclonal antibody has contributed immeasurably to the armamentarium of experimental and diagnostic procedures available to the dermatopathologist. Currently, immunohisto-chemical (IHC) methods are available for the precise identification of an array of extracellular, cellular and ultrastructural components. Subtyping of dermal lympho-cyte and mast cell populations along with enumeration of Langerhans' cell popula-tions is now routine.[7-13] The recent introduction of 5-bromo-2′-deoxyuridine (BrUdR) as an immunogenic thymidine analogue has provided a convenient method for the identification and quantification of proliferating epithelial cells (i.e., cells actively synthesizing DNA) following toxic insult. BrUdR-labeled cells can be readily identified in paraffin-embedded tissue sections by anti-BrUdR antibodies and the avidin-biotin immunoperoxidase method. The major advantages of this IHC pro-cedure over conventional tritiated thymidine methods are that no radioactive nucleo-sides are used and the prolonged incubation period for autoradiographic develop-ment is eliminated.[14]

When IHC procedures are anticipated, careful attention must be given to the use of appropriate tissue fixatives and the selection of antibody which is either species specific or demonstrates satisfactory cross-reactivity with the target antigen. B-5, Bouin's, and Zenker's fixatives as well as 10% BNF have been used with good success for many procedures, although OCT embedded frozen sections are required for some IHC methods. Unless species specific antibody is available, immuno-globins raised in nonhomologous systems must be thoroughly tested to ensure that adequate cross-reactivity is present. Because a broad variety of commercially avail-able anti-human antibodies is available, Smith[15] has extensively characterized non-homologous antibodies which demonstrate satisfactory cross-reactivity for canine and rodent antigens.

D. Autoradiographic Procedures

Either *in vivo* on *in vitro* exposure of viable skin sections to tritiated thymidine has proved to be a reliable index of proliferating epithelial cells. Experimentally, this procedure may establish subtle changes in the mitotic activity of skin sections exposed to toxic agents. Because autoradiographic procedures to identify cells labeled with tritiated thymidine often require a 2 to 6 week incubation period, experimental results are not immediately available. Otherwise autoradiographic procedures associated with dermatopathologic investigations usually provide excel-lent results. Paired samples (treated and control) are essential to distinguish slight changes. Figure 1 illustrates positive labeling of epidermal cells following incuba-

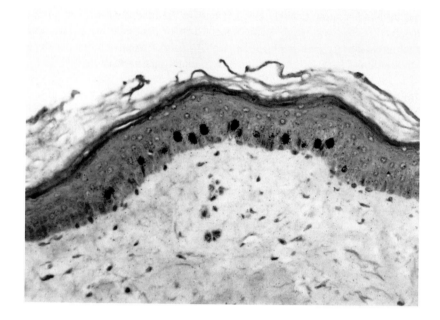

Figure 1.   Basilar cells engaged in active DNA synthesis are labeled with tritiated thymidine and are identified by dark granules in this skin section following autoradiographic development. Both tritiated thymidine and 5-bromo-2′-deoxyuridine can be used to establish proliferative indices for skin section in dermatotoxicologic studies.

tion of skin biopsies in tritiated thymidine culture medium for approximately 2 h postcollection. Autoradiographic procedures for skin specimens have been outlined by Fukuyama.[16]

E. Ultrastructural Procedures

Electron microscopic examinations have contributed greatly to the understanding of pathogenic mechanisms associated with systemic toxins. In dermatotoxicologic investigations, however, ultrastructural evaluations have been largely restricted to characterizing cell populations or identifying subtle connective tissue changes, especially those related to phototoxicity or radiation exposure. The introduction of immunoelectron microscopic procedures may assist in the accurate anatomic localization of many toxicant-induced subcellular events.[17,18] General electron microscopic procedures for examining skin specimens have been reviewed by Eady.[19]

F. Macroscopic and Microscopic Examinations

Ideally, all gross and microscopic examinations should be conducted by a dermatopathologist familiar with the species and reactivity of the suspected toxicant. In the experimental setting, cutaneous observations should be supplemented with complete histopathologic examination of all major organs along with appropriate clinical chemistry and hematology evaluations. Percutaneous absorption of many toxins may result in systemic toxicity or immunoincompetence which can markedly influence dermatologic changes.

III. DICTIONARY OF DERMATOPATHOLOGY TERMINOLOGY

A. General Terms

Dermatosis (pl. dermatoses) — A nonspecific term denoting any skin disease. In this text, it is used to identify skin changes not associated with significant inflammatory events.

Dermatitis (pl. dermatitides) — Any skin disease characterized by inflammatory changes. May be subclassified based upon the types of inflammatory cells which are present and their corresponding patterns of distribution. Histopathologic patterns of inflammation have proved useful in classifying skin diseases and may include the following categories — either singularly or in several combinations. The works of Ackerman [20] and Muller, Kirk, and Scott [21] should be consulted for expanded details and explanations. Basic patterns include: (1) perivascular dermatitis, (2) nodular to diffuse dermatitis (including granulomatous dermatitis), (3) intraepidermal vesicular and pustular dermatitis, (4) perifolliculitis, folliculitis, and furunculosis, (5) panniculitis; and (6) vasculitis.

1. **Perivascular dermatitis** exists when inflammatory cell infiltrates are predominately distributed in circumvascular patterns. Four variants are recognized. These include: (1) "perivascular dermatitis" involving the superficial or deep dermal blood vessels; (2) "interface dermatitis" where inflammatory cell infiltrates closely follow the dermoepidermal junction as well as perivascular distributions; (3) "spongiotic dermatitis" which is characterized by intercellular edema in the epidermal layer which may progress to form intraepidermal vesicles and bullae; and (4) "hyperplastic dermatitis" which is a variant associated with chonic irritation and varying degrees of epidermal hyperplasia and hyperkeratosis. Any of these four variants of perivascular dermatitis might be indicative of contact with noxious agents, food or drug allergies or several additional infections, or toxic or physical agents. Interface dermatitis may be associated with "toxic epidermal necrolysis", drug reactions and autoimmune skin diseases. Spongiotic dermatitis is perhaps the most common initial change following contact with a dermal irritant.

2. **Nodular to diffuse patterns** of dermal inflammation may be associated with many chemical, physical, or infectious agents. With nodular patterns, inflammatory cell infiltrates form relatively distinct and separate dermal collections whereas diffuse patterns are characterized by evenly dispersed cellular infilitrates. A particular variant of these patterns is that of **granulomatous inflammation**. This variant is indicative of chonic insult, especially involving microorganisms or foreign materials which cannot be readily eliminated by host

defense mechanisms. "Granulomas" usually include central accumulations of macrophages and possibly multinucleated giant cells surrounded by proliferating fibroblasts, lymphocytes, plasma cells and occasional granulocytic leukocytes.

3. **Intraepidermal vesicular and pustular dermatitis** is frequently a sequelae to contact with external toxicants and begins with intraepidermal edema (spongiotic dermatitis) which coalesces to form fluid-filled vesicles, etc. Often, vesicles may be filled with neutrophils in which case the term pustular dermatitis is appropriate. Epidermal and subepidermal vesicular dermatitides are common features of autoimmune skin diseases (e.g., pemphigus) and many viral infections.

4. **Folliculitis, perifolliculitis, and furunculosis** all pertain to inflammatory diseases centered around hair follicle structures. Furunculosis is a common sequelae to folliculitis where there is rupture of the hair follicle and escape of highly irritating sebum, keratin, and hair shaft material into the subcutis. Although follicular diseases may be associated with toxicant exposure (e.g., halocarbons), many follicular changes observed clinically are associated with bacterial pathogens or dermatophytes.

5. **Panniculitis** can be a perplexing disease syndrome with multiple etiologic factors. Simply stated, the term means "inflammation of the subcutaneous fat layer". Causes may remain entirely obscure or be identified as penetrating foreign bodies or infectious organisms. Autoimmune diseases have been incriminated in some cases.[22]

6. **Vasculitis** is appropriately diagnosed when inflammatory cells and degenerative signs are noted within the blood vessel wall as well as within the surrounding tissue. These changes may be noted with percutaneous exposure to some toxic chemicals; however, subendothelial or intramural deposition of circulating immune complexes in autoimmune or chonic infectious diseases frequently is identified as a cause of vasculitis.

B. Macroscopic Pathology Terminology

Abscess — Usually a soft, fluctuant lesion filled with pus.

Acral — Pertaining to the extremities.

Alopecia — Hair loss.

Atrophy — Reduction in the size of an organ due to reduced numbers of cells. (confirmed microscopically).

Bullae — A fluid-filled vesicle greater than 1 cm in diameter.

Comedo — A hair follicle filled and distended with keratinized debris and degenerating cells.

Crust — A "scab" consisting of dried exudates, blood, and cellular debris.

Erythema — A red macula usually due to dermal vascular congestion and dilatation.

Erosion — Loss of superficial epithelium without discontinuity of the basal lamina (basement membrane).

Eschar — A skin slough following contact with a harsh irritant.

Excoriation — A self-induced traumatized area.

Fissure — A linear crack in the skin.

Gangrene — Necrosis of tissue, often following disruption of the blood supply.

Lichenification — Thickened skin with exaggerated marking and furrows.

Macula — A circumscribed, flat, discolored spot <1 cm in diameter.

Nodule — A small, solid, elevated, well-circumscribed mass which is >1 cm in diameter.

Papule — A small, solid, elevated, well-circumscribed mass which is <1 cm in diameter.

Patch — An irregular macula that is >1 cm in size.

Plaque — A flattened, but slightly elevated skin mass which is larger than a papule.

Pustule — A small intracutaneous abscess.

Scale — A small, thin, dry flake of keratin debris.

Squama — Essentially same as scale.

Tumor — Any enlarged, abnormal mass. Often, but not invariably, associated with neoplastic growth.

Ulcer — An erosion of the epithelium which breaches the basal lamina.

Wheal — A well-circumscribed, white-to-pink elevated skin lesion due to subcutaneous edema. Characteristic of acute allergic reactions or insect bites.

Vesicle — A well-demarcated, dome-shaped, thin-walled intraepidermal cyst filled with a clear, thin fluid.

C. Microscopic Pathology Terminology
1. Epidermis

Acantholysis — Loss of cohesion of cells in the stratum spinosum layer often leading to vesicle/bullae formation.

Acanthosis (hyperplasia) — An increase in epidermal thickness due to increased numbers of cells in the stratum spinosum.

Ballooning Degeneration — A degenerative change in epidermal cells characterized by swollen, eosinophilic, vacuolated cytoplasm often associated with acantholysis.

Dyskeratosis — Premature, faulty keratinization of individual epithelial cells.

Epidermitis — Inflammation restricted to the upper (epidermis) skin layer.

Erosion — Loss of the superficial epidermal layers without discontinuity of the basal lamina.

Exocytosis —The migration of leukocytes and erythrocytes through the epidermal layers to the exterior.

Folliculitis — An inflammatory reaction involving hair follicles.

Follicular Keratosis — Plugging and distention of hair follicles with disorganized keratin debris.

Furunculosis — Folliculitis associated with rupture of the hair follicle and the escape of irritating sebum, keratin, and hair shaft material into the dermis.

Hydropic Degeneration — Vacuolization (intracellular edema) of the cells of the deep epidermal strata.

Hypergranulosis/Hypogranulosis — An increase or decrease in the thickness of the stratum granulosum.

Hyperkeratosis/Hypokeratosis — An increase or decrease in the thickness of the stratum corneum often associated with chonic irritation and hyperplasia.

Hyperpigmentation/Hypopigmentation — An increase or decrease in the amount of melanin pigment in the epidermis.

Hyperplasia/Hypoplasia — An increase or decrease in the thickness of the noncornified epidermis. Often accompanied by the formation of rete ridges.

Intercellular Edema — (Spongiosis) early separation of the cells of the epidermis by extracellular fluid. May coalesce to form vesicles/bullae.

Intracellular Edema — Vacuolar degeneration of the epidermal cells.

Necrosis — The pathologic death of cells in the epidermis or dermis.

Subcorneal pustules — Intraepithelial microabscesses immediately subjacent to the stratum corneum.

Ulcer — Erosion of the epidermis with disruption and loss of the subjacent basal lamina.

2. Dermis

Collagen Degeneration — Replacement of normal collagen fibrils with eosinophilic, glassy to granular substance which may be disorganized and fragmented (collagenolysis). Most collagen changes are not consistent with specific toxicant exposure.

Edema — Infiltration and separation of dermal connective tissue spaces with pink, protein-rich extravasated fluid. Often associated with allergen/toxin-mediated degranulation of mast cells.

Fibroplasia and Sclerosis — Increased and more densely packed collagen fibers in the dermis. Regarded as hallmarks of chonic irritation or inflammation. Changes are often nonspecific.

Inflammatory Cell Infiltrates — Mixed populations of mononuclear and granular leukocytes frequent the subepidermal connective tissue with many skin diseases. Leukocytes may be attracted by chemotatic factors associated, percutaneous movement of toxic chemicals

IV. DERMATOPATHOLOGIC CHANGES FOLLOWING TOXICANT EXPOSURE

A. Pathologic Changes Associated with Contact Irritants

Cutaneous irritation following contact with xenobiotic chemicals and materials is the most common form of toxicant-induced skin disease. Pathologic changes associated with contact irritants range from "subjective irritation" where no gross or histomorphic changes can be identified (e.g., burning or stinging following topical application of astringents) to severe ulcerative skin lesions with life-theatening consequences. Following contact with many skin irritants, dermatopathologic changes usually progress along a continuum of nonspecific alterations, although two fundamental patterns of reactivity frequently can be recognized.

The first pattern typifies exposure to toxic chemicals or materials with sufficient potency to cause an active dermatitis. The hallmarks of this dermatopathologic response are signs of active inflammation. Redness, swelling, heat, or pain may be detected clinically, and, microscopically, there is evidence of increased vascular permeability, inflammatory cell infiltration and varying degrees of tissue insult. Classes of chemicals which usually produce these lesions include strong acids, alkalis, or other corrosives.

At the other end of the spectrum are patterns associated with irritants of low toxic potential. With these toxicants, gross lesions may consist of mild to moderate dermal thickening (lichenification) or pigmentary changes. Microscopic findings are usually restricted to proliferative responses such as epidermal hyperplasia and hyperkeratosis along with dermal fibrosis. Inflammatory cell infiltrates are essentially absent in this pattern of response, and correspondingly, changes may be referred to as a dermatosis rather than a dermatitis (technically, a dermatosis is *any* pathologic condition of the skin). Examples of contact irritants which might produce a dermatosis include petroleum greases, vegetable oils, or lubricants when applied to occluded skin for a prolonged period.

Because of the importance in distinguishing irritant-induced contact dermatitis and dermatosis from other dermatotoxic phenomena, histopathologic changes associated with these two skin diseases will be considered further.

1. Histopathologic Findings in Contact Irritant Dermatitis

Following contact with irritants which cause an active dermatitis, initial changes usually include dermal congestion and edema along with epidermal spongiosis and ballooning degeneration of spinous epithelial cells (Figure 2). Depending on the nature and continued presence of the irritant, these mild changes may spontaneously and uneventfully regress or they may progress to epidermal vesicle formation with subsequent erosion and ulceration. Superficial dermal blood vessels invariably display cuffs of lymphocytes and histiocytes along with varying populations of neutrophils and eosinophils. Features which may serve to distinguish contact irritants from contact allergens (next section) are the presence of superficial epidermal necrosis, (with possible erosions or ulcerations), neutrophilia, and subcorneal or intraepidermal abscesses in irritant reactions.[20,23] Additionally, ultrastructural and immunopathological studies have suggested that Langerhans' cells may respond differently following exposure to contact irritants when compared with allergens. In this regard, Ferguson et al.[24] observed that epidermal Langerhans' cells were greatly reduced following exposure to irritants, but remained within normal limits following challenge by contact allergens. Marks et al.[25] demonstrated that 1 to 14 days postexposure to allergens, Langerhans' cell populations shifted from the epidermis to assume perivascular dermal locations. In contrast, following irritant exposures, Langerhans' cells appeared in diffuse dermal locations before returning to the epidermis between 14 to 21 days after contact. It should be emphasized that some dermatopathologists have been unable to demonstrate significant differences between the responding cell types or the sequence of histomorphic changes when comparing contact irritant dermatitis with allergic contact dermatitis in some species.[26,27]

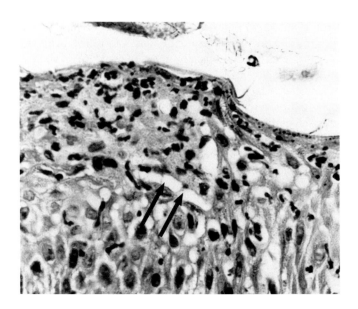

Figure 2.   Marked spongiosis (intercellular edema), with early vesicle formation (double arrows), along with hydropic degeneration and superficial necrosis of keratinocytes are common histopathologic findings associated with contact irritants. Exocytosis of leukocytes through the epithelium is generally present, and may result in intracutaneous pustules, especially in the rabbit.

In cases where contact irritation persists, microscopic findings often include marked epidermal hyperplasia, hyperkeratosis, and ulceration with severe dermal fibroplasia and mixed inflammatory cell infiltrates (Figure 3). Occasionally, proliferating fibroblasts in areas of chronic inflammation undergo neoplastic transformation and develop into an invasive fibrosarcoma.

In considering experimentally induced contact dermatitis in laboratory animals, it should be recognized that skin responses in the **rabbit** may differ somewhat from the "generic" histomorphic changes described previously. This distinction is important because the New Zealand white rabbit is widely used to evaluate contact irritants. In this species, exposure to cutaneous irritants may produce classic gross signs of erythema, edema, and eschar formation, but histopathologic changes are frequently dominated by subcorneal (intraepidermal) pustules filled with intense accumulations of heterophils. Spongiosis of the deeper epidermis and dermal inflammatory infiltrates may be relatively slight (Figure 4). If the irritation persists, the rabbit may respond with striking epidermal hyperplasia (Figure 5a,b). It should be recognized that rabbit heterophil (neutrophil) possesses intense numbers of eosinophilic granules and should not be mistaken for an eosinophil.

In the **guinea pig**, histopathologic changes which develop following contact with strong irritants have been charted by Fisher et al.[28] and Hunziker.[29] Following an

Figure 3. Transition from normal epidermis (left) through hyperplasia and hyperkeratosis (double arrows) to frank ulceration (right) with thick crust formation consisting of keratin, necrotic keratinocytes, serum, and inflammatory exudates, and often, colonies of bacteria. The dermis exhibits early fibroplasia.

epicutaneous application of 1-chloro-2, 4-dinitrobenzene (DNCB) in olive oil, serial biopsies revealed that significant tissue changes were delayed until approximately 6 h postexposure. By 12 h, most specimens exhibited moderate to severe epidermal necrosis with associated dermal edema and mixed populations of polymorphonuclear and mononuclear inflammatory cells. At 24 h, sections displayed separation of superficial, necrotic epithelial cells by basilar to suprabasilar clefts filled with intense accumulations of polymorphonuclear leukocytes (i.e., intraepidermal abscesses). In cases where frank ulceration supervened, collagen degeneration and fragmentation was present in the superficial dermis. By the third day postexposure, repair mechanisms had progressed to include proliferation of epidermal cells from the upper portion of hair follicles to cover denuded basal lamina and connective tissue. In the guinea pig model, responses to contact irritants were dominated by superficial epithelial necrosis and infiltration by polymorphonuclear leukocytes.

2. Histopathologic Findings in Irritant-induced Dermatosis

In contrast to dermatologic changes characterized by inflammatory signs, exposure to many chemicals or materials may result in a nonspecific, proliferative dermatosis which is largely restricted to the epidermal compartment. Epidermal hyperplasia and hyperkeratosis may be expected when the skin is exposed for prolonged periods to chemicals of low toxic potential. In these cases, histopathologic

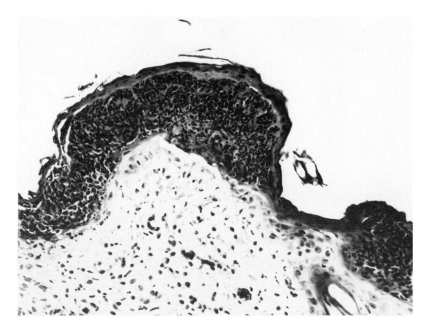

Figure 4.   Subcorneal pustules in a rabbit following exposure to a strong contact irritant. Pustules, filled with degenerating heterophils, may form without significant spongiosis or vesicle formation as a preceding event. Progression to ulceration is common.

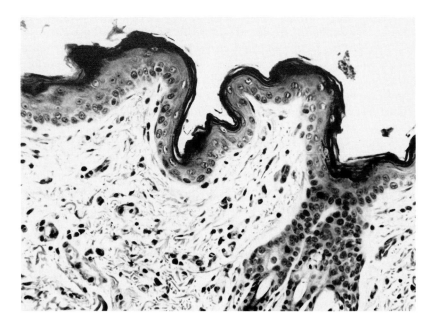

Figure 5a.   Normal rabbit epidermis for comparison with sections illustrating hyperplasia (Figure 5b). (b) Marked epidermal hyperplasia in rabbit skin treated with a petroleum-based ointment and covered with an occlusive dressing for 21 days.

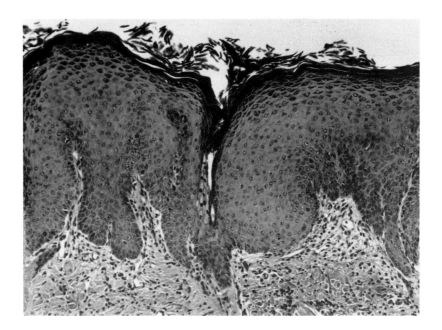

Figure 5b.

changes may be very subtle and dermatopathologic evaluations can be complicated by the regional variability of cutaneous histomorphology (i.e., thick skin is normal on the plantar surface of the foot, whereas in other locations a diagnosis of hyperkeratosis would be appropriate). Therefore, when evaluating very mild dermatoses, it is usually helpful to compare exposed skin specimens with control samples from an identical anatomic site. This may be undesirable in the clinical setting; however, under experimental conditions, comparisons of treated vs. untreated skin are entirely practical and may be absolutely essential in determining that a subtle dermatosis exists. Comparative procedures may be greatly facilitated by the use of computerized morophometric measurements which precisely quantify small changes in the dermal or epidermal compartments. Additional methods which may confirm slight epidermal proliferative responses include the use of tritiated thymidine or 5-bromo-2′-deoxyuridine (BrUdR) to label basilar cells that are undergoing active DNA synthesis (see Sections II.C and II.D). Once mitotically active cells have been identified, treated vs. nontreated specimens may be compared to establish proliferative indices.

B. Pathologic Changes Mediated by Immune Processes

Historically and practically, pathologic evaluations of *immune-mediated* dermatotoxic responses have been based on macroscopic skin changes (e.g., erythema and edema) along with systemic manifestations of allergic reactions. Because of the nonspecific nature of many histopathologic findings associated with dermal hypersensitivity, routine microscopic examinations have not been incorporated into standard clinical or experimental procedures (e.g., the Draize test, etc.). As refined diagnostic methods have been implemented, however, several pathologic changes have been identified which may assist in differentiating allergic contact dermatitis from other dermatotoxic responses. Some of these pathologic features, both gross and microscopic, are as follows.

Clinically, contact allergies represent either immediate (anaphylactic) or delayed hypersensitivity reactions. Erythema, and especially edema are the cardinal signs of immediate hypersensitivity while eczematous, pruritic lesions characterize delayed reactions. Excoriations resulting from pruritis and self trauma may dominate later gross lesions.

Microscopically, immediate allergic reactions are dominated by congestion and edema, without changes which are pathognomonic for specific etiologic factors. With contact allergens which evoke delayed, cell-mediated immunity, microscopic changes may include marked spongiosis which can progress to vesicles containing lymphocytes or eosinophils. Edema is usually present in the superficial dermis, along with perivascular cuffs of lymphocytes, histiocytes, plasma cells, and eosinophils. In contrast to skin lesions associated with contact irritants, necrotic keratinocytes are rare, and usually, epidermal erosions and ulcerations are absent. Additionally, deeper epidermal strata may be affected first with allergic reactions.[23]

In the sensitized guinea pig model, sequential changes which occur following challenge with a contact allergen have been documental by Fisher et al.[28] Using serial biopsies, these investigators determined that by 3 h post-challenge, significant vascular-oriented changes were present in the upper to middle dermis which included congestion, edema, and perivascular cuffs of mononuclear cells which often displayed "linear" migration toward the overlying epidermis. Polymorphonuclear leukocytes were essentially absent. By 6 h postexposure, moderate dermal populations of leukocytes extended to the dermo-epidermal junction and into the overlying epidermis. Between 6 and 12 h, the epidermis increased in thickness from 2 to 4 cells to approximately 3 to 6 cells and displayed intercellular edema and the formation of microvesicles. Dermal changes reached a maximum by 24 h and were characterized by collagen degeneration, edema, mononuclear cells, and lesser populations of polymorphonuclear leukocytes. Epidermal changes were most severe at 48 h and consisted of extensive vacuolar degeneration and marked inflammatory cell infiltrates along the dermo-epidermal junction and within the epidermis. Repair activities were well advanced by 5 days postexposure and consisted of a new, hyperplastic epidermis and resolving inflammation.

As an important extension of Fisher's studies. Medenica et al.[30] and Robinson et al.[31] have noted that reactive cell populations in contact sensitized guinea pigs usually include significant numbers of dermal basophils. Because increased numbers of basophils are not a feature of contact irritation reactions, this cell marker has found promising utility in the **cutaneous basophil hypersensitivity (CBH)** test for distinguishing weak contact allergens from cutaneous irritants. Dermatopathologic procedures for this test require the quantitative assessment of dermal basophil populations in Giemsa-stained thin sections (1 μm) following glycol methacrylate embedding procedures. In normal guinea pig skin, dermal basophil populations average less than 10 (per 400 leukocytes counted), whereas counts of 25 to 50 are not uncommon following challenge with weak allergens. Strong allergens may result in counts in excess of 200 per 400 cells counted. Basophil populations in irritation reactions should approximate normal skin.

In considering other dermatopathologic changes requiring immune interactions, it should be recognized that recent advances in immunobiology have identified sophisticated interactions between the integumentary and lymphoreticular systems. Examples include percutaneous antigen processing and presentation by Langerhans' cells, lymphokine production by keratinocytes, and immunomodulatory effects of ultraviolet irradiation.[7,10,12,32-35] These modalities, combined with the well-characterized role of the immune system in generalized hypersensitivity reactions, suggest that experimental evaluations of immune-mediated dermatotoxic phenomena should include pathologic analysis of all tissues which might contribute to a better understanding of toxicologic mechanisms or immune competency. In this connection, *complete histopathologic examination* of laboratory animals used for testing contact allergens or other dermatotoxic agents may identify spontaneous or chemically induced systemic lesions which could greatly modify immune responses. Thymic and lymphoid atrophy, aplastic anemia, and greatly modified local inflammatory responses are distinct examples of histopathologic changes which may occur following cutaneous or systemic exposure to xenobiotic chemicals or pharmaceuticals. In addition to documenting changes indicative of immunomodulation, expanded microscopic examination may assist in confirming the allergic basis for select dermatologic changes. With immediate hypersensitivity reactions, for example, multisystem gross and microscopic examination may reveal laryngopulmonary edema, splanchnic congestion, or other lesions compatible with generalized anaphylaxis.

C. Pathologic Changes Induced by Light

From a mechanistic standpoint, light-induced dermatopathologic changes can be divided into phototoxic and photoallergic categories. Phototoxic skin damage results from the direct interaction of irradiation with subcellular targets while photoallergic reactions pivot around immunomodulation of cutaneous photoreactivity. Both variants require initiation by exogenous light, but subsequent cytopathologic mechanisms may be substantially different.

With phototoxicity, light may originate directly from exogenous sources such as the sun, artificial lighting, or photodynamic topical chemicals, or it may emanate from endogenous sources such as photodynamic drugs or chemicals following activation or excitation by percutaneous irradiation. Subcellular targets have not been completely characterized, but may include the formation of thymine dimers, DNA-protein cross-links, or photodependent oxidations. Immunologic processes are not involved in this form of photosensitivity.[36-38]

With photoallergic reactions, cytopathologic events are believed to be even more complex than with direct phototoxicity. Although many mechanistic features remain obscure, fundamental concepts include the photoactivation of endogenous or xenobiotic haptens so that they combine with cellular proteins and form a complete antigen. Subsequent immunologic reactions especially, cell-mediated hypersensitivity, complete the sensitivity process.[38,40]

Although precise cytopathologic mechanisms have not been established for many photosensitivity reactions, clinical and pathological features have been extensively documented. The following outline describes key diagnostic findings which serve to differentiate photosensitivity reactions from other dermatologic phenomena.

1. Phototoxicity

Clinical Findings — Clinical manifestations of phototoxicity include erythema, heat, pain, and sometimes edema in irradiated areas of the skin. These initial changes may be followed by accelerated exfoliation of cornified layers and hyperpigmentation. Sunburn is the most prevalent form of phototoxicity in humans, while veterinary patients often exhibit photosensitization following the ingestion of photodynamic plants. In animals, lesions are usually restricted to sparcely haired, nonpigmented skin regions, and occasionally, may be so severe that full-thickness necrosis exposes large areas of dermal connective tissue. White or hairless laboratory animals (mice and guinea pigs) may serve as excellent models for predicting the photosensitizing potential of unknown chemicals or drugs.[39-41] In this regard, the albino mouse ear assay system has gained wide acceptance in evaluating potentially phototoxic chemicals. Figure 6a-d illustrate both gross and microscopic findings in the pinna of CFW mice following ingestion of photodynamic plant material.[42]

Histopathologic Findings — It should be emphasized that microscopic changes resulting from phototoxicity share many features with contact irritation dermatitis. With both conditions, there may be hydropic degeneration of basilar cells, epidermal edema (spongiosis), and perivascular lymphohistiocytic infiltrates in the superficial dermis. Two key features which may serve to differentiate irritant from photoinduced lesions include: (1) photosensitivity reactions are restricted to light-exposed areas; and (2) contact irritants often induce superficial epidermal necrosis with subcorneal abscess formation.

In humans with sunlight-induced phototoxicity (sunburn), individual, necrotic keratinocytes may be present in the upper epidermis subjacent to a normal stratum corneum. These necrotic keratinocytes are termed "sunburn cells" and exhibit round, intensely eosinophilic cytoplasm and a pyknotic nucleus.[43] Sunburn cells may be difficult to distinguish in veterinary patients.

2. Photoallergy

Clinical Findings — Usually, but not invariably, dermatologic lesions are restricted to light-exposed areas. Changes may vary from urticaria to papular and eczematous eruptions with subsequent exfoliation and lichenification. Microscopically, it is very difficult to distinguish photoallergic reactions from nummular eczemia, atopia dermatitis, eczematous drug eruptions, and especially, allergic contact dermatitis.[20,23,44]

Histopathologic Findings — Generally, microscopic findings do not provide an adequate basis for separating photoallergic reactions from the eczematous drug eruptions and allergic contact dermatitides previously discussed. Salient features include spongiosis with lymphocytic exocytosis, mild dermal edema, and mild to

Figure 6a. Photosensitization in albino CFW mice following ingestion of a photodynamic plant and exposure to UV irradiation. When
compared with the control mouse (upper left), the ears of the mouse
fed the photodynamic material are swollen, erect, and erythematous.
Microscopic changes are illustrated in Figures 6b to 6d. Photographs
courtesy of Dr. L.D. Rowe et al., Agricultural Research Service, College
Station, TX. With permission from *Am. J. Vet. Res.*

moderate dermal perivascular cuffing consisting of lymphocytes, histiocytes, and
varying numbers of eosinophils. A feature which may distinguish photoallergy from
contact allergy in human skin is that inflammatory cell infiltrations in light-induced
allergic reactions may be both superficial and deep within the dermis, whereas with
contact allergy they tend to be limited to the superficial dermis.[20]

D. Pathologic Changes Associated with Systemic Toxicants

Dermatopathologic changes associated with systemic toxicants can be divided

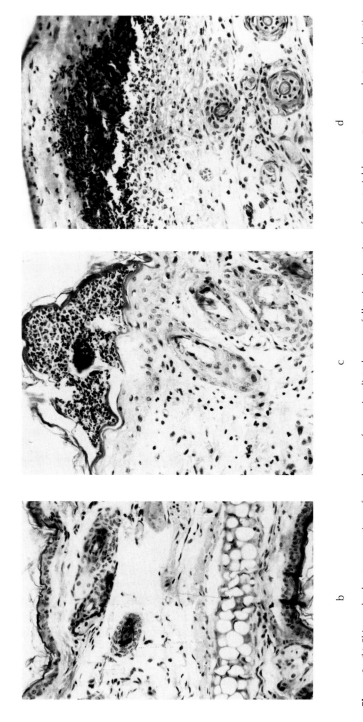

d

c

b

Figure 6. (b) Skin and subcutaneous tissue from the ear of a nonirradiated mouse following ingestion of a material known to cause photosensitization. Note the general absence of microscopic changes. (c) Focal epidermal necrosis with pustule formation in the ear of a mouse exposed to UV irradiation following ingestion of a diet containing 7.8% photodynamic material. (d) Diffuse necrosis and epidermal desquamation in the skin of the ear of a mouse exposed to UV irradiation following ingestion of a diet containing 30% photodynamic material. Photographs courtesy of Dr. L.D. Rowe et al., with permission from *Am J. Vet. Res.*

into several broad categories: (1) toxicants which directly damage integumentary structures following systemic distribution, (2) systemic drugs or chemicals which induce skin lesions via secondary mechanisms (e.g., allergic reactions), (3) contact agents which directly damage skin *and* internal organs following percutaneous absorption, (4) contact agents which fail to induce skin changes, but cause significant damage to internal organs. These four mechanisms, either alone or in combination, offer an almost unlimited array of dermal-systemic toxicologic interactions. It is not feasible to address all possible toxicologic relationships between the skin and other organs. Instead, the purpose of this section is to provide key examples of dermal-systemic toxicant interactions where pathologic findings in one or more tissues provide helpful diagnostic or mechanistic data.

1. Toxicants Which Can Directly Damage Skin Following Systemic Distribution

Thallium toxicosis — Skin lesions associated with thallium ingestion have largely disappeared in the U.S. since the use of this rodenticide has been terminated. Thallium is directly toxic to epidermal cells, resulting in disseminated parakeratotic hyperkeratosis, degeneration of keratinocytes and ulceration. Alopecia following follicular keratosis is also prominent.[21]

Halogen Acne (Chloracne) — Dating from the era when the ingestion of highly chlorinated naphthalenes was recognized as the cause of "Hyperkeratosis X" disease in cattle, systemic exposure to polychlorinated and polybrominated compounds has been incriminated as the causative factor for skin and systemic lesions in many animals.[45,46] Histopathologic changes usually consist of widespread epithelial hyperplasia and hyperkeratosis with follicular keratosis and squamous metaplasia of glandular epithelia. Liver changes, when present, may consist of persistently swollen hepatocytes with homogeneous, eosinophilic, finely granular cytoplasm. Because skin lesions may be associated with toxicant-induced hypovitaminosis A, it is possible that some halocarbon-induced dermatoses should be placed in the following mechanistic category.[47,48]

2. Systemic Agents Which Induce Skin Lesions Via Secondary Mechanisms

Toxic epidermal necrolysis (TED) — TED is a rare vesiculobullous to ulcerative skin disease characterized microscopically by hydropic degeneration of basilar cells and full-thickness necrosis of the epidermis. Inflammatory cell infiltrates may be sparse.[49] Although precise mechanisms have not been established, hypersensitivity reactions to various pharmaceuticals probably contribute to this dermatologic disease.[50]

3. Contact Agents Which Damage Skin and Other Organs Following Percutaneous Absorption

A classic example of this variant is "flea collar dermatitis" in domestic animals wearing vinyl collars impregnated with the organophosphorous insecticide, dichlorovos (DDVP). Not only may DDVP induce contact irritation, it may also cause neurotoxicity via cholinesterase inhibition.[21,51]

4. Contact Agents Which Fail to Induce Skin Changes but Cause Significant Toxicity to Other Organs

A large number of systemic toxicants are rapidly absorbed without significant skin damage. Noteworthy are organophosphorous and chlorinated hydrocarbon insecticides which may readily pass though the intact skin and cause fatal neurologic perturbations.[51,52]

V. CUTANEOUS NEOPLASIA INDUCED BY CHEMICAL OR PHYSICAL AGENTS

The carcinogenic potential of topical agents was recognized over 80 years ago when Sir Percival Pott described soot-induced scrotal carcinomas in London chimney sweeps. Following this landmark discovery, a variety of chemical and physical agents has proved to be either initiators or promoters of skin cancer. Polycyclic aromatic hydrocarbons, aromatic amines, nitrosoamines, and azo dyes head the list of chemical carcinogens, while ionizing and nonionizing radiation are also known to induce cutaneous neoplasia.[37] All of these carcinogens, and many more, have been extensively evaluated in laboratory animals, and an exhaustive body of literature now exists which addresses skin cancer induced by environmental agents. The purpose of this section is to describe pathologic changes in the skin of laboratory animals following exposure to initiators or promoters of cancer, and to identify laboratory species which may be used as models for skin cancer.

Because of their simplistic management and relatively short life spans, mice and rats have served as the principal research models for cutaneous neoplasia. More importantly, selective breeding has established strains which are hypersusceptible to chemically induced, two-stage skin carcinogenesis. The SENCAR mouse, which was derived from crossing Charles River CD-1 mice with "skin tumor-sensitive" mice (STS) has proved to be extremely susceptible to many tumorigenic chemicals.[52,54-56] Additionally, SKh:hairless mice have been used extensively to evaluate the ability of sunscreens and other chemicals to prevent photoinduced skin cancer.[5] Rats have been less widely used in studies to evaluate topical carcinogens, but have provided exhaustive data concerning the ability of systemic chemicals to induce cutaneous neoplasia.[57,58]

In considering skin tumors, both epithelial (epidermal) and connective tissue (dermal) neoplasms must be addressed. Epithelial tumors, both spontaneous and induced, generally can be divided into three groups: (1) squamous cell; (2) basal cell; and (3) sebaceous tumors. The separation is not absolute because many tumors may be composed of mixed cell types. In the dermal compartment, fibromas and fibrosarcomas originating from transformed fibrocytes are the predominant skin-associated tumors. Melanomas, mast cell tumors, and other cutaneous lymphoreticular neoplasms are recognized in laboratory animals, but usually these neoplasms are spontaneous rather than induced.

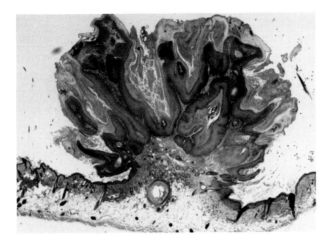

a

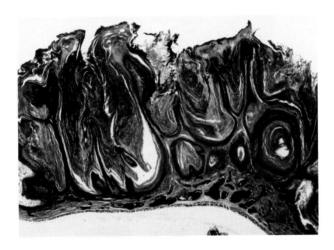

b

Figure 7.   (a) A benign, pedunculated papilloma from the muzzle area of a Fischer 344 rat. (b) A sessile papilloma. Distinction from keratoacanthoma (Figure 9b) may be difficult in some cases.

Squamous neoplasms may be either benign papillomas (Figure 7a,b) or malignant carcinomas (Figure 8) which actively invade juxtaposed tissue and may metastasize to distant sites. While metastasis to regional lymph nodes does occur, it is not common with most rodent epithelial neoplasms. A variant of the squamous cell tumor is the benign keratoacanthoma which often consists of dense, lamellated keratin protruding from a crater-like skin depression. Keratoacanthomas or papillo-

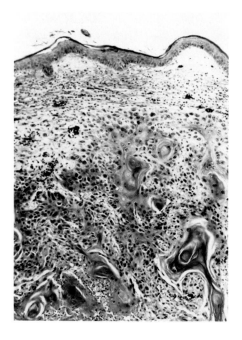

Figure 8. A malignant squamous cell carcinoma in the skin of a SENCAR mouse following skin painting with a known carcinogen. Although highly invasive, keratinocytes in this neoplasm exhibit distinct squamous differentiation. Some squamous carcinomas, in contrast, are poorly differentiated, and may resemble spindle cell tumors.

mas may be responsible for "cutaneous horn" formations in some animals (Figure 9a,b).

Basal cell tumors originate from the basilar germinal cells of either the follicular or interfollicular epithelium and are locally invasive. Only very rarely do they metastasize. Microscopically, they usually exhibit ribbons and festoons of basilar cells which may differentiate into squamous, sebaceous, or follicular structures (Figure 10).

Sebaceous tumors may consist of a benign proliferation of sebaceous gland cells which grow by local expansion only, or they may demonstrate active invasion or metastasis typical of a carcinoma (Figure 11).

Select adnexal glands may also respond to chemicals which induce skin neoplasia in rodents. Noteworthy are modified sebaceous glands associated with the external ear canal and the subcutaneous tissue of the inguinal region. These glands, referred to as Zymbal's glands, and the preputial (male)/clitoral (female) glands, respectively,

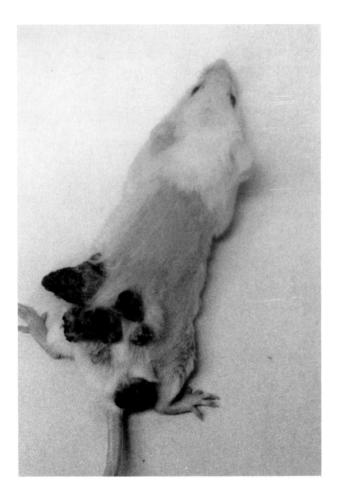

Figure 9a.   Multiple keratoacanthomas (and sessile papillomas) in a SENCAR mouse following skin painting with a known carcinogen.

may demonstrate neoplastic transformation following exposure to known cutaneous carcinogens.[59-61] Correspondingly, they should be included with the integumentary system when evaluating the tumorigenic potential activity of a test chemical or material.

Fibrosarcomas originating in subcutaneous connective tissue may be observed with many experimental manipulations or chemical exposures. Noteworthy is the ability of many relatively inert substances to induce fibrosarcomas when implanted in the subcutaneous tissues of laboratory animals.[62] Neoplastic transformation of fibrocytes may also occur in areas of chronic inflammation or ulceration. Usually, fibrosarcomas proliferate by local invasion rather than distant metastasis.

Because of the widespread use of laboratory rodents in dermal bioassay proce-

Figure 9b.   A keratoacanthoma from the skin of a SENCAR mouse. Note the subgross appearance of a "cup-like" crater filled with stacks and concentric lamellae of keratin. Differentiation from a sessile papilloma may be difficult in some cases.

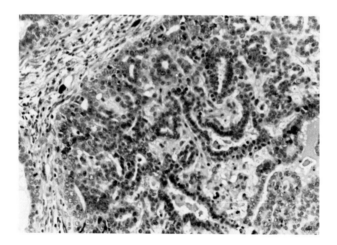

Figure 10.   A basal cell epithelioma from the subcutaneous tissue of a Fischer 344 rat. Note the ribbons and festoons of proliferating basilar cells. These neoplasms may exhibit squamous, glandular, or sebaceous differentiation.

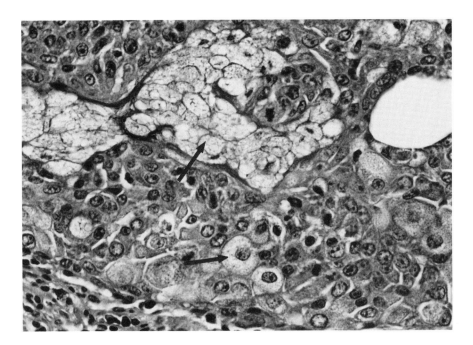

Figure 11.   Proliferating neoplastic cells display sebaceous differentiation (arrows) in this invasive carcinoma from the skin of a Fischer 344 rat.

dures, the U.S. Environmental Protection Agency (EPA) convened several workshops (April 1987 and May 1988) to identify the most suitable test species and establish uniform procedures for dermal carcinogenesis testing.[63,64] Although no obligatory guidelines were promulgated by the workshops, the following observations and recommendations were made relative to the use of mice, rats, and hamsters in skin cancer testing.

A. Mice

Eight strains were identified as being susceptible to tumorigenesis by the dermal route. They include: BALB/C, B6C3F1, CD-1, C3H, C57Bl/6, Swiss (ICR), SENCAR and possibly SKh/hr (hairless). Of these eight strains, spontaneous lesions or genetic defects may reduce the usefulness of four stocks. Specifically, SENCAR mice display problems with genetic drift and may not be readily available from commercial breeders. C3H mice have a high spontaneous incidence of mammary tumors. C57Bl/6 mice frequently develop ulcerative skin lesions of poorly defined etiology, and aging SKh/hr mice develop striking numbers of follicular cysts which may rupture and incite severe granulomatous dermatitis. Regardless, mice appear to be the best laboratory rodents for skin carcinogenesis testing because they respond to small doses of carcinogens and absorption/pharmacokinetic data are often available. Benign papillomas and keratoacanthomas along with malignant squamous cell carcinomas are the most common forms of skin cancer in mice.[63,64] Of critical importance in mouse "skin painting" (carcinogenesis) studies is the selection of high dose concentrations which can be tolerated by the host without significant epidermal damage. Dose levels which exceed the maximum tolerated dose (MTD) usually result in epidermal necrosis, erosion, and/or ulceration with possible compromise of the test system via protein loss, electrolyte imbalances, or secondary infectious agents. Additionally, treatment-induced epidermal damage may stimulate chronic proliferative (repair) activity which, in itself, may act as a tumor promoter or enhance spontaneous neoplastic transformation of replicating keratinocytes. As a consequence, the selection of MTDs should be based upon range-finding studies where histopathologic assessments confirm that significant epidermal insult has not occurred.

Appendix I outlines a histomorphology worksheet which may be used to establish MTDs based upon composite severity scores for select epidermal changes. Additionally, this worksheet may be used to rank-order proliferative changes in skin specimens based upon subtotal and composite severity scores. This worksheet has proven to be highly useful in distinguishing reactive changes induced by test materials of low toxicity or irritancy. For purposes of standardization, most diagnostic parameters listed on this worksheet have been illustrated in an *Atlas of Dermal Lesions* (2OT-2004, August 1990) published by the U.S. Environmental Protection Agency (TS-796), Washington, D.C. 20460.[65]

B. Rats

The workshop did not identify the most suitable strain of rat for skin testing, but cited the extensive tumorigenesis data base which has been developed by the National Toxicology Program for the Fischer-344 rat. When compared with mice, greater doses are usually required to induce skin cancer in rats, and the most frequent tumor is the basal cell carcinoma.[63,64]

C. Hamsters

The Syrian golden hamster may be more susceptible than the rat to the development of both dermal and systemic neoplasia following skin exposures; however, little data exist concerning dermal absorption in hamsters and husbandry for these rodents may be difficult. Squamous cell carcinomas and melanomas are the most common chemically induced skin tumors in hamsters.[63,64]

The EPA workshops concluded that the mouse was to be recommended as the first test species for skin carcinogenesis and the rat should be selected as the second test species. Special circumstances may justify the use of other species (e.g., hamsters) based upon physiologic data and susceptibility to the test chemical.

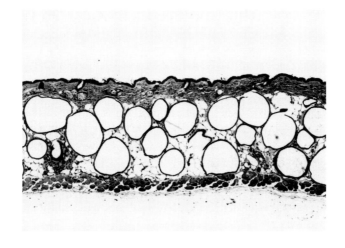

a

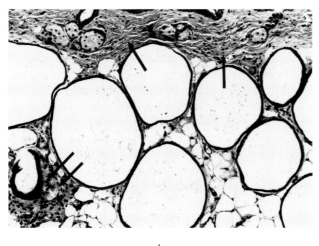

b

Figure 12. (a) Skin and subcutaneous tissue of a 10-month-old SKh hairless mouse illustrating spontaneous cystic dilatation of dysplastic hair follicles. Follicles are empty, suggesting that *in vivo* contents were dissolved by aqueous fixatives or fat solvents used in tissue processing. (b) Higher magnification of Figure 12a illustrating dermal fibrosis (single arrows) and early granulomatous inflammation associated with follicular rupture (double arrows). The changes may become severe and compromise the SKh strain as a useful model in long-term dermatotoxicologic studies.

VI. SPECIAL ANATOMIC STRUCTURES ASSOCIATED WITH THE SKIN OF LABORATORY ANIMALS

In addition to Zymbal's glands and the preputial/clitoral glands (Section V), laboratory rodents possess several unique integumentary structures or functions which deserve consideration. Noteworthy are the "pigmented flank organs" of hamsters and the "caudal glands" of guinea pigs which may be mistaken for melanomas or treatment-induced lesions. The flank organs of hamsters are located bilaterally in the dorsal flank area and consist of a dense collection of sebaceous glands with abundant melanin pigment. Guinea pig caudal glands occur over the coccyx and have a similar histologic appearance. Experimentally, the flank organs have proved to be valuable in testing sebaceous gland responses to exogenous sex hormones.[66] In routine skin patch testing, however, they may reduce the area of skin available for treatment.

The use of nude or hairless rats, mice, and guinea pigs has gained wide acceptance in skin research. Especially beneficial has been the contribution of athymic nude mice in transplantation studies and other skin changes requiring interaction with the immune system. Another valuable mouse strain in skin studies has been the SKh hairless mouse. In this strain, normal hair follicles are present in neonates, but follicles become markedly dysplastic by 30 to 35 days of age and fail to produce significant external hair shafts. With increasing age, deep follicular structures become dilated resulting in multiple, cystic hair follicles in the subcutis. Not infrequently, these cysts rupture and release highly irritating sebum and keratin into the subcutis resulting in granulomatous inflammation and dermal fibrosis, (Figure 12a,b). These spontaneous changes must be distinguished from treatment-related effects.[5]

An additional feature of the rodent integument which may go unrecognized in dermatotoxicologic investigations is the importance of the tail as a thermoregulatory organ. In this connection, amputation or disruption of the heat conduction functions of the tail may contribute to fatal hyperthermia. Furthermore, rats subjected to low environmental humidity may demonstrate reflex annular constriction of the tail with gangrenous necrosis of the tail tip. This condition, known as "ring tail", is more common in juvenile rats.[67]

VII. SYSTEMIC DISEASES OR CHANGES WHICH MIGHT COMPROMISE DERMATOTOXICOLOGIC STUDIES IN LABORATORY ANIMALS

It is apparent that a plethora of systemic changes, either functional or anatomic, may occur following the percutaneous absorption of toxicants. This section describes important spontaneous and induced systemic lesions in laboratory animals which might mask or complicate the evaluation of experimental results. Because of the widespread use of rabbits in skin toxicity studies, these animals deserve special consideration.

A. Rabbits

Infectious Diseases — Unexpected morbidity and mortality in rabbits used for percutaneous toxicity testing may be attributed to several infectious diseases. Although specific pathogen free animals may be purchased which are largely free of these diseases, rabbits from most commercial sources are infected. Losses may be so great that studies are hopelessly compromised.

1. **Enterotoxemia** — Several variants of this highly fatal disease "complex" may occur in young rabbits following stress, dietary changes, or experimental manipulation. Causative organisms may include toxigenic strains of *Escherichia coli* or *Clostridium* sp. Clinical and pathological findings range from acute death with mild intestinal hyperemia as the only pathologic change to a more protracted course characterized by anorexia, weight loss, and "mucoid enteropathy". In most cases, treatment is of limited value.[68-70]

2. **Pasteurellosis** — Infection with the organism *Pasteurella multocida* usually begins as a mucopurulent rhinitis ("snuffles") and progresses to involve regional lymph nodes, middle and inner ear structures, and the lungs. Clinically, torticollis (head tilt) is observed when the organism reaches the inner ear, and death rapidly ensues following pleuropulmonary dissemination. Chemotherapy is generally useless.[70]

3. **Encephalitizoonosis** — This insidious disease, which is caused by the microsporidian parasite, *Encephalitozoon cuniculi,* may result in significant lesions in the kidney, central nervous system (CNS), and the hepatobiliary system. Although replication of this parasite in the renal tubular epithelium frequently causes a severe tubulointerstitial nephritis, the presence of parasite-induced microgranulomas in the CNS is of more concern in dermatotoxicology studies. Granulomas may be widespread throughout the brain and spinal cord and mask or potentiate neurologic signs associated with the test article.[71]

1. Spontaneous or Induced Lesions in Rabbits

Testicular Degeneration - Degeneration of the testicular germinal epithelium is a common, spontaneous lesion in laboratory rabbits restrained for skin testing procedures.[72] Precise causative factors remain obscure, but probably relate to caging and restraining methods. Microscopic findings include attenuation of spermatogenic activity with the formation of spermatid giant cells in many tubules (Figure 13). When adequate numbers of control animals are omitted from the test protocol, testicular changes might be erroneously attributed to the test substance.

Topical Corticosteroids — Studies have indicated that the rabbit may be highly sensitive to the topical application of corticosteroids followed by occlusive dressings. Percutaneous absorption of glucocorticoids in this species has been associated with severe thymic and adrenocortical involution (Figures 14a,b and 15a,b). Correspondingly, these systemic effects may abrogate immune or inflammatory responses within the skin or other organs.

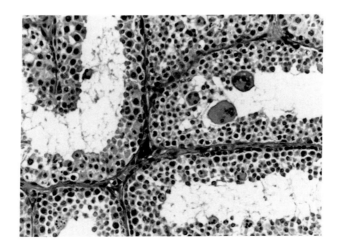

Figure 13. Testicular seminiferous tubules from a control New Zealand white rabbit used in percutaneous toxicity studies. There is diffuse attenuation of spermatogenic activity and the formation of spermatid giant cells within the tubular lumina. These degenerative changes may occur spontaneously in control and treated rabbits, and they should not be regarded as an exposure-related event without careful evaluation of all test results.

Figure 14a. Thymic lobule from a control New Zealand white rabbit exhibiting a normal medulla and a cortical mantle richly populated with lymphocytes. (b) Severe thymic involution with marked depletion of cortical lymphocytes in a 4-month-old rabbit treated with topical applications of a 1% corticosteroid ointment for approximately 1 month.

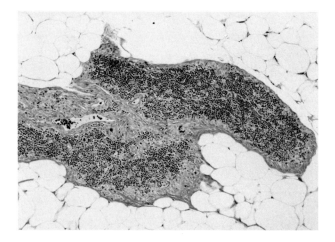

Figure 14b.

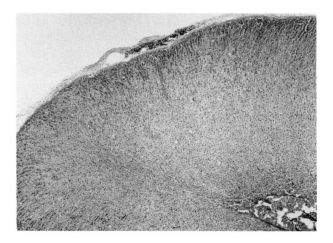

Figure 15a. Normal adrenal gland from a 4-month-old New Zealand white rabbit. Note the relative thickness of the zona fasciculata (middle cortical layer) which secretes glucocorticoids. The medulla is identified by an arrow. (b) Severe adrenocortical involution in an age-matched (see Figure 15a) rabbit treated with topical applications of a 1% corticosteroid ointment for approximately 1 month.

Figure 15b.

B. Other Laboratory Animals

Adjuvant-induced systemic lesions — Immunomodulating agents, such as Fruend's adjuvant, have gained widespread acceptance in studies to characterize contact allergens. These immune boosters may cause multisystem granulomas, amyloidosis, or severe arthritis in many laboratory species (Figure 16a,b). These secondary lesions may reduce the longevity of test species.

Gastric ulcers and bleeding — Experimental manipulation of rats along with restricted feeding often results in stress ulcers of the stomach. These ulcers, in turn, may result in fatal gastric hemorrhage.

Retinal degeneration and lenticular cataracts —Exposure of albino rats and mice to high intensity lighting typical of phototoxicity experiments may induce severe ocular changes including retinal degeneration and cataractous changes in the lens. These lesions eventually result in functional blindness.[67]

Systemic lesions associated with skin ulcers — In some strains of mice, especially C57Bl/6, spontaneous skin ulceration may occur with great frequency and completely compromise investigational results. This is especially problematic when mice are gang caged, and there is extensive alopecia and ulceration associated with barbering (trichophagia) or fighting. It should be remembered that many strains of mice are persistently infected with oncornaviruses, resulting in a host of systemic abnormalities or neoplastic diseases.

Hepatocellular fatty change is a common finding in liver cells of rodents with chronic ulcerative skin disease. Cutaneous ulcers provide a gateway for pathogenic organisms to enter the body and for loss of serum proteins and electrolytes. Additionally, significant renal disease and pneumonitis has been associated with skin ulceration.

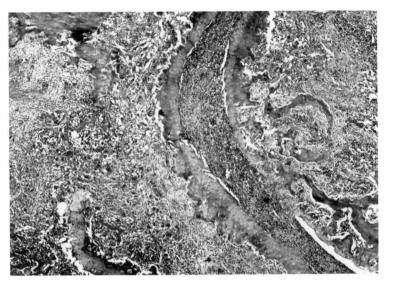

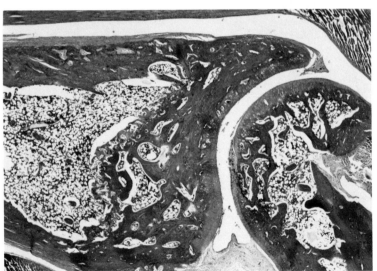

Figure 16. (a) Normal tibiotarsal joint from a Sprague-Dawley rat serving as a control animal in a pharmacology experiment. (b) Severe osteoarthritis and degenerative joint disease in a rat receiving a single subcutaneous injection of modified Freund's adjuvant (MFA) 2 months previously. The use of MFA to modulate immune responses in contact allergy studies may result in severe secondary lesions in many organs.

VIII. CONCLUSIONS

This chapter has surveyed toxicologic pathology as related to the largest organ system in the body — the skin. The methods and interpretations presented herein should assist toxicologists and pathologists in selecting appropriate procedures for examining toxicant-exposed skin specimens, and they should provide guidance for evaluating histomorphic changes. Although efforts were made to incorporate some of the latest technicologic advancements into the text, dermatologists should be alert for future pathology breakthroughs which might enhance dermatotoxicology investigations. Additionally, it is important to recognize that toxicant-induced systemic changes may markedly influence skin responses, and where possible, dermatotoxic studies should include complete clinical and anatomic pathology evaluations.

ACKNOWLEDGMENTS

The author extends deep appreciation to Ms. Vickie Phillips and Ms. Cathy Sutton of Pathology Associates, Inc. for their patience and skill in the preparation of this manuscript. Thanks are also extended to The Procter and Gamble Company, Cincinnati, Ohio, and to the Toxic Hazards Division, Air Force Aerospace Medical Research Laboratory, Wright-Patterson, AFB, Ohio, for illustrative material.

APPENDIX I
HISTOMORPHIC ASSESSMENT OF MOUSE SKIN SPECIMENS
PATHOLOGY ASSOCIATES, INC.

DAYS ON TEST: 23

Group No. ___1___   Species: SENCAR   Test Article: Compound X

ANIMAL NUMBER	1	2	3	4	5	6	7
EPIDERMIS							
MTD Parameters							
1. Erosion	2	1	0	1	0	0	1
2. Ulceration	1	0	0	0	0	0	2
3. Crust Formation	2	1	0	0	0	1	1
4. Epidermitis	3	1	1	2	1	1	1
5. Necrosis/Acantholysis	0	0	0	0	0	0	0
6. Spongiosis/Hydropic Degen.	3	2	1	2	2	1	1
Sub-Total MTD Score	11	5	2	5	3	3	6
Proliferative Changes							
7. Orthokeratosis	1	2	2	2	2	2	2
8. Parakeratosis	3	3	1	2	1	0	0
9. Hyperplasia	4	4	3	4	4	3	3
10. Dyskeratosis	0	0	0	0	0	0	0
11. Epidermal Thickness (µm) *	85	83	49	57	63	52	41
12. Epi. Cells per 100 µm (Ave.)	61	69	41	53	45	42	37
13. Mitoses Per 4mm	10	11	8	16	10	7	8
14. _____							
Sub-Total Prolif. Score	164	172	104	134	125	106	91

DERMIS							
14. Inflammatory Infiltrates	4	3	3	3	3	3	3
15. Edema	1	0	0	0	0	0	0
16. Vasodilatation	2	1	0	1	1	0	1
17.							
Sub-Total Dermal Score	7	4	3	4	4	3	4
COMPOSITE SCORE	182	181	109	143	132	112	101

SCORE: 0 = Regular/Normal, 1 = Minimally Irregular/Abnormal, 2 = Mildly Irregular/Abnormal,
3 = Moderately Irregular/Abnormal, 4 = Marked Irregular/Abnormal, 5 = Severely Irregular/Abnormal.
MTD = Maximum Tolerated Dose (MTD Score ≤5 = below MTD; 6-8 = borderline; >8 = exceed MTD)
* Epidermal thickness and Epi. cells per 100μm are averages of five micrometer-based measurements per specimen.

REFERENCES

1. W. F. Lever and G. Schaumberg-Lever, Eds., *Histopathology of the Skin,* 5th ed., J. B. Lippincott, Philadelphia (1975).
2. L. G. Luna, Ed., *Manual of Histologic Staining Methods of the Armed Forces Institute of Pathology,* 3rd ed., McGraw-Hill, New York (1968).
3. D. C. Sheehan and B. B. Hrapchak, Eds., *Theory and Practice of Histotechnology,* 2nd ed., Battelle Press, Columbus (1980).
4. W. C. Johnson, Proc. Armed Forces Inst. Pathol. Course Appl. Histochem. Pathol., Washington, D.C., Jan. (1977).
5. L. H. Kligman, F. J. Akin, and A. M. Kligman, *J. Invest. Dermatol.* 78, 181 (1982).
6. E. Gomez, O. J. Corrado, D. L. Baldwin, A. R. Swanston, and R. J. Davies, *J. Allerg. Clin. Immunol.* 78, 637 (1986).
7. J. A. Ashworth, M. L. Turbitt, and R. Mackie, *Clin. Exp. Dermatol.* 11, 153 (1986).
8. J. D. Bos, I. Zonneveld, P. K. Das, S. R. Krieg, C. M. Van der Loos, and M. L. Kapsenberg, *J. Invest. Dermatol.* 88 (5), 569 (1987).
9. J. Ferguson, J. H. Gibbs, and J. S. Beck, *Contact Dermatitis* 13, 166 (1985).
10. P. C. Kung, *J. Cutaneous Pathol.* 10, 457 (1983).
11. A. I. Lauerma, K. Visa, M. Pekonen, L. Forstrom, and S. Reitamo, *Arch. Dermatol. Res.* 279, ISS 6, 379 (1987).
12. J. A. K. Patterson, *J. Cutaneous Pathol.* 10, 425 (1983).
13. N. S. Penneys, *J. Cutaneous Pathol.* 10, 431 (1983).
14. A. deFazio, J. A. Leary, D. W. Hedley, and M. H. N. Tattersall, *J. Histochem. Cytochem.* 35, 5, 571 (1987).
15. R. A. Smith, Proc. Natl. Soc. Histotechnol., Louisville, KY, October 9-14 (1988).
16. K. Fukuyama, Autoradiography, in *Methods in Skin Research.* D. Skerrow and C. J. Skerrow, Eds. John Wiley & Sons, Chichester (1985).
17. C. A. Holden and D. M. MacDonald, *J. Cutaneous Pathol.* 10, 448 (1983).
18. G. Kolde and J. Knop, *J. Invest. Dermatol.* 90 (3), 320 (1988).
19. R. A. J. Eady, Transmission Electron Microscopy, in *Methods in Skin Research,* D. Skerrow and C. J. Skerrow, Eds. John Wiley & Sons, Chichester (1985).
20. A. B. Ackerman, *Histologic Diagnosis of Inflammatory Skin Diseases,* Lea & Febiger, Philadelphia (1978).

21. G. H. Muller, R. W. Kirk, and D. W. Scott, *Small Animal Dermatology*, 3rd ed., W. B. Saunders, Philadelphia (1983).
22. J. K. Billings, S. S. Milgram, A. K. Gupta et al., *Arch. Dermatol.* 123, 1662 (1987).
23. A. B. Ackerman, J. Niven, and J. M. Grant-Kels, Eds. *Differential Diagnosis in Dermatopathology*, Lea & Febiger, Philadelphia, (1982).
24. J. Ferguson, J. H. Gibbs, and J. Swanson Beck, *Contact Dermatitis* 13, 166 (1985).
25. J. G. Marks, Jr., R. J. Zaino, M. F. Bressler, and J. V. Williams, *Int. J. Dermatol.* 26 (6), 354 (1987).
26. H. C. Maguire, Jr. and D. Cipriano, Allergic Contact Dermatitis in Laboratory Animals, in *Cutaneous Toxicity,* V. A. Drill and P. Lazar, Eds., Raven Press, New York (1984).
27. C. M. Willis, E. Young, D. R. Brandon and J. D. Wilkinson, *Br. J. Dermatol.* 115, 305 (1986).
28. J. P. Fisher and R. A. Cooke, *J. Allergy* 29(5), 411 (1958).
29. N. Hunziker, in *Advances in Modern Toxicology, Vol 4, Dermatotoxicology and Pharmacology*, F. N. Marzulli and H. I. Maibach, eds., Hemisphere Publishing, Washington, D.C., 373 (1977).
30. M. Medenica and A. Rostenberg, Jr., *J. Invest. Dermatol.* 56(4), 259 (1971).
31. M. K. Robinson, E. R. Fletcher, G. R. Johnson, W. E. Wyder, and J. K. Maurer, *J. Invest. Dermatol.* 94(5), 636 (1990).
32. W. Aberer, N. Romani, A. Elbe, and G. Stingl, *J. Immunol.* 136, 1210 (1986).
33. M. Orita, *Br. J. Dermatol.* 117, 721 (1987).
34. G. Stingl, C. A. Stingl, W. Aberer, and K. Wolff, *J. Immunol.* 127, 1707 (1981).
35. S. Sullivan, P. R. Bergstresser, R. E. Tigelaar, and J. W. Streilein, *J. Immunol.* 137, 2460 (1986).
36. C. Carraro and M. A. Pathak, *J. Invest. Dermatol.* 90 (3), 267 (1988).
37. E. A. Emmett, Toxic Responses of the Skin, in *Casarett and Doull's Toxicology,* 3rd ed., C.D. Klaassen, M.O. Amdur and J. Doull, Eds., Macmillian, New York (1986).
38. A. Kornhauser, W. Warner, and A. Giles, Jr. Light-induced Dermal Toxicity: Effects on the Cellular and Molecular Level, in *Dermatotoxicology*, 3rd ed., F. N. Marzulli and H. I. Maibach, Eds., Hemisphere Publishing, Washington, D.C. (1987).
39. L. C. Harber, R. B. Armstrong, and H. Ichikawa, *J.N.C.I.* 69(1), 237 (1982).
40. L. C. Harber, A. R. Shalita, and R. B. Armstrong, Immunologically Mediated Contact Photosensitivity in Guinea Pigs, in *Dermatotoxicology*, 3rd ed., F. N. Marzulli and H. I. Maibach, Eds., Hemisphere Publishing, Washington, D. C. (1987).
41. P. A. Giudici and H. C. Maguire, Jr., *J. Invest. Dermatol.* 85, 207 (1985).
42. L. D. Rowe, J. O. Norman, R. E. Corrier, S. W. Casteel, B. S. Rector, E. M. Bailey, J. L. Schuster, and J. C. Reagor, *Am. J. Vet. Res.* 48 (11), 1958 (1987).
43. I. Willis, Photosensitivity, in *Dermatology, Vol. I*, S.L. Moschella, D.M. Pillsbury, and H.J. Hurley, Jr., Eds., W. B. Saunders, Philadelphia (1975).
44. R. J. Barr, Spongiotic Dermatitis, in Proc. 38th Annu. Meet. Am. Coll. Vet. Pathol., Monterey, CA (1987).
45. T. F. Jackson and F. L. Halbert, *J. A.V. M. A.* 165(5) 437 (1974).
46. E. E. McConnell, J. A. Moore, B. N. Gupta, A. H. Rakes, M. I. Luster, J. A. Goldstein, J. K. Haseman, and C. E. Parker, *Toxicol. Appl. Pharm.* 52, 468, 1980.
47. T. C. Jones and R. D. Hunt, Eds., *Veterinary Pathology*, 5th ed., Lea & Febiger, Philadelphia (1983).
48. M. J. VanRafelghem, D. R. Mattie, R. H. Bruner, and M. E. Andersen *Fund. Appl. Toxicol.* 9, 522 (1987).
49. J. C. Guillaume, J. C. Roujeau, J. Reuaz, D. Penso, and R. Touraine, *Arch. Dermatol.* 123, 1166 (1987).
50. J. Revuz, D. Penso, J. C. Roujeau, J. C. Guillaume, C. R. Payne, J. Wechsler, and R. Touraine, *Arch. Dermatol.* 123, 1160 (1987).
51. A. Curley, R. E. Hawke, and R. D. Kimbrough, et al., *Lancet* (2), 296 (1971).
52. R. D. Radeleff, *Veterinary Toxicology*, Lea & Febiger, Philadelphia (1964).
53. E. Hecker, *Toxicol. Pathol.* 15(2), 245 (1987).
54. P. T. Strickland, *Environ. Health Perspect.* 68, 131 (1986).

55. T. J. Slaga, *Environ. Health Perspect.* 68, 27 (1986).
56. J. M. Ward, S. Rehm, D. Devor, H. Hennings, and M. L. Wenk, *Environ. Health Perspect.* 68, 61 (1986)
57. J. D. Burek, Ed., *Pathology of the Aging Rat*, CRC Press, West Palm Beach, FL (1978).
58. D. G. Goodman et. al., Chemically Induced and Unusual Proliferative and Neoplastic Lesions in Rats, Registry of Veterinary Pathology, Armed Forces Institute of Pathology, Washington, D.C. (1984).
59. G. Reznik and J. M. Ward, *Vet. Pathol.* 18, 228 (1981).
60. J. C. Seely, *Lab Animal* May/June, 25 (1986).
61. G. B. Pliss, Tumors of the Auditory Sebaceous Gland, in *Pathology of Tumors in Laboratory Animals*, Vol. 1 (pt. 1), V. S. Tursov, Ed. International Agency for Research on Cancer, Lyon (1973).
62. R. L. Carter, F. J. C., Roe, and R. Peto, *J.N.C.I.* 46, 1277 (1971).
63. Environmental Protection Agency, Summary of the EPA Workshop on Carcinogenesis Bioassay via the Dermal Route, April 28-29, 1987, Washington, DC (EPA 560/6-89-002), 1989.
64. Environmental Protection Agency, Summary of the Second EPA Workshop on Carcinogenesis Bioassay via the Dermal Route, May 18-19, 1988, Research Triangle Park, NC (EPA 560/6-89-003), 1989.
65. Environmental Protection Agency, Atlas of Dermal Lesions (2OT-2004), Washington, DC (1990).
66. A. Weissmann, J. Bowden, B. L. Frank, S. N. Horwitz, and P. Frost, *Arch. Dermatol.* 121, 57 (1985).
67. J. G. Fox, B. J. Cohen, and F. M. Loen, Eds., *Laboratory Animal Medicine,* Academic Press, New York (1984).
68. N. M. Patton, H. T. Holmes, R. J. Riggs, and P. R. Cheeke, *Lab. Anim. Sci.*, 28 (5), 536 (1978).
69. J. E. Rehg and S. P. Pakes, *Lab. Anim. Sci.*, 32 (3), 253 (1982).
70. S. H. Weisbroth, R. E. Flatt, and A. L. Kraus, Eds., *The Biology of the Laboratory Rabbit*, Academic Press, New York (1974).
71. J. A. Shadduck and M. J. Geroulo, *Lab. Anim. Sci.*, 29 (3), 330 (1979).
72. D. Morton, S. E. Weisbrode, W. E. Wyder, J. K. Maurer, and C. C. Capen, *Vet. Pathol.* 23, 176 (1986).

3

Considerations in the Design and Conduct of Subchronic and Chronic Dermal Exposure Studies with Chemicals

WILLIAM C. EASTIN, JR.
National Institute of Environmental Health Sciences
Research Triangle Park, North Carolina

I. INTRODUCTION

At the present time, subchronic and chronic toxicology and carcinogenesis studies are performed using laboratory animal models to provide information that can be used to serve common public health concerns. Subchronic (e.g., 14-day and 90-day) studies serve to provide data on the short-term toxicity and disposition of a chemical or chemical mixture and to aid in setting doses for chronic (e.g., 2-year) studies. Two-year chronic exposure studies using laboratory animal models continue to be the most definitive means for the identification of chemical carcinogens. Historically, subchronic and chronic toxicity studies like those conducted by the National Toxicology Program (NTP) have used skin as the route of exposure less frequently relative to the number of studies performed by dosed feed, dosed water, and gavage routes. This is due in part to the intensive technical effort required for preparation of the test animals and application procedures. However, subchronic and chronic studies conducted under conditions that are most relevant to human exposure situations should provide as much information as possible to more completely characterize toxicity and there is an increasing emphasis on conducting animal studies using the most likely route of human exposure.[1-5]

The likelihood of the skin being exposed to toxic chemicals is extremely high; it is exposed in the occupational setting, while using chemicals at home, and to a wide range of cosmetic and toiletry formulations and drugs in a variety of vehicles. Thus, skin exposures are either repetitive or may occur over a long period of time; conducting subchronic and chronic toxicity studies using skin as the route of exposure is appropriate.

Chronic studies are designed to observe effects of chemicals on animals exposed for a major portion of their lifetime. As these studies are performed at many

111

laboratories throughout the world, the conditions for conducting the studies have been extremely variable. In 1979, an international committee of scientists convened to discuss flaws in study design and data interpretation and to set basic requirements for the design, conduct, analysis, and reporting that would improve the quality and comparability of the results of these studies.[4]

Skin exposure studies have inherent practical problems that make them costly to perform, such as a requirement to physically treat each animal, the need to prevent chemical cross-contamination of animals, and routine hair clipping. For these reasons, fewer subchronic and chronic skin exposure studies have been conducted and methods for these studies are not as well defined as in the protocols for the more frequently performed dosed-feed, gavage, and dosed-water studies.

The objective of this chapter is to present some important considerations for designing subchronic and chronic toxicity studies using skin as a route of exposure. For the purpose of this chapter, it will be assumed that (1) local and systemic toxicity depend on a chemical penetrating the skin, and (2) application of a chemical or chemical mixture to the skin requires penetration through the epidermal layer before reaching the dermis and entering the systemic circulation. In reviewing the literature, the terms "skin exposure", "skin painting", and "dermal studies" have often been used interchangeably to describe studies using the skin as an exposure route. While the phrase "dermal studies" is not quite appropriate, it is ingrained in the literature and will be used in the following discussion to mean skin exposure studies.

II. ANIMAL MODELS

A. Species

Many subchronic and chronic toxicity studies utilizing exposure routes other than the skin (i.e., intravenous, subcutaneous, intramuscular, oral, and inhalation) are designed to examine systemic effects of chemical exposures and assume that some or most of the chemical enters the body. In contrast, many dermal studies use short-term designs to examine the toxic effects or determine irritation potential of a chemical only at the site of application. Subchronic and chronic chemical exposure studies designed to study both local and/or systemic toxic effects of chemical applied to the skin are less common. The species most frequently used in dermal toxicity/carcinogenicity studies are rats, mice, Syrian hamsters, and less often, rabbits and guinea pigs. The selection of the species to be used is usually based on practical considerations such as a comparatively short life span, small body size, and availability.[1,6,7] Rats and mice have been used extensively in toxicology and carcinogenicity studies and large databases have been compiled for these species.[2,8,9] From these studies, a knowledge of the physiology, anatomy, genetics, husbandry, nutrition, spontaneous diseases, spontaneous tumor incidences, and susceptibilities to tumor induction has been acquired that provides a more reliable basis to assess the results of toxicology studies with these species.[9] The best comparison is with control data collected under similar study conditions, i.e., using the same route of exposure. However, the number of dermal toxicity studies conducted to date is relatively small and a comparison of the results to a database established using other routes of exposure is usually unavoidable. Therefore, because of the large database established with other routes of exposure rats and mice are most frequently used in prechronic and chronic dermal studies. There are special dermal studies in which other species are more commonly used. For example, hamsters, guinea pigs, and rabbits are routinely used in studies to determine irritation and sensitization potential of chemicals found in topically applied drugs and cosmetics.[6,10]

B. Strains

The nature and severity of a toxic response is not only a function of the sensitivity of the animal species used in the study, but also may be related to the strain used within a given species. In selecting an appropriate animal strain, genetics (i.e., whether the animals are inbred, hybrid, or outbred) is usually an important consideration. Many laboratories have experience with outbred strains because they are more readily available and these strains are commonly more disease resistant than inbred animals. However, randomly bred strains are subject to genetic drift and this can produce considerable variation in response. On the other hand, inbred strains are single genotypes and may not be representative of the species. F1 hybrids are a uniform genotype, but they have a level of heterozygosity more closely resembling the outbred animals.[1,3]

Other genetic factors are also important in strain selection for dermal studies. The differential response of the skin of certain strains of mice and hamsters to certain carcinogens has been used to study tumorigenesis.[11] Mutant hairless strains have been used to study photocarcinogenesis (hairless mouse)[12] and as a model for human percutaneous absorption (fuzzy rat).[13] In studies with haired species where exposure will continue for more than a few weeks, the site of application will need to be clipped regularly to insure consistent application of chemical. Thus, it is also important to consider the inherent excitability of the species.

Mice — A number of strains of mice have been used in dermal studies, but few of these studies measured both local (site of application) and systemic effects. In toxicology studies involving other routes of exposure, several strains (e.g., B6C3F$_1$, CD-1, and C3H, etc.) have been commonly used and a considerable data base on the general biology (including tumor incidences) on these strains exists. There is also a large database on biology and chemical effects on skin of strains used in the mouse skin initiation/promotion studies.[2,3,14-16]

Rats — Rats are commonly used animal models in toxicology studies using other routes of exposure.[4,6,7,10,21] However, because the rat has not been shown to be a good model for dermal initiation/promotion studies, the database from toxicology studies using skin as the route of exposure is limited with this species. Nevertheless, the large historical database that has been established for some rat strains from studies using routes of exposure other than skin favors the selection of rat as a test animal in dermal studies (e.g., F344/N at NTP and Osborne-Mendel from early NCI and FDA studies).[17,18]

Hamsters — Hamsters are a good model to use in dermal initiation/promotion protocols to study melanoma induction. Toxicity data on hamsters have also been obtained from studies on respiratory tract carcinogenesis.[18-20] However, hamsters are less frequently used in toxicology/carcinogenesis dermal studies because of their size and difficulty in handling.

Guinea Pigs — This species has been used in subchronic dermal studies and in irritation/sensitization studies; adequate background information is available in the literature .[6,10]

Toxicologic studies are generally conducted for extrapolation to humans, but no animal model is known to respond identically to man in every respect. Therefore, it has generally been recommended that toxicity/carcinogenicity studies be performed in at least two mammalian species to increase the extrapolative power of the results.[6,10,17,21] Both sexes are usually included in subchronic and chronic studies in order to examine the possibility of sex-dependent toxicity. Using species with an established biological database (e.g., the Fischer 344 rat and B6C3F1 mouse used extensively in toxicity/carcinogenesis studies by the NTP) has obvious advantages.

C. Site of Application

The usual areas of chemical application are the ear and the lateral and dorsal areas of the body. The skin of the ear has been used in some species, e.g., rabbits, but these studies are usually conducted to examine the potential for dermal irritation.[6,10] The skin on the lateral surface of the body is the site commonly used in studies to determine irritation and allergic responses to chemicals.[10] Irritation and allergic response studies are usually short term and the sites may be covered by a protective device or the animal may be restrained to eliminate the possibility of external contact with the dose site.[6,10] The most commonly used site in subchronic and chronic dermal studies is the interscapular area.[6,17,21] It is more difficult for the animal to reach and clipping and dose application at this site is less difficult to perform. Because skin thickness is not uniform over all body locations, percutaneous penetration may vary at different application sites.[22,23] Therefore, whatever the site, the dose should be applied to a fixed standard area for consistency.

Subchronic studies are usually conducted to determine local and systemic toxic effects, as well as to establish doses for long-term studies. In order to maximize the potential to observe a toxic effect in dermal studies, a larger dose volume can be applied to a larger area of skin to increase the total absorption of the dose. When the dose volume is increased, skin from the interscapular region extending posteriorly to near the base of the tail can be used. In most cases, however, the application site does not exceed 10% of the total body surface area.[6,10,21]

D. Maintenance

1. Housing

Polycarbonate or stainless steel, wire mesh cages are used in large scale toxicology studies. Cages should be chemical specific and destroyed or decontaminated after the study is completed to reduce the potential of cross-chemical contamination between studies. Polycarbonate cages are less drafty and temperature is less variable.

Stainless steel, wire mesh cages allow for air to circulate, an important consideration to help reduce an unintentional inhalation exposure if the chemical or vehicle solvent is volatile.

Animals may be housed singly or in groups. There are guidelines which limit the number and size of animals of a given species which may be grouped together.[24] There is some evidence that individual housing of animals alone may, in itself, produce biologic changes [15,25-27] and an increase in spontaneous tumors in some species.[28] Group-housed animals groom each other and may inflict wounds by fighting. These factors increase the potential for unwanted exposure orally and through wounds. Male mice are particularly prone to fighting and individual caging is recommended. Fighting may also significantly affect survival and incidence of disease in a study. Individual caging eliminates these problems, but it is more costly, especially when large numbers of animals are involved. If the objective is to study the systemic effects of dermal exposure, minimizing the potential for other exposure routes by individual housing is recommended. Animal care and environmental conditions should follow the recommendations set forth in the U.S. Department of Health and Human Services, *Guide for the Care and Use of Laboratory Animals.*[24]

III. STUDY DESIGN

A. Disposition and Absorption

It is important to know the magnitude of absorption and metabolism and disposition of chemicals before they are administered via any route of exposure in toxicology studies. This information provides a better estimate of the potential systemic effects of a chemical and helps to determine the need to conduct further studies. The skin is a formidable barrier to penetration and it is especially important to determine the extent and factors which can affect percutaneous absorption (see Chapters 7, 8, and 9). The study of chemical-specific metabolism is also useful to define the metabolites and tissue accumulation, and to determine if species differences in metabolism will be a factor in the interpretation of the results. Absorption, metabolism, and disposition data should be obtained before the design of toxicology studies to help select doses, vehicles, frequency of exposure, etc. The results from percutaneous absorption/disposition studies may, in fact, indicate that a selected test chemical is poorly absorbed and that it would be inappropriate to use the dermal route to determine systemic toxicity.

B. Vehicle Selection

Test chemicals may be applied neat or diluted with or dissolved in a vehicle. If applied neat or diluted, the dose volume may be adjusted to achieve a constant (e.g., mg/kg body wt) exposure as the animals grow. However, in some studies the objective is to keep the dose volume constant to expose a fixed area of skin. The test agent concentration may be increased in order to achieve a constant exposure (mg/kg body wt), but it should be noted that as the animal grows, the skin will be in contact with increasing concentrations of chemical and local irritation could develop with time. Whether the test chemical is a solid or a liquid, the selection of a vehicle should be decided before toxicity studies begin so that the effects on percutaneous absorption can be determined. Dimethylsulfoxide (DMSO) has been used as a vehicle in some studies, but DMSO is known to alter the integrity of the skin.[22] There may be several solvents of choice depending on the chemical characteristics of the test agent. Selection of an appropriate vehicle, the vehicle effect (enhancement or inhibition) on absorption, and the toxicity of the test chemical should be considered. Although water is a good solvent for many chemicals, surface tension may cause this vehicle to "bead" and run off when applied to the skin of some species. Solvent mixtures (e.g., the combination of ethanol with water) may be used to decrease the surface tension of water and to achieve a better exposure. Ethanol has been used as a vehicle, but there is some question about the toxicity of this alcohol when applied to skin.[22,29] Acetone is commonly used because it is a good solvent for many chemicals and because it has a low surface tension and vapor pressure.[17,21] This vehicle spreads easily and evaporates quickly from the skin, but with frequent use, acetone tends to dry the skin at the site of application. If a vehicle is to be used, the effect of the vehicle alone (vehicle control) should be included as a test group within the study.

C. Dose Selection

Dose selection is usually based on the biological effects and the chemical characteristics of the test agent in question. Results of absorption studies should indicate the degree of percutaneous penetration. These data indicate the amount of chemical absorbed per unit skin area relative to time and allow the investigator to estimate the percent of the applied dose that will enter the circulation. It is assumed that compound applied in excess of this value will remain on the skin.

Subchronic studies are usually done prior to long-term studies to establish appropriate doses and results of short-term studies should provide data on both local and systemic effects. When selecting dose concentrations for future toxicity studies, it is important to know at what concentrations skin irritation occurs and the maximum percutaneous absorption. Since a local effect is an important toxicological endpoint in dermal studies, dose selection usually includes at least one dose close to the lowest concentration predicted to cause a histopathologic effect. It is also relevant to select a dose that is estimated to be at the level of human exposure.

D. Chemical Application
1. Site Preparation

In fur-bearing species, shaving has the potential to cause damage to the skin and use of a depilatory agent can alter skin permeability. The use of clippers to remove the haircoat and to provide access to skin at the application site is usually preferred because the hair can be rapidly removed with minimal chance of damaging the skin. The hair remaining after clipping is short and is not thought to present an appreciable barrier to skin exposure; however, clipping needs to be done more frequently than with other methods of hair removal. The extent of hair removal varies depending on the technique and dose volume. Some investigators clip only the site of application, e.g., the interscapular region, while others remove the hair from the entire back. The amount of hair removed usually depends on the quantity of chemical to be applied. While leaving most of the hair may lessen the effect of hair removal on an animal's ability to thermoregulate, chemicals applied to the application site may run into hair remaining around the dose site and lose contact with the skin. Since there is little difference in time between clipping interscapular region vs. the entire back, clipping an area somewhat larger than the site of application is recommended.[6,10,21] This also allows for ease of application of larger volumes of a chemical and better visibility of the toxic effect of a chemical on the skin. If hair is removed from the site of application of the animals to receive test chemical, hair should also be clipped from the control groups.

2. Method of Chemical Application

It is recommended that the exposure site be limited to a maximum of 10% of the body surface area; about 35 cm^2 in rats and 6 cm^2 in mice.[6,10,21] The dose volume, frequency, and duration of dosing are a function of the total surface area available for exposure, toxicity, solubility of chemical, chemical characteristics, and effects of the vehicle. Some study designs require treating the specified area throughout the period of exposure and with fixed volumes (e.g., of 30 and 10 µl) to maintain a constant concentration/exposed area relationship. When exposure is to cover the entire 10 % of the body surface area, doses of 300 µl for rats and 100 µl for mice are considered the optimal dose volumes. Experience has shown that sometimes dose volumes may be excessive; for example with some vehicles, uncontrolled spreading of application volume will occur with one vehicle and not with another. To achieve a desired dose (mg/kg body wt) with chemicals having a low solubility in the vehicle may also require a larger volume than recommended. In these cases, it may be necessary to divide the dose and to apply several, sequential, smaller doses and to allow time for drying in between applications. Doses can be applied to the skin in a variety of ways. Some older studies used an artist's brush to "paint" the chemical onto the site of application. Hence, the term "skinpaint" was used to indicate the method of application. Although a brush is rarely used in contemporary studies, skinpaint still commonly appears in the literature to indicate that the chemical was applied to the skin. More recent studies use applicators that control the volume and distribution of the dose better than paint brushes. If the dose is a free-flowing liquid, it can be applied conveniently with a micropipette or pipettor with a disposable tip. More viscous chemicals may require the use of a positive displacement pipette or syringe. The frequency of application can range from one to seven times per week for subchronic and chronic studies depending on chemical-specific characteristics. Application frequency should take into consideration the dose concentrations to be applied and the absorption rate. The duration (e.g., length of time for skin exposure) is also related to the chemical characteristics and sometimes to the manner in which humans are exposed to the chemical. Thus, the chemical may remain in contact with the skin until the next subsequent exposure or it may be physically and/or chemically removed after every exposure to simulate human use (e.g., cosmetics). In some studies, the design calls for occluding the site of application after exposure in order to mimic the exposure of humans (e.g., chemicals in contact with skin under clothing, or in dose-maximized irritation and sensitization studies).

E. Observations
1. Clinical

Observations are important in skin exposure studies to obtain an accurate measure of toxicity. Observations are usually made during the time of chemical application and findings are recorded.[6,7,10,21] In NTP studies, scheduled observations of clinical signs are routinely made and recorded for all animals in all study groups once a week for subchronic studies, and once a week for the first 13 weeks and every 4 weeks thereafter in chronic exposure studies. These data are often captured electronically and include, among other things, body weight, food consumption, and overt signs of toxicity. The skin is a target site and close observation is required to help determine effects on skin that may decrease with continued exposure and therefore, not be detected during the microscopic evaluations. It is useful to create a map of the skin exposure area to track the first appearance, development, and locations of any tumors or lesions that may occur and to describe the progression of development of signs of toxicity.[21]

2. Gross

A complete gross necropsy is usually performed on exposure groups and controls. This includes an external examination of the animal body surfaces and orifices and of the major organs and tissues. Because the method of chemical application is dermal, the toxic effects on the skin should have been recorded throughout exposure and a final map should be charted at necropsy for correlation of the in-life observations with the microscopic lesions. Tumors and other lesions are identified and selected for histopathologic evaluation and their apparent state of development since first appearance in the chemical record should be described in a pathology narrative. Also, at the time of necropsy, photographs (e.g., color slides, color prints, etc.) to show details of representative tumors or lesions may provide graphic examples of the verbal description. In this way, the history of skin lesions examined microscopically is traceable from first appearance during the in-life portion of the study.

Gross observations that are frequently recorded for skin at the site of application are often categorized as those lesions that can be tolerated (nonlife threatening) and those that are considered excessively toxic (potentially life threatening). Erythema, scaling, subcutaneous edema, alopecia, and thickening are gross signs of chemical effects that are usually considered nonlife threatening; whereas, ulcers, fissures, exudate and crust-formation, and necrosis are examples of chemical effects that are regarded as life threatening.

3. Histopathologic

Histologic sections of selected organs and tissues are prepared in the same way as for other routes of exposure. In skinpaint studies, skin samples should be taken from exposed and nonexposed sites. It is also important to prepare skin samples for histology to be able to identify the orientation and site on the animal from which the sample was taken. In this respect, it is very useful to cut the anterior (cranial) border with an arrow shape. Excised skin may be laid on a piece of labeled index card and placed in a fixative; this will keep the skin flat. Tumors submitted for histopathologic evaluation should include, at a minimum, those identified at the gross observation. Also, samples of normal-appearing skin should be selected from the site of application, if possible, and from a designated site away from the site of application. These samples permit comparisons to be made between treated skin with gross lesions, treated skin without gross lesions, and untreated skin. Care should be taken when removing selected samples from the excised skin for histology to retain the remainder of the excised skin intact for reference. The skin sample is usually trimmed from anterior to posterior (cranial to caudal) and approximately 1 to 2 cm in length for slides. In trimming tumors, it is important to include adjacent nontumorous skin in the section. The nontumor section is important for assessing non-neoplastic changes (e.g., dermatitis, hyperplasia, etc.) compared with skin from control animals. Histopathologic evaluation of skin not from the site of application in dosed animals is important to assess whether there has been a systemic effect of chemical treatment on untreated skin.

Examples of common histopathologic findings are inflammation; mild spongiosis; minimal degeneration and necrosis; epidermal hyperplasia, hyperkeratosis, parakeratosis and dyskeratosis; dermal edema; fibrosis; atrophy or hyperplasia of adnexa. These are generally signs of tolerated exposure concentrations. With increased toxicity, microulcers, marked spongiosis, degeneration, and mild to moderate necrosis, marked inflammation, and edema that destroy the integrity of the skin may be observed.

A complete evaluation of chemical toxicity in dermal studies includes, as in studies with other routes of exposure, observations made at necropsy and microscopically on individual organs, and all significant clinical observations that have been made during the study. In contrast to other routes of exposure, dermal studies allow for a continuous description of the effects of the test article on the primary target organ, skin, to be made from the first day of exposure. This history of development of skin lesions is especially useful when the chemical causes skin tumors and is the basis for the use of mouse skin in initiation/promotion studies.

F. Initiation/Promotion Studies

A modified skinpaint protocol is widely used to study the stages in tumorigenesis. Carcinogenesis has been defined as a multistage process operationally described as initiation, promotion, and progression and each of these stages can be affected by different agents through various mechanisms.[4] It has been shown that the skin of some strains of mice will develop tumors after exposure to certain chemicals and temporal combinations of chemicals.[29-31] The obvious advantage to using mouse skin is the ability to observe the development of tumors over time without sacrificing the animal. The mouse skin initiation/promotion protocol is the most thoroughly studied procedure used to investigate promotion.[32-35] In this model, skin tumors are induced by the application of a subthreshold dose of a carcinogen (initiation stage) followed by repetitive treatment with a known or suspected tumor promoter (promotion stage) and tumor development is monitored over time.[33,36] This system can be used not only to determine the tumor-initiating and -promoting activities of a compound, but also, to determine if it is a complete carcinogen, i.e., if it has both tumor-initiating and -promoting activities.[36]

The mouse skin protocol is commonly used to study tumorigenesis and it is has been used by the NTP [37] as an optional study to obtain additional data to provide a more complete toxicologic characterization of a chemical.[38] The standard protocol uses the sequential application of a subthreshold dose of a carcinogen (initiation) followed by repetitive treatment with a noncarcinogenic promoter (promotion).[36] Mouse strain selection for these studies is usually based on the initiation/promotion literature and on the objectives of the study. In the past, selection was from readily available strains that were known to be responsive, e.g., Swiss CD-1 and HA/ ICR.[39,40] More recently, the SENCAR mouse, especially bred for sensitivity to 7,12-dimethyl-benzanthracene-induced tumor initiation and 12-O-tetradecanoylphorbol-13-acetate-induced promotion, has been gaining acceptance as the strain of choice.[11,41]

Although the basic design is quite similar to skinpaint studies designed to evaluate both local and systemic test agent effects, there are some differences. For example, it is important for the hair growth cycle to be in the resting phase for optimal initiation and therefore, mice should be approximately 55 days of age at the time of first dosing.[42] Since the development of tumors on the skin is usually the only endpoint in initiation/promotion studies, complete necropsies are not routinely performed and only the skin is examined microscopically. Because the endpoint is tumor appearance and development on the skin, tumor maps should be maintained for each animal to record the location and to follow tumor growth with time. Time of first appearance of tumors, number of tumors per mouse, number of animals with tumors, number of papillomas and number of carcinomas per mouse, etc., are all standard initiation/promotion study experimental data critical for evaluation of relative carcinogenic potential. Therefore, criteria need to be established to be able to collect consistent observations and to be able to compare the results with the literature data. For example, if a tumor is present for 14 days, at least 2 mm in diameter, and not attached to the subcutis it may be recorded as a papilloma, whereas tumors necrotic in appearance and attached to the subcutis may be recorded as carcinomas. Recording the time of first tumor appearance for individual animals provides important data to be able to estimate a "mean time to tumor" for mice in that exposure group.[43] The underlying concept in initiation/promotion studies is that animal models that display a high degree of responsiveness to carcinogens provide a tool for identifying some of the key steps involved in tumorigenesis and the mouse skin initiation/promotion protocol has been frequently used to distinguish chemicals that are initiators from promoters.[38]

IV. DISCUSSION

It has been estimated that 20 to 30 years can elapse between the onset of exposure to a carcinogen and the development of the neoplastic disease in the human population.[44] Therefore, it is important to devise methods to identify these agents so that human exposure can be minimized. Subchronic and chronic chemical toxicity studies are most frequently conducted to address a public health concern, as well as for the development and validation of new and better integrated test methods. Data from these animal studies are provided to or used by federal and state agencies and private sector organizations to help respond to issues relevant to the effect of chemical substances in the environment. While there are limitations in using data from animal studies to extrapolate risk to humans, animal models are still the most practical way to evaluate subchronic and chronic toxicity, and laboratory animal studies offer the best compromise between reliability, rapidity, and economy.

Toxicology and safety assessment are emerging disciplines and the methodologies associated with these activities are under constant development and change. Variables in study design, if not properly controlled, will profoundly influence the outcome of long-term animal studies. Regulatory agencies recognize that acceptable standards for the conduct of animal studies need to be modified and updated as knowledge of the science of toxicology expands. In recent years considerable attention has been focused on design aspects of toxicity studies, including appropriateness of the animal species and strain, the level and duration of dosing, duration of testing, and the choice of exposure route.

The use of prechronic and chronic study data to predict chemical effects on humans demands the most appropriate study design. To provide the most reliable data, studies must be designed to control, as much as possible, all of the extraneous variables that could confound the interpretation of the results. The large number of subchronic and chronic studies conducted using the more routine, dosed-feed, dosed-water, and gavage routes of exposure has evolved well-defined protocol designs for those studies and the conduct has become routine. Now, the greater interest in conducting toxicity studies by the most common route for human exposure will require the same kind of considerations for the design of skin exposure studies.

REFERENCES

1. D.P. Rall, M.D. Hogan, J.E. Huff, B.A. Schwetz, and R.W. Tennant, *Annu. Rev. Publ. Health* 8, 355 (1987).
2. J.K. Haseman, J.E. Huff, E. Zeiger, and E.E. McConnell, *Environ. Health Perspect.,* 74, 229 (1987).
3. J.E. Huff, E.E. McConnell, J.K. Haseman, G.A. Boorman, S.L. Eustis, B.A. Schwetz, G.N. Rao, C.W. Jameson, L.G. Hart, and D.P. Rall, *Ann. N.Y. Acad. Sci.* 534,1 (1988).

4. International Agency for Research on Cancer Monographs. Long-term and short-term screening assays for carcinogens: A critical appraisal. Suppl. 2, Lyon, France, 1980.

5. Report of the NTP ad hoc panel on chemical carcinogenesis testing and evaluation. Prepared for the National Toxicology Program's Board of Scientific Counselors, August 17, 1984.

6. U.S. Environmental Protection Agency, Office of Pesticides and Toxic Substances, Health Effects Test Guidelines. EPA-560/11-82-002 (1982).

7. U.S. Food and Drug Administration, Bureau of Foods, Toxicological principles for the safety assessment of direct food additives and color additives used in food. 1982.

8. J.K. Haseman, J. Huff, and G.A. Boorman, *Toxicol. Pathol.* 12, 26 (1984).

9. J.K. Haseman, J.E. Huff, G.N. Rao, J.E. Arnold, G.A. Boorman, and E.E. McConnell, *J. Natl. Cancer Inst.* 75, 975 (1985).

10. The Organization for Economic Cooperation and Development (OECD) guidelines for testing of chemicals. 1981.

11. T.J. Slaga, A.J.P. Klein-Szanto, R.K. Boutwell, D.E. Stevenson, H.L. Spitzer, and B. D'Motto, *Skin Carcinogenesis: Mechanisms and Human Relevance.* Vol. 298, Alan R. Liss, New York (1989).

12. J.H. Epstein, Photocarcinogenesis: A Review, in International Conference on Ultraviolet Carcinogenesis, NCI Monograph 50, 13 (1978).

13. C.R. Behl, N.H. Bellantone, and G.L. Flynn, Influence of Age on Percutaneous Absorption of Drug Substances, in *Percutaneous Absorption.* R.L. Bronaugh and H.I. Maibach, Eds. Dermatology, Vol. 6, Marcel Dekker, New York, 183 (1985).

14. U.S. Department of Health and Human Services, National Institute of Environmental Health Sciences, The SENCAR Mouse in Toxicological Testing. *Environ Health Perspect.*, Vol. 68 (1986).

15. F. Stënback, *Acta Pharmacol. Toxicol.* 46, 89 (1980).

16. G.N. Rao, L.S. Birnbaum, J.J. Collins, R.W. Tennant, and L.C. Skow, *Fund. Appl. Toxicol.* 10, 385 (1988).

17. J.H. Weisburger and E.K. Weisburger, Tests for Chemical Carcinogens, in *Methods in Cancer Research,* Vol 1, H. Busch, Ed. (1967).

18. S.P. Sher, *Crit. Rev. Toxicol.* 10, 49 (1982).

19. R.A. Adams, J.A. DiPaolo, and F. Homburger, *Cancer Res.* 38, 3642 (1988).

20. P. Nettesheim, *Prog. Exp. Tumor Res.* 116, 185 (1972).

21. U.S. Department of Health and Human Services, National Toxicology Program General Statement of Work for the Conduct of Toxicity and Carcinogenicity Studies in Laboratory Animals. (Revised April 1987). (Available from NTP Information Office, NIEHS, P.O. Box 12233, Research Triangle Park, NC 27709).

22. R.J. Scheuplein, Permeability of the Skin, in *Handbook of Physiology, Section 9, Reactions to Environmental Agents.* American Physiological Society, Bethesda, MD, 299 (1977).

23. R.L. Bronaugh, R.F. Stewart, and E.R. Congdon, *J. Soc. Cosmet. Chem.,* 34, 127 (1983).

24. U.S. Department of Health and Human Services, National Institutes of Health, Guide for the Care and Use of Laboratory Animals, NIH Publication No. 86-23 (1985).

25. J.R. Lindsey, M.W. Conner, and H.J. Baker, Physical, Chemical, and Microbial Factors Affecting Biologic Response, in *Laboratory Animal Housing, II. The Animal Environment.* National Research Council, Institute of Laboratory Animal Resources. Washington, D.C., 31 (1985).

26. D.E. Davis, Social Behavior in a Laboratory Environment, in *Laboratory Animal Housing.* National Academy of Science, Washington, D.C. (1978).

27. G.N. Rao, Significance of Environmental Factors on the Test System, in *Managing and Conduct and Data Quality of Toxicology Studies,* Princeton Scientific, Princeton, NJ, 173 (1986).

28. M. Chvédoff, M.R. Clarke, J.M. Faccini, E. Irisarri, and A.M. Monro, *Arch. Toxicol.* Suppl. 4, 435 (1980).

29. E.R. Cooper, Vehicle Effects on Skin Penetration, in *Percutaneous Absorption,* R.L. Bronaugh and H.I. Maibach, Eds. Dermatology, Vol. 6, 525 (1985).

30. I. Berenblum, *Cancer Res.*, 1, 807 (1941).
31. J.C. Mottram, *J. Pathol. Bacteriol.*, 56, 181 (1944).
32. I. Berenblum and P. Shubik, *Br J. Cancer*, 1, 383 (1947).
33. R.K. Boutwell, *Prog. Exp. Tumor Res.*, 4, 207 (1964).
34. R.K. Boutwell, The Function and Mechanism of Promoters of Carcinogenesis, *CRC Crit Rev. Toxicol.*, 2, 419 (1974).
35. B.L. Van Duuren and S. Melchionne, *Prog. Exp. Tumor Res.*, 11, 31 (1969).
36. T.J. Slaga, Mechanisms Involved in Two-Stage Carcinogenesis in Mouse Skin, in *Mechanisms of Tumor Promotion. Vol. II Tumor Promotion and Skin Carcinogenesis*, T.J. Slaga, Ed. CRC Press, Boca Raton, FL, 1 (1984).
37. National Toxicology Program Technical Report on the Toxicology and Carcinogenesis Studies of *ortho*-Phenylphenol in Swiss CD-1 Mice, NTP TR 301, March 1986.
38. M.A. Pereira, *J. Am. College Toxicol.*, 1, 47 (1982).
39. T.J. Slaga, S.M. Fischer, C.E. Weeks, and A.J.P. Klein-Szanto, *Curr. Probl. Dermatol.*, 10, 193 (1980).
40. F.J. Burns, M. Vanderlaan, E. Snyder, and R. Albert, Induction and Progression Kinetics of Mouse Skin Papillomas, in *Carcinogenesis: A Comprehensive Survey, Vol. 2. Mechanisms of Tumor Promotion and Cocarcinogenesis.* T.J. Slaga, R.K. Boutwell, and A. Sivak, Eds., Raven Press, New York, 91 (1978).
41. T.J. Slaga and S.M. Fischer, *Prog. Exp. Tumor Res.*, 26, 85 (1983).
42. F.G. Bock, *Prog. Exp. Tumor Res.*, 26, 169 (1983).
43. W.C. Eastin, Jr., Developing Design Standards for Dermal Initiation/Promotion Screening Studies, in *Skin Carcinogenesis: Mechanisms and Human Relevance.* T.J. Slaga, A.J.P. Klein-Szanto, R.K. Boutwell, D.E. Stevenson, H.L. Spitzer, and B. D'Motto, Eds. Progress in Clinical and Biological Research, Vol. 298, Alan R. Liss, New York, 295 (1989).
44. D.P. Rall, Laboratory toxicity and carcinogenesis: testing underlying concepts, advantages and constraints. Living in a Chemical World. *Ann. N.Y. Acad. Sci.* 534, 78 (1988).

4

Evaluation of the Dermal Irritancy of Chemicals

CARL T. OLSON
Battelle Memorial Institute
Columbus, Ohio

I. INTRODUCTION

Regulatory agencies, manufacturers, and users of commercial products and chemicals require that potential user safety hazards be evaluated prior to marketing. One aspect of this safety testing involves determining the effects when compounds are applied directly to the skin. Adverse effects which may be produced by such testing include: signs and symptoms of systemic toxicity due to percutaneous absorption of the test agent with distribution to target organs; dermal irritation due to direct chemical interactions in the skin; the development of allergic responses due to immunologically mediated influences; and photodermatitis due to the absorption of light energy by the agent or its metabolites. Dermal irritation is usually assessed by the degree of inflammatory response produced. Inflammation is defined in *Dorland's Medical Dictionary* as "a localized, protective response elicited by injury or destruction of tissues, which serves to destroy, dilute, or wall off (sequester) both the injurious agent and the injured tissue.[1] It is characterized in the acute form by the classical signs of pain (dolor), heat (calor), redness (rubor), swelling (tumor), and loss of function (*functio laesa*). Histologically, it involves a complex series of events, including dilatation of arterioles, capillaries, and venules, with increased permeability and blood flow; exudation of fluids, including plasma proteins; and leukocytic migration into the inflammatory focus." This chapter discusses existing methods and potential alternatives for determining the degree of acute irritation — the local, reversible inflammatory response of normal skin to direct injury caused by a single application of a toxic substance, without the involvement of an immunologic mechanism.

Terms such as "contact irritation" or "primary irritation" are commonly used to distinguish inflammation due to the direct, irritating properties of compounds in intimate contact with skin from allergic, immunologically linked effects or photosensitization. Substances are sometimes classified as either "irritant" or "nonirritant". This is unsound in that all substances are irritating for someone under some conditions, although extreme exposures may be necessary. Water is perhaps the ultimate

125

example. Maintaining moist skin by application of impermeable tape for 2 to 3 weeks and then removing the occlusion often results in erythema and a scaling rash, even when microbial growth is controlled.[2] Substances with weak irritating potential also may become injurious when used in conjunction with other mild irritants or "nonirritants".

Expressions of mild irritation may be vague and inconspicuous, sometimes little more than a subtle change in surface texture or reflectance. With many irritants, earliest changes may be entirely invisible, yet functionally serious.[2] Morphologic indications of damage encompass the entire dermatologic spectrum, from barely discernible inflammatory changes (eczema, wheals, purpura, and blisters) to corrosion (ulcers, connective tissue scars, granulomas, etc.).

The factors that may affect the degree of skin inflammation produced by a chemical are many, including: animal species and strain, sex, age, nutritional and disease status, individual animal variations, properties of the chemical (pH, volatility, polarity, etc.), dose, concentration, vehicle, interactions of chemical with vehicle or skin, skin condition (health, excoriations, hydration, temperature, occlusion), site of application, climate or season, single or repetitive dosing, mechanism(s) of action to create damage, and skin metabolism. A number of laboratory animals have been proposed for use in identifying potential skin irritants for man, and many articles written comparing the responses to skin irritants of man and animals. McCreesh and Steinberg[3] reported the degree of dermal responsiveness of animals, in decreasing order of responsiveness, to be rabbits, guinea pigs, albino rats, humans, and swine. Other authors switch the order of rats and guinea pigs.[4-6] The dermal responsiveness of the mouse is similar to that of the rat. The percutaneous permeability of rhesus monkeys appears reasonably similar to that of man.[5]

The irritancy of a substance does not depend solely on its inherent ability to injure cell membranes or cellular constituents, but also on its ability to penetrate (and its rate of penetration) the stratum corneum, and reach its target. The stratum corneum, as described in another chapter, is the outer layer of the epidermis and consists of dried, flattened keratinocytes which have no metabolic activity and represent the nonviable end product of the synthetic activity of the lower layers.[5] This layer is the main barrier to skin penetration of water and most chemicals. Although the principles governing penetration of the stratum corneum are the same for all mammals, there are substantial differences between species in the amounts of a compound reaching a particular tissue compartment.[7] Loss of barrier function of the stratum corneum may occur through disease or damage. Stripping of layers of the stratum corneum by applying and removing cellophane tape a number of times is a commonly used technique in dermatotoxicology studies to increase skin permeability. Stripping of the stratum corneum in this manner increased the penetration of hydrocortisone in humans fourfold.[8] Occluding the hydrocortisone with a plastic film during the first 24 h caused a 10-fold increase in absorption over that of unoccluded skin. Occlusion causes hydration of the skin and an increased temperature, and prevents evaporation of test compounds. The skin of newborn and elderly people undergoes changes which can affect permeability.

The greatest potential for percutaneous absorption occurs when a high concentration of compound is spread over a large surface area of the body.[8] Nonpolar compounds (lipophilic) are generally better skin penetrators than polar (hydrophilic) compounds. Vehicles such as dimethylsulfoxide (DMSO) readily permeate skin, and can greatly enhance absorption of test chemicals. Repetitive dosing of a chemical can also greatly increase absorptivity over a single application, even if the same total dose is applied.[8]

There are appreciable interspecies differences in the metabolic handling of a compound, and in the enzyme content in the skin, and the location of these enzymes within the layers of the skin.[5,9] The skin is metabolically active, and metabolic enzymes in the skin may be induced. Metabolites of applied chemicals may be more or less irritating than the parent compound.

Differences in skin irritability occur among strains or individuals of the same species. A good example of this is man. Blacks are far less susceptible to skin irritation, and darkly pigmented whites from the Mediterranean area are a good deal more resistant than light-complexioned individuals. People of Celtic extraction (Welsh-Scottish-Irish) are the best subjects for assaying mild irritants because they have the most sensitive skin.[2] Sun-sensitive skin is also sensitive to other irritants.

At least in man, the climatic conditions can greatly affect skin sensitivity to irritation. A soap and a general-purpose detergent were tested for irritancy in the same individuals both in warm weather and again in cold weather.[10] The warm weather test was conducted in early June when the mean temperature was 55°F and the mean dew point 47°F. The cold weather test was performed in mid-February when the mean temperature was 22°F and the mean dew point 21°F. Both products gave much stronger reactions in cold weather than in warm. In studies grading the skin of hands of a group of housewives every 3 months over a period of 1 year, the difference created by the use of the mildest vs. the least mild of the major washing products, even in winter when the differences were maximum, were less than one half as large as the seasonal effects.[10]

The body site tested may also greatly affect the irritation produced by a compound. As a rule, irritability and permeability are directly related. In man, the face is very susceptible — its stratum corneum is not very sturdy and there is a high density of follicles that act as shunts to bypass the stratum corneum barrier.[2] The leg is less easily irritated than the forearm because it is less permeable. The back is more permeable, and therefore more easily irritated than the forearm.[2,8]

The skin of other animals is not like that of man. There are anatomical, physiological, and pharmacological differences. It is the general consensus that the skin of common laboratory animals is more easily irritated than human skin. This is not, however, always true. Nixon et al.[11] found that hypochlorite bleach caused only slight visible irritation in rabbits and guinea pigs, but produced serum leakage and eschar reactions on intact skin sites in four of seven human subjects. A mixture of primarily octanoic and decanoic fatty acids produced only very slight reactions in guinea pigs and moderate to severe erythema in rabbits, but, again, caused serum leakage and eschar formation on intact skin of two of seven humans. Marzulli and

Maibach,[12] testing irritation potential of sunscreen ingredients, also found one (sulisobenzone) of 12 tested substances which appeared to be significantly more irritating to human skin than to rabbit skin. The degree of skin reaction in laboratory animals often correlates poorly with the response in humans. Strong irritants will usually evoke reactions in laboratory animals, but the rank order of severity is often very different from that noted in humans. Furthermore, animal skin lacks the capacity to differentiate among substances which in human testing cover a range from very mild to moderately irritating. Rank order differences disappear; the skin of laboratory animals reacts the same — generally erythema with some scaling.[2] Mechanisms by which irritants produce an inflammatory response are not always the same.[12,13] Skin thickening may occur in the absence of a comparable increase in erythema if mainly epidermal rather than dermal changes are involved, and is caused primarily by an increase in cellular layers (acanthosis) rather than edema.[12]

Soaps, detergents, cosmetics, pesticides, petroleum products, and a multitude of occupational and environmental agents are recognized as potential causes of nonimmunologic, chemically induced skin inflammation, and the importance of this has long been known by regulatory and industrial toxicologists.[13] Manufacturers of household products, cosmetics, pesticides and other chemical products bear the responsibility of ensuring that their products are safe for their intended uses and are not unduly hazardous when misused in reasonably foreseeable ways.[11] Regulations requiring safety testing and delegating responsibility for its enforcement derive from federal laws including the Federal Food, Drug and Cosmetic Act (FDCA), the Federal Insecticide, Fungicide and Rodenticide Act (FIFRA), the Federal Hazardous Substances Act (FHSA), and the Toxic Substance Control Act (TSCA). In the U.S., four regulatory agencies, the Consumer Product Safety Commission (CPSC), the Environmental Protection Agency (EPA), the Food and Drug Administration (FDA), and the Occupational Safety and Health Administration (OSHA), agreed in 1977 to work together to reform and standardize the regulatory processes, and formed the Interagency Regulatory Liaison Group (IRLG). In 1979, the Food Safety and Quality Service (FSQS) of the Department of Agriculture joined the IRLG. These agencies recognized that although they often regulated the same chemicals, toxicity testing guidelines were not always uniform. A Testing Standards and Guidelines Work Group was established to resolve existing differences in methodology. The IRLG guidelines developed for irritancy testing were coordinated with the European Organization for Economic Cooperation and Development (OECD), and were published in 1981 as the Recommended Guideline for Acute Dermal Toxicity Testing.[14]

The recommended procedures currently used in regulatory submissions in the U.S. for evaluating dermal irritancy are based on a protocol developed by the FDA in response to a request for assistance from Great Britain in the mid-1940s for testing eye irritants proposed for use in chemical warfare.[15] The results of an earlier experimental study on the acid-base tolerance of the corneas of rabbits were reported in 1944 by Friedenwald et al.[16] For quantification of the lesions, the authors developed a numerical scoring scheme that graded the significant lesions of different components of the eye (cornea, iris, conjunctiva) according to their relative importance to visual function. This work formed the basis for the paper of Draize et al.[17] in which they described methods and a grading system for the effects on skin and mucous membranes caused by compounds or preparations intended for therapeutic, cosmetic, or other topical use.[18] This skin irritancy test was a patch technique using albino rabbits. An area of the back was clipped free of hair and a fixed amount of test material was applied followed by a standard square patch of gauze. The test material was applied both to normal and excoriated skin. The scratches were through the stratum corneum, but not deep enough to produce bleeding. The patches were secured with tape and covered in an occlusive fashion. The dressing and patch were removed at 24 h and the resulting reaction evaluated. The reaction was reevaluated after a further 48 h. The end result was a numerical irritancy score. Testing could also be performed to evaluate effects of various substances on the eye.

This type of skin testing has been performed in humans. However, ethical, legal, financial, practical, and scientific aspects must be considered in determining whether definitive skin irritation testing should be conducted in man. For compounds about which essentially nothing is known of its irritancy potential, it would not be advisable to initiate testing with humans, but trials with humans are necessary for many compounds prior to marketing the product. With this form of dermatologic assay, numbers are assigned according to individual judgment of the severity of lesions in an attempt to make a qualitative test quantitative. As a result of slightly different procedures or interpretation of effects, analyses of irritancy of compounds vary greatly between laboratories, and even within the same laboratory.[19]

The indiscriminate use of definitive LD_{50} tests, i.e., determination of the dose of a particular compound necessary to kill 50% of exposed animals within a certain time under specified conditions, has been sharply criticized for needless waste of animal lives to obtain data with unwarranted precision and of limited value. Probably the second most commonly criticized procedure is the rabbit eye and skin irritation testing, the Draize test.[15,20-25] An obvious answer to decreasing the number of animals used for dermal irritancy testing is to develop *in vitro* techniques. *In vitro* techniques have been, and will continue to be, developed, and are discussed in Chapter 9. With *in vitro* testing, however, one must still correlate the response measured with the effect seen *in vivo,* and the correlation determined for each chemical, or family of chemicals, may well not be the same for all compounds of interest. The same problem exists, obviously, for extrapolating results in laboratory animals to man. There is definitely a need for improved procedures to quantitatively and objectively assess the irritation potential of chemicals, and various ways in which such testing may be developed will be presented.

II. STANDARD METHODS OF SKIN IRRITANCY TESTING

Several regulatory authorities have issued guidelines on the conduct of animal tests for skin irritation. Information from such tests is required for registration, classification, and labeling purposes. Guidelines have been issued by the EPA (1978, 1979, 1982),[26] the CPSC (1981, 1986),[27] the IRLG (1981),[14] the OECD (1981),[28] and

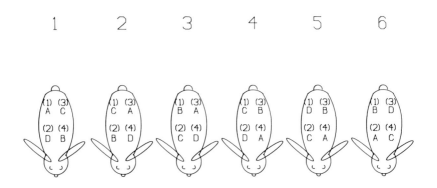

Figure 1. Four preparations, A, B, C, and D are assigned to the four areas, 1 , 2 , 3, and 4 as illustrated in the six rabbits. Areas 1 and 4 are intact skin and areas 2 and 3 have four epidermal abrasions.

the European Economic Community (EEC, 1979),[29] as well as by the National Academy of Sciences-National Research Council (NAS-NRC, 1977).[6] The "standard procedure" for determining skin irritancy of chemicals, as mentioned in the introduction, was described in a paper written by J. H. Draize, G. Woodard, and H. Calvery entitled *Methods for the Study of Irritation and Toxicity of Substances Applied Topically to the Skin and Mucous Membranes* and published in the *Journal of Pharmacology* in 1944.[17] The authors, working in the Division of Pharmacology of the Food and Drug Administration, had been investigating chemical toxicity to skin for some time, and were well aware of potential bias, of the importance of controlling variables, and the need for accurate and statistically valid interpretation. They divided the overall effects into distinct elements for grading, as previously described, for the measurement of injuries to rabbit eyes by Friedenwald et al.[18]

The procedures they recommended for determining contact irritation of a test compound applied to the skin were as follows. Intact and abraded skin of albino rabbits is used. The hair is clipped from the back and flanks. Four areas on the back, approximately 10 cm apart, are designated for the position of patches on each of 6 rabbits, as indicated in Figure 1. Areas 2 and 3 are abraded by making four epidermal incisions (two perpendicular to two others in the area of the patch). The patches consist of two layers of light gauze cut in squares, 2.5 cm to a side. These are secured to the area by thin bands of adhesive tape, and the material to be tested introduced under the patch (0.5 ml, if liquid or 0.5 g, if solid). The entire trunk of the rabbit is wrapped in rubberized cloth to help hold the patches in position and retard evaporation of volatile test substances. Four compounds are tested per series of six rabbits, with three applications of each to intact skin and three to abraded skin, as assigned in Figure 1.

The animals are placed in restraint devices prior to dosing, and held there for the 24-h exposure period. Upon removal of the patches, skin reactions are evaluated on

Table 1
Evaluation of Skin Reactions

Erythema and eschar formation:	
Very slight erythema (barely perceptible)	1
Well-defined erythema	2
Moderate to severe erythema	3
Severe erythema (beet redness) to slight eschar formation (injuries in depth)	4
Total possible erythema score	4
Edema formation	
Very slight edema (barely perceptible)	1
Slight edema (edges of area well defined by definitive raising)	2
Moderate edema (area raised approximately 1 mm)	3
Severe edema (raised more than 1 mm and extending beyond area of exposure)	4
Total possible edema score	4
Total possible score for primary irritation	8

the basis of a scale of weighted scores, as shown in Table 1. Readings are made again 48 h later (72 h from the start of exposure), and the final score is an average of the 24- and 72-h readings. If signs of irritation are still detectable at 72 h, further observations, up to 14 days postexposure, are recommended.

The table of scores gives no values for agents which are severe vesicants or severe escharotics. When such agents are encountered, dilute solutions in a bland, nonirritating solvent are tested. Sufficient dilution is made so that the responses elicited may be graded by the scale of Table 1.

As a general rule, the authors found that substances which produce a score of 2 or less at 24 h become negative, i.e., lesions are completely healed, before the 72-h reading. The combined average was referred to in their laboratory as "the primary irritation index". Compounds producing a primary irritation index of 2 or less were considered only mildly irritating, whereas those with indices of 2 to 5 were labeled moderate irritants and those above 6 considered severe irritants.

For over 40 years the "Draize primary irritation skin test" has remained essentially unchanged, and the recognized standard, although Draize in 1959 did discuss the desirability "in the case of solids to attempt solubilizing in an appropriate solvent and to apply the solution as for liquids".[30] The 1986 CPSC Code of Federal Regulations,[27] legislated under the provisions of the FHSA, prescribes procedures virtually identical to those described by Draize. The NAS-NRC guidelines stated the necessity of including a comparison standard, i.e., a compound similar to the test substances in composition and properties and with a known human skin irritancy. NAS-NRC guidelines also recommended that the exposure period of the Draize test be shortened to 4 h unless longer exposure periods could be justified by intended use criteria. A 4-h skin exposure period had previously been recommended by the FDA.[31] The NAS-NRC also recommended the deletion of the requirement to test abraded skin, stating that it was difficult to standardize abrasion techniques, and that researchers[11] had shown that classification of irritancy based only on intact skin was not usually different from that using abraded skin. The 1978 guidelines proposed by EPA stated that substances with a pH of 1 to 3 or 12 to 14 need not be tested on animals since these compounds would predictably be corrosive. The newer (1982) EPA guidelines, as well as the OECD guidelines, exempt skin irritancy testing of compounds with a pH below 2.0 or above 11.5. The OECD guidelines recommend using "at least three healthy adult animals", suggesting the use of the albino rabbit, although "additional animals may be required to clarify equivocal responses". The OECD guidelines also recommend a 4-h exposure period with a 30 to 60 min postexposure scoring period, and deleted any requirement for testing on abraded skin. The IRLG and EEC guidelines are not as specific, but tend to follow the procedures recommended by Draize.

Weil and Scala[19] performed an oft-cited study of intra- and interlaboratory variability in the results of rabbit eye and skin irritation testing using the Draize method. Twenty-two different laboratories, governmental or industrial, voluntarily provided results of skin irritancy testing using 12 standard reference materials. This collaborative testing clearly indicated that some laboratories consistently rated materials either more or less irritating than the majority, and there was a good deal of variation within certain laboratories. Not only was there great variability in the scoring of irritation, which determined whether a material was deemed an irritant or not and the severity of its irritancy, but the relative ranking of irritancy of the 12 compounds differed considerably between laboratories. Certain materials were rated as the most irritating tested by some laboratories and, contrariwise, as the least irritating by others. The all-or-none, irritant or nonirritant, skin rating of the reference samples was determined quite differently in different laboratories. Two laboratories were arbitrarily selected and agreed to rerun the skin tests on fresh aliquots of the same samples. An independent, trained observer participated in the retests in both laboratories. A technician from each laboratory treated rabbits, as did the observer, and both scored the effects 24 h later. The conclusion of this follow-on study was that the major differences in these two laboratories, which scored at opposite extremes in the original tests, was not primarily produced by differences in the stocks of rabbits used, but was largely the result of different grading by the technicians. It is noteworthy, however, that the independent observer obtained different readings in the two laboratories with these tests. There was also some variability in interpretation and performance of the procedures between laboratories, which added to the discrepancies in scoring. The authors of the article recommended that courses or clinics in methodology be conducted to decrease variability between laboratories, and that, in the meantime, procedures currently recommended by federal agencies for use in testing irritancy of materials not be recommended in any new regulations.

Various researchers have suggested modifications to the Draize test as originally proposed. Most suggested modifications deal with choice of animal species, site of application, duration of exposure, deletion of abraded skin testing, the use of occlusive or semiocclusive dressings, the time of scoring of lesions, the procedures to use when testing solid substances, i.e., to moisten or dissolve or apply as a solid, and the use of solvent or negative, and known positive control substances. There have been recommendations to use repeat application testing with animals, as is done in man.[12,32,33] Repeated testing may demonstrate the production of irritation by mild irritants, but rather than due strictly to a true contact dermatitis, there may be immunologically mediated responses. The precise reasons for the irritation, however, may be of little concern to the consumer. The same immunologically mediated response could be possible in research animals if they have been previously exposed to a test compound or one of its components.

Meyer et al.[9] studied the skin of the domestic cat, dog, and pig for use as models of humans, and concluded that the pig was the only domestic animal having morphological and functional skin characteristics comparable to human skin. Brown[34] compared the effect of surfactants on the skin of rabbits, guinea pigs, rats, and human volunteers, using a variety of test methodologies. He found that, given the same test conditions, there were no apparent differences in the reactivity of rabbit and albino guinea pig skin with the surfactants tested. He reported that, based on a much more varied range of chemicals than covered in his paper, the guinea pig and rabbit tests are complementary rather than alternatives. He also reported good correlation between results obtained with the surfactants tested under occluded skin patches on formalin-pretreated abdominal rat skin and those on rabbit skin without pretreatment with formalin, but stated that he had been less successful in discriminating between a series of concentrations of sodium lauryl sulfate using the formalin-treated rat skin method. There was poor correlation in his results, however, obtained by different methods. For example, the results from one surfactant fluctuated between most irritant, intermediate, and least irritant for both human skin and animal skin. Nixon et al.[11] found that a number of materials caused vastly different reactions when 24 household materials or industrial chemicals were applied for 4 h to intact and abraded skin of rabbits, guinea pigs, and humans. Some caused tissue damage at abraded sites in the animals, but caused only mild reactions on either intact or abraded sites of humans. Hypochlorite bleach caused severe reactions on intact human skin, but considerably less reaction on both intact and abraded skin of rabbits and guinea pigs. From their results, they concluded that, (1) neither rabbit nor guinea pig skin should be relied upon exclusively to identify potentially hazardous irritants to human skin, and (2) that the testing of abraded skin is unnecessary and can be misleading in the interpretation of results, making the animal tests less, rather than more, reliable. They reported that the rabbit and guinea pig appear to be similarly predictive as models for preliminary screening of household products and their chemical components for skin irritancy. Roudabush et al.[35] compared rabbits and guinea pigs in their primary irritation response to various compounds, and concluded that unabraded abdominal skin of guinea pigs is as sensitive as, or more sensitive than, rabbit skin in eliciting dermal reactions. The authors concluded that the intact guinea pig skin test serves as a useful alternative for the rabbit test and that there is no scientific basis for preferring the rabbit as a test animal. Griffith and Buehler[36] found that for weak irritants, guinea pig skin reactions are more predictive of human response than rabbit skin reactions. They also found, however, that guinea pig skin is more tolerant of strong irritants and corrosive materials than is human skin, and the converse is true for the rabbit, suggesting that the best animal model for man is dependent on the type of material to be tested. They also found that irritancy data from tests on abraded skin lead to substantially the same classification of materials as data from intact skin tests, and that a 4-h semiocclusive patch test was a more realistic exposure than one of 24 h. Gloxhuber and Schulz[37] wrote that since the skin of rabbits, or of other experimental animals, reacts differently than human skin, it would be advisable to run analogous experiments with different species, such as guinea pigs, rats, and mice, and that the resulting trend can be used as a prediction for humans. They also expressed the desirability of including, for the sake of comparison, a similar substance or commercial product of known irritancy. Marzulli and Maibach[12] reported that hairless animals have increased stratum corneum thickness and skin permeability characteristics closely approximating those of man. In studies, however, they found hairless rats and hamsters to show little response to skin irritants. Vinegar[38] stated that some laboratories prefer to use the abdomens of rabbits in routine primary skin irritation testing, rather than the back; he found in experiments using the same test material that there was a clear tendency for the primary irritation indices to be higher on abdominal skin than on the back, but that neither erythema nor edema indices were significantly higher, indicating that the higher primary irritation indices on abdominal skin were due to a combination of both factors, not to erythema or edema alone. The conclusion was that, since the dorsal rabbit skin is considered a very conservative model for predictive testing in humans, using the abdominal skin in rabbits further prejudices the results. The relationship of dermal penetration in man and other animal species is variable, and no single animal species can be used to predict dermal penetration of all varieties of compounds.[39] The degree of inflammation elicited in man with known irritants varies markedly among individual test sites.[40]

Potokar et al.[41] investigated the effects of varying the exposure time and the extent of occlusion of rabbit skin in evaluating the corrosiveness of industrial chemicals, and found that exposures of 1 h with semiocclusion led to realistic assessments of corrosiveness. Four-hour exposures led to unrealistic hazard assessments, and results using the semiocclusive method did not usually differ from results using occlusive testing. Gilman et al.[42] found that the form and concentration of test material, as well as the length of contact with the skin and the degree of patch occlusivity, are prime factors influencing the degree of dermal irritation. Using granular detergents, the most severe skin reactions were obtained when the products were applied as a paste, and left in contact with the skin for 24 h using an occluded patch. The authors believed it more relevant to perform the primary rabbit dermal irritation assay using a granular product as a dry powder since it would more likely be in that form in the case of misuse or accidental exposure, and to limit exposure time to 4 h using a semiocclusive patch. Sullivan et al.[43] demonstrated the differences in results obtained in irritation studies with six commercial powder detergent products applied to the intact skin of rabbits in the dry or moistened state. Products were administered to unmoistened skin in the dry form, to premoistened skin, to skin covered with a moist gauze pad, in the dry form with distilled water injected through the patch, or as a solution or slurry. In all cases, a gauze pad was used and held securely in place with adhesive tape. Observations were made at 4, 24, and 72 h for each application and the treated sites scored for erythema and edema and observed for necrosis. The resultant observations ranged from no response to necrosis of the skin, clearly demonstrating the adverse effects of moistening the skin. Gad et al.[44] evaluated 72 compounds and mixtures for both skin and eye irritation to correlate ocular and dermal irritancy. They found that primary dermal studies reproducibly evaluate and classify the potential for materials to cause ocular irritation. They also found that, as a general rule, abrading skin does not significantly enhance its sensitivity to irritation when compared to intact skin, and either abraded or intact skin may be more responsive to particular compounds. Other techniques suggested for modification of the Draize methodology include the postexposure injection of dyes (e.g., trypan blue) to help discriminate areas of acute inflammation due to extravasation of the dye into the area, and varying the concentration of test substances to determine a "threshold irritation concentration", i.e., the lowest concentration of a compound which elicits an observable inflammatory response.[45]

What we now have is an over 40-year-old "standard procedure" which has been independently modified in many different ways, which cannot be accurately replicated in different laboratories or by different individuals, which cannot always be relied upon to give an accurate estimation of irritancy of a compound to man, but is required by law for the testing of many household goods, cosmetics, and industrial chemicals. Nevertheless, the procedure has been of use in ranking relative irritancy of chemicals and in calling attention to the importance of primary dermal irritation and the need for further testing. Although the Draize testing method has limitations, in its over 40 years of existence, no one has developed a better, more universally accepted, alternative method. As it is currently performed, the test usually predicts severe human irritants, but fails to separate mild and moderate skin irritants and, in most cases, the animal tests tend to overrate irritancy potential.[2]

III. POTENTIAL ALTERNATIVE *IN VIVO* METHODS FOR IRRITANCY TESTING

Other than modifications to the Draize test, there have been a number of alternative methods proposed for determining dermal irritancy of compounds *in vivo*. As research continues and results in new findings, and instrumentation and techniques improve, it now appears likely that more refined, accurate, objective, and replicable testing methods will be developed and will be accepted.

Different methods advocated include the use of skin chambers or immersion of test animals, measurements of skinfold thickness, ear swelling, cutaneous blood flow, transepidermal water loss, thermometry, skin staining and reflectance studies, and analysis of blood or skin perfusates for enzymes, eicosanoids, or reactive proteins. For example, hairless guinea pigs or mice can be restrained and their bodies immersed in a 37°C bath with the test compound at various concentrations for 1 h daily, for 5 days, and examined for skin reactions or alterations (erythema, edema, necrosis, etc.).[37] Five animals at a time are usually exposed in a single bath, and skin reactions compared to those of other animals exposed to other compounds concurrently. This is similar to arm immersion techniques used with man. Gloves containing known quantities of test compounds such as detergents are also used with man; after a 30-min period or longer, solutions in the glove are examined for extracted protein and amino acids.

A. Skin Chambers

Kligman[2] believed patches to be archaic devices that can slip or fold and allow spreading of test material beyond the confines of the patch. He replaced the use of patches with aluminum chambers. The chambers are occlusive and prevent loss of test material through spreading or evaporation. The use of chambers is not new — Rokstad introduced a celluloid chamber in 1940, but it was awkward and difficult to seal, and the celluloid was often attacked by test agents. In 1975, Pirila introduced the Finn chamber, an aluminum chamber designed with an elevated flange around the rim to prevent leakage. Its capacity of 0.020 ml for a 50 mm^2 surface area proved to be a handicap, however, and the Duhring chamber, an enlarged version of the Finn, was developed.[46] The Duhring chamber is made from 18 mm disks of 0.3 mm thick aluminum punched with a special die to form an elevated flange 2 mm high. The inner diameter of the chamber is 12 mm, providing an exposure area of 113 mm^2, with a maximum capacity of 0.250 ml. Creams, ointments, and powders are loaded directly into the chamber while liquids are applied to a double disk of nonwoven cotton cloth cut to fit snugly into the chamber. The filled chambers are firmly fixed to the skin with porous tape. The flange indents the skin, confining the test substance, and producing a pressure ring on the skin, visible when the chamber is removed, which indicates a good seal. Nixon[47] performed studies using Hill Top chambers, relatively flat chambers made of a flexible polyethylene-vinyl acetate copolymer, with no added plasticizers, which has been approved by the FDA for food and beverage (alcoholic and nonalcoholic) container use. A nonwoven, absorbent, non-reactive cotton material is used inside the chambers, which can be obtained in 19 mm diameter (capacity 0.20 ml) or 25 mm diameter (capacity 0.30 ml) sizes. There is an inner and outer ring of plastic which forms a double seal with the skin, and tape is used to hold the chamber in place. The flexibility of the chamber allows it to follow the contours of the body, maintaining the seal without discomfort to the wearer. Its occlusivity has been demonstrated.[48] Nixon, using these chambers, was able to reduce the problems associated with gauze pads, *viz.*, confining the dose to the intended treatment site, and interactions which can occur between the test substance and tape used to secure the patch. Nixon concluded that these chambers, or an equivalent, in rabbit skin irritation testing can meet federal requirements yet reduce the number of animals needed for screening compounds since up to eight sites per animal can be used. Chambers can be used on man or on laboratory animals. Effects of the test compound on the skin are usually evaluated in a manner similar to that used for the Draize method, a scoring system of erythema and edema. Vickers[49] used a small chamber, containing Triton X® (a nonionic detergent used as an emulsifier or dispersing agent), fixed to the skin. A test compound was applied to the surface and, after a period of time, the solution collected and the desquamated corneocytes counted in a hemocytometer. In addition to total cell counts, the size of clumps of corneocytes and their fine morphology could be determined. Cell size and nuclear remnants in the cell are measures of the age and rate of development of corneocytes, and altered epidermopoesis may be a manifestation of an irritant effect. Vickers believed this technique to be a very sensitive measure of scaling, but large numbers of experiments were necessary to obtain statistically valid results.

B. Skin Thickness Measurements

One method used to quantify the edematous response of skin to a topically applied irritant is the measurement of a skinfold thickness. In some cases, skin thickening may occur in the absence of a comparable increase in erythema if epidermal rather than dermal changes are manifested.[12] Change in skinfold thickness is primarily a measure of edema, but in some instances this may be due to an increase in cellular layers (acanthosis) rather than fluid accumulation. This would be expected to occur only in cumulative irritation testing in which a test substance is applied daily over a period of weeks. Marzulli and Maibach[12] applied test substances to the clipped backs of albino rabbits daily, except for weekends, for 14 days, using four sites per rabbit and applying 0.05 ml of test substance and leaving the site uncovered. Each test site was scored visually, every 24 h, for erythema by comparing it with a centrally located control site. Sites were also measured, to the nearest 0.1 mm, for skinfold thickness, using a caliper with a constant tension of 10 g/mm^2, and making readings within a few seconds to avoid compression of the soft tissue. Results obtained with 11 substances demonstrated a high degree of correlation ($r = 0.86$; $p < 0.001$) between the erythema scored and the measured thickness of skin. When a second control site was compared with the first control site, there was virtually no difference in skinfold thickness. The authors stated that a mean increase in skinfold thickness above 0.4 mm may serve as a warning that the test substance may be irritating when applied repeatedly to human skin.

Wahlberg[50] evaluated the response of humans, rabbits, and guinea pigs to topically applied sodium lauryl sulfate, an anionic surface active agent, and nonanoic acid using skinfold thickness as the measurement of irritancy. He found that the laboratory animals had a normal increase in skinfold thickness with time, probably related to increasing age and weight; in the guinea pig, an average increase of 30% in 1 month was observed. He believed an unexposed control site for each series of tests was obligatory, as well as a pretreatment period of 4 to 5 days to check stability of the readings. Another drawback with the rabbit was intermittent and unpredictable hair growth, which interfered with measurements. In guinea pig testing he found the flanks to be suitable sites, whereas the skin of the back adhered too strongly to underlying tissue to be appropriate for skinfold thickness measurements. Although increasing the concentration of test substances caused an increase in fluid accumulation and earlier responses, thus establishing a dose-response relationship, the guinea pig was found to be less responsive than the rabbit, and, for 5% sodium lauryl sulfate, less responsive than man.

Tan et al.[51] discussed techniques for measuring skin thickness in man using xerography or high frequency pulsed ultrasound. Skin thickness was found to increase linearly with age up to 20 years, and then to decrease linearly. Differences in skin thickness between the sexes and between different body sites were demonstrated. Although a limiting factor in using the ultrasound technique may be changes in acoustic velocity through diseased skin, the authors believe that this would not differ to a significant degree from normal skin, and that the ultrasound technique could be useful in measuring alterations in skin thickness. Similarly, Marks and Kingston[7] used an ultrasound device to determine skin thickness and believed that it offered a simple, accurate, and reproducible measure of the degree of edema present.

Bucher et al.[52] determined the irritant effects of solutions with unphysiological pH values, using the abdominal skin of juvenile white mice. Mice weighing between 13 and 14 g were depilated, using a commercial cream, on a 3 to 4 cm^2 area of abdominal skin just below the xiphoid process. Prior to administering test solutions, the "tenderness" of the skin was determined by reefing the skin a few times with thin pincers, the tips of which were 7 mm apart, and thus evaluating the maximum number of wrinkles (skinfolds) which could be obtained. The skin was reefed a few times, purposely somewhat altering the geometry of the position of the tips of the pincers and the pressure against the skin to obtain the maximum number of wrinkles possible. The solutions tested were then injected intracutaneously, in a volume of 0.01 ml, forming a weal. At periods of 0.5, 1, 1.5, 2, 2.5, 3, 4, 5, and 6 h after injection, the maximum number of wrinkles obtainable was again determined. A pH 7 control solution was also used, and showed an initial drop in the number of wrinkles due to mechanical disturbance created by the weal fluid; after 1 to 2 h, the number of wrinkles returned to preinjection values. The authors believed this procedure to be a simple and reliable test method for irritancy of solutions with pH values between 3 and 11, and could be used as a standard test for topical irritancy. The same authors published another paper[53] describing the same test for identifying irritant substances. Because of differences in man and laboratory animals, and the difficulty in extrapolating from animal to man, they injected test substances intracutaneously, essentially removing any barrier effect of the skin, to determine if there would be any acute risk of contact of test substances with mucous membranes. They expressed their desire, by developing this assay, to make the Draize rabbit ocular irritation studies superfluous in most cases. Walz and Bucher[54] and Walz[55] wrote about an algorithm developed which transformed the number of wrinkles recorded, just before as well as 2.5 and 6 h after injection, into a measure of the intensity of the edematous reaction. They found the results obtained with ten mice per solution tested to be reliable and reproducible, and to conform to dose-response relationships, and hence quantified local irritancy.

Gloxhuber and Schulz[37] had previously advocated the intradermal injection of 0.05 ml of test substances into depilated abdominal skin of mice with evaluation of the subcutaneous surface of the skin after sacrifice of the animals at 24 h. Gloxhuber and Kastner[56] further defined this intradermal testing in mice. While they cautioned against directly extrapolating results in mice to that expected in man, they believed by using this intradermal test that a new chemical could be compared with reference substances, and determine if it would be better, or equally, or less well tolerated by man than the reference chemical.

Pharmacologists have developed assays, based on ear swelling, in which inflammation is quantitated by less subjective and more reproducible means than those used in Draize testing. The activity of anti-inflammatory drugs has been evaluated by applying topically to ears of rats a solution containing an irritant and the anti-inflammatory drug being investigated, and applying to the ears of control rats only the irritant. In the original assay of this test, the degree of inflammation was quantitated as the differences in wet and dry weights of treated and untreated ears. The technique was later refined by measuring ear thickness with a micrometer to assess degree of inflammation.[13] Other investigators have used the ears of laboratory mice in similar testing. Patrick et al.[13] wrote that a rodent ear assay procedure using ear thickness to quantify the degree of inflammation would effectively circumvent many problems of the more commonly used skin irritation assays: it is not necessary to clip hair from the site, which can cause skin damage; the site is not abraded; the test area is not occluded; and, the response can be evaluated by nonvisual measurements in which the units are linear.

C. Cutaneous Blood Flow Measurements

The use of laser Doppler flowmetry to estimate the flow of erythrocytes in the cutaneous microvessels offers a safe and accurate method of measuring skin blood flow, and thereby estimating erythema in man quantitatively.[57-59] The operating principle of the flowmeter is based on the fact that monochromatic light from a laser is scattered in moving erythrocytes, broadening the spectral frequency of reflected radiation, while beams in static tissue structures are reflected essentially unchanged in frequency. Light from a helium-neon (He-Ne) laser is guided by an optical fiber in a probe to the body surface, where it permeates the skin to a depth of 0.5 to 1 mm. A portion of the back-scattered light is transmitted by a pair of fibers to photodetectors, and changes in magnitude and frequency related to cutaneous blood flow are recorded. Using a differential detector system and a processing arrangement incorporated in flowmeters, a low-noise output signal that is linearly related to blood flow for low and moderate flow rates is obtained. The analog output signal can be fed to a recorder for continuous flow recording, the signal expressed in relative and dimensionless blood flow values. Marks and Kingston[7] reported that, for assessment of mild reactions with minimal edema, the laser Doppler measurement of skin blood flow provided a very sensitive method of detecting a positive irritation reaction. Van Neste et al.[60] stated that cutaneous blood flow values, as measured by laser Doppler flowmetry, only reflect secondary, delayed inflammatory changes. The authors made a distinction between primary chemical insult, identified clinically as roughness and functionally as increased transepidermal water loss associated with alterations of barrier function of the skin, and secondary delayed inflammatory response, clinically recognized as erythema and functionally by increased cutaneous blood flow. The use of laser Doppler flowmetry in laboratory animals may be of somewhat limited value due to differences in skin vascularity between man and other animals. Skin vasculature in humans is quite vast, due primarily to the thermoregulatory function, whereas cutaneous blood flow under fur is principally related to nutrition of the hair follicles, and does not seem to be increased in heat stress.[61] The stratum corneum is also thinner in most animal models, frequently compensated by a relatively thick hair cover, and consequently shaving or clipping the hair coat of animals probably creates a much more sensitive testing system.

Another method of measuring cutaneous blood flow is the ^{133}Xe washout technique.[62,63] At a given time after exposure to an irritant or test substance, a known amount of radioactive ^{133}Xe, an inert gas, in solution is injected intradermally or applied on the skin. Radioactive washout, the decrease in the emission of the low energy gamma radiation, is measured with a sodium iodide crystal coupled with a scintillation system. The period required to reduce the radioactive counts by one half (a half-life or $T_{1/2}$) is calculated, and is a function of blood flow to the area.

Vickers[49] reported this method to be sensitive enough to measure the reduction in skin blood flow produced by the vasoconstriction of topically applied corticosteroids. The technique could be developed to provide a definitive measurement of skin blood flow which is comparable to erythema, at least in man, but it suffers from the necessity of using a radioisotope.

D. Transepidermal Water Loss Measurements

Transepidermal water loss (TEWL), determined by evaporimetry, is used to assess alterations of the barrier function of skin. The probe of an evaporimeter contains two sets of thermistors and capacitive thin-film transducers located 3 and 9 mm from the tip of the probe. Temperature and humidity of air flowing freely through the probe are determined at both sites. The partial water pressure is determined at each level and the TEWL derived from the pressure gradient and displayed in g/m^2/h on a central unit. Excessive turbulence of air flow entering the probe can be prevented by a small windscreen around the site. There is usually a lag time of 20 to 40 s prior to achieving stable readings, and this necessitates a minimum 60 s contact time of the probe with the skin.[60] With damaging substances such as sodium lauryl sulfate, it is easy to demonstrate that TEWL begins to increase well before the development of a visible reaction. Kligman[2] found that hydrophilic ointment USP, a very useful vehicle for many drugs, contained 1% sodium lauryl sulfate and that occlusive application of this ointment for a few days produced no visible change, yet measurement of TEWL showed a 100% increase within 24 h and a 3- to 4-fold increase after 2 to 3 days. The same order of change was also observed after washing forearm skin twice daily for 2 days with a popular hand soap and the enhanced permeability, measured as TEWL, preceded signs of redness and scaling by about 1 week. Van Neste et al.[60] found a significant positive correlation between clinical signs of skin irritation and increased TEWL values. The use of TEWL measurements in laboratory animals to predict human responses, however, may be of limited value due to the same reasons cited for cutaneous blood flow measurements, i.e., the skin vasculature and thermoregulatory functions of human skin are not similar to those of other animals.

E. Cutaneous Thermometry

Thermometry, the measurement of temperature, was discussed by Marks and Kingston.[7] Using human volunteers, they studied the irritancy of dithranol (anthralin), an antipsoriatic. Dithranol stains the skin brownish-purple and erythema due to irritancy cannot be accurately determined. Various concentrations of the drug were applied occlusively in Finn chambers for a 24 h period to the flexor aspects of forearms of ten healthy subjects per concentration. In order to contrast the irritant reactions due to dithranol with reactions due to other irritants, 20% benzoyl peroxide and 7% sodium lauryl sulfate aqueous solutions were applied to other sites on the forearms. At the end of the 24 h exposure period, skin thickness was determined with calipers and the skin surface temperature measured with a thermocouple device. Biopsies were also taken, and sections were fixed in formalin and stained with hematoxylin and eosin, and examined microscopically. The skin exposed to dithranol showed a concentration-dependent rise in temperature, but skin exposed to 20% benzoyl peroxide demonstrated a minimal temperature increase, and skin exposed to 7% sodium lauryl sulfate showed no increase in temperature, although increases in skin thickness were evident and edema and inflammation present on microscopic examination of the biopsy specimens of skin exposed to both positive control substances. The only skin temperature increase which met the predefined positive reaction, an increase of 0.5°C over surrounding tissue, was created by the highest concentration of dithranol. Vickers[49] reported that there have been many attempts to measure alteration in skin temperature. He stated that early attempts with thermistors were no more successful in the long run than more sophisticated attempts with liquid crystals or thermoscan camera thermography, and that such measures have been abandoned in favor of other techniques.

Miller and Wildnauer[64] used thermomechanical analysis and differential scanning calorimetry to examine skin biopsies from human volunteers treated locally with a retinoic acid product (with occlusive therapy) petrolatum, propylene glycol-ethanol, or physiologic saline. Shifts were observed in softening transitions (due to lipid and protein alterations with heat) of treated skin from untreated skin, implying possible alterations in higher-level organization of keratin in the stratum corneum.

F. Cutaneous Morphometry

Changes in skin surface in the form of roughness and brittleness are common complaints of users of irritant substances. Gloxhuber and Schulz[37] report research of others accomplished to determine a method of measuring these effects. The affected skin is stained with Primulin O and examined under a fluorescence microscope.

Roughness of the surface can be well observed with this method, and a rating scale developed. Other dyes can also be used on the skin, with the quantity of dye retained being an indicator of the roughness. Reflectance measurements can also be made to quantitate the degree of roughness.

Holzle and Plewig[65] investigated the effects of increased or decreased epidermal turnover on the morphology of human corneocytes. The desquamating portion of the stratum corneum was sampled with a detergent scrub technique using Triton® X-100, and cells evaluated for quantity, size, shape, nuclear inclusions, and trabeculae. Corneocytes from skin of patients with allergic contact dermatitis differed from those of normal skin in that they were smaller, of irregular shape, and had asymmetrical trabeculae; more cells were nucleated and greater numbers of cells were collected. Topically applied steroids greatly improved all parameters. The authors believed such a bioassay permitted sensitive measurements of corneocyte morphology in conditions with altered epidermal cellular kinetics, and provided a method to evaluate the effects of corticosteroids on these parameters.

G. Measurement of Inflammatory Indices

Analyses for indicators of inflammation in blood or tissue of laboratory animals have been proposed for quantitating the inflammatory response to irritation. Different indicators and different procedures have been advocated. It is highly desirable to be able to easily obtain and objectively measure an indicator of skin inflammation. Analysis of body fluid, such as blood, or exudate or transudate collected from blisters or chambers over sites of inflammation, for the presence of such an inflammatory indicator would be ideal. Potential indicators include enzymes, eicosanoids (derivatives of arachidonic acid, including prostaglandins and leukotrienes), and acute phase reactant proteins.

Enzymes are protein catalysts of biological origin, a catalyst being defined as a substance which increases the rate of a particular chemical reaction without itself being consumed or permanently altered.[71] Knowledge of enzymes and their actions goes back at least 150 years. Clinical enzymology, the application of the knowledge of enzymes to the diagnosis and treatment of disease processes, is a still rapidly developing field in contemporary clinical chemistry. Measurements of digestive enzyme activity in body fluid as an aid to diagnosis date back to the early 1900s. Measurement of enzyme activity in serum began in the 1920s and 1930s with studies on alkaline phosphatase in bone and liver diseases, and acid phosphatase in prostatic cancer. In 1943, Warburg and Christian observed increased activities of glycolytic enzymes in sera of tumor-bearing rats, and, in 1955, LaDue et al. reported a transitory rise of glutamic-oxaloacetic transaminase (SGOT; now called aspartate transaminase, AST) activity in serum following acute myocardial infarction.[71]

Over 700 enzymes have been isolated, each capable of catalyzing a specific organic or inorganic reaction. Some enzymes are found only in specific organs, while others are found in most every cell of the body. Enzymes may also be quite specific to certain compartments or organelles within the cells, such as cytosol, lysosomes, peroxisomes, membrane border, nucleus or nucleolus, endoplasmic reticulum, Golgi bodies, or mitochondria. Enzymes common to many or most organs of an individual animal may exist in different forms, which demonstrate dissimilar patterns on electrophoresis. Different forms of the same enzyme are called isozymes, or, more preferably, isoenzymes. Levels of enzymes in body fluids usually have a "normal" range which often varies with species, age, sex, etc., and may vary from one individual animal to another. Using blood, one or more baseline levels may be obtained prior to any procedure. Increased levels of enzymes may be due to leakage from injured or dying cells, or due to increased synthesis as a result of damage or tissue insult or an increase in the number or activity of cells producing the enzyme.

Marks and Kingston,[7] in addition to using thermometry, measurements of skin thickening, and histopathology to assess the irritancy of dithranol, also used suction to form skin blisters, applied dithranol on the roofs of the blisters, and subsequently analyzed blister fluid for the enzymes lactate dehydrogenase and beta-glucuronidase, and for protein. Protein concentration and beta-glucuronidase activity were not increased in blister fluid after irritation of the skin, but lactate dehydrogenase activity increased in blister fluid after dithranol application either to blister roofs or to blister sites prior to blister formation. Lactate dehydrogenase activity increased after doses of dithranol that did not normally elicit observable clinical responses. Lactate dehydrogenase (LDH or LD) is an enzyme which catalyzes the reversible lactate to pyruvate reaction, and is found in all body tissue.[72] Marks and Kingston referenced another study in which dimethyl sulfoxide (DMSO) created similar results. They postulated that irritants made epidermal cells "leaky" without causing irreversible cell damage, and that as the intensity of the stimulus increased, irreversible cell damage and cell death occurred. Hamami and Marks[73] found that a 50% solution of ammonium hydroxide and sodium lauryl sulfate (concentration not given) also increased LDH activity in blister fluid. Middleton[74] advocated measuring enzyme activity in suction blister fluid as a measure of skin irritation. Activities of LDH, malate dehydrogenase (MDH), AST, and glutamate dehydrogenase (GLDH) were increased in blister fluid of skin exposed prior to blister formation to tributlytin, at doses and at times when there were no comparable increases in plasma enzyme activity. Lazarus et al.[75] wrote that acid phosphatase, acid proteinases (Cathepsin D and Cathepsin B_1), and neutral proteinases of lysosomes in the skin were found in increased levels in inflamed areas. They postulated that this was due to lysosomal lysis due to irritants such as UV light.

Lewis,[76] in the 1960s, catheterized femoral lymph vessels of cats prior to heating, cooling, or creating a local ischemic condition. His reasoning was that inflammation is a local reaction, and changes in interstitial fluid content, rather than that of blood, should be examined. Since interstitial fluid itself is inaccessible, lymph, the fluid drainage of the interstitial spaces, should be analyzed. He measured LDH, AST, alanine transaminase (ALT), acid phosphatase, and beta-glucuronidase levels in the lymph of the injured leg, of the contralateral limb, and in venous blood from the limb. Burns caused rapid and marked increases in the concentration of ALT, AST, and LDH, but only small increases in acid phosphatase and beta-glucuronidase, in lymph collected from the injured leg. Only minor changes were observed in levels of these enzymes in blood or lymph from the contralateral limb. Freezing a limb caused greatly increased levels of all enzymes measured, but ischemia did not cause a significant increase in any enzymes. The author discussed the intracellular distribution of enzymes (LDH and most of the AST being cytoplasmic, ALT and a small amount of AST being mitochondrial, and acid phosphatase and beta-glucuronidase being mainly in lysosomes) and believed his results demonstrated various degrees of cellular damage, resulting in changes in enzyme levels in lymph.

McCreesh and Steinberg[3] report on the work of those who examined test animals for stress-induced biochemical changes by measuring phosphogluconate dehydrogenase, glucose-6-phosphate dehydrogenase, acid phosphatase, alkaline phosphatase, monoamine oxidase, and succinic dehydrogenase. No changes attributable to irritation testing could be found.

More recently, analyses of other enzymes have been advocated for measurement of cutaneous inflammation. Bradley et al.[77] and Kensler et al.[78] reported on the quantitation of myeloperoxidase for assessing dermal inflammation. The accumulation of neutrophils (polymorphonuclear leukocytes) is a prominent feature in many inflammatory conditions. An ability to estimate the quantity of neutrophils in inflamed tissue might prove useful in judging the intensity of inflammation or the effect of experimental or therapeutic regimens in altering the inflammatory response. Myeloperoxidase is a plentiful constituent of neutrophils, amplifies the reactivity of superoxide anions and hydrogen peroxide generated by polymorphonuclear leukocytes, and is located within the primary granules. Myeloperoxidase activity has been shown to increase 15-fold in the skin by 16 h after treatment with a phorbol diester.[78] To completely extract myeloperoxidase from neutrophils or skin, hexadecyltrimethyl-ammonium bromide is used to solubilize the enzyme, which is then quantitated spectrophotometrically. The enzyme level is directly related to the number of neutrophils present. In addition to peroxidases, it is possible one could also measure changes in skin superoxide dismutase, an enzyme found in tissue, including red blood cells, which catalytically scavenges the toxic free radical superoxide anion (O_2^-).

Canonico et al.[79] wrote that sterile inflammatory lesions cause a nonspecific anabolic response of liver and nonparenchymatous liver cells (such as macrophages) resulting in the synthesis of a class of glycoproteins collectively termed acute-phase globulins. These proteins are thought to function by restricting tissue damage through inhibition of proteases released at sites of inflammation, by aiding in wound healing, by modulating blood clotting, by activating phagocytes, by promoting phagocytosis of necrotic tissue, and by modulating the immune response. These authors proposed that increased activity of glycosyltransferases in serum represented a nonspecific response to inflammation created by an increased demand for glycosylation of newly synthesized, acute-phase globulins. Serum levels of sialyltransferase, galactosyltransferase, a_2-fucosyltransferase, and a_3-fucosyltransferase were measured using scintillation spectroscopy after various tissue insult, including bacterial infection and skin burns. Variable, multifold increases in serum enzyme levels were observed, depending upon the type of tissue injury. The authors concluded that certain patterns of serum glycosyltransferase levels may be recognized as having diagnostic significance.

Volden et al.[80] studied lysosomal, cytosolic, and plasma membrane enzymes in epidermis, dermis, and suction blister fluid after UV irradiation. The authors measured arylamidase, beta-galactosidase, alpha-mannosidase, beta-acetylglucosaminidase, acid phosphatase, lactate dehydrogenase, and phosphodiesterase I in epidermal and dermal homogenates and in blister fluid of humans exposed locally to UV radiation. Slightly lower lysosomal enzyme activities were found in epidermis 18 h after irradiation, while values were invariably higher in the dermis of inflamed skin. The activities of the plasma membrane and cytosol enzymes were higher in normal epidermis and dermis than in inflamed skin. Higher values were also found in suction blister fluid raised on inflamed skin, suggesting destruction of the plasma membrane during the inflammatory process.

Prostaglandins (first isolated from the prostate gland) were discovered in the 1930s at about the same time as the essential fatty acids, but there was nothing to connect the two groups until, in the early 1960s, the structures of two prostaglandins (PGE$_1$ and PGF$_{1\alpha}$) were discovered to be cyclized fatty acids.[81] It is now known that arachidonic acid (C20:4; 5,8,11,14-eicosatetraenoic acid) is the major source of prostaglandins in mammalian tissues.[82] With the exception of some blood cells, virtually all mammalian tissue which is under the influence of cyclooxygenase, a widely distributed enzyme, can synthesize prostaglandins. Cyclooxygenase activity is normally initiated through the release of arachidonic acid by injury or stimulation of tissue. Prostaglandin release has been demonstrated in acute inflammation induced by irritants in laboratory animals, and elevated prostaglandin synthesis occurs in man with inflammatory skin diseases, psoriasis, and allergic eczema.[81,83] Prostaglandins fall into several main classes, designated by letters, and distinguished by substitutions on the common cyclopentane ring. The main classes are further subdivided in accord with the number of double bonds in the side chains, which are indicated by subscripts. Prostaglandin E$_2$ (PGE$_2$) is the predominant prostaglandin found as a result of inflammatory conditions of the skin, but prostaglandin F$_{2\alpha}$ (PGF$_{2\alpha}$), prostaglandin I$_2$ (PGI$_2$; prostacyclin), prostaglandin D$_2$ (PGD$_2$), and thromboxane A$_2$ (TXA$_2$) have also been reported.[63,81,84] Arachidonic acid may also be acted upon, under the influence of lipoxygenase, to form the lipid peroxide 12-hydroperoxyeicosatetraenoic acid (12-HPETE), which subsequently forms 12-hydroxyeicosatetraenoic acid (12-HETE) or a complex group of compounds known as leukotrienes (see Figure 2). Leukotriene B$_4$ (LTB$_4$) has been detected by radioimmunoassay (RIA) in inflammatory exudates obtained following the implantation of irritant-soaked polyester sponges in rats.[85] LTB$_4$ attracts leukocytes to sites of inflammation and causes degranulation of these cells, contributing to the inflammatory response.[85-87]

Marks and Kingston[7] reported on research which measured increased levels of pro-inflammatory prostaglandins in perfusates from primary irritant dermatitis reactions, but stated that their role in the genesis of the irritant reaction is uncertain. Yoshino et al.[88] formed subcutaneous pouches by injection of 20 ml of sterilized air

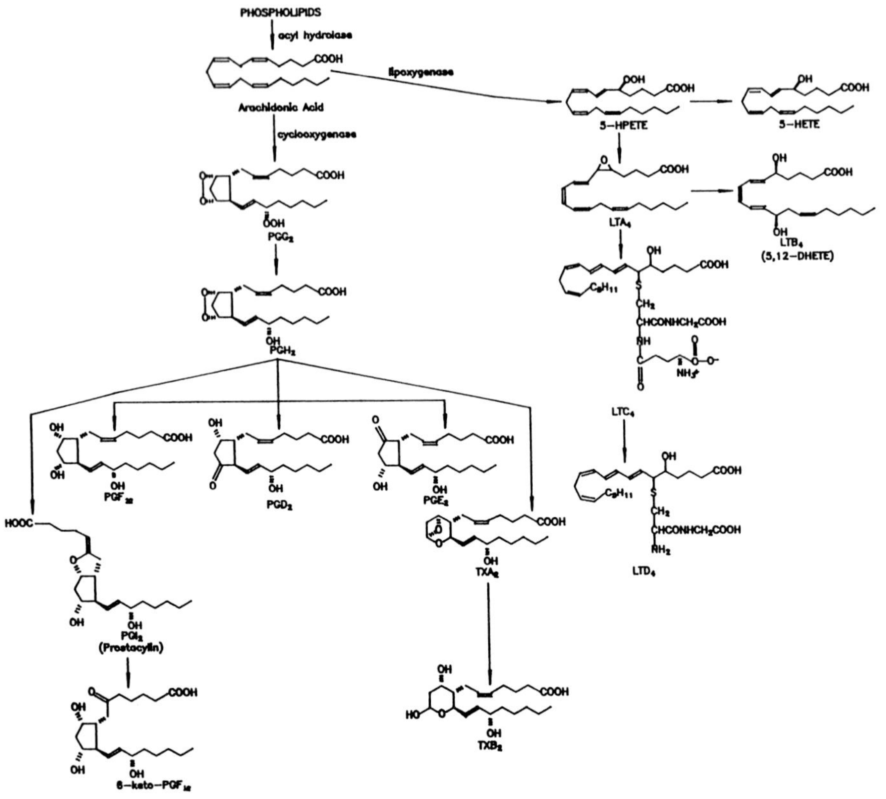

Figure 2. Products of arachidonic acid.

on the dorsum of rats and followed this with injection of an additional 10 ml of air 7 days later. Streptococcal cell wall fragments, suspended in phosphate-buffered saline, were then injected into the pouch to create an inflammatory condition. The arachidonic acid metabolites, PGE_2 and LTB_4, were measured in the pouch fluid and concentrations served as indicators of the degree of inflammation induced. Analyses of local prostaglandin levels have been used to evaluate the anti-inflammatory efficacy of experimental therapeutic compounds. Changes in plasma levels of PGE_2, TXB_2 (the stable degradation product of TXA_2, which has a very short chemical half-life), and LTB_4, after an acute inflammatory condition of the skin had been created, were not measured using commercially available radioimmunoassay kits (Olson et al., unpublished data). Changes in pre- and postirritation levels of plasma arachidonic acid, likewise, have not been found to be significantly different.

Marzulli and Maibach[12] found no differences in the PGE content of irritated skin and that of control skin. Although increases in arachidonic acid derivatives in tissues at local sites of inflammation have been demonstrated, when assaying plasma, no observable changes from preirritation levels have been seen. Biologically active prostaglandins have short life spans. The half-life of TXA_2 is approximately 30 s, and more than 95% of infused PGE_2 is inactivated after one circulation through the lungs.[86] This complicates the measurement of prostaglandin levels. Some prostaglandins have more stable degradation products (e.g., PGI_2 to 6-keto-$PGF_{1\alpha}$ and TXA_2 to TXB_2), which can be measured to indirectly determine changes in concentration of parent compounds. The use of a skin chamber or blister with analysis of local changes in concentrations of prostaglandins or leukotrienes would allow multiple dosing sites per animal, and could become a sensitive, quantitative, and efficient measure of irritancy of test compounds in the future.

Another potential method of quantitating irritancy using inflammatory mediators is the measurement of acute phase reactant proteins. Tissue injury and inflammation are potent stimuli for the increased synthesis of certain plasma proteins collectively known as acute phase reactants, which are nonspecific and rapid participants in inflammation and tissue repair.[89] About 20% of the approximately 100 isolated and well-characterized human plasma proteins are classified as acute phase proteins, which are present in various quantities and with little known functional relationship among them.[90] Some belong to the coagulation (e.g., fibrinogen, factor VII) or complement (e.g., C2, C3, C4, C5, C9, B) systems or are transport proteins (e.g., haptoglobin, ferritin, ceruloplasmin) or proteinase inhibitors (e.g., $alpha_1$-antitrypsin, $alpha_1$-antichymotrypsin, C1-inactivator), while others have unknown functions. The majority of acute phase proteins are glycoproteins, but two acute phase proteins in humans, C-reactive protein (CRP) and serum amyloid-A (SAA), are pure polypeptides.

In 1930, Tillett and Francis observed that sera obtained from patients during acute febrile illnesses precipitated with an extract of the pneumococcus, first designated Fraction C and later C-polysaccharide.[91] The substance responsible for this reaction was termed C-precipitin, but as further study demonstrated C-precipitin was a protein, it was designated C-reactive protein or CRP. CRP shares with immunoglobulins the ability to initiate multiple effector functions that are associated with the inflammatory response. CRP and SAA changes in serum are evident within a few hours of initiation of inflammation and reach peak values, which, depending upon the severity of inflammation, may be more than 1000-fold normal, after 24 to 48 h.

The majority of acute phase proteins is synthesized in hepatocytes, under the influence of circulating mediators released in association with inflammation and necrosis, and, during acute inflammatory processes, the extent of the acute phase response has a quantitative relationship with the degree of inflammation.[90] The acute phase response is observed not only in man, but in all mammals, as well as birds, although the major acute phase proteins vary between species. While in the human, CRP and SAA represent the major acute phase proteins, CRP is the major acute phase protein in rabbits, SAA and serum amyloid-P (SAP), which has remarkable structural similarity to CRP, are important in the mouse, and $alpha_1$-acid glycoprotein and $alpha_2$-macroglobulin are predominant in the rat. As immunoassay techniques and instrumentation improve, the measurement of acute phase proteins in blood or local tissue may well serve as the most sensitive and quantitative indicator of inflammation, and become routinely used for this purpose.

H. Other Techniques

Some other less commonly reported techniques used in an attempt to quantify skin irritancy include photoacoustics,[66] electrical resistivity,[67] and measurements of intracorneal cohesion,[68] and vascular permeability.[69,70]

IV. CONCLUSIONS

Various *in vivo* techniques to assess dermal irritancy have been discussed in this chapter, but the standard Draize test, or one of the many modifications of it, is still routinely used, in many cases because it is required by government regulation. It has served the purpose of identifying severe irritants relatively well, but has come under criticism due to its potential to elicit pain and due to requirements for large numbers of animals. There are also difficulties with the Draize test in accurately identifying or ranking moderately or minimally irritating compounds. To date, however, a technique has not been developed which has received universal acceptance as an alternative. *In vitro* techniques of assessing dermal irritancy are being studied, and may eventually be used as an adjunct to or replacement for *in vivo* procedures. Proper validation of these *in vitro* tests with *in vivo* results will be necessary, and the complete replacement of *in vivo* with *in vitro* testing will probably not be possible. For example, the well-known Ames *in vitro* assay for carcinogenicity received acclaim and relatively rapid acceptance due to the need to lower the time and cost requirements for the testing of potential carcinogens, but the relatively large numbers of false positive and negative results has resulted in the use of the test as a mutagenicity screen only. Another problem is the extrapolation of testing results in laboratory animals to what is to be expected in humans. In this chapter, differences in sensitivity between animal species, sexes, individuals, etc., have been briefly discussed. These differences will have to be addressed if a viable alternative testing strategy is to be developed.

Epidemiologic studies can be invaluable in evaluating risks from dermal irritants, but often such studies are inconclusive and must be performed after the problem is manifest in the human population. How often are individuals exposed to only one chemical or compound? It is difficult, if not impossible, to accurately identify a causative agent for many effects, and it may actually be due to a combination of chemicals. For dermal irritancy, after initial *in vitro* and/or animal studies, with products shown not to be exceedingly irritating, it may be possible, in some cases, to use human volunteers, especially those of Celtic origin, who appear to be most sensitive to dermal irritation. It would probably be advisable in such cases to determine a threshold dose for irritancy, rather than using a single selected dose, and determining the mechanism(s) of irritation would be very helpful in developing less irritating substitutes.

In vivo testing for dermal irritancy will probably never be completely abandoned, but it should be possible to improve techniques to more objectively and quantitatively assess irritancy and reduce the number of animals used for this purpose. As more is learned about the mediators of inflammation and endogenous anti-inflammatory substances, and as instrumentation is improved and better techniques of extraction and detection developed, more sensitive, more objective, and more replicable methods for assessing skin irritation will be developed and, hopefully, accepted.

REFERENCES

1. *Dorland's Illustrated Medical Dictionary,* 25th ed. W. B. Saunders, Philadelphia (1974).
2. A. M. Kligman, in *Principles of Cosmetics for the Dermatologist.* P. Frost and S. N. Horowitz, Eds. C. V. Mosby, St. Louis, 265 (1982).
3. A. H. McCreesh and M. Steinberg, in *Dermatotoxicology,* 2nd ed. F. N. Marzulli and H. I. Maibach, Eds., Hemisphere, New York, 147 (1983).
4. DHSS, *Guidelines for the Testing of Chemicals for Toxicity.* Committee on Toxicity of Chemicals in Food. Consumer Products and the Environment. Report on Health and Social Subjects. no. 27.9.43. HMSO, London, (1982).
5. E. A. Emmett, in *Casarett and Doull's Toxicology,* 3rd ed. C. D. Klaassen, M. O. Amdur, and J. Doull, Eds. 412 (1986).
6. National Academy of Sciences, *Principles and Procedures for Evaluating the Toxicity of Household Substances.* Washington, D.C.; a report prepared by the Committee for the Revision of NAS Publication 1138, under the auspices of the Committee on Toxicology, National Research Council, National Academy of Sciences (1977).
7. R. Marks and T. Kingston, *Food Chem. Toxicol.* 23, 155 (1985).
8. R. C. Wester and H. I. Maibach, *Dermatotoxicology,* 2nd ed. F. N. Marzulli and H. I. Maibach, Eds. Hemisphere, New York, 131 (1983).
9. W. Meyer, R. Schwarz, and K. Neurand, *Skin, Drug Application and Evaluation of Environmental Hazards. Current Problems in Dermatology, Vol. 7.* G. A. Simon, Z. Paster, and M. A. Klingberg, Eds. S. Karger, Basel, 39 (1978).
10. R. O. Carter and J. F. Griffith, *Toxicol. Appl. Pharmacol.* 7(Suppl.) 60 (1965).
11. G. A. Nixon, C. A. Tyson, and W. C. Wertz, *Toxicol. Appl. Pharmacol.* 31, 481 (1975).
12. F. N. Marzulli and H. I. Maibach, *Food Cosmet. Toxicol.* 13, 533 (1975).
13. E. Patrick, A. Burkhalter, and H. I. Maibach, *J. Invest. Dermatol.* 88 ISS 3 Suppl., 24s (1987).
14. IRLG, *Recommended Guideline for Acute Dermal Toxicity Test: Testing Standards and Guidelines Work Group.* Washington, D.C. (1981).
15. A. N. Rowan, *Acute Toxicity Tests: Possible Alternatives,* F. W.van der Kreek, Ed. Proc. Symp. Utrecht Dutch Soc. Protect. Animals, Dutch Soc. Toxicol., Anti-Vivisection Found. (1981).
16. J. S. Friedenwald, W. F. Hughes, Jr., and H. Herrmann, *Arch. Ophthal.* 31, 279 (1944).
17. J. H. Draize, G. Woodard, and H. O. Calvery, *J. Pharmacol.* 82, 377 (1944).
18. E. Bosshard, *Food Chem. Toxicol.* 23, 149 (1985).
19. C. S. Weil and R. A. Scala, *Toxicol. Appl. Pharmacol.* 19, 276 (1971).
20. J. R. Nethercott, E. R. Maraghi, and D. Andrews, *Occup. Health Chem. Ind., Proc. 11th Int. Congr.* 427 (1984).
21. C. Schlatter and C. A. Reinhardt, *Food Chem. Toxicol.* 23, 145 (1985).
22. R. Sharpe, *Food Chem. Toxicol.* 23, 139 (1985).
23. P. J. Simons, *The Use of Alternatives in Drug Research.* A. N. Rowan and C. J. Stratmann, Eds. Macmillan, London, 147 (1980).

24. A. M. Sincock, *Acute Toxicity Tests: Possible Alternatives*. F. W. van der Kreek, Ed. Proc. Symp. Utrecht Dutch Soc. Protect. Animals, Dutch Soc. Toxicol., Anti-Vivisection Found. (1981).

25. L. J. Vinson, *Detergency, Theory and Test Methods. Part II*. W. G. Cutler and R. C. Davies, Eds. Marcel Dekker, New York, 679 (1975).

26. Office of Toxic Substances, Environmental Protection Agency, Primary Dermal Irritation, in *Health Effects Test Guidelines*. (August 1982).

27. CPSC. CFR Title 16, Subchapter IIC — Federal Hazardous Substances Act Regulations, Secs 1500.40, 1500.41, and 1500.42 (1986).

28. OECD. Acute Dermal Irritation/Corrosion, in *OECD Guidelines for Testing of Chemicals*. Section 4, no. 404. OECD, Paris (1981).

29. EEC. Council Directive 79/831/EEC(18 September). Annex V, Part B: Methods for the Determination of Toxicity. Annex VII: 4.1.5 Acute Toxicity, Skin Irritation. Off. J. Europ. Commun. 22, (L259), 10 (1979).

30. J. H. Draize, *Appraisal of the Safety of Chemicals in Foods, Drugs and Cosmetics*, Association of Food and Drug Officials of the United States, 1959, Fourth Printing (1979).

31. USFDA: Hazardous Substances: Proposed Revision of Test for Primary Skin Irritants, *Fed. Reg.,* 37 (244), 27635 (1972).

32. R. L. Bronaugh and H. I. Maibach, *Principles of Cosmetics for the Dermatologist*, P. Frost and S. N. Horwitz, Eds. C. V. Mosby, St Louis, 223 (1982).

33. L. Phillips, M. Steinberg, H. I. Maibach, and W. A. Akers, *Toxicol. Appl. Pharmacol.* 21, 369 (1972).

34. V. K. H. Brown, *J. Soc. Cosmet. Chem.* 22, 411 (1971).

35. R. L. Roudabush, C. J. Terhaar, D. W. Fassett, and S. P. Dziuba, *Toxicol. Appl. Pharmacol.* 7, 559 (1965).

36. J. F. Griffith and E. V. Buehler, *Prediction of Skin Irritancy and Sensitizing Potential by Testing with Animals and Man*. Proc. 3rd Conf. Cutaneous Toxicity, Academic Press, New York, 155 (1977).

37. C. Gloxhuber and K. H. Schulz, *Detergency, Theory and Test Methods. Part II*. W. G. Cutler and R. C. Davies, Eds. Marcel Dekker, New York, 695 (1975).

38. M. B. Vinegar, *Toxicol. Appl. Pharmacol.* 49, 63 (1979).

39. R. E. Grissom, Jr., C. Brownie, and F. E. Guthrie, *Bull. Environ. Contam. Toxicol.* 38, 917 (1987).

40. M. V. Dahl, F. Pass, and R. J. Trancik, *J. Am. Acad. Dermatol.* 11, 474 (1984).

41. M. Potokar, O. J. Grundler, A. Heusener, R. Jung, P. Murmann, C. Schobel, H. Suberg, and H. J. Zechel, *Food Chem. Toxicol.* 23, 615 (1985).

42. M. R. Gilman, R. H. Evans, and S. J. DeSalve, *Drug Chem. Toxicol.* 1, 391 (1978).

43. J. B. Sullivan, J. C. Strausburg, and R. W. Kapp, Jr., *Toxicol. Appl. Pharmacol.* 33, 165 (1975).

44. S. C. Gad, R. D. Walsh, and B. J. Dunn, *J. Toxicol. Cutaneous Ocul. Toxicol.* 5(3), 195 (1986).

45. A. M. Kligman and W. M. Wooding, *J. Invest. Dermatol.* 49, 78 (1967).

46. P. J. Frosch and A. M. Kligman, *Contact Dermatitis* 5, 73 (1979).

47. G. A. Nixon, Human and Environmental Safety Division, Procter & Gamble, Cincinnati, OH, personal communications (1987).

48. R. A. Quisno and R. L. Doyle, *J. Soc. Cosmet. Chem.* 34, 13 (1983).

49. C. F. H. Vickers, *Int. J. Cosmet. Sci.* 1, 363 (1979).

50. J. E. Wahlberg, *Contact Dermatitis* 9, 21 (1983).

51. C. Y. Tan, B. Statham, R. Marks, and P. A. Payne, *Br. J. Dermatol.* 106, 657 (1982).

52. K. Bucher, K. E. Bucher, and D. Walz, *Agents Actions* 9, 124 (1979).

53. K. Bucher, K. E. Bucher, and D. Walz, *Agents Actions* 11, 515 (1981).

54. D. Walz and K. E. Bucher, *Agents Actions* 12, 552 (1982).

55. D. Walz, *Food Chem. Toxicol.* 23, 199 (1985).

56. C. Gloxhuber and W. Kastner, *Food Chem. Toxicol.* 23, 195 (1985).

57. G. A. Holloway, Jr. and D. W. Watkins, *J. Invest. Dermatol.* 69, 306 (1977).

58. G. E. Nilsson, U. Otto, and J. E. Wahlberg, *Contact Dermatitis* 8, 401 (1982).

59. J. E. Wahlberg, *Scand. J. Work Environ. Health* 10, 159 (1984).
60. D. Van Neste, G. Mahmoud, and M. Masmoudi, *Contact Dermatitis* 16, 27 (1987).
61. E. J. Calabrese, *Principles of Animal Extrapolation*. John Wiley & Sons, New York (1983).
62. J. K. Kristensen and S. Wadskov, *J. Invest. Dermatol.* 68, 196 (1977).
63. S. B. Palder, W. Hurval, S. Lelcuk, F. Alexander, D. Shepro, J. A. Mannick, and H. B. Hechtman, *Surgery* 99, 72 (1986).
64. D. L. Miller and R. H. Wildnauer, *J. Invest. Dermatol.* 69, 287 (1977).
65. E. Holzle and G. Plewig, *J. Invest. Dermatol.* 68, 350 (1977).
66. A. Rosencwaig and E. Pines, *J. Invest. Dermatol.* 69, 296 (1977).
67. S. D. Campbell, K. K. Kraning, E. G. Schibli, and S. T. Momii, *J. Invest. Dermatol.* 69, 290 (1977).
68. R. Marks, S. Nicholls, and D. Fitzgeorge, *J. Invest. Dermatol.* 69, 299 (1977).
69. R. Cummings and A. W. Lykke, *Br. J. Exp. Pathol.* 51, 19 (1970).
70. M. C. Middleton and I. Pratt, *J. Invest. Dermatol.* 68, 379 (1977).
71. J. F. Kachmar and D. W. Moss, *Fundamentals of Clinical Chemistry*. N. W. Tietz, Ed. W. B. Saunders, Philadelphia (1976).
72. L. A. Kaplan and A. J. Pesce, *Clinical Chemistry Theory, Analysis, and Correlation*. C. V. Mosby, St. Louis, MO (1984).
73. I. Hamami and R. Marks, *J. Invest. Dermatol.* 82, 557 (1984).
74. M. C. Middleton, *J. Invest. Dermatol.* 74, 219 (1980).
75. G. S. Lazarus, V. B. Hatcher, and N. Levine, *J. Invest. Dermatol.* 65, 259 (1975).
76. G. P. Lewis, *J. Physiol.* 191, 591 (1967).
77. P. P. Bradley, D. A. Priebat, R. D. Christenses, and G. Rothstein, *J. Invest. Dermatol.* 78, 206 (1982).
78. T. W. Kensler, P. A. Egner, K. G. Moore, B. G. Taffe, L. E. Twerdok, and M. A. Trush, *Toxicol. Appl. Pharmacol.* 90, 337 (1987).
79. P. G. Canonico, J. S. Little, M. C. Powanda, K. A. Bostian, and W. R. Beisel, *Infect. Immun.* 29, 114 (1980).
80. G. Volden, G. Kavli, H. F. Haugen, and S. Skrede, *Br. J. Dermatol.* 109(Suppl 25), 68 (1983).
81. G. A. Higgs, *Prog. Lipid Res.* 25, 555 (1986).
82. S. Moncada, R. J. Flower, and J. R. Vane, *Goodman and Gilman's The Pharmacological Basis of Therapeutics*. 7th ed. A. G. Gilman, L. S. Goodman, T. W. Rall, and F. Murad, Eds. Macmillan, New York (1985).
83. J. Sondergaard, M. W. Greaves, and H. P. Jorgensen, *Arch. Dermatol.* 110, 556 (1974).
84. M. Rampart and T. J. Williams, *Am. J. Pathol.* 124, 66 (1986).
85. P. M. Simmons, J. A. Salmon, and S. Moncada, *Biochem. Pharmacol.* 32, 1353 (1983).
86. A. J. Higgins, *J. Vet. Pharmacol. Ther.* 1 (1985).
87. J. A. Salmon, *Prostaglandins* 27, 364 (1984).
88. S. Yoshino, W. J. Cromartie, and J. H. Schwab, *Am. J. Pharm.* 327 (November 1985).
89. R. F. Dyck and S. L. Rogers, *Clin. Invest. Med.* 8, 148 (1985).
90. H. G. Schwick and H. Haupt, *Behring. Inst. Mitt.* 80, 1 (1986).
91. H. Gewurz, C. Mold, J. Siegel, and B. Fiedel, *Adv. Intern. Med.* 27, 345 (1982).

5

Dermal Hypersensitivity: Immunologic Principles and Current Methods of Assessment

GERRY M. HENNINGSEN
Division of Biomedical and Behavioral Sciences
National Institute for Occupational Safety and Health*
Robert A. Taft Laboratories
Cincinnati, Ohio

I. INTRODUCTION

A. Background

Clinicians and research scientists evaluate allergenicity in the skin for two main reasons: to diagnose human or animal allergies, and to predict dermal sensitization by new chemicals and products.[1-3] Scientists first reported the disease of allergic contact dermatitis (ACD), resulting from chemical exposure, at the turn of the 20th century. Researchers later recognized that ACD from chemical exposure was similar to delayed-type cutaneous hypersensitivity induced by microbial antigens.[4] Epidemiologists found that skin disease comprised about half of the reported industrial diseases and days lost at work in the U.S., with contact dermatitis as the most frequent diagnosis.[5,6] The U. S. Bureau of Labor Statistics published in the 1987 Annual Survey of Occupational Injuries and Illnesses that noninjurious dermatological conditions accounted for 28% of the 190,400 reported occupational illnesses.[7] Currently, the National Institute for Occupational Safety and Health (NIOSH) categorizes "dermatological conditions" as one of the ten leading work-related diseases.[8]

Estimates of the prevalence of contact dermatitis in the general population lie between 1.5 and 5.4%; however, these rates include both primary irritant and ACD.[9] The North American Contact Dermatitis Group (NACDG) reported an incidence as high as 12% reactivity to nickel sulfate in clinical contact allergy tests of 1200 patients.[10] When considering these rates, one must understand the important distinction between "primary irritants" and "contact allergens". Primary irritants corrode or inflame and injure the skin upon contact, while contact allergens induce delayed

---

* The views contained in this chapter belong solely to the author and do not reflect any official policy by NIOSH, CDC, PHS, or the DHHS of the federal government.

cutaneous hypersensitivity in susceptible individuals at nonirritating doses. The clinical appearance of irritant and ACD, as well as atopic dermatitis, is similar in that most patients have varying degrees of eczema (erythema, edema, papules, vesicles, and/or exudate with secondary bacterial infection). In one retrospective study on nearly 8000 eczema patients, dermatologists diagnosed contact dermatitis as the cause in about 16% of the cases.[11] Another estimate places immune sensitization as the cause of 20 to 25% of all cases of occupational dermatitis.[12]

The foregoing prevalence rates may underestimate actual occurrences of contact dermatitis, since consumers usually solve (and do not report) many dermal reactions to cosmetics by trial and error use of alternate products.[13] In contrast, about 1.5% of the general population are dermally anergic, or unable to produce skin contact allergies.[14] The proclivity or tendency for dermal sensitization probably follows a somewhat normal population distribution, whereby hypersensitive (atopic) and hyposensitive (anergic) persons are at opposite tails of the distribution curve, and individual susceptibility to the effects of allergen exposure depends on genetic and environmental factors.[3,9]

In the U.S. the five most common causes of ACD have been reported to be *Toxicodendron* (*Rhus*) plant resin (urushiol from poison ivy, poison oak, and poison sumac), nickel, paraphenyldiamine, rubber compounds, and ethylenediamine.[14] It is estimated that 50 to 70% of the American population are sensitized to *Toxicodendron* and would react positively to provocative testing, while about 30% are assumed to have an antigen-specific genetic tolerance to this contact sensitizer.[15] The Compositae family of plants, especially the genera *Ambrosia* and *Chrysanthemum*, are other common plant contact sensitizers, due principally to the presence of sesquiterpene lactones.[16] Many other contact sensitizers are chemicals which act as haptens, or incomplete antigens with low molecular weights that bind to macromolecules in the body to become allergenic.

B. Definitions

Dermal hypersensitivity reactions are inflammatory reactions of the skin that either defend the host against pathologic agents or damage host tissue and cause disease. The protective effects of hypersensitivity are a desirable part of host "immunity", while the detrimental effects arise from immune-mediated lesions defined as "immunopathologic" disease.[17] The terms "allergy" and "hypersensitivity" commonly denote deleterious immune reactions, which involve the pathophysiologic interaction of *antigens* (substances that induce an immune response) with specific *antibodies* (gamma globulin proteins) or with sensitized T lymphocytes.[18] The immunopathologic consequences of these interactions are quite diverse, and will be discussed later. The term "allergy" generally designates immediate or humoral antibody reactions, while "hypersensitivity" usually signifies delayed cellular immune reactivity.[17,19] The preferred term for delayed dermal hypersensitivity to contact allergens is "allergic contact dermatitis", clinically known as dermatitis venenata.[13]

C. Scope

Although this chapter presents an overview of current methods and immunologic aspects of both immediate and delayed hypersensitivity in the skin, it emphasizes delayed contact hypersensitivity reactions, or ACD. This focus on predictive aspects of ACD supports the main theme of this book on dermal toxicology. However, other facets of delayed dermal hypersensitivity in the areas of dermal immunology, immunotoxicology/pharmacology, and assessment of cell-mediated immunity in humans and animal models will be discussed. The purpose of this chapter is not to provide an exhaustive or in-depth review of the mechanisms and many procedures to assess dermal hypersensitivity, since several recent and excellent review articles[9,20] and textbooks[21-23] examine these expansive areas. Rather, the intention here is to provide some fundamental background and to familiarize the reader with (1) the concept of the skin as an immune organ, (2) selected procedures currently used to assess dermal hypersensitivity, (3) an awareness of the problems with interpretation and confounding factors of tests used to measure ACD, (4) toxicologic research needs in the area of contact allergy, and (5) new or future developments in assessing and controlling ACD.

II. SKIN AS AN IMMUNE ORGAN

A. Background

The primary role of the immune system is defense of the host against disease. Immunity involves both innate and adaptive mechanisms (Table 1). The epithelial surfaces of the body serve as major nonspecific barriers to infections, in the structural form of external skin and internal mucous membranes. Components within these structures also play a major role in adaptive resistance to disease. Adaptive immunity manifests the ability to distinguish between self and nonself antigens and to mount a specific host-defense response against foreign molecules. The adaptive immune system consists of a complex network of immunocytes (primarily lymphocytes and macrophages) and regulatory cytokines (soluble biomolecules secreted from cells) which act in unison to generate an immune response.[24] Activation of this system requires antigen presentation by macrophages, and differentiation and clonal proliferation of lymphocytes (Figure 1).

The two major arms of the immune system are humoral immunity and cell-mediated immunity. Specific antibodies from B lymphocytes mediate humoral immunity, whereas specifically sensitized T lymphocytes produce cell-mediated immunity. Various cells (macrophages) in the mononuclear phagocytic system (MPS) process and present antigens on their surface membranes in context with class II major histocompatibility (MHC) antigens. These antigen-presenting cells enable specific lymphocytes to recognize the foreign antigen and to generate an immune response against the antigenic substance. The cell in the skin with this antigen presentation role is primarily the Langerhans' cell.[25] T lymphocytes function as two main subpopulations which express different surface markers, the helper/effector T

Table 1
Comparison of Innate and Adaptive Immunity

Characteristic	Innate Resistance	vs.	Adaptive Resistance
Specificity of response	Indiscriminate		Specific, memory, discriminating
Mechanical barrier	Epithelium		Granuloma
Humoral factors	pH, lysozyme, serum proteins		Gamma globulins
Cellular factors	Leukocytes		Sensitized lymphocytes
Induction of response	Constitutive		Active immunization

Adapted from S. Sell, in *Immunology, Immunopathology, and Immunity*, 4th ed., Elsevier, New York, 3 (1987).

cells (T_H) and the suppressor T cells (T_S). Human T_H cells express CD4$^+$ and T_S cells express CD8$^+$ surface antigens.[17]

Immunohormones (cytokines) primarily regulate the active immune response through signals transmitted by receptors on the cell surfaces of immunocytes and other inflammatory cells. The major regulatory cytokines are the monokine, interleukin-1 (IL-1) and the lymphokines, interleukin-2 (IL-2) and gamma interferon (IFN).[17] Other immune-modulating regulatory molecules include various interleukins, alpha-IFN, neurotransmitters, hormones, cyclic nucleotides, chemotactic factors, and vasoactive amines.[26,27] In turn, immunocytokines can also modulate a variety of physiological activities, such as neural and endocrine functions, to help maintain homeostasis through local and systemic feedback mechanisms.[28]

When this delicate network of immunocytes and regulatory cytokines operates in a healthy balance, then the host is immunocompetent and able to ward off infectious diseases or cancer and neutralize foreign antigens. However, a multitude of environmental and genetic factors can compromise the immune system and lead to disease susceptibility.[17,29] Although the immune response usually succeeds in its normal homeostatic role of host resistance to disease, it can produce immunopathologic disorders such as autoimmunity or allergies.[17] Some authorities on hypersensitivity believe that inappropriate and deleterious dermal allergic reactions to environmental substances are the evolutionary price we pay to enable our immune systems to ward off complex parasites and cancer.[17,30]

B. Immunologic Structure of the Skin

Skin possesses cellular structures and physiologic functions that qualify it as a secondary immune organ, both in terms of innate resistance to infections and acquired specific immunity to foreign antigens. The concept of tissue-associated lymphoid tissue (TALT) portrays related immune structures and responses within an

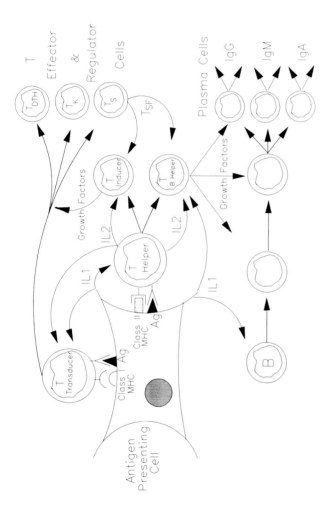

Figure 1. Schematic representation of activation of the immune response. Cooperation of macrophages, T cells, and B cells in the induction of antibody formation to haptens. Macrophages (antigen-presenting cells or dendritic cells) nonspecifically process antigen and provide activation signals (IL-1) to T and B lymphocytes. T cells recognize the antigenic molecule in association with class II MHC markers. The IL-1-activated T cells secrete IL-2 which stimulates the production of other growth factors by T lymphocytes to support B cell proliferation or T effector cell differentiation and division. (Adapted from S. Sell, in *Immunology, Immunopathology, and Immunity*, 4th ed, Elsevier, New York, 202, 1987).

organ, and encompasses "skin-associated lymphoid tissue" (SALT).[31] This specialized set of lymphoid tissues in the skin accounts for epidermotrophism of sensitized
T lymphocytes, antigen presentation by Langerhans' cells, and actions of other
dermal cells that may contribute to skin immunity: mast cells; keratinocytes; and
possibly dendritic epidermal cells.

1. Immunocyte Distribution in Skin

Bos and colleagues recently characterized the distribution and immune pheno-
types of lymphocyte subpopulations in normal human skin, and redefined the
complex immunocyte composition as the "skin immune system" (SIS).[32] T cells were
clustered around postcapillary venules of the papillary vascular plexus, and included
nearly equal numbers of T_H and T_S cells. Most T_H cells expressed a CD4+, 4B4+
"helper-inducer" phenotype, and rarely exhibited CD4+, 2H4+ "suppressor-inducer"
surface markers. Also, most of the perivascular T cells displayed HLA-DR and IL-
2 surface receptors, indicating they were activated. The authors suggest that prefer-
ential perivascular localization of activated T cells is characteristic of normal human
skin, and may involve endothelial cell interactions. Immunocytes circulate throughout the body via the blood and lymphatic vascular systems.[17] Epidermotrophism, or
migratory homing and selective extravasation of immune cells into the skin, is
important for dermal cell-mediated immunity.[31,33] Likewise, dendritic cells migrate
to regional lymph nodes within 24 h of contact sensitization, and are able to
adoptively transfer contact hypersensitivity to naive nude mice.[34] Recently, a common antigenic determinant on lymphocytes and Langerhans' cells in guinea pigs was
identified with a mouse monoclonal antibody designated MSgp 2.[35] Such an antigen
is postulated to play a role in cell migration.

Langerhans' cells play a critical role by processing antigens in the skin for the
induction of both humoral IgA immunity[36] and cutaneous cell-mediated immunity or
ACD.[4] Located mostly in the epidermis, the Langerhans' cell extends dendrites
between keratinocytes and serves as a defensive "outpost" of the immune system.[37]
A Langerhans' cell system has been described to explain the pathophysiologic
processes of subpopulations of the MPS within the skin. The MPS cells of this
"system" include the Langerhans' cells, indeterminate epidermal cells, lymphoid
interdigitating cells, and the lymph-veiled cell.[38] Langerhans' cells develop from
bone marrow macrophage precursors, and may reconstitute themselves by pheno-
typic transformation of dermal phagocytic macrophages under certain conditions.[25]
Keratinocyte-derived IL-1 (epidermal thymocyte activating factor) along with gran-
ulocyte/macrophage colony-stimulating factor (GM-CSF) were required for maturation of Langerhans' cells into potent sensitizing dendritic cells in mice.[39]

The Langerhans' cell is usually not visualized on hematoxylin and eosin (H & E)
stained histological sections, but can be distinguished by certain antigenic markers
(immunocytochemistry) and by the presence of Birbeck granules seen by electron
microscopy.[17] Human Langerhans' cells stain ATPase-positive and exhibit HLA-DR
(Ia in mice) and other surface antigens, depending on species and conditions.[40-42]
Langerhans' cells appear to have 50 to 100 times more surface HLA-DR antigens on

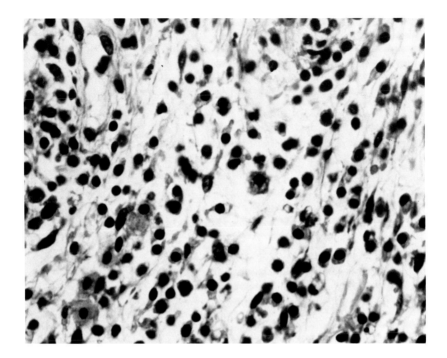

Figure 2. Mononuclear cell infiltrate in the footpad of a sensitized rat injected with a protein antigen 24 h earlier. Macrophages and T lymphocytes are the predominant inflammatory cells that characterize a type-IV delayed hypersensitivity reaction.

their cell membranes than do peripheral blood monocytes, which correlates with high antigen presentation capability.[43] The population density of Langerhans' cells varies in different regions of the skin and decreases with age, but these quantitative differences appear to have variable effects on disease susceptibility and cutaneous reactivity.[44,45]

2. Histocytologic Characterization of Skin

Immunopathologic lesions in dermal hypersensitivity reactions are the result of inflammation, which can result in clinical dermatitis such as eczema.[46]

The inflammatory process is a general response to injury and, as such, it is not usually reflective of the various types of dermal hypersensitivity when examined by histopathology alone.[47] Instead, morphologic characterization of dermal inflammation effectively differentiates acute, subacute, and chronic lesions based on cellular infiltrates and other pathologic changes in tissues. For example, immediate hypersensitivity produces acute inflammation characterized by polymorphonuclear leukocyte infiltrates, whereas delayed hypersensitivity is characterized by perivascular infiltrates of mononuclear cells (Figure 2).[48] Both of these types of hypersensitivity reactions can proceed simultaneously to different degrees, which confounds the histopathologic picture.[49] Therefore, different mechanisms of immunopathology, rather than histopathologic characterizations, are most frequently used to classify the various types of immediate and delayed hypersensitivity.[48] However, studies on the microscopic and ultrastructural characterizations of delayed-type (vs. immediate) hypersensitivity skin reactions in man[50,51] and animals[52,53] have generally recognized numerous perivascular mononuclear cells (T lymphocytes and macrophage subsets), with fewer numbers of mast cells and neutrophils depending on experimental conditions.

Through continued advances in histocytochemistry, researchers have more and better immunocytochemicals to identify cell surface markers, which promise to produce a better understanding of the cellular kinetics, distribution patterns, and unique activities/interactions that occur during dermal hypersensitivity reactions. *In situ* studies of immunophenotypes of Langerhans' cells (T6), interdigitating cells (RFD1), keratinocytes (Ia1), monocytes (M1), and lymphocytes (T3, T4, T8, T11) have provided extensive pathogenic information on ACD.[54-56]

The immunophenotypic differences between irritant dermatitis and ACD have been recently compared in human skin samples. As expected with delayed-type hypersensitivity, mostly T lymphocytes were apposed to Langerhans' cells, and few polymorphonuclear leukocytes infiltrated reaction sites during either irritant dermatitis or ACD in one study.[57] In other studies damaged and fewer Langerhans' cells were observed during irritant reactions, while more and translocated (from epidermis to dermis) Langerhans' cells were seen during ACD.[58-60] Natural killer cells and B cells were usually absent, and T helper/inducer cells predominated over T suppressor/cytotoxic cells for both types of reactions.[60] Keratinocytes expressed HLA-DR antigens only during ACD and not during irritant contact dermatitis in another related study.[61]

C. Immunologic Function of the Skin
1. Disease Resistance

Immunity provides living organisms with protection from disease.[24] Mammals are protected by both innate and adaptive resistance to disease-causing agents. Innate resistance against foreign substances in the skin is provided by mechanical barriers (stratum corneum), secreted products (sweat and sebaceous glands), and inflammatory cells (phagocytes). Adaptive resistance, or acquired immunity, is normally quiescent until stimulated by a specific antigen, and is mediated by specific antibodies or sensitized T lymphocytes. Cell-mediated immunity is one of the two major arms of acquired immunity, and is mediated by sensitized T (thymus-derived) lymphocytes and activated macrophages. The other arm is humoral immunity which is mediated by specific antibodies that are produced by mature B (bursa-equivalent, bone marrow-derived) lymphocytes termed as plasma cells. Resistance to disease requires the involvement of either or both of these arms of the immune system.[17] Cell-mediated immunity is protective against viral infected cells, many protozoal and helminth infections, mycobacteria, fungal infections, and certain tumor cells in an

Table 2

Classification of Hypersensitivity Reactions Based on Immunopathologic Mechanisms

Reaction type	Immunologic mechanism	Reaction time	Predominant immunocyte	Gamma globulin	Primary cytokine	Tissue injury
Type I	Anaphylaxis asthma	Immediate (10-20 min.)	Mast cell basophil	IgE, (IgG)[1]	Histamine, SRS, kinins	Smooth muscle contraction
Type II	Ig-dependent cytotoxicity	Variable	K cells (p complement)	IgG, IgM	None	Cell destruction
Type III complex	Immune (6-18 hours)	Intermediate (complement)	Polymorphs IgM	IgG, enzymes	Lysosomal	Basement membrane damage
Type IV	Cell-mediated	Delayed	T_{DTH} lymphocyte	None	Lympho-kines	Granuloma, dermatitis
Type V (II?)	Biomolecular binding	Variable	None	IgG IgM	None	Stimulation/ inactivation

[1] 7S IgG_1 in guinea pigs and mice.

immune surveillance (anti-tumor) role.[62] Humoral immunity protects against many bacteria and some viruses and protozoa.

2. Immunopathology

The same immune mechanisms that furnish both resistance against infections and so-called "immune-surveillance of neoplasia" can also produce immunopathologic disease (a "double-edged sword" effect).[17,63] Cell-mediated immunity becomes harmful during delayed hypersensitivity reactions that cause tissue damage, such as in ACD, graft-vs.-host disease, certain autoimmune disorders, and larger space-occupying granulomas.[18] Likewise, humoral immunity can also damage host tissues during immediate hypersensitivity reactions, such as in anaphylaxis, immune-complex disease, and cytotoxicity of host cells. As with disease resistance, hypersensitivity can involve either or both of these arms of acquired immunity, depending on a multitude of variables.[17,18,48,49]

Because the simple division of hypersensitivity reactions into immediate vs. delayed categories does not adequately describe the many diverse clinical manifestations, four general classifications have been proposed (originally by Gell and Coombs[64]) based on immunopathogenic mechanisms of tissue damage (Table 2).[18,49] The first three types involve humoral hypersensitivity, while the last type describes cell-mediated (type IV) hypersensitivity. This classification scheme has proved useful but slightly constraining; therefore, some investigators propose its expansion to better characterize and segregate reactions such as granulomas (classically type IV), and antibody binding of biomolecules with subsequent neutralization or activation (classically type II).[17,18] The humoral types I, II, and III of hypersensitivity cause immediate-onset (minutes to hours) reactions, and can be transferred via the serum to naive recipients. Delayed type-IV hypersensitivity develops later (hours to days) and can be transferred by cells or by a lymphocytic lysate known as "transfer factor".[65] Although ACD is predominantly a type-IV hypersensitivity reaction, any one or all of the other hypersensitivity reactions may sometimes be present to varying extents.[48,49]

Thus, depending on the species tested and exposure scenarios, the four types of immunopathologic hypersensitivity classifications are sometimes inadequate or overlapping. Furthermore, they do not account for simultaneous types of dermal hypersensitivity nor for cutaneous basophilic hypersensitivity (the Jones-Mote reaction).[18,30,66-68] Cutaneous basophilic hypersensitivity has a delayed onset and involves the release of vasoactive amines from tissue mast cells that results in less induration than clinically observed in type-IV hypersensitivity. In guinea pig models it can be mediated by either T cells, B cells, IgG_1, or IgE.[66,67] Mice may also generate a mast cell reaction that has an early T-dependent phase with considerable neutrophil infiltration that is typical of murine delayed hypersensitivity.[68,69] However, other investigators dispute the asserted importance of the role of mast cells in eliciting delayed hypersensitivity in mice, since mast cell-deficient mice produced reactions as great as their heterozygous littermate controls.[70] Mast cells themselves are functionally variable depending on tissue location.

Only skin mast cells expressed functional receptor sites for neuropeptides and other basic compounds, which resulted in their activation by nonimmunologic stimuli.[71] The functions of granulocytic leukocytes in delayed hypersensitivity appear to vary considerably among species and often depend upon conditions of induction and elicitation of hypersensitivity reactions.[44,45,72,73]

D. Variations in the Expression of ACD

The clinical expression and incidence of ACD has long been known to vary considerably among individuals, species, regions of the body, with exposure conditions, and with differences in age, gender, race, and physiological conditions (pregnancy, menstrual cycle, etc.) in humans.[23] For example, the neck, eyelids, and genitalia are generally more sensitive to ACD than are other areas of the human body. Various specialties in the medical profession must also contend with mucosal hypersensitivities such as allergic contact stomatitis or vaginitis in sensitive patients exposed to alloys or other sensitizers.[23,74] Although there is probably no difference in susceptibility to contact sensitizers due to gender, frequency of contact allergies do relate to intensities of exposures that can differ greatly between males and females. Predisposition to ACD from exposure to certain plants may be an exception, since women seem to react more frequently to primrose and men more frequently to ragweed.[15] Moreover, sensitivity of regional body sites does seem to vary between genders, with women having more sensitive oral mucosa and men having more sensitive popliteal fossae.[75] The most common site of contact allergy in humans has been reported to be the lower leg.[75]

Children have only about one eighth the incidence of ACD compared to older age groups, and they do not have a full capability to react to contact sensitizers until 3 to 8 years of age.[23,76] Furthermore, noneczematous dermal lesions can occasionally dominate eczema, especially in sensitized elderly persons who also tend to have more persistent and intractable cases of ACD.[23] Hyporesponsiveness to contact sensitizers in aged animals has been observed to correlate with significant decreases in numbers of Langerhans' cells and reduced T cell function.[77] Finally, atopic individuals are predisposed to ACD, presumably due to a lowered threshold for irritation.[78]

E. Immunologic Basis of Dermal Hypersensitivity

Hypersensitivity reactions are induced by specific antigens, and are elicited by either specific antibodies from sensitized B lymphocytes (antibody-secreting plasma cells) or by sensitized T lymphocytes in the case of cell-mediated immunity, including ACD.[9,13] Most contact sensitizers induce hypersensitivity through antigenic stimulation in the form of a "hapten", which is an incomplete antigen since it requires a suitable carrier molecule in order to become antigenic. A contact-sensitizing hapten can be organic or inorganic, generally has a low molecular weight of <600 Da, and is able to penetrate the skin and covalently bind with amino acids in biological proteins.[79] This hapten-protein conjugate must be of sufficient molecular size to be recognized as a foreign antigen, and a specific antibody and/or specific cell-mediated immune response ensues which sensitizes the skin immune system to the hapten molecule. Upon re-exposure of the skin to the sensitizing chemical hapten, a dermal hypersensitivity reaction may be elicited. This inflammatory reaction is generally delayed-onset type-IV hypersensitivity which stems from a cell-mediated immune response (T lymphocytes and macrophages), but it can sometimes involve humoral immunity (IgE, IgG, or IgA) and leukocytes (which can phagocytize debris or degranulate and release vasoactive amines and oxidative free radicals).[17,36,67] For example, in type I hypersensitivity there is specific IgE produced that binds to mast cells which degranulate when the surface-bound IgE binds to an allergen. There are also certain chemicals, such as codeine, which can directly cause nonspecific release of leukocytic inflammatory mediators that leads to urticaria (wheel and flare reactions) without prior immunologic induction.

F. Allergic Contact Dermatitis Mechanisms

ACD appears after repeated contact of allergenic haptens with the skin of sensitized individuals. The sensitization or induction phase is characterized by an activation (differentiation and proliferation) of allergen-specific T-effector lymphocytes (T_{DTH}), which requires presentation of the hapten in association with class II MHC antigens on the surface of Langerhans' cells (Figure 3). T cell activation is modulated by IL-1, IL-2, γ-IFN, prostaglandins, and other immunoregulatory cytokines. The elicitation, or challenge, phase of delayed contact hypersensitivity results from a reaction of activated T_{DTH} lymphocytes with the eliciting antigen (Figure 4). The antigen may be processed and presented by apposing macrophages. The T_{DTH} cells then release lymphokines which serve as chemotactic, migratory inhibition, and

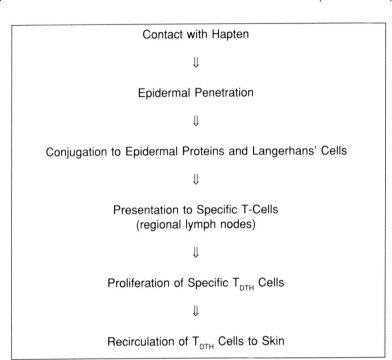

Figure 3.   Steps in the induction of contact sensitization.

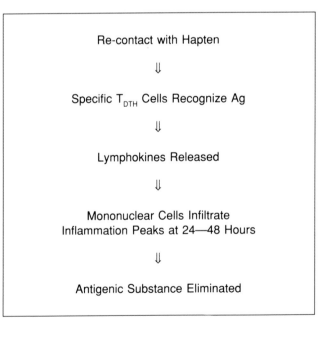

Figure 4.   Steps in the elicitation of contact sensitization.

amplification factors for mononuclear and granulocytic leukocytes. These infiltrating cells release additional vasoactive substances and other inflammatory chemical mediators which produce the typical ACD reaction.[13,79] T-suppressor cells, and anti-idiotypic antibodies, can provide further modulatory control of this cell-mediated immune response.[80]

The Langerhans' cell appears to be the crucial cell for presentation of contact allergens to appropriate T lymphocytes. Migration of some T cells and Langerhans' cells to regional lymph nodes is necessary for induction of hypersensitivity.[13] Normally, Langerhans' cells are the only cells in the epidermis that express immune-response associated antigens (Ia in mice or HLA-DR in humans).[61] However, during ACD, local keratinocytes in the epidermis are also induced to express these antigens in an amount that directly corresponds to the intensity and duration of the contact hypersensitivity.[61,81] ACD has been viewed as a wound-healing response without a visible initiating wound, whereby keratinocytes proliferate and express HLA-DR in response to IL-2-induced activated T-cell production of γ-IFN. Hyperproliferation of keratinocytes may be prevented by higher levels of γ-IFN that are released from increasing numbers of activated T cells at the height of the contact hypersensitivity reaction.[82]

Only certain haptenated cells (including Langerhans' cells, splenic dendritic cells, and thymic epithelial cells) can experimentally induce allergic contact hypersensitivity following intravenous immunization of mice.[83] However, similar studies in the guinea pig showed that haptenated Langerhans' cells were not essential to induce contact sensitivity, although they can be involved in eliciting contact sensitivity in a sensitized animal.[42] The elicitation phase of contact sensitivity in mice may involve sequential responses of T effector lymphocytes. The first step appears to occur within 2 h of antigen challenge and depends on isotypic-like antigen-binding T cell factors which initiate vasoactivity. The second step occurs as T_{DTH} cells emigrate into the challenge site and release chemotactic lymphokines that attract mononuclear cells over the next 24 to 48 h to produce the typical delayed-type hypersensitivity response.[84]

G. Atopic Dermatitis Mechanisms

Atopic dermatitis is one of the most common eczematous dermatoses seen clinically. It also has several distinguishing features when compared to ACD that enable its definitive diagnosis.[85] Atopic dermatitis is characterized by severe itching, onset in infancy, inheritance, and associations with elevated IgE, multiple positive immediate-type skin tests, allergic rhinitis, and bronchial asthma. The mechanisms remain unknown, and the pathogenesis does not fit any of the Gell-Coombs classifications.[64] It is controversial whether IgE is causally related to atopy, since patients respond to allergen exposure by skin prick tests with wheel and flare reactions that are IgE mediated; however, cutaneous antigen exposure does not elicit typical atopic eczema.[17] Elevated plasma histamine has been significantly correlated with increased basophil counts and IgE levels in atopic children.[86] A recent study also showed that patients with atopic dermatitis developed both positive immediate prick test and delayed cutaneous patch test reactions to aeroallergens, suggesting a combined mechanistic role for IgE- and cell-mediated hypersensitivity.[87] Hence, the unique presence of IgE molecules on epidermal Langerhans' cells in patients with atopic dermatitis may also account for the observed high frequency of positive patch test reactions to inhaled allergens.[88]

Atopic patients have been reported to frequently have impaired cell-mediated immunity, with decreased lymphoblastogenesis and increased delayed-type hypersensitivity. Evidence of corresponding, significant reductions of T suppressor cells could explain these immune function abnormalities and possibly account for the atopy.[85] Atopic dermatitis is usually differentiated clinically from ACD on the basis of distribution of the eczematous eruptions. Atopic dermatitis commonly involves the antecubital fossae and popliteal spaces, which are uncommon sites for eczematous ACD.

III. IMMUNE SYSTEM DYSFUNCTION: EFFECTS ON DERMAL HYPERSENSITIVITY

A. General Considerations

Alterations in immunological control of hypersensitivity can have a significant effect on the resulting degree of inflammation. Tolerance, or hyporesponsiveness, can be genetically or environmentally related. Enhanced hypersensitivity can also result from various genetic or environmental influences on the immune system. Some examples of immune-mediated effects and their consequences for dermal hypersensitivity are outlined in the following sections that describe the actions of various physical, chemical, and biological agents. Much of this work has centered on the mechanisms of immune modulation caused by various drugs and other chemicals, including carcinogens, and by specific wavelengths of UV radiation.

B. Immunologic Tolerance

Tolerance to contact sensitizers has been observed under a variety of experimental and environmental conditions. High levels of exposure to sensitizers during induction, or exposure to sensitizers by parenteral routes, are both known to result in lowered delayed-type hypersensitivity.[17] Mouse skin that was treated with the immunosuppressant and carcinogen 7,12-dimethylbenzanthracene (DMBA) could not generate an ACD response to 2,4-dinitrofluorobenzene (DNFB) in the treated area that became depleted of Langerhans' cells, whereas ACD could ensue if DNFB was applied elsewhere. Also, suppressor cells specific for DNFB were generated at the DMBA-treated sites which inhibited an allergic response.[89] In another study mice actually developed contact hypersensitivity to DMBA and benzo(a)pyrene, an outcome which could potentially modify the immunotoxicity and carcinogenicity of these polyaromatic hydrocarbons (PAHs) through their immunologic inactivation.[90]

Mouse skin that was treated with UVB (290 to 320 nm) radiation, and topical arachidonic acid or PGE_2 (but no other tested prostaglandins), decreased the number of Langerhans' cells and induced antigen-specific tolerance to DNFB. Oral indomethacin, a cyclooxygenase inhibitor, abrogated the suppression of ACD to

DNFB. This study suggests that PGE_2 is one of the biochemical mediators that suppresses contact hypersensitivity, but the mechanism of its action remains unclear.[91] UV radiation-induced suppression or tolerance to contact sensitizers could also be prevented by the treatment of the irradiated site with glucocorticoids.[92] Genetic-based tolerance to contact allergens is also observed in certain animal strains and in certain human populations. Depigmenting C57Bl/Ler-vit/vit mice, and humans with vitiligo, lose their ability to exhibit contact hypersensitivity. This tolerance could be reversed with cyclophosphamide treatment in the mice.[93]

C. Immunopharmacologic and Immunotoxicologic Compounds
1. Biomolecules

IL-1 was able to enhance plasma levels of corticosterone and to depress elicitation of dermal hypersensitivity to contact allergens in mice. This suppression of DTH by IL-1 may be mediated by PGE_2, since treatments with PGE_2 or arachidonic acid produced a comparable suppression that was inhibited by indomethacin (a potent PGE_2 inhibitor). Alpha-melanocyte-stimulating hormone also inhibited these effects of IL-1, but did not affect PGE_2 levels, suggesting that this endogenous neuropeptide may additionally function as a regulator of the immunologic and inflammatory activity of IL-1.[94] In addition, IL-1 and tumor necrosis factor (TNF) or other cytokines are capable of directly causing the release of histamine from human basophilic leukocytes.[95]

Certain acute-phase reactive proteins also possess immunoregulatory properties; for instance, serine protease displayed chemotactic activity for macrophages during delayed-type hypersensitivity reactions in the skin of guinea pigs.[96] IL-1 treatment stimulated the production of murine serum amyloid P, which tends to counter the pyrogenic and inflammatory activities of IL-1.[94] Transferrin can inhibit the release of histamine from mast cells in vivo and in vitro, and therefore may play an important role in regulating histamine release during allergic and inflammatory reactions.[9]

A complement (C5)-derived macrophage chemotactic factor with a high molecular weight has been described in guinea pig sera that was distinct from C5a-like anaphylatoxin.[98] IL-2 and γ-IFN were able to restore contact hypersensitivity in aged mice through the increased numbers of Ia^+ Langerhans' cells by IL-2 treatment and through γ-IFN-mediated T cell restoration.[77] Keratinocyte-derived IL-1, along with GM-CSF (IL-3), may be critical in the sensitization phase of cell-mediated immunity in mice through their maturation effects on Langerhans' cells.[39]

2. Drugs

Corticosteroids are well known for their potent immunosuppressive effects, including their ability to suppress both the induction and elicitation phases of ACD. This anti-inflammatory activity in humans has been shown to be due to a decreased number of $HLA-DR/T6^+$ Langerhans' cells and a decrease in Langerhans' cell-dependent T lymphocyte activation.[99]

Cyclophosphamide is a potent immunosuppressive drug that is an alkylating mustard compound which interferes with cell proliferation. Murine T_S cells, which regulate induction and the early 2-h hypersensitivity elicitation phase, seem to be sensitive targets of cyclophosphamide. Their inhibition can result in enhanced contact hypersensitivity.[84] Cyclophosphamide has anti-inflammatory action against both specific (oxazalone-induced) ACD and nonspecific (croton oil) irritant contact dermatitis in guinea pigs. Methotrexate and azathioprine have similar but less potent effects on dermal inflammation; whereas, cyclosporin A has no effect on nonspecific inflammation, but it is one of the more potent suppressors of ACD in guinea pigs.[100]

3. Chemicals

A number of chemicals have been shown to alter (most often as suppression) delayed-type hypersensitivity in animal models.[63] This list includes lead, selenium, tetrachlorodibenzodioxin (TCDD), tetrachlorodibenzofuran (TCDF), methylcholanthrene, carbofuran, 2,4-dichlorophenol, polychlorinated biphenyls (PCB) and organotins. PCB exposure has also produced decreased DTH reactions in people. The extent of effects and mechanisms of immunotoxic chemicals have been the subject of recent reviews and chapters.[29,63,101,102] Selected examples of immunotoxic chemical effects on ACD are given below.

Several complete (second stage) tumor promoting chemicals, but not incomplete (first stage) promoters or nonpromoters, were shown to suppress murine contact hypersensitivity to dinitrochlorobenzene (DNCB) only when the promoters were topically administered at the site of induction prior to sensitization. The suppression was prevented by co-application of promoters with phospholipase or lipoxygenase inhibitors or the steroid fluocinolone. The complete promoters, phorbol myristate acetate and mezerein, were also able to greatly enhance contact hypersensitivity, but only when applied at 24 h prior to challenge. The authors suggested that these findings support the possibility that second-stage promotion may involve suppression of a tumor-specific immune response.[103] This theory seems to be supported by other studies noting induced tolerance by carcinogens (previous section) or UV radiation (next sections) and the role of specific immunocyte and cytokine responses.

TCDD suppresses cell-mediated immunity and dermal hypersensitivity in both animals and humans.[104] Dextran sulfate prohibits lymphoblasts from extravasating into dermal sites of delayed hypersensitivity, and suppresses T_{DTH} cell function.[105] An uncommon, life-threatening disorder that can be provocated by certain drugs, other chemicals, viruses, and bacteria is known as toxic epidermal necrolysis. The disease results in extensive shedding of the skin, and has features of autoimmunity possibly directed against keratinocytes or Langerhans' cells.

Cases from contact are very rare, and include dermal exposures to paraffin, oil fuel, erythromycin, and a chemical intermediate of tetramisole.[106]

D. Immunomodulating Physical Agents and Disease
1. UV Radiation

UV radiation produces some of the more profound effects on dermal hypersensitivity when compared to other physical agents. Its mechanism of action differs significantly from that of low frequency or gamma radiation. Low frequency radiation primarily alters immunity and DTH through its thermal effects. Gamma irradiation results in endothelial cell edema and microvascular occlusions within lymphocyte-receptive areas of murine lymph nodes, which may account for the decreased contact hypersensitivity observed due to a diminished ability of lymphocytes to localize into skin sites with antigen deposition.[107]

Sunlight and its component ultraviolet UVB (295 to 320 nm) radiation have been found to persistently suppress contact hypersensitivity to allergens applied subsequently to nonirradiated skin in mice, through the generation of suppressor T cells. Visible ($\geq$400 nm) radiation produced only slight suppression of ACD; whereas, UVA (320 to 400 nm) radiation was shown to enhance ACD.[108] Recent studies on contact hypersensitivity to DNFB in C3H mice showed that ATPase/Ia positive staining antigen-presenting cells (APC) are also suppressed by UV radiation, but only transiently.[109] Indomethacin pretreatment restored contact hypersensitivity in parallel to APC recovery, which indicated that prostaglandins may be involved in the mechanisms of UV-induced immune suppression. Other studies have shown a genetic predisposition for the UVB effects on contact hypersensitivity. While UVB produced profound depletion of Langerhans' cells in all strains of tested mice, some strains were still able to respond with vigorous contact hypersensitivity to DNFB.[110] Likewise, studies on effects of UV radiation in humans show that wavelengths differ from those in mice that cause decreased numbers of Langerhans' cells and depressed contact hypersensitivity. The UV induction of a non-Langerhans' cell with the phenotype of a T6-DR+ epidermal cell at 3 days postexposure may be involved with the induction of specific T_s cells and hyporesponsiveness.[111]

Mouse skin that was exposed to low doses of UV radiation was unable to induce delayed-type hypersensitivity to haptens at the site of irradiation; however, sensitization at a skin site that was distant from the irradiated site produced a decreased delayed-type hypersensitivity response that could be returned to normal by pretreatment with indomethacin. Indomethacin also enabled normal elicitation of delayed-type hypersensitivity at irradiated skin sites, implicating prostaglandins in the pathogenesis of the observed immunosuppression. Mice sensitized at nonirradiated sites had hypersensitivity effector cells in their draining lymph nodes, while mice that were contact sensitized through irradiated skin had no effector cells. Furthermore, immunomodulating drugs may potentiate a suberythemal, mildly suppressive dose of UV radiation in mice.[112] On the other hand, UVB-induced hyporesponsiveness of contact hypersensitivity has actually proven beneficial in reducing the pathogenicity of cutaneous leishmaniasis in mice.[113]

2. Tumorigenesis

Interference by carcinogens and tumors with dermal immunology and hypersensitivity, and interference by dermal immunocytes with carcinogens and tumorigenesis, have provided interesting scenarios on the possible mechanisms of their interactions. The ability of transplanted murine tumor cells to induce anergy, as measured by delayed-type hypersensitivity, has been known for some time. For example, attempts to sensitize mice to DNFB 10 days following tumor transplantations with allogeneic melanomas, syngeneic lymphomas, or fibrosarcomas failed to produce delayed-type hypersensitivity reactions.[114] Mice that were repeatedly exposed to UV radiation developed suppressor T lymphocytes (T_S) that facilitated the growth of UV-induced tumors by blocking the generation of helper T lymphocytes (T_H). These results suggest that T_H cells may play a central role in the immunological rejection of such tumors.[115]

Mice have developed contact hypersensitivity specifically to the carcinogens DMBA and benzo(a)pyrene (BaP), an outcome which could potentially lower the immunotoxicity and carcinogenicity of these PAHs through their immunologic consumption.[116] Conversely, potential immunotoxicity and carcinogenicity of some PAHs could actually increase under certain conditions that alter the metabolic conversion of PAHs. Exposure of both murine skin microsomes (induced with 3-methylcholanthrene) and BaP to 365 nm UV radiation resulted in three to eight times more BaP metabolism, due to the synergistic effects of UV radiation on both BaP (photoactivation) and skin microsomes (induction).[117]

Morphologically altered Langerhans' cells were observed in basal cell carcinomas, giving rise to the possibility that these tumors could preferentially arise in areas with abnormal Langerhans' cells or, alternatively, the tumor cells could affect the Langerhans' cell morphology.[118]

Murine Thy-1$^+$ dendritic epidermal cells are bone marrow derived, have T cell lineage, and exhibit *in vitro* cytotoxicity to tumor cell lines that is opposite that seen with natural killer cells. Their location in the epidermis and ability to destroy UV-induced skin tumors was postulated to be evidence of a possible immune surveillance role against skin cancer.[62]

3. Infectious Organisms and Other Diseases

As an organ of host defense, the skin is more susceptible to disease in the presence of local or systemic immunodeficiency.[36] Insulin deficiency in experimental diabetes has differential effects on distinct subpopulations of T lymphocytes. In studies of contact hypersensitivity in mice, a lack of insulin affected only the late-acting lymphokine-producing T cells, while the early-phase factor-producing T cells were not affected.[120] Severe uremia is another metabolic disorder that can produce changes in delayed-type hypersensitivity.[121] Age-related changes in epidermal immunity may contribute to increased susceptibility to cutaneous infections and neoplasms in geriatric individuals.[122]

IV. CURRENT METHODS TO ASSESS DERMAL HYPERSENSITIVITY

A. Overview

Diagnosis of dermal hypersensitivity requires a complete history and physical examination to eliminate other possible causes of reactions that are observed. Dermal hypersensitivity reactions may be used for either diagnostic purposes to identify aeroallergens, food allergens, or contact sensitizers, or the reaction may be used to predict potential sensitization of substances (haptens) which could act as contact allergens. Patch tests are frequently used as diagnostic tools for human cases of allergic dermatitis. The objectives of the major types of tests, generally described below, are to either diagnose or to predict dermal allergies.

B. Immediate Hypersensitivity Tests
1. Human Tests

Tests for immediate type I hypersensitivity in the skin are used to help diagnose some lung and gastrointestinal diseases through the identification of specific allergens that elicit an IgE-mediated (reaginic) inflammatory reaction.[21] New guidelines have been issued by the American College of Physicians regarding allergy testing.[123] Major recommended tests include:

1. Skin prick test (SPT) — a drop of allergen is placed on the skin and a needle is repeatedly inserted through the drop into the epidermis. A weal ≥4 mm indicates a positive result. The test is simple, reliable, economical, relatively painless, and usually successful in identifying IgE-mediated allergies.
2. Intradermal test — a small amount of extract is injected into the skin, and a weal ≥4 mm indicates a positive test. This test has a greater chance, compared to the SPT, of producing an unwanted anaphylactic reaction to an allergen. A variety of acceptable techniques can be used to detect allergenicity by this method.[124]
3. Skin test titration — only used to help determine immunotherapeutic doses in patients undergoing desensitization treatments (see Reference 123 for details).
4. Provocation tests — bronchial or oral provocation tests are rare, specialized tests to assist with diagnosis of asthma or food allergies.

Also, evaluation of total IgE and allergen-specific IgE is useful for diagnosing immediate type I hypersensitivity (asthma or rhinitis) in humans and pets, and for predicting the allergenicity of products in experimental animals.[2,21,23,85,119,125]

Skin prick test (SPT) — The SPT is the clinical diagnostic test of choice for attempting to identify allergen-specific IgE-mediated allergies. Results can be evaluated using several different endpoints, such as mean weal diameter (MWD) and allergen histamine weal ratio (AHWR). Varying the weal diameters from different allergens had variable correlations to sera radioallergosorbent tests (RAST) positivity in humans.[126] New allergen-coated lancets have been tested and used as a simpler, standardized method of performing the SPT.[127] The SPT was evaluated for precision and dose response relationships to allergens, histamine, and histamine releasers. Correlations between dose and areas of weals best fit a log-log relationship, and responses produced parallel slopes within patients for allergens, histamine, and histamine releasers. Coefficients of variation for weal sizes ≥10 mm^2 were about 40% for allergens and 25% for histamine. Histamine concentrations ≥1mg/ml provided the best positive control for the SPT.[128]

Radioallergosorbent tests (RAST) — RAST procedures measure specific IgE levels and have been used successfully to diagnose immediate type I hypersensitivity. RAST tests have shown good correlation with skin tests for human reaginic hypersensitivity to potent antigens such as castor bean.[129]

Serum IgE — Determinations of total serum IgE levels are indicated in children to differentiate atopic individuals and in adults to help diagnose allergies when the SPT is equivocal. Normal nonatopic total IgE serum levels average about 25 U/ml, while levels exceeding 400 U/ml usually designate atopy. Several good commercial kits are available to determine total and specific IgE levels.[130] Specific-allergen IgE tests are less frequently indicated, being performed only after total IgE levels are ascertained, and conducted when skin testing is contraindicated or unreliable, such as in:

1. Very young children
2. Eczematous skin
3. Dermographism: false-positive skin tests
4. Antihistamine medication: false-negative skin tests
5. Anaphylactic sensitivity
6. Food allergies and elevated IgE: false-negative skin tests
7. Insect sensitivities: false-positive skin tests
8. Discordance between clinical history, skin tests, and total IgE levels

Leukocyte histamine release assay — This basophil degranulation method is useful for measuring noncirculating cell-bound IgE. Improvements and standardizations have been recently made for this immunoassay.[131]

2. Animal Tests

A variety of animal tests are available to measure type I (immediate) hypersensitivity reactions.[21,119] However, since these tests are more diagnostic in nature, rather than predictive of dermal disease, they are beyond the scope of this chapter and will not be described.

C. Delayed-Type Hypersensitivity Tests
1. Human Diagnostic Tests

Patch testing and provocative use tests are the usual methods used clinically to diagnose delayed contact hypersensitivity in humans. These tests are briefly described as follows.

Patch testing — This test is most commonly used to determine the cause of ACD.[123] Clinical diagnosis, as well as direct predictive testing, of ACD in humans relies upon the patch test.[13,23] The principle behind patch testing is simply to place a concentration of test substance that is neither irritating nor sensitizing on the skin, using a suitable vehicle such as white petrolatum.[132] Nonirritancy of the substance is established by testing the compound on a large control population. Generally, suitable test concentrations are established through trial and error, and only a limited number of substances are properly standardized and validated. The test substances are usually applied to normal skin on the back underneath an occlusive bandage for 48 h. Readings are made from 24 to 96 h after the patch is removed.[21,133,134] Unique problems, such as allergic contact stomatitis, can also be tested for offending sensitizers through an intraoral patch test.[74] Patch tests are also employed for children and use the same allergen concentrations as those used in adults; however, positive results are infrequently obtained in this age group.[135]

Use tests — So-called "use" or provocation tests are usually conducted to better simulate the mode and frequency of actual human exposures. They can verify allergic (vs. irritant) results of patch tests or help interpret negative patch test results in allergic patients. The repeated open application test (ROAT) has sometimes proven useful in this role.[136] Delayed-type hypersensitivity to foods is best evaluated by patch testing, but sometimes scratch-chamber tests or ROATs may perform better in detecting some ingested allergens.[137]

2. Human Predictive Tests

Patch testing of humans is used as a predictive test for ACD potential of new commercial or environmental products when prior testing produced negative delayed-type hypersensitivity results in the guinea pig model or when there is intended use of a sensitizer at less than sensitizing levels.[138] Ingredients of concern as well as the entire formulation can generally be tested with a repeated insult patch test in several hundred subjects at different times. The vehicle is usually aqueous, vs. ethanol or acetone used in guinea pigs. Induction for this test involves application of the nonirritating concentrations of materials by an occlusive patch three times per week for 3 weeks to the forearm. Challenge is performed 2 weeks later by applying nonirritating concentrations of the test substance on both the original site of application and on a naive site on the other arm; rechallenges are recommended within another 1 or 2 weeks to aid in interpreting results. The use test may also be incorporated to better predict the sensitizing potential of a substance.[139]

3. Human Clinical Assessments

Testing for delayed-type hypersensitivity is one of the simplest and the most important methods for clinical assessment of the cellular immune response in patients.[140] For instance, patch or intradermal testing for *Candida albicans* antigen offers the ability to evaluate cell-mediated immunity to a specific ubiquitous recall antigen.[141] Under certain circumstances, delayed-type hypersensitivity testing can provide prognostic information on patients with diseases such as cancer.[142] Investigators at New York Medical College have employed a new skin window test which is under development for measuring a patient's immune reactivity against their own cancer.[143] Delayed-type hypersensitivity to foods is best evaluated by patch testing, but sometimes scratch-chamber tests or ROAT may perform better at detecting these allergens.[144]

4. Animal Models

ACD — Predictive tests for ACD are most commonly performed in the guinea pig by a variety of methods.[9,145] As a model for human contact hypersensitivity, guinea pig skin is suitable but has several structural differences that must be considered, for example, more numerous hair shafts, a smoother dermal/epidermal junction, and the

Table 3
Contact Sensitization Tests with Guinea Pigs — Selected Examples

	Buehler	GPMT	Maguire
Induction			
Route	Topical	i.d. & topical	Topical
Applications (days)	0, 7, 14	0 (FCA), 7	0, 2, 4 (FCA), 7
Duration (patch)	6 h	48 h	48 h
Challenge			
When	Day 28	Day 21	Day 21
Route	Topical	Topical	Topical
Duration (patch)	6 h	24 h	Open
Readings	24 & 48 h	24 & 48 h	24 & 48 h
Scoring	Erythema	Sensitization	Erythema/edema
Grade	(0—3)	(% rate: I—V)	(+/–)

Note: Abbreviations used = i.d.: intradermal; FCA: Freund's complete adjuvant; h: hours.

presence of the panniculus carnosus.[146] Guinea pigs also react to contact sensitizers with mostly erythema, while human responses are generally more complex and can involve vesicle formation.[138]

CMI — Dermal delayed (type IV) hypersensitivity reactions are commonly used to assess cell-mediated immunity (CMI) by *in vivo* measurements, but CMI is sometimes assessed with graft vs. host or skin graft reactions.[147] The dermal hypersensitivity assays often employ protein-antigen injections which may differ somewhat from those assays which sensitize with haptens applied by topical routes, but the immunological basis is similar enough to serve as a model for CMI or delayed contact hypersensitivity.[148] Several models of CMI in rodents or other mammals exist which can be used effectively to evaluate immunotoxicity of various chemicals.[2,72,149]

D. Specific Procedures for Delayed Allergic Contact Dermatitis
1. Guinea Pig Predictive Models

Because no satisfactory *in vitro* tests are yet available to reliably predict ACD, predictive tests for safety assessment of new products still require the use of intact mammals with functioning immune systems.[132] Hartley or Pinbright strains of albino guinea pigs, aged 1 to 3 months and weighing 300 to 500 g, are most often preferred for testing. Epicutaneous (topical) administration is the preferred route of exposure, since it is more realistic and relevant to human exposures than procedures using other routes of exposure. The use of Freund's Complete adjuvant may be administered to increase immunological activity when testing weak sensitizers. After a period of rest for a couple of weeks following induction, the animals are challenged with the test material and responses are compared with those produced in animals from appropriate control groups. Skinfold thickness was one of the most precise, while intensity of erythema was one of the easiest, endpoints to be measured in the guinea pig models.[150] Some of the more common test procedures are outlined in Table 3.[151] They have been further divided according to their use of immunopotentiating adjuvants. Satisfactory predictive tests for ACD in animal models should be able to identify weak sensitizers, be correlated with known human responses, and be reproducible, efficient, and economical. Each method has its own inherent merits and disadvantages as a predictive test for human ACD. Several investigators have compared the various methods, but no test has proven ideal for all scenarios. The Beuhler test attempts to simulate realistic environmental exposure conditions, and thus avoids adjuvants.[145] The Maguire split adjuvant test was found to be most suited for use in ACD assessments by one toxicology group.[152] Both of these tests will be discussed later in detail. Correlations of the human Draize test with various guinea pig skin hypersensitivity tests have ranged from 0.24 to 0.69.[153]

a. Nonadjuvant Test Methods

Buehler topical patch technique — Buehler and colleagues devised and modified a screening method in guinea pigs where single occlusive patches containing solubilized test material are repeatedly applied to the shaved (or depilated, with a chemical depilatory agent) backs of guinea pigs at weekly intervals for 3 weeks.[154,155]

- Ten to 20 animals per group are used.
- Test groups receive 0.5 ml of various concentrations of the potential allergen; negative controls receive 0.5 ml of vehicle only; and positive controls receive 0.5 ml of 0.05% dinitrochlorobenzene (DNCB).
- Patches are applied topically for 6 h on days 0, 7, and 14 for induction. 80% ethanol is usually used as the vehicle for induction. The negative control group receives patches only at challenge.
- On day 28, after a 2-week resting period, challenges are made with the highest nonirritating concentration of the test material on naive skin. Acetone is generally used as the vehicle for challenges. The results are read at 24 and 48 h after application of the occlusive challenge patch, and are compared with results from an appropriate challenge control group.
- The challenge sites are graded 0 through 3 to denote the degree of erythema. A single guinea pig within a test group that responds with a higher skin grade than controls is considered to be positive and the test material is assumed to be allergenic in this species.
- Rechallenge is performed within a couple of weeks after the primary challenge (day 42), at different sites to aid in interpretation of results.

This model uses six patch application sites on the backs of guinea pigs. One site is used to induce sensitization, and the remaining five sites are used for various combinations of challenge and rechallenge.[155] The major strengths of this model are its ability to simulate human exposures and to detect most sensitizers under the proper experimental conditions.[145] Currently, this basic method requires about 38 animals to test contact hypersensitivity at a cost of about $100 per animal; while about ten additional controls are needed for rechallenge studies at a cost of about $60 each.[156]

Open epicutaneous test — Klecak proposed the use of the open epicutaneous test (OET) in combination with the Freund's complete adjuvant test (FCAT), to screen and identify contact allergens with clinical relevance. The OET employs repeated topical applications for induction and elicitation under open conditions of exposure.[157] A preinduction, nonirritating concentration is determined by applying 0.025 ml dilutions of uncovered test material on the shaven backs of guinea pigs. Irritancy is assessed 24 h later. Four concentrations, two with minimal irritation and two maximum levels without irritation, are tested at challenge days 21 and 35 of the assay. The test is considered positive if a nonirritating challenge concentration produces a reaction 24 h later.

b. Adjuvant Test Methods

Maximization test — The guinea pig maximization test (GPMT), developed by Magnusson and Klingman, employs topical applications on disturbed skin.[158] This assay is performed over a 3-week period with test material concentrations at the highest tolerated (nonirritating) levels applied to shaven backs.

- 20 to 25 animals are used per group.
- Groups receive either: (1) 0.1 ml of Freund's complete adjuvant (FCA); (2) 0.1 ml of potential allergen; (3) 0.1 ml of allergen in FCA; (4) 0.1 ml vehicle only; and (5) 0.1 ml of vehicle in FCA.
- On day 0, duplicate intradermal injections of the above materials are made over the shoulder of the guinea pigs on both sides of the spine.
- On day 7, hair is removed from the same areas over the day 0 injections, and test material in petrolatum is placed under an occlusive patch for 48 h. If the contact produces no irritation in preliminary tests, then 10% sodium lauryl sulfate should be applied just before applying the patch on day 7 to ensure adequate irritation occurs.
- On day 21, after the animals have rested for about 2 weeks, challenge is performed with an occlusive patch for 24 h. The test material is applied to one side of the spine, while the petrolatum vehicle is applied to the opposite side, above the original application sites.
- Intensity and duration of reactions are measured at 24 and 48 h after application. However, frequency of reactions was determined to be the most important measurement to grade sensitizing materials. Five grades of sensitization rate (%) are used to classify contact allergic potential of test materials.

Sensitization Rate (%)	Grade	
0 to 8	I	Weak
9 to 28	II	Mild
29 to 64	III	Moderate
65 to 80	IV	Strong
81 to 100	V	Extreme

Split adjuvant technique — The Maguire split adjuvant test uses the maximum nonirritating concentration of test material, determined by preliminary tests conducted on the shaved flanks of guinea pigs.[159,160]

- The back is shaved and chemically depilated above the forelegs on day 0.
- 0.1 ml of the test material is applied under an occlusive patch for 48 h and scored for inflammation upon removal.
- On day 2, another identical application is made at the same site for 48 more hours.
- On day 4, the site is scored and 0.2 ml of FCA is injected in two sites (0.1 ml each) adjacent to the insult site; then another application of 0.1 ml of test material is made under a new patch at the same site.
- On day 7, the site is scored and another application of 0.1 ml of test material is made under a patch at the same site. After 48 h this patch is removed and the animals are scored (day 9) and rested for about 2 weeks. It may be necessary to tape the animals' toes on the rear feet to prevent them from scratching the irritated areas.
- On day 21, both flanks are shaved and one flank is challenged with the same test material as used during the earlier induction period, except no occlusive patch is used. Vehicle is applied to the opposite flank.
- The challenged sites are measured for erythema and/or edema 24 and 48 h later. Positive responders show greater inflammation at the site challenged with test material compared to the site of vehicle challenge.

Freund's complete adjuvant test — The FCAT uses intradermal injections to induce contact hypersensitivity and uses topical applications to challenge the animals for elicitation of ACD.[161,162] A test and control group of eight to ten animals are used. Half the maximum nonirritating concentration is injected (0.1 ml) with FCA intradermally every other day over a 10-day period, while controls only receive injections of 0.1 ml FCA. On days 21 and 35 of the assay, challenge is carried out with four predetermined concentrations (two minimally irritating and two maximum nonirritating doses) of 0.025 ml of the test material. Like the OET described above, the test is considered positive if a nonirritating challenge concentration produces a reaction 24 h later.

Other tests — The U. S. Environmental Protection Agency (EPA) cites seven ACD methods as being acceptable for use in toxicity testing under the Toxic Substances Control Act (TSCA).[163] They include, in addition to the five tests described above, the Mauer optimization test and the footpad technique in guinea pigs. Besides the EPA Good Laboratory Practices (GLP) reporting requirements, the EPA specifies that dermal sensitization test reports supply the following: method; positive control information; number and sex of animals; species and strain; individual animal weights at the start and finish of the study; the grading system used; and the reading made for each animal. Delayed contact hypersensitivity of the vagina to sensitizing haptens can be readily produced and evaluated in the guinea pig through histological, rather than gross pathological, observations.[164] Readers are referred to a recent review by Andersen and Maibach for more detailed comparisons of the various guinea pig assays.[20]

2. Other Animal Models
a. MEST Assay

The mouse ear swelling test (MEST) has been validated as an alternative animal model that may be a more accurate, sensitive, and efficient predictor of human contact hypersensitivity than are the traditional albino guinea pig ACD tests. Numerous strains of mice, ages, induction scenarios, vehicles, and durations were evaluated along with different species (rat and guinea pig), sexes, sites, and adjuvant effects. The MEST test outperformed the guinea pig tests under the proper set of conditions: (1) prior determination of dermal irritation or toxicity of test materials; (2) measuring ear swelling at both 24 and 48 h postexposure; and (3) using appropriate vehicles and concentrations of test substances. In the validation study, the MEST test correctly identified the contact allergenic potential of 71 of 72 materials which were known to be either positive or negative sensitizers in humans. This method has additional advantages of requiring less labor, resources, animals, and costs.[165] A brief outline of this optimized and validated procedure is provided.

- Healthy mice with normal intact ears are randomly selected for study. 6- to 8-week-old female CF-1 mice are preferred for testing.
- On day 0, abdominal fur is shaved and the skin is "tape-stripped" to remove the outer epidermal cells prior to induction. Two intradermal injections with 0.05 ml each of FCA are administered in the abdominal induction sites. Test groups of 10 to 15 mice receive 0.1 ml of test material applied topically to the prepared abdominal site, while 5 to 10 control mice receive 0.1 ml of vehicle only. The site is quickly dried with an electric hair dryer.
- On days 1, 2, and 3, the abdominal sites are again tape-stripped, 0.1 ml of the material is applied over the site, and the area is rapidly dried.
- On day 10, 0.02 ml of test material is topically applied to the left ears of all mice, and an equal volume of vehicle only is applied to the right ear, followed by rapid drying. At both 24 and 48 h later, the animals are lightly anesthetized to permit accurate measurements of ear thickness with a micrometer.
- Mice are considered positive responders if the challenged ear thickness is ≥120% that of the contralateral control ear thickness. Results can also be reported as group mean relative thickness of challenged ears.
- A nonstandard design can be used whereby three every-other-day patch applications replace the four sequential daily topical applications, but no performance advantages were observed with this modification. The CF-1 and BALB/c strains, between 5 and 13 weeks old, gave superior responses when compared to other strains or ages.

The MEST test was recently used to evaluate dermal sensitizing potencies of four diisocyanates.[166] The magnitude of ear swelling at 24 h postchallenge was compared to the dose of isocyanate, and three relationships were observed that showed a no-effect region, a dose-response region, and a region with reduced responses at highest doses. Potency of the sensitizers was expressed as a SD_{50} value (dose needed to sensitize 50% of the mice in each group) in mg/kg body weight, and cross-reactivity with chemical analogs could be discerned with assay. Advantages of the MEST were its ease, effectiveness, and use of a smaller and less expensive animal model.

A caution should be pointed out for how one actually measures swelling in dermal tissues, such as the ear or footpad, since a significant amount of variation in absolute measurements can be produced by the type of instrument used. Micrometers and calipers are common tools used to measure tissue swelling; however, some that are spring-loaded employ greater pressure (thus, giving smaller readings) at the tissue site of swelling. Those instruments that exert greater pressure may not be able to detect early phase edematous reactions, but they perform adequately in detecting the induration caused by cellular infiltration.[167] Replicate readings with high-precision electronic digital calipers have given superior results in another study.[72]

3. In Vitro Screening Methods

Some common *in vitro* tests for cell-mediated immunity include lymphoprolifera-tion to mitogens or allogeneic leukocytes in mixed lymphocyte cultures (MLC), and lymphokine production.[101] Enzyme-linked immunosorbent assays (ELISA) and lymphoproliferation assays have also shown utility in diagnosis of ACD under certain conditions.[168] For instance, ELISAs with high sensitivity to chemical-protein conjugates are able to detect humoral anti-hapten responses that may be involved in dermal hypersensitivity reactions.[169,170]

In vitro tests to detect induction of sensitizing potential of chemicals for delayed-type hypersensitivity have made little progress.[125] Still one of the more promising approaches is to treat antigen-presenting Langerhans' cells with antigen and then co-culture with lymphocytes to measure lymphocyte receptor expression or synthesis of IL-2.[171] Other prospective tests include (1) leukocyte procoagulant activity of periph-eral blood mononuclear cells as an indicator of contact hypersensitivity;[172] (2) a combined immunogen-Langerhans' cell binding, Langerhans' cell activation/migra-tion, and autologous lymphocyte blastogenesis assay;[173] and (3) chromatin activation of blood lymphocytes detected by polarization microscopy and cytophotometry[174]

4. Interpretation of Test Results

Since there are many factors that influence the interpretation of results from dermal hypersensitivity tests, guinea pig allergy test results cannot stand alone.[20] Results should be interpreted with available human patch and use tests, literature on case reports of adverse reactions and retrospective epidemiology studies, U.S. De-partment of Health and Human Services monitoring programs for adverse reactions, and diagnostic test results from dermatology clinics.[1,20] The aim of the selected guinea pig test often defines its limitations; for instance, the guinea pig maximization test is used to establish the potential of a substance to act as a contact sensitizer.

Table 4
Experimental Variables in Detecting Inherent Allergenicity

Genetic factors	Biological properties
Humans: resistant or sensitive	Immunotoxic agent
Guinea pigs: strains 2 or 13	Physiologic status
Route of Exposure	Vehicles
Injection	Solubility/penetration
Topical	Exposure
Occlusive patch	Concentration
Skin	Contact time
Morphology	Number of episodes
Area	Grading

Adapted from E. V. Buehler, H. L. Ritz, and E. A. Newmann, *Reg. Toxicol. Pharmacol.*, 5, 46 (1985).

Therefore, a positive test result does not necessarily preclude the use of the substance, since the test conditions are exaggerated compared to normal conditions of use and a considerable margin of safety would be assumed.[20] Likewise, a negative result may mean the substance is unlikely to sensitize, but does not mean that the substance will never sensitize anyone under the proper conditions of exposure. Moreover, a positive hypersensitivity result in a nonadjuvant guinea pig test is cause for relatively more concern about the substance's potential contact sensitizing capacity under typical conditions of use.

In practical terms, although nearly all compounds have dermal sensitizing potential, their use at low levels presents an extremely low risk of induction of clinical dermatoses. Prospective dose-response studies in humans with repeated insult patch tests have shown apparent threshold levels, below which there was no sensitization, and relatively safe use of the material could be presumed.[175] Some investigators have attempted to rank sensitizers from weak to extreme according to their sensitization rate,[158] but this scheme ignores exposure levels and types of use as well as host factors which all determine to varying extents whether or not clinical ACD will occur.[145]

Variation — Proper interpretation and extrapolation of experimental contact hypersensitivity results have been the subject of several reviews.[138] Adequate exposure histories and proper selection of matched controls are vital in human studies to eliminate false positive results and accurately predict the incidence of dermal hypersensitivity caused by a test substance.[176] Some variables that can influence ACD are listed in Table 4. Factors that contribute to variability and should be considered when interpreting test results or planning experimental designs are further described below.

Acidic or alkaline test materials are sometimes too irritating to be tested at sufficient levels to determine their ACD potential. Buffering the pH to between 4 and 9 allowed successful testing of strong acids and bases with patch tests.[177] The time of reading may affect dermal reactivity results for certain allergens if one only looks at either 24 or 48 h dermatitis; therefore, it is recommended that readings be performed at both times.[178] Cross-reactivity of allergens, especially certain metals (such as nickel and cobalt) or organic analogs, may be responsible for some false positive results.[181] Different vehicles can influence the outcome of patch testing by apparently modifying the quantity of allergen released into the skin.[179] Vehicle-allergen interactions have resulted in different dermal reactivity to the same hapten applied in different vehicles.[180]

Lesional skin provides a different immunological microenvironment from normal skin which can enhance delayed hypersensitivity reactions by predisposing that site for further immunological reactions. Therefore, a negative patch test on normal skin does not preclude the possibility that the exposure of damaged or diseased skin to an allergen could provoke ACD.[182] One should also be aware that a lower threshold dose of dermal sensitization or elicitation may result from tests on more susceptible damaged skin.[183] This phenomenon has been observed with "excitatory" or "angry" skin syndrome.[23] Also, passive transfer of some haptens through skin can increase by 1000 times if the skin is diseased or damaged, apparently through a loss of the skin's natural barrier function.[184] Despite the immunosuppressive effects of UV sunlight on experimentally induced contact hypersensitivity, a retrospective study of 8000 Belgian patients found no influence of UV sunlight on patch test results.[185]

Positive, though somewhat weak, correlations have been made between several predominant sites of eczema and specific sensitizing chemicals; for instance, lower leg dermatitis and contact allergy to lanolin.[75] One report described different absolute reactions at duplicate sites of dermal hypersensitivity on the same individual; however, relative potencies of the allergenic extracts remain similar among sites and patients.[124]

Several types of medications are capable of interfering with intradermal skin testing; therefore, withdrawal of the medicine for the proper length of time is required before ACD tests can be accurately evaluated. These medications include glucocorticoids, antihistamines, epinephrine, tranquilizers, opiates, and mast cell stabilizers.[186] Endogenous neuroendocrine effects may also influence ACD. Acknowledging the regulatory role of the central nervous system on immunity, investigators have begun looking at effects of behavior and hypnotism on delayed-type hypersensitivity.[187]

False positive or negative results have been attributed to improper preparation, storage, and administration of antigen, demographic influences, nonspecific release of histamine, and various medications.[140] For example, the tuberculin skin test failed to identify up to 20% of patients with active tuberculosis.[188] Failure of a hapten to penetrate the skin can also lead to false negative error. A sixfold difference in the range of skin absorption of a model allergen, paraphenylenediamine, was observed between various patch test systems. The Hill Top Chamber design produced the most percutaneous absorption of the allergen when compared to other systems.[189] The foregoing examples demonstrate that an investigator or clinician should carefully scrutinize test results in view of all available information before drawing conclusions regarding dermal allergenicity.

V. FUTURE DIRECTIONS

A. Standardization of Allergens and Reagents

It is obvious that environmental and experimental conditions of exposures can have dramatic effects on delayed-type hypersensitivity and incidences of contact dermatitis. Therefore, it becomes critical to define and understand the importance of such factors as allergen concentration, total amount of test substance applied, site and size of application area, and vehicle effects when comparing test results in guinea pigs and humans.[190,191] In recent years, standardized panels of allergens for patch testing have been recommended by expert groups, such as the North American Contact Dermatitis Group[23] and the International Contact Dermatitis Group[10]. However, accumulated knowledge of reactivity with these suggested allergens could be used to further refine the panels, and would provide benefit for improved or promising new skin testing assays by utilizing the most appropriate test antigens.[192]

Efforts have been made to equilibrate biological activity of allergen preparations for SPT, whereby the concentration of allergen giving the same size weal as 1 mg/ml histamine HCl in the median sensitive patient would be defined as 1000 biological units/ml. However, use of 10 mg/ml histamine for the SPT, in some studies, has remained a better reference concentration when compared to 1 mg/ml.[193] SPT was used to detect atopy in 938 patients with allergic airway diseases, and investigators found that a limited number of allergens could reliably predict asthma and allergic rhinitis. The allergenic extracts of timothy, birch, and ragwort accounted for positive reactions in 98% of allergic rhinitis patients. House dust mite, cat dander, and timothy extracts caused positive skin prick tests in 85% of atopic patients with asthma or allergic rhinitis.[194]

Results of clinical laboratory analyses for total and allergen-specific IgE were recently reviewed and guidelines for testing were issued to assist in providing standardized results. Many commercial kits and various standard antigens or antibodies are available for IgE determinations. Minimal sampling rates, a common IgE standard (British Standard BS 75/502), and a coefficient of variation of <10% on three IgE concentrations are among the major recommendations to improve IgE testing.[130] The above tests remain the mainstays for differential diagnosis and clinical management of allergic diseases.[125]

Other delayed-type hypersensitivity tests, such as the tuberculin skin test, suffer from nonstandardized antigens that are crudely prepared and contribute to varying responses that enable only a semiquantitative measure of reactivity.[188] The National Academy of Sciences recently convened a workshop on biomarkers of immunotoxicology, and recommendations for standardization of both test allergens and methods were one goal and anticipated product of the expert panel.

B. New Animal Models

Because no satisfactory *in vitro* tests are yet available to reliably predict ACD, predictive tests for safety assessment of new products still require the use of intact mammals (primarily guinea pigs) with functioning immune systems. However, different strains and species offer the prospect for improved safety assessments which may provide better extrapolation and prediction of human dermal hypersensitivity. In addition, new biotechnology tools, such as a transplanted human immune system in mice developed by the Medical Biology Institute at La Jolla, CA, offer the prospect of unraveling the pathogenesis of ACD in humans.

The cynomolgus monkey (*Macaca fascicularis*) has been proposed as a useful animal model to study respiratory and immunological consequences of experimentally induced immediate hypersensitivity diseases such as asthma, and for use in studying allergic disease processes in human populations.[111]

Hairless strains of guinea pigs offer advantages of having skin that is anatomically closer to that of humans for which they model. Hardy, euthymic strains of hairless guinea pigs are now available, and appear to be suitable alternatives to the standard Hartley strain.[195] The cutaneous morphology and pharmacology of these animals is being characterized.[196]

A mutant strain of mouse, the C57Bl/Ler-vit/vit, exhibits a loss of epidermal pigment cells and a selective cell-mediated deficiency to epicutaneously administered allergens. This observation is consistent with humans with vitiligo, a disorder where pigment cells are destroyed and frequent autoimmunity occurs, along with a loss of contact hypersensitivity.[197] Since other immune components were normal and contact hypersensitivity responses could be restored with skin transplants, this vit/vit strain of mouse could serve in an excellent role for the investigation of various aspects of contact hypersensitivity reactions.

C. Quantitative Measurements

Improved dermal delivery systems for standardized or customized application of test substances with occlusive bandaging have been developed. Such systems, such as the Hill Top Chamber, provide uniform and reproducible exposures of skin.[156] Also, the degree of allergic reactivity may be quantitated by titration of the response with varying doses of antigens during skin testing and inhalation challenges. This capability is improving with the incorporation of appropriate *in vitro* tests and the standardization of extracts, some of which may be in the form of recombinant allergens.[198]

Several newer bioengineering techniques have been applied to assess patch test results with the intent of providing a less invasive but more objective, accurate, and quantitative measurement of dermal reactivity; however, most of these techniques are time consuming and many require specialized equipment and training.[199] These techniques include: laser doppler velocimetry to measure cutaneous blood flow;[200] skin reflectance measurements;[201] colorimeters; ultrasonic measure of skin thickness;[202,203] transepidermal water loss by evaporimetry to measure the water-barrier function of the stratum corneum;[204] electrical resistance measurements; polysulfide rubber replicas of papulovesicular lesions;[204] and transcutaneous oxygen tension measurement.[205]

Mathematical models have been used in an attempt to calculate a quantitative assessment of skin sensitization potential by analyzing structure-activity relationships using a relative alkylation index.[206] Quantitative structure-activity methods show some promise by being able to interpret various allergic contact hypersensitivity results.[207] A common slope for the dose-response relationship of an allergen and histamine can be used to estimate skin sensitivity.[128] Dose-response curves are important to accurately assess the sensitizing potential of chemicals, since tolerance, or decreased responsiveness, can occur at elevated levels of exposure (inverse dose-response).[166] Recently, probit analyses of allergen dose-response curves have been suggested as a useful approach to evaluate skin sensitivity in epidemiologic surveys.[208] The effective dose (ED_{50}) of skin reactivity to allergen extracts was able to distinguish symptomatic from asymptomatic reactors using the probit analysis.

D. *In Vitro* Dermatoimmunology Assays

Although several *in vitro* tests such as lymphocyte transformation tests and macrophage migration inhibition tests sometimes correlated with delayed-type hypersensitivity in selected cases of ACD,[168] there are as yet no such *in vitro* tests with sufficient reliability for routine diagnostic use.[132] Likewise, no satisfactory *in vitro* tests are yet available to reliably predict ACD; therefore, all such predictive tests still require the use of intact mammals with functioning immune systems. However, progress is being made with *in vitro* methods due to rapid advances in understanding immunological mechanisms and application of new biotechnology.

The field of immunohistocytology has grown greatly in importance through the development of tools used to identify and characterize skin diseases. Recently, a full-color atlas of dermatoimmunohistocytology was compiled to illustrate principles, examples of normal and abnormal states, and clinical applications.[209]

The incorporation of monoclonal antibodies into enzyme and fluorescent immunoassays has produced new diagnostic and research tools that are sensitive and specific for molecules involved in dermal hypersensitivity reactions.[168] For instance, EIA and FIA that are specific for histamine have been developed using monoclonal antihistamine antibodies.[210]

Leukocyte histamine release assays or basophil degranulation tests, have become commercially available. Also, new glass microfiber-based histamine spectrofluorometric method uses small blood volume samples and is sensitive enough to detect nanogram levels of histamine that are released from peripheral basophilic leukocytes. This method compares well with histamine release assays, SPTs, bronchial provocation, and RAST; and it is recommended for pediatric cases where: (a) a positive SPT does not correspond with case history; (b) the BP is considered too hazardous; or (c) confirmation is needed for a negative/inconclusive SPT or RAST.[211]

E. Risk Assessment

Risk-benefit analyses are appropriate when interpreting and extrapolating results from guinea pig allergy tests. No substantial risk may be accepted for a potential sensitizer used as a cosmetic ingredient; whereas, the risk may be acceptable in a beneficial therapeutic drug.[20]

Biological Monitoring — Because the immune response has the exceptional qualities of sensitivity and specificity, it can be manipulated as a biological indicator of exposure to a variety of chemicals, including allergens. The utility of an immune response to detect exposure to contact sensitizers and aeroallergens was demonstrated in BALB/c mice.[212] Dose-dependent changes were observed in immune responses which could serve as an indicator of recent exposure to a specific allergenic chemical. The responses were highly specific (few false-positives) due to specific immunologic recognition of a chemical allergen moiety.[213] In addition, specific serum antibodies can serve as qualitative indicators of prior exposures to a chemical. Such allergenic chemicals usually elicit an immunologic response against their form as a hapten, which is recognized by the circulating immunoglobulin during subsequent exposures in a sensitized host.[119]

Epidemiology — The patch test can be effectively used in prevalence or incidence studies of contact sensitizers in populations.[9] A recent retrospective study evaluated the trend of allergic contact sensitivity over 7 years in a population of nearly 12,000 patients who were patch tested annually.[192] The investigators reported a doubling in positive reactions to nickel from about 6 to 12%, as well as other significant increases and decreases to specific allergens over this time. Continuous revision of the standard series of test allergens was recommended to better account for changes in the spectrum of reactivity.

F. Immunotherapy

Effective preventive measures are currently limited to avoidance of contact with allergens that have high sensitizing potential. This tends to be problematic, as many contact sensitizers are so ubiquitous that avoidance is nearly impossible. Because ACD is self-limiting in the absence of antigen, therapy should be simple, safe, conservative, and aimed at relieving pruritus. Wet, cold dressings provide relief of acute dermatitis, and steroidal treatment provides relief from pruritus. Topical medications that contain antihistamines, local anesthetic esters, or zirconium should not be administered since these substances can be potent sensitizers in damaged skin.[15] Several available therapeutic agents are described below which can provide some relief of the symptoms and signs of inflammatory responses in the skin.

Glucocorticoids, such as triamcinolone acetonide, budesonide, or beclomethasone, are effective suppressors of ACD and can penetrate the skin to have some systemic action. A newer 21-oic acid methyl ester derivative of triamcinolone acetonide exerts its immunosuppressive activity only at the site of application, since it is rapidly inactivated by plasma esterases.[92] Topical cyclosporin A inhibited skin patch test responses in some patients who were sensitized to nickel, and no side effects of the cyclosporin treatment were noted in a preliminary study.[214]

Several experimental inhibitors of delayed-type hypersensitivity have shown effectiveness in animal models. Clonidine is an alpha$_2$-adrenergic receptor agonist which is able to suppress the early phase of dermal hypersensitivity in guinea pigs by decreasing mast cell degranulation and infiltration of neutrophils and eosinophils. This effect was obtained mainly in immediate hypersensitivity, minimally in delayed-type hypersensitivity, and did not occur in irritant contact hypersensitivity.[215] A calmodulin inhibitor, trifluoroperazine, was able to block the expression phase of delayed contact hypersensitivity in mice that received topical applications.[216]

Chronobiological influences of circadian and circannual rhythms are being studied for their effects on allergic reactions. In one epidemiologic study, peak times of allergic symptoms occurred during early morning and during January to April.[217] Knowledge of endogenous rhythmic influences by metabolic, immunologic, and endocrine systems on allergenicity should aid in determining varying susceptibilities to allergens or in providing more effective treatment of allergy symptoms.

VI. SUMMARY

The skin can be an important target site of injury from sensitizing chemicals. Dermal hypersensitivity arises from a myriad of immunopathologic mechanisms which are influenced by many genetic and environmental confounding factors. Although great advances are being made in biotechnology, dermatology, and immunology to further our understanding of dermal hypersensitivity, much remains to be learned before the manifestations of this skin disorder can be effectively controlled or prevented.

There are critical issues and needs that must be addressed along these lines, some of which have been discussed above. Basic toxicology (dermatologic and immunologic) research is needed in order to achieve sufficient understandings of the pathogenesis of ACD and to develop better *in vitro* tests. Applied toxicologic studies are required to enable better understanding of chemical structure-activity relationships of contact sensitizers, contributing factors (genetic and environmental) that result in increased individual susceptibility to development of clinical disease, and improved standardized tests that employ the best available animal model for predicting human contact sensitization by accurate extrapolations. Improved epidemiologic surveillance and prospective studies are needed for detecting population changes in prevalence of ACD. Political and social decisions are also needed to allow acceptable identification and protection of hypersensitive individuals in order to minimize ACD that arises from occupational and other environmental sources.[218-220]

In summary, current information has been presented to emphasize the fact that the skin is an active and complex **immune organ**.[221] Much is being learned about the skin's normal and diseased states through the application of modern biotechnical tools. Although the use of predictive animal models for ACD is absolutely necessary, alternative *in vitro* tests may be able to significantly augment *in vivo* tests after further successful developments. Toxicologists who study contact hypersensitivity need to be familiar with the biological and chemical interactions that eventually comprise the ACD response in an individual. They also must be aware of the relevant features of predictive or diagnostic tests, and the confounding factors which may influence the interpretation and use of their experimental results. A multidisciplinary approach is encouraged in order to evaluate all relevant aspects of ACD when designing and conducting studies on the dermal sensitization potential of chemicals.

REFERENCES

1. F. N. Marzulli and H. I. Maibach, *Dermatotoxicology,* 3rd ed. F. N. Marzulli and H. I. Maibach, Eds. Hemisphere, Washington, D.C., 319 (1987).
2. G. H. Muller and R. W. Kirk, *Small Animal Dermatology,* 2nd ed, W. B. Saunders, Philadelphia, 413 (1976).
3. G. Klecak, *Allergic Diseases, Diagnosis and Management,* 3rd ed. R. Patterson, Ed. J. B. Lippincott, Philadelphia, 227 (1985).
4. R. L. Baer, *Contact Dermatitis,* 3rd ed. A. A. Fisher, Ed. Lea & Febiger, Philadelphia, 9 (1986).
5. E. G. Bagnulo, *Indust. Med.* 9, 23 (1940).
6. M. L. Johnson, A. E. Burdick et al., *J. Invest. Dermatol.* 73, 395 (1979).
7. Annual Survey of Occupational Injuries and Illnesses, U.S. Bureau of Labor Statistics, Department of Labor, Washington, D.C. (1987).
8. J. D. Millar, *Am. J. Indust. Med.* 13, 223 (1988).
9. K. E. Andersen, C. Benezra, D. Burrows, J. Camarasa, A. Dooms-Goosens, G. Ducombs, P. Frosch, J. Lachapelle, A. Lahti, T. Menne, R. Rycroft, R. Scheper, I. White, and J. Wilkinson, *Contact Dermatitis* 16, 55 (1987).
10. E. J. Rudner, W. E. Clendenning, and E. Epstein, *Arch. Dermatol.* 108, 537 (1973).
11. N. K. Veien, T. Hattel, O. Justesen, and A. Nirholm, *Contact Dermatitis* 17, 35 (1987).
12. D. J. Birmingham, *Occup. Med. State Art. Rev.* 3, 511 (1988).
13. H. J. Eiermann, W. Larsen, H. I. Maibach, and J. S. Taylor, *J. Am. Acad. Dermatol.* 6, 909 (1982).
14. R. G. Slavin, *Allergic Diseases, Diagnosis and Management,* 3rd ed. R. Patterson, Ed. J. B. Lippincott, Philadelphia, 662 (1985).
15. K. F. Lampe, *Cutaneous Toxicology,* V. A. Drill and P. Lazar, Eds., Raven Press, New York, 229 (1984).
16. B. Singh, *Immunologic Considerations in Toxicology,* Vol. 1, R. P. Sharma, Ed. CRC Press, Boca Raton, FL, 123 (1981).
17. S. Sell, *Immunology, Immunopathology, and Immunity,* 4th ed, Elsevier, New York (1987).
18. J. D. Lakin and R. A. Strecker, *Allergic Diseases, Diagnosis and Management,* 3rd ed. R. Patterson, Ed. J. B. Lippincott, Philadelphia, 1 (1985).
19. J. L. Turk and D. Parker, *Allergic Diseases, Diagnosis and Management,* 3rd ed. R. Patterson, Ed. J.B. Lippincott, Philadelphia, 191 (1985).
20. K. E. Andersen and H. I. Maibach, *Curr. Probl. Dermatol.* 14, 263 (1985).
21. R. Patterson, Ed., *Allergic Diseases, Diagnosis and Management,* 3rd ed. J. B. Lippincott, Philadelphia, (1985).
22. F. N. Marzulli and H. I. Maibach, Eds., *Dermatotoxicology,* 3rd ed., Hemisphere, Washington, D.C. (1987).
23. A. A. Fisher, Ed., *Contact Dermatitis,* 3rd ed. Lea & Febiger, Philadelphia (1986).
24. P. H. Bick, *Immunotoxicology and Immunopharmacology,* J. Dean, M. I. Luster, A. E. Munson, and H. Amos, Eds. Raven Press, New York, 1 (1985).
25. G. F. Murphy, D. Messadi, E. Fonferko, and W. W. Hancock, *Am. J. Pathol.* 123, 401 (1986).
26. R. G. Coffey and J. W. Hadden, *Fed. Proc.* 44, 112 (1985).
27. G. Jancs'o, F. Ob'al, I. T'oth-K'asa, M. Katona, and S. Husz, *Int. J. Tissue React.* 7, 449 (1985).
28. A. del Ray and H. O. Besedovsky, *Hormones and Immunity,* I. Berczi and K. Kovacs, Eds. MTP Press, Boston, 215 (1987).
29. L. D. Koller, *Toxicol. Pathol.* 15, 346 (1987).
30. E. B. Mitchell and P. W. Askenase, *Clin. Rev. Allerg.* 1, 427 (1983).
31. J. W Streilein, *J. Invest. Dermatol.* 85, 10 (1985).
32. J. D. Bos, I. Zonneveld, P. K. Das, S. R. Krieg, C. M. van der Loos, and M. L. Kapsenberg, *J. Invest. Dermatol.* 88, 569 (1987).
33. G. J. Spangrude, B. A. Araneo, and R. A. Daynes, *J. Immunol.* 134, 2900 (1985).
34. S. C. Knight, J. Krejci, M. Malkovsky, A. Colizzi, A. Gautman, and G. L. Asherson, *Cell. Immunol.* 94, 427 (1985).

35. D. G. Healey, D. Baker, S. E. Gschmeissner, and J. L. Turk, *Int. Arch. Allerg. Appl. Immunol.* 82, 120 (1987).

36. P. Fritsch, G. Schuler, and H. Hintner, Eds. *Immunodeficiency and Skin, Curr. Prob. Dermatol.* 18, 1 (1989).

36a. G. M. Halliday and H. K. Muller, *Immunol. Cell. Biol.* 65, 71 (1987).

37. T. Bieber, *Hautarzi* 37, 424 (1986).

38. S. Barbey and C. Nezelof, *Pathol. Biol.* 34, 259 (1986).

39. C. Heufler, F. Koch, and G. Schuler, *J. Exp. Med.* 167, 700 (1988).

40. P. A. Hall, C. J. O'Doherty, and D. A. Levison, *Histopathology,* 11, 1181 (1987).

41. E. Tschachler, M. Tani, T. R. Malek, E. M. Shevach, W. Holter, W. Knapp, K. Wolff, and G. Stingl, *J. Immunol.* 137, 155 (1986).

42. D. Baker, D. Parker, D. G. Healey, and J. L. Turk, *Immunology,* 62, 659 (1987).

43. L. R. Breathen, *Dermatol. Beruf. Umwelt.* 35, 58 (1987).

44. E. Sprecher and Y. Becker, *J. Virol.* 61, 2515 (1987).

45. K. L. Choi and D. N. Sauder, *Mech. Ageing Dev.* 39, 69 (1987).

46. A. B. Ackerman, *Contact Dermatitis,* 3rd ed. A. A. Fisher, Ed., Lea & Febiger, Philadelphia, 46 (1986).

47. P. M. Ward, *Principles of Pathobiology,* 3rd ed. R. B. Hill and M. F LaVia, Eds. Oxford University Press, New York, 112 (1980).

48. M. F. LaVia, *Principles of Pathobiology,* 3rd ed. R. B. Hill and M. F LaVia, Eds. Oxford University Press, New York, 163 (1980).

49. R. G. Thompson, *General Veterinary Pathology,* W. B. Saunders, Philadelphia, 230 (1978).

50. A. M. Dvorak, C. Martin, and H. F. Dvorak, *Lab. Invest.* 34, 179 (1976).

51. H. F. Dvorak, M. C. Mihm, Jr., and A. M. Dvorak, *J. Invest. Dermatol.* 67, 391 (1976).

52. H. P. Godfrey, M. E. Phillips, and P. W. Askenase, *Int. Arch. Allerg. Appl. Immunol.* 70, 50 (1983).

53. Y. Sonoda, S. Asano, T. Miyazaki, and S. Sagami, *Arch. Dermatol. Res.* 277, 44 (1985).

54. J. D. Bos, I. D. van Garderen, S. R. Krieg, and L. W. Poulter, *J. Invest. Dermatol.* 87, 358 (1986).

55. A. I. Lauerma, K. Visa, M. Pekonen, L. Förström, and S. Reitamo, *Arch. Dermatol. Res.* 279, 379 (1987).

56. G. Kaplan, A. Nusrat, M. D. Witmer, I. Nath, and Z. A. Cohn, *J. Exp. Med.* 165, 763 (1987).

57. C. M. Willis, E. Young, D. R. Brandon, and J. D. Wilkinson, *Br. J. Dermatol.* 115, 305 (1986).

58. D. J. Gawkrodger, E. McVittie, M. M. Carr, J. A Ross, and J. A. Hunter, *Clin. Exp. Immunol.* 66, 590 (1986).

59. J. Ferguson, J. H. Gibbs, and J. S. Beck, *Contact Dermatitis,* 13, 166 (1985).

60. J. G. Marks, R. J. Zaino, M. F. Bressler, and J. V. Williams, *Int. J. Dermatol.* 26, 354 (1987).

61. A. Scheynius and T. Fischer, *Contact Dermatitis,* 14, 297 (1986).

62. H. Okamoto, K. Itoh, E. Walsh, J. Trial, C. Platsoucas, C. Bucana, and M. L. Kripke, *J. Leukocyte Biol.* 43, 502 (1988).

63. L. D. Koller, *Immunol. Allerg. Pract.* 7, 13 (1985).

64. P. G. H. Gell and R. R. A. Coombs, *Clinical Aspects of Immunology,* Blackwell, Oxford (1963).

65. C. H. Kirkpatrick, *J. Allerg. Clin. Immunol.* 81, 803 (1988).

66. F. M. Graziano, L. Gunderson, L. Larson, and P. W. Askenase, *J. Immunol.* 131, 2675 (1983).

67. D. Mahapatro and R. C. Mahapatro, *Am. J. Dermatol.* 6, 483 (1984).

68. P. W. Askenase and H. Van Loveren, *Immunol. Today* 4, 259 (1983).

69. R. G. Titus and J. M. Chiller, *J. Immunol. Methods* 45, 65 (1981).

70. S. J. Galli and I. Hammel, *Science* 226, 710 (1984).

71. M. A. Lowman, P. H. Rees, R. C. Benyon, and M. K. Church, *J. Allerg. Clin. Immunol.* 81, 590 (1988).

72. G. M. Henningsen, L. D. Koller, J. H. Exon, P. A. Talcott, and C. A. Osborne, *J. Immunol. Methods* 70, 153 (1984).

73. I. Tizard, *Veterinary Immunology,* 3rd. ed. W. B. Saunders, Philadelphia, 319 (1987).

74. L. A. van Loon, P. W. van Elsas, T. van Joost, and C. L. Davidson, *Contact Dermatitis* 14, 158 (1986).

75. B. Edman, *Contact Dermatitis* 13, 129 (1985).
76. S. Hurwitz, Ed. *Clinical Pediatric Dermatology,* W. B. Saunders, Philadelphia, 39 (1981).
77. D. V. Belsito, R. M. Dersarkissian, G. J. Thorbecke, and R. L. Baer, *Arch. Dermatol. Res.* 279 Suppl., S76 (1987).
78. E. Shmunes, *Occupational Medicine State of the Art Reviews,* Vol. 1, R. M. Adams, Ed. Hanley & Belfus, Philadelphia, 219 (1986).
79. J. Knop, *Immun. Infekt.* 13, 171 (1985).
80. L. Polak, *Occupational and Industrial Dermatology,* H. I. Maibach and G. A. Gellin, Eds. Year Book Medical Publishers, Chicago, 39 (1982).
81. L. K. Roberts, G. J. Spangrude, R. A. Daynes, and G. G. Krueger, *J. Immunol.* 135, 2929 (1985).
82. V. B. Morhenn, *Immunol. Today* 9, 104 (1988).
83. J. S. Britz, J. M. Jason, W. Ptak, C. A. Janeway, and R. K. Gershon, *Clin. Immunol. Immunopathol.* 30, 227 (1984).
84. W. Ptak, M. Bereta, M. Ptak, and P .W. Askenase, *J. Immunol.* 136, 1554 (1986).
85. D. Sparks, *Allergic Diseases, Diagnosis and Management,* 3rd ed. R. Patterson, Ed. J. B. Lippincott, Philadelphia, 491 (1985).
86. T. Fujisawa, M. Komada, K. Iguchi, and Y. Uchida, *Ann. Allerg.* 59, 303 (1987).
87. A. D. Adinoff, P. Tellez, and R. A. F. Clark, *J. Allerg. Clin. Immunol.* 81, 736 (1988).
88. C. Bruynzeel-Koomen, D. F. van Wichen, J. Toonstra, L. Berrens, and P. L. Bruynzeel, *Arch. Dermatol. Res.* 278, 199 (1986).
89. G. M. Halliday and H. K. Muller, *Cell. Immunol.* 99, 220 (1986).
90. J. C. Klemme, H. Mukhtar, and C. A. Elmets, *Cancer Res.* 47, 6074 (1987).
91. L. A. Rheins, L. Barnes, S. Amornsiripanitch, C. E. Colliuns, and J. J. Nordlund, *Cell. Immunol.* 106, 33 (1987).
92. P. M. Ross, J. A. Walberg, and H. L. Bradlow, *J. Invest. Dermatol.* 90, 366 (1988).
93. S. Amornsiripanitch, L. M. Barnes, J. J. Norlund, L. S. Trinkle, and L. A. Rheins, *J. Immunol.* 140, 3438 (1988).
94. R. A. Daynes, B. A. Robertson, B. H. Cho, D. K. Burnham, and R. Newton, *J. Immunol.* 139, 103 (1987).
95. M. Haak-Frendscho, C. Dinarello, and A. P. Kaplan, *J. Allerg. Clin. Immunol.* 82, 218 (1988).
96. T. Kawaguchi, K. Ueda, T. Yamamoto, and T. Kambara, *Am. J. Pathol.* 115, 307 (1984).
97. W. Gross-Weege, K. Theobald, and W. Kronig, *Agents Actions* 19, 10 (1986).
98. I. Kukita, T. Yamamoto, T. Kawaguchi, and T. Kambara, *Inflammation* 11, 459 (1987).
99. J. Ashworth, J. Booker, and S. M. Breathnach, *Br. J. Dermatol.* 118, 457 (1988).
100. C. Anderson, *Acta Dermatol. Venereol.* 116, 1 (1985).
101. J. H. Dean, M. J. Murray, and E. C. Ward, *Casarett and Doull's Toxicology,* C. D. Klaassen, M. O. Amdur, and J. Doull, Eds. Macmillan, New York, 262 (1986).
102. M. I. Luster and J. A. Blank, *Annu. Rev. Pharmacol. Toxicol.* 27, 23 (1987).
103. B. Czernieki, G. Witz, C. Reilly, and S. C. Gad, *J. Appl. Toxicol.* 8, 1 (1988).
104. A. P. Knutsen, S. T. Roodman, R. G. Evans, K. R. Mueller, K. B. Webb, P. Stehr-Green, R. E. Hoffman, and W.F. Schramm, *Bull. Environ. Contam. Toxicol.* 39, 481 (1987).
105. A. Bellavia, I. Brusca, V. Marino, S. M. Peri, P. Di Fiore, and A. Salerno, *Immunopharmacology,* 13, 173 (1987).
106. R. Valsecchi, P. Cassina, and T. Cainelli, *Contact Dermatitis* 16, 277 (1987).
107. W. E. Samlowski and C. L. Crump, *Blood* 70, 1910 (1987).
108. W. L. Morison, R. A. Pike, and M. L. Kripke, *Photodermatology* 2, 195 (1985).
109. B. D. Jun, L. K. Robets, B. H. Cho, B. Robertson, and R. A. Daynes, *J. Invest. Dermatol.* 90, 311 (1988).
110. J. W. Streilein and P. R. Bergstresser, *Immunogenetics,* 27, 252 (1988).
111. O. Baadsgaard, H. C. Wulf, G. L. Wantzin, and K. D. Cooper, *J. Invest. Dermatol.* 89, 113 (1987).
112. G. E. Kelly, A. Schreibner, and A. G. Sheil, *Immunol. Cell Biol.* 65, 153 (1987).
113. M.S. Giannini, *Infect. Immun.* 51, 838 (1986).
114. J. M. Jessup, M. H. Cohen, M. M. Tomaszewski, and E. L. Flex, *J. Natl. Cancer Inst.* 57, 1077 (1976).

115. C. A. Romerdahl and M. L. Kripke, *Cancer Res.* 48, 2325 (1988).
116. J. C. Klemme, H. Mukhtar, and C. A. Elmets, *Cancer Res.* 47, 6074 (1987).
117. W. B. Peirano and D. Warshawsky, *Proc. Am. Assoc. Cancer Res. Annu. Meet.* 28, 129 (1987).
118. E. Azizi, C. Bucana, L. Goldberg, and M. L. Kripke, *Am. J. Dermatopathol.* 9, 465 (1987).
119. R. E. Biagini, W. J. Moorman, J. B. Lal, J. S. Gallagher, and I. L. Bernstein, *Lab. Anim. Sci.* 38, 194 (1988).
120. W. Ptak, M. Rewicka, A. Gryglewski, and J. Bielecka, *Int. Arch. Allerg. Appl. Immunol.* 81, 136 (1986).
121. R. F. Gagnon, *Nephron* 43, 16 (1986).
122. D. N. Sauder, *Dermatol. Clin.* 4, 447 (1986).
123. P. P. VanArsdel, Jr. and E. B. Larson, *Ann Intern. Med.* 110, 304 (1989).
124. S. Dreborg, G. Nilsson, and O. Zetterstrom, *Ann. Allerg.* 58, 33 (1987).
125. H. A. Homburger, *Crit. Rev. Lab. Sci.* 23, 279 (1986).
126. P. L. Paggiaro, E. Bacci, D. L. Amram, O. Rossi, and D. Talini, *Clin. Allerg.* 16, 49 (1986).
127. N. I. Kjeliman, S. Dreborg, and K. Falth-Magnusson, *Allergy* 43, 277 (1988).
128. S. Dreborg, M. Holgersson, G. Nilsson, and O. Zetterstrom, *Allergy* 42, 117 (1987).
129. R. Panzini and S. G. Johansson, *Clin. Allerg.* 16, 259 (1986).
130. R. Fifield, A. G. Bird, R. H. Carter, A. M. Ward, and J. T. Whicher, *Ann. Clin. Biochem.* 24, 232 (1987).
131. J. Saint-Laudy, *J. Immunol. Methods* 98, 279 (1987).
132. E. A. Emmett, *Casarett and Doull's Toxicology,* C. D. Klaassen, M. O. Amdur, and J. Doull, Eds. Macmillan, New York, 423 (1986).
133. N. Hjorth, *Dermatotoxicology,* 3rd ed. F. N. Marzulli and H. I. Maibach, Eds. Hemisphere, Washington, D.C., 307 (1987).
134. K. E. Malten, J. P. Nater, and W. G. van Ketel, Eds. *Patch Testing Guidelines,* Dekker & van de Vegt, Nijmegan (1976).
135. I. Pevny, M. Brennennstuhl, and G. Razinskas, *Contact Dermatitis* 11, 201 (1984).
136. M. Hannuksela and H. Salo, *Contact Dermatitis,* 14, 221 (1986).
137. A. Niinimaki, *Contact Dermatitis,* 16, 11 (1987).
138. E. V. Buehler, H. L. Ritz, and E. A. Newmann, *Reg. Toxicol. Pharmacol.* 5, 46 (1985).
139. F. N. Marzulli and H. I. Maibach, *Dermatotoxicology,* 3rd ed. F. N. Marzulli and H. I. Maibach, Eds. Hemisphere, Washington, D.C., 319 (1987).
140. N. J. Doll, B. E. Bozelka, and J. E. Savaggio, *Environ. Occup. Med.* 49, (1983).
141. H. Tagami, S. Urano-Suehisa, and N. Hatchome, *Br. J. Dermatol.* 113, 415 (1985).
142. H. M. Anthony, G. H. Templeman, K. E. Madsen, and M. K. Mason, *Cancer* 34, 1901 (1974).
143. OT Briefs, *Oncol. Times* 10, 31 (1988).
144. A. Niinimaki, *Contact Dermatitis* 16, 11 (1987).
145. E. V. Buehler and H. L. Ritz, *Immunotoxicology and Immunopharmacology,* J. Dean, L. I. Luster, A. E. Munson, and H. Amos, Eds. Raven Press, New York, 123 (1985).
146. E. V. Buehler, *Curr. Probl. Dermatol.* 14, 39 (1985).
147. M. I. Luster, J. H. Dean, and G. A. Boorman, *Environ. Health Perspect.* 43, 31 (1982).
148. J. L. Turk, *Delayed Hypersensitivity,* Elsevier/North Holland, Amsterdam, 40 (1980).
149. J. H. Exon, L. D. Koller, P. A. Talcott, C. A. O'Reilly, and G. M. Henningsen, *Fund. Appl. Toxicol.* 7, 387 (1986).
150. B. Przybilla, J. Ring, and J. G. Schmid, *Contact Dermatitis* 11, 229 (1984).
151. M. R. Gilman, *Principles and Methods of Toxicology,* A. W. Hayes, Ed. Raven Press, New York, 214 (1982).
152. J. R. Horton, J. D. MacEwen, and E. H. Vernot, AFAMRL-TR-81-131, WPAFB, OH (1981).
153. F. Marzulli and N. C. Maguire, Jr., *Dermatotoxicology,* Hemisphere, Washington, D.C., 237 (1983).
154. E. V. Buehler, *Arch. Dermatol.* 91, 171 (1965).
155. H. L. Ritz and E. V. Buehler, *Current Concepts in Cutaneous Toxicity,* Academic Press, New York, 25 (1980).

156. The Hilltop Companies, P. O. Box 429501, Cincinnati, OH 45242.
157. G. Klecak, *Advances in Modern Toxicology,* F. N. Marzulli and H. I. Maibach, Eds. Hemisphere, Washington, D.C., 305 (1977).
158. B. Magnusson and A. M. Klingman, *J. Invest. Dermatol.* 52, 268 (1969).
159. H. C. McGuire and M. W. Chase, *J. Invest. Dermatol.* 49, 460 (1967).
160. H. C. McGuire, *J. Soc. Cosmet. Chem.* 24, 151 (1973).
161. G. Klecak, *Advances in Modern Toxicology,* F. N. Marzulli and H. I. Maibach, Eds. Hemisphere, Washington, D.C., 305 (1977).
162. M. R. Gilman, *Principles and Methods of Toxicology,* A. W. Hayes, Ed. Raven Press, New York, 214 (1982).
163. Health Effects Test Guidelines, EPA 560/6-82-001, Office of Pesticides and Toxic Substances, USEPA, Washington, D.C. (1982).
164. E. A. Newmann, E. V. Buehler, and R. D. Parker, *Fund. Appl. Toxicol.* 3, 521 (1983).
165. S. C. Gad, B. J. Dunn, D. W. Dobbs, C. Reilly, and R. D. Walsh, *Toxicol. Appl. Pharmacol.* 84, 93 (1986).
166. P. S. Thorne, J. A. Hildebrand, G. R. Lewis, and M. H. Karol, *Toxicol. Appl. Pharmacol.* 87, 155 (1987).
167. H. Van Loveren, K. Kato, R. E. Ratzlaff, R. Meade, W. Ptak, and P. W. Askenase, *J. Immunol. Methods* 67, 311 (1984).
168. C. Babiuk, K. L. Hastings, and J. H. Dean, *Fund. Appl. Toxicol.* 9, 623 (1987).
169. J. H. Yeung, J. W. Coleman, and B. K. Park, *Biochem. Pharmacol.* 34, 4005 (1985).
170. R. E. Biagini, S. Klincewicz, G. M. Henningsen, B. A. MacKenzie, J. Gallagher, L. Bernstein, and D. Bernstein, *Life Sci.* 47, 897 (1990).
171. W. E. Parish, *Food Chem. Toxicol.* 24, 481 (1986).
172. R. D. Aldridge, J. I. Milton, and A. W. Thompson, *Int. Arch. Allerg. Appl. Immunol.* 76, 350 (1985).
173. J. D. Bos, *Med. Hypotheses* 15, 103 (1984).
174. J. M. Banga and I. Pfeiffer, *Anal. Quant. Cytol. Histol.* 8, 63 (1986).
175. C. W. Cardin, J. E. Weaver, and P. T. Bailey, *Contact Dermatitis* 15, 10 (1986).
176. S. D. Prystowsky, A. M. Allen, R. W. Smith, J. H. Nonomura, R. B. Odum, and W. A. Akers, *Arch. Dermatol.* 115, 959 (1979).
177. M. Bruze, *Contact Dermatitis* 10, 267 (1984).
178. K. Kalimo and K Lammintausta, *Contact Dermatitis* 10, 25 (1984).
179. A. Y. Mendelow, A. Forsyth, A. T. Florence, and A. J. Baillie, *Contact Dermatitis* 3, 29 (1985).
180. B. Bjorkner and B. Niklasson, *Contact Dermatitis* 11, 268 (1984).
181. K. Lammintausta, O. Pitkanen, K. Kalimo, and C. T. Jansen, *Contact Dermatitis* 13, 148 (1985).
182. J. Smolle and H. Kresbach, *Z. Hautkr.* 62, 1681 (1987).
183. J. P. Arlette and M. J. Fritzler, *Contact Dermatitis* 11, 31 (1984).
184. R. Scheuplin and J. Blank, *Physiol. Rev.* 51, 4 (1971).
185. A. Dooms-Goossens, E. Lasaffre, M. Heidbuchel, M. Dooms, and H. Degreef, *Contact Dermatitis* 19, 36 (1988).
186. J. L. Barbet and R. L. Halliwell, *JAVMA* 194, 1565 (1989).
187. S. E. Locke, B. J. Ranzil, N. A. Covino, J. Toczydlowski, C. M. Lohse, H. F. Dvorak, K. A. Arndt, and F. H. Frankel, *Ann. N. Y. Acad. Sci.* 496, 745 (1987).
188. J. A. Sbarbaro, *Semin. Respir. Infect.* 1, 234 (1986).
189. H. O. Kim, R. C. Wester, J. A. McMaster, D. A. W. Bucks, and H. I. Maibach, *Contact Dermatitis* 17, 178 (1987).
190. O. B. Christensen, M. B. Christensen, and H. I. Maibach, *Contact Dermatitis* 10, 166 (1984).
191. M. C. Anderson and H. Baer, *Clin. Rev. Allerg.* 4, 363 (1986).
192. R. Gollhausen, F. Enders, B. Przybilla, G. Burg, and J. Ring, *Contact Dermatitis* 18, 147 (1988).
193. S. Dreborg, A. Basomba, L. Berlin, S. Durham, R. Einarsson, N. E. Eriksson, A. B. Frostad, O. Grimmer, R. Halvorsen, and M. Holgerssson, *Clin. Allerg.* 17, 537 (1987).
194. N. E. Eriksson, *Allergy* 42, 189 (1987).

195. Charles River Laboratories, Inc. 251 Ballardville Street, Wilmington, MA 01887.
196. D. F. Woodward, *Models in Dermatology,* Vol. 4, H. I. Maibach and N. J. Lowe, Eds. S. Karger, New York (1989).
197. S. Amornsiripanitch, L. M. Barnes, J. J. Nordlund, L. S. Trinkle, and L. A. Rheins, *J. Immunol.* 15, 3438 (1988).
198. S. Yoshida, G. Halpern, and M. E. Gershwin, *Allergol. Immunopathol.* 15, 335 (1987).
199. E. Berardesca and H. I. Maibach, *Contact Dermatitis* 18, 3 (1988).
200. P. Olsson, A. Hammarlund, and U. Pipkorn, *J. Allerg. Clin. Immunol.* 82, 291 (1988).
201. A. Y. Mendelow, A. Forsyth, J. W. Feather, A. J. Ballie, and A. T. Florence, *Contact Dermatitis* 15, 73 (1986).
202. J. Serup, B. Staberg, and P. Klemp, *Contact Dermatitis* 10, 88 (1984).
203. S. Brazier and S. Shaw, *Contact Dermatitis* 15, 199 (1986).
204. K. Peters and J. Serup, *Acta Dermatol. Venereol.* 67, 491 (1987).
205. E. P. Prens, T. van Joost, and J. Steketee, *Contact Dermatitis* 16, 142 (1987).
206. B. F. J. Goodwin and D. W. Roberts, *Food Chem. Toxicol.* 24, 795 (1986).
207. D. W. Roberts, *Contact Dermatitis* 17, 281 (1987).
208. G. M. Corbo, A. Foresi, S. Morandini, S. Valente, S. Matolli, and G. Ciappi, *J. Allerg. Clin. Immunol.* 81, 41 (1988).
209. H. Ueki and H. Yaoita, Eds., *A Color Atlas of Dermato-Immunohistocytology,* CRC Press, Boca Raton, FL (1989).
210. J. L. Guesdon, D. Chevrier, J. C. Mazire, B. David, and S. Avrameas, *J. Immunol. Methods* 87, 69 (1986).
211. H. Nolte, O. Schultz, and P. S. Skov, *Allergy* 42, 366 (1987).
212. M. H. Karol, P. S. Thorne, and J. A. Hillebrand, *Occupational and Environmental Chemical Hazards,* V. Foa, E. A. Emmett, M. Maroni, and A. Columbi, Eds. Ellis Horwood, London, 87 (1987).
213. J. C. Stadler and M. H. Karol, *Toxicol. Appl. Pharmacol.* 78, 445 (1985).
214. R. D. Aldridge, H. F. Sewell, G. King, and A. W. Thompson, *Clin. Exp. Immunol.* 66, 582 (1986).
215. C. D. Anderson, B. R. Lindgren, and R. G. Anderson, *Int. Arch. Allerg. Appl. Immunol.* 83, 371 (1987).
216. G. Antonelli, M. Santiano, P. Romano, V. Colizzi, and F. Dianzani, *Int. J. Immunopharmacol.* 9, 237 (1987).
217. A. Reinberg, P. Gervais, F. Levi, M. Smolensky, L. Del Cerro, and C. Ugolini, *J. Allerg. Clin. Immunol.* 81, 51 (1988).
218. U.S. Congress, Office of Technology Assessment, Identifying and Controlling Immunotoxic Substances—Background Paper, OTA-BP-BA-75, U.S. Government Printing Office, Washington, D.C. (1991).
219. National Research Council, Commission on Life Sciences, Board on Environmental Studies and Toxicology, Committee on Biologic Markers, *Biologic Markers in Immunotoxicology,* National Academic Press, Washington, D.C., in press.
220. R.M. Adams, *Occupational Skin Disease,* 2nd ed. W.B. Saunders, Philadelphia (1990).
221. J.D. Bos, *Skin Immune System,* CRC Press, Boca Raton, FL (1990).

6

Cutaneous Photosensitization

STAN W. CASTEEL
Veterinary Medical Diagnostic Laboratory
College of Veterinary Medicine
University of Missouri
Columbia, Missouri

I. INTRODUCTION

Pathological skin changes that occur because of the interaction of light with some chemical not normally present in this tissue are commonly grouped together and called cutaneous photosensitization reactions. In a classical photosensitization reaction, neither the foreign chemical nor light exposure alone is capable of inducing pathological changes in the skin. At the molecular level, photosensitization is characterized as a reaction in which light absorption by a photoreactive molecule results in the structural modification of a constituent molecule of a biosystem. Sensitization of the skin to light usually is due to a photoreaction involving certain drugs, industrial chemicals, cosmetic, or plant-derived compounds. Consideration of these interactions is important because both humans and animals have opportunities for extensive exposure both to sunlight and to a wide variety of potentially photoreactive compounds.

The specific factors necessary for a cutaneous photosensitization reaction include: (1) an adequate concentration of a photoreactive chemical in the cutaneous circulation, or as a contact reactor; (2) the delivery of light energy of appropriate wavelength, duration, and intensity; (3) the presence of susceptible cutaneous tissue; (4) and depending on the specific photochemical reaction mechanism, molecular oxygen.

The term "photosensitization" encompasses both phototoxic and photoallergic mechanisms which may be clinically indistinguishable. Both have the general appearance of an exaggerated sunburn. Phototoxic reactions are usually dose related and can be induced in most individuals who are exposed to adequate amounts of the chemical and radiation of appropriate wavelength, duration, and intensity. These reactions do not require prior sensitization as in the case of photoallergic mechanisms. The phototoxic reaction is clinically evident within 5 to 18 h following exposure to the sun and is usually maximal by 36 to 72 h. In contrast, photoallergic reactions involve an immune-mediated mechanism. Light initiates a photochemical reaction between the photoactive chemical and cutaneous proteins, resulting in the formation of a photoantigen. Each of these phenomena will be discussed in turn as to the mechanisms involved and the methods used for their specific evaluation. In addition, photobiologic principles relevant to cutaneous photosensitization will be discussed under a separate heading to enhance the understanding and uniqueness of this disease entity.

II. PHOTOBIOLOGIC PRINCIPLES

Photobiology is the study of the responses of living systems to irradiation with light. The responses of biological systems to light are based on chemical reactions initiated by the absorption of light by molecules in the system. This is consistent with the first law of photochemistry, which states that light must be absorbed for a photochemical event to occur.[1]

The sun is the principal environmental source of electromagnetic radiation to which humans and animals are exposed. Only a small portion of light energy from the sun actually reaches the earth. The character and amount of such radiation varies with the seasons and with changing atmospheric conditions. Absorption by the ozone layer and molecular scattering of shorter wavelengths impinging on the upper atmosphere serve to diminish light reaching the earth's surface and to remove wavelengths of solar radiation shorter than 290 nm. Carbon dioxide and water vapor further modify the spectral distribution of solar radiation reaching the earth by absorbing red and IR wavelengths. The sunlight that does reach the earth's surface contains wavelengths from about 290 to 4000 nm. This light can be further categorized according to wavelength into UV, visible, and IR bands.[1]

The UV portion of the electromagnetic spectrum can be subdivided into UVA, UVB, and UVC bands. These bands have certain physical and toxicological characteristics related to their interaction with biological systems. The UVC (200 to 280 nm) radiation from the sun is effectively screened out by the atmosphere's ozone layer and does not reach the earth's surface. Artificial sources of this band are used effectively for their germicidal capabilities in a number of sterilization procedures and can induce skin erythema in moderate doses. Ordinary sunburn results from excessive exposure of the skin to UV rays in the UVB (280 to 320 nm) radiation band. UVB induces skin erythema that is more persistent (lasting up to 120 h) than that induced by UVC wavelengths (lasting 12 to 36 h). These sunburn-producing wavelengths are completely filtered out by ordinary window glass and to a great extent by smoke and smog. Clinical signs of ordinary sunburn appear in humans 1 to 24 h after exposure and usually peak in 72 h. The UVA (320 to 400 nm) region of the spectrum is responsible for the majority of photosensitization reactions, but does not, by itself, induce erythema when applied in moderate doses. This wavelength band also is called "near" or blacklight UV.

The visible portion of the spectrum is in the 400 to 700 nm range and is involved infrequently in photosensitivity reactions. Visible radiation lacks the energy to induce photochemical reactions in most proteins and in DNA, but is effectively absorbed by colored compounds in general and by rhodopsin in the eye and chlorophyll in plants.

Responses of biological systems to light are based on chemical reactions initiated by molecular absorption of light. Absorption of light by specific molecules in the system results in the promotion of electrons to higher energy states. These "excited-state" molecules possess the necessary energy of activation required to undergo a photochemical reaction. The absorbing molecule may photochemically react in a reagent manner with adjacent biomolecules and thus be exhausted in the reaction or it may resemble a catalytic mechanism causing a modification of a biomolecule without itself undergoing a permanent chemical change. The latter mechanism allows the photosensitizing molecule to return to the unexcited (ground) state and thus be available to absorb another quantum of light and repeat the photochemical reaction cycle again.

The integument has certain intrinsic protective barriers to minimize damage from light exposure. Skin is an optically inhomogeneous medium that serves to modify the radiation that reaches deeper structures via the mechanisms of reflection, refraction, scattering, and absorption. The stratum corneum reflects 5 to 10% of incident solar radiation. In addition, the dead cells of the stratum corneum are composed primarily of keratin, a fibrous protein that absorbs significantly in the UVB and UVC spectral bands and also scatters most visible radiation because of its structural characteristics. The DNA present in these outer epidermal layers will absorb incident UV radiation allowing only 5 to 10% of solar radiation in the 290 to 310 nm band to actually penetrate to the basal epidermal cell layer and superficial dermal vasculature. However, wavelengths above 330 nm readily penetrate the epidermis to reach the deeper photosensitive dermal layers. In Caucasian skin, as much as 50% of incident UVA radiation may be transmitted to the basal cell layer and dermis. There is a significant reduction in UVA transmission in blacks, and those individuals with very dark skin may have as little as 5 to 10% of the incident UVA penetrate the epidermis. This phenomenon is due to the quantitative differences in the light-absorbing pigment, melanin, which serves to further protect the skin from light-induced damage. A more thorough review of these principles is available in other sources.[1]

A. Action Spectrum

An important concept in photobiology is the "action spectrum", which is the relative response of a system to a scan of impinging wavelengths (Figure 1). With respect to photosensitization, an action spectrum is a description of the relative effectiveness of photons at each wavelength in inducing this specific photobiologic response. It is an approximation of the absorption spectrum (Figure 2) of the responsible chromophore (light-absorbing moiety) in the test system. In an action spectrum, peaks represent the most efficient wavelengths for producing the response. Differences between the action spectrum and the absorption spectrum of the effective chromophore frequently arise due to differences in the absorption characteristics of the chromophore *in vivo* vs. that in an *in vitro* test system. Many times the wave-

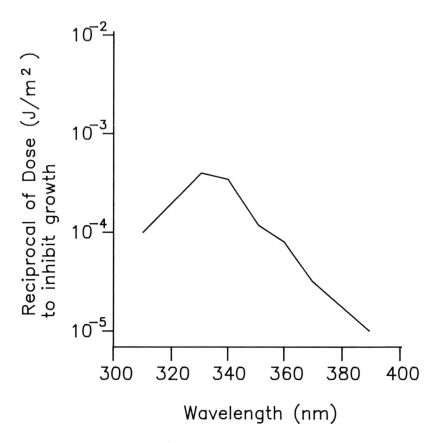

Figure 1.   Action spectra for *C. albicans* growth inhibition by 10 mM 4-methyl *N*-ethyl pyrrolo(3,2)coumarin.

length distribution of radiation actually reaching the target chromophores in tissue will be modified by the optical properties of interposed layers.

III. CHARACTERISTICS OF PHOTOSENSITIZING COMPOUNDS

There are a large number of compounds capable of inducing photosensitization in both man and animals. Chemicals with known photosensitizing potential include a variety of compounds used in medical practice, industry, agriculture, and the home. Molecular requirements of most phototoxic agents include a relatively low molecular weight (200 to 500 Da), and a planar, tricyclic, or polycyclic configuration that is highly conjugated. With respect to the immune-mediated photosensitizers, increases in molecular size and complexity are associated with increased immunogenicity and therefore macromolecular therapeutic agents such as protein or peptide hormones and dextrans are antigenic as such. The majority of drugs are small organic molecules with molecular weights of <1000 Da. Small molecules require the capabability

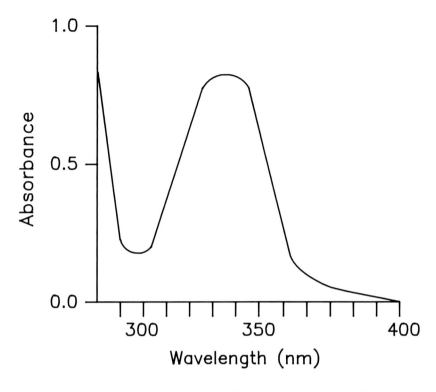

Figure 2.   Absorption spectra of 10^{-5} M ethanolic solution of 4-methyl *N*-ethyl pyrrolo(3,2)coumarin.

to function as haptens to elicit an immune response. This capability depends on their being able to form stable covalent bonds with tissue proteins, thus imparting full antigenic character. Fortunately, most drugs have little or no ability to form covalent bonds with tissue components. Table 1 is a partial list of common photosensitizers.

IV. MECHANISM OF ACTION

Photosensitization reactions involve phototoxic reactions and the much less common photoallergic reactions. The majority of common photosensitizers have action spectrums in the long-wave UV (UVA) range. In a mechanistic sense, photosensitization is a change initiated by the impingement of appropriate light on a reaction system containing biologic material and a photosensitizer (PS). When this system is irradiated, the light-absorbing PS induces another component of the biologic system to react to the absorbed radiation and to produce detectable changes. The PS may or may not be consumed in the reaction, but it significantly lowers the energy required to induce similar changes that radiation alone may be capable of evoking at significantly higher doses.[2]

Table 1
Common Photosensitizers Categorized by Their General Mechanisms[a]

Phototoxic Reaction Mechanism

Coal tar derivatives: acridine, anthracene, naphthalene, phenanthrene, thiophene
Dyes: eosin, rose bengal, orange red, toluidine blue, anthraquinone
Essential oils: oil of cedar, oil of bergamot, oil of lime, oil of lavender, oil of sandalwood
Furocoumarins: 8-methoxypsoralen, 5-methoxypsoralen, trimethoxypsoralen
Nonsteroidal anti-inflammatory drugs: benoxaprofen, carprofen, naproxen, tiaprofenic acid, piroxicam
Phenothiazines: chlorpromazine, promazine
Plants of the Umbelliferae and Rutaceae families: contain furocoumarins and include figs, parsley,
  parsnips, caraway, dill, and celery infected with the parasitic fungus, *Scelerotina*
Tetracylines: doxycycline, democlocycline, chlortetracycline

Photoallergic Reaction Mechanism

Bithionol
Fentichlor
Halogenated salicylanilides: tetrachlorosalicylanilide, trichlorosalicylanilide, tribromosalicylanilide,
  dibromosalicylanilide
Hexachlorophene, dichlorophene
Musk ambrette
6-Methylcoumarin
p-Aminobenzoic acid and its esters
Sulfanilamide[a]

[a]   Some compounds may elicit photosensitization via both mechanisms.

A. Phototoxic Reactions

Phototoxicity is a light-induced, nonimmunologic, skin response to a photoactive chemical. The route of exposure to the photoactive chemical may be by direct application to the target tissue or indirectly via the circulatory system following ingestion or parenteral administration. Phototoxic reactions resemble primary irritation reactions in that they may be elicited following a single exposure, in contrast to photoallergic reactions which require an induction period prior to elicitation of the response.

Perhaps the most widely studied group of phototoxic compounds are the psoralens. These are derivatives of the naturally occurring tricyclic furocoumarins present in a variety of plants including several members of the Umbelliferae and Rutaceae families. Some of these compounds, such as 8-methoxypsoralen and 4,5'8-trimethylpsoralen, possess significant therapeutic potential in combination with light (photochemotherapy) for the treatment of various skin disorders, such as vitiligo and psoriasis, and for certain tumors.[2] The precise mechanism by which psoralens induce phototoxicity is not completely understood. The manner by which they induce cellular injury involves several cellular locations simultaneously including DNA, RNA, enzymes, mitochondria, and cell-membrane lipids. Psoralens participate in two types of reactions when exposed to UVA light.[3,4] An oxygen-independent reaction (type I) whose primary site of cellular damage is DNA, involves the direct reaction of photosensitizer with a cellular constituent. Psoralens intercalate between DNA base pairs in the absence of light and the subsequent absorption of light results in the formation of photoadducts with pyrimidine bases. Absorption of a second quantum of light by the appropriate adduct may result in a second cycloaddition reaction with a pyrimidine base on the opposite strand of DNA giving rise to an interstrand cross-linkage. The monofunctional adducts and interstrand cross-linking of the double helix of DNA severely inhibit cellular function. An oxygen-dependent reaction (type II) involving the formation of reactive oxygen species such as super-oxide anion, singlet oxygen, and free radicals also will induce significant cellular damage. This is an indirect reaction in the sense that the activated photosensitizer reacts first with molecular oxygen and not directly with cellular constituents (Figure 3). These reactive chemical species damage arterioles, venules, and keratinocytes.

The interaction of phototoxic chemicals with biomolecular systems is mediated by several distinct photochemical reaction mechanisms. With respect to specific cellular target sites, photomodification of biological membranes is an important mode of action for certain PS. The association of the PS with membranes is an important determinant of the potency of photomodification.[5] In an aqueous environment binding of the PS to cell membranes increases the chances of a direct or oxygen-independent type I reaction. Here, the excited triplet state of the PS reacts directly with membrane components to form photoproducts. Spatial separation of the PS and membrane favors indirect or oxygen-dependent type II reactions. In these reactions the excited triplet state of the PS reacts first with molecular oxygen, generating an active oxygen intermediate such as singlet oxygen, superoxide anion, or hydroxy radicals, that subsequently reacts with adjacent substrates.

Phenothiazine derivatives photosensitize cleavage of DNA and form covalent photoadducts with DNA and nucleotide bases.[6] Phototoxicity of the tetracyclines have been demonstrated by *in vitro* cell culture methods using selected tetracyclines plus UVA radiation to inhibit polymorphonuclear leukocyte migration.[7] Phototoxicity induced by the nonsteroidal anti-inflammatory drug (NSAID), benoxaprofen, is attributed to membrane photosensitization. Specifically, mast cell degranulation due to interaction of the mast cell membrane with benoxaprofen and light, is responsible for the urticarial lesions seen in clinical cases. Piroxicam, another NSAID, is responsible for both photoallergy and phototoxicity. Testing results suggest that piroxicam photosensitivity may involve the formation of a metabolite that is preferentially formed or accumulated in the skin.[8] Likewise, sulfanilamide appears to induce both photoallergic and phototoxic reactions in skin by free radical mechanisms.[9] The photoinduced binding of sulfanilamide to protein and DNA also has been demonstrated.[10]

B. Hepatogenous or Secondary Photosensitization in Ruminants

Secondary photosensitization (hepatogenous) is considered to be the most common type in ruminants, especially cattle and sheep.[11] In this case the liver fails to

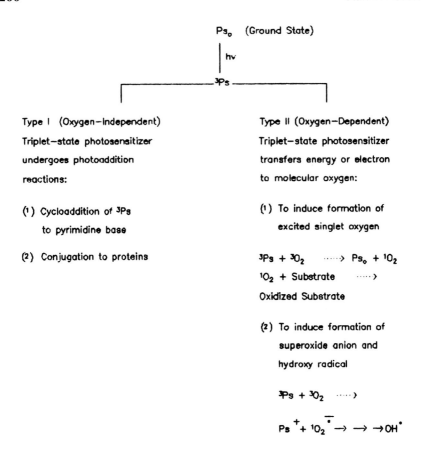

Figure 3.   Schematic representation of photochemical reactions. Ps_o = ground state of the photosensitizer. ^3Ps = excited triplet state of the photosensitizer.

detoxify and excrete a normally harmless dietary metabolite, the chlorophyll derivative phylloerythrin (Figure 4), which results in phylloerythrin reaching the dermal circulation in a concentration adequate to incite phototoxic damage. Only a small minority of toxic hepatic illnesses are accompanied by photosensitization and the animal must be on a diet containing sufficient amounts of chlorophyll as a prerequisite for disease development. In any case, the degree of photosensitization reflects the inability of the liver to detoxify phylloerythrin.

Skin lesions in photosensitized animals are restricted to white or lightly pigmented areas exposed to sunlight. Susceptible sites in cattles include the udder, teats, scrotum, muzzle, nostrils, eyes, eyelids, vulva, escutcheon, insides of the legs, and the dorsal midline. Clinical signs intensifying upon exposure to sunlight include restlessness, photophobia, and pruritis which may induce cows to kick at their udders. Epiphora and keratitis may be evident, with progressive corneal edema sometimes resulting in blindness. The range of changes in affected skin begins with

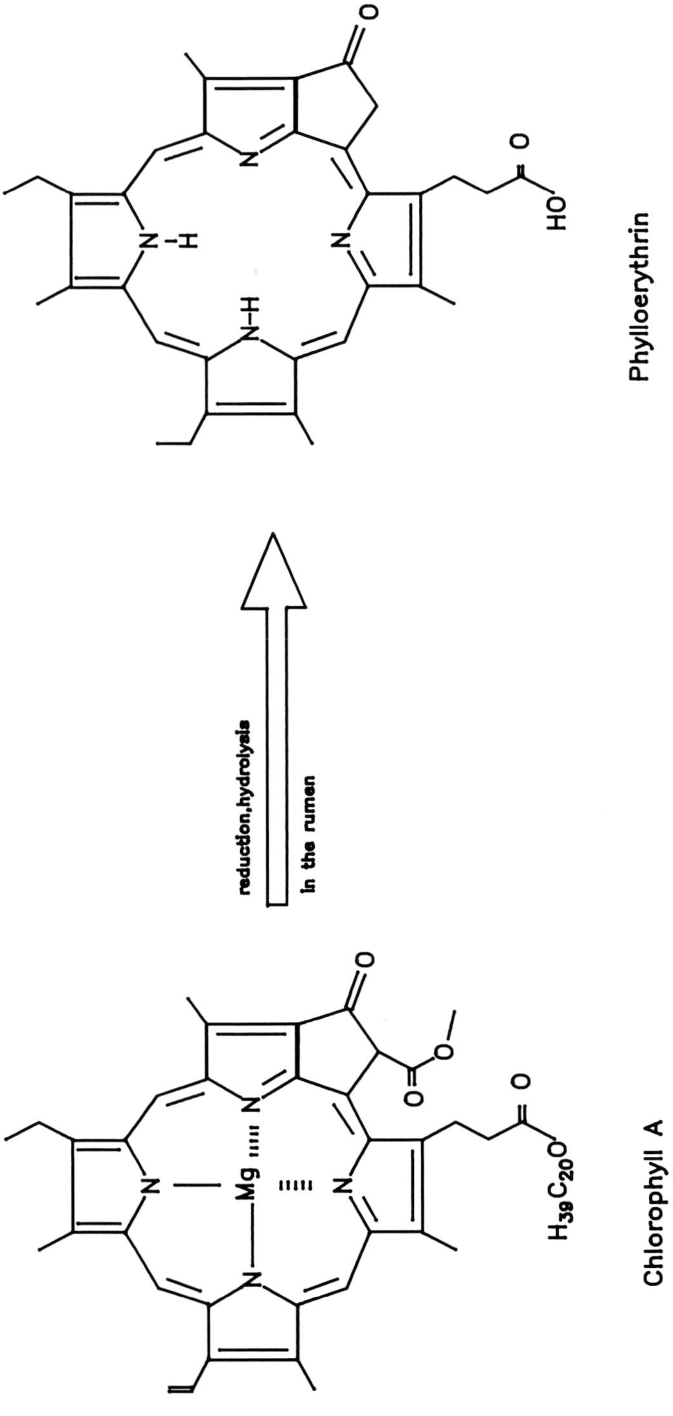

Phylloerythrin

Chlorophyll A

$H_{39}C_{20}O$

reduction, hydrolysis

In the rumen

Figure 4.   Formation of phylloerythrin from chlorophyll A in the anaerobic environment of the rumen.

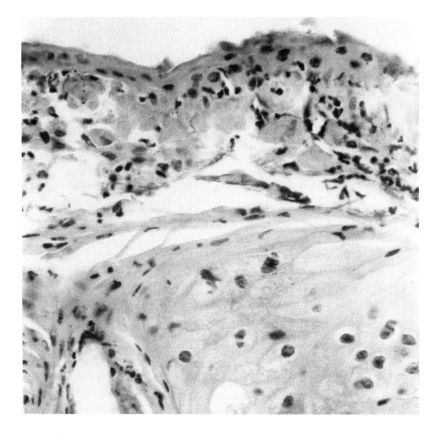

Figure 5.   Ballooning degeneration of stratum spinosum with neutrophilic exudate in parakeratotic stratum corneum, skin of bovine neck. H & E stain, magnification × 100.

erythema, followed by edema, serum exudate, crust formation, necrosis, cracking, fissuring, and sloughing of the epidermal layers.[11] Microscopic lesions seen in skin include a superficial crusting, with accumulation of keratin and purulent debris (Figure 5). Purulent keratitis is a typical histopathologic finding seen in bovine cornea (Figure 6).

C. Photoallergic Reactions

Photoallergy is an acquired immunologically mediated reaction to a chemical initiated by the formation of photoproducts. The occurrence of a photoallergic response to a chemical is sporadic and highly dependent upon the specific immune reactivity of the host. Photoallergic responses are thought to be cell-mediated hypersensitivity reactions involving two distinct mechanisms. In the first reaction type, light initiates the conversion of the hapten (synonymous with photosensitizer) to a complete allergen. Animal studies suggest that the photoreactive chemical in the skin

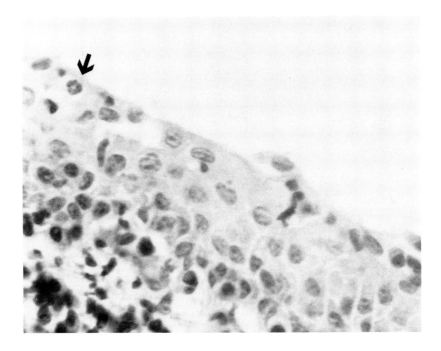

Figure 6.   Accumulation of neutrophils and plasma cells in the substantia propria of the bovine cornea with an occasional neutrophil (arrow) in corneal epithelium and distortion of the basal layer of epithelium by exudates. H & E stain, magnification × 100.

absorbs light and is converted to a photoproduct that subsequently binds to tissue proteins producing a complete antigen.[12-14]

$$\text{PS} \quad \rightarrow \quad \underset{\text{(Hapten)}}{\text{Photoproduct}} \quad + \quad \text{Tissue Protein} \rightarrow \quad \underset{\text{(Antigen)}}{\text{Allergen}}$$

Halogenated salicylanilide photoproducts are believed to be formed in this fashion. In the second type of reaction, light absorbed by the photosensitizer results in its conversion to a photoproduct that is a more potent allergen than the parent compound.

$$\underset{\text{(Weak Allergen)}}{\text{PS}} \quad \rightarrow \quad \underset{\text{(Strong Allergen)}}{\text{Photoproduct}}$$

The photoproduct of sulfanilamide is thought to be formed by this second pathway in which the parent sulfanilamide compound is converted by UV light to the potent allergic sensitizer p-hydroxyaminobenzene sulfonamide.[15] Patients with this type of photoallergy have demonstrated an allergic reaction to sulfanilamide in the dark.

An important complication of some of the chemicals inducing photoallergic responses is the development of persistent light reactions in which a marked sensitivity to light persists despite the apparent termination of exposure. Removal of the offending photoallergen in these cases does little to abate the condition and the action spectrum broadens to include the UVB as well as the UVA bands. As the phrase implies, this condition is long lived and troublesome.[16] This particular problem validates the importance of developing and utilizing screening tests for photoallergenicity to prevent exposure of a susceptible population of people to chemicals with this potential.

V. ASSESSMENT OF PHOTOSENSITIZATION

Photosensitivity reactions account for a very small percentage of the total number of undesirable effects from environmental chemicals. However, the increasing incidence and severe disability resulting from these types of skin changes, particularly when the photosensitivity response is of the persistent light reactor mechanism, suggest that additional photobiologic research efforts are needed. Predictive testing is an obvious approach used to assess the photosensitizing potential of new chemicals entering the commercial market. These methods make it possible to identify and possibly minimize or eliminate exposures to those compounds demonstrating risk-benefit ratios that are undesirable for the general population or especially sensitive individuals.

In vitro and *in vivo* methods with predictive value for estimating the photosensitizing potential of new compounds have developed rapidly to meet the demanding requirements of today's society. *In vitro* methods for assessing photosensitization are desirable because they are usually rapid and inexpensive and therefore allow screening of a large number of compounds (see also Chapter 11). Many of these methods are not very specific, however, and will generate a greater percentage of false-positive results than *in vivo* tests using animal or human models. Complex *in vitro* test systems appear to be useful in identifying the site and mechanism of action in certain situations.[6] Continued evolution of *in vitro* methodologies will add to the understanding of the photosensitization mechanism as better correlation is established with *in vivo* studies.

A. Assessment of Phototoxicity

A number of assay systems for phototoxic substances exist. They include the use of biochemical tests, cell cultures and suspensions, microorganisms, nonhuman mammalian skin, and human skin. This is an area in which the use of *in vitro* testing will have significant predictive value. However, replacement of animal tests with alternative methods is likely to lead to a reduction in predictive precision, just as animal model usage has led to enhanced precision.

The distribution of phototoxic reactions has changed significantly since 1981, with benoxaprofen and other NSAIDs causing the majority of problems.[17] As a result, phototoxicity predictive tests have become increasingly important in the development of new NSAIDs. This type of testing, in contrast to irritation testing, is not a routine procedure for all new chemicals entering the market and therefore, much less experience is apparent in this area of skin testing. Accordingly, it is necessary to ascertain what screening tests are available and how sensitive they are, as well as to identify the reference standards that are to be used to determine the specificity of a phototoxicity test.[17]

B. *In Vitro* Testing

Different endpoints can be used to evaluate the phototoxic potential of test compounds in *in vitro* systems. Growth inhibition of the common yeast *Candida albicans* originally was developed in 1965 and has since undergone several modifications.[18] Table 2 presents a compilation of some of the published variations and results of this widely used method. Differences in the sensitivity of various strains and species of *Candida* were investigated in an assessment of sensitivity to psoralens, but no essential differences between the various strains were detected.[19] Not all compounds known to be phototoxic in humans have induced reactions in this test system. For example, demethylchlortetracycline was negative in this particular assay.[18,20] The *Candida* test has been recommended by others, with the reservation that false-positive and false-negative results are always a possibility.[21]

A general description of the performance of the *Candida* method is as follows:

1. Specimens to be tested are applied as pure crystals, plant material, or filter paper disks impregnated with test compounds, to the agar surface of Sabouraud-dextrose media plates previously seeded with 48-hour-old cultures of *Candida albicans* suspended in Sabouraud broth.
2. Antimicrobial agents such as penicillin-streptomycin and kanamycin are added to the agar medium at a concentration of about 100 mg/ml to inhibit bacterial growth.
3. All samples with a zone of growth inhibition after 24 to 48 h of UVA irradiation at 25°C are rerun, accompanied by a non-UVA-irradiated control.
4. A compound is considered to have phototoxic activity when inhibition of growth of *Candida albicans* is greater around the disk on the UVA-irradiated plate than that around the corresponding disk on the nonirradiated plate. Positive control PS such as the psoralens can be included as a further validation of the testing procedure.

A convenient source for most phototoxic compounds activated by light in the UVA spectral band is the blacklight fluorescent lamp. These lamps provide a broad band output in the UVA region centered around 350 nm. A pane of glass interposed between the source and the target will eliminate the small amount of UVB emitted by these lamps. Several lamps separated by 5 to 10 cm and mounted 15 to 20 cm above the upper surface of the agar are required for effective radiation intensity. A variety of instruments are available for measuring the intensity of the radiation at the agar surface. This method is especially sensitive to the psoralen compounds and coal

Table 2
Variations of the *Candida* Test of Daniels[18-20,22-24]

		Results	
Investigators	Organism	Positive	Negative
Daniels (1965)	*Candida albicans*	Coal tar derivatives, psoralens	Sulfanilamides, demethylchlor-tetracycline, rose bengal
Ison and Davis (1969)	*C. albicans*	8-Methoxy-psoralen, chlorpromazine	Demethylchlor-tetracycline, griseofulvin, tolbutamide
Lohrisch et al. (1980)	*C. albicans*	Acridine, 8-methoxypsoralen	
Serrano et al. (1984)	*C. albicans*		Piroxicam
Knudsen (1985)	*C. albicans* 12 strains *Candida,* 15 species	Methoxypsoralen	
Ljunggren (1985)	*C. albicans*	Propionic acid derivatives	

tar derivatives used in dermatology. It also can be used for the presumptive identification of plants causing phytophotodermatitis.

Certain phototoxic compounds such as the porphyrins can damage cell membranes when excited by light. This reaction forms the basis of the red blood cell photohemolysis test. Since erythrocytes are anuclear, chemicals that react primarily with DNA cannot be evaluated with this technique. In this method, red blood cells (RBCs) are prepared from heparinized fresh mammalian blood by washing three times with phosphate-buffered saline (PBS; 0.01 M in phosphate). Packed RBCs are stored and used within 1 week.[25] Lysis of RBCs is monitored by light scattering at 600 nm. RBC suspensions (1.5×10^8 cells/ml) are incubated for 30 min with the test compound in PBS. Phototoxic membrane damage may occur by either oxygen-dependent or oxygen-independent mechanisms.[26] To evaluate the participation of oxygen in membrane damage, a comparison is made of the relative rates of photohemolysis of RBCs in air-saturated and nitrogen-purged cell suspensions. The oxygen concentration is reduced in samples by purging with nitrogen via syringe needles in rubber septum-capped tubes for 15 min prior to irradiation or incubation in the dark.[25] Half of the samples are then opened and purged with air before irradiation. Air- and nitrogen-saturated samples are irradiated in parallel and the percent hemolysis is determined at various time intervals.

The photohemolytic action of chlorophyll metabolites such as phylloerythrin may also be investigated *in vitro*.[27] To study the mechanism of phototoxic action of

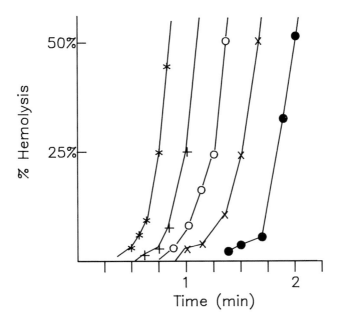

Figure 7. Effect of different concentrations of phylloerythrin on light-induced hemolysis time: ● = 5.4 μ*M*/l; x = 8.1 μ*M*/l; ○ = 10.8 μ*M*/l; + = 13.5 μ*M*/l; * = 16.2 μ*M*/l. (Modified from *Veterinary Pharmacology and Toxicology,* Ruckebusch, Toutain, and Koritz, Eds. MTP Press, Falcon House, Lancaster, England, 1983.)

phylloerythrin, hemolysis of goat RBCs subsequent to UV exposure is studied. Phylloerythrin from sheep feces is isolated by ether extraction, followed by methylation of the carboxyl group and recrystallization from methanol-chloroform (3:1, v/v). Before use the ester is hydrolyzed and free phylloerythrin is dissolved in a maximum ethanol concentration of 5% and added to washed goat RBCs. Incubation with erythrocytes is performed in tubes placed horizontally in a rotating rack under a UV lamp with emission maximum around 360 nm. Hemolysis is determined by centrifuging a sample and measuring the hemoglobin concentration in the supernatant with a spectrophotometer at 540 nm. There is a concentration-dependent lag phase for phylloerythrin, after which hemolysis proceeds quickly (Figure 7). A linear relationship exists between the hemolysis time and the logarithm of the phylloerythrin concentration in the range of 5 to 20 μ*M*/l. The time required to reach 50% hemolysis (HT_{50}) can be extrapolated from a time vs. log concentration graph (Figure 8).

More detailed studies can be undertaken to evaluate the molecular mechanism of action of PS and the participation of active oxygen species. Singlet oxygen and hydroxy radicals attack cell membrane constituents and both have been implicated as the active oxygen species produced by many PS.[28-30] The hydroxy radical (OH·)

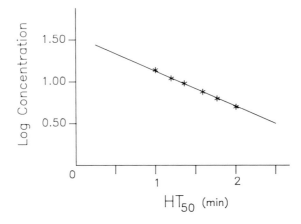

Figure 8. Relationship between the time required to reach 50% hemolysis (HT_{50}) and the log concentration of phylloerythrin. (Modified from *Veterinary Pharmacology and Toxicology*, Ruckebusch, Toutain, and Koritz, Eds. MTP Press, Ltd., Falcon House, Lancaster, 1983.)

is formed from superoxide anion ($O_2^{-}\cdot$) in water. Sodium azide at a concentration of 30 to 300 mM can be used to quench singlet oxygen, and superoxide dismutase at 50 µg/ml is an effective quencher of superoxide anion. Both of these agents can be added to photoreactive mixtures prior to irradiation to evaluate the participation of active oxygen species in reaction mechanisms.

Capacity is known in cell biology as the ability of a host cell to support virus growth. The inactivation of this cellular function can result from treatment of the cell with DNA-toxic chemicals. Herpes simplex virus is especially useful for mammalian cell capacity investigations because it can be propagated in a wide variety of cells and has a rapid growth cycle. A description of the effects on capacity for herpesvirus growth by chemical photosensitization of host cells follows. Inactivation of the capacity of monkey kidney cells to support herpesvirus plaque formation by phototoxic treatment of the cells with photosensitizer and light has been demonstrated.[31] Confluent monolayers of hepatocytes or other cells of interest are incubated with different concentrations of photosensitizer, exposed to radiation from fluorescent lamps, filtered with glass to remove UVB and UVC bands, and infected with herpesvirus. Virus replication is measured by plaque formation. Appropriate negative controls with photosensitizer and without light, and with light and without photosensitizer, are included in the study design.

Another *in vitro* cell culture technique involves the induction of SV40 tumor virus by photoreactive chemicals of SV40-transformed cells with photosensitizer and light.[32,33] Monolayers of clone E cells are treated with different concentrations of photosensitizer, irradiated with glass-filtered light from fluorescent lamps, and incubated 3 days for virus expression. The amount of SV40 induced is determined by measurement of the SV40 infectivity of cell extracts in permissive CV-1 monkey kidney cells. Both cell-culture testing systems have the disadvantage of requiring virology laboratory experience and facilities. In addition, differences between responses of virus-infected cells and normal cells to PS may exist and limited validation of the tests have been made with known phototoxic compounds. Neither is it known whether the systems would react to treatments causing primarily cell membrane damage. For screening purposes, it is important to use several different concentrations of chemical, light exposure times, and wavelengths. The light absorption spectrum for each compound is useful in selection of the appropriate wavelengths of light for irradiation.

A photobiological assay has been described for a DNA-repair-deficient *Escherichia coli* bacterium in which growth inhibition is the response evaluated.[34] According to the authors of this study, use of an organism deficient in normal DNA repair improves the sensitivity of detection compared to methods from previous similar experiments.

Other *in vitro* methods exist for evaluating the phototoxic potential of various compounds. They can be used to screen chemicals in a rapid and relatively inexpensive way; however, they do have their limitations. Methods employing most microorganisms are not very specific and they generate a relatively large number of false-positive responses compared to *in vivo* tests. The use of these types of methods will be refined with time and their value will be better realized when correlations to data from *in vivo* systems are effectively validated.

C. *In Vivo* Testing

Various *in vivo* model systems have been developed to investigate the phototoxicity of compounds likely to come in contact with humans through therapeutic, cosmetic, and environmental avenues. Most of these assays are either qualitative or semiquantitative and are time consuming and relatively expensive to perform. Several species of animals, including mice, rabbits, guinea pigs, swine, and monkeys, have been used in an attempt to mimic the human situation. Until additional experience is obtained, new compounds developed for widespread use in humans also can be examined on human skin. Because of a common response in humans, testing can be done with minimal hazard in small test areas in a few subjects. Important features of any system for these types of investigations include availability and ease of performance, reproducibility, as well as a reasonably quantifiable phototoxicity measurement. Factors affecting phototoxicity at a particular skin site include:

1. The quantity and location of the photosensitizer on the skin
2. Capacity of the chemical for percutaneous absorption as well as exposure of traumatized areas such as an abrasion
3. The effect of vehicle on percutaneous absorption
4. The pH, the presence of enzymes, and solubility conditions at the exposure site
5. The duration and intensity of activating light

6.   Penetration depth of activating light
7.   Humidity and ambient temperature
8.   Thickness of the stratum corneum of the selected site
9.   Degree of skin pigmentation

D. Animal Testing

Currently, there are no regulatory guidelines or requirements for testing the phototoxic potential of new chemicals, so it is not surprising that no standardized test system is available.[17] It is difficult to compare the different experimental methods used to evaluate phototoxicity because of the large number of variables involved in these procedures. To reiterate, some of these variables include species, light source, intensity and duration of light exposure, route of chemical administration, and the time interval between chemical exposure and light irradiation. These factors must be kept in mind when comparisons are to be made between available methods. Specific consideration has been afforded the effects of these variables in a recognized testing laboratory.[17] The guinea pig and mouse are the principal species used. Xenon and fluorescent light are the most common irradiation sources. Duration of light exposure is limited by its erythemagenic potential and a variation of 0.5 to 100 J/cm^2 has been measured for UVA lamps. The human exposure situation dictates the choice of a systemic or topical route of exposure to the test compound and the interval between chemical application and irradiation is dictated primarily by the route chosen.

A series of experiments designed to investigate the effects of these variables on animal testing revealed the following results.[17] The interval between exposure to 8-methoxypsoralen and irradiation required to induce a positive response in mice was dependent on the vehicle used. It may be necessary to define a specific time interval based on pharmacokinetic parameters and their relationship to the physicochemical characteristics of the photosensitizer. Pharmacokinetic considerations are especially important when various chemicals are administered systemically and when topical application is the delivery mode for testing. Models of uptake from skin require estimates of the area involved and blood flow to the contact area. These same principles apply to systemic as well as local administration. Accurate control of lamp output can be monitored using a radiometer with detectors for UVA and UVB bands. Because erythema is not a feature conducive to quantitative assessment in mice, measurement of edema formation is the method of choice. This can be quantitated as a weight increase in the skin or tail, or caliper measurement of thickness of an ear or skinfold. Evaluation of time-course studies is facilitated by thickness measurements. Comparing ear, back, and tail positions on mice, the back skin appears to be the most suitable. Ear thickness measurements are more dependent on the precise location of measurement on the ear pinna, resulting in a greater variation in values compared to back measurements. The magnitude of increase in back-skin thickness also appears to be larger than that for ears.

Murine ear swelling is a response used to assess not only photosensitization but also immunologic reactions.[35] This model meets the criteria of ready accessibility, and reasonable reproducibility and ease of quantitation. Use of this method involves

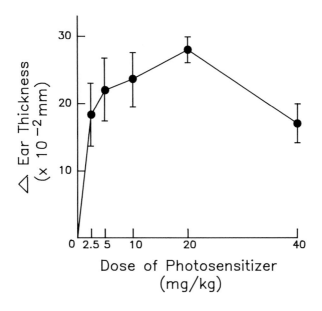

Figure 9.   Effect of varying doses of photosensitizer (hematopor-phyrin derivative) on the ear swelling response of C3H mice irradiated with 1.0 J/cm² 405 nm radiation. Panels of mice were inoculated with doses of photosensitizer ranging from 2.5 to 40 mg/kg. Values repre-sent mean ± SEM of four mice. (Modified by C. W. Hawkins, D. R. Bickers, H. Mukhtar, and C. A. Elmets, Cutaneous porphyrin photosen-sitization: murine ear swelling as a marker of acute response, *J. Invest. Dermatol.* 86, 638, 1986.)

administration of the photosensitizer orally or parenterally followed by exposure to a UVA light source after an appropriate time interval to allow for distribution to the test site. Irradiation must occur at a time when there is maximum uptake of the compound by the ear. Duration of radiation delivery may be somewhat empirical but is generally in the range of 2 to 4 h. Mice may be placed in rotating cages to ensure uniform exposure to the light source. The amount of ear swelling that occurs after irradiation may be determined at various times using a thickness gauge micrometer. These measurements are compared to those made prior to irradiation according to the following formula:[35]

> Ear thickness = ear thickness after irradiation (imme-diately or at 12 or 24 h) - ear thickness prior to irradia-tion (see Figure 9).

Histologic evaluation of the photosensitivity response is performed on the ears of mice by staining sections of excised tissue with hematoxylin and eosin. The micro-scopic examination of the immediate response usually is characterized by marked dermal edema and vascular engorgement. No epidermal changes or cellular infiltrates are present in the dermis at this particular stage of the study.[35] In this experiment and others, microscopic examination carried out on ear sections taken more than 24 h after irradiation reveal dermal edema accompanied by mild to moderate infiltration of neutrophils and occasional mononuclear cells.[36] Phototoxic reactions in mice also can be evaluated by grading the edema, erythema, and necrosis of the ears and tail according to other variations.[37-39] An additional method that measures the increase in wet weight of the mouse tail can be used to measure the phototoxic response.[40] This particular method is more labor intensive and the requirement of tail removal eliminates the option of monitoring serial changes in the same animals.

Albino guinea pigs and miniature pigs (minipigs) also are utilized in photosensitization studies.[41] This method involves using a depilatory preparation to remove hair from the backs of guinea pigs 24 h prior to chemical administration. The PS is injected intradermally at different doses, followed by UVA (20 J/cm^2) irradiation 6 h after chemical treatment. An erythematous response is graded 24 h postirradiation according to the following scale: 0 — no reaction; 1+ — minimal erythema with sharp borders; 2+ — bright erythema without edema; 3+ — marked erythema with edema; 4+ — violaceous erythema with vesiculation. Minipigs can be handled and graded in a similar manner except that sedation with xylazine and ketamine and acclimation to a sling are needed for restraint.

E. Human Testing

Phototoxicity has been experimentally induced in humans. The phototoxicity of certain tetracyclines was demonstrated following intradermal injection and exposure to sunlight.[42] In another study, phototoxic reactions to several clinically recognized PS were induced following intradermal injection, as well as topical application to stripped skin.[43] A somewhat improved method exists for evaluating systemic photosensitizing chemicals using intradermal injections in humans.[44] The untanned lower back of white healthy adults serves as the test site in this method. A UVA-VIS light source is used and intensity measurements are made with a calibrated thermopile. Exposure to UVA and solar simulated UV is carried out separately. UVA exposure should begin with about 30 J/cm^2 and increased up to 40 if necessary. A volume of 0.1 ml is injected intradermally with a 27-gauge needle and attached tuberculin syringe. The test compounds are dissolved in physiologic saline solution when possible. Agents insoluble in this vehicle are given as a suspension in saline with the aid of an ultrasonification unit. Pilot studies using rabbits or guinea pigs may be necessary to determine optimal nonirritating concentrations, time intervals, and light doses. Controls consist of an unirradiated drug-injected site and an irradiated site injected with solvent. Test sites are irradiated 15 min after injection and responses are graded 24 h later and compared with control sites.

Another method has been developed for investigating the phototoxic potential of NSAIDs in humans.[45] The method involves patients taking the appropriate drug orally for at least 1 week at the time of phototesting. The light source consists of a 900 W xenon arc lamp optically coupled to a single grating monochromator. Light is transmitted from the exit slit of the monochromator through a liquid-filled light guide to permit irradiation of a limited area of skin (0.5 cm diameter) in a controlled manner. Alternatively, banks of four to six fluorescent lamps may be used to irradiate relatively large areas of skin. Protective goggles should be worn by both patient and investigator to shield the eyes from short-wave UV light. The back of each patient is used as the phototesting site because of the high sensitivity and more uniform response to UV radiation than most other locations. Subjects are irradiated with a 20 nm bandwidth centered around 320 nm. Each individual is given a series of radiation doses ranging from 0.5 to 4 J/cm^2 in a geometric fashion equivalent to doubling alternate exposures. Test sites are examined following exposure and the minimum radiation dose inducing the immediate reactions of urticarial wheals, erythema, and flaring is recorded. Subjective sensations of itching and burning also are noted.

Phototoxicity is one of the easiest forms of toxicity to detect because of the relative ease of performance of predictive assays *in vitro* and *in vivo*. The high incidence of expression in susceptible populations facilitates detection of chemicals with the potential to cause widespread problems. With the inevitable increase in reliability of more innovative techniques, the likelihood of exposing susceptible individuals to phototoxic compounds in commerce eventually will become a thing of the past. Development of standardized testing certainly also will facilitate inter-laboratory comparisons.

F. Assessment of Photoallergy

Photoallergy is considered to be far less common than phototoxicity; however, it is generally thought to be the more serious of the photosensitization reactions due to the persistent light reactor problem previously mentioned. Photoallergic contact dermatitis resembles allergic contact dermatitis clinically and usually is limited to light-exposed areas of the skin, although extension beyond these sites does occur. Testing methods designed to identify potential photoallergens evolved in the wake of the photoallergic contact dermatitis outbreak caused by the antimicrobial halogenated salicylanilides in the early 1960s. The following criteria are useful in testing and diagnostic situations to help define and characterize the photoallergic response:[46]

1. Demonstration of an incubation period and spontaneous flare response when the condition is induced experimentally
2. Irradiation of a distant site that induces flares at a previously exposed site
3. Subsequent exposures resulting in a more severe reaction than initial exposure
4. Demonstrable involvement of the immune system by passive transferability
5. Photoallergy can be induced by topical exposure to chemicals followed by irradiation and probably by systemic exposure to certain compounds and light

So as not to overstate the reliability of testing specifically for photoallergy, it must be kept in mind that some photosensitizing chemicals can induce both photoallergic and phototoxic reactions, resulting in clinical and laboratory confusion.[47]

G. *In Vitro* Testing

Induction of the photoallergic response involves the photochemical binding of the photoallergen to a protein as the initial step.[12-14] A number of compounds with widely differing structures have been incriminated as photoallergens, but relatively few examples of protein conjugation have been reported. The following screening procedure is designed to detect potential photoallergens based on protein conjugate formation[48]

1. Chemicals are selected for testing based on their ability to absorb the appropriate wavelengths of UV or visible light. For chemicals with unknown action spectra, it is necessary to use a light source with a broad emission spectrum in the UV-VIS region.

2. Stock solutions of photoallergens to be tested are prepared according to their solubility in water or ethanol and then are added to solutions of human serum albumin (3 to 4×10^{-5} M) in 0.1 M Tris-HCl buffer pH 8.1 using a microsyringe. The molar ratio of protein to photoallergen should be 1.0 and the concentration of ethanol in appropriate samples should not exceed 1%.

3. Samples are contained in a 3 ml quartz cuvette and directly irradiated for 30 min with a medium pressure Hg-arc lamp fitted with Schott UG-5 (2 mm; UV-transmitting) and WG-310 (1 mm) filters, resulting in 58% transmission at 313 nm.

4. Samples are then passed through a column of Sephadex$^{®}$ G-10 (5 cm^2 × 33 cm) equilibrated with 0.05 M ammonium bicarbonate to separate the protein from the unbound photoallergen.

5. Spectrophotometry is used to obtain UV spectra from each sample before and after irradiation and after the chromatographic procedure.

6. Unirradiated solutions of human serum albumin and photoallergen are passed through the Sephadex$^{®}$ column and their spectra are recorded to determine if there is any binding of photoallergen to human serum albumin in the absence of light exposure.

7. The spectra of other samples are monitored to determine if there is binding of photoproducts to human serum albumin by placing photoallergen in buffer, irradiating this solution for 30 min, letting it stand for 2 h (to allow time for decay of photoactivated species), incubating for 2 h with human serum albumin, and passing the solution through the Sephadex$^{®}$ column.

8. Irradiated samples passed through a column of Sephadex$^{®}$ G-10 will produce spectra with contours identical to those obtained prior to the column treatment, but with contours different from those of the spectrum of human serum albumin alone. These spectral changes suggest photochemical binding of the photoallergen molecules to the protein. Or, alternatively, there may be photochemical modification of the protein by the photoallergen without binding. This possibility can be ruled out since the absorption spectra taken before and after the column treatment can be expected to be different.

The above-mentioned *in vitro* test demonstrates good correlation between compounds known to be photoallergens and their ability to form covalent conjugates with protein. A weakness of this method lies in the fact that it is strictly a qualitative procedure based on UV spectroscopy. This qualitative limitation is the result of a lack of knowledge of the extinction coefficients of the protein conjugates formed. This method may give rise to false-positives in situations where noncovalent binding of photoallergens or photoproducts cannot be distinguished from covalent binding. If more detailed information of the stoichiometry of the photochemical reaction is needed, it may be necessary to use radiolabeled photoallergens according to other available methods.[49,50]

H. *In Vivo* Testing

There are two phases in the evaluation of photoallergic reactions in animal models. The first, the induction phase, is an attempt to induce an immune response in the test animal and the second phase is the elicitation phase, in which the animal is challenged with the same test compound used in the induction phase. This second phase indicates whether the induction phase was successful. Guinea pigs frequently are used because of their reactivity to many of the same compounds that affect humans. They are used to demonstrate photosensitivity of the delayed hypersensitivity type in several reported animal studies.[51-53] Hartley strain albino guinea pigs are used routinely. Females are preferable because they are less likely to scratch the phototest site.[12]

To eliminate compounds with irritant potential, an irritation screen should be conducted on the backs of about ten animals, using three concentrations of test solution, applied in duplicate on opposite sides of the back. One side is covered with an opaque material to test for irritant-contact dermatitis, while the other side is irradiated with UVA to test for phototoxicity. Sites are evaluated at 24, 48, and 72 h after application of test compounds. This step is helpful in selecting appropriate concentrations of test solutions required for the definitive identification of photoallergens.

The induction site is prepared by removing hair, usually from the nuchal region of the guinea pig. This is accomplished by shaving the animal, followed by chemical depilation. Other procedures, such as stripping of the stratum corneum with cellophane tape, may be used to remove further barriers to chemical penetration. The test chemical is applied in solution using acetone, ethanol, or water as the usual vehicle. The concentration and volume of test solutions are determined empirically but usually 0.1 ml of a 0.1 to 1.0% concentration is utilized. A lag time is allowed for penetration of the chemical prior to irradiation of the site. Again, this is somewhat empirical in that the ability of the compound to cross biological membranes must be considered. The depth of penetration required has not been established, but logically it must at least reach viable epidermal tissue. Irradiation of the site must be timed so that an adequate concentration of the chemical is present at the induction site to obviate the possibility of a false-negative result. This very problem has been documented as a cause of false-negative results using 6-methyl-coumarin.[54] In general, a

30-min lag time between application of the test chemical and irradiation has proved adequate for most compounds tested with the guinea pig model. Irradiation with 10 to 20 J/cm^2 of UVA light has proven effective for all of the recognized photoallergens. If deemed necessary, an *in vitro* spectral scan of the test compound is an aid in selecting the appropriate bands of light. These steps are repeated at least five times during a 2-week period with this phase ending 2 weeks after the last induction procedure. There are no standardized methods for this phase and a review of the literature will reveal many variations on this basic theme.

The elicitation phase is performed to confirm the success of the induction phase. Because of the potential phototoxicity of some of these compounds, elicitation may be tested on the back to avoid a possible false-positive result induced by chemical retained at the induction site. Hair on the back is removed by clipping and depilation. The test compound is applied at a concentration below that which induces phototoxic reactions in previously unexposed animals. Test solution is applied symmetrically to sites on the back with one of two sites being covered with an opaque bandage. The covered site serves as a control to detect simple irritation potential of the test solution. Another control site may be selected to monitor the erythemagenic potential of the light source alone. Irradiation of the uncovered site with 10 to 20 J/cm^2 of UVA follows a 30-min lag time. Animals are examined after 24 h and the amount of erythema and edema visualized is recorded. Two conditions must be satisfied to verify that a photoallergic reaction was successfully induced: (1) erythema and edema at the irradiated test site and (2) no detectable reaction at any of the control sites. If a clinical response is noted at either of the control sites, a photoallergic reaction may not be solely responsible for the reaction at the test site.[55] The number of positive reactors in the test group is compared statistically with the number of positive controls using the Fisher's Exact Test.[56]

Reaction to the more potent photoallergens can be detected in guinea pigs using the basic method outlined above. Modifications may be necessary to induce positive reactions to some of the weaker photoallergens. The use of cellophane tape to remove layers of the stratum corneum is one such modification.[52] An intradermal injection of Freund's complete adjuvant around the site and at the time of first induction also will enhance the number of positive reactors.[55] Induction involves shaving and depilation of the nuchal areas of 10 to 20 test animals as before. Areas of 6 to 8 cm^2 are defined by injection of 0.1 ml of Freund's complete adjuvant into four corners. Test compound is then applied to each area under an occlusive patch as defined by the injection sites. Following patch removal, the sites are irradiated and, with the exception of adjuvant injection, this procedure is repeated four to five times in 2 weeks. Negative control animals receive identical treatment concurrently, except that the sensitizer is not included. A maximum number of reactors may be observed with a combination of both enhancement techniques.[12,53]

Murine models for photoallergic contact dermatitis also have been developed.[57,58] In these models the systemic administration of cyclophosphamide or the intradermal injection of *Corynebacterium parvum* is used to alter the response by their immuno-modulating effects.

I. Human Testing

The photomaximization test is conducted in humans and is simply a repeated insult procedure that involves an exaggerated exposure to both chemical and UV light.[59] A continuous spectrum, 150 W, xenon arc solar simulator light source is used. The test compound at a 5% concentration in an appropriate base, such as a hydrophilic ointment, is delivered to the skin at a concentration of 10 $\mu l/cm^2$ with a tuberculin syringe. The material is spread evenly using a glass rod and the sites are covered with cotton cloth and adhered to the skin with occlusive tape. The patches are removed 24 h later and the sites are exposed to three minimal erythema doses (MED) of light from the solar simulator. The MED is determined for each subject beforehand by exposing the skin sites to 25% increments of radiation. The dose required to induce minimal uniform erythema with a distinct border 24 h after exposure is the MED. Following a rest interval of 48 h, a similar application is made to the same site for another 24 h and followed again by 3 MED of light. These steps are repeated for a total of six exposures during a period of 3 weeks. Following a rest period of 14 days, the subjects are challenged by a single exposure to a fresh skin site. An occlusive application is made with a 1.0% concentration of the test compound for 24 h, followed by exposure to 4.0 J/cm^2 of UVA light. The UVA band is obtained from the light source by filtering the radiation through a 2-mm Schott WG345 filter or a pane of ordinary window glass. Test sites are evaluated 48 and 72 h after irradiation. Several layers of opaque adhesive tape are used to cover the sealed, unirradiated, control sites. Development of the typical reaction of erythema, edema, and possibly vesicles in the irradiated, but not the unirradiated, sites is evidence that a photoallergic contact dermatitis has been induced. Each compound is usually evaluated in 25 subjects.

Photopatch testing is used as a diagnostic procedure when it is suspected that sunlight is involved in the elicitation of some form of contact dermatitis. This procedure involves exposing the individual to the suspected compound alone and to radiant energy alone to rule out simple contact dermatitis and sensitivity to sunlight, respectively. Another site is exposed sequentially to the compound plus a light source. When clinical signs compatible with the naturally occurring syndrome are induced only at this last site, the diagnosis of photoallergic contact dermatitis is confirmed. The following procedure requires several visits by patients with this problem.[60]

Step 1 — UVB irradiation of eight 1-cm squares of skin on the lower back is performed in a stepwise series of doses. The lowest dose is one third the estimated MED. Doses in subsequent squares progress geometrically in increments of 40%. Sites are evaluated immediately and 30 min following irradiation for the presence of a weal. Concurrently, compounds suspected of causing a reaction are applied in duplicate and covered with opaque material and taped in place. Suspect chemicals are formulated in petrolatum at a concentration of 0.1 to 1.0%.

Step 2 — Exposed sites are examined 24 h later and the site receiving the lowest dose of UVB that resulted in detectable erythema filling the square with distinct borders is selected as the MED. Different squares are then exposed to large single doses of UVB (10 MED) and UVA (20 times the exposure period for UVB, but with radiation filtered through window glass). Photopatch sites are uncovered and evaluated for diagnostic clinical signs. One set is recovered immediately and the other is irradiated with UVA, then recovered. If the subject is very sensitive to UVA, a dose less than the MED is given, and if not, 10 J/cm^2 is delivered as the standard.

Step 3 — Photopatch sites are evaluated 48 h after Step 2 for the presence of a photoallergic response (erythema and edema). The multiple MED sites may show some changes.

Step 4 — All sites are re-evaluated 72 h after Step 3 for any additional delayed reactions. This procedure may be shortened by stripping of the stratum corneum from test sites with adhesive tape just prior to chemical application. Sites are irradiated 1 h following chemical application.

Experience with human testing is somewhat limited. Variables that influence induction and elicitation, such as concentration, vehicle, UV exposure and intensity, and the time interval between chemical application and light exposure, require further investigation and standardization of techniques.

VI. CONCLUSIONS

This chapter has attempted to describe the methods and complexity of *in vitro* and *in vivo* photosensitization testing and the various factors that influence each system, as well as an overall lack of standardization. Phototoxicity is the easiest form of photosensitization to detect. Photoallergy, however, is somewhat more involved and is very dependent on the immune status of the model selected for observation. We have seen that there are a variety of methods for screening the photosensitizing potential of chemicals. To facilitate the valid comparison of results between laboratories, it is imperative that testing reports include a detailed description of the methodology used. Testing should be performed with the goal of preventing photosensitization reactions from occurring in the general population.

No single test system currently available will predict the potential photosensitization capability of all compounds. It is important that one not be too preoccupied with the testing technique and its variances. One must be cognizant of the fact that the purpose is to test the compound and not the method. It is advisable to select a method of testing and refine and standardize the associated techniques.

REFERENCES

1. J. M. Megaw and L. A. Drake, *Photobiology of The Skin and Eye,* Marcel Dekker, New York (1986).
2. M. A. Pathak, *J. Natl. Cancer Inst.* 69, 163 (1982).
3. E. Ben-Hur and M. M. Elkind, *Biochim. Biophys. Acta* 331, 181 (1973).

4. M. A. Pathak, D. M. Kramer, and T. B. Fitzpatrick, Photobiology and photochemistry of furo-coumarins (psoralens), in *Sunlight and Man: Normal and Abnormal Photobiologic Responses*, M. A. Pathak, L. C. Harber, M. Seiji, and A. Kukita, Eds. University of Tokyo Press, Tokyo, 336 (1974).
5. D. P. Valenzeno, *Photochem. Photobiol.* 46, 147 (1987).
6. I. E. Kochevar, *Photochem. Photobiol.* 45, 891 (1987).
7. J. Glette and S. Sandberg, *Biochem. Pharmacol.* 35, 2883 (1986).
8. I. E. Kochevar, W. L. Morison, J. L. Lamm, D. J. McAuliffe, A. Western, and A. F. Hood, *Arch. Dermatol.* 122, 1283 (1986).
9. A. G. Motten and C. F. Chignell, *Photochem. Photobiol.* 37, 17 (1983).
10. B. K. Sinha, J. T. Arnold, and C. F. Chignell, *Photochem. Photobiol.* 35, 413 (1982).
11. S. W. Casteel, E. M. Bailey, J. C. Reagor, and L. D. Rowe, *Vet. Hum. Toxicol.* 28, 251 (1986).
12. L. C. Harber, R. B. Armstrong, and M. Ichikawa, *J. Natl. Cancer Inst.* 69, 237 (1982).
13. P. S. Herman and W. M. Sams, Jr., *J. Lab. Clin. Med.* 77, 572 (1971).
14. L. C. Harber, I. E. Kochevar, and A. R. Shalita, *Photomedicine*, Plenum Press, New York (1980).
15. I. E. Kochevar, *The Effects of Ultraviolet Radiation on the Immune System*, Johnson and Johnson Baby Products Co., Skillman, NJ (1983).
16. E. A. Emmett, *Casarett and Doull's Toxicology; The Basic Science of Poisons*, 3rd ed. Macmillan, New York, 424 (1986).
17. T. Maurer, *Food Chem. Toxicol.* 25, 407 (1987).
18. F. Daniels, *J. Invest. Dermatol.* 44, 259 (1965).
19. E. A. Knudsen, *Photodermatology* 2, 80 (1985).
20. A. E. Ison and C. M. Davis, *J. Invest. Dermatol.* 52, 193 (1969).
21. G. Kavli and G. Volden, *Photodermatology* 1, 204 (1984).
22. I. Lohrisch, R. Müller, J. Barth, and C. Schönborn, *Dermatol. Monatsschr.* 166, 156 (1980).
23. G. Serrano, J. Bonillo, A. Aliaga, E. N. Gargallo, and C. P. Pelufo, *J. Am. Acad. Dermatol.* 11, 113 (1984).
24. B. Ljunggren, *Photodermatology* 2, 3 (1985).
25. I. E. Kochevar, K. W. Hoover, and M. Gawienowski, *J. Invest. Dermatol.* 82, 214 (1984).
26. I. E. Kochevar, *J. Invest. Dermatol.* 76, 59 (1981).
27. B. J. Blaanboer and M. Van Graft, *Veterinary Pharmacology and Toxicology*, MTP Press, Boston, 671 (1983).
28. K. Kameda, T. Ono, and Y. Imai, *Biochim. Biophys. Acta* 572, 77 (1979).
29. M. J. Thomas and W. A. Pryor, *Lipids* 15, 544 (1980).
30. J. D. Spikes, *The Science of Photomedicine*, Plenum Press, New York, 113 (1982).
31. L. E. Bockstahler, T. P. Coohill, C. D. Lytle, S. P. Moore, J. M. Cantwell, and B. J. Schmidt, *J. Natl. Cancer Inst.* 69, 183 (1982).
32. L. E. Bockstahler and J. M. Cantwell, *Biophys. J.* 25, 209 (1979).
33. L. E. Bockstahler, *Prog. Nucleic Acid Res. Mol. Biol.* 26, 303 (1981).
34. M. J. Ashwood-Smith, G. A. Poulton, O. Ceska, M. Lin, and E. Furniss, *Photochem. Photobiol.* 38, 113 (1983).
35. C. W. Hawkins, D. R. Bickers, H. Mukhtar, and C. A. Elmets, *J. Invest. Dermatol.* 86, 638 (1986).
36. L. D. Rowe, J. O. Norman, D. E. Corrier, S. W. Casteel, B. S. Rector, E. M. Bailey, J. L. Schuster, and J. C. Reagor, *Am. J. Vet. Res.* 48, 1658 (1987).
37. K. Konrad, H. Honigsmann, F. Gschnait, and K. Wolff, *J. Invest. Dermatol.* 65, 300 (1975).
38. H. W. Lim and I. Gigli, *J. Invest. Dermatol.* 76, 4 (1981).
39. A. Ison and H. Blank, *J. Invest. Dermatol.* 49, 508 (1967).
40. B. Ljunggren and H. Moller, *J. Invest. Dermatol.* 68, 313 (1977).
41. J. L. McCullough, G. D. Weinstein, J. L. Douglas, and M. W. Berns, *Photochem. Photobiol.* 46, 77 (1987).

42. W. F. Schorr and S. Monash, *Arch. Dermatol.* 88, 440 (1963).

43. A. M. Kligman and R. Breit, *J. Invest. Dermatol.* 51, 90 (1968).

44. K. H. Kaidbey and A. M. Kligman, *J. Invest. Dermatol.* 70, 272 (1978).

45. B. L. Diffey and S. Brown, *Br. J. Clin. Pharmacol.* 16, 633 (1983).

46. J. H. Epstein, *Dermatotoxicology,* 3rd ed. Hemisphere, Washington, D.C., 441 (1987).

47. R. D. Granstein, *Dermatology in General Medicine,* 3rd ed. T. B. Fitzpatrick, A. Z. Eisen, K. Wolff, I. M. Freedberg, and K. F. Austen, Eds. McGraw-Hill, St. Louis, 1464 (1987).

48. M. D. Barratt and K. R. Brown, *Toxicol. Lett.* 24, 1 (1985).

49. D. M. Rickwood and M. D. Barratt, *Photochem. Photobiol.* 35, 643 (1982).

50. D. M. Rickwood and M. D. Barratt, *Photobiochem. Photobiophys.* 5, 365 (1983).

51. J. Griffith and R. D. Carter, *Toxicol. Appl. Pharmacol.* 12, 304 (1968).

52. I. E. Kochevar, G. L. Zalar, J. Einbinder, and L. C. Harber, *J. Invest. Dermatol.* 73, 144 (1979).

53. H. Ichikawa, R. B. Armstrong, and L. C. Harber, *J. Invest. Dermatol.* 76, 498 (1981).

54. R. T. Jackson, L. T. Nesbitt, Jr., and V. A. DeLeo, *J. Am. Acad. Dermatol.* 2, 124 (1980).

55. L. C. Harber and D. R. Bickers, *Photosensitivity Diseases. Principles of Diagnosis and Treatment,* W. B. Saunders, Philadelphia (1981).

56. S. Gad and C. S. Weil, *Statistics and Experimental Design for Toxicologists,* Telford Press, Caldwell, NJ (1986).

57. Y. Miyachi and M. Takigawa, *J. Invest. Dermatol.* 78, 363 (1982).

58. H. C. Maguire and K. Kaidbey, *J. Invest. Dermatol.* 79, 147 (1982).

59. K. H. Kaidbey and A. M. Kligman, *Contact Dermatitis* 6, 161 (1980).

60. J. D. Bernhard, M. A. Pathak, I. E. Kochevar, and J. A. Parrish, *Dermatology in General Medicine,* 3rd ed. T. B. Fitzpatrick, A. Z. Eisen, K. Wolff, I. M. Freedberg, and K. F. Austen, Eds. McGraw-Hill, St. Louis, 1501 (1987).

7

Principles of Skin Permeability Relevant to Chemical Exposure

RICHARD H. GUY
Departments of Pharmacy and Pharmaceutical Chemistry
University of California, San Francisco
San Francisco, California

AND

JONATHAN HADGRAFT
The Welsh School of Pharmacy
University of Wales, College of Cardiff
Cardiff, Wales

I. INTRODUCTION

There is no question that dermal exposure to toxic substances represents a major occupational hazard and that successful anticipation of potential risk could significantly reduce the incidence of this chronic health and environmental problem. Recent interest in the transdermal delivery of drugs to elicit systemic pharmacological effects has stimulated research into the mechanism(s) of percutaneous absorption and a detailed understanding of the barrier function of the skin. On the basis of this emerging information, it is now feasible to predict, with a reasonable degree of reliability, the systemic exposure of the body to a chemical following dermal contact. It should then prove possible to determine, on a rational basis, whether a toxicity problem is likely and, if so, what steps should be taken to minimize the risk. To understand dermal penetration and the factors which control this route of chemical entry into the body, it is first necessary to review the salient anatomical features of the skin that control the barrier to absorption. The objective of this chapter is to identify those biological and physicochemical parameters which determine the rate and extent of chemical penetration across human skin. Subsequently, we will discuss how these parameters interact with the physicochemical properties of the dermally contacting chemical to determine the kinetics and degree of penetration.

The skin is the largest organ of the body and covers a surface area of nearly 2 m^2 in the adult human. The basic structure of skin is a bilaminate membrane comprising the dermis and epidermis. Although the thicknesses of these layers differ from site to site on the body, the microscopic detail is remarkably constant. The major determinants of the barrier function of the skin are the stratum corneum and the viable tissue, the two regions which constitute the epidermis. The viable epidermis evolves from a basal endothelial cell layer. As the cells mature, they migrate toward the skin surface and undergo the process of differentiation. In so doing, a thin, completely keratinized cell layer, the stratum corneum, is formed as a 10 μm thick layer at the surface. The stratum corneum can be depicted as a "brick wall".[1] The keratin-filled corneocytes are the bricks and a complex mixture of apolar lipids[2] form the mortar and confer structural integrity. A principal function of the stratum corneum is to provide a barrier to the transepidermal loss of tissue water. By forming this resistance, it is consequential that the stratum corneum is also an excellent barrier to the inward movement of many dermally contacting materials.

The diffusion environment of the stratum corneum is primarily lipophilic (see below). The viable epidermis, on the other hand, is an essentially aqueous region. In addition, the epidermis is avascular, the microcirculation of the skin being confined to the dermis. For a topically applied chemical to reach the systemic circulation, therefore, requires that it transports through both lipophilic and aqueous regions. A schematic representation of the skin is shown in Figure 1.

In most cases, it is the lipoidal stratum corneum which provides the rate-limiting step in absorption.[3] However, for very lipophilic substances, slow partitioning from the stratum corneum into the less "attractive" viable tissue layer can assume overall control of the penetration process. If the latter situation prevails, then a reservoir of chemical can be established within the stratum corneum and can provide a slow release of the agent into the body over a prolonged period of time.[4]

II. ROUTES OF PENETRATION

Previous discussions concerning the routes of chemical penetration across the skin (and, in particular, the stratum corneum) have identified three possible pathways[5] (Figure 2): (1) transcellular; (2) intercellular; and (3) appendageal (primarily, follicular).

On the basis of a number of disparate observations over the last 10 to 15 years, it now appears that, for the majority of compounds, the intercellular route predominates. In man, transport via the hair follicles and sweat glands is unlikely on the basis of available surface area; in other words, except in isolated regions, humans are not very hairy nor does the skin contain a high number of sweat glands. The transcellular path, although maximizing the surface area parameter, requires that transport takes place through the densely packed corneocytes and that multiple partitioning steps between these cells and the intercellular lipids occur. It has been shown, for example, in experiments localizing the position of butanol during its passage across the stratum corneum, that the chemical is concentrated in the intercellular domains and is excluded from the interior of the corneocytes.[6] Earlier studies,[7] which investigated the passage of nicotinic acid esters across the skin, demonstrated that a transcellular route was physicochemically implausible and that the intercellular path was pre-

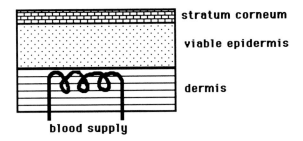

Figure 1.   A schematic representation of the structure of the skin.

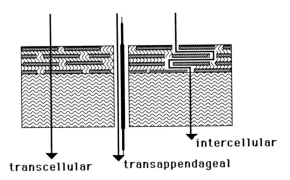

Figure 2.   Potential pathways of drug penetration through the skin.

ferred. More recently, the link between the physical state of the intercellular lipid and the status of the barrier function has been established.[8] There appears to be no question that fluidization of the lipid "mortar" is directly correlated with facilitated stratum corneum transport. It is appropriate, therefore, to view the stratum corneum as an essentially lipid membrane. One should also point out that the lipids are organized into broad lamellar sheets and that these structures can be visualized by careful microscopy.[9] A feature of the lipid composition comprising the intercellular region is the high fraction of ceramides and the virtual absence of phospholipids. It is possible that some of the ceramides act as "rivets" to hold together adjacent lamellae.[10]

Identification of the intercellular lipid domain as the transport path has a significant ramification from the standpoint of dermal exposure to toxic materials. Another key study in the determination of barrier properties involved solvent extraction of the stratum corneum.[11] When the skin is treated with volatile solvents, the barrier to chemical transport is reduced, presumably because of lipid extraction. Replacement of the lipid restores barrier function. This observation is highly relevant: many occupational exposures to toxic chemicals are mediated via the solvents in which the chemicals are dissolved. Hence, not only are these materials contacting the skin, they

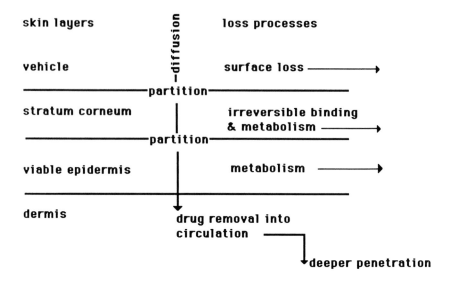

Figure 3.   The sequential physicochemical steps involved in drug transfer in percutaneous absorption.

are being "delivered" to the surface in a vehicle which itself can compromise the barrier.

III. PHYSICOCHEMICAL DETERMINANTS OF SKIN PENETRATION

The sequential steps involved in percutaneous absorption are schematically illustrated in Figure 3.

The initial process is the partitioning of the chemical into the lipid environment of the stratum corneum. The ease of this event will be determined by both the inherent attraction of the chemical for the lipids and the nature of the "vehicle" in which the chemical is delivered to the skin surface. If the chemical is in a solid state, e.g., a powder, then the particle size and polymorphism of the material may contribute to the kinetics of this first step in absorption. The material must be in solution before it can partition into the stratum corneum and hence the dissolution rate can contribute to the overall absorption.[12] If the agent is liquid or is in solution, the partition coefficient between the applied phase and the outer layers of the stratum corneum is important. A number of different partition coefficients are of importance in skin penetration; in order to differentiate them a subscript system is often invoked. Thus, the partition coefficient of an agent between the stratum corneum (s) and the applied vehicle (v) is $K_{s/v}$. Recent findings indicate that the solubility parameter of the chemical may be a factor that should be considered in predicting $K_{s/v}$.[13]

Having partitioned into the stratum corneum, the chemical must now diffuse through the intercellular lipids. The diffusional barrier of the stratum corneum is high and has been characterized by diffusion coefficients as low as 10^{-13} cm^2/s. However, these values are frequently based on a path-length of transport that ignores the tortuosity of the intercellular route. As a result, most diffusion coefficients quoted for stratum corneum transport are underestimated. Further, derivation of diffusion coefficients has usually involved *in vitro* measurement of a diffusional lag time.[14] Unfortunately, experimental determination of this parameter is subject to considerable variability. The use of lag times to calculate diffusion coefficients, therefore, is, at best, approximate and, more typically, unreliable. There is general agreement that the diffusional resistance of the stratum corneum increases (i.e., the diffusion coefficient decreases) as the molecular size (or weight) of the penetrant increases. The dependence on molecular weight (MW), however, and whether there is a "cut-off" in MW beyond which skin transport does not occur, has not been resolved. It has been suggested that diffusion through the stratum corneum is analogous to transport through a polymer network and that the diffusion coefficient depends exponentially on molecular volume.[15] In the light of the previous statements concerning the nature of the diffusional route, transport may be modeled better by consideration of a lipid array rather than a polymeric matrix. Therefore, the dependence of diffusion coefficient on molecular size would be less severe than that given by an exponential relation. In line with values obtained for diffusion in liquids, it is anticipated that the diffusion coefficient will vary as a function of $(MW)^{-b}$, where $b = 0.3$ to 0.6.[16] In predictions of the transdermal delivery of drugs, a cube root dependency has been found to be a satisfactory approximation.[17]

The next step in percutaneous penetration is the partitioning of the chemical from the lipid environment of the stratum corneum into the much more aqueous, viable epidermis. Chemicals which are extremely lipophilic will be severely rate limited by this process due to their low solubility in the viable tissue and the resultant slow interfacial transfer kinetics at the lipid-aqueous boundary. The degree of penetrant lipophilicity which leads to this change in transport-controlling step is not precisely defined. However, on the basis of simulations of the skin transport process[18] and from analyses of experimentally determined absorption data,[19] chemicals with a log (octanol/water) partition coefficient (log P) >3.0 may be expected to be at least partially rate controlled by the stratum corneum to viable epidermis transfer step. The sensitivity of this component of percutaneous penetration to the oil/water partition coefficient of the chemical is illustrated in the discussion of kinetic modeling below. As stated previously, an important ramification of this step being kinetically limiting is that a substantial reservoir of chemical can be established in the stratum corneum. For more water-soluble substances, of course, interfacial transport across the stratum corneum-viable tissue interface is a facile and rapid process that does not influence the overall absorption rate.

Having reached the viable epidermis, the penetrant is relatively free to diffuse deeper into the skin toward the cutaneous microcirculation in the upper dermis. The diffusion environment resembles an aqueous protein gel[20] and is characterized by diffusion coefficients of the order of 10^{-6} cm^2/s. As this value is much greater than those representative of the stratum corneum, it follows that this diffusion step is unlikely to determine the absorption rate unless the outer layer of the skin is damaged.

During the transfer processes the agent may also be subject to metabolic degradation. Many enzyme systems have been identified in the skin but their exact location and extent have not been fully documented. Until a greater understanding of skin metabolism is appreciated, it is difficult to quantify its overall significance; however, it is a subject that is being currently addressed and one that should not be ignored in estimating the total amount of an agent that penetrates the skin. It is also possible that an innocuous compound may, during skin transfer, become metabolized to a substance which possesses toxic characteristics.

Finally, the penetrant will encounter a blood vessel and gain entry to the systemic circulation. This will normally be a very efficient process and is one that has been characterized by a first-order rate constant[21] of approximately 10^{-3} s^{-1}. It is possible that certain chemicals may induce significant changes in cutaneous blood flow. Those which cause vasodilatation are unlikely to enhance the overall flux of penetrant in the body because of the general efficiency of the process in the unperturbed state. Vasoconstrictors, on the other hand, may impede their own clearance from the dermis and retard the rate of appearance of the chemical in the body. Deeper penetration into subcutaneous tissues (fat, muscle) may occur and lead to the formation of a long-lived depot.[22] The precise mechanism by which this occurs is not fully understood. However, simple calculations of transfer rates suggest that passive diffusion alone cannot be responsible for localization of topically applied agents in the deeper tissues of the skin.

At this point, it is possible to identify certain physicochemical parameters which can give an indication of a molecule's ability to penetrate skin. From the discussion above, it is clear that the relative lipophilicity of the penetrant is a key determinant. Ionized materials are poor penetrants unless they are able to form ion pairs. It can also be shown that skin penetration is inversely related to permeant melting point (MP). For example, in Figure 4, the steady state flux of an unrelated group of compounds across excised human skin *in vitro* is plotted as a function of penetrant MP;[23] a linear relationship is seen. Not unexpectedly, for this same series of compounds, flux increases proportionately with oil/water partition coefficient. However, as indicated above, penetration does not continue to rise with ever-increasing partition coefficient. At some point, the lipophilicity becomes high enough that transfer out of the stratum corneum is rate limiting. Hence, flux as a function of partition coefficient will plateau or, if the range of penetrants is sufficiently large, show a parabolic form[24] (see, for example, Figure 5). The generality of this observation, though, is not, at this time, fully established. In conclusion, the following basic rules can be deduced.

1.   Chemicals with a log P between 1 and 2 will be well absorbed.
2.   Poorly soluble substances are not well absorbed.
3.   The lower the MP of the agent, the better its absorption.
4.   Molecules which can hydrogen bond diffuse more slowly through the skin.

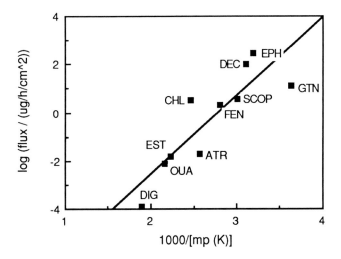

Figure 4. The relationship between the steady-state flux for a number of compounds across excised full thickness skin and their melting points. The compounds are ATR, atropine; CHL, chlorpheniramine; DEC, diethylcarbamazine; DIG, digitoxin; EPH, ephedrine; EST, estradiol; FEN, fentanyl; GTN, glyceryl trinitrate; OUA, ouabain; SCOP, scopolamine.

IV. STRUCTURE-ACTIVITY RELATIONSHIPS IN PERCUTANEOUS PENETRATION

There have been, on the whole, relatively few quantitative attempts to relate percutaneous absorption to the structure and physicochemical properties of the permeant. To do so requires that systematic evaluations of skin penetration be performed on several sets of homologous or analogous chemicals. These experiments are time consuming and do not necessarily address compounds of immediate significance to a particular therapeutic or toxicologic situation. Nevertheless, there is no question that, ultimately, these studies will form the cornerstone on which valid predictions of dermal exposure can be made. It is important, therefore, that this type of approach be pursued and encouraged. It is equally important that the experiments be conducted in as meaningful a fashion as possible; i.e., if a model system (animal, *in vitro*, etc.) is to be used, then some attempt to relate the data obtained to results in humans must be made. In this way not only may a useful structure-activity relationship be derived, but one may also gain insight as to the degree of extrapolation necessary to convert "model system" information into a risk assessment for man.

With respect to structure-transport studies in skin absorption, the following specific chemical classes have been investigated: n-alkanols,[25,26] phenols,[27,28] phenylboronic acids,[29] steroids,[30,31] nicotinic acid esters,[28,32] alkanoic acids,[33,34] polynuclear aromatics,[35] and nonsteroidal anti-inflammatory drugs.[36]

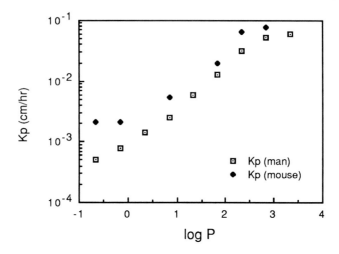

Figure 5.   The relationship between permeability and octanol-water partition coefficient or a series of n-alkanols. The similarity between human and hairless mouse data is demonstrated.

n-Alkanols — The steady-state permeability from aqueous solution, of the homologous series of n-alcohols through excised human epidermis *in vitro* has been measured.[25] The results, plotted as permeability coefficient (Kp) against log P, are shown in Figure 5. Also shown on this graph are the corresponding data for the same molecules permeating full-thickness hairless mouse skin.[26] The results for human skin are described by the equation[36]

$$\log K_p = 0.54 \log P - 2.88 \tag{1}$$

with a correlation coefficient of 0.98. A slightly improved correlation is obtained if the stratum corneum-water partition coefficient is used.[36] The hairless mouse data fit the equation

$$\log K_p = 0.50 \log P - 2.52 \tag{2}$$

with, again, r = 0.98. It follows that, in this case, a result from the animal model would be quite predictive of human skin absorption. Further discussion of animal to man extrapolation is presented below. One may also conclude from these results that, within this series of compounds, a log-linear relationship dependent upon partition coefficient alone is perfectly adequate to assess Kp. As will become apparent, however, it is not necessarily true that the equations above can be used to predict the Kp of a chemical which is not an alcohol. A final point is that the results for transport across human epidermis hint (and are frequently used to demonstrate) that a maximum Kp value has been reached. That is, with nonanol (the most lipophilic alcohol

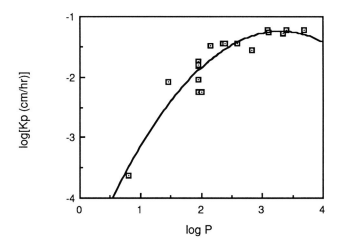

Figure 6.   The relationship between the permeability of human skin and the octanol-water partition coefficient for a series of phenols.

considered), the control of permeation is influenced by slow transfer out of the stratum corneum into the viable tissue.

Phenols — The permeability of a wide range of phenol derivatives across human epidermis was measured by Roberts et al.[27] The results, plotted as a function of log P, are reproduced in Figure 6. In this case, significant nonlinearity in the data is apparent and a parabolic dependence of log Kp on log P may be determined

$$\log Kp = -0.36(\log P)^2 + 2.39 \log P - 5.2 \tag{3}$$

The correlation coefficient is 0.94.

In an attempt to develop simple models for measuring percutaneous penetration, the permeability characteristics of organic liquid membranes in a rotating diffusion cell have been considered.[28] The lipid phases employed were isopropyl myristate (IPM) and tetradecane (TD) and were chosen to simulate the apolar nature of the intercellular domains of the stratum corneum. Again, a diverse range of phenols was considered (although there was not complete coincidence with the Roberts' compounds[27]) and this allowed an extensive span of lipophilicity to be evaluated. The permeabilities of the phenols through the two lipid models are presented as a function of log P in Figure 7. Quadratic fits to the results give the following equations

$$\log Kp \ [IPM] = -0.48(\log P)^2 + 2.32 \log P - 2.2 \qquad [r = 0.96] \tag{4}$$

$$\log Kp \ [TD] = -0.40(\log P)^2 + 2.55 \log P - 4.0 \qquad [r = 0.96] \tag{5}$$

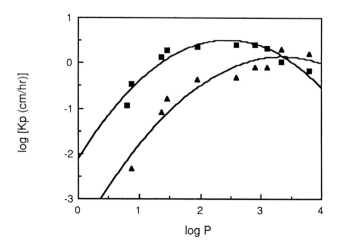

Figure 7.   The relationship between the resistance to transport across model skin lipid membranes and the octanol-water partition coefficient of a series of phenols. The two model membranes are: ▲ tetradecane; ■ isopropyl myristate.

As can be seen from Figures 6 and 7, the organic liquid models are more permeable than human epidermis. Nevertheless, the magnitude of the difference is quite constant and predictable (see section below on *Miscellaneous Compounds*). The functional dependence of log Kp on log P is very similar for the "real" and "model" systems (as can be deduced by the similarity of the coefficients pre-multiplying the log P and (log P)^2 terms in Equations 3, 4, and 5). Once again, a reasonable prediction of the Kp of a phenol through human skin may be obtained from an extrapolation based on log P or from a model *in vitro* experiment.

It is of interest, at this point, to ask the question: "can phenol permeability be predicted from the alkanol structure-transport relationship?" In Figure 8, the phenol permeability coefficients are compared to the dependency predicted by the alcohol results (Equation 1). It can be seen that the alkanol data, in general, overestimate phenol penetration and that the discrepancy is most marked in the range log P <2.0. It follows that the answer to the above question depends on the nature of the penetrant and that this situation is clearly not optimal.

Phenylboronic acids — The percutaneous absorption of meta- and para-substituted derivatives showed linear relationships with log P and with the log (benzene/water) partition coefficient.[29] However, because the measurement of skin penetration differed from a simple Kp determination, it is not possible to compare the structure-activity equations for this set of compounds with those discussed above.

Steroids — The steady-state permeability of 14 steroids was measured across human epidermis *in vitro*.[30] The Kp values are plotted as a function of four organic/aqueous partition coefficients in Figure 9. Linear regression parameters on the results

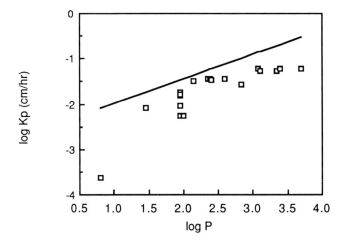

Figure 8.   The variation of the permeability of human skin with log P for a series of phenols. The linear relationship shows the predicted values based on the alcohol data described by Equation 1.

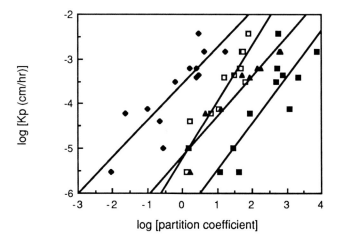

Figure 9.   The relationship between the permeability of human skin and various solvent-water partition coefficients for a series of steroids. The solvents are: ♦, hexadecane; ❑, amyl caproate; ▲, ether; ■, octanol. The linear relationships shown are given in Table 1.

Table 1
Relationships between Observed Steroid Permeabilities and their Partition Coefficients between
Various Solvents and Water

Solvent	α	β	R
Octanol	1.04	6.51	0.85
Hexadecane	0.81	3.55	0.86
Amyl caproate	1.26	5.21	0.93
Ether	0.89	5.17	0.96

Note: Linear regression analysis was applied to the data presented in Figure 9 using the equation log Kp
= αlog K(solvent) –β. R is the correlation coefficient.

are presented in Table 1. The slopes of the lines are quite consistent and the values
of the intercepts shift in the direction expected as the solvent dielectric constant
increases. Scheuplein et al.[30] also measured stratum corneum-water partition coefficients (K_{sc}) in this study. Linear regression of the permeability data with these values
gives[36]

$$\log Kp = 2.63 \log K_{sc} - 7.54 \tag{6}$$

with r = 0.93. Clearly, there is discrepancy between this relationship and those given
in Table 1 for the simple organic solvents. Although the exact reason for this
difference has not been identified, two possibilities may be suggested. First, the
technique used to measure K_{sc} requires that the tissue be removed from the remainder
of the epidermis; it is not known whether the isolation procedure alters, in any way,
the nature of the stratum corneum. It is also unclear how to separate partitioning from
binding. Secondly, as the intercellular lipids provide the permeation pathway, it is
appropriate that a volume correction be applied when calculating K_{sc}.

In a later study the steady-state flux of hydrocortisone and of a number of its 21-
esters (acetate through heptanoate) was measured across hairless mouse skin *in
vitro*.[31] The Kp values determined are plotted, as a function of the ether/water
partition coefficient ($K_{e/w}$), for each chemical in Figure 10. The linear relationship (r
= 0.95) is described by:

$$\log Kp = 0.56 \log K_{e/w} - 3.39 \tag{7}$$

In this case it would not be useful for extrapolate from the mouse data to man (see
Table 1), a conclusion in contrast to the situation for alkanols and, to a certain extent,
for phenols. This observation reflects the fact that an animal model, which is
applicable to various classes of penetrant, has not yet been identified for human skin
permeation prediction.

Nicotinic acid esters — These compounds are potent vasodilators when applied
topically to the skin. Their percutaneous penetration in man has been quantified,

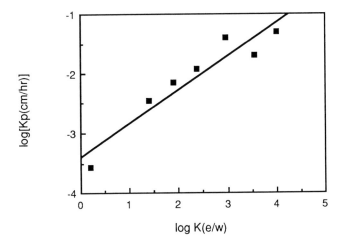

Figure 10.   The relationship between the permeability of hairless mouse skin and the ether-water partition coefficient for a series of 21-esters of hydrocortisone.

therefore, by the time to onset of erythema (skin reddening) following application. In an initial set of experiments using several nicotinates,[32] the threshold concentrations (C) necessary to induce visible erythema were determined. The reciprocals of these concentrations are plotted as a function of nicotinate ether/water partition coefficient ($K_{e/w}$) in Figure 11. A classic parabolic relationship is observed. The form of this dependency can be explained on the basis of a change in the rate-controlling step of penetration as discussed above (i.e., a switch from stratum corneum diffusion to slow partitioning at the stratum corneum-viable tissue interface). This structure-activity relationship must reflect transport rather than pharmacological effect since intradermal injection of the different nicotinates has shown that they are equipotent.[32]

These results may be compared to the permeabilities (Kp) of a similar range of nicotinates across a tetradecane model membrane.[28] In Figure 11 the two sets of data are juxtaposed and can be seen to be remarkably similar. This further supports the hypothesis proposed above, in which physical chemistry rather than pharmacology controls the structure-activity behavior observed. Quadratic fits to the data give the following:

$$\log[1/C(mM)] = -0.17 + 0.60 \log K_{e/w} - 0.30(\log K_{e/w})^2 \qquad (8)$$

$$\log[Kp(cm/h)] = -0.11 + 0.69 \log K_t - 0.27(\log K_t)^2 \qquad (9)$$

(where K_t is the tetradecane/water partition coefficient) with correlation coefficients of 0.90 and 0.99, respectively. Good coincidence in the coefficients is apparent. This agreement is particularly notable (and potentially valuable) because of the different

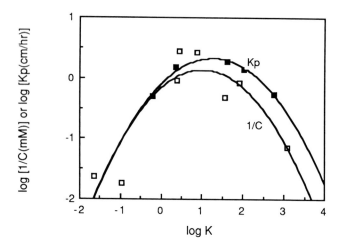

Figure 11. A parabolic relationship between the reciprocal threshold concentration (c) required to induce erythema and the ether-water partition coefficient for a series of nicotinates (▫). Also shown is a similar relationship for the permeabilities across a tetradecane model membrane (■), and the tetradecane-water partition coefficient. The ether-water partition coefficient for octyl nicotinate has been estimated by extrapolation of the values for the lower homologues. The value quoted in Reference 32 (log K = 1.49) seems implausible.

nature of the experimental procedures employed. In other words, the results suggest that a measurement of chemical permeation through a model membrane system may be predictive of an *in vivo*, nonsteady state, short-term exposure situation.

Alkanoic acids — The steady-state permeability coefficients of a number of n-alkanoic acids have been measured across excised porcine skin.[33,34] In one set of experiments, the acids were delivered in their pure state and the results are expressed, together with the corresponding melting points, in Figure 12 for the n = 4 to 8 homologues. As previously noted, there is an inverse relationship between Kp and MP.

Polynuclear aromatics — Recently, Roy et al.[35] have determined the *in vitro* percutaneous penetration of several compounds chosen to represent those typically found in refinery streams. The experimental procedure involved the use of excised rat skin (dermatomed to a thickness of 350 μm) and of a receptor phase, which contained a small (6%) concentration of nonionic surfactant to ensure adequate solubility of the lipophilic penetrants. The data (expressed as a percent of applied dose absorbed within 96 h of exposure) were analyzed by multivariate regression. The equation derived, which best characterized the data set, contained four independent variables: molecular surface area; molar refractivity; molecular moment of inertia; and molecular symmetry. The coincidence between observed and predicted

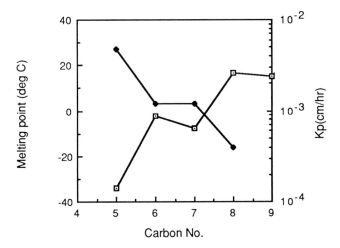

Figure 12.   The relationship between permeability (♦) of porcine skin and carbon number for a series of alkanoic acids. The inverse relationship between melting point (▫) and permeability is also demonstrated.

absorption is shown in Figure 13. With the exception of two outliers, there is good agreement between experiment and theory. The authors concluded from their interpretation of the results that skin permeation was not a particularly selective process and that the molecular descriptors identified would impact primarily on dissolution and diffusion phenomena. However, an alternative presentation of the results also implicates (as one might expect from the preceding discussion) partitioning as a key factor in the determination of percutaneous penetration. In Figure 14 skin permeation, again expressed as a percent of applied dose absorbed, is plotted as a function of log P. The data for the 3-ring polynuclear aromatics (PNA) are plotted separately from those for the 4- and 5-ring compounds. It can be seen that the two groups lie on separate lines and that, in combination, the effect is to produce an apparent parabolic structure-activity relationship. In constructing Figure 14, the data point for carbazole (a 3-ring PNA), for which absorption was 90% and log P = 3.51, has been ignored. Once again, there would appear to be evidence for a change in the rate-determining step of skin penetration as the lipophilicity of the permeant becomes large. For the PNAs, however, the maximum absorption occurs at a higher log P than that seen for the other molecular groups considered above. The reason for this difference is not understood. It is possible that the methodology employed may be a contributory factor. For example, rat skin is much more permeable than human skin and is less discriminatory, as a result, to permeant properties. The cumulative dose absorbed at 96 h is also an imprecise parameter to characterize penetration because it contains essentially no kinetic information. It is difficult, therefore, to extrapolate these results to man. Nevertheless, the pattern of behavior observed is consistent with

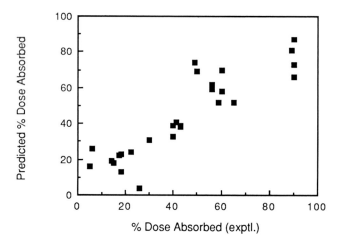

Figure 13. The percentage dose absorbed through rat skin within 96 h has been predicted for a series of polynuclear aromatics. The prediction includes four independent variables, molecular surface area, molar refractivity, molecular moment of inertia, and molecular symmetry. Good correlation with experimental data is shown.

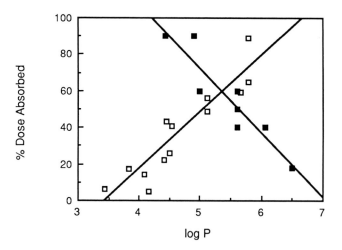

Figure 14. The relationship between percentage dose absorbed through rat skin within 96 h and log P for a series of polynuclear aromatic compounds. The substances have been separated into 3-ring compounds (□) and 4 and 5-ring aromatics (■).

that demonstrated by other chemical classes discussed in this chapter and there would appear to be no reason to predict anomalous structure-skin absorption relationships for the PNAs.

Nonsteroidal anti-inflammatory drugs (NSAIDs) — Yano et al.[36] measured the skin permeability of 18 NSAIDs *in vivo* in humans. The chemicals tested included a set of eight salicylate derivatives. Percutaneous absorption was quantified as the percent of the applied dose which was not recoverable by surface washing following a 4-h contact period. The chemicals (0.5 mg) were administered in 10 ml of either acetone or methanol. The results are plotted in Figure 15 as a function of the chemicals' log P values. The set of salicylates is highlighted in the figure but is seen to coincide completely with the general pattern of the (once again) parabolic structure-activity relation

$$\log[\% \text{ dose abs.}] = -0.42 + 1.14 \log P - 0.23(\log P)^2 \qquad (10)$$

with r = 0.96. The coefficients describing this set of experiments may be compared favorably with those characterizing the nicotinic acid ester data described above (Equations 8 and 9). Although the coincidence may reflect the similarity in structure between the two sets of compounds, the agreement is, nevertheless, remarkable given the disparate methodology employed in the three investigations.

Miscellaneous — The studies considered so far have focused upon chemical permeants of similar structure. This may not be typical of the occupational exposure situation in which contact with diverse materials may occur. An important question is whether absorption as a function of physicochemical properties can be predicted for a wide range of compounds. Quantitative and extensive information, which can be used to address this issue, is lacking. However, there have been two recent efforts[37,38] to measure percutaneous absorption of unrelated chemicals and to correlate the behavior seen with basic chemical properties.

In these experiments the steady-state fluxes of barbitone, phenobarbitone, butobarbitone, amylobarbitone, hydrocortisone, nicotine, salicylic acid, and isoquinoline were measured across excised human skin *in vitro* and across four lipid-impregnated model membranes in the rotating diffusion cell. It was found that IPM and TD again provided model membranes, the transport properties of which followed classical behavior: Kp values increasing with the corresponding lipid/water partition coefficient (see Figure 16). For IPM and TD the log Kp vs. log K data lie on a common line.

$$\log \text{Kp} = -0.71 \log \text{K} - 0.03 \qquad [r = 0.97] \qquad (11)$$

The span of log K values is smaller than those considered earlier and this explains why a parabolic dependence is not seen with these data. Also shown in Figure 16 are the Kp results for human skin transport plotted against the log K values for TD/water.

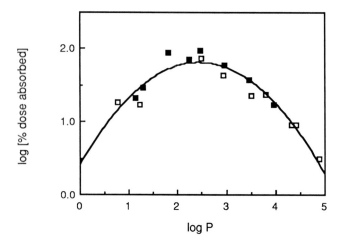

Figure 15.   The percutaneous absorption of a series of nonsteroidal anti-inflammatory agents, including salicylates (■), plotted as a function of log P.

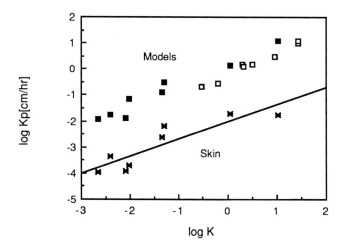

Figure 16.   The relationship between permeability and partition coefficient for a series of miscellaneous compounds. The permeability was measured across excised human skin (*), and membranes impregnated with isopropyl myristate (□) and tetradecane (■).

The resulting correlation

$$\log \text{Kp} = 0.66 \log \text{K} - 2.02 \qquad [r = 0.89] \qquad (12)$$

is strikingly parallel to the model membrane results and again implies the usefulness of, in particular, TD as a lipoidal representation of stratum corneum. As observed with the phenols and, as expected, the skin is more resistant to permeation than the simple organic liquid membrane. There would appear to be between a 10- and 100-fold difference in permeability coefficient. It is conceivable that this discrepancy results from the fact that stratum corneum transport involves permeation via the intercellular (lipid-filled) channels and that only a fraction of the skin surface can act, therefore, as "productive" area. The lipids in stratum corneum are known to occupy 5 to 15% of the total membrane volume;[39] given that, for the TD system, essentially all of the membrane is "available" for solute transport, then the magnitude of the difference between skin and model permeabilities is plausible.

V. CALCULATIONS OF BODY EXPOSURE FROM SKIN ABSORPTION DATA

There have been a number of investigations designed to evaluate percutaneous absorption in man.[40] The majority of these studies have involved topical application of the chemical in question to a small area on the ventral forearm. While these data are of considerable value, it is useful to know how (or whether) one can extrapolate the findings to a "real" exposure situation in an occupational or environmental setting. To do so requires two key pieces of information: first, and relatively easily, the area of contact between the chemical and skin must be known. If one assumes that the amount of chemical absorbed into the body is a linear function of contact area (an assumption for which no contrary evidence exists), then a simple correction can be made. Second, and more difficult, the relative permeability of the forearm compared to other anatomic sites must be understood. The latter information is available for only three cases, namely for hydrocortisone[41] and for the pesticides, malathion, and parathion.[42] The results from these studies indicate that the skin of the genitalia is, on average, 25 times more permeable than that of the forearm; that of the trunk is 2.5 times more permeable, while that of the face and head is approximately 5-fold more permeable. Absorption across leg skin is similar to that on the forearm.

Given these data, it is possible to calculate relative body exposures on the basis on forearm data alone or on the basis of differential penetration at specific skin sites.[43] For example, it can be shown that total body exposure based on forearm absorption data will underestimate actual exposure by a factor of 2. One can also demonstrate that an exposure limited to the hands and face could result in three- to fourfold higher absorption compared to that estimated from the single value of forearm permeability. While these calculations are usefully illustrative, it must be emphasized that they are based on a small amount of information from only three compounds. Therefore, before one can use the approach predictively for compounds, which are not closely similar to hydrocortisone, malathion, and parathion, it will be necessary to assess skin penetration, as a function of anatomic position, for a range of chemicals of diverse physicochemical characteristics.

VI. MATHEMATICAL MODELING OF SKIN PENETRATION

Percutaneous absorption is complex and involves many variables. To model this process mathematically, therefore, is a substantial task. Inevitably, one runs the risk either of formulating a model, which is simplistic and incapable of subtle prediction, or of producing a simulation too complex for testing in any reasonable experimental system. Hence, there is a need to tread a delicate path between these extremes if progress is to be made. An ideal model will be sufficiently flexible to incorporate the important variables of skin penetration and will be both sensitive to the biology and condition of the skin, and to the physicochemical characteristics of the permeant. The schematic shown in Figure 3 includes the key steps of percutaneous absorption and a number of additional processes which may impact upon the overall kinetics and extent of penetration. Clearly, a mathematical model which completely described the scheme as shown would be extremely complicated and could only be handled by an efficient computer. To explore the possible utility of a model, therefore, requires that the problem be broken down into more manageable pieces so that the relative significance of the constituent parts can be rationally assessed.

In the simplest case, the problem is treated using Fick's laws of diffusion and an estimate of total body burden following a dermal exposure is found. When skin is contacted with a chemical, the amount penetrating into the body as a function of time is shown in Figure 17. Two situations are illustrated: in the infinite dose case, a constant driving force is maintained on the skin surface and, following a lag-time (t_L), the amount of chemical crossing the skin increases linearly with time (the so-called steady-state situation). In the finite dose case, the skin is exposed to a limited amount of chemical. Initially, the amount reaching the body increases with time but then slows down as the chemical on the surface is depleted. There may be a region of the finite dose case which is coincident with the steady-state behavior following infinite dosing. However, it is not easy to determine where the two situations may overlap. Most *in vitro* evaluations of skin penetration have used an experimental design in which the infinite dose situation applies (see above). At steady state, the flux (J) of chemical across the skin is given by Fick's first law of diffusion

$$J = A*Kp*\Delta C \tag{13}$$

where A is the area of exposed skin, Kp is the permeability coefficient of the chemical across the skin and ΔC is the difference $(C_d - C_r)$ in chemical concentration between the donor solution on the skin surface and the receptor solution beneath the skin. In designing an infinite dose experiment, two criteria should be met: (1) C_d

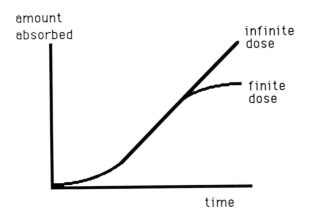

Figure 17.   A typical profile showing the amount of substance penetrating the skin as a function of time for infinite and finite dose application.

should be sufficiently large that significant depletion of chemical does not occur; and (2) C_r should not rise above 1/10 of the saturation solubility. Under these conditions

$$Kp = J/A*C_d \qquad (14)$$

With a knowledge of Kp, an estimation of body burden following a dermal exposure can be made. If an area A is contacted by chemical at concentration C for a time t, then the amount of chemical (M) which will enter the body is given by

$$M = A*Kp.*C*t \qquad (15)$$

Depending upon the anatomic site of contact, Equation 15 may need to be modified to take into account the site dependency of penetration as discussed above.

Some caveats to this approach should be noted. The infinite dose technique often produces complete hydration of the skin tissue. It is well documented that hydration of skin lowers the barrier to chemical penetration. Second, most occupational or environmental exposures do not involve an infinite dose of chemical. Typically, contact times will be relatively short and, following exposure, the chemical will be washed from the skin surface. It follows that a prediction of body burden by Equation 15 will overestimate the true exposure level. This latter contention is appropriate, however, only when the skin is intact. If the barrier is broken, then much freer passage is possible,[44] particularly for polar, water-soluble chemicals. If the skin is diseased, then there is also the potential for altered barrier function. The extent to which different skin diseases impact upon absorption, though, is not well understood and cannot be rationally modeled at this time. The straightforward approach detailed above does not address the complex sequelae of diffusion and partitioning illustrated

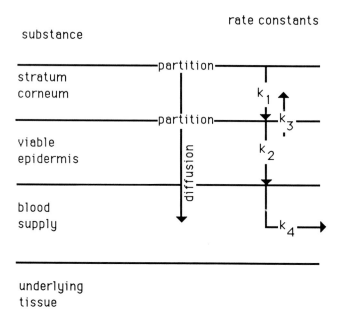

Figure 18.   A kinetic description of dermal absorption using first-order rate constants.

in Figure 3. To do so requires that nonsteady-state solutions to Fick's 2nd law of diffusion, with appropriate boundary conditions, be obtained. This sophistication has been reported and relationships applicable to short contact exposures have been derived.[45-47] The amount of chemical penetrating the skin can be shown to depend upon basic physicochemical parameters (diffusion coefficients, partition coefficients, transport path lengths, etc.) and the effect of changes in these parameters on the penetration kinetics can be explored. However, experimental determination of these key descriptors is difficult and simple testing of the model predictions requires significant approximations and assumptions. Furthermore, this more rigorous approach does not include other important processes. For example, a volatile chemical will be subject to concomitant evaporation and absorption, reducing thereby the total bioavailability. The skin is a metabolically active organ and activation, or detoxification, of absorbing molecules is possible.[48] Little is known of the magnitudes of the possible effects due to these "loss" processes. Again, mathematical modeling has been carried out[49,50] but validation of the conclusions deduced awaits appropriate experimentation.

Another approach, which is also based on the physicochemical properties of the penetrant, can be adopted to model skin absorption. In this case the diffusion and partitioning are approximated by a series of first-order rate constants[51,52] as depicted in Figure 18: k_1 describes diffusion through the stratum corneum. For intact skin this

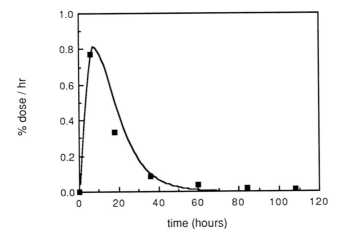

Figure 19.   The theoretical and experimental (■) urinary excretion data for diethyl-m-toluamide. The theoretical profile has been generated using the kinetic model and the physicochemical properties of the permeant.

rate constant will be relatively small and will depend upon the molecular size of the penetrant. This dependency has been suggested to follow a simple cube-root function,[51] although a more severe exponential function has been recently proposed:[15] k_2 describes chemical permeation through the viable epidermis. The rate constant is again size dependent but has a magnitude characteristic of diffusion through an aqueous protein gel.[20] The value of k_2 is therefore greater than that of k_1. k_3 measures the relative affinity of the chemical for the stratum corneum compared to the more aqueous in nature viable epidermis. The ratio k_3/k_2 may be regarded, therefore, as an effective "stratum corneum-viable epidermis" partition coefficient of the chemical. The larger the value of k_3, the higher the affinity of the chemical for the stratum corneum and the slower the transfer of molecules from stratum corneum to viable tissue. It has been shown that k_3 may be determined empirically from the octanol/water partition coefficient (determined at pH 7.4) of the chemical:[51] $k_3/k_2 = P/5$. k_4 characterizes the elimination half-life ($t_{1/2}$) of the chemical:

$$k_4 = \ln 2/t_{1/2}$$

The model can be used to predict the percutaneous absorption of a diverse range of chemicals on the basis of their molecular weights and octanol/water partition coefficients. Reasonably good agreement between the model and *in vivo* human skin penetration data has been found.[51] The correlation is adequate for the prediction of both plasma concentration[53] and urinary excretion rate data.[51] As an example, the coincidence between experimental data and model prediction for the insect repellant diethyl-m-toluamide is shown in Figure 19.

Other advantages of the kinetic model are apparent. In addition to the fact that it is a nonsteady state simulation, it can be adapted and expanded to take into account surface evaporation of penetrant,[54] cutaneous metabolism,[55] and different chemical input functions (zero order, first order, etc.).[18] Furthermore, the approach offers, at least, the potential to extrapolate data acquired in animals to man. It is anticipated that the value of k_1 will be the most species-dependent parameter since it will be a function of stratum corneum thickness, the number of cell layers, and the diffusional path length. It is unlikely that k_2 is significantly different between species and the partitioning process between lipophilic stratum corneum and hydrophilic viable tissue (common features of all animal models and man) should remain quite constant. Hence, if a scale of relative k_1 values for different species could be developed, then the absorption process in man might be deduced from an animal experiment. Of course, the value of k_4 may also depend upon the animal used but this parameter can be found with relative ease by experiment. Unfortunately, the ability to extrapolate from one species to another does not presently exist because of the paucity of comparative data across animals for a wide range of compounds. Although there have been a few studies[56-58] in which the penetration of a number of chemicals across different animal skins has been measured, the results are insufficient to formulate simple rules for human absorption prediction from animal data. Qualitatively, however, it is possible to draw the following conclusions about the predictability of absorption data from various animal models.[40]

1. Small "hairy" animals (e.g., rats, rabbits) usually yield penetration values much higher than those seen in man.
2. The rhesus monkey has been shown to be a valuable *in vivo* model for human percutaneous absorption.
3. Newer models, such as the weanling (or mini- or micro-) pig and the human-skin-grafted rat, offer alternatives of considerable promise.

VII. CONCLUSIONS

At present, our understanding of the skin permeation process has reached a reasonable level. General rules governing the penetration of chemicals have been established and the impact of structural and/or physical changes is being intensively studied. However, while correct qualitative statements about percutaneous absorption can be made, there is much less certainty when *quantitative* conclusions are required. Obviously, in the area of risk assessment and dermal exposure prediction, this is a serious drawback. In this chapter, we have attempted to highlight those areas in which progress has been made and those in which considerable further work needs to be done. The discussion on quantitative structure-activity relationships indicates a fruitful path for additional study. The power of this type of approach has been demonstrated in other fields of biological transport and pharmacology — one is optimistic that skin penetration will also prove amenable to this analysis. On the other hand, comprehension of the role of skin metabolism, for example, and the optimal use of animal model experiments remain problems of substance. There is no question, in our opinion, that the skin is a complex membrane, the transport properties of which deserve greater attention and considerable respect. Until these properties are unraveled, prediction of dermal exposure to toxic chemicals will be, at best, qualitative and, in less favorable circumstances, wrong.

ACKNOWLEDGMENTS

We thank the U.S. Environmental Protection Agency (CR-812474 and CR-816785) and the National Institutes of Health (GM-33395 and HD-23010) for financial support. R. H. G. was the recipient of a Special Emphasis Research Career Award from the National Institute of Occupational Safety and Health (K01-OH00017).

REFERENCES

1. A.S. Michaels, S.K. Chandrasekaran, and J.E. Shaw, *A.I.Ch.E.J.* 21, 985 (1975).
2. P.M. Elias, *J. Invest. Dermatol.* 80, 44s (1983).
3. B.W. Barry, *Dermatological Formulations — Percutaneous Absorption,* Marcel Dekker, New York (1983).
4. J. Hadgraft, *Int. J. Pharm.* 4, 229 (1979).
5. R.J. Scheuplein, *J. Invest. Dermatol.* 45, 334 (1965).
6. M.K. Nemanic and P.M. Elias, *Cytochemistry* 28, 573 (1980).
7. W.J. Albery and J. Hadgraft, *J. Pharm. Pharmacol.* 31, 140 (1979).
8. G.M. Golden, J.E. McKie, and R.O. Potts, *J. Pharm. Sci.* 76, 25 (1987).
9. L. Landmann, *J. Invest. Dermatol.* 87, 202 (1986).
10. P.W. Wertz and D.T. Downing, *Science* 217, 1261 (1982).
11. R.K. Winkelmann, *Br. J. Dermatol.* 81 (Suppl. 4), 11 (1969).
12. K. Al-Khamis, S.S. Davis, and J. Hadgraft, *Int. J. Pharm.* 40, 111 (1987).
13. C.D. Vaughan, *J. Soc. Cosmet. Chem.* 36, 319 (1985).
14. B.J. Poulsen and G.L. Flynn, *Percutaneous Absorption,* R.L. Bronaugh and H.I. Maibach, Eds. Marcel Dekker, New York, 431 (1985).
15. G.B. Kasting, R.L. Smith, and E.R. Cooper, *Skin Pharmacokinetics.* B. Shroot and H. Schaefer, Eds. S. Karger, Basel, 138 (1987).
16. C.R. Wilke and P.C. Chang, *A.I.Ch.E.J.* 1, 264 (1955).
17. R.H. Guy and J. Hadgraft, *Transdermal Delivery of Drugs, Vol. 3.* A.F. Kydonieus and B. Berner, Eds. CRC Press, Boca Raton, FL, 3, (1987).
18. R.H. Guy and J. Hadgraft, *J. Control. Rel.* 1, 177 (1985).
19. R.H. Guy, J. Hadgraft, and H.I. Maibach, *Toxicol. Appl. Pharmacol.* 78, 123 (1985).
20. R.J. Scheuplein, *J. Invest. Dermatol.* 48, 79 (1967).
21. W.J. Albery, R.H. Guy, and J. Hadgraft, *Int. J. Pharm.* 15, 125 (1983).
22. R.H. Guy and H.I. Maibach, *J. Pharm. Sci.* 72, 1375 (1983).
23. G. Ridout and R.H. Guy, Unpublished data.
24. T. Yano, A. Nakagawa, M. Tsuji, and K. Noda, *Life Sci.* 39, 1043 (1986).
25. R.J. Scheuplein and I.H. Blank, *Physiol. Rev.* 51, 702 (1971).
26. C.R. Behl, G.L. Flynn, T. Kurihara, N. Harper, W. Smith, W.I., Higuchi, N.F.H. Ho, and C.L. Pierson, *J. Invest. Dermatol.* 75, 346 (1980).

27. M.S. Roberts, R.A. Anderson, and J. Swarbrick, *J. Pharm. Pharmacol.* 29, 677 (1977).
28. J. Houk and R.H. Guy, *Chem. Rev.,* 88, 455 (1988).
29. W.E. Clendenning and R.B. Stoughton, *J. Invest. Dermatol.* 39, 47 (1962).
30. R.J. Scheuplein, I.H. Blank, G.J. Brauner, and D.J. MacFarlane, *J. Invest. Dermatol.* 52, 63 (1969).
31. B. Idson and C.R. Behl, *Transdermal Delivery of Drugs, Vol. 3.* A.F. Kydonieus and B. Berner, Eds. CRC Press, Boca Raton, FL, 85 (1987).
32. R.B. Stoughton, W.E. Clendenning, and D. Kruse, *J. Invest. Dermatol.* 35, 337 (1960).
33. Z. Liron and S. Cohen, *J. Pharm. Sci.* 73, 534 (1984).
34. Z. Liron and S. Cohen, *J. Pharm. Sci.* 73, 538 (1984).
35. T.A. Roy, J.J. Chang, and M.H. Czerwinski, *In Vitro Toxicology — Approaches to Validation.* A.M. Goldberg, Ed. Mary Ann Liebert, New York, 471 (1987).
36. E.J. Lien and G.L. Tong, *J. Soc. Cosmet. Chem.* 24, 371 (1973).
37. J. Hadgraft and G. Ridout, *Int. J. Pharm.* 39, 149 (1987).
38. J. Hadgraft and G. Ridout, *Int. J. Pharm.,* 42, 97 (1988).
39. P.M. Elias, *Int. J. Dermatol.* 20, 1 (1981).
40. R.H. Guy, J. Hadgraft, R.S. Hinz, K.V. Roskos, and D.A.W. Bucks, *Transdermal Controlled Systemic Medications.* Y.W. Chien, Ed. Marcel Dekker, New York, 179 (1987).
41. R.J. Feldmann and H.I. Maibach, *J. Invest. Dermatol.* 48, 181 (1967).
42. H.I. Maibach, R.J. Feldmann, T.H. Milby, and W.F. Serat, *Arch. Environ. Health* 23, 208 (1971).
43. R.H. Guy and H.I. Maibach, *J. Appl. Toxicol.* 4, 26 (1984).
44. H. Schaefer, A. Zesch, and G. Stuttgen, *Skin Permeability.* Springer-Verlag, Berlin, 1982.
45. W.J. Albery and J. Hadgraft, *J. Pharm. Pharmacol.* 31, 129 (1979).
46. R.H. Guy and J. Hadgraft, *Int. J. Pharm.* 6, 321 (1980).
47. R.H. Guy and J. Hadgraft, *J. Pharmacokinet. Biopharm.* 11, 189 (1983).
48. R.H. Guy, J. Hadgraft, and D.A.W. Bucks, *Xenobiotica* 17, 325 (1987).
49. R.H. Guy and J. Hadgraft, *Int. J. Pharm.* 18, 139 (1984).
50. R.H. Guy and J. Hadgraft, *Percutaneous Absorption.* R.L. Bronaugh and H.I. Maibach, Eds. Marcel Dekker, New York, 57 (1985).
51. R.H. Guy, J. Hadgraft, and H.I. Maibach, *Dermal Exposure Related to Pesticide Use.* R.C. Honeycutt, G. Zweig, and N.N. Ragsdale, Eds. ACS Symp. Ser. 273, American Chemical Society, Washington, D.C., 19 (1985).
52. R.H. Guy and J. Hadgraft, *J. Pharm. Sci.* 73, 883 (1984).
53. R.H. Guy and J. Hadgraft, *J. Control. Rel.* 4, 237 (1987).
54. R.H. Guy and J. Hadgraft, *J. Soc. Cosmet. Chem.* 35, 103 (1984).
55. R.H. Guy and J. Hadgraft, *Int. J. Pharm.* 20, 43 (1984).
56. M.J. Bartek and J.A. La Budde, *Animal Models In Dermatology.* H.I. Maibach, Ed. Churchill Livingstone, New York, 103 (1975).
57. R.C. Wester and H.I. Maibach, *Percutaneous Absorption.* R.L. Bronaugh and H.I. Maibach, Eds. Marcel Dekker, New York, 251 (1985).
58. W.G. Reifenrath, E.M. Chellquist, E.A. Shipwash, and W.W. Jederberg, *Fund. Appl. Toxicol.* 4, S224 (1984).

8

In Vivo Assessment of Dermal Absorption

GEORGE J. KLAIN

AND

WILLIAM G. REIFENRATH
Letterman Army Institute of Research
Division of Cutaneous Hazards
Presidio of San Francisco
San Francisco, California

I. INTRODUCTION

Evidence accumulated in past years has clearly shown that large numbers of toxic substances in the environment have a capacity to penetrate the skin and produce occupational disease and systemic toxicity. This fact not only dispelled the view that the skin is an impermeable barrier, but also stimulated scientific inquiries into the mechanisms and factors affecting skin permeability.

A perusal of the literature shows that the early history of skin permeability was somewhat involved. In the scientific era before 1877 a view was held that the skin was freely permeable only to gases.[1] From 1877 to about 1900 skin was considered essentially impermeable.[2] As most of the experimental work in that era was done *in vivo,* further progress in this endeavor depended upon the ability of the investigator to detect the test substance in the blood or urine. Thus, after 1900 it was concluded that the skin was permeable to some substances and that fat-soluble chemicals penetrated the skin more rapidly than water-soluble compounds[3]. A review published in 1928 stated that the skin absorbs all substances.[4] Subsequently, it has been realized that the skin is not a simple membrane and its permeability is not governed only by the laws of physics. Numerous studies have demonstrated that the skin is a living and dynamic tisssue, with its own metabolic characteristics which can markedly alter the absorption conditions.[5] In addition, the dermal absorption of chemicals may be affected by a variety of external factors, including the solvent used to deliver the test compound, physical condition of the skin, occlusion, air movement over the test site, physical state of the test compound (liquid vs. gas), and others.

Animal models are used extensively in dermal absorption studies to evaluate dermal toxicity of topically applied compounds since many of these studies cannot be performed using human subjects. The results obtained from animal studies are thus used to predict percutaneous absorption in man. However, there are marked differences in the dermal absorption of chemicals between man and animal species and even among different animal species, which makes it more difficult to extrapolate to human situations. Thus, it seems unlikely that we can hope to completely duplicate the dermal absorption characteristics of a large number of chemical compounds using a given animal model with those in man. We can only develop animal models with the highest possible degree of human correlation. Furthermore, in many cases *in vivo* measurements of dermal absorption can only be made by indirect methods. The selection of experimental techniques and an animal species will then depend upon the objective of a specific study. In this chapter we will present current methodology used in the assessment of dermal penetration, indicating advantages and disadvantages of the various experimental procedures described.

II. METHODOLOGY

A. Experimental Procedures

The following methods have been used in *in vivo* dermal absorption studies.

1. Indirect Method

In the indirect method the dermal penetration of the test compound is determined from the extent of excretion from urine or feces. The method was originally developed by Feldmann and Maibach[6,7] for use in experimental studies with human subjects. As the compound may be retained in the tissues, a correction must be made by administration of a single parenteral dose and a determination of the extent of excretion of the dose. The use of radiolabeled test compounds (^{14}C, ^{3}H, etc.) allows the investigator to rapidly measure excretion of the label and the identification and measurement of individual metabolites is not necessary. Adequate time must be allowed for the excretion of administered radioactivity, usually 5 days. However, radioactivity must be checked daily until it reaches background level. The following formula is used to determine the percent dermal absorption of a topical dose:

$$\frac{\text{Total \% excreted radioactivity after topical dose}}{\text{Total \% excreted radioactivity after parenteral dose}} \times 100\%$$

For example, if 15% of the radioactive dose of compound A was recovered in the urine after topical application, and 85% of the radioactive dose was recovered after subcutaneous injection, the percent absorption would be 15/85 × 100, or 18%. The indirect method can introduce more error as the recovery after subcutaneous injection becomes less. For example, if only 20% of compound A was recovered in the urine after subcutaneous injection, the percent absorption would be 75%, and small errors in the recovery after topical application are magnified fivefold.

An advantage of this method is that it may be used in human subjects if the test compound is not toxic. In animal experiments the animals need not be killed and can be used repeatedly after the radioactivity reaches the background level. On the other hand, this method does not lend itself to study the kinetics of absorption and does not account for the metabolism of test compound by skin.

A variation of the indirect method for the determination of dermal absorption is the measurement of the areas under the plasma concentration-time curves.[8,9] A good correlation has been found in bioavailability of topical nitroglycerin determined by the urinary excretion measurements (72.7%) and by the plasma concentration-time curve (77.2%).[9]

2. Direct Method

The direct method will permit the study of absorption kinetics and will provide a better approximation for estimating the absorption of substances with a slow excretion rate from the body. After application of the radioactive test compound to a group of animals, several animals are killed at predetermined time intervals, and blood, various tissues, the application site, and the remaining carcass are assayed for radioactivity. Expired carbon dioxide, urine, and feces are also collected for radioactivity determination (Table 1). This method is more time consuming than the indirect method and, obviously, cannot be used with human subjects. The direct method has been described and used by several investigators.[10-13] A good correlation between the two methods has been reported[11,14] (Table 2).

3. Disappearance of Test Compound from Application Site

The difference between the applied and residual concentration of a nonvolatile test compound at the application site may indicate percutaneous absorption. This method has several disadvantages, however, including the uncertainty of the full recovery of the compound from the skin and the requirement for highly sensitive analytical procedures. In addition, this method does not distinguish between the retention of the compound in a skin reservoir and its actual absorption. Nevertheless, a high correlation has been reported between the dermal absorption of several test compounds in hairless rats and the residual amount in the stripped stratum corneum.[15] (Figure 1).

4. Physiological Responses

If the test compound produces a biological response when it reaches the target tissue, the response can be used to determine absorption. An example would be the effect of topical organophosphates on the activity of blood acetylcholinesterase.[16] Another response would be the change in blood pressure, observed after topical application of nitroglycerine,[17] sweat gland secretion, vasodilatation, and vasoconstriction.[18]

5. Other

Methods for measuring dermal absorption may also include whole body autora-

Table 1

Disposition of Radioactivity Following Topical Application of Radiolabeled Parathion

	Recovery of radioactivity (% of applied dose)						
	Protective skin application						
Animal	Urine	Feces	Patch	Scrub[a]	Site[b]	Carcass[c]	Total
Athymic nude mouse	63 ± 17	3 ± 1	11 ± 2	0.4 ± 0.2	0.2 ± 0.2	5 ± 2	83 ± 11
Human skin grafted athymic nude mouse	25 ± 5	2 ± 2	47 ± 7	3 ± 2	4 ± 3	0.4 ± 0.2	81+6
Pigskin grafted athymic nude mouse	24 ± 8	1.3 ± 0.6	39 ± 15	4 ± 1	7 ± 2	0.9 ± 0.3	76 ± 4
Weanling pig	17 ± 2	0.5 ± 0.4	50 ± 14	11 ± 1	1.5 ± 0.3	—	80 ± 10
Hairless dog	20 ± 6	0.4 ± 0.4	64 ± 12	12 ± 5	—	—	96 ± 10

[a] 24 or 48 h postapplication. The skin surface was decontaminated with cotton swabs lightly soaked in ethanol.
[b] Radioactivity recovered from the skin and subcutaneous fat bounded by the application site.
[c] Recovery of radioactivity following assay of various organs and remaining carcass, minus the skin of the application site.

From W. G. Reifenrath, E. M. Chellquist, E. A. Shipwash, W. W. Jederberg, and G. G. Krueger, *Br. J. Dermatol.* 111(Suppl. 27), 123 (1984). With permission.

Table 2

Comparison of Indirect and Direct Method for Determining Percutaneous Absorption of Parathion in Rats

	Cumulative Percent of Recovered Dose[a]	
Hours	Indirect[b]	Direct[c]
4	19 ± 1	23 ± 1
8	21 ± 5	27 ± 4
12	50 ± 8	53 ± 8
24	67 ± 5	65 ± 6
48	90 ± 1	90 ± 2
120	95 ± 1	99 ± 1

[a] Mean and 1 SE of three replicates. Data derived from Reference 11.
[b] Data are corrected for incomplete urinary excretion as follows: raw values of percent recovery in the urine were multiplied by a factor of 1.15 (1/0.87), since 87% of the dose was recovered in the urine after intraperitoneal injection.
[c] Sum of recoveries in the urine, feces, blood, liver, and carcass minus the recovery from the application site (unabsorbed parathion).

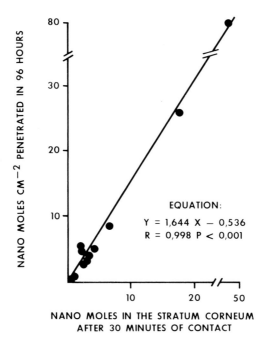

EQUATION:

$$Y = 1,644\ X - 0,536$$
$$R = 0,998\ \ P < 0,001$$

NANO MOLES IN THE STRATUM CORNEUM
AFTER 30 MINUTES OF CONTACT

Figure 1.   Correlation between the level of penetration and the concentration in the stratum corneum. (From A. Rougier et al., *J. Invest. Dermatol.* 81, 275 (1983). With permission.)

diography,[19] histology, fluorescence, and affinity of the test compound for a specific organ or tissue.[20,21]

B. Selection of Animal Models

In vivo dermal penetration studies have utilized small laboratory rodents, such as mice, rats, guinea pigs, and rabbits. These animal species are readily available, convenient to use, and easy to maintain. However, the penetration data obtained with these animal models sometimes do not correlate well with the data previously acquired for man.[14,22] In general, rabbit skin appears to be most permeable to topically applied compounds, followed closely by rat skin. The wide differences in the penetration characteristics among laboratory rodents is apparently due to factors such as the thickness of the stratum corneum, the number of hair follicles per unit surface area, the skin blood supply and others. Furthermore, before the test compound can be topically applied, the hair is first closely clipped and usually shaved. This procedure can damage the stratum corneum and enhance penetration. To overcome some of these problems, several investigators have used the Mexican hairless dog as an animal model for percutaneous absorption.[14,23,24] However, no correlation between the human and the hairless dog data was found[14] (r = 0.58, n = 9, $p > 0.05$), and this was explained by the histologic differences of the skin.[24]

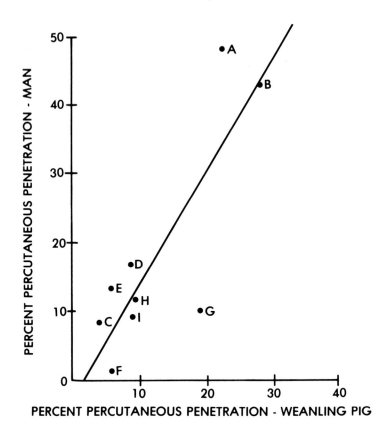

Figure 2. Correlation between pig penetration data and those reported for man. (A) caffeine, (B) benzoic acid, (C) malathion, (D) N,N,-diethyl-m-toluamide, (E) testosterone, (F) fluocinolone acetonide, (G) parathion, (H) progesterone, (I) lindane. (From W. G. Reifenrath et al., *Br. J. Dermatol.* 111(Suppl. 27), 123 (1984). With permission.)

Another animal model available for dermal penetration studies is the domestic pig. Pig skin has many physiologic, morphologic, and metabolic characteristics in common with human skin.[25,26] Several investigators working with a wide variety of test compounds reported a high correlation between *in vivo* pig penetration values and those reported for man[14,22,27,28] (Figure 2). Maintenance cost and facility requirements are, however, significantly higher than those for small laboratory animals.

Congenitally athymic nude mice have been used in dermal penetration studies.[29] This animal model readily accepts human skin grafts and skin grafts from other animal species. The skin grafts are used as application sites in dermal absorption studies. Indeed, a significant correlation exists between the penetration data obtained with the human skin-grafted nude mouse (man-mouse) and those reported for the man.[14] In contrast, no correlation was found between the data obtained from pigskin-

Table 3
Percutaneous Absorption of Increased Topical Doses of Several Compounds in the Rhesus Monkey
and Man (*In Vivo*)

Penetrant	Dose (μg/cm^2)	Percent dose absorbed	
		Rhesus	Man
Hydrocortisone	4	2.9	1.9
	40	2.1	0.6
Benzoic acid	4	59.2	42.6
	40	33.6	25.7
	2000	17.4	14.4
Testosterone	4	18.4	13.2
	40	6.7	8.8
	400	2.2	2.8

Data derived from Reference 63.

grafted nude mouse or from the nude mouse and human skin penetration data.[14] However, no significant differences in penetration values were observed when comparing the data from the man-mouse vs. the pig-mouse. Better agreement between penetration values for the man-mouse vs. man or for the pig-mouse vs. man was noted for the more hydrophilic compounds. It was suggested that the grafting of skin, in this case without a complete dermis, may remove a barrier to the penetration of more lipophilic compounds. Further testing of the man-mouse or pig-mouse is needed with additional compounds to better define the value of these models in percutaneous absorption studies.

Several investigators studied the percutaneous absorption of selected compounds in the monkey and compared the data to those reported for man.[28,30] In direct comparison studies the experimental procedures and application sites in the monkey and the man were similar.[31,32] The results indicate that dermal absorption in the monkey and in man is usually similar (Table 3). Thus, considering all the experimental evidence, it appears that the pig and the monkey are the animals of choice for use in *in vivo* percutaneous absorption studies. Disadvantages of employing these two animal models are the relatively high maintenance cost and the possible handling difficulties. Larger animals are advantageous in the execution of comparison studies with man, since the proportionality of dose, expressed as mg/cm^2 and mg/kg, can be more easily maintained. Monkeys have been successfully used in dermal absorption studies with pesticides.[33]

Other possible animal models include an athymic (nude) rat skin flap model,[34] hairless rats,[35,36] hairless mice,[35,37] and fuzzy rats.[37]

C. Preparation of Animals

The body weight of each animal should be recorded before experimentation. This serves essentially two purposes: (1) the loss in body weight at the end of the experimental period may indicate the degree of toxicity of the test compound and (2) the applied dose per unit area can be calculated as the quantity per unit weight.

Where applicable, fur on the back is closely clipped with an electric animal hair clipper about 24 h before experimentation. Shaving the clipped area is not recommended. To protect the application site from contact with the metabolism cage, and from the animal's reach and potential ingestion of the test compound, a protective nonocclusive patch is usually placed over the treated area.[22] A nonocclusive foam pad device has been used on rabbits, pigs, and dogs.[14,22]

A protective collar of rubber tubing placed behind the forelegs of rats also has been used.[22,38] In short-term studies, mice can be immobilized by taping a strip of exposed film, 4 × 1 in., around the torso and by inserting the hindquarters into a styrofoam drinking cup, and into a 1 cm diameter hole in the base of the cup.[39] The test compound is then applied to the protruding back of the mouse, immediately in front of the tail.

In the monkey, the ventral forearm has been used as the site of application. The site may be lightly clipped and there is no evidence that clipping enhances absorption.[40] Each monkey, preferably trained for metabolic studies, is placed in a metabolism cage and the arms are secured to the side of the chair. Thus, the application site on the ventral forearm is isolated from the fecal and urine collection area.[40]

D. Procedures for Application of Chemicals to the Skin
1. Liquids and Solids

Dermal absorption is affected by a variety of factors, including the amount of applied dose and surface area. With a constant surface area, dermal absorption increases with increasing chemical dose, but absorption as a percentage decreases (Table 4). Dermal absorption is also enhanced when a constant dose is applied to an increasing surface area.[41] In the study, a constant amount of liquid VX (11.0 mm³ × 10⁻³/ kg) was applied to rabbit skin with the only variable being the skin contact area. The data show that greater skin contact area resulted in greater total penetration (Table 5). The authors suggest that the small contact area may make for a slightly greater penetration rate, since skin contact time and skin contact area are interrelated. The entire process is maximized when a large quantity of test compound is applied to a large area of the body. To facilitate application, the test compound is generally dissolved in a volatile organic solvent (e.g., acetone, ethanol) and a small volume of the solution is applied to a designated area of skin. The dose is usually expressed as $\mu g/cm^2$. It must be realized that some vehicles may alter the integrity of stratum corneum and thus promote absorption. Most dermal penetration studies are usually conducted with a single application. A nonocclusive protective patch is then placed around the application site. The patch can be constructed from a foam pad (Reston® Foam Pad, 3M Company, Minneapolis, MN) from which the central portion has been removed using a cork borer or a blade. Nylon screen can be glued over the cut-out portion of the pad and the assembly is secured over the application site with a surgical tape.[14,22]

Occlusion of the application site prevents evaporation of volatile test compounds.

Table 4

In Vitro Skin Absorption of N,N-Diethyl-m-Toluamide as a Function of Dose

Dose (μg/cm²)	Percent absorption/50 h[a]	Amt absorbed/ 50 h (μg)
4[b]	6 ± 1	0.24
360[c]	13 ± 2	47
125,000[d]	0.6 ± 0.1	750

[a]   Mean value and 1 SD.
[b]   Data from Reference 62.
[c]   Value after only 24 h; however, absorption was essentially complete.
[d]   W. G. Reifenrath, unpublished data.

Table 5

Effect of Skin Surface Contact Area on Mean Penetration Rate and Mean Total Penetration of VX through Rabbit Skin

Skin contact area (mm²)	Penetration rate (mm³ × 10⁻³/mm²/min)	Total penetration (mm³ × 10⁻³/min)
7.1	0.07	0.521
38.5	0.05	1.797

Data from Reference 41.

However, there is evidence that occlusion enhances percutaneous absorption by increasing the hydration of the skin or altering other physical and chemical characteristics of the skin. Indeed, it was found that occlusion increases the penetration of lipid-soluble, nonpolar molecules but has smaller effect on the penetration of polar compounds.[42]

2. Vapors

Several methods are available for the determination of dermal absorption of volatile materials in experimental animals. One method employs a charcoal-containing stainless steel disk which is glued to the backs of experimental animals. The charcoal absorbs the portion of the test substance which would be lost by evaporation.[43] The charcoal is covered with a Teflon cap and the test compound is delivered through a guide needle to the isolated skin area (Figure 3). After application, the syringe and the guide needle are removed and the animal is placed into a metabolism cage. Another method employs specially designed face masks for rodents in order to avoid pulmonary uptake of volatile compounds in studies concerned with whole-body vapor exposure[44] (Figure 4). Experimental animals are kept in an exposure chamber and during the exposure period blood samples can be obtained from

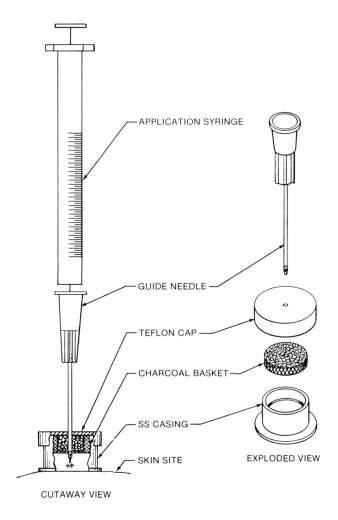

APPLICATION SYRINGE

GUIDE NEEDLE

TEFLON CAP

CHARCOAL BASKET

SS CASING

SKIN SITE

EXPLODED VIEW

CUTAWAY VIEW

Figure 3.   Device for percutaneous penetration of volatile compounds. (From A. S. Susten, B. Dames, and R. W. Niemeier, Proc. 15th
Annu. Conf. Environ. Toxicol., Wright-Patterson AFB, OH, 322, 1985.)

indwelling jugular catheters. A third method employs glass cups containing a filter-
paper circle.[45] A small volume of the volatile test compound is placed on the filter
paper and the cup is attached by means of a double-stick tape to the clipped back of
experimental animals (Figure 5). Each animal is placed in a metabolism cage to
collect metabolic products.

E. Collection of Biological Samples

Metabolism cages allow quantitative separate collection of urine, feces, expired

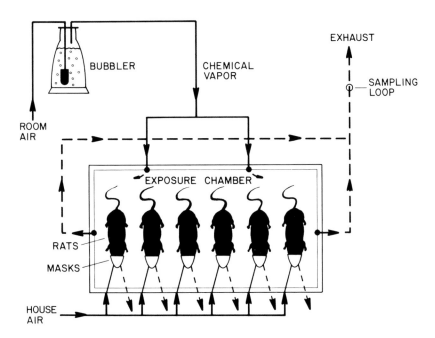

Figure 4.   Schematic of dermal vapor exposure system. (From J. N. McDougal, G. W. Jepson, H. J. Clewell, III, and M. E. Anderson, *Toxicol. Appl. Pharmacol.* 79, 150 (1985). With permission.)

carbon dioxide, and other volatile metabolites. The animal has free access to food and water. Metabolism cages for small laboratory rodents are usually constructed from Pyrex® glass or from various polymeric materials. For larger laboratory animals such as pigs or dogs, the cages are constructed from stainless steel. Because of a large volume of circulating air involved, these cages are not constructed to facilitate the collection of expired carbon dioxide. Urine and feces are collected daily to determine the completeness of excretion of the test compound. After removal of the fecal material, the collection area is rinsed with distilled water and ethanol to ensure a complete recovery of the urine. The volume of the urine plus the washings are recorded and kept in a refrigerator. The fecal material is frozen until analyzed.

If the test compound is labeled with ^{14}C it is desirable to determine what fraction of the applied dose has been oxidized and expired as $^{14}CO_2$. Therefore, CO_2-free air is drawn through the metabolism cage and through a 2% solution of $NaOH$[46,47] or an organic base, such as ethanolamine or Hyamine®. Other volatile metabolites in expired air may be trapped on a glass column containing Tenax® absorbent. The column will not retain $^{14}CO_2$.

Figure 5.   Glass cups for dermal absorption of volatile compounds.
(From P. Schmid, J. R. Jaeger, G. J. Klain, J. P. Hannon, C. A. Bassone,
and M. C. Powanda, Letterman Army Inst. Res. Rep. No. 230, Presidio,
San Francisco, 1987.)

F. Analysis of Biological Samples
1. Determination of Radioactivity

1.   Aliquots of urine, sodium hydroxide or organic base solutions are
      pipetted into scintillation vials and mixed with 10 ml of an aqueous
      counting solution (ACS).
2.   Tenax® samples are placed directly into scintillation vials and the
      columns are flushed with 10 ml of ACS. Radioactivity in the solu-
      tions is determined using a liquid scintillation spectrometer
      equipped with an automatic external standard.

Table 6
Effect of Anatomic Region on Absorption of ^{14}C-Hydrocortisone

	Total excretion[a]		
Anatomic region	Experiment	Forearm control	Ratio
Forearm (ventral)	1.04	1.04	1.0
Forearm (dorsal)	1.19	1.04	1.1
Foot arch (plantar)	0.17	1.27	0.14
Palm	0.74	0.94	0.83
Back	1.26	0.72	1.7
Scalp	4.41	1.23	3.5
Forehead	7.65	1.27	6.0
Jaw angle	12.25	0.94	13.0
Scrotum	36.2	0.86	42.0

[a] Urinary ^{14}C excretion expressed as percent applied dose.

Data from Reference 49.

3. Fecal samples are freeze-dried and thoroughly homogenized.
4. Treated skin sites are cut into small (50 mg) pieces.
5. Carcass, tissues, and organs are homogenized with an equal volume of water. Radioactivity in fecal, tissue homogenate (equivalent to about 150 mg tissue), blood, and plasma samples is then determined following processing using a tissue oxidizer.

2. Other Analytical Techniques.

The meaningful use of nonradioactive test compounds in *in vivo* dermal penetration studies depends on the sensitivity of analytical techniques to measure the concentration in blood and urine. Blood levels after topical application are usually very low due to dilution, tissue uptake, or rapid excretion. It may be difficult to completely extract the test compound from tissues. Appropriate analytical techniques include high performance liquid chromatography, gas-liquid chromatography with various detectors, bioassays, enzyme immunoassays, spectrophotometric, and spectrofluorometric methods. These are techniques of general acceptability and their use will depend on the design of each study.

G. Protocols

Ample evidence demonstrates that the rate of dermal absorption varies according to the anatomical site to which the test compound is applied.[48,49] This effect has been observed both in animal and human studies. In man, for example, a high absorption of hydrocortisone was observed on the forehead, followed by the scalp, back, forearms, palms, ankle, and the foot arch[49] (Table 6). These findings emphasize the importance of choosing a uniform regional application site in both animal and human dermal absorption studies. Otherwise, exaggerated variation in absorption may be expected. Regional absorption differences for chemicals have been attributed to the thickness of the stratum corneum and the hair follicle density. Marked differences in the thickness of the rat stratum corneum obtained from the back and abdomen have been found.[50] However, this finding does not apply to all situations. In the monkey, no regional site effects on dermal absorption of nitroglycerin have been found.[51] In contrast, dermal absorption of other drugs is site dependent in the rat.[52] However, there is evidence that there are marked differences in absorption among individuals even when the experimental conditions are identical.[53] In view of this biological variability, the selection of test sites should be carefully considered in dermal absorption studies. In particular, this is important in human studies or in studies that may be conducted with a small number of experimental animals because of high maintenance costs or limited availability. The data obtained from such studies may contain a large amount of experimental error and the investigator may not be able to draw any valid conclusions from the study. Thus, an appropriate experimental design, including an adequate number of animals per treatment group is essential for the meaningful execution of *in vivo* dermal absorption studies. There is no exact way to determine the number of subjects needed for a given experiment. The experience of the investigator is certainly important in this respect. However, to select an appropriate statistical number of subjects for a given experiment, the proposed experimental design should be subjected to a power analysis test which takes into account the variability of the measurements and the probability of a type I and a type II error.[54] As an example, suppose one had three different formulations (r = 3) of the pesticide parathion. From past experience, it is known that mean skin absorption of parathion from the formulations ranged from 20 to 30 µg under certain test conditions with a standard deviation (s) of 3 µg. If it is desired to detect a difference (d) of 5 µg between the formulations, Equation 2 is used to calculate the value of φ.

$$\phi = \frac{d}{s}\sqrt{\frac{1}{2r}} \qquad (1)$$

$$\phi = \frac{5}{3}\sqrt{\frac{1}{(2)(3)}} = 0.68 \qquad (2)$$

Assuming a probability of type I error of 0.05 (a = 0.05) and type II error of 0.20 (b = 0.20 or P = 0.80), from Table A-10 in Reference 54, we obtain a value of 8 for the number of replicates needed for each formulation.

III. ANALYSIS OF RESULTS

In vitro studies of percutaneous absorption typically measure cumulative absorption or absorption rate over time for a compound. A lag time, permeability coefficient, or a percent absoption can be calculated for the penetrant. A percentage absorption is typically derived from *in vivo* studies, based on recoveries of penetrant or metabolite(s) from urine, feces, expired air, application site, or carcass. In comparison studies, the Student's *t* test or analysis of variance techniques may be used for differences among the means of two or more populations, respectively. A significance level of 0.05 is common practice.

IV. COMPARISON OF *IN VITRO* AND *IN VIVO* PERCUTANEOUS ABSORPTION MEASUREMENTS

Percutaneous absorption studies have been done *in vitro* with excised skin for reasons of economy of animal use, greater experimental control, and ease of analytical measurement. In such *in vitro* studies, chemicals are placed in contact with the stratum corneum or donor side of the skin and their appearance in a receptor solution bathing the visceral side of the skin is monitored. *In vivo,* loss of chemicals from the skin surface can occur by evaporation, abrasion, or exfoliation of the stratum corneum. Air flow, temperature, and humidity can be factors affecting surface loss. Penetration and systemic uptake is governed in part by the microcirculation of the skin. Metabolic transformation of a chemical may occur before systemic uptake. Chemicals (e.g., irritants or vasoconstrictors) may exert pharmacological effects on the microcirculation that affect their own distribution and fate. It would be difficult or impossible to retain all these factors in an *in vitro* model without destroying the very advantages of making *in vitro* measurements. Fortunately, not all of the above-mentioned factors occur simultaneously in most cases and many can be controlled or taken into account in many *in vitro* models.

If the surface dose is made constant, by application of a very large dose (infinite dose) or if a rate-controlling membrane is used to meter the dose to the skin, evaporation is unimportant. For modeling finite doses of volatile compounds, evaporation-penetration cells have been designed[55] to allow a controlled flow of air over the surface to mimic *in vivo* evaporative loss (Figure 6). By conditioning the air prior to entry into the evaporation cell, skin surface temperature can be controlled. Use of freshly excised skin and a receptor fluid which supports viability will retain skin metabolism in the *in vitro* experiments.[56] Skin esterase activity can be preserved *in vitro* without these measures; however, this activity is sensitive to heat and can be destroyed by techniques used to separate epidermal membranes.[57]

In vivo, systemic uptake can occur via the microcirculation at the epidermal-dermal junction of the skin, while *in vitro* the compound must travel through the dermis (if present) to reach the receptor fluid. Full thickness skin can give results in good agreement with *in vivo* determinations for relatively hydrophilic compounds (Table 7). For lipophilic compounds (octanol/water partition coefficient >1000) the dermis can introduce an artifact by acting as a reservoir or barrier to diffusion into the receptor fluid. Better agreement for these compounds can be obtained by using split thickness skin, and considering residue in the dermis as penetration (Table 7). This was done by isolation and extraction of the dermis at the end of an experiment.[58]

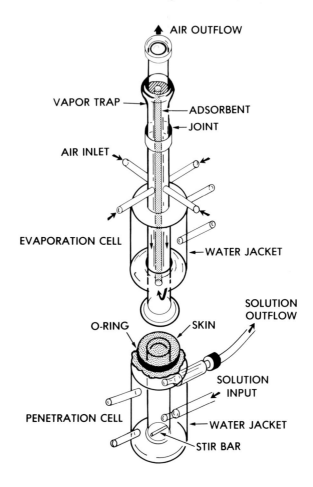

Figure 6.   Skin penetration and evaporation cells. (From G. S. Hawkins and W. G. Reifenrath, *J. Pharm. Sci. 75*, 378 (1986). With permission.)

The data in Table 7 also demonstrate the importance of air flow on *in vitro* penetration measurements. It was necessary to use a higher air flow (600 vs. 60 ml/min) through the evaporation cell to get better agreement for the range of compounds, specifically malathion and N,N-diethyl-m-toluamide, with *in vivo* values.

An alternative approach to achieve agreement between *in vitro* and *in vivo* penetration measurements for lipophilic compounds has been the use of receptor fluids containing surfactants.[37] These surfactants may have a direct effect on the permeability properties of the skin, as well as increasing the solubility of the penetrant in the receptor fluid.[59] *In vivo* percutaneous absorption of highly lipophilic compounds (e.g., DDT) may be complicated by poor excretion of the compound or

Table 7
Percutaneous Penetration of Radiolabeled Compounds on Pigskin *In Vivo* and Under Different *In Vitro* Conditions[a]

		Percent of applied radioactive dose[b]				
			Split-thickness skin			
Compound	Log P[c]	Receptor fluid (AF[d] = 60)	Receptor fluid (AF = 60)	Receptor fluid + dermal residue (AF = 60)	Receptor fluid + dermal residue (AF = 600)	*In vivo*
Caffeine	0.01	20 ± 2	18 ± 3	21 ± 4	20 ± 4	23 ± 9
Benzoic acid	1.95	20 ± 13	17 ± 6	21 ± 7	15 ± 8	28 ± 6
N,N-Diethyl-m-toluamide	2.29	14 ± 4	19 ± 13	21 ± 13	6 ± 1	9 ± 4
Fluocinolone acetonide	2.48	2 ± 1	1.1 ± 0.9	2 ± 2	4 ± 4	6 ± 1
Malathion	2.98	16 ± 11	21 ± 6	24 ± 7	10 ± 6	4.4 ± 0.3
Parathion	2.98	1 ± 1	12 ± 5	21 ± 5	10 ± 4	19 ± 2
Testosterone	3.31	3 ± 2	9 ± 4	13 ± 5	11 ± 7	6.0 ± 0.3
Lindane	3.66	1 ± 1	6 ± 2	9 ± 4	6 ± 2	8 ± 1
Progesterone	3.78	1 ± 1	5 ± 2	9 ± 4	12 ± 7	10 ± 1

a The chemical dose of all compounds was 4 μg/cm^2. Data from tables in References 27 and 55.
b Mean ± SD of at least three replicates. *In vivo* values are corrected for incompleteness of excretion.
c Logarithm of the octanol/water partition coefficient.
d Air flow through the evaporation cell in ml/min.

metabolites. Error in calculating percent percutaneous absorption is then magnified. If possible, a direct method should be utilized.

The receptor fluid should serve as a "sink" for penetrant leaving the skin. For hydrophobic compounds, the solubility in the receptor fluid may be limited; improved agreement between *in vitro* and *in vivo* results can sometimes be obtained by the use of "flow through" cells which effectively increases the volume of the receptor fluid as compared to a penetration cell with the static receptor fluid,[60] or by the use of more lipophilic receptor fluid. Certain studies have shown that the agreement between *in vitro* and *in vivo* penetration is dose dependent,[60,61] i.e., penetration is higher *in vivo* than *in vitro* at higher doses. Solubility in the receptor fluid may be one reason for this discrepancy.

V. CONCLUSIONS

The complexity of the percutaneous penetration, coupled with diverse physicochemical properties of test compounds and profound differences in skin penetration characteristics observed in man and experimental animals, do not permit the development and use of a "universal" system for the assessment of *in vivo* dermal penetration. We have presented general principles of the *in vivo* assessment of dermal penetration methodology and highlighted the problems that may exist. Many of these principles have been applied in our laboratory. A great deal of interest has been generated in recent years in dermal penetration studies of harmful or beneficial chemicals, including the transdermal delivery of drugs and prodrugs. Since the design and goals of each *in vivo* study are different, each researcher will use the outlined principles according to the specific needs.

REFERENCES

1. S. Rothman, Handbuch der Normalen und Pathologischen, *Physiologie* 4, 107 (1929).
2. R. Fleischer, *Habilitationsschrift*, Erlangen (1877).
3. A. Schwenkenbecher, *Arch. Anat. Physiol.* 121 (1904).
4. K. Stejskal, *Zentralbl. Haut. Geschlechtskr.* 26, 537 (1928).
5. B. W. Barry, *Dermatological Formulations — Percutaneous Absorption,* Marcel Dekker, New York, 127 (1983).
6. R. J. Feldmann and H. I. Maibach, *Arch. Dermatol.* 91, 661 (1965).
7. R. J. Feldmann and H. I. Maibach, *J. Invest. Dermatol.* 54, 399 (1970).
8. R. C. Wester and P. K. Noonan, *J. Invest. Dermatol.* 70, 92 (1978).
9. R. C. Wester and H. I. Maibach, Eds., *Dermatotoxicology,* 2nd ed., Hemisphere, New York, 131 (1983).
10. T.J. Franz, *J. Invest. Dermatol.* 64, 190 (1975).
11. P. V. Shaw and F. E. Guthrie, *J. Invest. Dermatol.* 80, 291 (1983).
12. H. Hofer and E. Hruby, *Food Cosmet. Toxicol.* 21, 331 (1983).
13. T. Nishiyama, Y. Iwata, K. Nakajima, and T. Mitsui, *J. Soc. Cosmet. Chem.* 34, 263 (1983).
14. W. G. Reifenrath, E. M. Chellquist, E. A. Shipwash, W. W. Jederberg, and G. G. Krueger, *Br. J. Dermatol.* 111(Suppl. 27), 123 (1984).
15. A. Rougier, D. Dupuis, C. Lotte, R. Roguet, and H. Schaefer, *J. Invest. Dermatol.* 81, 275 (1983).

16. J. Van Genderen and O. L. Wolthuis, *Skin Models,* Springer-Verlag, Berlin, 85 (1986).
17. G. S. Francis and A. D. Hagen, *Angiology* 28, 873 (1977).
18. B. W. Barry and R. Woodford, *Skin Models,* Springer-Verlag, Berlin, 103 (1986).
19. L. Bloomquist and W. Thorsell, *Acta Pharmacol. Toxicol.* 41, 235 (1977).
20. J. Tas and Y. Feige, *J. Invest. Dermatol.* 30, 193 (1958).
21. G. Cyr, D. Skauen, J. E. Christian, and C. Lee, *J. Pharm. Sci.* 38, 615 (1959).
22. M. J. Bartek, J. A. LaBudde, and H. I. Maibach, *J. Invest. Dermatol.* 58, 114 (1972).
23. J. J. Loux, P. D. Depalma, and S. L. Yankell, *J. Soc. Cosmet. Chem.* 25, 473 (1974).
24. N. Hunziker, R. J Feldman, and H. I. Maibach, *Dermatologica* 156, 79 (1978).
25. W. Meyer, R. Schwarz, and K. Neurand, *Current Problems in Dermatology,* Vol. 7, W. Mali, Ed. S. Karger, Basel, 39 (1978).
26. G. J. Klain, S. J. Bonner, and W. G. Bell, S*wine in Biomedical Research,* Vol. 1, Plenum Press, New York, 667 (1986).
27. W. G. Reifenrath and G. S. Hawkins, *Swine in Biomedical Research,* Vol. 1, Plenum Press, New York, 673 (1986).
28. M. J. Bartek and J. A. LaBudde, *Animal Models in Dermatology,* Churchill-Livingstone, New York, 103 (1975).
29. G. G. Krueger and J. Shelby, *J. Invest. Dermatol.* 76, 506 (1981).
30. R. C. Wester and P. K. Noonan, *J. Soc. Cosmet. Chem.* 30, 297 (1979).
31. R. C. Wester and H. I. Maibach, *Models in Dermatology,* Vol. 2, S. Karger, Basel, 159 (1985).
32. H. I. Maibach and L. J. Wolfram, *J. Soc. Cosmet. Chem.* 32, 223 (1981).
33. R. B. L. van Lier, *Am. Chem. Soc. Symp. Ser.* 273, 81 (1985).
34. Z. Wojciechowski, L. K. Pershing, S. Huether, L. Leonard, S. A. Burton, W. I. Higuchi, and G. G. Krueger, *J. Invest. Dermatol.* 88, 439 (1987).
35. M. Walker, P. H. Dugard, and R. C. Scott, *Human Toxicol.* 2, 561 (1983).
36. A. Rougier, C. Lotte, and H. I. Maibach, *J. Invest. Dermatol.* 88, 577 (1987).
37. R. L. Bronaugh and R. F. Stewart, *J. Pharm. Sci.* 75, 487 (1986).
38. P. V. Shaw, H. L. Fisher, M. R. Sumler, R. J. Monroe, N. Chernoff, and L. L. Hall, *J. Toxicol. Environ. Health* 21, 353 (1987).
39. D. N. Bailey, *Res. Commun. Substance Abuse* 1, 443 (1980).
40. R. C. Wester and H. I. Maibach, *Toxicol. Appl. Pharmacol.* 32, 394 (1975).
41. F. N. Marzulli and J. S. Wiles, Chem. Warf. Lab. Rep. 2153, Army Chem. Center, MD (1957).
42. C. R. Behl, G. L. Flynn, T. Kurihara, N. Harper, W. Smith, W. I. Higuchi, N. F. H. Ho, and C. L. Pierson, *J. Invest. Dermatol.* 75, 346 (1980).
43. A. S. Susten, B. Dames, and R. W. Niemeier, Proc. 15th Annu. Conf. Environ. Toxicol., Wright-Patterson AFB, OH, 322 (1985).
44. J. N. McDougal, G. W. Jepson, H. J. Clewell, III, and M. E. Andersen, *Toxicol. Appl. Pharmacol.* 79, 150 (1985).
45. P. Schmid, J. R. Jaeger, G. J. Klain, J. P. Hannon, C. A. Bossone, and M. C. Powanda, Letterman Army Inst. Res. Rep. No. 230, Presidio, San Francisco (1987).
46. G. J. Klain, W. G. Reifenrath, and K. E. Black, *Fund. Appl. Toxicol.* 5, S127 (1985).
47. G. J. Marco, B. J. Simoneaux, S. C. Williams, J. E. Cassidy, R. Bissig, and W. Muecke, *Am. Chem. Soc. Symp. Ser.* 273, 43, (1985).
48. R. C. Wester, P. K. Noonan, and H. I. Maibach, *Arch. Dermatol. Res.* 267, 229 (1980).
49. R. J. Feldmann and H. I. Maibach, *J. Invest. Dermatol.* 48, 181 (1967).
50. S. T. Horhota and H. L. Fung, *J. Pharm. Sci.* 67, 1345 (1978).
51. P. K. Noonan and R. C. Wester, *J. Pharm. Sci.* 69, 365 (1980).
52. R. L. Bronaugh, R. F. Steward, and E. R. Congdon, *J. Soc. Cosmet. Chem.* 34, 127 (1983).
53. H. I. Maibach, *Dermatologica* 152(Suppl. 1), 11 (1976).
54. J. Neter and W. Wasserman, Eds., *Applied Linear Statistical Models,* Prentice-Hall, Homewood, IL, 492 (1974).
55. G. S. Hawkins and W. G. Reifenrath, *Fund. Appl. Toxicol.* 4, S133 (1984).

56. J. Kao, F. K. Patterson, and J. Hall, *Toxicol. Appl. Pharmacol.* 81, 502 (1985).
57. M. Loden, *J. Invest. Dermatol.* 85, 335 (1985).
58. G. S. Hawkins and W. G. Reifenrath, *J. Pharm. Sci.* 75, 378 (1986).
59. R. T. Riley and B. W. Kemppainen, *Percutaneous Absorption: Mechanisms, Methodology, Drug Delivery,* Marcel Dekker, New York, 387 (1985).
60. B. W. Kemppainen, J. G. Pace, and R. T. Riley, *Toxicon* 25, 1153 (1987).
61. F. N. Marzulli, D. W. C. Brown, and H. I. Maibach, *Toxicol. Appl. Pharmacol.* Suppl. 3, 76 (1969).
62. W. G. Reifenrath, G. S. Hawkins, and M. S. Kurtz, *J. Am. Mosquito Control Assoc.* in press (1989).
63. R. C. Wester and P. K. Noonan, *Int. J. Pharm.* 7, 99 (1980).

9

In Vitro Assessment of Dermal Absorption

JOHN KAO
Department of Drug Metabolism and Pharmacokinetics
SmithKline Beecham Pharmaceuticals
King of Prussia, Pennsylvania

I. INTRODUCTION

The skin is a complex, multilayered organ which serves as a living, protective envelope surrounding the body. As a primary interface between the body and its external environment, the skin is constantly exposed to a variety of chemical agents, and is a surface upon which drugs and cosmetics are intentionally applied. While it forms an effective barrier against invasion of the body by microorganisms and prevents excessive loss of body water, it is becoming increasingly apparent that the skin is not a complete barrier, and that it is an important portal for the entry of chemicals into the systemic circulation. Consequently, studies on the dermal absorption of hazardous chemicals have become an important aspect of dermatotoxicology. From a dermatopharmaceutic standpoint, the recent increased interest in skin absorption is derived from the view that the permeability of the skin may provide an opportunity to use the transdermal route as a means of controlled delivery of potent drugs for systemic therapy. The initial success of transdermal nitroglycerin, and subsequently with scopolamine, clonidine, and estradiol have provided much of the impetus for this development.

Skin absorption is a complex phenomenon. It can be viewed as the translocation of surface-applied substances through the various layers of the epidermis and dermis to a location where they can enter the systemic circulation via the dermal microvasculature and lymphatics, or remain in the deeper layers of the skin. This transport of substances through the skin involves complex diffusional and metabolic processes and is influenced by the interactions of a variety of physiochemical, biophysical, and biochemical factors. Dermal absorption is the net result of the penetration outward margin, cutaneous metabolism, binding and permeation of a topically applied chemical into and through the different strata of the skin (Figure 1). The fundamental concepts in percutaneous absorption have been extensively reviewed in the literature, and noteworthy reviews found in the recent literature include the in-depth treatises of Schaefer et al.[1] and Barry.[2]

267

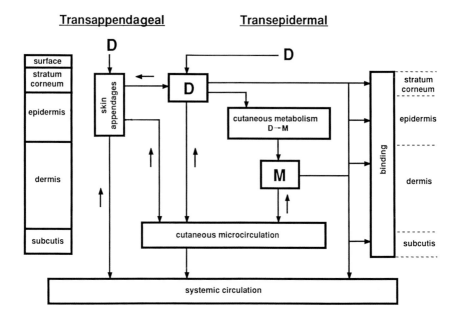

Figure 1. Schematic showing the percutaneous fate of a drug (D) following topical application. M denotes cutaneous metabolite(s) of D.

There are a wide variety of experimental approaches which have been developed to assess dermal absorption; however, it should be noted at the outset that, at present, there are no generally accepted techniques for assessing skin absorption. Debates and conflicting opinions continue to revolve around the various factors that are important in influencing percutaneous absorption; consequently, the rationales by which experimental models are selected and developed are continually being modified and revised.

Fundamentally, the questions most often encountered in dermal absorption investigations are concerned with how much, how fast, and what are the modulating factors that may influence the penetration and percutaneous fate of the topically applied agents. To answer these questions the primary methods available are based upon *in vivo* and *in vitro* studies. Both approaches have their advantages, limitations, and weaknesses. The purpose of this chapter is to review the current knowledge concerning *in vitro* methods for assessing skin absorption, and to focus on the issues relating to the practical aspects in experimental design and conduct of *in vitro* studies. The techniques and pitfalls of the *in vivo* methodology and its application in assessing dermal absorption were discussed in the previous chapter.

II. METHODOLOGY

A. General Principles
It is generally recognized that the systematic characterization of the rate-limiting

barrier in skin began with pioneering investigations in the early 1960s,[3-6] and the classic treatises of Scheuplein and co-workers[7-14] on the mechanisms of skin absorption, have provided the theoretical bases for much of the *in vitro* experimental strategies currently employed in studies of skin penetration. A detailed discussion on the mathematical analysis of percutaneous absorption is beyond the scope of this review; however, for interested readers, excellent reviews may be found elsewhere in the extensive literature published on the subject.[1,2,14-20]

Diffusion forms the basis of *in vitro* studies, and the problems of skin absorption are often simplified to a number of diffusion equations and mass transfer coefficients. The justification of such methodology centers upon the generally accepted assumption that the stratum corneum is the principal barrier limiting skin permeability.[5,14,21] Since this outermost layer of the skin is composed essentially of nonliving tissue, it is therefore reasoned that biochemical processes cannot influence the diffusional characteristics of the rate limiting membrane, and hence, *in vitro* diffusion studies will accurately predict *in vivo* skin penetration and absorption.

Unfortunately, by considering the skin to be merely a diffusional membrane, concepts relating to biochemical factors which may influence the percutaneous fate of topically applied substances have received little attention. This diffusional membrane assumption has persisted despite the fact that the skin is known to be an organ active in many essential biochemical activities, including those involved in the metabolism of xenobiotics.[22-26] Since the skin is capable of metabolizing xenobiotics, chemicals that are applied to the surface of the skin, once they penetrate the stratum corneum, are necessarily exposed to any available biotransformation system that exists in the skin. Consequently, inactivation, activation, and interaction with tissue components may occur during translocation of the chemicals across the skin. Although investigations in this area are still in their infancy, recent *in vitro* studies have demonstrated that skin metabolism can influence the percutaneous fate of certain chemicals.[27-31] It is evident that the principles and practices of *in vitro* skin absorption studies are complex and multifaceted; nevertheless, with careful consideration, *in vitro* methodology can be an extremely useful tool for assessing dermal absorption and biotransformation.

In general, *in vitro* skin absorption studies are conducted using diffusion chambers in which the skin preparation is the diffusion membrane. Typically, in an *in vitro* experiment the diffusion chambers are assembled with the skin preparation separating the donor and receptor compartments. An appropriate fluid is placed into the receptor chamber and the chemical under investigation, usually radiolabeled, in an appropriate vehicle is placed into the donor compartment; the recovery of radioactivity, or the chemical of interest, with time in the receptor fluid then provides an estimate of skin penetration. For the purposes of the following discussion, *in vitro* skin permeation studies can be divided into the following components: (1) diffusion chambers; (2) membrane preparation; (3) receptor fluid; (4) dose application; (5) data presentation; and (6) checks and balances. These elements will be examined from a standpoint of experimental design and relevance in skin absorption investigations using *in vitro* methodology.

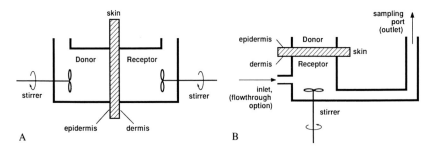

Figure 2.   Schematics of generic chamber designs for (A) "infinite dose" and (B) finite dose techniques.

B. Diffusion Chambers

There are two distinctly separate approaches to the experimental design of *in vitro* skin absorption studies and they are summarized schematically in Figure 2. The conventional strategy is to reduce the problems of skin absorption to a series of diffusion equations. Experiments are performed with rigorous compliance to the principles of diffusion, and at steady-state flux conditions the diffusional character- istics of a given compound, in a given vehicle through a defined membrane prepa- ration can be characterized in terms of its permeability coefficient.[1,2,13,14] This ap- proach, variously described as the "steady-state flux" or "infinite dose tech- nique",[2,18,32] has provided some of the most definitive insights into the physiochemi- cal parameters that govern membrane diffusion, and has provided much of our knowledge concerning the diffusional mechanisms controlling percutaneous pene- tration. In the second strategy experiments are designed to mimic the *in vivo* situation and this has become known as the "finite dose technique".[2,32] This approach has recently received a great deal of attention, since it is potentially a very powerful tool for studying factors that influence percutaneous absorption and may provide a means for assessing dermal absorption of toxic agents.

Both approaches use diffusion chambers, and over the years various penetration chambers have been described that are suitable for use in absorption studies. These chambers were usually designed and fabricated for specific applications; in general, their designs differ only in nonessential details and the basic components of the chambers are essentially identical.[2] A detailed discussion on the design of diffusion chambers was published recently; it concentrated on the calibration of the hydrody- namic characteristics of various *in vitro* diffusion chambers, and indicated that these characteristics may have significant influence on *in vitro* permeation measure- ments.[33] Ideally, chambers should be made from inert material and constructed so that the contents of the chamber are visible. The chambers should be robust, easy to assemble, and suitable for use with a variety of membranes. Furthermore, they should provide a precisely defined area of tissue exposure, and the incorporation of devices which will permit the thorough mixing of the chamber fluid is absolutely essential. Also desirable is the ability to precisely regulate the temperature of the chambers. Today, penetration chambers are readily available commerically (e.g., Vangard International, Inc., Neptune, NJ and Laboratory Glass Apparatus, Inc., Berkeley, CA); they are easy to use, and are well designed, having incorporated many of the desirable features of diffusion chambers described by investigators with many years of research experience in the field of skin absorption.

In operation, the essential features of the chamber used in steady-state flux conditions require that the skin be mounted as a membrane between two fluid-filled chambers (see Figure 2A). The chamber fluid is usually an aqueous solution. The chambers are well stirred and the temperature appropriately controlled. The study compound is added to solution on the donor side and absorption is determined by serially sampling and assaying its concentration on the receptor side.[7,34-38] This type of diffusion chamber is frequently used in studies for determining the physical constants, such as permeability coefficient, diffusion coefficient, and partition coefficient that are pertinent in membrane diffusion as they relate to skin absorption. Modifications of this basic design have also been described which allow for determining the diffusion of vapors through the skin.[39]

The salient features of chambers used to mimic *in vivo* conditions are that the unstirred donor compartment is open to the atmosphere (see Figure 2B), and the skin preparation is mounted horizontally, epidermal side up, over the well-stirred, fluid-filled receptor compartment.[40,41] The study compound in an appropriate formulation is applied to the surface of the skin as a small finite dose and permeation is determined by the recovery of the compound in the receptor fluid. It should be pointed out that when the applied dose is large and absorption is minimal, infinite dose conditions may prevail.[2] A chamber with this basic design constructed with glass and containing a built- in water jacket for temperature regulation was described by Franz;[42] variations of this cell form the basis of many of the chambers that are currently in general use. One of the advantages of this design is that it permits the application of compounds on the skin as a solid deposited from a volatile solvent, as a liquid, as a semisolid (e.g., ointment gels and creams), as a film, or in a transdermal drug delivery device.[33,43] Furthermore, since the donor compartment is open, the applied compound may be removed during the permeation experiment; equally, more compound may be added. Also, the environmental conditions above the skin may be controlled. The design is flexible in that the donor compartment may be closed to mimic the occluded situation; it may be open to the ambient or defined environmental conditions, such as controlled humidity; and chamber design with glass towers which provide a mean for assessing the evaporation of the applied compound or the vehicle from the surface of the skin have also been described.[44] A further refinement in the chamber design to mimic *in vivo* conditions is the use of a flow-through system for the receptor fluid. This has the obvious advantage in automation of the sample collection process, and since the receptor fluid is replaced continuously, sink conditions may be maintained; also movement of the receptor fluid through the cell may be sufficient to achieve adequate mixing and to mimic the subcutaneous blood flow. Flow-through designs for *in vitro* permeation chambers have been described and used successfully by a number of investigators.[28,45,46]

The choice of diffusion chamber used for a particular *in vitro* study will depend on the research strategy and desired information. The attractions for the use of the infinite dose approach are that steady-state experiments are easy to control and interpret. They will provide information on flux, lag times, and relevant mass transfer coefficients that are pertinent to steady-state diffusion; however, this approach may be of limited value as a predictive model for assessing dermal absorption *in vivo*. The most obvious liability of the methodology is the extensive hydration of the skin preparation in the fluid-filled chambers. Hydration of the stratum corneum is known to result in drastic alteration in skin permeability.[12,47-49] Also, the routine use of aqueous solutions in the donor chamber fails to take into consideration the influence of the vehicles in the skin absorption process. A detailed discussion concerning the limitations of the infinite dose technique and the advantages of the finite dose technique have been reported; the indications are that the finite dose methodology may be more reflective of dermal absorption *in vivo*.[32]

C. Membrane Preparation

Whatever the source of the skin, the fundamental assumption for *in vitro* absorption studies is that the excised tissue retains its barrier functions.[50] The integrity of this membrane is of major importance, but the type of skin preparations used can greatly influence the outcome of the *in vitro* study. Therefore, regardless of the design and conduct of the *in vitro* study, preparation of the skin membrane is probably the most critical step.

The rate-limiting barrier is generally believed to be the stratum corneum, and it is often assumed that diffusion through the epidermis and the papillary dermis is much higher, and therefore, will not significantly influence overall dermal absorption. While this may be a valid assumption for substances such as aqueous alcohol solutions,[7,8] it is becoming increasingly evident that the stratum corneum is not the sole source of diffusion resistance in the skin.[1] Indeed, it has been suggested that for certain lipophilic chemicals the lipid-rich stratum corneum may not be a barrier; rather it may act as a reservoir and function more as a sponge, capable of absorbing a quantity of lipophilic material limited only by the solubility of the substance in the sebaceous and intrinsic epidermal lipids. For such hydrophobic compounds the largely aqueous epidermal and dermal tissues may be the more important barrier,[27,28] and therefore the thicknesses of the skin preparations can have an important influence on the transit rate determined *in vitro*.

It is generally assumed that the diffusion process terminates at the site of vascular entry, and the dermal microvasculature acts as a "sink", removing the penetrants.[1,2] In the common laboratory animals and in man, the dermis (approximately 2 to 3 mm) compared to the epidermis (approximately 50 to 100 μm) is a relatively thick tissue.[1,2] By virtue of its relative volume, the dermis may retain permeating substances so that compounds which penetrate the stratum corneum easily may appear to permeate moderately through the skin due to their long residence time in the dermis. Since the cutaneous microcirculation is situated within the region immediately beneath the epidermis, the use of skin preparations with a full-thickness dermis may not be representative of the *in vivo* situation. However, it should be noted that evidence is accumulating which indicates that the cutaneous microvasculature does not always remove all of the permeating substances; it is not always a perfect "sink", and a fraction of the absorbed material may be delivered to the subcutaneous fat and underlying connective and muscular tissues.[51] Appreciation of some of these issues would greatly assist in choosing the appropriate skin preparations for *in vitro* investigations.

Full-thickness skin preparations, however, are often used, particularly when the source is a laboratory animal whose skin is relatively thin; for example the skin of the mouse is approximately 1 mm thick.[52] In preparing the skin from animals for use as a membrane in *in vitro* penetration studies, the hair on the animal should be lightly shaved with an electric clipper some time prior to harvesting the skin. Chemical depilatories are sometimes used, but these can have deleterious effects on the skin and may result in enhanced skin absorption.[53] The skin should be removed from the body immediately following sacrifice. Upon excision the subcutaneous fat should be carefully removed as far as possible to reduce the artificial accumulation of penetrating substances in the skin, especially if they are lipophilic in nature. For mouse skin this is a relatively simple procedure requiring only gentle scraping with curved forceps;[54] for other mammalian skin, careful dissection may be necessary if one is to remove the fat without causing fundamental damage to the dermis.[30]

Because the use of full-thickness skin in diffusion studies may not be reflective of the *in vivo* situation, various techniques have been advanced to isolate the epidermis and stratum corneum so as to better define the functional barrier for *in vitro* studies. Split-thickness skin preparations have been obtained by a variety of physical and chemical methods. Chemical vesication with occlusive application of cantharidin,[55] ammonium hydroxide,[56] or formic acid[57] *in vivo*, and subsequent excision of the blister tops has been one of the techniques used for obtaining the diffusional membrane. Subcutaneous injection with staphylococcal epidermolytic toxin[58-60] has also been employed to achieve separation of the epidermis and dermis. Epidermal sheets have been prepared by exposure of whole skin to ammonia fumes for 30 min,[61] and for hairless skin a short exposure (60 s) of whole skin to mild heat (60°C) has been a convenient way for preparing an epidermal membrane.[55] For haired skin this is not a useful method because hairs remain in the dermis during the separation and leave holes in the epidermal membrane when it is peeled away. An alternative method is to treat the skin with concentrated salt solutions. Soaking whole skin in 1.5 *M* sodium bromide or 30% quanidine hydrochloride[62] for several hours will often separate epidermis from dermis such that hair will be removed with the epidermis. Controlled incubation of skin in dilute trypsin solution has also been used successfully to separate the stratum corneum from the epidermis.[55]

Perhaps the most frequently used and arguably the easiest method to obtain split-thickness skin is simply to use a dermatome.[2,46,63-65] However, for skin sections prepared with a dermatome, the thickness of the slices can be influenced by the angle of the cutting blade, the pressure applied, and the speed of cutting;[66] therefore, a high degree of skill and practice may be necessary in order to obtain reproducible preparations. Also, depending on the desired thickness chosen, usually 500 µm or less, a fraction of the papillary dermis will be part of the membrane. Since thickness of the membrane is an important parameter in the diffusion process, an assessment of the thickness and reproducibility of the skin preparation should be an integral part of an *in vitro* study. Thicknesses are usually determined microscopically and ideally, measurement should be taken from frozen sections to overcome the well-known destructive effects of conventional embedding and fixative procedures on the stratum corneum.[67]

A general consensus among investigators in percutaneous absorption is that human skin is preferred and should be used for *in vitro* assessment of dermal absorption,[68-71] but it is also recognized that a major liability of human skin as a research tissue *in vitro* is its notoriously high variability in barrier properties.[72] The source of human skin is frequently from cadavers, and since the investigator often has little or no control over the source and characteristics of the donor skin, the high variability observed with human skin preparations is to be expected. Characteristics such as treatment of the cadaver, elapsed time from death to harvest of tissue, skin site, age, health, sex, race, and skin care habits are examples of variables which may bias the *in vitro* penetration studies. Also, when skin samples are derived from elective surgery, the preoperative procedures such as scrubbing with antimicrobial disinfectants, the surgical manipulations, and the manner in which the membrane is prepared from the excised tissue are important details of concern. Again, these variables may influence the *in vitro* penetration observations. It has been recommended that where possible, in an *in vitro* study with human skin, such information should be routinely collected and carefully documented.[71]

Although human skin is the tissue of choice, its limited accessibility to many investigators, and the variability experienced with human skin have led many researchers to explore skin from various animals as models for skin absorption. However, species differ considerably in the structure and function of their skin and it is unlikely that animal skin will have barrier properties that are identical to that of human skin. Nevertheless, animal skin is routinely used for evaluating dermal toxicity and percutaneous absorption. Histological evidence and physiochemical studies have concluded that animal skin can provide reasonable percutaneous absorption models that approximate human skin; however, the debate concerning the appropriate animal model continues.[25,52,69] Numerous comparative studies, both *in vivo* and *in vitro*, have been conducted to identify the ideal animal model; from the results obtained thus far, it would appear that the choice of animal model will depend on the preference of the investigators and the compound under investigation. The pig, the monkey, hairless mice, and more recently the fuzzy or hairless rat have been described as species with the potential to be good candidates as predictive models of skin absorption in man.[73-75]

Because physical diffusion is assumed to be the principal determinant in skin absorption, and since opinions concerning the selection of an appropriate animal model remain in conflict, artificial barrier systems have been explored as potential models for evaluating absorption in human skin.[76] These systems offer some advantages over biological models in that they are reproducible, easily prepared, and the composition of the membrane can be readily manipulated; such membranes offer a defined matrix with which basic physical concepts regulating permeation may be examined. Various materials have been used in the construction of artificial membranes, and they include chemicals such as cellulose acetate, isopropyl myristate, mineral oil, and dimethyl polysiloxane.[2] Materials such as collagen[77] and eggshell membrane,[78] which are of biological origin, have also been used. In general, the construction of these artificial membranes attempts to mimic the stratum corneum barrier, and their use in diffusion studies have provided some useful information on the underlying mechanisms governing the physiochemical properties of chemicals and the relative abilities of the chemicals to diffuse through lipid membranes.[2,79] However, use of artificial membranes in dermal absorption assessment has been limited, but they have been used as models for evaluating potential drug formulations during the development of topical preparations and transdermal delivery systems.[2,80] Nevertheless, the extent to which artificial membranes may serve as useful surrogates for skin in dermal absorption studies has yet to be established.

Ideally, freshly prepared and viable skin samples should be used for *in vitro* permeation measurements. Unfortunately, this is not always possible. Because of the intermittent nature of the supply, and the variability of the source, skin preparations of human origin used for *in vitro* studies are often stored either under reduced temperature or in a dessicated state. The influence of storage conditions on the *in vitro* permeability properties of the skin is therefore a cause of concern. Based on the permeability of water, it has been reported that frozen storage over extended periods does not affect the barrier properties of the skin.[81-84] Penetration of phenol in rat skin was not affected by frozen storage.[85] Freezing (–20°C for 8 h) pig skin appeared to have no effect on the percutaneous penetration of a number of compounds including, benzoic acid, caffeine, N,N-diethyl-m-toluamide, fluocinolone acetonitrile, and lindane.[86] On the other hand, increased permeability of a chromone acid (6,7,8,9,-tetrahydro-5-hydroxy-4-oxo-propyl-4H-naptho (2,3-b)-pyran-2-carboxylic acid) was observed in excised human cadaver skin following frozen storage (–17°C). When the skin samples were dried under controlled humidity, at ambient temperature, and stored refrigerated (1°C), permeability to the chromone acid was unchanged following appropriate rehydration.[87] Permeability of T_2 toxin in human and monkey skin stored frozen at –60°C has also been reported to be greater than in the skin preparations that are stored refrigerated at 4°C.[88] The penetration and evaporation of the insect repellant N,N-diethyl-m-toluamide in excised pig skin was also affected by frozen storage. Pig skin frozen (–80°C) for longer than 1 week was more permeable than the freshly excised skin.[64] From the limited number of studies it is evident that storage conditions can influence the barrier functions of the skin and may vary with compounds; however, the extent of this effect on the results of *in vitro* studies awaits a more systematic examination of the problem. When circumstances dictate that the skin preparations must be stored, the effect of storage on the permeation of the specific compound under investigation should be determined.[71]

It is clear from the discussion thus far that the diffusional integrity of the skin preparation is of primary importance. However, observations from a limited number of studies have implied that skin metabolism may play a role in dermal absorption. Cutaneous first pass metabolism of compounds such as benzoyl peroxide[89] and nitroglycerin[90] has been suggested, and evidence showing the metabolism of topical T_2 toxin[91] and aldrin[92] by skin has been reported. *In vitro* penetration of benzo(a)pyrene was shown to be determined primarily by the epidermal viability,[27] and the permeation was accompanied by extensive cutaneous metabolism.[28,29] Studies with freshly prepared skin samples also demonstrated that factors which modulate the activities of drug-metabolizing enzymes, such as inducers and inhibitors, can also influence the permeation of benzo(a)pyrene.[30] Furthermore, cutaneous first pass metabolism was also observed to accompany the percutaneous permeation of topically applied steroids such as testosterone, estrone, estradiol, and cortisol.[31] From these recent studies it is evident that the metabolic status of the tissue can have a major influence on the percutaneous fate of topically applied agents. Therefore, in designing *in vitro* dermal absorption studies the relevance of biochemical viability of the excised tissue needs to be considered. It is perhaps worth noting that techniques for the cryopreservation of viable mammalian skin have been described in the literature.[93-96] Although the criteria used in these reports for demonstrating biological viability vary considerably, systematic assessments of these methods as potential means for "banking" skin preparations may provide useful opportunities for *in vitro* dermal absorption studies.

D. Receptor Fluids

Assessment of skin absorption by *in vitro* methods is dependent upon the assumption that the recovery of the compound of interest in the receptor compartment provides an accurate measure of penetration and permeation. The process governing the validity of this assumption relates to the partitioning of the compound of interest from the skin preparation to the receptor fluid. Consequently, the nature of the *in vitro* measurements will depend considerably on the appropriateness of the receptor fluid chosen. Therefore, caution must be exercised in the interpretation of *in vitro* observations. Consideration must be given to the partitioning criteria and the limitations of the receptor fluid used.

Historically, the selection of receptor fluid is essentially empirical. Its choice largely reflected the views of individual investigators as to the structure and function of the skin as an interface in dermal absorption. In the pioneering studies distilled water was utilized as the receptor fluid.[7,8] These classic studies demonstrated the significance of laws of mass action and diffusion in dermal absorption, and are the bases of our current knowledge on skin permeability. The concepts developed in these investigations also governed much of the *in vitro* methodology used in skin absorption investigations.

The emphasis of *in vitro* skin absorption studies has been concerned traditionally with the ability of chemicals to diffuse across the nonviable skin barrier. Until recently, issues regarding the possible influence of cutaneous metabolic transformations on diffusion and permeation through the tissue have received negligible attention. It is assumed that metabolic activities in the skin contribute little to the percutaneous absorption process. The justifications for this viewpoint have come not only from studies demonstrating the importance of diffusion laws to percutaneous absorption, but also from some favorable *in vivo* and *in vitro* comparisons of dermal absorption. Using *in vitro* conditions in which metabolic viability of the skin would not be expected to be maintained, reasonable correlations between *in vivo* and *in vitro* permeation values for a range selected of compounds has been reported.[42,63,74,97,98]

However, an *in vitro* approach for studying the absorption and cutaneous metabolism of topically applied substances using skin maintained as short-term organ cultures was reported recently.[28,29] Fundamentally, the elements of the two culture systems described, a static and a flow-through design, are based essentially on that of the finite dose technique. *In vitro* viability and morphological integrity of the excised skin were maintained under defined conditions with the use of an appropriate culture medium. The culture medium, which was minimal essential medium containing fetal calf serum (10% v/v) and gentamicin (0.1 mg/ml) also serves as the receptor fluid. Using this approach it was demonstrated that skin penetration and the percutaneous fate of compounds such as benzo(a)pyrene and steroids, in a variety of mammalian skin samples, were influenced by epidermal viability and metabolic status of the skin.[30,31] With these observations, the importance of cutaneous metabolism during skin absorption has received new emphasis, and viability of the excised tissue may have important relevance for *in vitro* skin absorption studies.

Until recently, the overwhelming choice of receptor fluid has been isotonic saline, although buffered, usually pH 7.4, isotonic salt solutions have also been employed.[1,2] However, increasingly there are concerns regarding the suitability of these aqueous salt solutions as receptor fluid in absorption studies of lipophilic compounds with limited water solubility. Also of concern is the ability of these salt solutions to maintain tissue viability when *in vitro* studies are conducted over an extended period of time. For hydrophobic compounds the limiting step may not be penetration of the stratum corneum but rather partitioning from the skin into the aqueous receptor. This inability of the water-insoluble chemical to freely partition from the excised skin into the aqueous receptor fluid would seriously compromise the permeation measurement. When applied to the skin *in vivo*, these hydrophobic chemicals may readily penetrate the stratum corneum, diffuse through the skin, and, because of the solubilizing and emulsifying abilities of biological fluid, be taken away by the blood in the dermal micovasculature. Therefore, permeability of these compounds determined *in vitro* will appear to be considerably lower than *in vivo*. The inability of hydrophobic compounds to partition from the skin into the receptor fluid is a serious liability which may limit the value of the *in vitro* methodology. This problem has been alluded to by other investigators,[42,97] and what constitutes an appropriate receptor fluid therefore requires careful examination.

A twofold increase in penetration of hexachlorphene in isolated stratum corneum was observed when albumin (3% w/v) in a physiological salt solution was used in place of isotonic saline as the receptor fluid.[100] Enhanced *in vitro* penetration of parathion in pigskin was also observed when swine serum was used as the receptor fluid.[99] The incorporation of aqueous alcohol into receptor fluids, to facilitate partitioning, in some *in vitro* studies was practiced by some investigators,[101] and in a recent study a nonionic surfactant (poloxamer 188) was employed to facilitate the solubility of linoleic acid.[102] Thus, it is evident that increased *in vitro* permeation can be achieved by improving the lipophilicity of the receptor fluids.

An examination of the effect of various receptor fluids on the permeation of two fragrance compounds, cinnamyl anthranilate and acetyl ethyl tetrametyltetralin, with hydrophobic properties, in rat skin was reported recently.[97,101] In this study full-thickness and dermatomed (350 μm) skin preparations were used, and various concentrations of nonionic surfactants (polyethylene glycol 20 oleyl ether, octoxynol, poloxamer 188, polysorbate 80) in water, bovine serum albumin (3% w/v) in water, mixtures of glycerol and water (50:50) and ethanol and water (40:60), together with normal saline and rabbit serum were evaluated as receptor fluids. The effect of the receptor fluids on the integrity of the skin preparation was monitored by assessing the permeation of reference compounds, such as cortisone, urea, and water simultaneously with the fragrance chemicals. The results of these studies indicated that replacing normal saline with any of the other solutions improved the extent of *in vitro* skin permeation of the two fragrance compounds. However, altered barrier function with respect to cortisone penetration was observed for receptor fluids containing organic solvents and some nonionic surfactants. Although physiologically based fluids, such as the use of serum or plasma proteins in buffered salt solutions, may be more appropriate; their use, however, did not substantially increase the *in vitro* absorption of the fragrances. It was determined that a 6% solution of polyethylene glycol 20 oleyl ether in water provided the best improvements in permeation. This nonphysiological solution had no apparent effects on the integrity of the barrier, and it was selected as the receptor fluid of choice. However, despite the obvious improvements in permeation, estimates of skin absorption of the two hydrophobic compounds by the *in vitro* methodology was still significantly lower than the corresponding *in vivo* determinations. It was evident that the thicknesses of the skin preparations contributed significantly to the anomalies noted in the *in vivo* and *in vitro* comparisons. In subsequent studies the authors suggested that a better correlation with *in vivo* observations for a number of hydrophobic compounds was obtained when thinner sections of microtomed skin (<300 μm) were used in conjunction with receptor fluids containing polyethylene glycol 20 oleyl ether.[65]

It is evident that the ideal receptor fluid has yet to be formulated; indeed, what constitutes an ideal fluid remains to be established. As described, the different approaches employed by various investigators to manipulate the lipophilicity of the receptor fluid may provide the basis for a rational approach to the development of such a fluid. Depending on the desired information, it is clear that maintaining the viability of the excised tissue should be considered as important, particularly in investigations where extensive cutaneous metabolism is suspected. An ideal receptor fluid should therefore incorporate the appropriate physiochemical and biochemical properties that are desirable for *in vitro* percutaneous absorption and metabolism investigations. Unfortunately, what these desirable properties may be remains to be characterized, but it is suggested that physiologically based fluids would be more appropriate. Biological fluids consisting of lipid- and water-soluble components in a homogeneous mixture, such as milk, plasma, and serum, supplemented with culture media and proteins, may provide the desirable properties of an appropriate receptor fluid. However, the chemical composition of biological fluids is not constant. For example, diet, disease, and pregnancy are some of the factors known to alter the concentrations of various serum proteins which bind hormones, drugs, and lipids.[103] Therefore, good quality control will be essential in order to avoid inconsistencies in receptor fluid composition prepared with supplements of biological origin. Receptor fluids with constant chemical composition prepared with nonionic surfactants, organic solvents, and other excipients in culture media, at concentrations that do not compromise viability of the tissue, may also provide the means to facilitate the *in vitro* dermal absorption of the compound of interest. Combinations of such solutions deserve evaluation not only as a receptor fluid, but also as a medium for maintaining the structural integrity and biological viability of the skin preparation. Until an ideal receptor fluid has been identified, choosing the optimum receptor fluid for a particular compound under investigation remains empirical. Therefore, in evaluating *in vitro* observations of dermal absorption, it is well to bear in mind the key words *partition* and *metabolism*.

E. Dose Application

How the chemical of interest is applied to the skin is often dependent upon the strategy of the *in vitro* approach used and the nature of the investigations to be performed. In general, dose application in dermal absorption studies conducted *in vitro* is relatively straightforward. For studies using the conventional steady-state diffusional (infinite dose) procedure to determine mass transfer coefficients, a well-stirred donor solution of the compound of interest, at a defined constant concentration, is used to deliver the penetrant across the skin preparation, and the penetrant is received by a well-stirred receptor. The important design features of these studies are that the quantity of compound that penetrates the membrane must be kept small relative to the total amount available and that there is no appreciable reduction in the concentration of the compound in the donor compartment. Because of the biological variability, up to a 10% depletion of the donor phase, and correspondingly, in the receptor a buildup of 10% is permissible without significantly violating zero-order flux conditions.[2] Traditionally, aqueous donor solutions are used,[7,8,13] but compounds with limited water solubility present significant problems, and as discussed by others, hydration of the diffusional membrane can have a major influence on the permeability determinations.[12,47-49]

For the *in vivo* mimic (finite dose) technique, the compound of interest is prepared in an appropriate vehicle which may be liquid or semisolid. An appropriate amount of the dosing preparation is then applied uniformly onto the surface of the skin. The amount of compound applied is often determined empirically, and is often based on doses reflective of *in vivo* applications. For a liquid vehicle containing the test compound, aliquots, usually 5 to 10 $\mu l/cm^2$ of skin, can be applied using a micropipette. For semisolid or nonliquid preparations application can be achieved by gently rubbing with a small stirring rod onto the surface of the skin; the amount of material applied can be determined by weighing the rod before and after application.[32] Uniformity of the application is an important aspect in these studies; however, assessing these criteria may be difficult and often uniformity is assumed without supporting evidence. Recently, *in vitro* dermal absorption studies have also been conducted in which the chemical was applied using transdermal drug delivery devices.[104]

These relatively simple dosing procedures often overshadow important issues that can have a significant impact on the *in vitro* assessment of dermal absorption. In diffusion terms relative to percutaneous absorption, flux is dependent upon the concentration gradient across the barrier membrane and the partition of the penetrant between the vehicle and the stratum corneum;[2,13,15] consequently, solubility of the penetrant in the vehicle used could influence the outcome of the permeation determinations. The role of the vehicle as a factor that may influence skin absorption has been discussed in detail by others[1,2,98] and is not the objective of this section. However, it is worth noting that some solvents are chosen as vehicles because they act as penetrant enhancers, as in the case of, for example dimethylsulfoxide and azone,[105] while others, such as some lipid solvents and surfactants,[14,106] may increase skin permeation by destroying the barrier properties of the stratum corneum. Thus, caution must be exercised in selecting the appropriate vehicle for dose application. Postapplication loss of volatile components in the vehicle can alter the permeation characteristics of chemicals,[107] and when highly volatile vehicles are used, for example acetone, it may result in the compound of interest being deposited as a thin film of solid onto the surface of the skin. A nonvolatile vehicle, such as an ointment, may have occlusive properties and change the degree of hydration in the stratum corneum. Both of these situations can greatly influence the extent of percutaneous permeation.[1,2]

In most toxicologic and pharmacologic investigations, the administered dose is precisely defined and dose response relationships are usually carefully evaluated. Unfortunately, this is not always the case in dermal absorption studies, and in the percutaneous absorption literature a great deal of information of questionable relevance exists, since the dose applied was frequently not clearly defined or reported. Available evidence indicated that the amount of absorption is greatly dependent upon the concentration of the applied dose and the surface area of exposure.[68,108] It has been reported that increasing the concentration of the applied dose may alter the percent of the applied dose being absorbed, but total absorption is increased.[109] Furthermore, increasing the surface area of exposure will also result in increases in the extent of absorption.[110] Therefore, in defining the dose applied, one must consider not only the amount of chemical applied per unit area, but also the total surface area of application, and the the total dose applied. Additionally, the frequency of application and the duration of exposure can influence the extent of skin absorption.[108,111,112]

Examination of the interrelationship of the various parameters pertaining to dose application in dermal absorption is only just beginning. How these parameters may affect the extent of dermal absorption is being explored,[108] and it is clear from that discussion that our current knowledge in this area is far form complete; therefore, in assessing the dermal absorption and dose response of a compound using *in vitro* techniques, defining the dose for investigation requires careful consideration.

F. Data Presentation

A characteristic of *in vitro* skin absorption studies is that results are frequently associated with a high degree of variability. For the purpose of the following discussion it is assumed that a sufficient number of replicate experiments have been conducted, and that the results of the studies were appropriately analyzed statistically.

The way in which the results of *in vitro* dermal absorption studies are presented is, of course, dependent on the type of experiment performed. In an "infinite dose" experiment, performed with strict compliance to the principles of diffusion, three parameters in theory may be monitored as a function of time.[2] These are the amount of drug or chemical entering the skin membrane, the amount permeating the skin and appearing in the receptor compartment, and the amount of drug remaining in the skin membrane. In practice, however, the amount of chemical recovered in the receptor fluid with time is the parameter which is routinely monitored.

The conventional approach to presenting data from this type of study is to plot the cumulative amount of drug reaching the receptor as a function of time (Figure 3). From the linear portion of this plot, we obtain the most important piece of information, i.e., steady-state flux of the compound across the skin membrane. This value is generally normalized with respect to the area of the skin membrane, and is usually expressed as amount of drug per unit area per unit time. The intercept on the x-axis, obtained by extrapolating the linear part of the curve, gives a measure of the time required to establish a linear concentration gradient across the skin membrane, and is referred to as the lag time. From these two parameters it is relatively simple, using the diffusion equations, to calculate the permeability coefficient, and derive the other mass transfer parameters such as the diffusion coefficient of the drug across the skin, the partition coefficient of drug between the skin and the receptor fluid, and the diffusional thickness of the membrane.[2,15] The validity of using diffusion equations in skin penetration is based on a number of important assumptions.[9,10,13,14] However, because the skin is not an isotropic membrane, and may offer multiple layers of diffusional resistance, these assumptions are often violated in *in vitro* experiments. Also, given the known complexity of the skin absorption process, the utility of these derived parameters may be of limited value. The amount of drug that enters and remains in the skin membrane during steady-state absorption may be of value, and this parameter may be obtained from analysis of the skin membrane after the flux

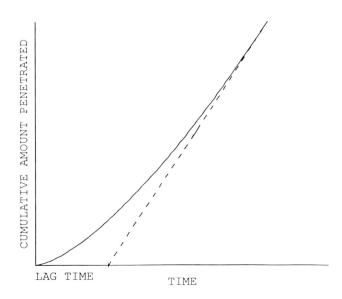

Figure 3.   A typical plot of cumulative amount of drug permeating
the skin as a function of time — "infinite dose".

data have been obtained. However, to determine the amount of drug entering the skin
during the approach to steady state is labor intensive,[1] and has not been conducted
to any significant degree. Therefore, in an "infinite dose" experiment the most
reliable and relevant parameters generated are probably the steady-state flux and
permeability coefficient values.

A plot of cumulative amount of drug appearing in the receptor fluid as a function
of time is again the simplest way to present results from an *in vivo* mimic, "finite
dose" experiment (Figure 4). Frequently, the amount of drug permeating is expressed
as a percent of the applied dose. The initial part of this curve resembles that for the
"infinite dose" situation, and parameters such as lag time and flux may be determined
in a similar manner. In theory, because the amount of drug in the donor compartment
is depleted with time, the plot becomes sigmoidal and plateaus once the drug in the
donor compartment is exhausted. In practice, this is only observed when the dose in
the donor compartment is small and there is extensive skin permeation. Where the
dose is high and permeation is small so that there is negligible depletion in the donor
compartment, the experimental conditions approach that of the "infinite dose" situation.

The results of *in vivo* mimic experiments can also be presented in the form of rate
of appearance in the receptor fluid as a function of time. Typically, this involves a
plot of the percent of the dose recovered in the receptor per hour vs. time, and for
a readily permeable compound a typical absorption curve is observed (Figure 5). The
advantage of this form of data presentation is that the information may be used to

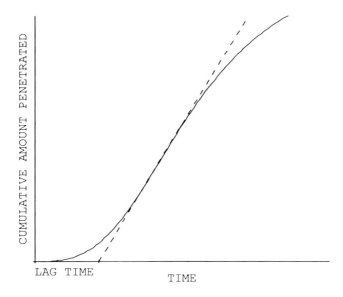

Figure 4. A typical plot of cumulative amount of drug permeating the skin as a function of time — "finite dose".

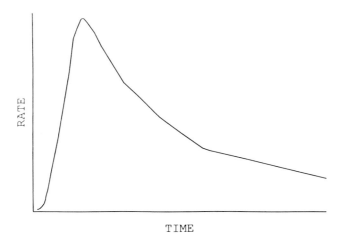

Figure 5. A typical plot of rates of permeation as a function of time — "finite dose".

evaluate and compare penetration rates at different times following drug application under different experimental conditions. The results from these studies can be applied to various kinetic models of skin absorption.[113-121] From these models it may be possible to derive kinetic parameters that characterize the skin absorption process, and using these kinetic parameters, make predictions concerning the extent of skin absorption *in vivo*. Unfortunately, these are only theoretical predications, and despite the advances made, the status of our current understanding of skin absorption is such that *in vivo* measurements invariably will be required for conformation.

G. Checks and Balances

The review, thus far, has focused primarily on the various elements that contribute to the design of an *in vitro* study of skin absorption. In this section some of the necessary checks and balances that need to be considered in the conduct of an *in vitro* investigation will be discussed.

1. Barrier Integrity

Establishing the integrity of the diffusional barrier of the skin preparation is one of the critical checks that should be performed prior to every *in vitro* investigation. This is particularly important in the preparation of very thin epidermal sections, especially from haired animals. Holes in the membrane corresponding to hair follicles can be a major problem. This problem may be reduced somewhat if the dermatome sections of epidermal membrane are prepared so as to include the hair roots; however, the increase in the thickness of the membrane may compromise the subsequent permeation measurements. This is a dilemma that individual investigators will have to rationalize based on the nature of the study and the compound being investigated. In a recent study it was reported that about half of the dermatome sections of 300 μm epidermal membranes obtained from normal rats were deemed to have been damaged during preparation,[65] although the authors also reported that 200 μm thick dermatome sections from human and the fuzzy haired rat could be prepared routinely, without damage.

The integrity of the membrane should also be assessed at the termination of the *in vitro* permeation measurement. This is necessary because it is possible that the intrinsic toxic potential of the chemical under investigation may interfere with the barrier functions of the membrane and influence the skin permeability of the compound, especially following extended contact during a lengthy penetration study. In a study on the *in vitro* absorption of glycol ethers through human skin, it was reported that contact with some of the ethers resulted in slight irreversible effects on the barrier function of the skin;[37] however, the nature of the damage or how the damage may have altered the permeability of the skin to individual glycol ethers could not be judged from the experiments reported.

Holes due to hair shafts or other gross defects in the membrane may be detected by simple visual inspection of the skin preparation. The diffusional integrity of the skin preparations is often assessed by determining their permeability to tritiated water. This is often performed as the first part of an *in vitro* study. Generally, penetration of tritiated water is measured for several hours, and if the extent of permeation is within a preselected value, the tissue is deemed acceptable and may be used in subsequent permeation studies.[37,65] In a slight variation to this approach, the permeation of a control compound, ^{3}H-cortisone, was monitored simultaneously as the ^{14}C-labeled test compound by dual-label technique and the integrity of the skin barrier under different experimental conditions was correlated to permeability constants determined for the control compound.[65] The permeability of other compounds, known not to be metabolized by the skin and whose penetration is dependent only on diffusion, can also serve as functional references.

2. Biological Integrity.

When biological integrity of the tissue is an important part of the *in vitro* skin absorption study, as in the case where skin metabolism of the compound is a special concern, it is necessary to establish that the *in vitro* conditions are capable of maintaining the metabolic viability of the tissue during the course of the investigation. The desirable *in vitro* conditions employed can be established in separate experiments prior to the permeation investigations. Various parameters used to demonstrate and to assess viability and structural integrity of the excised skin preparations have been previously described.[29,54] For example, histological examination of the skin preparation, appropriately stained with hematoxylin and eosin, by light microscopy may be used to establish the structural integrity of the cultured tissue. The *in vitro* incorporation of radiolabeled precursors into tissue macromolecules is another example of a technique which has been used to demonstrate the viability of the cultured skin preparations. Incorporation experiments are usually performed by incubating the cultured tissue for a defined short period of time in culture medium containing the appropriate radiolabeled precursors. Viability is demonstrated based on the extent of incorporation of ^{3}H-thymidine and ^{14}C-leucine into epidermal DNA and proteins, respectively, and is determined following isolation and purification of the appropriate macromolecules. Tissue viability can also be assessed based on the ability of the excised and cultured skin preparation to metabolize radiolabeled glucose to radiolabeled carbon dioxide.[30,94] In using a flow-through penetration chamber with a bicarbonate-buffered culture medium as the receptor fluid, a simple qualitative demonstration of *in vitro* viability is to stop the flow of receptor fluid at the end of the study, and monitor visually the change in the color of the phenol red indicator in the culture medium. Viable tissues will continue to carry out intermediary metabolism and produce acidic byproducts, such as lactic acid, resulting in the straw-colored appearance of the culture medium within the receptor chamber. The color of the medium from nonviable tissue under these condition will remain red.[122]

3. Toxicity

As alluded to earlier, skin contact with a toxic chemical, either as the vehicle or the compound of interest, can result in local effects that may affect its barrier properties. Since the stratum corneum is the major barrier involved, its physical integrity is of primary importance. Where there is obvious damage to the surface of the skin, enhanced permeability is to be expected in many instances.[14] However, biochemical effects in response to biologically active chemicals may also influence the functional capabilities of the excised skin. Therefore, an assessment of local toxicity following topical application should be a factor for consideration in both *in vivo* and *in vitro* skin absorption studies. For *in vitro* studies, the biochemical and histological parameters described may be used to evaluate possible tissue toxicity.[29,54,123] In addition, the leakage of intracellular enzymes into the receptor medium from the excised tissue following *in vitro* topical exposure will provide another indicator for monitoring cellular injury.[54,123-125] In this regard the activities of lactate, malate, and glutamate dehydrogenases, and glutamic-oxalactate transaminase in the receptor fluids can be readily determined using commerically available diagnostic kits. However, as very little is known concerning the potential influence of local toxicity on cutaneous metabolism and skin absorption, it is suggested that whenever possible a "no-effect" level of the compound should be used in studying its *in vitro* dermal absorption and metabolism,[20,30] thus avoiding unnecessary complications. In any case, some knowledge of the physical and biological integrity of the skin preparation before, during, and after the experiments would greatly facilitate the interpretation of the *in vitro* observations.

4. Environmental Conditions

The environmental conditions of the *in vitro* studies can have a major influence on the outcome of dermal absorption investigations, especially when open diffusion chambers are used to mimic the *in vivo* situation. The temperature the skin and of the air can both affect the measured flux, and it has been estimated that a change of 10°C will change the flux by approximately a factor of 2.[9,126-129] However, in controlled systems where temperature variations are rarely more than 1 to 2°C, temperature effects are likely to be small compared to the other variables of *in vitro* absorption discussed in this chapter. Nevertheless, the skin temperatures selected for *in vitro* studies should reflect the *in vivo* situation, and it has been recommended that the surface temperature of the skin should be maintained at 32 ± 1°C.[71] Generally, this can be achieved when the temperature of the receptor fluid is kept at 37°C and the air temperature is maintained at 25°C.[119] However, the ambient temperature under which *in vitro* studies are normally conducted is in the range of 25 to 30°C, and if the upper part of the chamber is not maintained at a constant temperature, the skin temperature will vary.[64] This is normally not a major concern, but it is especially problematic in studies involving potentially volatile penetrants or volatile delivery systems.

Volatile chemicals such as solvents and insect repellents, which are designed to evaporate slowly from the skin surface, present additional challenges to both *in vivo* and *in vitro* skin absorption methodologies. A volatile compound, when applied topically to the skin, will penetrate the skin and be absorbed. However, at the same time a portion of the compound will be lost due to evaporation from the skin surface. Consequently, the rate of evaporation and the relationship of evaporation to penetration will affect the rate of penetration and the quantity of chemical absorbed dermally.[64,130-133] The extent of evaporation from the skin surface is a function of the dose, air flow, and temperature of the skin surface.[44] Therefore, the design of *in vitro* methods to determine the dermal absorption of volatile compounds must include techniques for controlling the variables that are associated with evaporation, and diffusion chambers have been developed in which these variables may be monitored and their influences on both evaporation and penetration have been assessed simultaneously.[44,64]

5. Loss and Recovery of Compound

It is evident that loss of chemicals under investigation, as in surface evaporation, may have serious implications in the assessment of dermal absorption, and raises the issue of the importance of mass balance and dose accountability determinations. Such determinations are often not reported or conducted, even though a low recovery of the applied material may compromise the validity and interpretation of both *in vivo* and *in vitro* skin absorption studies. The recovery of the material of interest in the receptor fluid traditionally serves as the basis for determining skin permeation *in vitro*. Consequently, much of the interests have focused on the receptor fluid, and recovery of material in the skin itself has received only limited attention. However, the distribution, metabolism, and binding of the agent of interest in the skin are integral parts of the skin absorption process,[1,2] and estimates of the chemical in the tissue should be included as part of any dermal absorption study. Investigations where the emphasis is on the localization of material in the different layers of the skin and its use as measures of percutaneous absorption and local bioavailability have been published, and the merits of these investigation extensively discussed.[1] The horizontal slicing technique[134] advocated in these studies brings an additional dimension to *in vitro* dermal absorption studies that should be explored. In any event, an assessment of the amount of drug present in the skin or in the various layers of the stratum corneum, epidermis, and dermis would be desirable assets in an *in vitro* dermal absorption investigation. Such analysis of the skin would contribute to establishing mass balance and dose accountability. This is another one of the important checks that should be determined routinely in an *in vitro* investigation.

Radiolabeled chemicals are routinely employed in dermal absorption investigations, and frequently liquid scintillation spectrometry is the only analytical procedure used for the detection of the penetrating substances. The chemical identity of the penetrant is assumed to be the applied material. However, evidence is accumulating to indicate that, in many incidences, chemical transformations accompany the permeation and partition of the topically applied material into the receptor fluid, and these transformations are often associated with metabolic processes.[28-31,135] Consequently, the recovery of radioactivity in the receptor fluid reflects the combined permeation of both the parent compound and metabolites. Therefore, in assessing the dermal absorption profile of a particular chemical it may be necessary to develop additional analytical methodologies, such as HPLC, which would help to differentiate the various components of the penetrant. In view of the possible involvement of cutaneous first pass metabolism in skin absorption and toxicity of topically applied chemicals, such analyses are increasingly being performed as an integral part of *in vitro* dermal absorption investigations. However, caution must be exercised in assessing the significance of metabolism and chemical transformations during absorption based on observations from *in vitro* investigations.

As alluded to earlier regarding the use of aqueous receptor fluids in the *in vitro* permeation of hydrophobic chemicals, it has been suggested that under the conditions employed where the water-insoluble applied compound is unable to freely partition from the skin to the aqueous receptor fluid, the role of a metabolically viable epidermis is simply to convert a water-insoluble compound to a water-soluble metabolite capable of partitioning into the receptor fluid. Consequently, the recovery of significant amounts of metabolites in the receptor may be artifactual, and therefore the influence of skin metabolism on dermal absorption may be overestimated. In this regard much criticism has been directed at the metabolism and permeation studies reported for benzo(a)pyrene, but not for similar studies with testosterone.[29-31] However, for hydrophobic chemicals, it is possible that the predominantly aqueous viable epidemis in fact may be the rate-limiting barrier to skin absorption and cutaneous metabolism simply facilitates this process. In either case, additional experiments will be necessary in order to rationalize these concerns.

Compounds applied topically will encounter the resident microorganisms of the skin. Such organisms have been shown to be capable of metabolizing drugs such as betamethasone-17-valerate[136] and glyceryl trinitrate,[137] and indeed the metabolism of topical benzoylperoxide to benzoic acid has been solely attributed to skin microorganisms.[89] However, experience in our laboratory with metabolism studies in viable human skin preparations which have been thoroughly scrubbed with antimicrobial disinfectants, has led us to the view that it is cutaneous rather than microbial metabolism which is primarily responsible for the observed biotransformations. In support of our conclusions, it has been reported that during the development of methods for assessing pregraft viability for skin based on the ability of the excised tissue to metabolize various substrates, the metabolic contributions due to skin microorganism were determined to be negligible.[94] Nevertheless, the extent to which microbial metabolism may be involved is as yet unresolved and what role skin microorganisms may play in the fate of topically applied chemicals remains to be investigated.

It is generally assumed that the metabolites found in the receptor fluid are the products of cutaneous metabolism, formed during the permeation of the parent compound through the viable skin. However, it has been suggested that these metabolites may result from transformations in the receptor fluid after skin permeation. These transformations may be simple hydrolysis or they may be mediated chemically due to the inherent instability of the parent compound in the receptor fluid. Alternatively, as has been advocated, these transformations may be mediated by active enzymes that have leaked into the receptor fluid from the culture tissue. Consequently, the metabolites formed may be artifacts of the *in vitro* methodology. This is a serious limitation and there are reports to support this contention.[138,139] It is conceivable that soluble enzymes, such as nonspecific skin esterases,[140] may leak into the receptor fluid from the excised tissue, and be responsible for hydrolyzing the parent esters. This can be considered to be an example of passive metabolism.[1] In contrast, active metabolism is exemplified by, for example, the oxidative transformations of polycyclic hydrocarbons and steroids. Such active metabolic reactions are mediated by microsomal enzymes and are dependent upon the availability of cofactors and the generation and utilization of energy. In the absence of added cofactors and a regenerating system in the receptor fluid it is unlikely that such biotransformations would proceed. Furthermore, in a metabolically viable and structurally intact tissue leakage of membrane-bound enzymes is expected to be negligible. Nevertheless, in assessing the significance of metabolism in dermal absorption, a clear determination of the origin of the metabolites found in the receptor fluid may be necessary. This may involve separate experiments to determine the stability of the penetrant in the culture fluid, and to assess the ability of the receptor fluid, which has been exposed to the excised tissue, to support transformation of the compound of interest. Such studies are important in providing information useful in developing appropriate remedial strategies in the design of the *in vitro* experiments.

From the above discussion it is evident that the principles of *in vitro* methodology for the assessment of dermal absorption are well defined; in practice, the design, conduct, and interpretation of *in vitro* studies require careful consideration. Appropriate use of the *in vitro* methodology can be an extremely powerful tool in evaluating and understanding dermal absorption, and there are substantial theoretical and practical experiences to guide research projects in this area. Given the available options, there are many advantages to conducting experiments with viable skin preparations as the diffusional membrane. This would greatly increase the confidence that well-designed and -executed in vitro experiments will produce information that has acceptable *in vivo* relevance. Furthermore, by considering the skin not just as a barrier to absorption, but as an organ capable of many metabolic and toxicologic interactions, the *in vitro* results may contribute to our understanding of the complicated process of skin absorption and disposition of topical agents.

III. CONCLUSIONS

In reviewing the literature on percutaneous absorption, it is fair to say that studies using excised skin mounted in diffusion chambers have provided much of our current understanding on the mechanism of dermal absorption. Indeed, the *in vitro* approach has contributed significantly to defining the fundamental physiochemical parameters underlying dermal absorption, and has led to important advances in the area of dermatopharmaceutics as evidenced by the success of transdermal delivery of drugs. In the area of cutaneous toxicology, *in vitro* methodologies are receiving much attention as techniques for assessing and predicting dermal absorption of toxic agents. The popularity and success of the *in vitro* approach stem from the fact that the methodology used in the *in vitro* studies is relatively simple to follow. Compared to *in vivo* studies, *in vitro* experiments generally provide the investigator with the ability to manipulate and control the experimental conditions. The *in vitro* methodology provides the investigator with the unique opportunity to monitor the rate and extent of percutaneous absorption in skin tissue removed from the rest of the body, and in general, *in vitro* results can also be obtained relatively quickly.

The relevance of *in vitro* absorption results is largely assumed, and a great deal of information has been generated using *in vitro* models. Probably because they are difficult to perform, few studies have been designed specifically to correlate results from *in vitro* studies with observations from *in vivo* studies.[32,42,97,98] In general, these studies, albeit with only a limited number of compounds, have demonstrated that there were good qualitative correlations between *in vivo* and *in vitro* skin permeation. Quantitatively, the correlations, in terms of the percentage of the applied dose absorbed, were somewhat variable, and frequently the cause of the poor correlation was believed to be due to technical limitations of the *in vitro* methodology. Some of these limitations, such as the appropriate choice of receptor fluids and thickness of the epidermal membrane, were discussed in preceding sections. In reviewing the literature on *in vivo-in vitro* comparisons of dermal absorption it is interesting to note that the *in vivo* results are often the standards by which the validity of the *in vitro* techniques are judged. The validity of the *in vivo* results is assumed, despite the fact that *in vivo* absorptions are often determined indirectly (see Chapter 8). The methodology used is based on excretion data. Frequently, corrections are necessary to account for the fraction of the absorbed dose that was not excreted, and these corrections are dependent on the validity of unsubstantiated assumptions.[70,141] Both the *in vivo* and *in vitro* methodologies for measuring dermal absorption are evolving, and are constantly being developed and refined; therefore, in assessing the value of such comparisons, caution should be exercised. Meaningful *in vivo-in vitro* comparisons should be made only when the experimental parameters of the *in vitro* studies closely resemble those of the *in vivo* studies. Some of the parameters that need to be considered include concentration of the applied dose, surface area of application, the total dose, duration of exposure, occlusion, vehicle, and the length of the *in vivo* and *in vitro* studies.

Percutaneous absorption is a complex process. In the past, many of the *in vitro* evaluations of skin permeation were conducted with skin preparations that were incapable of respiration and devoid of any active biochemical processes. Therefore, the reported recovery of material diffusing through such skin preparations may only be of limited value. Since both diffusional and metabolic processes are potentially important in determining the percutaneous fate of topically applied agents, the use of nonviable skin preparations and inert membrane may only provide information relative to the diffusional aspects of skin penetration. The net result of the diffusional and metabolic processes on the fate of surface-applied substances may be obtained by the use of metabolically viable and structurally intact skin maintained in organ culture. The relative importance of these processes, however, will depend on the physiochemical properties of the compounds and the metabolic capabilities of the epidermal cells toward the compounds in question.

Some of the important issues relating to *in vitro* skin permeation and metabolism studies have been discussed in this chapter. Others, such as the potential importance of the dermal microvasculature in skin absorption also require attention. New techniques, such as those involved with the isolation and creation of skin flaps with defined and accessible vasculature, are being developed and validated as models to further our understanding of percutaneous absorption (see Chapter 10). In the meantime, questions concerning differences in skin site, skin appendages, skin condition, age, sex, and species, together with the influence of various biochemical and physiochemical factors that affect dermal absorption, can be addressed using *in vitro* skin organ culture systems. Such studies, coupled with studies using the second generation of *in vitro* skin absorption models, such as the skin flap models, will provide the means whereby differences in skin absorption and metabolism between species, including man, may be investigated. With this ability to directly compare the percutaneous fate of topical xenobiotics in mammalian skin under defined conditions, it should be possible to establish a basis for extrapolation and provide a predictive estimate for human skin absorption and bioavailability following topical exposure. When skin contact with a chemical results in local effects, pathological changes in the skin may be expected to affect its barrier properties, and hence influence the percutaneous fate of surface-applied chemicals. The stratum corneum is the major barrier involved; therefore its integrity is of primary importance. However, biochemical changes in the skin in response to topical exposure to biologically active chemicals may also influence the metabolic capabilities and metabolic status of the skin, and thereby modulate the cutaneous disposition of topically applied substances. Studies have demonstrated that the metabolizing activities of the skin readily respond to modulation by inducers and inhibitors of drug-metabolizing enzymes.[22-26,29-30,142-144] Where significant cutaneous metabolism is anticipated, such modulation of the skin could have important implications on the outcome of dermal absorption of xenobiotics.

The skin, in addition to being a portal of entry for a variety of topically applied chemicals, is also a drug-metabolizing organ, and a target organ for local toxicity. Thus, knowledge of the processes involved in the translocation of chemicals through the skin into the systemic circulation, coupled with the response of the skin to such chemicals and their effects on the physiological disposition and availability of topical agents, are important aspects of skin toxicology and pharmacology. Research in this area is in its infancy and offers many opportunities. Mechanistic and functional approaches to dermal absorption need to be developed. Such development will, hopefully, result in a better understanding of the interplay between penetration and cutaneous metabolism, and consequently, their relevance in skin toxicity and availability of topical agents. Indeed, to what extent this interrelationship may be controlled and modulated remains to be established, and exploitation of such knowledge would greatly facilitate the continual development of new strategies in topical therapy and transdermal delivery of drugs and prodrugs.

ACKNOWLEDGMENTS

The author gratefully acknowledges Liz Graichen for the graphics and Marge Schnellen for the expert secretarial assistance in the preparation of this manuscript.

REFERENCES

1. H. Schaefer, A. Zesch, and G. Stuttgen, *Skin Permeability,* Springer-Verlag, Berlin (1982).
2. B.W. Barry, *Dermatological Formulation: Percutaneous Absorption,* Marcel Dekker, New York (1983).
3. I.H. Blank, *J. Invest. Dermatol.* 43, 415 (1964).
4. I.H. Blank, *J. Invest. Dermatol.* 45, 249 (1965).
5. F.N. Marzulli and R.T. Tregear, *J. Physiol.* 157, 52 (1961).
6. F.N. Marzulli, *J. Invest. Dermatol.* 39, 387 (1962).
7. R.J. Scheuplein, *J. Invest. Dermatol.* 45, 334 (1965).
8. R.J. Scheuplein, *J. Invest. Dermatol.* 48, 79 (1967).
9. R.J. Scheuplein and I.H. Blank, *Physiol. Rev.* 51, 702 (1971).
10. R.J. Scheuplein and I.H. Blank, *J. Invest. Dermatol.* 60, 286 (1973).
11. R.J. Scheuplein, I.H. Blank, G.J. Brauner, and D.J. MacFarlane, *J. Invest. Dermatol.* 52, 63 (1969).
12. R.J. Scheuplein and L. Ross, *J. Invest. Dermatol.* 62, 353 (1974).
13. R.J. Scheuplein, *Handbook of Physiology — Reaction to Environmental Agents,* American Physiological Society, Washington, D.C., 299 (1977).
14. R.J. Scheuplein and R.L. Bronaugh, *Biochemistry and Physiology of the Skin,* Vol. 2, L.A. Goldsmith, Ed. Oxford University Press, New York, 1255 (1983).
15. P.H. Dugard, *Dermatotoxicology,* 2nd ed. F.N. Marzulli and H.I. Maibach, Eds. Hemisphere, New York, 91 (1981).
16. A.S. Michaels, S.K. Chandrusekaran, and J.E. Shaw, *AIChE J.* 21, 285 (1975).
17. G.L. Flynn, *Percutaneous Absorption,* R.L. Bronaugh and H.I. Maibach, Eds. Marcel Dekker, New York, 17 (1985).
18. B.J. Ponlsen and G.L. Flynn, *Percutaneous Absorption,* R.L. Bronaugh and H.I. Maibach, Eds. Marcel Dekker, New York, 431 (1985).
19. R.H. Guy and J. Hadgraft, *Percutaneous Absorption,* R.L. Bronaugh and H.I. Maibach, Eds. Marcel Dekker, New York, 3 (1985).
20. B. Berner and E.R. Cooper, *Transdermal Delivery of Drugs,* Vol. 2, A.F. Kydonieus and B. Berner, Eds. CRC Press, Boca Raton, FL, 41 (1987).
21. F.N. Marzulli, D.W.C. Brown, and H.I. Maibach, *Toxicol. Appl. Pharmacol.* Suppl. 3, 76 (1969).
22. A. Pannatier, P. Jenner, B. Testa, and J.C. Elter, *Drug Metab. Rev.* 8, 319 (1978).
23. P.K. Noonan and R.C. Wester, *Dermatotoxicology,* 2nd ed., F.N. Marzulli and H.I. Maibach, Eds. Hemisphere, New York, 71 (1983).
24. D.R. Bickers, *Biochemistry and Physiology of the Skin,* Vol. 2, L.A. Goldsmith, Ed. Oxford University Press, New York, 1169 (1983).
25. R.C. Wester and H.I. Maibach, *Progress in Drug Metabolism,* J.W. Bridges and L.F. Chasseand, Eds. Taylor & Francis, London, 95 (1986).
26. R.J. Martin, S.P. Denyer, and J. Hadgraft, *Int. J. Pharm.* 23 (1987).
27. L.H. Smith and J.M. Holland, *Toxicology* 21, 47 (1981).
28. J.M. Holland, J. Kao, and M.S. Whitaker, *Toxicol. Appl. Pharmacol.* 72, 272 (1984).
29. J. Kao, J. Hall, L.R. Shugart, and J.M. Holland, *Toxicol. Appl. Pharmacol.* 75, 289 (1984).
30. J. Kao, F.K. Patterson, and J. Hall, *Toxicol. Appl. Pharmacol.* 81, 502 (1985).
31. J. Kao and J. Hall, *J. Pharm. Exp. Ther.* 241, 482 (1987).
32. T.J. Franz, *Curr. Probl. Dermatol.* 7, 58 (1978).
33. K. Tojo, *Transdermal Controlled Systemic Medications,* Y.W. Chien, Ed. Marcel Dekker, New York, 127 (1987).
34. F.J. Nugent and J.A. Wood, *Can. J. Pharm. Sci.* 15, 1 (1985).
35. G.L. Flynn and E.W. Smith, *J. Pharm. Sci.* 60, 1713 (1971).
36. N.K. Petal and N.E. Foss, *J. Pharm. Sci.* 53, 94 (1964).
37. P.H. Dugard, M. Walker, S.J. Mawdsley, and R.C. Scott, *Environ. Health Perspect.* 57, 193 (1984).

38. D. Southwell and B.W. Barry, *J. Invest. Dermatol.* 80, 507 (1983).
39. B.W. Barry, S.M. Harrison, and P.H. Dugard, *J. Pharm. Pharmacol.* 37, 84 (1985).
40. M.K. Samitz, S. Katz, and J.D. Shrager, *J. Invest. Dematol.* 48, 514 (1967).
41. M.F. Goldman, B.J. Poulsen, and T. Higuchi, *J. Pharm. Sci.* 58, 1098 (1969).
42. T.J. Franz, *J. Invest. Dermatol.* 54, 190 (1975).
43. P.R. Keshary, Y.C. Huang, and Y.W. Chien, *Drug Dev. Ind. Pharm.* 11, 1213 (1985).
44. W.G. Reifenrath and T.S. Spencer, *Percutaneous Absorption,* R.L. Bronaugh and H.I. Maibach, Eds. Marcel Dekker, New York, 305 (1985).
45. S.A. Akhter, S.L. Bennett, I.L. Walker, and B.W. Barry, *Int. J. Pharm.* 21, 17 (1984).
46. R.L. Bronaugh and R.F. Stewart, *J. Pharm. Sci.* 74, 64 (1985).
47. M.K. Polano, *Arch. Dematol.* 112, 675 (1976).
48. I.H. Blank, *Percutaneous Absorption,* R.L. Bronaugh and H.I. Maibach, Eds. Marcel Dekker, New York, 97 (1985).
49. C.R. Behl, G.L. Flynn, T. Kurihara, N. Harper, W.M. Smith, W.I. Higuchi, N.F.H. Ho, and C.L. Pierson, *J. Invest. Dermatol.* 75, 346 (1980)
50. R.T. Tregear, *Physical Functions of Skin,* Academic Press, New York, (1966).
51. J.P. Marty, R.H. Guy, and H.I. Maibach, *Percutaneous Absorption,* R.L. Bronaugh and H.I. Maibach, Eds. Marcel Dekker, New York, 469 (1985).
52. R.L. Bronaugh, R.F. Stewart, and E.R. Congdon, *Toxicol. Appl. Pharmacol.* 62, 481 (1982).
53. K.E. Anderson, H.I. Maibach, and M.O. Anjo, *Br. J. Dermatol.* 102, 447 (1980).
54. J. Kao, J. Hall, and J.M. Holland, *Toxicol. Appl. Pharmacol.* 68, 206 (1983).
55. A.M. Kligman and E. Christopher, *Arch. Dermatol.* 88, 702 (1963).
56. P. Frosch and A.M. Kligman, *Br. J. Dermatol.* 96, 461 (1977).
57. P. Lehmann and A.M. Kligman, *Br. J. Dermatol.* 109, 313 (1983).
58. P.M. Elias, H.M. Hermyer, G. Tuppeiner, P. Fritsch, and K. Wolff, *J. Invest. Dermatol* 63, 467 (1974).
59. P.M. Elias, H.M. Hermyer, P. Fritsch, G. Tuppeiner, and K. Wolff, *J. Lab. Clin. Med.* 84, 414 (1974).
60. P.M. Elias, E.R. Cooper, A. Korc, and B.E. Brown, *J. Invest. Dermatol.* 76, 297 (1981).
61. M.S. Roberts, R.A. Anderson, and J. Swarbrick, *J. Pharm. Pharmacol.* 29, 677 (1977).
62. E.R. Cooper and B. Berner, *Methods in Skin Research,* D. Skerrow and C.J. Skerrow, Eds. John Wiley & Sons, New York, 407 (1985).
63. R.L. Bronaugh and H.I. Maibach, *Dermatotoxicology,* 2nd ed., F.N. Marzulli and H.I. Maibach, Eds. Hemisphere, New York, 117 (1983).
64. G.S. Hawkins and W.G. Reifenrath, *Fund. Appl. Toxicol.* 4, S133 (1984).
65. R.L. Bronaugh and R.F. Stewart, *J. Pharm. Sci.* 75, 487 (1986).
66. M.C. Middleton and R. Hasmall, *J. Invest. Dermatol.* 68, 108 (1977).
67. E.O. Bernstein, D.W. Coble, and E.O. Kairinen, *Scanning Electron Microsc.* 3, 347 (1979).
68. R.C. Wester and P.K. Noonan, *Int. J. Pharm.* 7, 99 (1980).
69. R.C. Wester and H.I. Maibach, *Transdermal Delivery of Drugs,* Vol. 1, A.F. Kydonieus and B. Berner, Eds. CRC Press, Boca Raton, FL, 61 (1987)
70. R.H. Guy, A.H. Guy, H.I. Maibach, and V.P. Shah, *Pharm. Res.* 3, 253 (1986).
71. J.P. Skelly, V.P. Shah, H.I. Maibach, R.H. Guy, R.C. Wester, G.L. Flynn, and A. Yacobi, *Pharm. Res.* 4, 265 (1987).
72. D. Southwell, B.W. Barry, and R. Woodford, *Int. J. Pharm.* 18, 299 (1984).
73. H.I. Maibach, *Animal Models in Dermatology: Relevance to Human Dermatopharmacology and Dermatotoxicology,* Churchill Livingstone, New York (1975).
74. R.C. Wester and H.I. Maiback, *Models in Dermatology,* H.I. Maiback and F.N. Marzulli, Eds. Marcel Dekker, New York, 159 (1985).
75. A. Rougier, C. Lotte, and H.I. Maibach, *J. Invest. Dermatol.* 88, 577 (1987).
76. Nacht S. and D. Yeung, *Percutaneous Absorption,* R.L. Bronaugh and H.I. Maibach, Eds. Marcel Dekker, New York, 373 (1985).

77. M. Nakano, A. Kuchila, and T. Arita, *Chem. Pharm. Bull. (Tokyo)* 24, 2345 (1976).
78. M. Washitaka, Y. Takashima, S. Tanaka, T. Anmo, and I. Tanaka, *Chem. Pharm. Bull (Tokyo)* 30, 2885 (1980).
79. J. Hadgraft and G. Ridout, *Int. J. Pharm.* 39, 149 (1987).
80. D. Yeung, P. Walter, and S. Nacht, *Transdermal Delivery of Drugs,* Vol. II, A.F. Kydonieus and B. Berner, Eds. CRC Press, Boca Raton, FL, 19 (1987)
81. G.E. Burch and T. Windsor, *Arch. Int. Med.* 74, 437 (1944).
82. G.S. Berenson and G.E. Burch, *Am. J. Trop. Med. Hyg.* 3, 842 (1951).
83. J.P. Astley and M. Levine, *J. Pharm. Sci.* 65, 210 (1976).
84. S.M. Harrison, B.W. Barry, and P.H. Dugard, *J. Pharm. Pharmacol.* 36, 261 (1984)
85. M.S. Roberts, C.D. Shorey, R. Arnold, and R.A. Anderson, *Aust. J. Pharm. Sci.* 3, 81 (1984).
86. G.S. Hawkins and W.G. Reifenrath, *J. Pharm. Sci.* 75, 378 (1986).
87. J. Swarbrick, G. Lee, and J. Brom, *J. Invest. Dermatol.* 78, 63 (1982).
88. B.W. Kemppainen, R.T. Riley, J.G. Pace, and F.J. Hoerr, *Food Chem. Toxicol.* 24, 211 (1986).
89. S. Nacht, D. Yeung, J.N. Beasley, M.D. Anjo, and H.I. Maiback, *Am. Acad. Dermatol.* 4, 31 (1981).
90. R.C. Wester, P.K. Noonan, S. Smeach, and L. Kosobud, *J. Pharm. Sci.* 72, 745 (1983).
91. B.W. Kemppainen, R.T. Riley, J.G. Pace, F.J. Hoerr, and J. Joyave, *Fund. Appl. Toxicol.* 7, 367 (1986).
92. M.J. Grahm, P.M. Williams, and M.D. Rawlins, Abst. IUPHAR 9th Int. Congr. 138P (1984).
93. S.R. May and F.A. DeClement, *Cryobiology* 17, 33 (1980).
94. S.R. May and F.A. DeClement, *Cryobiology* 19, 362 (1982).
95. S.R. May, R.M. Guttman, and J.F. Wainwright, *Cryobiology* 22, 205 (1985).
96. S.R. May and J.F. Wainwright, *Cryobiology* 22, 18 (1985).
97. R.L. Bronaugh, *Percutaneous Absorption,* R.L. Bronaugh and H.I. Maibach, Eds. Marcel Dekker, New York, 267 (1985).
98. R.L. Bronaugh and T.J. Franz, *Br. J. Dermatol.* 115, 1 (1986).
99. D.W.C. Brown and A.G. Ulsamer, *Food Cosmet. Toxicol.* 13, 81 (1985).
100. R.T. Riley and B.W. Kemppainen, *Food Chem. Toxicol.* 73, 67 (1985).
101. R.L. Bronaugh and R.F. Stewart, *J. Pharm. Sci.* 23, 1255 (1984).
102. A. Hoelgaard and B. Mollgaard, *J. Pharm. Pharmacol.* 34, 610 (1982).
103. W.M. Pardridges, *Biological Transport of Radiotracers,* L.G. Colombetti, Ed. CRC Press, Boca Raton, FL, 189 (1982).
104. Y.C. Huang, *Transdermal Controlled Systemic Medication,* Y.W. Chien, Ed. Marcel Dekker, New York, 159 (1987).
105. E. Cooper, *Percutaneous Absorption,* R.L. Bronaugh and H.I. Maibach, Eds. Marcel Dekker, New York, 525 (1985).
106. E. Cooper, *Transdermal Controlled Systemic Medication,* Y.W. Chien, Ed. Marcel Dekker, New York, 83 (1987).
107. M.F. Coldman, B.J. Poulsen, and T. Higuchi, *J. Pharm. Sci.* 58, 1098 (1969).
108. R.C. Wester and H.J. Maibach, *Percutaneous Absorption,* R.L. Bronaugh and H.I. Maibach, Eds. Marcel Dekker, New York, 347 (1985).
109. R.C. Wester and H.I. Maibach, *J. Invest. Dermatol.* 67, 518 (1976).
110. P.K. Noonan and R.C. Wester, *J. Pharm. Sci.* 69, 385 (1980).
111. R.C. Wester, P.K. Noonan, and H.I. Maibach, *Arch. Dermatol. Res.* 113, 620 (1977).
112. R.C. Wester, P.K. Noonan, and H.I. Maibach, *Arch. Dermatol. Res.* 267, 299 (1980).
113. H.Y. Ando, N.F.H. Ho, and W.I. Higuchi, *J. Pharm. Sci.* 66, 1525 (1977).
114. J.L. Fox, C.D. Yu, W.I. Higuchi, and N.F.H. Ho, *Int. J. Pharm.* 2, 41 (1979).
115. J. Hadgraft, *Int.J. Pharm.* 4, 229 (1980).
116. R.H. Guy and J. Hadgraft, *Int. J. Pharm.* 20, 43 (1984).
117. R.H. Guy, J. Hadgraft, and H.I. Maibach, *Int. J. Pharm.* 11, 119 (1982).
118. R.H. Guy and J. Hadgraft, *J. Pharm. Sci.* 73, 883 (1984).

119. K. Sato, T. Oda, K. Sugibayashi, and Y Morimoto, *Chem. Pharm. Bull.* 36, 2624 (1988).
120. R.H. Guy and J. Hadgraft, *Int. J. Pharm.* 24, 267 (1985).
121. E. Nakashima, P.K. Noonan, and L.Z. Benet, *J. Pharmacokinet. Biopharm.* 15, 423 (1987).
122. J. Kao, Unpublished observations.
123. R.G. Helman, J. Hall, and J. Kao, *Fund. Appl. Toxicol.* 7, 94 (1986).
124. M.C. Middleton, *J. Invest. Dermatol.* 74, 219 (1980).
125. M.C. Middleton, *Testing for Toxicity,* J. W. Gorrad, Ed. Taylor & Francis London, 275 (1981).
126. N.H. Creasey, J. Battensby, and J.A. Fletcher, *Curr. Probl. Dermatol.* 7, 95 (1978).
127. H. Durrheim, G.L. Flynn, W. Higuchi, and C.R. Behl, *J. Pharm. Sci.* 69, 781 (1980).
128. Y.W. Chien and K.H. Valia, *Drug Dev. Ind. Pharm.* 10, 575 (1984).
129. Z. Liron and S. Cohen, *J. Pharm. Sci.* 73, 534 (1984).
130. T.S. Spencer, J.A. Hill, H.I. Maibach, and R.J. Feldman, *J. Invest. Dermatol.* 72, 317 (1979).
131. W.G. Reifenrath and P.B. Robinson, *J. Pharm. Sci.* 71, 1014 (1982).
132. A.S. Susten, B.L. Dames, and R.W. Niemecier, *J. Appl. Toxicol.* 6, 43 (1986).
133. R.H. Guy and J. Hadgraft, *J. Soc. Cosmet. Chem.* 35, 103 (1984).
134. W. Schalla and H. Schaefer, *Percutaneous Absorption,* R.L. Bronaugh and H.I. Maibach, Eds. Marcel Dekker, New York, 281 (1985).
135. B.W. Kemppainen, R.T. Riley, and S. Biles-Thurlow, *Food Chem. Toxicol.* 25, 379 (1987).
136. F.L. Brookes, W.B. Hugo, and S.P. Denyer, *J. Pharm. Pharmacol.* 34, 61P (1982).
137. S.P. Denyer, W.B.Hugo, and M. O'Brien, *J. Pharm. Pharmacol.* 36, 61P (1984).
138. C.D. Yu, J.L. Fox, N.F.H. Ho, and W.I. Higuchi, *J. Pharm. Sci.* 68, 1347 (1979).
139. H. Bundgaard, A. Hoelgaard, and B. Mollgaard, *Int. J. Pharm.* 15, 285 1983.
140. W. Montagna, *J. Biophys. Biochem. Cytol.* 1, 13 (1955).
141. R.H. Guy, J. Hadgraft, R.S. Hinz, K.V. Roskos, and D.A.W. Bucks, *Transdermal Controlled Systemic Medications,* Y.W. Chien, Ed. Marcel Dekker, New York, 179 (1987).
142. M.J. Finnen, M.L. Herdman, and S. Shuster, *J. Steroid Biochem.* 20, 1169 (1984).
143. H. Mukhtar, B.T. Deltito, M. Das, E.P. Chemiack, A.D. Chemiack, and D.R. Bickers, *Cancer Res.* 44, 4233 (1984).
144. S.E. Rattie, F. Williams, and M.D. Rawlin, *Xenobiotica* 10, 255 (1986).

10

Isolated Perfused Skin Flap and Skin Grafting Techniques

J. EDMOND RIVIERE
Cutaneous Pharmacology and Toxicology Center
College of Veterinary Medicine and Toxicology Program
North Carolina State University
Raleigh, North Carolina

AND

MICHAEL P. CARVER
Colgate-Palmolive Company
Piscataway, New Jersey

I. INTRODUCTION

Numerous *in vitro* and *in vivo* experimental models exist to assess percutaneous absorption of xenobiotics and drugs in animals and man. However, there are limitations present in both types of systems which preclude either approach from being optimal. The major impetus for development of the alternative animal models reviewed herein focuses on the anatomical and physiological limitations inherent to most *in vitro* systems presently in use. For the most part, there is concern over the lack of cutaneous viability and absence of a functional microcirculation in simple diffusion models. The approach taken to circumvent these basic limitations in technology involves creating either *in situ* or *in vitro* perfused skin preparations. The second major concern of many animal models, *in vitro* and *in vivo*, is that they are obviously not human and thus the extrapolation to humans is questionable. The current solution to this problem involves human-xenograft models which allow compound absorption through human skin to be assessed in a laboratory animal, wherein other variables of drug disposition and metabolism can be controlled. When human skin cannot be utilized, pigskin has been employed as an acceptable surrogate.

It is the purpose of this section to review the development and application of both of these experimental methods as they apply to percutaneous absorption. As will

Table 1
Variable Processes Influencing Percutaneous Absorption

1. Release of drug to the skin surface
2. Partitioning of available drug into the stratum corneum
3. Diffusion through the stratum corneum
4. Partitioning of the drug from the stratum corneum into the viable epidermis
5. Diffusion through the viable epidermis into the dermis
6. Biotransformation of the compound and/or intracellular uptake and sequestration of drug
7. Diffusion through the dermis and uptake into the cutaneous microcirculation
8. Delivery of absorbed drug into the systemic circulation

become obvious, these approaches are rather recent additions to the toxicologist's procedural repertoire and thus large "data gaps" still exist in the literature relative to the database for more classical techniques. However, significant progress has been made in the development and validation of these models and thus an overview of this work is essential if their application is ever to be fully realized.

II. JUSTIFICATION FOR NEW EXPERIMENTAL MODELS

When xenobiotics or drugs are applied topically onto skin, percutaneous absorption may be affected by the processes outlined in Table 1. Earlier chapters in this text (7, 8, and 9) and published reviews[1-6] present an overview of the mechanisms operative in xenobiotic percutaneous absorption and should be consulted for a more detailed perspective on the physiochemical processes involved in penetration through the epidermal layer of skin. Each of the variables listed in Table 1 are important in predicting the total penetration of a compound. Ideally, in order to quantitate the total extent and rate of a compound's penetration, all of these factors should be accounted for in any test system utilized. Moreover, numerous experimental variables may influence any one, or a combination, of several of these events. For example, the formulation of compound used (vehicle, solvent, enhancers, surfactants) and the method of application (occlusive vs. nonocclusive) may interact with the first five processes listed. Anatomical differences in the structure of skin either within or between species (thickness, presence or absence of adnexial structures, relative cutaneous blood flow), can affect all of these variables. The condition of the skin used (hydrated, diseased, adult vs. neonatal) may affect penetration by interacting with factors three and five through eight. Likewise, environmental conditions (temperature, relative humidity, surface air flow) could influence the first four factors or processes.

The study of percutaneous absorption is facilitated by selecting *in vitro* and *in vivo* approaches which are most sensitive to the specific processes of interest. Those steps which primarily involve physicochemical interactions between the chemical and stratum corneum are probably best studied using simple, *in vitro* test systems, in which *in vivo* confounding variables are minimized. Artificial membranes or sheets of stratum corneum may be used to investigate partitioning phenomena, especially formulation problems. Although *in vitro* systems are adequate to study some aspects of percutaneous absorption, they are not designed to investigate the total phenomenon. For example, these more simplified *in vitro* systems are not optimal for examining the net effect of these processes on the total flux of absorbed compound presented to the systemic circulation. Their major limitation is the inability to model processes five through eight in Table 1.

In vivo, whole animal studies are well suited to study the total process of percutaneous absorption because the skin is anatomically and physiologically intact. However, *in vivo* systems cannot directly measure the cutaneous flux of drug presented to the systemic circulation. When venous blood samples are assayed for topically absorbed drug, two major confounding factors are present. First, the absorbed drug is diluted and then distributed to the rest of the animal following cutaneous venous efflux of drug into the systemic circulation. Since the transcutaneous flux of compound is often very slow and only a small fraction of applied compound is absorbed, especially for lipid-soluble xenobiotics, chemical concentrations in the systemic circulation are often near or below the limits of analytical detection. Second, because of this circulatory mixing, cutaneous biotransformation is confounded with systemic (primarily hepatic) biotransformation of the absorbed parent compound. It is also likely that some of the absorbed cutaneous metabolites are further transformed systemically. Thus, *in vivo* studies cannot quantitatively examine all aspects of percutaneous absorption.

The utility of *in vivo* studies lies mainly in determining the total bioavailability of a topically applied compound. In most cases, radiolabeled chemicals are used and a parallel intravenous study is conducted to correct for kinetic differences between routes of administration. Two approaches are then taken to determine percutaneous absorption. First, the area under the curve (AUC) of the blood concentration time profile is calculated for each route of administration and the ratio of the two is considered equal to systemic bioavailability (F). Alternatively, the excretion of radiolabeled chemicals in the urine and feces (optimally in expired air), and absorbed body burden are determined for both routes and the percentage of the applied dose excreted via each route is compared in a similar manner as the AUC described above. This latter technique is the one most often utilized.[3,7-14] However, in humans and large laboratory animals (pigs), radiolabel excretion in expired air is often difficult if not impossible to obtain, and calculations must be based on urinary and fecal excretion alone. In many cases, only urine is available and there is evidence that this can seriously affect the interpretation of results.[15-20] In all cases, differentiation between cutaneous and systemic biotransformation is usually impossible.

Despite the limitations of whole animal studies, the ability to estimate overall systemic bioavailability after topical application onto viable skin in a laboratory animal renders these studies valuable. The problem which arises is how to extrapolate results obtained in a laboratory animal species to man. Human skin xenografts onto athymic (immunodeficient) rodents, originally developed for immunological investigations, have now been adopted for use in percutaneous absorption trials in an attempt to overcome this difficulty. Another niche exists in percutaneous model development which requires a preparation that could directly assess the cutaneous venous flux of an absorbed chemical independent of confounding systemic effects. Such a model should also maintain skin in as close an anatomical and physiological state to the *in vivo* situation as possible. Perfused skin preparations would appear to satisfy these needs. Because these experimental models have shown promise as alternatives to traditional methods, past efforts and recent advances in xenograft and perfused skin models will be reviewed in detail below.[21]

III. SKIN GRAFTING AND ATHYMIC RODENT MODELS

A. Introduction

The development of skin graft models for investigating percutaneous absorption is driven by the demonstrated utility of a whole animal preparation for realistically assessing topical bioavailability, coupled with the desire to utilize human skin wherever possible. The techniques utilized are those of reconstructive surgery applied to laboratory animals, primarily in rats and mice. A full discussion of the immunobiology, wound healing, and surgical techniques involved is far beyond the scope or intent of this chapter. However, there are common features present in all skin grafting techniques utilized in cutaneous pharmacology and toxicology research which will be addressed.

The first problem encountered in grafting the skin of one species to another is that of cell-mediated immune rejection of the xenograft. This natural reaction can be avoided by using congenitally athymic (nude) rodents, which are inherently unable to launch a graft rejection response due to the absence of the thymus. Studies have demonstrated that nude mice are capable of maintaining viable skin xenografts obtained from humans, as well as from a virtual zoological menagerie which includes other rodents, hamsters, rabbits, pigs, pigeons, chickens, lizards, and grass snakes.[22-26] The congenitally athymic nude mouse is the most common host species employed and the phenotype arises from an autosomal recessive gene (nu locus of the VII linkage group). Through repeated back-crossing, this mutation can be transmitted into inbred mice, resulting in a number of commercially available nude mouse strains (BALB/C-nu/nu, NIH Swiss, BALB/C/A/BOM, NMRI/BOM-nu/nu, CBA, ONU). Homozygotes (nu nu) are relatively hairless, have abnormal keratinization of hair follicles, a sulfhydryl group deficiency and retarded growth. In addition, they possess only a rudimentary thymus and have abnormally depressed leukocyte numbers in the systemic circulation, spleen, lymph nodes, and Peyer's patches.[22,23,27-31] A hairless, athymic rat model has also been developed, although supplemental treatment with the immunosuppressant drug cyclosporine (10 mg/kg/day) is required for long-term maintenance of human grafts.[32] Finally, normal euthymic rats treated with an even greater dose of parenteral cyclosporine (12.5 to 25 mg/kg/day) rarely reject human skin grafts.[33] This latter approach could also be applied to other animal species.

A problem in using athymic nude rodents is their susceptibility to infectious disease. This requires that microbiologically "clean" environments be provided.

Cages, bedding, food, water, and any other material in contact with the animals must be sterilized before use. Likewise, laboratory and animal care personnel must follow strict sanitary guidelines such as the use of sterile gowns and gloves, masks, and shoe covers, etc. Finally, investigators must remain vigilant about control of bacterial and viral disease in other animals housed in the facility, since the athymic animals are at greater risk. Detection and isolation of asymptomatic carriers is especially important. Familiarity with the specialized animal husbandry requirements of immunodeficient animals is an important consideration in attempting to use these experimental models.

B. Techniques

There are a number of surgical techniques available for grafting human skin to the athymic mouse.[25,34-38] It must be reiterated that all procedures should be conducted under a biological isolation hood, using sterile instruments and techniques. Human donor skin may be obtained from a number of possible sources, subsequent to obtaining informed consent. Sources include skin harvested in the course of facial, breast, or abdominal surgery, foreskin obtained during circumcisions, or fresh cadaver skin. A brief medical history of the patient should be obtained to provide baseline data to control for "unexpected" results. Skin and subcutaneous tissue samples are cut into conveniently sized strips and placed on gauze pads premoistened with tissue culture media, such as RPMI 1640 or Eagle's MEM®. In general, these skin strips are dermatomed (≥ 0.2 to 0.4 mm) to obtain a split-thickness graft, cut into discs with a sterile biopsy punch (size dependent upon animal, usually 2 cm in diameter), and stored in the above medium at 4°C until use. The storage period should not exceed 3 to 4 days. If a nonhuman donor species is used, skin may be obtained directly by dermatome sectioning of the animal (aseptic, anesthetized procedure) and immediately grafted onto the recipient, thereby minimizing tissue handling.

After anesthesia is induced in healthy nude mice, a disc of skin (same size as graft above), which includes subcutaneous tissue and underlying panniculus muscle, is surgically removed from the lateral thorax or other site. The dermatomed xenograft is placed on this area and secured in place with sterile tape or sutures. The tape or sutures should be loosened ventrally after 2 to 3 days and removed after 2 weeks, if not already shed. A scale or crust may form over the graft and will be lost in 2 to 3 weeks, allowing a successful graft to last for the remainder of the animal's life. Numerous variations on thickness, including the use of full-thickness grafts, have been employed. In some preparations, adnexial structures may be retained. A variation of this technique is the subcutaneous approach, which utilizes an incision into the skin over the thorax of the mouse and insertion of the graft into the orifice created.[34,39] The incision is then closed and the mouse skin is removed 3 weeks later, leaving a viable xenograft. The major advantage of this latter technique is that thicker grafts may be used.

A final variation is to induce a granulating wound in the mouse, place a Teflon® collar inside this wound, and transplant a human epidermal cell culture into this tissue bed.[40] The result of this preparation is an epidermal cell population from a well-defined cell line. Recent advances in skin organ culture may also yield reproducible "sheets" of human epidermal cells which could be grafted onto athymic animals. Such a system should produce a more homogeneous epidermal graft, and reduce variability in percutaneous absorption studies. An exciting area of research using this approach involves retroviral transfer of the human growth hormone gene into cultured keratinocytes. These are then grafted onto athymic nuce mice to result in systemic delivery of growth hormone.[41] The possibilities for application of this technology to the development of novel percutaneous chemical or drug delivery systems are obvious.

Studies have demonstrated that transplanted human skin retains some histological and biochemical properties of normal human skin, including barrier functions and susceptibility to chemically induced proliferation. Grafted, diseased skin (psoriasis, itchthyosis) has been shown to retain its pathological appearance and biochemical characteristics. However, certain graft-related lesions (central necrosis, acanthosis, epidermal thickening) have also been noted.[26,35,38] A recent histological study on the response of human-xenografted nude mouse skin to topical arsenicals, reported that the histological appearance of nonexposed xenografted human skin differed considerably from normal.[37] Differences included blunted or absent rete ridges, sparse appendages, foreign body granulomas, and an abnormal dermis which consisted of undifferentiated fibroblastic proliferation with an increase in connective tissue fibers. The possible interference of these background histological lesions to interpretation of dermal toxicology studies must be assessed.

C. Applications

Petersen and colleagues used the human-xenografted athymic nude mouse to investigate the percutaneous penetration of a number of radiolabeled cosmetic ingredients.[42] Reifenrath and co-workers have used the human skin-grafted nude mouse to study the percutaneous absorption of a series of compounds, providing the most comprehensive database currently available.[43,44] This work compared absorption of compounds through both human and pigskin grafts on nude mice to *in vivo* absorption in pigs and humans. The best predictions were obtained within species. The correlation between human-grafted nude mice and *in vivo* human data for caffeine, benzoic acid, malathion, N,N-diethyl-m-toluamide (DEET), testosterone, fluocinolone acetonide, parathion, progesterone, and lindane was good ($r = 0.74$, $p = 0.05$), although the regression line did not pass through the origin. Similar results were obtained with porcine xenografts on nude mice compared to the *in vivo* absorption in pigs. Surprisingly, the strongest correlation to *in vivo* human percutaneous absorption occurred in comparison with *in vivo* weanling pigs ($p = 0.83$, $p = 0.05$). These results suggest that the xenografting procedure introduces a consistent experimental error in extrapolating data collected in the nude mouse xenograft to the *in vivo* donor animal. This is consistent with the histological findings noted above and although human epidermal cells and some adnexal appendages are intact, the graft is perfused by mouse capillaries which may have introduced unknown variables. This limitation of xenograft models has not been adequately explored.

In conclusion, it appears that the human-xenografted athymic nude mouse may be a useful experimental model. Its major strength, one which has not yet been adequately exploited, appears to be in the study of compounds which undergo significant first-pass cutaneous biotransformation. Human skin would be the optimal tissue for these studies. However, like other *in vivo* models, it is difficult to separate cutaneous from systemic metabolism in a whole animal preparation. Additionally, comparisons between investigators using various sources of skin is problematic due to differences in thickness, age, distribution of adnexial structures, and pathophysiologic state. The *in situ* skin flap system described below attempts to address this limitation.

IV. SKIN FLAPS AND PERFUSED SKIN MODELS

Blood supply to the skin involves two primary types of vasculature: musculocutaneous and direct cutaneous arteries. The venous return generally runs parallel to the arterial supply. Musculocutaneous arteries supply relatively small areas of skin, after penetrating the underlying muscle and subcutaneous tissue, and would not be amenable to direct perfusion. In contrast, direct cutaneous arteries traverse directly to skin and supply much larger surface areas, making them ideally suited for perfusion. Furthermore, the skin of the ventral abdomen of many species is supplied by the superficial epigastric artery and vein which is surgically accessible for cannulation and perfusion. Two approaches have been taken in skin perfusion studies; isolated and *in situ*. Isolated perfusion studies involve surgically lifting the area of skin perfused by a direct cutaneous artery and maintaining this organ preparation *in vitro* in a controlled environment, much as isolated perfused liver and kidney studies are conducted. *In situ* approaches, in contrast, require the skin flap to remain on the animal to allow sampling the venous return from the perfused skin area.

The isolated perfusion of cat[45] and dog[46-50] skin has been previously reported using skin flaps perfused by the saphenous artery. These preparations perfuse both the skin and underlying musculature. The cat flap was utilized to demonstrate that drug-induced release of cutaneous stores of histamine resulted in cutaneous vasoconstriction and edema. Biochemical pathways of carbohydrate, nucleic acid, and lipid cutaneous metabolism were investigated in the perfused dog skin flaps. Although viability could be maintained for periods of 18 h, these models were never further developed for percutaneous absorption studies.

The isolated perfusion of human skin using a groin flap preparation has been reported, although percutaneous absorption studies are also lacking in this model.[51,52] This approach uses a flat skin preparation in which the dermis is exposed to the ambient environment, necessitating perfusion at 100% humidity in an oxygenated environment. Effusate from the exposed dermal capillary beds is then collected to measure "venous flux". The problem of an exposed dermis could theoretically be avoided by using an axial pattern, tubed skin flap preparation, created such that the skin is perfused by an artery traversing along the axis of the flap. By creating a tube or a sandwich flap, the underlying dermis is protected from the ambient environment, allowing considerable flexibility in experimental design. The direct application of this invasive surgical procedure to humans is not possible due to ethical considerations. Two models have been developed for percutaneous absorption studies using this latter approach and thus merit special attention: the *in situ* rat-human skin flap system (RHSFS) developed by Dr. Gerald Kreuger and associates at the University of Utah and the isolated perfused porcine skin flap developed in our laboratory at North Carolina State University, in collaboration with Drs. Nancy Monteiro and Karl Bowman.

V. THE RAT-HUMAN SKIN FLAP SYSTEM

The RHSFS utilizes microvascular surgical techniques to create a rat/human-xenografted sandwich flap supplied by the superficial epigastric artery on nude rats.[53,54] Briefly, a human split-thickness skin graft (0.3 to 0.5 mm) is grafted to the subcutaneous surface (dermis to dermis) of a skin flap created on the ventral abdomen of a rat. After successful growth of this graft, the sandwich flap and its associated vasculature are transferred to the back of the rat through a subcutaneous tunnel (Figure 1). Since athymic nude rats are immunodeficient, cyclosporine must be administered (approximately 10 mg/kg/day) to prevent xenograft rejection throughout the procedure. In addition to xenografting human skin onto the host rat flap, split-thickness skin from syngenetic rats has also been utilized as a control on the effects of grafting alone. Cutaneous blood flow has been well characterized using both fluorescent dyes and laser Doppler velocimetry, the average flow being between 1.5 and 3 ml/min.[54]

The profile of benzoic acid absorption across the RHSFS is depicted in Figure 2. In these studies, total absorption after 4 h ranged from 3.5 to 20% and was reduced by vasoconstriction resulting from phenylephrine iontophoresis. In addition, time of peak absorption rate was also delayed. A decrease in body temperature produced a 40% decrease in blood flow and lowered benzoic acid flux by a factor of 3 in other studies. For both phenylephrine- and temperature-induced decreases in blood flow, a rebound phenomenon of enhanced benzoic acid penetration occurred after normal blood flow as resumed.[55,56]

The percutaneous absorption of caffeine has also been investigated and flux through the RHSFS was much greater than in *in vitro*. The hypothesis that early caffeine binding in the dermis saturated a potential dermal reservoir and resulted in rapid absorption through the proximate microcirculation in the RHSFS was thought to be consistent with the lack of a microcirculation in diffusion chamber studies *in vitro,* where the larger dermal reservoir results in a decrease in net absorption into the receptor fluid.[57] The transdermal absorption and cutaneous biotransformation (deamination) of the antiviral drug vidarabine has also been studied in the rat-rat skin flap system.[58] The influence of the cyclosporine on percutaneous absorption has been investigated in the RHSFS, as well as in rat- and pig-xenografted flaps.[59] Cy-

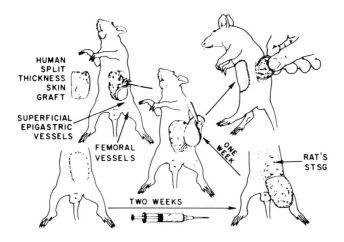

Figure 1. Surgical procedure used to generate the RHSFS. (From Krueger et al., *Fund. Appl. Toxicol.* 5, S112 (1985). With permission.)

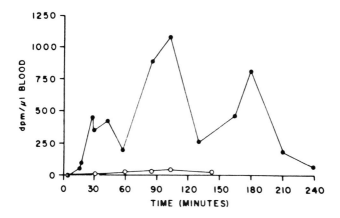

Figure 2. Percutaneous absorption of 14-C benzoic acid in the RHSFS. 200 μl of benzoic acid was added to phosphate buffered saline (pH = 6.0) in an occluded well in contact with the skin surface for 180 min. (From Krueger et al., *Fund. Appl. Toxicol.*, 5, S112 (1985). With permission.)

closporine was shown to cause a fourfold increase in percutaneous absorption of compounds. Transepidermal water loss was not affected by cyclosporine treatment, although fluoresceine retention in the treated skin occurred. The authors concluded that a significant cyclosporine effect was present which altered drug partitioning and skin permeability, but not its barrier function. This is an important line of investigation, since the strength of this model lies in the ability to directly assess percutaneous absorption through human skin. If a significant interaction occurs with the obligatory cyclosporine treatment, then the utility of this preparation may be compromised.

As can be appreciated, the RHSFS is capable of assessing all aspects of percutaneous absorption outlined in Table 1. Its strengths are that absorption across both human and rat skin can be evaluated, vascular access is available, cutaneous blood flow to the flap can be monitored, the skin is viable and served by endogenous nutrients and mediators, and the preparation is reusable. Moreover, factors which are important in determining percutaneous flux of a compound (viability, metabolism, blood flow) which cannot be studied in other experimental models, can be examined using the RHSFS. The main limitations are that the procedure is labor intensive, costly, and requires a great deal of technical skill, a relatively small area of skin is available for study (approximately 2 cm^2 or less), and blood samples are limited in volume (50 µl) and difficult to obtain. In addition, cyclosporine appears to enhance the absorption of agents applied to grafted rat skin. The mechanisms of the xenograft-related effects on percutaneous absorption noted for the nude mouse studies needs further investigation.

VI. THE ISOLATED PERFUSED PORCINE SKIN FLAP (IPPSF)

A. Introduction

The pig is well accepted as an animal model for percutaneous absorption studies.[13,44,60-64] Pig and human skin have a similar gross appearance, a sparse hair coat, a relatively thick epidermis compared to other laboratory animal species, a similar microstructure, comparable microcirculation, and similar arrangements of dermal collagen and elastic fibers. Differences in morphology primarily relate to the nature and distribution of the sweat glands.[65-72] Biochemically, cutaneous enzyme histochemistry, carbohydrate and fatty acid metabolism, and lipid composition are also comparable.[73-77] The rate of epidermal turnover in both species ranges from 26 to 30 days.[78] Of critical importance to skin flap techniques, the skin of the ventral abdomen of pigs is supplied by the superficial epigastric artery, making this species ideally suited for developing an isolated skin flap model.

B. Techniques

The IPPSF is a single pedicle, axial pattern tubed skin flap on the ventral abdomen of weanling swine.[79-81] Two flaps per animal, each lateral to the ventral midline, can be created in a single surgical procedure. As depicted in Figure 3, the procedure involves two steps: creation of the flap in stage I and harvest in stage II. Briefly, pigs weighing approximately 20 to 30 kg are premedicated with atropine sulfate and xylazine hydrochloride, induced with ketamine hydrochloride, and inhalational anesthesia is maintained with halothane. Each pig is prepared for routine aseptic surgery in the caudal abdominal and inguinal regions and a 4 × 12 cm area of skin, known from previous dissection and *in vivo* angiography studies to be perfused primarily by the caudal superficial epigastric artery and its associated paired venae comitantes, is demarcated. Following incision and scalpel dissection of the subcuta-

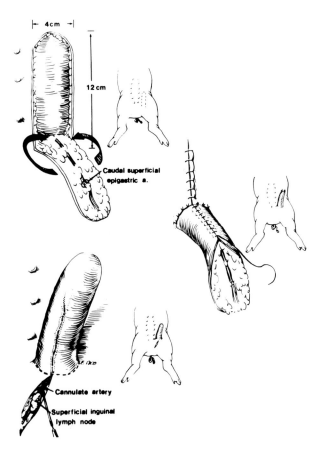

Figure 3. Surgical procedure for creating and harvesting porcine skin flaps.

neous tissue, the caudal incision is apposed and sutured and the tubed skin flap edges trimmed of fat and closed. The remaining deep and superficial subcutaneous tissues and skin incision are then closed as well. Two days later a second surgical procedure is used to cannulate the caudal superficial epigastric artery and harvest each of these skin flaps. The 2-day period between flap creation and harvest was determined to be optimal from the standpoint of lack of overall flap leakiness, normal histologic appearance (minimal variation in epidermal thickness), normal vascularization, and animal housing economics. The IPPSF is then transferred to the perfusion chamber described below. The remaining wound is flushed and allowed to heal and pigs are returned to the housing facility.

The isolated perfused organ apparatus, depicted in Figure 4, is a custom-designed temperature- and humidity-regulated chamber made specifically for this purpose (Diamond Research, Raleigh, NC). A computer monitors perfusion pressure, flow,

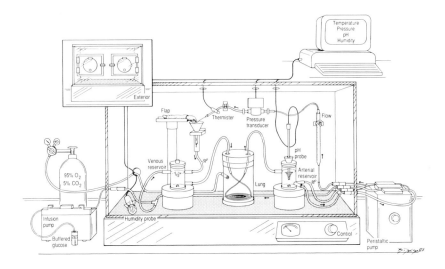

Figure 4. Isolated skin flap perfusion chamber.

pH, and temperature (±0.1°C). Flexibility is afforded in the experimental design by allowing both temperature and relative humidity to be maintained at specific set points (normally 37°C and 40 to 100% RH). Media is gassed with 95% oxygen and 5% carbon dioxide by passage through a silastic oxygenator. Normal perfusate flow through the skin flap is maintained at 1 to 2 ml/min/flap (3 to 7 ml/min/100 g) with a mean arterial pressure ranging from 30 to 70 mmHg. These values are very consistent with *in vivo* values reported in the literature.[82-85] Both recirculating and nonrecirculating configurations are possible. In the recirculating mode (Figure 4), perfusate is constantly shunted between an "arterial" reservoir, the silastic oxygenator, and the "venous" reservoir at a higher flow rate to maintain adequate mixing. In addition, each reservoir rests on a magnetic stirrer. The pH is measured in the arterial reservoir and maintained by infusion of appropriate buffers. A separate circuit, with ports for both arterial and venous sampling, delivers oxygenated arterial perfusate at the regulated flow rate into the arterial cannula implanted in the IPPSF. The skin flap sits on a cradle which collects venous drainage from the preparation and returns it to the venous reservoir.

The perfusion media is a Krebs'-Ringer bicarbonate buffer (pH 7.4, 350 mOsm/kg), containing albumin (45 g/l) and supplied with glucose (80 to 120 mg/dl) as the primary energy source. Since the IPPSF is not a sterile organ preparation, antimicrobials (penicillin G and amikacin) are included to prevent bacterial overgrowth from the microflora normally present on the skin surface. Heparin is included to prevent coagulation in the skin flap's vasculature from residual formed blood elements. In the nonrecirculating mode, perfusate is pumped through the oxygenator to the arterial reservoir and then to the IPPSF. A fraction collector can be used to automatically collect venous perfusate over defined time intervals. This single-pass system

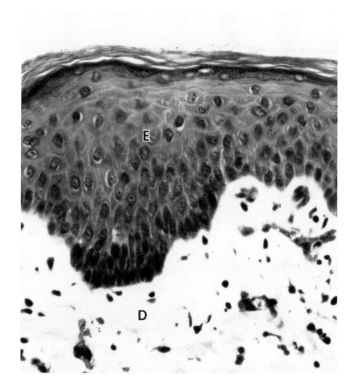

Figure 5. Light micrograph showing viable (E) epidermis and (D) dermis in an IPPSF perfused for 12 h. H & E, magnification × 396.

has advantages for assessing topical bioavailability and metabolite profiles when recirculation would confound the analysis.

C. Tissue Viability and Biochemistry

To date, our laboratory has perfused over 1200 IPPSFs. Biochemical function of skin has been assessed in the IPPSF by monitoring glucose utilization (arterial and venous glucose extraction), lactate production, lactate dehydrogenase (LDH) leakage, perfusate flow, pH, and pressure. Because of the high concentration of LDH normally found in skin, leakage into the perfusate should be an excellent marker of epidermal integrity.[86,87] Terminal LDH concentrations are usually less than 10 IU/L, increasing an order of magnitude in nonviable preparations. In addition, samples are collected for light and transmission electron microscopy after the completion of an experiment. Light microscopy on samples taken at the end of a perfusion experiment demonstrate essentially normal appearing porcine skin (Figure 5). Note the remarkable similarity of this photomicrograph of perfused porcine skin to that of normal human breast skin presented in Chapter 1 of this text (Figure 2). Similarly, detailed, electron microscopic studies have confirmed this finding.[79,80] Glucose utilization, flow, pressure, and pH have been determined to be sufficient for routine viability assessment during an experiment. Average glucose utilization is approximately 20 to 40 mg/h for a typical 30 g skin flap, dropping to <10 mg/h in nonviable preparations. Lactate production is linearly related to glucose consumption at a molar ratio of approximately 1.7. This finding is in agreement with numerous studies on cutaneous glucose utilization which suggest that 70 to 80% of cutaneous glucose is metabolized via glycolysis with lactate as the primary endproduct.[87,88]

The previously described isolated human groin flap utilizes a similar perfusate and experimental design.[51,52] However, the cannulated skin is perfused in a 100% humidified, fully oxygenated atmosphere and, since the groin flap is not tubed, "venous" drainage is actually dermal capillary effusate which is collected from the exposed dermis. Glucose is the primary energy source in both experimental models. If viability criteria, normalized to equivalent masses, surface areas, and perfusion periods are compared, the IPPSF has a glucose consumption (0.5 mg/cm^2/h) which is 20 times that of the groin flap preparation (0.024 mg/cm^2/h). Terminal LDH concentrations in the IPPSF are one tenth those of the human flap preparation.[89] Although it is difficult to make interspecies and interlaboratory comparisons, based on IPPSF criteria, these human groin flaps would not be considered viable.

D. Percutaneous Absorption in the IPPSF

In order to conduct percutaneous absorption studies, skin flaps are allowed to equilibrate for 1 to 2 h prior to compound application. At this time glucose utilization is used to determine whether the experiment will continue. If the skin passes this viability test, the IPPSF is then removed from the chamber and compound applied in the middle of the dorsal surface (opposite to suture line). This area, up to a maximum of approximately 10 cm^2, has been shown to possess adequate capillary perfusion. The method of application is dependent upon the nature of the compound studied. If volatile vehicles or solvents are involved, a dosing area (generally 5 cm^2) is demarcated with a flexible, inert, plastic border. If a transdermal delivery system is used, the patch is placed directly over the intended site. A major advantage of the IPPSF is that human prototype patches are applicable to the flap, thus eliminating the need for future scale-up studies. Experiments designed to model uptake of drug from the perfusate into the skin (systemic distribution, outward transdermal migration, "inverse" penetration, etc.) are conducted by adding drug to the arterial reservoir, so that it can be infused into the IPPSF rather than applied topically.

Figure 6 depicts the venous flux-time profiles of topical ^{14}C-radiolabeled parathion and malathion in the IPPSF.[16,90,91] These data, which are representative of other compounds tested, demonstrate the reproducibility of the model and its ability to differentiate between compounds by comparing the characteristic shapes of the venous flux profiles. Based on accepted anatomical and physiological constraints and pharmacokinetic principles described primarily by Hadgraft et al.[92-97] (see Chapter 7), linear compartmental models have been adapted to simulate percutaneous absorption rates in the IPPSF.[91] The first such pharmacokinetic model developed

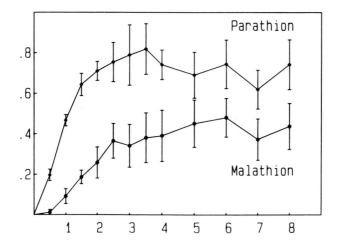

Figure 6.   Parathion and malathion percutaneous absorption in the IPPSF. All compounds were applied nonoccluded in ethanol (40 µg/cm²). Each data point represents the mean (±SE) of approximately four flaps per compound. Ordinate = % dose/h; abscissa = time in hours.

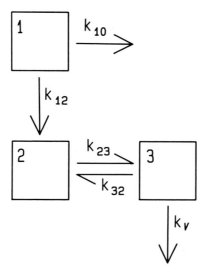

Figure 7.   Linear compartmental pharmacokinetic model used to simulate IPPSF flux profiles seen in Figure 8.

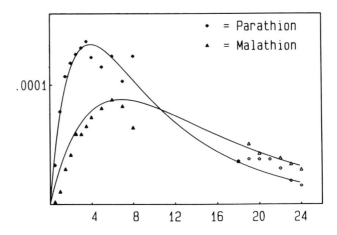

Figure 8.   Predicted vs. measured venous xenobiotic flux in an IPPSF. Simulations (curves) were based on pharmacokinetic model depicted in Figure 7. Ordinate = fraction of dose/min; abscissa = time in hours.

is depicted in Figure 7. This basic model contains three compartments: (1) the skin surface; (2) the stratum corneum and viable epidermis; and (3) the capillary perfusate. Rate constants represent surface loss (K10), penetration rate (K12), intracutaneous diffusion constants (K23, K32), and a flux function (Kv) dependent upon measured cutaneous blood flow. This model has been iteratively fit to perfusate flow and arterial and venous xenobiotic concentrations obtained in the IPPSF, with the aid of a computer modeling program (CONSAM). The percutaneous absorption profiles of most compounds tested is accurately simulated by this basic model, as exemplified by the fits for parathion and malathion seen in Figure 8. It must be stressed that this modeling approach was the first to be applied. (Significantly improved predictions have recently been obtained using a physiologically relevant pharmacokinetic model to study either drug distribution to skin[98] or percutaneous absorption.[98a]) However, this approach illustrates how different chemical entities can be studied using a relatively simple pharmacokinetic approach.

In order to further validate these models, *in vivo* absorption of topically applied caffeine (C), benzoic acid (B), malathion (M), parathion (P), testosterone (T), progesterone (R), and DFP (D) was determined. Percutaneous absorption estimates were based on radiolabel excretion in urine and feces over 6 days, with an intravenous correction experiment.[15,16] Pharmacokinetic model predictions to time infinity in the IPPSF (*in vitro*) are compared to the *in vivo* data in Figure 9.[91] The *in vivo* to *in vitro* correlation was excellent ($r = 0.94$, $p < 0.002$) and the linear regression line had a slope >1, suggesting that the IPPSF consistently overestimates percutaneous absorption *in vivo*. This may be partially explained by differences in ambient temperature (37°C *in vitro* vs. 25°C *in vivo*) or the greater relative humidity at which

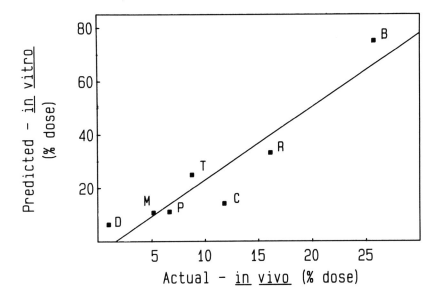

Figure 9.   Correlation of IPPSF (*in vitro*) to *in vivo* absorption using seven compounds applied at 40 μg/cm² nonoccluded in an ethanol vehicle in both systems. Caffeine (C), benzoic acid (B), malathion (M), parathion (P), testosterone (T), progesterone (R), and DFP. (D) R = 0.94; *p* <0.002.

the IPPSF studies were conducted (2x that *in vivo*), since the effects of temperature and hydration on skin absorption are well known.[99-106] The correlation of perfusate flow in the IPPSF to blood flow *in vivo* also may contribute to this finding. (When more physiologically based pharmacokinetic models are used to predict *in vivo* absorption, the regression slope is closer to 1.0 and the correlation is improved [R = 0.97].[98a,b] These models are similar to that depicted in Figure 11 with absorption and surface loss parameters added.)

Because the IPPSF is an intact organ, the effects of altered cutaneous function on the venous flux profile can be studied. A number of studies have been conducted on the transdermal flux of lidocaine administered by iontophoresis.[107,108] In Figure 10, the venous flux profile of lidocaine absorption after a 60 min episode of iontophoresis is presented along with that seen when the vasoconstrictor norepinephrine is co-iontophoresed. This figure clearly illustrates one of the major strengths of the IPPSF as an experimental tool, in that the venous flux profile observed is very dependent upon the functional state of the microcirculation. The interflap variability in these experiments is also minimal (C.V. from 10 to 30%). The net effect is similar to that previously discussed in the RHSFS above. Norepinephrine-induced vasoconstriction in the IPPSF is demonstrated by the reduced peak flux and the prolongation of the elution phase, as retained drug is slowly released from the cutaneous microcirculation. This phenomenon cannot be directly assessed in other, more traditional *in vitro* models and can only be inferred from *in vivo* studies. The IPPSF allows this

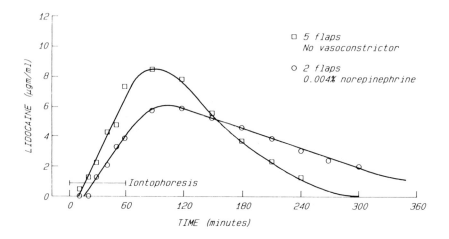

Figure 10.   IPPSF venous flux profile of 1% lidocaine administered by iontophoresis (5 cm², 2 mA).

effect to be quantitated and compartmental, deconvolution, or physiological modeling approaches provide a mathematical basis for extrapolating this *in vitro* skin absorption profile to the whole animal.

The IPPSF is also well suited to study drug distribution from the systemic circulation into the skin. As mentioned earlier, drugs may be infused into the afferent artery and the resulting efferent venous drug flux monitored as a function of time. The IPPSF allows experimental manipulation of variables such as perfusate flow, pH, temperature, co-administered drugs, vasoactivity, etc. to assess their effects on cutaneous uptake of drug. Current efforts have focused on assessing the effects of local hyperthermia (43°C) on the uptake of the platinum cancer chemotherapeutic agents cisplatin (CDDP) and carboplatin (CBDCA).[98,109] Again, pharmacokinetic models derived from the data become a means to test theories on cutaneous drug disposition, by providing parameters describing an input function into the IPPSF and an efflux function which is compatible with physiological models. The hybrid physiological/compartmental model depicted in Figure 11 was constructed to predict cutaneous drug distribution and efflux in these experiments.[98] The intercompartmental rate constants are as described in the earlier model except that a "deep" compartment is defined, with K23 corresponding to either intracellular uptake of drug, tissue binding, or cutaneous biotransformation. Distribution to the deep compartment may be reversible, however in the relatively short time frame of an IPPSF experiment (8 to 10 h), equilibrium will not have been achieved because K23 >> K32.

It must be stressed that this model is not the same as the basic compartmental model shown in Figure 7, since transit through the capillaries is expressed as a function of perfusate flow and a drug input function. This allows efflux, which is NOT a constant over time, to serve as the systemic input function into a physiological pharmacokinetic model. Additionally, delivery to the IPPSF can be defined as a

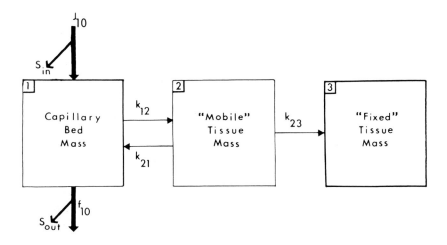

Figure 11. Hybrid pharmacokinetic model used to simulate drug uptake into the IPPSF after intra-arterial infusion. K_{12}, K_{21}, and K_{23} are transfer rate constants, f_{10} is a mass transfer function from compartment 1 to the venous effluent, J_{01} is the infusion flux which is the product of flow (Q) and concentration (C_{01}), and S_{in} and S_{out} are experimental sample ports.

function which mirrors the arterial concentration-time profile of a systemically administered drug. The model parameters in this model have also been proven mathematically to be uniquely identifiable, which allows correlation of disposition parameters between drugs and experimental conditions.[98] The rate constants in Figure 11 are not the same as those used in a traditional physiological pharmacokinetic model, which are solely dependent on physicochemical considerations. However, the motivation for utilizing this model is to study the effects of altered skin physiology or co-administered drug on the rate and extent of cutaneous drug uptake. The unique parameters in this pharmacokinetic model provide estimates of disposition which can be manipulated for this purpose. From these rate constants, partition coefficients, extraction ratios, volumes of distribution, and residence times can be calculated.

Figure 12 depicts the cutaneous uptake of cisplatin, carboplatin, tetracycline, and doxycycline predicted from this model. The adequacy of the model fits to the observed data in the IPPSF is clearly demonstrated. Model parameters obtained from these experiments are consistent with expected differences in skin disposition of these compounds based on physicochemical considerations, such as the enhanced tissue distribution of the more lipid-soluble doxycycline over tetracycline[110,111] and the increased tissue extraction and reactivity of cisplatin over carboplatin.[108,112] Quantitation of these processes in a viable skin preparation, such as the IPPSF, can help determine the underlying mechanisms operative. This hybrid model served as the foundation for a similar model describing transdermal drug delivery which incorporates cutaneous blood flow and disposition in skin.[98a,b] This will allow for the effects of various drug characteristics, dosage formulations, active delivery systems,

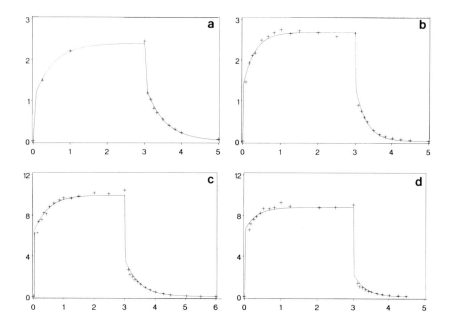

Figure 12.   Model-predicted (from Figure 11) and observed venous flux profiles from IPPSF infusions. Ordinate = μg/min; abscissa = time in hours. Cisplatin (a), carboplatin (b), tetracycline (c), and doxycycline (d).

and pathophysiologic conditions on cutaneous drug delivery to be tested. In addition, since extrapolations to the whole animal are possible, the delivery of active drug to the skin can be maximized.

The final application of the IPPSF to percutaneous absorption studies has been in characterizing the cutaneous biotransformation of xenobiotics.[16,113] Eight hours after topical application of ^{14}C-parathion in a recirculating system, approximately 70% of the radiolabel recovered in the venous perfusate was paraoxon, with 5% p-nitrophenol. When the flap is pretreated with 1-aminobenzotriazole, a nonspecific cytochrome monooxygenase inhibitor, paraoxon metabolism was blocked. Topical application of chlorobenzilate also results in significant cutaneous metabolism. These studies illustrate the utility of the IPPSF to identify cutaneous metabolites and to study factors which modulate this process. Use of a nonrecirculating system would allow development of pharmacokinetic models to quantitate the rate and extent of first-pass biotransformation reactions in the skin.

In conclusion, the IPPSF appears to be a useful model to investigate all phases of percutaneous absorption. To date, 1200 skin flaps have been perfused, some of these for periods of up to 18 h. If longer time periods are desired, compounds may be applied *in situ* and flaps harvested at later time points, as demonstrated using the organophosphates (Figure 8). As previously discussed, pigskin is generally perceived to be similar to human skin for studying the percutaneous absorption of many compounds, and in whole animal studies, has even been better than human xenografts for many compounds.[43,44] The advantages of the IPPSF are that absorption across viable epidermis and dermis occurs, a functional microcirculation is present, cutaneous metabolites can easily be detected without systemic interference, experimental conditions (temperature, humidity, perfusate composition, drug interactions) can be controlled or manipulated, large perfusate samples may be collected over a relatively long time period, and a large surface area of skin is available for testing human drug delivery system prototypes. The process of drug distribution into skin from systemic exposure may also be investigated. Topical IPPSF studies would thus directly determine in a single experimental protocol: bioavailability, absorption kinetics, and the presence or absence of cutaneous biotransformation or compound-induced vasodilation.

Since the IPPSF is also responsive to direct cutaneous injury, toxicologic processes interacting with xenobiotic absorption may be studied. This was demonstrated by biochemical and morphological changes seen after exposure of IPPSF to sodium fluoride administration.[79,80] (Recent studies have also demonstrated the IPPSFs utility to study cutaneous toxicity, including the ability to form blisters when exposed topically to vesicants.[118,119] These uses are extensively reviewed elsewhere.[98b]) Another advantage of the system is that since two IPPSF's can be raised from a single pig, a control preparation for the individual animal is always available. Last, but certainly not least, all absorption studies are humanely conducted *in vitro* without requiring the donor animal to be sacrificed.[114] After harvest of the skin flap, the pig may be returned to its prior existence. This aspect of the model is especially attractive when toxic compounds are under investigation. The major limitation is that the surgical procedures are somewhat labor intensive, although previously untrained laboratory technicians can be taught all stages of the procedure. There is a finite limit on the length of viable perfusion, a problem which is not relevant for most short-term transdermal drug applications. Since the studies are performed on porcine skin, extrapolation to humans is still across species lines. However, the advantages obtained by the ability to study specific mechanisms of percutaneous absorption in an anatomically intact and physiologically functioning preparation are considerable.

VII. ADVANTAGES, LIMITATIONS, AND FUTURE DIRECTIONS

Table 2 presents a relative comparison of the three model systems reviewed. Such comparisons are helpful, but obviously biased according to the needs and orientations of the investigator. It must be acknowledged that all of the experimental models described in this review are more expensive and technically demanding than classic *in vitro* diffusion cell approaches.

Perfused skin preparations may offer new insights into the mechanisms of xenobiotic percutaneous absorption and metabolism. They will not replace other approaches currently in use,[115-117] but fit well into the hierarchy of test systems, falling between *in vitro* diffusion models and *in vivo* whole animal studies. Screening for percutaneous absorption and initial formulation studies are probably best conducted

Table 2

Comparison between Available Animal Models

	Xenografts	RHSFS	IPPSF
Cost	++	+++	+++
Technical expertise required	+	+++	++
Special animal housing	+++	+++	+
Uses human skin	+	+	−
Control over study variables	−	+	+++
Ability to separate cutaneous from systemic biotransformation	−	+	+
Pharmacokinetic models can be utilized	+	++	+++

Note: (+) Are the relative advantages and (−) the disadvantages of selected factors between model systems.

in vitro, with confirmatory studies performed in whole animals. Perfused models are best suited to (1) studying mechanisms of absorption which involve interactions between the various steps involved in percutaneous penetration, (2) for quantitating cutaneous metabolic reactions, (3) for studying the effects of altered skin physiology on these processes, (4) for investigating the effects of altered blood perfusion on absorption, and (5) for designing transdermal drug delivery systems which require an accurate prediction of the cutaneous "input profile" into the systemic circulation.

Human xenograft models allow one to assess whether penetration and metabolism in human skin is significantly different from other species. If further systemic metabolism of the compound occurs after percutaneous absorption, extrapolation to the *in vivo* human situation may be affected. The rapid advances being made in skin organ culture and transplantation technology will undoubtedly result in another generation of more realistic animal models being developed. These may eventually be combined with isolated perfused preparations and result in significantly improved hybrid systems. As novel controlled and active drug delivery systems are developed, it will become essential to utilize models which are physiologically capable of responding in a fashion similar to the intact animal. This is especially true if alterations in microcirculation are involved, if cutaneous biotransformation occurs, or if depot formation is dependent upon cutaneous blood flow. Similarly, since most skin disease syndromes affect more than one of the parameters outlined in Table 1, experimental model systems must be capable of predicting these responses.[120]

Consistent with this complexity, another factor which needs to be addressed is the amount of variability often seen in many percutaneous absorption studies. A major source is the epidermal barrier itself which is controlled *in vitro* by using defined dermatome sections. However, with perfused preparations and whole animals, epidermal variability is further confounded by differences in epidermal metabolic activity, blood flow, and overall physiological status. A complete quantitative model should take into account all of the steps outlined in Table 1. Although at first this diversity appears to be a limitation, if controlled and quantitated, this natural variability is a major strength of perfused preparations for it allows one to assess which factors are dominant in determining absorptive flux. It is important that for whatever model system utilized, quantitative pharmacokinetic techniques be developed for extrapolation purposes. Early use of these mathematical models should also expedite the drug development process without greatly increasing the number of animals utilized, while reducing the number of human subjects required in early testing. For systematically toxic compounds where human testing is impossible, perfused skin studies using pharmacokinetic models may be the only alternative.

ACKNOWLEDGMENTS

We would like to thank the following for financial support: U.S. Army Medical Research and Development Command (DAMD-17-84C-4103), National Institute of Environmental Health Sciences (ES 07046 and ES 00044), National Cancer Institute (CA 42745), and the Becton Dickinson Corporation.

REFERENCES

1. R.C. Wester and H.I. Maibach, *Drug Metab. Rev.* 14, 169 (1983).
2. R.J. Scheuplein and R.L. Bronaugh, *Biochemistry and Physiology of the Skin,* Vol. 2, L.A. Goldsmith, Ed. Oxford University Press, New York, 1255 (1983).
3. R.C. Wester and H.I. Maibach, *J. Toxicol. Environ. Health* 16, 25 (1985).
4. M. Katz and B.J. Poulson, *Handbook of Experimental Pharmacology,* 1st ed. B.B. Brodie and J. Gillette, Eds. Springer-Verlag, New York, 103 (1971).
5. B. Idson, *J. Pharm. Sci.* 64, 901 (1975).
6. R.H. Guy, A.H. Guy, H.I. Maibach, and V.P. Shah, *Pharm. Res.* 3, 253 (1986).
7. W.G. Reifenrath and G.S. Hawkins, *Swine in Biomedical Research,* M.E. Tumbleson, Ed. Plenum Press, New York, 673 (1986).
8. R.J. Feldman and H.I. Maibach, *Arch. Dermatol.* 91, 661 (1965).
9. R.C. Wester and H.I. Maibach, *Cutaneous Toxicity,* V.A. Drill and P. Lazar, Eds. Academic Press, New York, 111 (1977).
10. A.M. Kligman, *Drug Develop. Ind. Pharm.* 9, 521 (1983).
11. R.C. Wester, P.K. Noonan, S. Smeach, and L. Kosobud, *J. Pharm. Sci.* 72, 745 (1983).
12. P.V. Shah, R.J. Monroe, and F.E. Guthrie, *Toxicol. Appl. Pharmacol.* 59, 414 (1981).
13. M.J. Bartek, J.L. LaBudde, and H.I. Maibach, *J. Invest. Dermatol.* 58, 114 (1972).
14. M.E. Andersen and W.C. Keller, *Cutaneous Toxicity.* V.A. Drill and P. Lazar, Eds. Academic Press, New York, 9 (1984).
15. M.P. Carver and J.E. Riviere, *Fund. Appl. Toxicol.* 13, 714 (1989).
16. J.E. Riviere, M.P. Carver, N.A. Monteiro, and K.A. Bowman, *Proc. Med. Chem. Def. Biosci. Rev.* 6, 763 (1987).
17. R.C. Wester, D.A.W. Bucks, H.I. Maibach, and J. Anderson, *J. Toxicol. Environ. Health* 12, 511 (1983).
18. K.E. Andersen, H.I. Maibach, and M.D. Anjo, *Br. J. Dermatol.* 102, 447 (1980).
19. P.V. Shah, M.R. Sumler, Y.M. Ioannou, H.L. Fisher, and L.L. Hall, *J. Toxicol. Environ. Health* 15, 623 (1985).

20. C.L. Sanders, C. Skinner, and R.A. Gelman, *J. Environ. Pathol. Toxicol. Oncol.* 7, 25 (1986).
21. L.K. Pershing and G.G. Krueger, *Skin Pharmacokinetics,* B. Shroot and H. Schaefer, Eds. S. Karger, Basel, 57 (1987).
22. E.M. Pantelouris, *Nature* 217, 370 (1968).
23. E.M. Pantelouris, *Immunology* 20, 247 (1971).
24. D.D. Manning, N.R. Reed, and C.F. Shaffer, *J. Exp. Med.* 138, 488 (1973).
25. N.D. Reed and D.D. Manning, *Proc. Soc. Exp. Biol. Med.* 143, 350 (1973).
26. J. Rygaard, *Acta Pathol. Microbiol. Scand. Sec. A* 82, 105 (1974).
27. S.P. Flannigan, *Genet. Res.* 8, 295 (1966).
28. M.A.B. de Sousa, D.M.V. Parrot, and E.M. Pantelouris, *Clin. Exp. Immunol.* 4, 637 (1969).
29. B. Kindred, *Eur. J. Immunol.* 1, 59 (1971).
30. H.H. Wortis, *Clin. Exp. Immunol.* 8, 305 (1971).
31. G.J. Eaton, *Transplantation* 22, 217 (1976).
32. A. Gilhar, Z.J. Wolciechowski, M.W. Piepkorn, G.J. Spangrude, L.K. Roberts, and G.G. Krueger, *Exp. Cell Biol.* 54, 263 (1986).
33. C. Biren, R. Barr, J. McCullough, K. Black, and C. Hewitt, *J. Invest. Dermatol.* 86, 611 (1986).
34. K.E. Black and W.W. Jederberg, in *Models in Dermatology,* H.I. Maibach and N.J. Lowe, Eds. S. Karger, Basel, 228 (1985).
35. G.G. Krueger, D.D. Manning, J. Malouf, and B.E. Ogden, *J. Invest. Dermatol.* 64, 307 (1975).
36. R.A. Briggaman and C.E. Wheeler, *J. Invest. Dermatol.* 67, 567 (1976).
37. E.L. McGown, T. van Ravenswaay, and C.R. Dumlao, *Toxicol. Pathol.* 15, 149 (1987).
38. Haftek, J.P. Ortonne, M.J. Staquet, J. Viac, and J. Thivolet, *J. Invest. Dermatol.* 76, 48 (1981).
39. D.E. Barker, *Arch. Pathol.* 32, 426 (1941).
40. P. Worst, I. MacKenzie, and N. Fusenig, *Cell Tiss. Res.* 225, 65 (1982).
41. J.R. Morgan, Y. Barrandon, H. Green, and R.C. Mulligan, *Science* 237, 1476 (1987).
42. R.V. Petersen, M.S. Kislalioglu, W.Q. Liang, S.M. Fang, M. Emam, and S. Dickman, *J. Soc. Cosmet. Chem.* 37, 249 (1986).
43. W.G. Reifenrath, E.M. Chellquist, E.A. Shipwash, and W.W. Jederberg, *Fund. Appl. Toxicol.* 4, S224 (1984).
44. W.G. Reifenrath, E.M. Chellquist, E.A. Shipwash, W.W. Jederberg, and G.G. Kreuger, *Br. J. Dermatol.* 111 (Suppl. 27), 123 (1984).
45. W. Feldberg and W.D.M. Paton, *J. Physiol.* 114, 490 (1951).
46. A.R. Kjaersgaard, *J. Invest. Dermatol.* 22, 135 (1954).
47. R.L. Bell, R. Lundquist, and K.M. Halprin, *J. Invest. Dermatol.* 31, 13 (1958).
48. K.M. Halprin and D.C. Chow, *J. Invest. Dermatol.* 36, 431 (1961).
49. V.R. Wheatley and D.W. Sher, *J. Invest. Dermatol.* 36, 169 (1961).
50. V.R. Wheatley, D.C. Chow, and F.D. Keenan, *J. Invest. Dermatol.* 36, 237 (1961).
51. H. Hiernickel, *Br. J. Dermatol.* 112, 299 (1985).
52. H. Hiernickel, H. Merk, and G.K. Steigleder, *Clin. Exp. Dermatol.* 11, 316 (1986).
53. G.G. Krueger, Z.L. Wojciechowski, S.A. Burton, A. Gilhar, S.E. Huether, L.G. Leonard, U.D. Rohr, T.J. Petelenz, W.I. Higuchi, and L.K. Pershing, *Fund. Appl. Toxicol.* 5, S112 (1985).
54. Z.J. Wojciechowski, L.K. Pershing, S. Huether, L. Leonard, S.A. Burton, W.I. Higuchi, and G.G. Krueger, *J. Invest. Dermatol.* 88, 439 (1987).
55. L.K. Pershing, R.L. Conkling, and G.G. Krueger, *Clin. Res.* 34, 418A (1986).
56. Z.J. Wojciechowski, S.A. Burton, T.J. Petelenz, and G.G. Krueger, *Clin. Res.* 33, 696A (1985).
57. L.K. Pershing, R.L. Conkling, and G.G. Krueger, *Clin. Res.* 34, 773A (1986).
58. S.A. Burton, Z. Wojciechowski, U. Rohr, G.G. Krueger, and W.I. Higuchi, *Clin. Res.* 33, 628A (1985).
59. L.K. Pershing, W.J. Jederberg, R.L. Conkling, and G.G. Krueger, *J. Invest. Dermatol.* 90, 597 (1988).
60. M.J. Bartek and J.A. LaBudde, *Animal Models in Dermatology,* H.I. Maibach, Ed. Churchill Livingstone, New York, 103 (1975).
61. R.C. Wester and H.I. Maibach, in *Models in Dermatology,* Vol. 2, H.I. Maibach and N.J. Lowe, Eds. S. Karger, Basel, 159 (1985).

62. R.L. Bronaugh, R.F. Stewart, and E.R. Congdon, *Toxicol. Appl. Pharmacol.* 62, 481 (1982).
63. F.N. Marzulli, D.W.C. Brown, and H.I. Maibach, *Toxicol. Appl. Pharmacol.* Suppl. 3, 79 (1969).
64. G.S. Hawkins and W.G. Reifenrath, *J. Pharm. Sci.* 75, 378 (1986).
65. N.A. Monteiro-Riviere, in *Swine in Biomedical Research,* Vol. 1, M.E. Tumbleson, Ed. Plenum Press, New York, 641 (1986).
66. N.A. Monteiro-Riviere and M.W. Stromberg, *Anat. Histol. Embryol.* 14, 97 (1985).
67. R.L. Bronaugh, R.F. Stewart, and E.R. Congdon, *Toxicol. Appl. Pharmacol.* 62, 481 (1982).
68. P.D. Forbes, in *Hair Growth, Advances in the Biology of Skin,* Vol. 9, W. Montagna and R.L. Dobson, Eds. Pergamon Press, New York, 419 (1969).
69. D.L. Ingram and M.E. Weaver, *Anat. Rec.* 163, 517 (169).
70. W. Meyer, K. Neurand, and B. Radke, *Arch. Dermatol. Res.* 270, 391 (1981).
71. W. Meyer, K. Neurand, and B. Radke, *J. Anat.* 134, 139 (1982).
72. N.A. Monteiro-Riviere, D.G. Bristol, T.O. Manning, R.A. Rogers, and J.E. Riviere, *J. Invest. Dermatol.* 95, 582 (1990).
73. G.J. Klain, S.J. Bonner, and W.G. Bell, in *Swine in Biomedical Research,* Vol. 1, M.E. Tumbleson, Ed. Plenum Press, New York, 667 (1986).
74. W. Meyer and K. Neurand, *Lab. Anim.* 10, 237 (1976).
75. G.M. Gray and H.J. Yardley, *J. Lipid Res.* 16, 434 (1975).
76. N. Nicolaides, H.C. Fu, and G.R. Rice, *J. Invest. Dermatol.* 51, 83 (1968).
77. C.L. Hedberg, P.W. Wertz, and D.T. Downing, *J. Invest. Dermatol.* 90, 225 (1988).
78. G.D. Weinstein, in *Swine in Biomedical Research,* L.K. Bustad, R.O. McClellan, and M.P. Burns, Eds. Pacific Northwest Institute, Richland, WA, 287 (1966).
79. J.E. Riviere, K.F. Bowman, N.A. Monteiro-Riviere, L.P. Dix, and M.P. Carver, *Fund. Appl. Toxicol.* 7, 444 (1986).
79a. K.F. Bowman, N.A. Monteiro-Riviere, and J.E. Riviere, *Am. J. Vet. Res.* 52, 75 (1991).
80. N.A. Monteiro-Riviere, K.F. Bowman, V.J. Scheidt, and J.E. Riviere, *In Vitro Toxicol.* 1, 241 (1987).
81. J.E. Riviere, K.F. Bowman, N.A. Monteiro-Riviere, *Swine in Biomedical Research,* Vol. 1, M.E. Tumbleson, Ed. Plenum Press, New York, 657 (1986).
82. W.J. Tranquili, M. Manohar, C.M. Parks, J.C. Thurmon, M.C. Thlodorakis, and G.J. Bensen, *Anesthesiology* 56, 369 (1982).
83. J.K. Kristensen and S. Wadskov, *J. Invest. Dermatol.* 68, 196 (1977).
84. T.J. Ryan, *Biochemistry and Physiology of the Skin,* Vol. 2, L.A. Goldsmith, Ed. Oxford University Press, New York, 817 (1983).
85. M.B. Meyers and G. Cherry, *Plast. Reconstr. Surg.* 38, 49 (1966).
86. K.M. Halperin and A. Ohkawara, *J. Invest. Dermatol.* 47, 222 (1966).
87. T.A. Johnson and R. Fusaro, *Adv. Metab. Disord.* 6, 1 (1972).
88. R.K. Frienkel, *Biochemistry and Physiology of the Skin,* Vol. 2, L.A. Goldsmith, Ed. Oxford University Press, New York, 328 (1983).
89. J.E. Riviere, K.F. Bowman, and N.A. Monteiro-Riviere, *Br. J. Dermatol.* 116, 739 (1987).
90. M.P. Carver, N.A. Monteiro-Riviere, R.A. Rogers, K.F. Bowman, and J.E. Riviere, *Toxicologist* 7, 244A (1987).
91. M.P. Carver, P.L. Williams, and J.E. Riviere, *Toxicol. Appl. Pharmacol.* 97, 324 (1989).
92. W.J. Albery and J. Hadgraft, *J. Pharm. Pharmacol.* 31, 129 (1979).
93. J. Hadgraft, *Int. J. Pharm.* 2, 265 (1979).
94. R.H. Guy and J. Hadgraft, *Int. J. Pharm.* 6, 321 (1980).
95. R.H. Guy and J. Hadgraft, *J. Pharmacokin. Biopharm.* 11, 189 (1983).
96. R.H. Guy, J. Hadgraft, and H.I. Maibach, *Toxicol. Appl. Pharmacol.* 78, 123 (1985).
97. R.H. Guy and J. Hadgraft, *J. Soc. Cosmet. Chem.* 35, 103 (1984).
98. P.L. Williams and J.E. Riviere, *J. Pharm. Sci.* 78, 550 (1989).
98a. P.L. Williams, M.P. Carver, and J.E. Riviere, *J. Pharm. Sci.* 79, 305 (1990).
98b. J.E. Riviere and N.A. Monteiro-Riviere, *Crit. Rev. Toxicol.* 21 (1990).
99. T.F. Barkve, K. Langseth-Manrique, J.E. Bredesen, and K. Gjesdal, *Am. Heart J.* 112, 537 (1986).

100. F.M. Craig, E.G. Cummings, and V.M. Sim, *J. Invest. Dermatol.* 68, 357 (1977).

101. A. Danon, S. Ben-Shimon, and Z. Ben-Ziv, *Eur. J. Clin. Pharmacol.* 31, 49 (1986).

102. S.K. Chang and J.E. Riviere, *Fund. Appl. Toxicol.* 17, 1991, in press.

103. R.C. Spear, W.J. Poppendorf, W.F. Spenser, and T.H. Milby, *J. Occup. Med.* 19, 411 (1977).

104. J.D. Middleton, *Br. J. Dermatol.* 80, 437 (1968).

105. T.S. Spenser, C.E. Linamen, W.A. Akers, and H.E. Jones, *Br. J. Dermatol.* 93, 159 (1975).

106. L.J. Vinson, E.J. Singer, W.R. Koehler, M.D. Lehman, and T. Masurat, *Toxicol. Appl. Pharmacol.* 7 (Suppl. 2), 7 (1965).

107. J.E. Riviere, B. Sage, and N.A. Monteiro-Riviere, *Cutan. Ocular Toxicol.* 9, 493 (1989/90).

108. J.E. Riviere, B. Sage, and P.L. Williams, *J. Pharm. Sci.* 80, 615 (1991).

109. P.L. Williams and J.E. Riviere, *Int. J. Hyperthermia,* 6, 923 (1990).

110. M. Schach von Wittenau, *Chemotherapy (Basel)* (Suppl. 13), 41, (1968).

111. J.L. Riond and J.E. Riviere, *Vet. Human Toxicol.* 30, 431, 1988.

112. J.E. Riviere, R.L. Page, R.A. Rogers, S.K. Chang, M.W. Dewhurst, and D.E. Thrall, *Cancer Res.* 50, 2075 (1990).

113. M.P. Carver, P.E. Levi, and J.E. Riviere, *Pest Biochem. Physiol.* 38, 245 (1990).

114. J.E. Riviere, *Humane Innovations and Alternatives in Animal Experimentation: A Notebook,* Vol. 1, E.M. Bernstein, Ed. Psychologists for the Ethical Treatment of Animals, Saranac Lake, NY, 7 (1987).

115. J. Kao, J. Hall, L.R. Shugart, and M.J. Holland, *Toxicol. Appl. Pharmacol.* 75, 289 (1984).

116. J. Kao, F.K. Patterson, and J. Hall, *Toxicol. Appl. Pharmacol.* 81, 502 (1985).

117. J. Kao and J. Hall, *J. Pharmacol. Exp. Ther.* 241, 482 (1987).

118. N.A. Monteiro-Riviere, *Fund. Appl. Toxicol.* 15, 174 (1990).

119. J.R. King and N.A. Monteiro-Riviere, *Toxicol. Appl. Pharmacol.* 104, 167 (1990).

120. J.E. Riviere, *Cosmet. Toiletries* 105, 85 (1990).

11

In Vitro Alternative Methods for the Assessment of Dermal Irritation and Inflammation

DAVID W. HOBSON

AND

JAMES A. BLANK
Battelle Memorial Institute
Columbus, Ohio

I. INTRODUCTION

Several different types of toxic interactions are possible between a chemical or chemical mixture and the skin. These toxic interactions can be broadly classified as being corrosive, irritative, or sensitive in nature. In each case, the pathology usually includes dermal inflammation and a variable amount of necrosis. Historically, *in vivo* test procedures have been used to assess the dermal toxicity of chemicals. Even though the precise mechanism(s) by which many dermatotoxicants produce their effects is still unknown, these *in vivo* tests are based on an appreciation for the fact that chemical interactions with the skin are often complex in nature (i.e., they may simultaneously involve various degrees of corrosion, irritation, and sensitization) and are probably best studied in a system capable of responding in a complex fashion. The procedures and rationale used for the *in vivo* assessment of dermal irritation and sensitization have been presented in detail in Chapters 4 and 5, respectively. Due to the availability of relatively extensive historical data from assessment of dermal toxicity *in vivo* with various different types of chemicals, it has recently become possible to attempt the development of *in vitro* models to predict some of the dermatotoxic effects observed *in vivo*. Rationale for developing such *in vitro* methods is (1) to evaluate specific aspects of the irritation or sensitization process related to a chemical or group of chemicals, (2) to supplement *in vivo* evaluation by serving as a first tier screen test to minimize *in vivo* testing and as a decision point on chemical/product development, or (3) to replace an *in vivo* procedure. In this chapter *in vitro* approaches to the assessment of dermatotoxicity, especially methods

323

proposed for predicting chemically induced dermal irritation and corrosion, will be presented.

Traditionally, tests to evaluate the dermal irritant or corrosive potential of chemicals have been performed *in vivo* (see Chapters 4 and 5 for further details) using various laboratory animal species. These *in vivo* tests have been the standard used to evaluate such effects for almost half a century. However, these procedures have been increasingly challenged as to their validity for human extrapolation and, more recently, as to their animal use requirements. As a result, toxicologists have responded by directing an increasing amount of attention toward the development of *in vitro* alternative approaches to such testing which might be both more predictive of the human response and more conservative in their need for animal subjects[1,2]. It is hoped that these endeavors will lead to the development and adoption of valid *in vitro* procedures to either supplement, augment, or replace the existing *in vivo* methods.

In Chapters 4 and 5, it has been shown that the *in vivo* responses of mammalian skin to irritants or allergic sensitizing agents involve complex processes which occur at the site of skin contact and also involve the responses of several other organ systems. This complexity is shown for dermal irritation and corrosion in Figure 1. Because of this complexity, it is an extremely difficult matter to develop *in vitro* models which completely simulate such responses within the skin. The development of *in vitro* models which mimic the systemic interactions occurring between the skin and various other organ systems is even more difficult. For this reason, current attempts to develop *in vitro* tests for the evaluation of the dermal irritancy and corrosive effects of chemicals and chemical formulations are being directed toward

- Examining the *in vitro* vs. *in vivo* correlation between measurable chemical effects on skin or skin surrogate interactions *in vitro* with known *in vivo* results
- The development of more representative *in vitro* dermal models
- The improvement of methods to quantify skin/chemical interactions *in vitro* for incorporation with *in vitro* models to improve their accuracy and sensitivity

The net result of this research with *in vitro* dermal models will be that our knowledge of chemical interactions with the skin will increase significantly over the next few years. From this, *in vitro* skin models which have excellent predictive characteristics should begin to emerge and we should begin to have a fairly complete mechanistic understanding as to the actual factors involved in producing the predictive strengths and weaknesses of such models.

This chapter is primarily concerned with discussing state-of-the-art *in vitro* dermal irritation and corrosion testing procedures. Because this is such a rapidly developing area of scientific endeavor, the objectives of this chapter are (1) to present examples of several current methods which have been proposed for the *in vitro* evaluation of dermal irritants and corrosive agents, (2) to discuss emerging procedures and technologies which may improve several aspects of the *in vitro* models relative to the current methods, and (3) to discuss approaches toward the

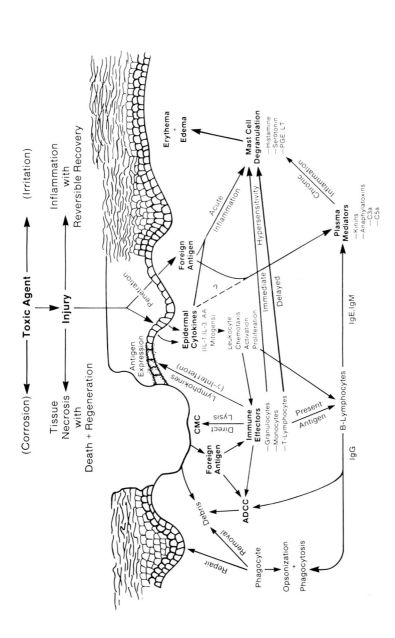

Figure 1. Illustration of cellular and humoral mechanisms of dermal irritation and corrosion. Abbreviations: IL-1, IL-3 = interleukins 1 and 3; AA = arachidonic acid metabolites; PGE = prostaglandins of E series; LT = leukotrienes; CMC = cell-mediated cytotoxicity; ADCC = antibody-dependent cellular cytotoxicity; Ig = immunoglobulin. (From S. T. Boyce, J. F. Hansbrough, and D. S. Norris, *Alternative Methods in Toxicology*, A. M. Goldberg, Ed. Mary Ann Liebert, New York (1988). With permission.)

validation of the new and current *in vitro* methods. The discussion is intended to be as objective as possible. It should also not be construed that statements made with regard to any of the methods described are intended to either endorse or refute any particular procedure.

II. TYPES OF ALTERNATIVE DERMAL IRRITATION MODELS

In recent years, several different types of *in vitro* alternative models to supplement, augment, or replace the current *in vivo* methods for the evaluation of dermal irritation have been proposed. Some of these models are *in vitro* models in the strictest sense (conducted with the use of living material outside of the target species) and others require either no living material at all or use lower vertebrate or invertebrate organisms as an alternative to the target species. In order to discuss these models in an organized fashion, it is convenient to classify them into five general categories: (1) structure-activity; (2) biochemical; (3) cellular; (4) tissue; and (5) lower vertebrate or invertebrate.

This chapter will be concerned primarily with discussing current *in vitro* cell or tissue culture approaches to the assessment of dermal irritation. However, examples of structure-activity and biochemical approaches will be mentioned briefly because, like other *in vitro* models, some of these models have been proposed and seriously evaluated as adjuncts or alternatives to the use of routine *in vivo* test procedures. Lower vertebrate or invertebrate models will not be discussed because, although proposed, such models appear to be giving way to other model types. In order to provide a logical framework for presentation, the different models will be discussed in their order of biologic complexity.

III. STRUCTURE-ACTIVITY MODELS

As we learn more about the nature of chemical-biological interactions, it appears likely that there are some recurring patterns in these interactions attributable to the structural characteristics of particular chemicals or structural classes of chemicals. With the advent of modern digital computers, it has become possible to study the correlation between the structure of a chemical and its biological activity to an extent not possible only a decade ago. Although their potential use in predicting mutagenicity and teratogenicity for some chemicals has been clearly demonstrated,[3] structure-activity models still have not received widespread acceptance. Obviously, the validity of such computerized predictive models are necessarily limited by the extent and quality of their informational base (database). Since most toxicologists are still trying to completely understand the chemical-biological interactions involved in the production of dermal inflammation, it is also not likely that such predictive approaches will gain widespread acceptance as a replacement or alternative to *in vivo* skin testing until more is known about these interactions.

Nevertheless, such models can be used to screen new chemical structures to assess the general likelihood that they might cause dermal irritation/sensitization relative to the existing database for similar structures. In this manner, decisions can be made early in the premanufacturing process as to whether a chemical warrants consideration as a potential dermal irritant/sensitizer or if the data are insufficient and indicate the necessity for more preliminary testing.

One example of this type of approach, as applied to mutagens and teratogens, is that of Einslein et al. which is based on the results obtained from numerous *in vivo* tests with various chemical types.[3] This approach appears to have remarkable predictive power for some chemical classes. It is likely that structure-activity approaches will be increasingly used in pre-production decision making as a direct result of our increasing knowledge concerning the nature and classification of dermal irritant and sensitization reactions. In fact, at least one approach to the prediction of dermal sensitizing agents using a structure-activity model has already been proposed.[4] Increased utilization of structure-activity models (or any other predictive model type for that matter) will depend upon demonstrating the validity of such models. The predictive power for such models is inseparably linked to the extent and quality of the database from which they are derived. As the knowledge of the toxicologic mechanism(s) giving rise to dermal irritation and sensitization is increased, it is certain that the database of *in vivo* structure-activity responses will be significantly expanded and available for the development of *in vitro* structure-activity models. Also, the development of valid biochemical, cellular, and tissue based *in vitro* dermal irritation/sensitization models should generate additional information, which will help establish the predictiveness of structure-activity models. For a good description of the uses for and problems involved in applying these models in toxicology, the reader is referred to a 1984 publication by Enslein.[5]

IV. BIOCHEMICAL MODELS

There are many different types of biochemical events that can occur to initiate dermal inflammation following percutaneous exposure to a chemical irritant. Such events may lead to the loss or alteration of viability in keratinocytes, dermal fibroblasts, glandular epithelial cells, or some other cell type within the skin. Some of the biochemical processes involved in the initiation of dermal inflammation and/or sensitization reactions following chemical exposure were discussed in some detail in Chapters 4 and 5, respectively. From this discussion it can be seen that, in a most rudimentary way, such reactions can be visualized as an interaction between a chemical agent and some sort of biochemical or biomolecular target. Obviously, this is not a new concept as its basis comes from the development of the theory of chemically receptive substances by Ehrlich, Langley, and others at the beginning of the 20th century.[6-8] More recently, the term "receptor" in pharmacology and toxicology has been commonly used to describe an interaction between a chemical substance and a biomolecule which results in a characteristic biologic effect.[9,10] The fact that chemically induced dermal inflammation probably arises, in many cases, from some sort of chemical-biochemical target interaction is precisely what developers of biochemical assays to predict the dermal irritancy of a chemical hope to exploit.

In order to develop a biochemical model for chemically induced dermal inflammation, it is important that the interaction between the inflammatory agent(s) and the biochemical target(s) or receptor(s) responsible for the initiation of inflammation be identified and understood. One approach to predict the initiation of dermal inflammation might be to identify and isolate receptor biomolecules specific for a chemical or chemical class and then develop a predictive biochemical test based on the *in vitro* measurement of that interaction. This has been the case with some organic arsenical compounds such as dichloro-2-chlorovinyl arsine (Lewisite) which are thought to produce their vesicating effects on the skin via interactions with important sulfhydryl-containing biomolecules such as the pyruvate dehydrogenase enzyme complex. It might be considerably more difficult, however, to develop a generalized biochemical test to predict the inflammatory potential of chemicals with a higher degree of structural variation as there are probably many different types of biochemical receptors for chemical irritants due to the chemical structural diversity of possible biomolecular targets. Nevertheless, the process of dermal inflammation, once initiated, is known to involve a fairly routine course of events which might involve some biochemical process common to many or all of the initiating reactions. Thus, the successful development of a valid biochemical test to predict the dermal inflammatory potential of chemical irritants will likely depend upon the identification of a biochemical initiating event which is common to chemicals within a structural class or the identification of a biochemical inflammation process which is common to chemicals across structural classes. Once identified, the system can be isolated and developed for routine testing purposes.

Unfortunately, there has been little work performed with the specific goal of determining the association between potential biochemical targets and various chemical structures known to cause dermal inflammation. Nevertheless, as mentioned above, there is increasing interest in the development of structure-activity models to predict dermal irritant and sensitizing chemicals which may help to identify common chemical structural correlates for known sensitizing and irritant agents using computerized databases. It is also likely that these efforts will be helpful in identifying gaps in our knowledge concerning the relationships between chemical structural correlates and their potential biochemical targets.

Although there are currently almost no biochemically based assays which have been proposed for the assessment of dermal inflammatory agent effects, the Skintex™ assay[11] is one example of such an approach for evaluating dermal toxicants. The Skintex™ assay was developed as a possible adjunct or alternative to *in vivo* dermal irritation testing. The assay is based on the premise that the initiation of toxic injury to the skin involves the penetration of a chemical through the stratum corneum followed by (1) the formation of "irritant complexes" with various cellular components (i.e., enzymes, DNA, receptors, cofactors, or membranes); (2) alteration of the physicochemical environment of skin molecules; or (3) by direct reactivity with intra- or intercellular skin components. In the Skintex™ model, the permeation and adsorptive characteristics of the stratum corneum are simulated by a keratin matrix barrier component and the direct reactivity of skin molecules with the test chemical are simulated by a protein matrix component. The permeation of the barrier component and direct reactivity with the protein matrix component are quantitated spectrophotometrically by monitoring the release of an indicator dye from the system in response to interactions between the test chemical and either the barrier and/or the protein matrix components.

The keratin matrix barrier component is prepared from a pH 8.0 buffered salt solution containing 10% keratin, 1% collagen, and saffranin bound to cellulose acetate by incubation with 0.1% glutaraldehyde. An indicator dye is attached to the resulting keratin/collagen barrier matrix membrane in order to detect changes in the structure or integrity of the matrix. The secondary, or protein matrix, component of the assay consists of a proprietary reagent solution consisting of globulin complexes, soluble collagen, glycosaminoglycans, amino acids, phospholipids, and free fatty acids in a buffered salt solution.

The basic protocol for the assay involves five steps:

1. Test samples containing an irritant, inflammatory, or corrosive agent are prepared as for *in vivo* dermal dosing.
2. Test samples and calibrators (consisting of materials whose *in vivo* dermal irritation scores are known in the form of Primary Dermal Irritation Index or PDII values) are applied at doses of between 30 and 100 μl to 0.25 cm^2 barrier matrix membrane patches. PDII values range from 0.0 to 8.0 and correspond with primary irritation scores of between 0 and 8 obtained from Draize rabbit skin tests.[12]
3. Barrier matrix patches containing test samples, calibrators, and any required blanks are inserted into cuvettes containing individual aliquots of the secondary reagent solution and incubated for 2 h.
4. Optical density (OD) readings at 580 nm are obtained from cuvettes containing the test samples, calibrators, and blanks.
5. Using the OD readings obtained from the calibrators, a scoring plot of OD readings vs. PDII values is created for the assay. PDII estimates are then obtained from the scoring plot using the corresponding OD readings (multiplied by 1000) obtained from each test sample. An example of the scoring plot is shown in Figure 2. A skin irritant is defined as a test sample which produces a PDII estimate >2.0. Estimates of 0.0 to 2.0 indicate minimal or mild irritation, estimates of approximately 4.0 indicate moderate irritation, and estimates of approximately 6.0 or greater indicate severe irritation.

The Skintex® assay is produced commercially in kit form. Further details concerning the development, validation, scoring, conduct, cost, or availability of the assay can be obtained from its commercial distributor (Ropak Corporation, Fullerton, CA).

It is also possible to develop *in vitro* biochemically based assays for the purpose of assessing specific toxicants or the effectiveness of candidate antidotes to counter the effects of specific toxicants. Because the skin contains a multitude of potential biochemical targets, the relative success of such assays depends upon the develop-

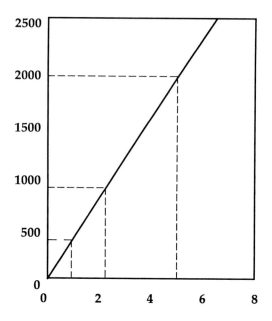

Primary Dermal Irritation Index (PDII)

Figure 2.   The primary dermal irritation index (PDII) of the Skintex®
Scoring System is estimated as a function of test system optical density.
(Provided courtesy of Ropak, Inc., Fullerton, CA.)

ment of a knowledge of the types of target interactions likely for particular structural classes of toxicants as well as the potential for a particular type of interaction to either result directly in, or be predictive of, the dermal toxic response of interest. Some current examples of assays which focus on specific biochemical targets found within the skin or other organs include methods to measure alterations in membrane lipid structure and metabolism by surfactants, the effects of organochlorine compounds on $Mg^{2+}ATPase$ activity, phospholipids, and proteins; the alkylation of cutaneous deoxyribonucleic acid by sulfur mustards; and the inhibition of pyruvate dehydrogenase activity by various arsenicals.[13-17] At present, the development of such assays is limited to the need to evaluate specific characteristics of the dermatotoxicity of specific chemicals or to examine the therapeutic efficacy of new candidate pharmaceuticals toward a specific target. Nevertheless, it would seem that the identification and systematic evaluation of specific biochemical target-toxicant interactions which may be important in the production of dermal toxicity might be of great benefit in the development of structure-activity models and in the refinement of biochemically oriented generalized models such as the Skintex™ assay. For this reason, it is hoped that more attention be given toward development of this important type of *in vitro* assay.

V. CELLULAR MODELS

Cell cultures have been used over the past 30 years in areas such as pharmacology, toxicology, immunology, and bacteriology for use in evaluating cancer chemotherapeutic compound and antimicrobial compound efficacy, various immunological parameters, compound mutagenicity through the Ames test, and the contamination of biotechnology products.[18] Cell culture methods are also being examined to serve as *in vitro* supplements and alternatives to *in vivo* dermal irritancy testing. The selection of cell types for use in the development of *in vitro* dermal irritant tests is difficult because the skin is a complex tissue, being composed of different cell types including keratinocytes, fibroblast, mast cells, Langerhans' cells, and leukocytes. The skin is also multilayered, which contributes to its complexity. Skin also possesses an acellular layer (stratum corneum) covering the skin surface which serves to minimize water loss and to protect the body from the external environment.

Recent efforts have been focused on the use of *in vitro* cytotoxicity tests as *in vitro* supplements and alternatives for *in vivo* ocular and dermal irritant tests. Thus far, the development of alternatives to the Draize ocular toxicity test has received the most attention in this regard. In this respect, good correlations between ocular irritancy and *in vitro* cytotoxicity tests have been observed with several classes of chemical compounds.[19-21] In these studies, *in vivo* and *in vitro* comparisons using established cell lines show that while there are differences in sensitivity of the different cell types, the ranking of chemical-induced cytotoxicity in the different cell lines are quite similar. [19] Other assays, such as functional and cytotoxic assays using neutrophils, macrophages, mast cells, and fibroblasts as the target cells, have also been examined as possible model systems.[22,23] Serum-free cultures of human epidermal keratinocytes without fibroblast feeder layers have recently been receiving increasing interest for use as potential *in vitro* models for the testing of potential dermal irritants.[24,25] The next section of this chapter describes the preparation of primary human epidermal keratinocyte cultures and some assays which have been used to assess cellular viability or cytotoxicity following chemical exposure.

A. Isolation and Culture of Human Keratinocytes

Several laboratory procedures have been published for the isolation and culture of epidermal keratinocytes.[26-28] More recently, human epidermal keratinocytes and the serum-free media to culture these cells have become commercially available (e.g., Clonetics Corporation, San Diego, CA). These cells can be obtained as either primary or secondary passaged cells. Figure 3 is a flowchart for the isolation of human keratinocyte as performed by Boyce and Ham.[29] The procedure is described below in more detail.

1.   Immediately after excision, the human skin tissue is placed in a sterile solution

consisting of 30 mM HEPES-NaOH buffer, 10 mM glucose, 3 mM KCl, 130 mM NaCl, 1 mM Na$_2$HPO$_4$ · 7H$_2$O, 0.0033 mM phenol red, pH 7.4. This solution, referred to as solution A, is supplemented with 100 U/ml penicillin, 100 µg/ml streptomycin, and 0.25 µg/ml Fungizone®. All procedures are performed under sterile conditions.

2. The skin sample is placed into a petri dish of sterile solution A and the pieces of subcutaneous connective tissue are removed from the dermis.

3. The tissue is washed for 15 to 30 s in a disinfectant solution such as 5% Dettol® (Reckitt & Coleman Pharmaceutical) dissolved in solution A. The tissue is then rinsed by dipping into three separate containers of sterile solution A.

4. The skin tissue is cut into pieces approximately 4 mm². This is performed with the tissue submerged in sterile solution A.

5. Place the tissue segments into 15 ml of MCDB 153 media containing Bovine Pituitary Extract (BPE; 10% V:V), 0.1 mM ionized calcium, 625 U/ml collagenase (Worthington). MCDB can be prepared following the procedure of Boyce and Ham[30] or can be purchased from Sigma Chemical Company or Grand Island Biological Company (GIBCO). BPE can be purchased from Sigma Chemical Company or Clonetics Corporation. Transfer equal volumes of the suspension to two 60-mm petri plates and incubate at 37°C under a 5% CO$_2$ and saturated humidity atmosphere for 90 to 120 min or until the epidermis can be removed from the dermis.

6. At this time prepare the sterile culture flask and media for keratinocyte culture. Culture media is prepared by making trace element supplemented MCDB 153 as described by Boyce and Ham.[30] This media can also be purchased from Sigma Chemical Company or GIBCO. After sterile filtering through a 0.22 µm sterile filter, the MCDB 153 is then supplemented with sterile reagents as follows: 10 ng/ml epidermal growth factor (EGF; Bethesda Research Laboratories), 5 µg/ml insulin (Sigma; bovine crystallized), 0.5 µg/ml hydrocortisone, 0.1 mM ethanolamine, 14 µg/ml BPE, and a 1:100 dilution of penicillin, streptomycin, and Fungizone® solution (GIBCO 100 X solution). Murine EGF and recombinant human EGF, which are commercially available through Boerhinger Mannheim, Calbiochem, and Sigma Chemical Company, may be used. Once EGF and insulin have been added, the media should not be filtered due to potential problems with protein adherence to the filters. It is suggested that the MCDB supplements are added as sterile stocks and added so that only a 2% or less dilution of the MCDB 153 occurs. MCDB with trace elements and supplements will be referred to as culture media. KGM media, which contains all of the above supplements, can be purchased from Clonetics Corporation. The shelf life of the culture media is short, being approximately 2 to 3 weeks from the date of preparation.

7. Flask preparation for culture involves adding culture media to flasks as follows: 25 ml media to 150 cm² flasks, 15 ml media to 75 cm² flasks, or 5 ml

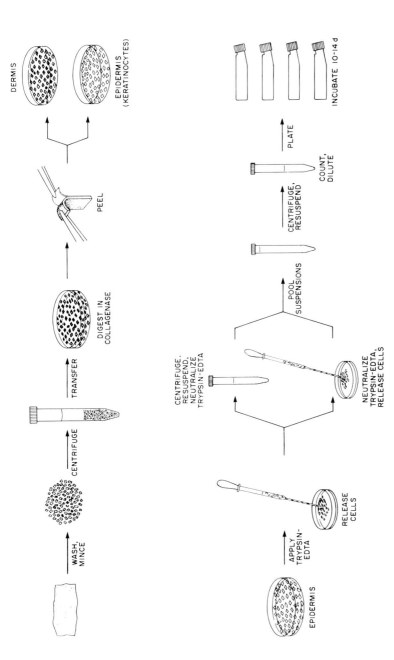

Figure 3. Flow diagram of a procedure for the isolation of human epidermal keratinocytes. (From S. T. Boyce and R. G. Ham, *In Vitro Models for Cancer Research*, M. M. Webber and L. I. Sekely, Eds. CRC Press, Boca Raton, FL (1986). With permission.)

media to 25 cm² flasks. The flasks are placed in an 37°C incubator (5% CO_2 and saturated humidity) and the flask lids are loosened so that the atmosphere and media in the flasks can equilibrate with that of the incubator.

8.   Carefully and aseptically aspirate the collagenase solution away from the tissue sections and add solution A to the plates. Remove the epidermis from the dermis for each tissue section using sterile forceps. The epidermis, which should detach as an intact sheet, is placed into a 60-mm petri plate containing solution A at room temperature. Between tissue sections, sterilize the forceps by dipping in ethanol, burning off the ethanol, and allowing to cool.

9.   Remove the solution A from the epidermal sections and add 6 ml of solution A containing 0.025% (w:v) trypsin (Sigma; type III bovine, chromatographically purified) and 0.01% ethylenediamine tetraacetic acid (w:v). Gently reflux the tissue sections with a sterile pasteur pipette for 3 to 4 min. The solution, which should contain keratinocytes, is removed from the remaining tissue sections and placed into a sterile centrifuge tube. The cells are pelleted by centrifugation at $250 \times g$ for 5 min.

10. 6 ml of MCDB 153 containing 0.1% (w:v) soybean antitrypsin (Sigma; type I-S) and 10% (w:v) Vitrogen 100 (Collagen Corporation) is added to the tissue sections left in the petri plate. The tissue sections are again refluxed for 5 min with a sterile pipette and the media (void of tissue sections) is then combined with the pellet from the centrifugation step that had been gently resuspended in the same solution. The cells are pelleted by centrifugation at $250 \times g$ for 5 min.

11. The supernate is removed from the cell pellet and the pellet is gently resuspended into a single cell suspension in 2 to 3 ml of culture media. The cells are counted and placed into the pre-equilibrated culture flask at a density of 3000 to 4000 cells per cm².

12. The media in the flask is mixed to evenly disperse the cells. The flasks are cultured at 37°C under a 5% CO_2 and humidity saturated atmosphere with the flask caps loosened to allow gas exchange. After 48 h of incubation, the media over the cells is gently removed and replaced with fresh, pre-equilibrated culture media. After 5 to 7 days of incubation, the media is carefully removed, the cells washed two times with pre-equilibrated solution A, and fresh, pre-equilibrated media is added. This last step is repeated at 5 to 7 days later or 1 day before the primary culture is harvested.

13. The culture is ready for subculture when the cell layer is approximately 50% confluent.

14. After removal of culture media, the cells are washed two times with solution A.

15. The cells are covered with solution A containing 0.25% trypsin (w:v) and 0.01% EDTA (w:v). After 30 to 60 s, the solution is removed and the flask caps are put firmly in place. After 3 to 4 min at room temperature, the cells are examined under an inverted microscope. If less than half of the cells have detached, continue the incubation for an additional 3 min and recheck the cells.

When more than half the cells have detached, the flask is rapped against a hard surface to detach cells from the flask surface.

16. At this point, the cells can be prepared for cryopreservation or subcultured. Initiate subculture in the same fashion that was utilized for primary culture of the cells. Conditions for subculture and growth in 96-well microtiter plates are discussed in the following section.

(Procedure reproduced with permission of S. Boyce.[29])

The reader is referred to the referenced publications of Boyce and Ham[29,30] for more detailed descriptions of the reagents reagent preparation, cell isolation procedure, and keratinocyte cryopreservation procedure.

B. Keratinocyte-Based Test Systems

Different cell types have been used for *in vitro* cytotoxicity testing including fibroblasts, keratinocytes, and cell lines of different types. With the commercialization of human epidermal keratinocytes, current research testing has centered around the use of keratinocytes. There are many culture and test conditions that should be considered when developing *in vitro* test protocols. These conditions include passage of cell, culture confluency, length of toxicant exposure, and the length of the culture period. Passage of cell and culture confluency are important issues as cell growth rates may vary from keratinocytes which are of second to fifth passages or of different confluency stages.

In an attempt to standardize cytotoxicity testing using human keratinocyte cultures, Clonetics Corporation has developed a protocol using tertiary cultured cells as the target and neutral red vital dye uptake as an endpoint. This protocol attempts to standardize confluency, length of toxicant exposure, and length of the culture period. This protocol also includes a positive control compound which is essential to monitor and control assay performance and to make interlaboratory comparisons. The design, as described by Triglia et al.,[31] uses secondary passaged keratinocytes seeded into 96-well microplates at 2500 cells per well in a 250 μl volume. After a 3-day incubation at 37°C and under a 5% CO_2 saturated humidity atmosphere, media is aspirated from the cultures and then media (untreated control), the positive control compound, or various dilutions of the test chemical are added. After the 2-day exposure period, neutral red dye uptake is examined and the concentration of toxicant required to reduce neutral red content by 50% is estimated (NR_{50}). In this manner, ranking of chemical toxicants is made according to the NR_{50} values and also may be compared to the reference standard compound value. The neutral red dye uptake assay is described in greater detail below.

Standardization of culture and assay conditions, as well as the measured endpoint, should maximize the benefit of the datasets generated. Before this occurs, protocols must be rigorously tested to evaluate their predictiveness of the *in vivo* situation and to define assay limitations. Such evaluations have already been initiated.[32,34] A more detailed description of assay design considerations and assay validation is presented in a later section.

C. Keratinocyte-Viability Assays

A host of different assays have been utilized to examine cellular proliferation, viability, and membrane integrity. These assays include, ^{3}H-thymidine incorporation, ^{3}H-uridine incorporation, cellular DNA measurements, cellular protein measurements, neutral red vital dye stain, MTT dye reduction, release of cellular enzymes such as lactate dehydrogenase and beta-glucuronidase, nonviable dye methods such as trypan blue retention, propidium iodide incorporation, arachidonic acid oxidation products, and glucose utilization. Described below are the procedures that have been used for several of the aforementioned assays.

1. Neutral Red Dye Uptake Assay

Neutral red (3-amino-7-dimethylamino-2-methyl phenazine hydrochloride) is a weakly cationic, water-soluble dye that accumulates in cellular lysosomes by electrostatic and hydrophobic interactions.[33] Alterations in cell membrane or lysosomal membranes lead to a decrease in the uptake and binding of the dye.[34] Cellular viability and proliferative assays have been developed based upon the retention of neutral red by viable cells and loss of dye by nonviable or damaged cells.[34-36] After incubation of the cells with neutral red, the cultures are washed, and the cells are solubilized releasing the neutral red, which is then quantified by measuring the optical density at 540 nm. This assay utilizes adherent cell types such as fibroblasts, keratinocytes, macrophages, hepatoma cells, and epithelial cells, which makes the wash and measurement steps easy and fast.

The neutral red method is relatively sensitive as it can detect as few as 1000 cells and provides a linear relationship between cell number and neutral red content up to 40,000 3T3 fibroblast cells.[35] This method does not require the use of radioactive material and is also fast, inexpensive, and can be easily automated. Data can be electronically captured by interfacing the microtiter plate spectrophotometer to a computer. In addition to making the assay quicker, electronic data capture simplifies data analyses and enhances quality control. This assay would be difficult to use with nonadherent cell types as it requires a wash step.

The procedure for the neutral red assay described below is essentially that previously described by Borenfreund and Puerner.[35] The procedure is one that has been used for examining the potential cytotoxicity of chemical agents.

1. Cells are seeded into 96-well microtiter plates prior to toxicant addition.
2. 24 h following toxicant addition, the media is removed and 0.2 ml of media containing 50 µg/ml neutral red is added. The neutral red solution is prepared the day before use and incubated at 37°C overnight. Prior to use the solution is centrifuged at 1500 × g for 10 min to precipitate undissolved neutral red.

3. After a 3-h incubation at 37°C, media containing the neutral red is removed and the cells are washed with a 4% formaldehyde, 1% calcium chloride solution prepared in deionized water to remove extracellular neutral red and promote cellular adhesion to the plate.

4. The wash solution is removed from the cells within 2 to 3 min of solution exposure. 200 µl of 1% acetic acid in a 50% ethanol-deionized water solution is then added to the wells. Following a 20 min room temperature incubation, the wells are gently mixed and the OD at 540 nm is determined.

5. To assess the effect of toxicant, the NR_{90} value, which is the concentration of toxicant that decreases neutral red content to 90% of controls, is determined. The NR_{90} value is useful in the testing process for the rank ordering of test compounds. In the neutral red protocol used by Clonetics Corporation, the NR_{50} value, which is the concentration of toxicant required to decrease neutral red content to 50% of controls, is determined.[24]

2. MTT Dye Uptake and Reduction Assay

A method based upon uptake and reduction of 3-(4,5-dimethylthiazol-2-yl)-2,5-diphenyl tetrazolium bromide (MTT) was developed by Mosmann in 1983 as a method that could be used to determine the number of viable cells present.[37] The method is based upon the findings that tetrazolium salts, such as MTT, are substrate for dehydrogenase enzymes located in cellular mitochondria.[38] MTT, which absorbs at 400 nm, is hydrolyzed in mitochondria of viable cells, forming a blue formazan product that absorbs at 570 nm.[37,39] The method is dependent upon the presence of active mitochondria and the viability and activational status of cells. Red blood cells, which do not contain mitochondria, and cells which have been lysed in the presence of complement and antibody to membrane antigens, do not appreciably form the measured product.[38] The activational status of the cell is important, as mitogen or factor-stimulated lymphocytes demonstrate enhanced formation of formazan product.[38] This has made the assay useful in various bioassays such as the detection of lymphotoxin,[40] growth factors,[44] interleukin,[42] and interferons.[43]

The MTT method is sensitive as it has been reported to detect as little as 300 cells and the formation of formazan product is linear up to 50,000 El_4G^- mouse lymphoma cells and Wehi-164 mouse fibrosarcoma cells.[38,44] This method is also fast, inexpensive, and easily automated. The MTT method does not require the use of radioactive material or the harvesting or washing of cells radioactive material. Data can be electronically captured by interfacing the microtiter plate spectrophotometer to a computer. In addition to making the assay faster, this simplifies data analyses and enhances quality control.

The MTT procedures can be performed in tissue culture media containing phenol red and serum. Phenol red, which is present in most tissue culture media, does not interfere with absorbance readings at 570 nm when the pH is maintained below 5.5. This is the purpose for acidification of the isopropanol and SDS-DMF solubilization solutions described below.[44] The MTT procedure described below is essentially that of Mosmann.[38]

1. MTT (Sigma catalog # M2128) is dissolved in phosphate buffered saline (PBS) at 5 mg/ml, filtered through a 0.45 μm filter to sterilize and remove insoluble residue.
2. Following the desired length of cellular exposure to growth factor or toxicant, 10 μl of the 5 mg/ml MTT solution is added per 100 μl of medium. The test plates are then incubated at 37°C for 4 h.
3. 100 μl of acidic isopropanol (0.04 N HCl) is added to all wells and mixed thoroughly to dissolve the formazan crystals. Following a 10 to 15 min incubation at room temperature, the OD at 570 nm is determined using a reference wavelength of 630 nm. The plates should be read within 1 h of adding the acidic isopropanol.

Problems with serum protein precipitation have been reported using the acidic isopropanol solution as described above.[39] This problem can be circumvented by changing the formazan product solubilization solution to one that includes sodium dodecyl sulfate (SDS).[39,44] Modifications to the method as described by Hansen and co-workers are reported to decrease assay variability and to increase assay sensitivity.[44] These modifications include the following:

1. Add 25 μl of the 5 mg/ml MTT stock per 100 μl of medium instead of 10 μl.
2. Replace the acidic isopropanol with an SDS-N,N-dimethyl formamide (DMF) mixture. The mixture is prepared by dissolving a 20% w/v solution of SDS in a 50% DMF-deionized water solution. The pH of the resulting solution is adjusted to 4.7 by the addition of 2.5% of 80% acetic acid and 2.5%of 1 N hydrochloric acid.
3. After a 2-h 37°C incubation with MTT, 100 μl of the acidified SDS-DMF solution is added. The plates are then incubated overnight at 37°C and the optical densities at 570 nm determined. This procedure does not require mixing before measuring OD.

For cell types not previously used in this assay, it is suggested that the following assay parameters be defined for better control of the assay:

1. Time course for formation of the formazan product — the length of incubation with MTT should be such that maximal formazan product is produced.
2. Linear relationship between cell number and formation of formazan product which absorbs at 570 nm; to examine cytotoxicity, the number of cells plated should be in the upper end of the linear part of this curve.
3. Metabolic status of the cells — since the metabolic status of cells affects the amount of formazan product form, the metabolic status of cells as well as potential effect of test chemicals on cellular metabolism must be taken into consideration when using this assay. Usefulness of this assay and other things to be taken into consideration are the use of test chemicals which inhibit the dehydrogenase responsible for producing the formazan product, interact with

MTT, or absorb at 570 nm. Microbial contamination can reduce MTT and thus may produce artifacts in contaminated cultures.[44] Iodoacetamide, which inhibits MTT hydrolysis, has been used as a positive control for this assay[31].

3. Cellular Release of Lactate Dehydrogenase

The release of cellular proteins (e.g., lactate dehydrogenase (LDH), beta-glucuronidase, or alkaline phosphatase) can be used as an endpoint for cytotoxicity assays.[45,46] This type of assay can quantify cytotoxicity by determining the fraction of enzyme released to total culture enzyme (solubilized or lysed cells). This assay has been performed primarily using adherent cell types but is easily adapted for nonadherent cells using a centrifugation step. LDH is measured through a series of redox reactions in which the measurement of the reduction of a tetrazolium salt is monitored spectrophotometrically. The assay of LDH as described by Korzeniewski and Callewaert is shown below.[46]

1.   Centrifuge test plates containing cultured cells at $300 \times g$ for 5 min at the end of the incubation period.
2.   Transfer 100 µl of supernate to an 96-well flat bottom assay plate for LDH analysis.
3.   Add 100 µl of the following: 0.054 M L(+) lactate, 0.66 mM 2-p-iodophenyl-3-p-nitrophenyl tetrazolium chloride, 0.28 mM phenazine methosulfate, and 1.3 mM nicotinamide adenine dinucleotide (NAD) in 0.2 M Tris buffer (pH 8.2).
4.   Monitor changes in the absorbance at 490 nm at 3 to 5 min intervals in each well of the 96-well plate using a plate reader; the change in absorbance with time (ΔA/min) is then calculated.
5.   The percent cytotoxicity is calculated by the following equation:

$$\frac{\text{Experimental} - \text{Spontaneous Release}}{\text{Maximal} - \text{Spontaneous Release}} \times 100$$

where: Experimental = ΔA/min of supernate from test chemical treated cultures.
Spontaneous Release = ΔA/min of supernate from nonexposed cultures.
Maximal = ΔA/min of supernate from sonicate or detergent permeabilized cultures.

The assay has been shown to detect the LDH content in as few as 2000 K-562 tumor cells and is linear up to the LDH content in 50,000 cells.[46] For other cell types, the cellular release of other enzymes such as beta-glucuronidase, alkaline phosphatase, or N-acetyl-glucosaminidase may provide a more sensitive indicator than LDH of membrane damage or cytotoxicity. Although effective, there are potential disadvantages to enzyme release methods. Corrosive agents may inactivate the enzyme or prevent its release from cells by acting as a fixative, and other chemicals may act as enzyme inhibitors.[22]

4. Measures of Cellular Metabolism

Cellular protein — For most cell cultures, there is a linear correlation between cell number and protein content. Changes in this correlation following chemical exposure can be reflective of an inhibition of protein synthesis. Several semi-automated assays for the determination of the total amount of protein present in culture exist, including a modified Coomassie blue dye method, a Kenacid blue dye method, and the BioRad protein assay.[47-49] Each method can be performed on samples contained in 96-well microtiter plates and is reported to be semi-automated, reproducible, fast, and economical.

Incorporation of uridine into RNA — This assay is based upon the incorporation of ^{3}H-uridine into RNA of the cultured cells. It has been reported to be a rapid means of measuring sublethal cytotoxicity and to provide a good correlation with *in vivo* damage as judged by the Draize test.[50,51]

Glucose utilization — The rate of glucose depletion from the culture media can be used as an index of cellular metabolic status. This assay is especially effective for the assessment of oxidative metabolic function in cell types with high glucose requirements. This method has been shown to be a sensitive indicator of keratinocyte exposure to cellular toxicants.[52]

D. Considerations on the Use of Human Keratinocyte Cultures

Thus far, the primary focus of keratinocyte cultures has been to measure cytotoxicity resulting from chemical exposure. While this is probably sufficient to identify many dermal irritants and corrosives, it is also somewhat limiting to only examining cell death as an endpoint. Other endpoints to examine cellular proliferation or differentiation, either used alone or used in conjunction with cytotoxicity tests as a battery of assays, may provide a more complete evaluation of the dermal toxicity of a chemical.

Keratinocyte cultures do not contain the stratified cellular layer that serves as a barrier *in vivo*. This could have an impact on the observed sensitivity and/or ranking of chemical irritants as assessed by the *in vitro* vs. *in vivo* system. In an attempt to overcome this limitation, collagen-agar matrix have been placed over the cultured cell monolayer.[22,23] Culture techniques have been published which allow alteration of the status of cellular differentiation.[53-55] Thus, stratified differentiated keratinocyte cultures could possibly be used and are discussed in more detail in following sections.

Once past the stratum corneum, chemicals may act as irritants, corrosives, or sensitizing agents through interaction with epidermal keratinocytes, mast cells, fibroblasts, or immunological cells. While the ability of human keratinocyte cultures to predict irritancy looks promising for certain classes of chemical compounds, its usefulness in predicting immunologically mediated reactions and reactions involving other cell types has not been demonstrated.[19-21] These aspects prevent the use of keratinocyte cultures as a complete replacement for the *in vivo* test. However, the keratinocyte culture appears very promising for use as a first-tier screening test to minimize chemical testing *in vivo* and for use in mechanistic studies to identify potential biochemical targets or receptors for dermatotoxicants. Other model systems which may be considered for use as replacements are systems involving co-cultures of cell types or a dermal equivalent system comprised of different cell types.

E. Photosensitivity Tests Using Cell Culture Systems

Photochemical reactions induced by UV rays may be enhanced as a result of chemical present in skin tissue from either dermal or systemic exposure. These reactions may also be classified as either phototoxic or photoallergic. Chemicals which enhance photochemical reactions are known as photosensitizers. Photochemical reactions may result in erythema, edema, papules, macules, epidermal thickening, proliferative changes in epidermal cells, suppression of T lymphocytes, or the development of precancerous and cancerous conditions.[56] Examples of photosensitizers include select sulfonamides, tetracyclines, nalidixic acid, sulfonylureas, thiazides, phenothiazines, furocoumarins, and coal tars.[57]

In vitro assays have been developed to test compound phototoxicity.[58-60] The most recent assay involves the use of human peripheral blood lymphocytes or a human T lymphoblastoid cell line and the measurement of cellular proliferation or cellular viability after exposure to test substance and UV rays.[61] The methodology in this assay involves exposing cells in culture to test chemical followed by exposure of cultures to UV light. After a 48 h incubation for the lymphoblastoid cell line or 72 h incubation of the lymphocytes with a T lymphocyte mitogen, lymphoblastoid viability or ^{3}H-thymidine incorporation in lymphocytes are examined. Control cultures consist of cells exposed to test compound and cells exposed to test compound and visible light.[61] As with other *in vitro* systems, the metabolites of a test chemical could be included in this system by the addition of the chemical metabolizing system such as microsomal enzyme hepatic microsomal preparations or other enzymatic systems. Human keratinocytes or a dermal equivalent system may also prove to be a useful model system for identifying phototoxic chemicals. With the commercial availability of melanocytes, it should be possible to co-culture melanocytes and keratinocytes yielding a system that more closely simulates the normal target. For a more complete discussion of cutaneous phototoxicity testing, both *in vivo* and *in vitro*, see Chapter 6.

VI. TISSUE MODELS

Animal models used to assess dermal irritancy, sensitization, and corrosion are accepted as valid due to their ability to serve as a basis to predict human response and because dermal inflammation, immunoresponse, and ulceration involve complex processes which are highly integrated and regulated *in vivo*. Questions concerning the validity of test results obtained from such models are generally related to differences in response or the ability to extrapolate test results between different species and not the presence or absence of mechanism(s) involved in response elicitation. Because different species may, in fact, be differentially responsive to a given chemical, the ideal would be to always evaluate chemicals using the target species. This would eliminate the need to extrapolate results from one species to another. Unfortunately, this is not always possible when man is the target species, which is often the case, and it becomes necessary to select some type of model system for the routine testing of chemicals. Biochemical and cell culture models can be used to study some interactions between specific classes of chemicals and potential dermal targets, but they often fall short in their ability to respond predictively to the broad spectrum of chemical structures with proven *in vivo* effect. In some cases this may be due to the lack of a particular target or combination of targets in the *in vitro* model. If the evaluation of a large number of different chemical classes is required, it will likely be necessary to develop more complex, tissue-based *in vitro* models for dermal toxicity testing for such purposes.

Tissue models offer some significant advantages over isolated cell culture models for the assessment of dermal toxicity *in vitro*. For example, tissue models are more highly integrated and have a greater degree of cell-to-cell interaction than cell culture models. These models have an intact stratum corneum and also provide increased diversity in the type of biomolecular targets available to interact with a test chemical. The individual cell types within the tissue generally appear morphologically more similar to cells of the same type found *in vivo*. In some tissue models, chemicals can be applied directly to membranes open to the ambient air, thereby, more accurately simulating *in vivo* exposure and eliminating the problem of test chemical degradation or binding loss in a cell culture medium. The ability to apply a chemical to a surface that requires penetration to effect toxicity also tends to increase the comparability of the tissue model with *in vivo* processes involved in chemical toxicity. Such models also provide a means to study some of the biochemical and biophysical mechanisms involved in the initiation of a toxic response isolated from responses and interactions occurring in the rest of the body. Thus, the data obtained using *in vitro* tissue models may provide an overall increased ability to accurately simulate events occurring in dermal tissues exposed to a toxicant *in vivo* relative to cell culture models.

There are currently several different types of tissue models available for the development of *in vitro* tests to assess dermal toxicity. The principal differences between each of these models is the type of tissue used in each model. The most obvious tissue source is viable, intact, epidermis obtained from a laboratory animal or human donor. Laboratory cultured, stratified, and differentiated human or animal keratinocyte membranes have also been developed for use in toxicity testing.

A. Skin Organ Preparations

Viable skin organ preparations have been developed from human or animal donors using either an epidermal slice technique or an isolated skin flap with an intact blood supply. Chapter 10 discusses the isolated skin flap technique in detail.

Epidermal slices obtained from human or animal donors has been examined as a possible means of evaluating the dermal toxicity of chemicals *in vitro*. In such evaluations similar procedures for skin culture have been developed by several investigators.[62,63] In most cases keratome slices containing stratum corneum, basement membrane, and fragments of dermis are placed in some sort of holding device with the dermal side in contact with cell culture medium and the surface of the stratum corneum open to the air to enable application of a test chemical to the corneum. Several different endpoints can then be used to evaluate the effects of the test chemical on the model. This approach is exemplified by the results of several recent studies.[64-66] In these studies, enzyme release (acid phosphatase, neutral protease, and lactate dehydrogenase) into the culture medium, incorporation of radiolabeled amino acids (^{14}C-labeled lysine and isoleucine), ^{51}Cr release, and histological/histochemical alterations were examined as endpoints in cultured human and laboratory animal skin slices exposed *in vivo* to *Clostridium perfringens* type C necrotizing toxins and various other chemical irritants and corrosive agents.

Parish summarized the results from several *in vitro* studies utilizing epidermal slices exposed to various classes of chemical insult (i.e., weak irritants, moderate irritants, bacterial toxins, and corrosive chemicals) and discussed the relevance of various *in vitro* endpoints (enzyme release, histology/histochemistry and the incorporation of radioisotope-labeled amino acids) to *in vivo* observations.[22,23] It was found that enzyme release was greatest following exposure to weak irritants and moderately irritating acids. Alkalis and bacterial toxins produced noticeably less enzyme release and corrosive agents resulted in no detectable release. The failure to detect enzyme release with corrosive agents was conjectured to be the result of destruction or denaturation of the enzyme prior to its release, or the fixation of cell membranes, thereby preventing release. Histochemistry and histological findings from *in vitro* preparations were reported to be relatively inconsistent with *in vivo* activity, whereas radioisotope incorporation appeared to be somewhat inversely correlated to the degree of insult and demonstrated more promise in the prediction of *in vivo* activity. The latter technique, however, was found to require controls and demonstrates that the test chemical did not destroy the radioisotope-labeled amino acids and was thought to be too complex for routine use. Epidermal slice techniques thus appear to provide endpoints which are quantifiable and potentially useful for the prediction of both the occurrence and the intensity of weak to moderate chemical irritants *in vivo*.

Corrosive agents, however, appear to completely eliminate the detection of endpoints such as enzyme release, histochemical/histological changes, and radiolabeled amino acid incorporation using epidermal slice techniques. Their detection may require the development of different endpoints. For example, Oliver et al. developed a model specifically for the detection of corrosives using epidermal slices which is based on changes in electrical resistance of the skin slice.[78,79] In this model, resistance across a skin slice disc is measured relative to an external circuit and chemical exposures which reduce resistance below the experimentally determined value of 4 kΩ per disc are predicted to be corrosive. In a correlation study using human and rat skin slices, 59 chemicals (composed of 43 corrosives and 16 irritants based on the results from *in vivo* rabbit studies) were evaluated using the *in vitro* skin slice resistance technique.[80] The results from this study indicate that the correlation between human and animal skin using this technique is better for corrosives than for irritants and the results obtained with rat skin slices appear to correlate more highly with *in vivo* rabbit findings than those for human skin slices. The lower correlation of the human findings was attributed to a lower susceptibility of human skin to chemical corrosive action than animal skin and indicated that full validation of the *in vitro* test for use in predicting chemical corrosives in human skin required confirmation by *in vivo* patch tests using human subjects.

The measurement of electrical resistance changes across the skin has been more commonly termed transepithelial resistance (TER). The successful use of TER measurements to evaluate corrosive agent effects or to supplement the use of other endpoints in *in vitro* models has been documented. Devices with which to make TER measurements have also been described in detail in the literature and are also available commercially in the form of a culture plate insert system developed by Millipore Corporation (Bedford, MA). With this type of technology so readily available, it is likely that such measurements will become more commonplace as *in vitro* endpoints if there use appears to routinely yield valuable information.

Epidermal slices used for studies such as those described above are prepared and maintained in much the same fashion used for *in vitro* penetration tests with viable skin (see Chapter 9). Samples are prepared as dermatome slices from the pelts of humanely killed laboratory animals or from human donor tissue. The depth of cut setting for the dermatome varies with the type of skin used and the application, but usually ranges from 0.4 mm thick for rodent or rabbit skin up to 1.0 mm thick for pig or human skin. Full-thickness skin samples containing a full depth of dermis may also be used after removal of the subcutaneous fat with a scalpel blade or other suitable instrument. These full-thickness samples have the advantage that they are skin samples with somewhat more complete integrity than the more uniform thickness dermatome preparations. Because the preparation characteristics may vary in several ways from one laboratory to another, epidermal slice methods may be difficult to standardize. The similarity to models used for *in vitro* dermal penetration studies is obvious, however, and there is some possibility that endpoints can be developed to measure the penetration rate of a chemical simultaneously with its effects on the integrity, viability, and proliferative capacity of the skin.

B. Skin Equivalents

In addition to *in vitro* tissue systems based on natural skin or membranes, differentiated keratinocyte cultures grown on various substrata have been recently developed and proposed for use as skin test systems. These so-called "skin equivalents" appear to hold some promise for use as models for *in vitro* skin research. Such models would have several advantages with respect to their use in dermal toxicity evaluations over both keratinocyte monolayer cultures and natural skin in that they (1) provide a relatively uniform surface above the cell culture medium for the topical application of test materials, (2) are derived using methods that produce a product somewhat similar to natural skin which can be controlled for consistent quality, and (3) can be manipulated to provide unique characteristics or cell types in order to optimize the model to produce a desired response.[81,82] Skin equivalents demonstrate many physiological and histological characteristics which more closely resemble natural skin than keratinocyte monolayer cultures grown submerged in culture media. Submerged monolayer cultures limit the determination of the effects of toxic agents to biochemical endpoints, cellular viability, and proliferation. However, differentiated models can be used to determine the toxic effect(s) of chemical agents on the attachment, proliferative activity, differential viability, growth, and differentiation of various cutaneous cell types. Such models would, therefore, allow for more complete *in vitro* evaluations of the potential toxic interactions between test chemicals and the skin.

Differentiated keratinocyte cultures may be grown from human- or animal-derived primary cultures. The cells required to establish primary keratinocyte cultures may be obtained either commercially or from the isolation of keratinocytes from epidermal tissues as previously described. Keratinocytes from such primary cultures are then allowed to attach to some type of substratum submerged in a suitable medium. As the attached culture is gradually raised to the air-medium interface, the keratinocytes begin to differentiate and stratify so as to resemble epidermal characteristics *in situ*. The procedures used to obtain such cultures as well as the relative success in the use of various substrata and media components to stimulate differentiation have been described in detail.[83,85] Figure 4 shows a comparison between viable human epidermis and a typical skin equivalent system.

Generally, differentiated keratinocyte cultures are most successfully grown on substrata which include some mesenchymal elements (fibroblast cultures, collagen gels, etc.).[84,85] There has, however, been some success reported in the growth of keratinocytes on synthetic membranes (i.e., nylon).[84] The evaluation of synthetic membranes has been pursued in order to hopefully eliminate some of the inherent variability between models which have highly variable substrata characteristics. For example, collagen and fibroblast preparations can vary significantly from lot to lot, whereas synthetic membranes are nearly identical. Regardless of substrata, a principal goal in developing nearly all differentiated keratinocyte culture models has been to produce a relatively consistent stratified model with an air-exposed epidermal surface which can be used in different laboratories to collectively evaluate the effects of various chemicals on the skin.

The following is a generalized procedure describing the preparation of a typical skin equivalent system:

1. Establish primary keratinocyte cultures. Prepare by trypsinization of excised skin sections from the species of choice as described above for the preparation of keratinocyte suspensions or monolayer cultures. Remove fibroblasts and other debris by centrifugation through a Ficoll gradient or other suitable cell separation procedure.
2. Select and prepare substrata for the skin equivalent system. The substrata

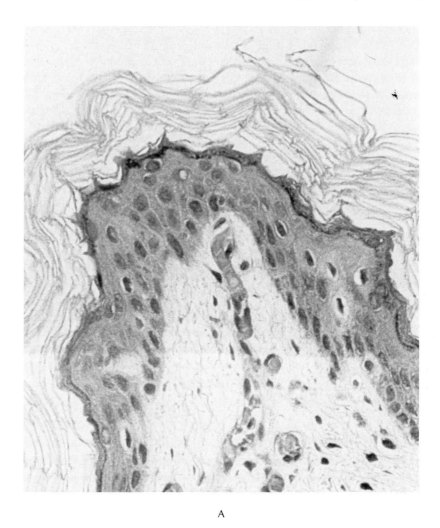

A

Figure 4. Cross-sectional comparison of human skin with a three-dimensional skin equivalent. (A) is a cross-section of human neonatal skin. Foreskin was fixed with 10% phosphate-buffered formalin, then dehydrated and embedded in paraffin. (B) is a cross-section of skin equivalent consisting of human fibroblast surrounded by a naturally secreted network of extracellular matrix proteins and human keratinocytes. Evident are the basal keratinocyte layer which possesses a cuboidal morphology and the multiple suprabasal layers. Culture techniques can also be implemented to increase the differentiation of the epidermal layer. Shown above are cross-sectional samples that were stained with hematoxylin and eosin, magnification × 400. (Photographs are courtesy of Marrow-Tech, Inc., La Jolla, CA.)

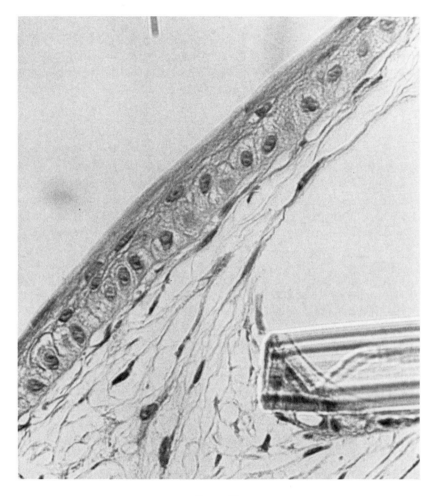

Figure 4B.

selected may be obtained in a variety of ways such as (1) the use of a commercially available filter membrane or nylon mesh such as selected and prepared as described by Vaughan et al.,[83] (2) formation of a collagen membrane derived from the culturing of fibroblasts with or without an underlying nylon mesh,[84] (3) the formation of a collagen gel which condenses to form a membrane.[85] The substrata are usually prepared in culture plates or other suitable containers.

3. Application of keratinocytes to the prepared substrata. Keratinocytes are suspended in a complete growth medium (usually not serum free) and seeded onto the prepared substrata at a predetermined concentration depending on the size of the desired membrane (e.g., 0.5 ml of a 5×10^5 cell/ml suspension per well for a 24-well culture plate).[83]

4. Attachment and growth of keratinocytes on substratum. Keratinocytes seeded onto the substrata are allowed to attach and form a monolayer in a submerged culture at 35°C in a 95% air: 5% CO_2 atmosphere with humidity maintained at 95%. The time required to form the attached monolayer is variable (i.e., 7 to 14 days) depending on the size of the membrane, the media used, and whether additional growth factors are employed in the culture.

5. Stratification and differentiation of attached keratinocytes. The procedures used to stimulate growth and differentiation of the attached keratinocytes to form an air/liquid interface and to begin the keratinization of the interface surface are also highly variable from one laboratory to another. Vaughan et al. describe a simple method using glass filter pads that they found successful in the cultivation of a skin equivalent membrane derived from rat keratinocytes which will be provided here as an example.[83] Membranes with attached keratinocyte monolayers are transferred to petri dishes containing glass filter pads saturated with growth medium. The dishes are covered and the cultures are grown in this fashion at the air/medium interface in the same ambient environment used for the submerged culture. Differentiation of the epidermal layer using this procedure is reported to take approximately 14 days to produce a stratified epidermal membrane containing 5 to 6 nucleated layers and 15 to 20 enucleated layers with normal markers of differentiation (i.e., keratohyalin-like granules, desmosomes, tonofilaments, etc.) being evident.

As noted above, the procedures used in the preparation of skin equivalent systems may vary significantly from one laboratory to another and there is much yet to be learned about the optimal concentrations of growth factors required to stimulate and maintain these systems. However, other than the source of the cells used to derive the model, the principal difference between one skin equivalent model and another is usually the source of cells and the type of substratum used. Since human cells are fast becoming the source of choice due to their increasing availability, cell source appears to be becoming less of an issue, however, and more attention is being focused on substrata differences. Some systems employ synthetic membranes as described above, whereas other procedures use collagen matrices formed from the contraction of collagen-containing gels or deposited by fibroblast cultures. Other systems even make use of a combination of artificial membrane and fibroblast culture. Thus, the endproduct is considered a skin equivalent even though the substratum may or may not actually contain true mesodermal components or characteristics as would be found in the dermis *in situ*. Differences in substrata characteristics between systems are becoming cause for debate as to which substrata may be most appropriate for use in the commercial development of skin equivalent systems for routine toxicity evaluations. Such discussions, while perhaps useful for promotion of one model over another, may be practically irrelevant, since the validity of any of skin equivalent models to predict *in vivo* skin responses has yet to be firmly established. It may be that other factors, as yet undetermined, may prove to be more important in the establishment of such models as valid skin surrogates or replacements. Until the effects of different substrata on the outcome of toxicity evaluations are completely appreciated, it may be most informative to evaluate and to compare the effects of a test chemical on the complete skin equivalent system and the substratum (sometimes referred to as the dermal equivalent) separately in order to more completely understand the effects of the chemical on the principal components of the model. There are several key references which provide more complete descriptions of the methods and procedures used to establish skin equivalent systems.[83-87] The reader is referred to these sources for additional details.

Until recently, skin equivalent systems have been developed and used exclusively in laboratories equipped to perform cell culture. As a result, there has been little opportunity for toxicologists not having extensive cell culture experience and a properly equipped and staffed cell culture laboratory to obtain and evaluate the potential uses of skin equivalents for their particular applications. This situation is rapidly changing, however, due to the commercial availability of complete, ready-to-use, skin equivalent systems. Just like the systems developed in research laboratories, commercially prepared skin equivalents also vary from one supplier to another in their composition. For the most part, compositional differences between commercial skin equivalents occur principally with respect to the type of substratum and the state of keratinocyte differentiation. Other differences between such systems include: the degree of quality control, availability of customized preparations which include optional sources of cells as well as additional skin cell types, the means by which the skin equivalent system is contained, and the cost.

Regardless as to how a skin equivalent system is composed, there are many possible endpoints which can be measured following the exposure of these systems to toxic chemicals. Most, if not all, endpoints described above for use with cell monolayer culture systems can be adapted for use with skin equivalents. Because skin equivalent systems also offer a functional barrier as well as attached and differentiating keratinocytes, several other endpoints not possible with cell monolayer culture systems can also be evaluated. Examples of endpoints possible for use with skin equivalent systems are given in Table 1.

In addition to the above, novel optical or electronic systems have been developed which can be used to continuously assess the adverse effects of chemicals on specific aspects of cell culture and/or skin equivalent systems. One example is a recently developed instrument which has been shown to be effective in the continuous assessment of viability in keratinocyte monolayer cultures. This instrument, shown diagrammatically in Figure 5, has been shown to be useful in the assessment of the adverse effects of toxic agents to membranes.[88,89] Another example might be the development of a method for the routine monitoring of cell growth in culture using infrared spectroscopy.[90] It is likely that the routine availability of human keratinocytes and high quality skin equivalent systems will lead to the development of even more elaborate electronic systems for use in the *in vitro* assessment of dermal toxicity from chemicals and chemical formulations in the near future.

It can be concluded that differentiated keratinocyte cultures offer a quality controlled, complex, and relatively new *in vitro* model with which to study the effects

Table 1
Some Examples of Endpoints which may be Determined Using Skin Equivalent Systems and Procedures Available for their Measurement

Endpoint	Procedure(s)	Ref.
Cell proliferation	Neutral red assay, MTT assay, thymidine or uridine incorporation	67
Cell viability	Neutral red assay, LDH release, MTT assay	68,69
Cell differentiation	Histological, immunocytochemical	70
Cell attachment	Histological	
Inflammatory mediator release	Arachidonic acid release, leukotrienes, interleukins, prostaglandins	69
Cell metabolism	NAD content, glucose utilization, oxygen consumption, amino acid incorporation, xenobiotic metabolism	67,71-73
Cell-cell communication	Histological	74
Adduct formation	DNA adduct and protein binding assays	75
Dermal barrier integrity	Chemical penetration assays, electrical potentiometry	76,77

of toxicants on epidermal growth, differentiation, cell surface interactions, adhesion, and cellular metabolism. Such cultures are also being developed to evaluate the penetrative characteristics of toxic chemicals as well. As presented above, it is evident that state-of-the-art skin equivalent systems are somewhat rudimentary as compared to the relative complexity and cellular diversity of natural skin *in situ*. Even though skin equivalent systems are currently produced which contain viable melanocytes and can be cultivated with keratinocytes obtained from different anatomical locations from donors of different ages, sexes, and skin type, several other potentially significant effector cells (i.e., macrophages, neutrophils, mast, Merkel, and Langerhans' cells) are noticeably absent from these systems. Also missing from current skin equivalent models are glandular elements (glands and secretory products) and hair follicles. To conclude that the lack of these additional cell types invalidates the use of *in vitro* skin equivalents for consideration as useful skin models would be very premature and most likely incorrect. The models do provide a means by which to effectively isolate and study specific aspects of the skin without the presence of additional, possibly complicating, factors. Such evaluations are not possible with any other *in vitro* culture system. Thus, skin equivalent systems already provide a relatively new and useful tool with which to study the effects of toxicants applied to the skin in a fashion not routinely available until now. Significant increases in the frequency of utilization of and the demand for such systems for use in toxicity evaluations is expected to occur over the next few years.

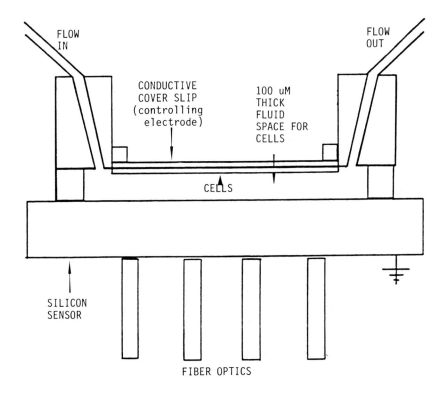

Figure 5. Cross-sectional view of silicon-based biosensor used for the continual monitoring of cell or tissue viability. Adherent cells grow either on a coverslip (as shown) or on the sensor surface. The chamber volume is approximately 10 μl/cm^2 area by 100 μM depth. Illumination of the underside of the sensor surface by light-emitting diodes (LED) via fiberoptics allows pH information to be gathered from four separate regions, each of 1 mm^2 or 100 nl within the chamber. Typically, a cell density of 10^5 cells per cm^2 in a confluent monolayer corresponds to 10^3 cells per site. The figure is not to scale and the diameter of the entire chamber is approximately 2 in. (Provided courtesy of Molecular Devices, Menlo Park, CA.)

VII. COMMON TECHNICAL PROBLEMS ASSOCIATED WITH *IN VITRO* ASSAY SYSTEMS

Although there are often many technical similarities between *in vitro* and *in vivo* procedures to evaluate the effects of dermatotoxicants, many *in vitro* test systems also require that additional, more technically complex operations be performed relative to their *in vivo* counterparts. As a result, there are many problems and sources of error which exist for such *in vitro* systems which are absent from *in vivo* test systems. Because several *in vitro* test systems have been developed, evaluated, and proposed for use as possible replacements of or adjuncts to various types of *in vivo* tests over the past 2 decades, there exists a relatively large and expanding body of knowledge concerning the nature and types of problems one can expect when developing or conducting *in vitro* toxicity tests. Although the technical problems associated with particular tests are routinely discussed in the literature, at scientific meetings or between individual scientists, very few publications are available which systematically present, discuss, and summarize many of the common technical problem associated with *in vitro* test systems. Recently, however, a technical report was issued by the Center for Alternatives to Animal Testing (CAAT) at the Johns Hopkins School of Hygiene and Public Health which discusses many of the technical problems which may be encountered in the development of cell culture assays to assess chemical toxicity. This report, although not intended to be a comprehensive guide to the identification and avoidance of such problems, does provide much useful insight into some of the more common technical pitfalls associated with the development and conduct of *in vitro* tests.[91] Although the report is principally focused on cell culture systems, much of the discussion provided is generally true for other types of *in vitro* test systems. Unfortunately, a complete discourse concerning problems which may occur during the conduct of *in vitro* toxicity assays is beyond the scope of this chapter and only a brief overview is possible. The reader is, therefore, advised to obtain a copy of the CAAT technical report for further information concerning this topic.

There are several factors which may confound the results obtained from *in vitro* tests. The CAAT report notes that these factors tend to fall into two general categories (1) those which affect biological components (i.e., cultured cells, enzymes, membranes, etc.) of the test, and (2) those which affect the dispersal and/or stability of the test material in the test medium. The first category includes factors which may unintentionally alter or influence the responsiveness of the biological component of the test to the inherent toxic effects of the test chemical such as

- pH changes in the test medium resulting from addition of the test chemical
- Alterations in test media osmolarity at high concentrations of the test chemical
- Alterations in the composition of test media resulting from reactions between test chemicals and media components (e.g., protein denaturation, alteration in the availability of essential substrates or cofactors, etc.)
- Confounding effects on the test system response occurring from the presence of chemicals (e.g., alcohols, dimethylsulfoxide, etc.) used as solvents for test chemicals
- Interactive effects of test chemicals and antimicrobial agents on test system response
- Complications due to evaporation of volatile test chemicals
- Interference with the endpoint measurement (e.g., the presence or formation of interfering colors in a colorimetric assay directly from test chemicals, interaction between an enzyme endpoint, and the test chemical)

- Influence of protein binding of test chemicals on the concentration-response curve
- Effect of test incubation time or temperature on the availability of essential nutrients
- Effect of test incubation time or temperature on the responsiveness of the test system to test chemicals
- Deterioration of test system purity and responsiveness as a result of microbial contamination
- Physical effects of insoluble particulates, formed as a result of interactions between test chemicals and components of the assay system, on the responsiveness of the test system

The second category of potential technical problems includes potentially confounding factors related to the dissolution of the test chemical in the test media. The avoidance of problems in this category involves the examination of specific items prior to testing. The elimination of any potential physical/chemical problems which may result from the incompatibility of the test chemical with the biologic components of the test is of prime concern. The fact that many test chemicals or chemical formulations are hydrophobic in nature and that most *in vitro* systems exist within an aqueous matrix in which such test chemicals are either insoluble or nondispersive is the cause of many problems of this type. Some of the most serious issues limiting the general usefulness of such *in vitro* systems stem from the fundamental problem of test chemical dispersal within the test matrix. For example, it is very difficult to establish the proper dose administered to a test system if the test chemical is incompletely dispersed within or degraded (e.g., hydrolyzed) by the matrix. It is also often difficult to determine the most appropriate means of physically introducing a test chemical into a test matrix. Aggregates of nondispersed chemicals or chemical mixtures can also have physical properties which cause mechanical damage to the test system rather than intrinsic toxicity (e.g., detergent or crystallization effects upon cellular membranes). Some specific issues of this nature identified at the CAAT workshop which must be addressed prior to the routine conduct of any *in vitro* toxicity test were

- What properties of test materials should be used to define incompatibility with the test system?
- What solvents or physical dispersants are acceptable for the introduction incompatible test materials into the test system?
- What separation or extraction techniques can be used to prepare test materials for the dosing of cell cultures?
- How will the test chemical exposure concentration be quantified?

Most of the potential problems noted above are not new to *in vitro* test procedures in general. In the development of *in vitro* assays to evaluate cutaneous toxicity, it is possible to make use of the past experience gained from the conduct of other *in vitro* test procedures (e.g., assays for the assessment of genetic, hepatic, cardiovascular, and renal toxicity, etc.) in order to identify and avoid some of the common problems to be encountered. For this reason, references describing previous experiences in the development of other *in vitro* toxicity tests should be of some value in the development and conduct of *in vitro* cutaneous tests.

The incompatibility of test chemicals with the test media is one of the most common problems encountered with *in vitro* dermal toxicity assays. This is probably because many *in vitro* tests require that the test chemical be introduced into a liquid test matrix (e.g., buffer, cell culture media, etc.), and degradation and/or binding of the test chemical in the test matrix can still occur even after pH, osmolarity, and solvent effects are controlled or eliminated. Table 2 gives examples of some of the possible effects which might occur as a result of interactions between different chemical classes of toxicants and various media components.

Based on the CAAT findings, several activities have been recommended to more fully characterize *in vitro* test systems to help avoid major technical difficulties. These recommendations are

1.   Characterize the biological components of *in vitro* systems as fully as possible.
2.   Document the effects of pH (ranging from 6.5 and 8.0) and osmolarity (ranging from that of the standard media to 350 mOsm) variation on the status of cells, or other biological component, and on the endpoint(s) being determined.
3.   Document concentration-response curves for solvents to be used (e.g., ethanol, dimethylsulfoxide, etc.) and any microbial/antifungal agents present in test media.
4.   During the tests, document interactions between test chemicals and test media. For example, the following information should be reported for at least the highest concentration of each test chemical: pH, osmolarity, turbidity, color change, precipitate formation, etc.
5.   Evaluate the stability of test chemicals in cell-free control media over the duration of the exposure period required by the test.

Of course no matter what technical problems are encountered, identified, and solved during assay development, the assay is only going to be used if it is capable of producing results with an acceptable degree of validity relative to expected *in vivo* results. While it is true that significant technical problems can and do often result in problems which affect the validity of such assays, the most difficult problems currently affecting the obtainment of valid results for most assays involve the means by which assays are designed and how determinations of validity are defined.

VIII. CONSIDERATIONS FOR *IN VITRO* ASSAY DESIGN AND VALIDATION

There is a growing number of *in vitro* systems which have been shown to be responsive in some fashion to known dermal irritants, inflammatory, or corrosive agents. Unfortunately, none of these systems has been found to be universally

Table 2

Possible Medium and Toxicant Interactions

Media component	Reactive functional group	Chemical class of toxicant	Cell effect of media component
Free amino acids	-NH$_2$ -COOH -OH -SH	Anhydrides Acid chlorides Organic alcohols Alkyl halides Oxidants Heavy metals	Synthesis of proteins Buffering capacity Energy sources Chelator of metals
Vitamins/ coenzymes	-SH	Heavy metals	Enzyme activity Cellular metabolism
Metals in salt solutions	Me^{++} (charge)	Chelators Glycols Diamines Sulfhydryl compounds	Cell morphology Enzyme activity Cell adhesion Growth rate
Serum proteins	Macromolecular structure (Denaturation/ polar absorption) -SH -COOH -OH -NH$_2$	As with free amino acids	Cell adhesion Availability of hormones, micronutrients, growth factors, fatty acid
Serum lipids	$-\overset{\mid}{C}=\overset{\mid}{C}-\overset{\mid}{C}=\overset{\mid}{C}-$	α, β-unsaturated carbonyl compounds Oxidants	Membrane fluidity

acceptable for use as a supplement or replacement for any *in vivo* procedure to evaluate such dermal effects. To some extent, this lack of acceptability appears to involve the inherent difficulty in obtaining any new technology intended to replace a generally accepted standard. There is, however, a strong and growing interest among toxicologists to utilize *in vitro* test systems whenever possible as replacements for or adjuncts to current *in vivo* test procedures. Unfortunately, there are several important scientific issues related to the design, operation, and validation of *in vitro* systems which must be resolved before any of these systems is found to be generally acceptable as an alternative to any *in vivo* test procedure. These issues vary from one proposed *in vitro* test system to another. Often a test system is identified which responds as expected to some test agents and then, upon more extensive evaluation, it is found to be unexpectedly unresponsive to others. The results obtained from a specific test may also differ significantly between different laboratories. One reason for this may be due to differences in the assay quality between replicates or between assays conducted in different laboratories. For example, cell culture systems using human keratinocytes may be derived from highly variable sources (different donors, anatomical regions, age groups, skin conditions, etc.) with cells from each source producing a differentiable result when used under identical test conditions. One of the first steps toward the development of universally acceptable assays, therefore, should be to perform studies to identify variables which must be controlled in order to establish an acceptable level of quality control.

Although quality control is a necessary component in any assay system, the establishment of adequate quality control procedures for *in vitro* assays is rarely a simple task. In addition to the use of different sources of cells mentioned above, such assays often use variable combinations of system components (purified enzymes, cell cultures, biochemicals, etc.) which are subject to significant lot to lot variation. This may result in statistically significant, periodic variations in the assay response which may be very difficult to eliminate. In some cases, periodic variation must be endured in the assay system and it is important that such variation be appropriately controlled in the statistical design of the assay.

A. Design Considerations

A complete discourse on the design of *in vitro* bioassays is beyond the scope of this chapter, however, some statistical considerations pertinent to the design and conduct of such assays can be found in Chapter 13 and much additional information can be found elsewhere.[92,93] Nevertheless, a brief discussion of the relative merits and shortcomings of some common designs is provided in the hope that such a discussion might stimulate interest in the development of improved approaches to the design of *in vitro* cutaneous assays.

There is, at present, no particular format for the conduct of *in vitro* assays designed to assess the dermal toxicity of chemicals. Such assays can be either of the direct or indirect type. In either type, the salient parameter is the relationship between the dose (concentration) of the test chemical to which the test system is exposed and the response of the test system to this exposure (dose-response relationship). Direct assays involve the direct determination of the doses (concentrations) of the test materials required to obtain a specified response from the *in vitro* test system. Indirect assays, alternatively, measure the response of the test system to specified doses of the test chemicals. An attempt may then be made to estimate doses (concentrations) of the test materials expected to produce a specified response by the test system using some type of dose-response model. At present, most *in vitro* assays to assess cutaneous toxicity are of the indirect type. This is probably because it is difficult and time consuming to obtain direct estimates concerning a specific response unless the response is rather broadly defined. Response measurements for direct or indirect assays may be either quantal (e.g., number of dead cells) or quantitative (e.g., enzyme activity levels). A standard material may be included in the assay and the ratio of the test material dose (concentration) response to that of the standard provides an estimate of the "potency" of the test material relative to that of the standard.

At present, *in vitro* cutaneous toxicity assays are designed to statistically compare test system responses obtained for one test material to another administered at fixed dose levels,[11,78] or they are designed to estimate the median effective concentration (EC_{50}) for each test material based on the median response level for a specific effect estimated using a variety of dose levels.[24,36,94] In either case, the measured response (e.g., cell viability, cell proliferation, indicator uptake or release, enzyme inhibition, etc.) may be quantal or quantitative. Procedures which use fixed dose levels may categorize responses and tabulate response data in each category to arrive at an overall response "score" with which to characterize the chemical's response in the assay,[11,78] or the comparisons may be completely quantitative and compare the data obtained for one chemical to that of another using standard parametric tests.[95,96]

In theory, designs which involve the estimation and comparison of EC_{50} values can be, from a statistical perspective, very powerful approaches to assay development. In some cases this approach may be preferred. Such a case occurs when it is desired to characterize and compare the concentration response characteristics of different test chemicals diluted in the same matrix relative to a standard chemical having a known EC_{50} response in the assay (e.g., determine which test chemicals have an EC_{50} significantly higher or lower than the standard). When a standard (or standards) is available and a database for each standard is compiled and maintained, the use of an EC_{50}-based design which makes use of this historical database can be one of the most statistically powerful assay designs possible. On the contrary, if no standard for comparison is used, the use of the EC_{50} endpoint may prove to be a problem. For example, it may not be possible to obtain an EC_{50} value for some test chemicals because the assay system does not respond sufficiently at any concentration to permit the estimation of an EC_{50} value and, without a standard value, the response is without reference. The interpretation of such responses must be categorized in some fashion in order to obtain some basis for comparison. One means of doing this is statistically defined classification limits in which a test chemical response may be included if it fails to produce an EC_{50} value at a concentration in excess of some limit value.

Once the assay design is established, the next essential step in its development involves the performance of initial trials to gain experience and confidence in its routine use. This step includes an examination of the reproducibility of the assay, the establishment of routine quality control procedures, and the development of a formal protocol describing all procedures to be followed. At this point, the assay is entering into the initial stages of the validation process.

B. Validation Considerations

Unfortunately, validation of new *in vitro* assays is a rather nondescript undertaking. This is because there are no generally accepted criteria for the validation of such assays and the concept of validation is currently subject to considerable interpretation among scientists. At present, validation for *in vitro* assays intended to evaluate some aspect of cutaneous toxicity must be established rather informally due to basic problems with the extrapolation of a particular *in vitro* test response to the myriad of possible responses which may occur following the exposure of skin *in situ* to a chemical. For example, some experts may define a valid *in vitro* dermal inflammatory assay as one which includes correlates for a multitude of *in vivo* factors before they can be considered valid.[97] While this approach to validating an *in vitro* assay may be very useful in defining important parameter estimates to consider for inclusion in the assay, it does not go very far toward the development of a practical strategy for assessing the degree of validity for such assays. This is because, aside from other factors, the most important determinants of *in vitro* assay validity should be the functional relationship with, and the predictive characteristics relative to, the intended *in vivo* response.

Validation, as it relates to the establishment of *in vitro* alternative assays to predict cutaneous toxicity, is a relatively unclear concept at present. In a recent review of the literature by Scala,[98] no formal scientific discussion on the topic of validation as it applies to *in vitro* alternative tests was found and it was concluded that this type of validation appears to "just happen". Fortunately, however, there are efforts underway which are attempting to develop a definition of validation as it applies to *in vitro* alternative assays and to establish guidelines for validation which are acceptable by scientific consensus.[99,100] Although the proposed approaches to validation differ somewhat in fine detail, there does appear to be significant agreement between them. The validation concepts and strategies presented here consist of some of the more common aspects of these proposed approaches.

"Validation" is a rather broad term used to describe a process which may include several synonymous actions such as to substantiate, to confirm, to sanction, to certify, to authenticate, to prove, to legalize, etc. It is derived from the latin root, *validus*, meaning strong. Given this, it appears to follow that in order to strongly confirm, sanction, or prove any scientific procedure or method, scientific data sufficient to provide conclusive evidence to establish as scientific fact that the procedure does what it is intended to do must exist. Thus, validation of an *in vitro* procedure to serve as an *in vivo* alternative must include the collection and evaluation of sufficient scientific evidence to establish that the procedure is, in fact, capable of serving in some fashion as an *in vivo* alternative. All validations should, therefore, have at least three components: (1) scientific data; (2) an objective for the validation; and (3) a concept as to how the criteria are sufficient to prove the objective.

Because the scientific evidence occurs in many forms, there is likely no singular or "best" approach to collection of the data required to validate an *in vitro* assay. All proposed approaches to validation of such procedures should probably, therefore, be tried and then viewed simply as recommended guidelines, until such time that a particular approach is proven to facilitate validation over all others. This does not imply that proposed approaches to validation are without merit or usefulness. On the contrary, they represent bold initiatives intended to assist in the establishment of properly validated *in vitro* tests.

Validation approaches appear to have at least two principal similarities (1) that valid procedures do have some common characteristics, and (2) that validation should be the result of a stepwise process rather than that of a single event. Once established, common characteristics of valid *in vitro* alternative approaches to *in vivo* tests should include the following:

- A valid *in vitro* procedure will usually have clearly recognizable qualitative and/or quantitative correlates to the *in vivo* test it is intended to supplement, augment, or replace.
- A valid procedure performs well quantitatively with respect to its specificity, sensitivity, predictive value, and overall equivalence toward predicting the responses of the *in vivo* test.
- Test validation occurs as the result of scientific consensus not by any other (e.g., political or financial) incentive.
- The results from valid test systems tend to enhance our understanding of the *in vivo* response.

Characteristics mentioned above such as specificity, sensitivity, predictive value, and equivalence have been quantitatively defined by Cooper et al.[101] These quantitative indices may be used to describe the performance characteristics of emerging *in vitro* assays. Thus, they may be considered as at least one acceptable means of quantitatively comparing the performance of several candidate bioassay procedures relative to another. The definitions and calculations used in the derivation of these quantitative indices are shown in Figure 6.

The steps recommended for validation of *in vitro* alternative assays vary somewhat from one source to another. In order to avoid overlooking the merits of one proposed validation scheme relative to another, a composite scheme consisting of various components from different schemes is provided. In this composite scheme, the validation of a new *in vitro* alternative test is treated as a series of sequential stages. Assay validation to the different stages may be dependent upon the intended use of the assay such as limited in-house testing of toxicants, generally accepted use, or regulatory agency approval. The stages included in this composite validation scheme are described as follows.

1. Stage 1: Preparation for Validation

This may also be termed the preliminary or developing test stage. This stage includes:

- Definition and refinement of the validation objective and the performance criteria necessary to meet the objective
- Selection of *in vitro* alternative tests which appear to have some potential to meet the validation objective
- Selection of standard test materials to be included in validation tests

In Vivo Outcome	Summary of In Vivo Skin Irritant Test Outcomes		
	Positive	Negative	Total
Positive	a	b	$a + b = n_3$
Negative	c	d	$c + d = n_4$
Number of Tests	$a + c = n_1$	$b + d = n_2$	$a + b + c + d = N$

Specificity $= d/n_2$ =   proportion of known in vitro non-irritants found negative in vitro.

Sensitivity $= a/n_1$ =   proportion of known in vivo irritants found positive in vitro.

Prevalence $= n_1/N$ =   proportion of test materials found positive in vitro among all test materials evaluated.

Predictive Value $= a/n_3$ =   proportion of test materials found postive in vitro among all known in vivo irritants tested.

Figure 6. Estimation of specificity, sensitivity, prevalence, and predictive value of an *in vitro* skin irritant test relative to known *in vivo* irritant outcome (from Cooper et al.). Each proportional term may be expressed as a percentage by multiplication × 100.

• Selection of experimental designs and statistical procedures to be used for the conduct of each validation experiment to be included in the entire validation scheme

This stage identifies tests which have working protocols, have some degree of quality control, and have been shown to be capable of the reproducible generation of results which appear to have some *in vivo* predictive value in preliminary studies. In other words, the test appears ready to begin the validation process and measures are in place to ensure that validation proceeds in a scientifically acceptable manner.

2. Stage 2: Informal Validation

This stage may also be called the intralaboratory evaluation stage. In this stage, for example, procedures for the initial estimation of the predictive value for a proposed *in vitro* dermal inflammation assay are performed by the assay developer to demonstrate the potential validity of the procedure. This might occur as exposures of the assay system to various concentrations of known dermal irritants to demonstrate the nature and consistency of the assay response. This type of validation is usually not very extensive with respect to the number or types of chemicals evaluated nor is it performed on large numbers of replicates. This validation may or may not be conducted, however, as a blind feasibility trial. The number, nature, and concentrations of the known irritants used should be left to the investigator performing the informal validation to decide, since it is possible that the assay may be intended for a specific use under a defined set of exposure conditions rather than for general application where the response limits for the assay have yet to be defined. The only rigid requirements are that there be a sufficient number of within and between assay replicates from each set of exposure conditions in order to statistically evaluate the overall variability of the assay response to each set of conditions. In most cases, the number of replicates performed should consist of at least three within assay replicates and be performed in at least three separate assays (between assay replicates) for each set of exposure conditions in order to obtain an initial estimate of the assay variability for each exposure condition presented in the informal validation. Although it is not very comprehensive, informal validation, nevertheless, should be considered an essential procedure in demonstrating the potential utility of the assay to other scientists. Assays presented without this type of validation data should be considered suspect for any proposed application until such data are obtained. Thus, informal validation of the assay is the first step toward determining whether further validation studies are warranted.

3. Stage 3: Formal Validation

Formal validation, sometimes termed "interlaboratory validation" or "macrovalidation", is the third stage in evaluating the predictive value of an assay. This validation procedure is an expansion of the informal validation procedure and should include at least four important steps: (1) define the intended use, (2) design a study to obtain the data required to evaluate the assay relative to its intended use, (3) perform the study as designed, and (4) statistically evaluate the predictive sensitivity of the *in vitro* assay response relative to the expected *in vivo* response. Formal assay validation procedures should be conducted in a blind fashion in different laboratories with single or multiple assays and may evaluate assay responses to a single or multiple chemical agents.

The first step in formal validation is to define the intended use for the assay(s) to be validated. It may be that the validation for use with a single chemical or defined class of chemicals is all that is necessary and it would be useless and costly to attempt validation beyond these applied limits. On the other hand, the desire may be to use the assay to evaluate a wide range of different chemical types and it then becomes important for the validation to examine the performance limits of the assay. In this case, a broad range of chemicals may be selected for the formal validation study. Another factor which should be considered during the intended use determination involves whether the assay may be intended for use in several laboratories or just a single laboratory. Defining the intended use for the assay should establish the intended performance requirements for the assay and should define the scope of the study for experimental design and data evaluation purposes.

The next step in formal validation is to design a validation study within the intended use requirements that extends the current database within these requirements up to or beyond the amount of data required for proper statistical evaluation. This statistical requirement is defined by the intended use and should be a major consideration in the study design.

After the validation study is designed, the remaining steps are to perform the study as closely as possible to its design and to then perform the statistical evaluations of the data obtained in order to quantitatively describe the relationship between the *in vivo* response and the *in vitro* response and to describe interlaboratory validation.

4. Stage 4: Scientific Review and Optimization

Also termed "test battery optimization", this stage involves steps toward scientific review and acceptance as well as refinement and optimization of the procedures used to conduct the assay. Some of the actions which occur in these steps are

- Presentation of results at scientific forums and in the literature
- Incorporation of suggestions and use of findings noted during the conduct of the previous validation studies to further refine and optimize the test procedure
- Documented agreement among scientists of the test as valid relative to the intended objective
- Dissemination of the assay protocol and successful establishment of the procedure in other laboratories
- Continuous review, refinement, and optimization of assay procedure
- Recognition by regulatory agencies of the potential validity of the test

5. Stage 5: Routine and Regulatory Use

Also known as "legal" or "regulatory" validation, this stage includes the actual routine use of the assay in several locations and the accepted use of the assay in regulatory schemes. This stage represents the current pinnacle of validation in that the *in vitro* assay has achieved general scientific recognition as attested by its routine utilization and its acceptance for use in regulatory decision making.

As shown above, a test may be considered to be validated *in,* rather than validated *by,* stages. For example, a test which has been validated at first stage might have the following characteristics: a working protocol with a clear definition of the validation objective for the test, a basic quality control procedure, a preliminary dataset showing that the test appears to be capable of producing results with some predictive value, and a list of standard test materials to be used for stage two validation of the test. Thus, each candidate *in vitro* alternative test might be considered validated to stages one, two, three, four, or five. To consider a test valid to evaluate a given set of test materials, it is only necessary that the test be validated to the stage required for scientifically credible evaluation of the results generated. In other words, if test results are to be used only in intralaboratory evaluations, it is only necessary that the test be validated to stage two. If the results are to be used for regulatory purposes, the test would need to be valid to stage five.

The validation of an *in vitro* assay relative to *in vivo* data appears relatively simple and straightforward. Unfortunately, this is rarely the case in practice because there are many factors which can complicate the validation comparison and render it difficult to interpret. Such factors may include:

- *In vivo* response data may be highly variable (especially if it originates from different laboratories and does not include any interlaboratory calibration control data)
- *In vitro* concentration/response relationships may have little or no direct relevance to *in vivo* concentration/response relationships
- Assays utilizing media logit responses (i.e., EC_{50}) may sometimes yield no response value, rather an upper or lower bound on the response range is all that is returned for some test chemicals
- Assay response may be evaluated at different times after exposure, resulting in the data having a time/response vector which must be considered in addition to the concentration/ response relationship which is of greatest importance

Unfortunately, despite a growing number of proposed tests, very few attempts have been made toward the formal validation of *in vitro* tests to predict dermal irritation *in vivo*. Most *in vitro* tests currently being proposed to supplement, augment, or replace current *in vivo* procedures generally do, however, include some type of informal data presentation or discussion to demonstrate the validity of the proposed assay. Whether or not there is a need for formalized validation of such assays depends upon the extent to which the assay will be used in multiple locations or might be broadly required to serve as a substitute for an *in vivo* procedure. For example, if an *in vitro* assay is intended to serve as a procedure conducted by a single manufacturer to screen a large number of alternative chemicals within a fairly defined structural class for their potential for producing inflammatory responses prior to the submission of the best candidates for *in vivo* testing, formal validation may not be extremely important. Results from an informal validation conducted on the widest range of chemicals within the structural class for which *in vivo* responses are known may be sufficient to demonstrate the validity of the procedure for limited screening applications. Conversely, an *in vitro* procedure which is intended to serve as a complete replacement for a widely accepted *in vivo* test procedure would require a much more formalized validation prior to its routine use.

IX. FUTURE TRENDS

Efforts currently being devoted to the development of supplemental and alternative *in vitro* assays to assess chemical dermatotoxicity are expected to continue into the near future. At present, it is difficult to predict the outcome of these efforts relative to their anticipated use as replacements for *in vitro* dermatotoxicity assays. However, the potential for *in vitro* assays to serve as supplements or adjuncts to *in vivo* tests is much more evident at this time. The initial role of *in vitro* dermatotoxicity assays is likely to be as preliminary screening procedures as to minimize the number of chemicals evaluated *in vivo*. Then, after considerable use as a supplement to *in vivo* tests, databases will exist to more widely assess the *in vivo* predictive characteristics of the *in vitro* procedures and these assays may then become accepted as replacements for *in vivo* tests.

There is a trend toward the development of *in vitro* dermatotoxicity assays with the objective of evaluating specific aspects of chemically induced dermal injury. This is likely the result of increasing appreciation among toxicologists for the complexity of events which occur during dermal inflammation and the difficulty of simulating the *in vivo* inflammatory process with *in vitro* models. As a result, some researchers appear to be shifting from attempts to develop all-inclusive *in vitro* models to the development of *in vitro* models which examine specific aspects of the inflammation process. This may lead to the subsequent development of a battery of *in vitro* tests that could be used to screen the inflammatory potential of chemicals from diverse chemical classes.

More emphasis needs to be directed toward the development and evaluation of biochemical models which can be used to routinely examine the effects of chemicals on specific biochemical targets. Such effort should be placed on the identification of biochemical targets commonly affected by different structural classes of dermal toxicants. Assays using isolated biochemical targets could then be developed into a controlled assay to routinely assess the effects of chemicals on these targets.

Serum-free culture of human epidermal keratinocytes has recently become possible. Efforts have recently been focused on the development of assays using these cultures which measure cellular viability and metabolism to predict the irritative potential of chemicals within a given chemical class. It may be of benefit to direct future efforts toward the development of endpoints to measure the release of inflammatory mediators such as platelet activating factor, complement products (e.g., C_{5a}), arachidonic acid oxidation metabolites, interleukins, or other agents acting as chemoattractants and activators of leukocytes. This may potentially increase the ability of *in vitro* systems to predict inflammation. Although cultured human keratinocytes may not be able to serve as a complete replacement for *in vivo* dermal testing, the model appears to be viable as a screen prior to *in vivo* testing. In order to examine the potential hypersensitivity of chemicals or chemical interactions with other skin components, it will be necessary to develop more sophisticated cell culture systems such as keratinocyte co-culture models or even to use dermal equivalent systems that contain several different cell types.

Skin equivalent test systems are rapidly emerging as viable alternative models to assess dermatotoxicity. These systems will likely undergo considerable evolution and improvement over the next few years as the commercial suppliers of these systems attempt to respond to the diverse needs of their clients. Presently there is a need for basic research directed toward increasing the complexity of skin equivalent models. Procedures for the incorporation of additional cell types and structural elements need to be developed so that these systems respond and are histologically more like natural skin. The addition of immunocytes to the culture media should also be considered. At present, these test systems appear to have the greatest potential of all proposed *in vitro* systems for emergence as a complete replacement for animal tests for the evaluation of chemically induced skin inflammation.

As the database on dermatotoxicant effects increases, the use of structure-activity models to predict toxicant potential within a given structural class should likewise increase. The increasing importance of structure-activity models for the prediction of dermal inflammatory agents is dependent upon several factors, including: our knowledge of chemically induced dermal inflammation mechanisms, the development and validation of *in vitro* skin models to investigate specific structure-activity correlates, and the availability of sufficient *in vivo* data with which to perform validation studies.

To date, no *in vitro* alternative assay to predict chemically induced dermal inflammation has been formally validated and proven acceptable for the replacement of any *in vivo* dermal inflammatory test. As such *in vitro* assays become more refined in the future, it is anticipated that some assays may gain acceptance as screening procedures and the data from these assays may be routinely included along with subsequent *in vivo* test data to provide additional evidence, or lack of evidence, of a potential dermal inflammatory hazard for a chemical. The way in which *in vitro* mutagenicity assays are currently used to screen chemicals prior to chronic *in vivo* testing is likely to be how most *in vitro* dermal inflammatory assays will be generally accepted for use in routine toxicity and hazard assessment procedures in the near future. Batteries of *in vitro* assays and skin equivalent systems may later emerge as *in vivo* replacement procedures, but not without the completion of extensive validation studies.

X. SUMMARY AND CONCLUSIONS

There are a number of different *in vitro* tests which have been proposed to evaluate the potential dermatotoxicity of toxic chemicals or chemical mixtures. These tests include the use of structure-activity, biochemical, cellular, and tissue models. The desire to develop valid *in vitro* methods is directed toward reducing or replacing laboratory animal tests that are currently used for chemical toxicant evaluation. Despite the increasing number of proposed *in vitro* tests, the acceptance of such tests by toxicologists for use as valid *in vivo* replacements has been cautious at best. This situation primarily appears to be the result of the need for more comprehensive and impartial validation studies involving a wide spectrum of chemicals with known *in vivo* dermal toxicity. Another factor that is involved is the need for the establishment of formalized criteria to define the circumstances for which an *in vitro* assay may be considered valid. As a result of the need for general acceptance of such *in vitro* assays, there exists a need for the funding of *in vitro* studies which are directed toward the establishment of a widely acceptable validation protocol and the formal validation of *in vitro* dermatotoxicity assays. Unfortunately, large-scale formal validation studies are currently not possible for many of the potential users of such tests due to funding problems and the amount of effort needed to organize such studies. It is anticipated, however, that such problems may soon be resolved due to the increasing desire among scientists, politicians, businesspersons, and consumers to identify and use suitable alternative tests. When and if this occurs, we may expect the use of *in vitro* alternative tests for the assessment of cutaneous toxicity to become more commonplace.

REFERENCES

1. Society of Toxicology, *Fund. Appl. Toxicol.* 13, 21 (1979).
2. A. M. Goldberg and J. M. Frazier, *Sci. Am.* 261(2), 24 (1989).
3. K. Einslein, T. R. Lander, M.E. Tomb, and W. G. Landis, *Teratogen. Carcinogen. Mutagen.* 3, 503 (1983).
4. G. Dupuis and C. Benezra, *Allergic Contact Dermatitis to Simple Chemicals: A Molecular Approach*, Marcel Dekker, New York (1983).
5. K. Einslein, *Pharmacol. Rev.* 36, 131 (1984).
6. J. N. Langley, *Proc. R. Soc.* B78, 170 (1906).
7. W. M. Fletcher, *J. Physiol.* 61, 1 (1926).
8. P. Ehrlich, *Lancet* 2, 445 (1913).
9. A. S. V. Burgen, *Annu. Rev. Pharmacol.* 10, 7 (1970).
10. H. P. Rang, *Nature* 231, 91 (1971).
11. V. P. Gordon, C. P. Kelly, and H. C. Bergman, *Toxicologist* 9, 6 (1989).
12. J. H. Draize, G. Woodward, and H. O. Calvery, *J. Pharmacol. Exp. Ther.* 82, 377 (1944).
13. J. Houk, C. Hansch, L. L. Hall, and R. H. Guy, Chemical structure — transport rate relationships for model skin lipid membranes, *Alternative Methods in Toxicology*, Vol. 5. A. M. Goldberg, Ed. Mary Ann Liebert, New York (1987).
14. B. Papirmeister, C. L. Gross, H. L. Meier, J. P. Petrali, and J. B. Johnson, *Fund. Appl. Toxicol.* 5, s134 (1985).
15. J. A. Faucher and E. D. Goddard, *J. Soc. Cosmet. Chem.* 29, 323, 1978.
16. K. S. J. Rao, *In vitro* enzymatic toxicity test model validity approach, *Alternative Methods in Toxicology*, Vol. 5. A. M. Goldberg, Ed. Mary Ann Liebert, New York (1987).
17. D. W. Hobson, T. H. Snider, M. J. Chang, and R. L. Joiner, *Toxicologist* 8, 20 (1988).
18. B. Ekwall, *Toxicology* 17, 127 (1980).
19. E. Borenfreund and O. Borrero, *Cell Biol. Toxicol.* 1, 55 (1984).
20. E. Borenfreund and C. Shopsis, *Xenobiotica* 15, 705 (1985).
21. J. Demetrulias and H. North-Root, Prediction of the eye irritation potential for surfactant-based household cleaning products using the SIRC cell toxicity test, *Alternative Methods in Toxicology*, Vol. 6. A. M. Goldberg, Ed. Mary Ann Liebert, New York (1988).
22. W. E. Parish, *Food Chem. Toxicol.* 23, 275 (1985).
23. W. E. Parish, *Food Chem. Toxicol.* 24, 481 (1986).
24. D. Triglia, P. T. Wagner, J. Harbell, K. Wallace, D. Matheson, and C. Shopsis, Interlaboratory validation study of the keratinocyte neutral red bioassay from Clonetics Corporation, *Alternative Methods in Toxicology,* Vol. 6. A. M. Goldberg, Ed. Mary Ann Liebert, New York (1988).
25. S. T. Boyce, J. F. Hansbrough, and D. S. Norris, Cellular responses of cultured human epidermal keratinocytes as models of toxicity to human skin, *Alternative Methods in Toxicology*, Vol. 6. A. M. Goldberg, Ed. Mary Ann Liebert, New York (1988).
26. S. Liu and M. Karasek, *J. Invest. Dermatol.* 71, 157 (1978).
27. A. E. Freeman, H. J. Igel, V. J. Herrman, and K. L. Koeinfeld, *In Vitro* 12, 352 (1988).
28. J. G. Rheinwald and H. Green, *Cell* 6, 331 (1975).
29. S. T. Boyce and R. G. Ham, *J. Tissue Culture Methods* 9, 83 (1985).
30. S. T. Boyce and R. G. Ham, Normal human epidermal keratinocytes, *In Vitro Models for Cancer Research*, M. M. Webber and L. I. Sekely, Eds. CRC Press, Boca Raton, FL (1986).
31. D. A. Swisher, M. E. Prevo, and P. W. Ledger, The MTT cytotoxicity test: correlation with cutaneous irritancy in two animal models, *Alternative Methods in Toxicology*, Vol. 6. A. M. Goldberg, Ed. Mary Ann Liebert, New York (1988).
32. H. Babich and E. Borenfreund, Structure-activity relationships of inorganic metals, organometals, and organic test agents determined in vitro with the neutral red assay, *Alternative Methods in Toxicology*, Vol. 6. A. M. Goldberg, ed., Mary Ann Liebert, New York (1988).
33. Z. Nemes, R. Dietz, J. B. Luth, S. Gomba, E. Hackenthal, and F. Gross, *Experientia* 35, 1475 (1979).

34. N. B. Finter, *J. Gen. Virol.* 5, 419 (1969).

35. E. Borenfreund and J. A. Puerner, *J. Tissue Culture Methods* 9, 7 (1984).

36. E. Borenfreund and J. A. Puerner, *Toxicol. Lett.* 24, 119 (1985).

37. T. Mosmann, *J. Immunol. Methods* 65, 55 (1983).

38. T. F. Slater, B. Sawyer, and U. D. Strauli, *Biochem. Biophys. Acta* 77, 383 (1963).

39. H. Tada, O. Shiho, K. Kuoshima, M. Koyama, and K. Tsukamoto, *J. Immunol. Methods* 93, 157 (1986).

40. L. M. Green, J. L. Read, and C. F. Ware, *J. Immunol. Methods* 70, 257 (1984).

41. F. Denizot and R. Lang, *J. Immunol. Methods* 89, 271 (1986).

42. K. Heeg, J. Reimann, D. Kabelitz, C. Hardt, and H. Wagner, *J. Immunol. Methods* 77, 237 (1985).

43. K. Berg, M. B. Hansen, and S. E. Nielsen, *APMIS* 98, 156 (1990).

44. M. B. Hansen, S. E. Nielsen, and K. Berg, *J. Immunol. Methods* 119, 203 (1989).

45. C. Korzeniewski and D. M. Callewaert, *J. Immunol. Methods* 64, 313 (1983).

46. A. Szekers, S. Pacsa, and B. Pejtsik, *J. Immunol. Methods* 80, 1 (1981).

47. C. S. Shopsis and B. Eng, *Toxicol. Lett.* 26, 1 (1985).

48. P. Knox, P. F. Uphill, J. R. Fry, J. Benford, and M. Balls, *Food Chem. Toxicol.* 24, 457 (1986).

49. S. L. Polizotto, P. E. Laux, H. E. Kennah, S. Hignet, and C. S. Barrow, *Toxicologist* 10, 79 (1990).

50. C. S. Shopsis, *J. Tissue Culture Methods* 9, 19 (1984).

51. C. S. Shopsis and S. Sathe, *Toxicology* 29, 195 (1984).

52. M. A. E. Mol, A. B. C. Van De Ruit, and A. W. Kluivers, *Toxicol. Appl. Pharmacol.* 98, 159 (1989).

53. H. Hennings, D. Michael, C. Cheng, P. Steinert, K. Holbrook, and S. H. Yuspa, *Cell* 19, 245 (1980).

54. S. T. Boyce, and R. G. Ham, *J. Invest. Dermatol.* 81, 33s (1983).

55. F. Bertolero, M. E. Kaighn, R. G. Camalier, and U. Saffiotti, *In Vitro Cell. Dev. Biol.* 22, 423 (1986).

56. E. A. Emmett, Toxic responses of the skin, *Casarett and Doull's Toxicology*, 3rd ed. C. D. Klaasen, M. O. Amdur, and J. Doull, Eds. Macmillan, New York (1986).

57. C. D. Klaasen, *The Pharmacological Basis of Therapeutics*, 6th ed. A. G. Gilman, L. S. Goodman, and A. Gilman, Eds. Macmillan, New York (1980).

58. R. G. Freeman, W. Murtishaw, and J. M. Know, *J. Invest. Dermatol.* 54, 164 (1970).

59. G. Kahn and B. Fleischaker, *J. Invest. Dermatol.* 56, 85 (1971).

60. W. L. Morison, D. J. McAuliffe, J. A. Parrish, and K. J. Block, *J. Invest. Dermatol.* 78, 460 (1982).

61. D. J. McAuliffe, T. Hasan, J. A. Parrish, and I. E. Kochevar, Determination of photosensitivity by an in vitro assay as an alternative to animal testing, *Alternative Methods in Toxicology*, Vol. 3. A. M. Goldberg, Ed. Mary Ann Liebert, New York (1985).

62. A. M. Dannenberg, Jr., K. G. Moore, B. H. Schofield, K. Higuchi, A. Kajiki, K. Au, P. J. Pula, and D. P. Bassett, Two new *in vitro* methods for evaluating toxicity to skin (employing short-term organ culture): I. Paranuclear vacuolization, seen in glycol methacrylate tissue sections; II. Inhibition of ¹⁴C-leucine incorporation, *Alternative Methods in Toxicology*, Vol. 5. A. M. Goldberg, Ed. Mary Ann Liebert, New York (1987).

63. G. J. A. Oliver and M. A. Pemberton, *Food Chem. Toxicol.* 23, 229 (1983).

64. W. E. Parish, *Food Chem. Toxicol.* 24, 481 (1986).

65. A. Janoff and J. D. Zeligs, *Science* 161, 702 (1968).

66. W. T. Gibson and M. R. Teall, *Food Chem. Toxicol.* 21, 581 (1983).

67. R. M. Scavarelli-Karantsavelos, S. Zaman-Saroya, F. L. Vaughan, and I. A. Bernstein, *Toxicologist* 10, 79 (1990).

68. R. J. Gay, M. Swiderek, T. Class, P. Kemp, and E. Bell, *Toxicologist* 10, 79 (1990).

69. E. Bell, R. Gay, M. Swiderek, T. Class, P. Kemp, G. Green, H. Haimes, and P. Bilbo, *NATO Advanced Research Workshop, Pharmaceutical Application of Cell & Tissue Culture to Drug Transport*, Bandol, France (1989).

70. L. Lewis, Y. Barrandon, H. Green, and G. Albrecht-Buehler, *Differentiation* 36, 228 (1987).

71. J. R. King and N. A. Monteiro-Riviere, *Toxicologist* 10, 80 (1990).

72. J. A. Blank, D. W. Hobson, G. S. Dill, and R. L. Joiner, *Toxicologist* 10, 331 (1990).

73. J. E. Storm, S. W. Collier, R. F. Stewart, and R. L. Bronaugh, *Toxicologist* 10, 256 (1990).

74. R. J. Ruch and J. E. Klaunig, *Toxicol. Appl. Pharmacol.* 94, 427 (1988).

75. S. W. Collier, N. M. Sheikh, J. L. Sakr, J. L. Lichtin, R. F. Stewart, and R. L. Bronaugh, *Toxicol. Appl. Pharmacol.* 99, 522 (1989).

76. L. L. Hall, H. L. Fisher, M. R. Sumler, S. P. Shrivastava, and P. V. Shah, *Toxicologist* 10, 257 (1990).

77. L. Bernstam, S. Jang, F. L. Vaughan, R. L. Bronaugh, and I. A. Bernstein, *Toxicologist* 10, 78 (1990).

78. G. J. A. Oliver and M. A. Pemberton, *Human Toxicol.* 2, 562 (1983).

79. G. J. A. Oliver, M. A. Pemberton, and C. Rhodes, *Food Chem. Toxicol.* 24, 507 (1986).

80. G. J. A. Oliver and M. A. Pemberton, *Food Chem. Toxicol.* 24, 513 (1986).

81. M. A. Karasek and M. E. Charlton, *J. Invest. Dermatol.* 56, 205 (1971).

82. E. Bell, S. Sher, B. Hull, C. Merrill, S. Rosen. A. Chamson, D. Asselineau, L. Dubertret, B. Coulomb, C. Lapiere, B. Nusgens, and Y. Neveux, *J. Invest. Dermatol.* 81, 2s (1983).

83. F. L. Vaughan, R. H. Gray, and I. A. Bernstein, *In Vitro Cell. Dev. Biol.* 22, 141 (1986).

84. G. Naughton, L. Jacob, and B. A. Naughton, *A Physiological Skin Model for In Vitro Toxicity Studies — A Symposium of the CAAT Technical Workshop of May 17-18, 1989*, Vol. 7. J. M. Frazier and J. A. Bradlaw, Eds. Johns Hopkins School of Hygiene and Public Health, Baltimore (1989).

85. E. Bell, L. Parenteau, H. B. Haimes, R. J. Gay, P. D. Kemp, T. W. FoFonoff, V. S. Mason, D. T. Kagan, and M. Swiderek, Testskin: a hybrid organism covered by a living human skin equivalent designed for toxicity and other testing, *Alternative Methods in Toxicology*, Vol. 6. A. M. Goldberg, Ed. Mary Ann Liebert, New York (1988).

86. H. Green, *Cell* 11, 405 (1977).

87. B. Coulomb, P. Saiag, E. Bell, F. Breitburd, C. Lebreton, M. Hesland, and L. Dubertret, *Br. J. Dermatol.* 114, 91 (1986).

88. D. G. Hafeman, J. W. Parce, and H. M. McConnell, *Science* 240, 1182 (1988).

89. K. M. Kercso, V. C. Muir, J. C. Owicki, and J. W. Parce, *Toxicologist* 9, 259 (1989).

90. T. B. Hutson, M. L. Mitchell, J. T. Keller, D. J. Long, and M. J. W. Chang, *Anal. Biochem.* 174, 124 (1988).

91. J. M. Frazier and J. A. Bradlaw, Eds., *Technical Report No. 1: Technical Problems Associated with In Vitro Toxicity Testing Systems — A Report of the CAAT Technical Workshop of May 17-18, 1989*, Johns Hopkins School of Hygiene and Public Health, Baltimore (1989).

92. Z. Govindarajulu, *Statistical Techniques in Bioassay*, S. Karger, New York (1988).

93. D. J. Finney, *Statistical Method in Biological Assay*, Griffin, London (1971).

94. P. A. Duffy, O. P. Flint, T. C. Orton, and M. J. Fursey, *Food Chem. Toxicol.* 24, 517 (1986).

95. D. W. Hobson, J. A. Blank, G. S. Dill, and R. L. Joiner, *Toxicologist* 10, 328 (1990).

96. V. DeLeo, J. Hong, S. Scheide, B. Kong, S. DeSalva, and D. Bagley, Surfactant-induced cutaneous primary irritancy: an *in vitro* model — assay system development, *Alternative Methods in Toxicology*, Vol. 6. A. M. Goldberg, Ed. Mary Ann Liebert, New York (1988).

97. L. A. Goldsmith, Skin toxicity, *Alternative Methods in Toxicology*, Vol. 6. A. M. Goldberg, Ed. Mary Ann Liebert, New York (1988).

98. R. A. Scala, Theoretical approaches to validation, *Alternative Methods in Toxicology*, Vol. 5. A. M. Goldberg, Ed. Mary Ann Liebert, New York (1987).

99. M. Balls, R. J. Riddell, S. A. Horner, and R. H. Clothier, The FRAME approach to the development, validation, and evaluation of *in vitro* alternative methods, *Alternative Methods in Toxicology*, Vol. 5. A. M. Goldberg, Ed. Mary Ann Liebert, New York (1987).

100. J. M. Frazier, The validation approach of the Johns Hopkins Center for alternatives to animal testing, *Alternative Methods in Toxicology*, Vol. 5. A. M. Goldberg, Ed. Mary Ann Liebert, New York (1987).

101. J. A. Cooper, R. Saracci, and P. Cole, *Br. J. Cancer* 39, 87 (1979).

12

Use of Ocular and Dermatotoxicologic Test Data in Product Development and in the Assessment of Chemical Hazards

SHAYNE COX GAD
G. D. Searle & Company
Skokie, Illinois

I. INTRODUCTION

The entire product safety assessment process, in the broadest sense, is a multistep process in which none of the individual steps is overwhelmingly complex, but the integration of the whole process involves fitting together a large number of pieces. This volume, as a whole, seeks to address the questions of the toxicology associated with agents that interact with the skin and eyes, i.e., the dermal and ocular routes of exposure. How the data generated by the various test systems and models described elsewhere in this volume are used by government and private enterprise to provide for a safe product life cycle is the subject of this chapter. As will be seen, this is a special case of the general product safety assessment problem, and it will be addressed by starting with the general case and progressing to the special case. Along the way, limitations of current models and places where testing and research data could be made more practically useful will be pointed out.

The single most important part of a product safety evaluation program is, in fact, the initial overall process of defining and developing an adequate data package on the potential hazards associated with the product life cycle (the manufacture, sale, use, and disposal of a product and associated process materials). To do this, one must ask a series of questions in a very interactive process, with many of the questions designed to identify and/or modify their successors. The first is — what information is needed?

This calls for understanding the way in which a product is to be made and used, and the potential health and safety risks associated with exposure of humans who will be associated with these processes. This is the basis of a hazard and toxicity profile. Once such a profile is established (as illustrated in Figure 1), the available literature is searched to determine what is already known.

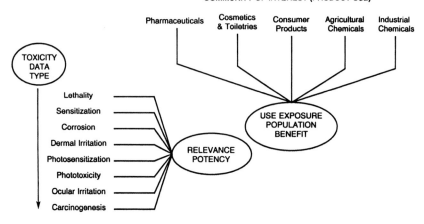

Figure 1.   Multidimensional matrix for ocular and dermal hazard assessment.

Taking into consideration this literature information and the previously defined exposure potential, a tier approach (Figure 2) is generally used to generate a list of tests or studies to be performed.* What goes into a tier system is determined by both regulatory requirements imposed by government agencies and the philosophy of the parent organization. How such tests are actually performed is determined on one of two bases. The first (and most common) is the menu approach: selecting a series of standard design tests as "modules" of data. The second is an interactive approach, where studies are designed (or designs are selected) based both on needs and what has been learned previously about the dermatotoxicity of the product.

The initial and most important aspect of a product safety evaluation program is the series of steps that leads to an actual statement of problems or of the objectives of any testing and research program. This definition of objectives is essential and, as proposed here, consists of five steps: (1) defining product or material use, (2) quantitating or estimating exposure potential, (3) identifying potential hazards, (4) gathering baseline data, and (5) designing and defining the actual research program. Later, we will look at the specific application of this to dermal and ocular toxicity cases, where the concept of *communities of interest* will become essential.

A. Product/Material Use

Identifying how a material is to be used, what it is to be used for, and how it is to be made are the first three essential questions to be answered before a meaningful dermatotoxicological risk assessment program can be performed. These determine, to a large extent, how many people are potentially exposed, to what extent and by

---

*   There is also the special case of pharmaceutical and pesticide products, where there are regulatorily mandated minimum test batteries.

Tier Testing	Mammalian Toxicology	Genetic Toxicology	Remarks
0	Literature review	Literature review	Upon initial identification of a problem data base of existing information and particulars of use of materials are established
1	Primary dermal irritation Eye irritation Dermal sensitization Acuted sytemic toxicity Lethality screens	Ames test In vitro SCE[a] In vitro cytogenetics Forward mutation/CHO[b]	R&D materials and low volume chemicals with severely limited exposure
2	Subacute studies Metabolism	In vivo SCE In vivo cytogenetics	Medim volume materials and/or those with
3	Subchronic studies Reproduction Teratology Chronic studies Specialty studies	—	Any material with a high volume or a potential for widespread human exposure or one which gives indications of specific long-term effects

[a]Sister chromatid exhange assay
[b]Chinese hamster ovary assay

Figure 2. Tier testing.

what routes they are exposed, and what benefits they perceive or gain from product use. The answers to these questions are generally categorical or qualitative, and become quantitative (as will be reviewed in the next section) at a later step in the risk assessment process (frequently long after acute data have been generated).

Starting with an examination of how a material is to be made (or how it is already being made), it generally occurs that there are several process segments, each representing separate problems. Commonly, much of a manufacturing process is "closed" (i.e., it occurs in sealed, airtight systems) thereby limiting exposures to (1) leaks with low level (generally) inhalation and dermal exposures to significant portions of a plant work force, and (2) maintenance and repair workers (generally short-term, higher level inhalation and dermal exposures to small numbers of personnel). Smaller segments of the process will almost invariably not be closed. These segments are most common either where some form of manual manipulation is required, or where the segment requires a large volume of space (such as when fibers or other objects are spun, formed, or individually coated with something) or where the product is packaged (such as powders being put into bags or polymers being removed from molds). The exact manner and quantity of each segment of the manufacturing process, given the development of a categorization of exposures for each of these segments, will then serve to help quantitate the identified categories.

Likewise, consideration of what a product is to be used for and how it is to be used should be of help to identify who, outside of the manufacturing process, may be exposed, by what routes, and to what extent.

The answers to these questions, again, will generate a categorical set of answers, but will also serve to identify which particular aspects of regulatory toxicity testing may be operative (such as those for the Department of Transportation [DOT] or

Consumer Product Safety Commission [CPSC]). If a product is to be worn (such as clothing or jewelry) or used on, or as, an environmental surface (such as household carpeting or wall covering), the potential for dermal exposure of a large number of people to the product (albeit at low levels in the case of most materials) is large. If the product is to be used on exterior portions of buildings, the potential for dermal exposure of a large number of individuals to the product would be smaller. Likewise, the nature of the intended use (say, as a true consumer product vs. an industrial product) may also affect the potential for the extent and degree of dermal exposure in overt and subtle manners. For example, a finish for carpets to be used in the home has a much greater potential for dermal and even oral (in the case of infants) exposure than such a product used only for carpeting in offices. Likewise, in general, true consumer products (such as household cleaners) have a greater potential for both accidental exposure and misuse.

B. Exposure Potential

The next problem (or step) is quantitating the exposure of the human population, both in terms of how many people are exposed by what routes (or means) and what quantities of an agent they are exposed to.

The process of identifying and quantitating exposure groups within the human population is beyond the scope of this chapter, except for some key points. Classification methods are the key tools, however, for identifying and properly delimiting human populations at risk and will be briefly discussed.

Classification is both a basic concept and a collection of techniques which are necessary prerequisites for further analysis of data when the members of a set of data are (or can be) each described by several variables. At least some degree of classification (which is broadly defined as the dividing of the members of a group into smaller groups in accordance with a set of decision rules) is necessary prior to any data collection. Whether formally or informally, an investigator has to decide which things are similar enough to be counted as the same and to develop rules governing collection procedures. Such rules can be as simple as "measure and record exposures only of production workers", or as complex as that demonstrated by the expanded classification presented below. Such a classification also demonstrates that the selection of which variables to measure will determine the final classification of data.

1. Which groups are potentially exposed?
2. What are segments of groups? Consumers, production workers, neonatal patients, etc.
3. What are routes of potential exposure?
4. Which group does each route occur in?
5. Which sex is exposed?
6. Which age groups are potentially exposed?

Data classification serves two purposes: data simplification (also called a descriptive function) and prediction. Simplification is necessary because there is a limit to both the volume and complexity of data that the human mind can comprehend and deal with conceptually. Classification allows us to attach a label (or name) to each group of data, to summarize the data (i.e., assign individual elements of data to groups and to characterize the population of the group), and to define the relationships between groups (i.e., develop a taxonomy).

Prediction is the use of summaries of data and knowledge of the relationships between groups to develop hypotheses as to what will happen when further data are collected (as when more production or market segments are expanded) and as to the mechanisms which cause such relationships to develop. Indeed, classification is the prime device for the discovery of mechanisms in all of science. A classic example of this was Darwin's realization that there were reasons (the mechanisms of evolution) behind the differences and similarities in species which had helped Linaeus to earlier develop his initial modern classification scheme (or taxonomy) for animals.

To develop a classification, one first sets bounds wide enough to encompass the range of data to be considered, but not unnecessarily wide. This is typically done by selecting some global variables (variables all data have in common) and limiting the range of each so that it just encompasses all the cases on hand. One then selects a set of local variables (characteristics that only some of the cases have, such as certain skin types, enzyme activity levels, or dietary preferences). These local variables then serve to differentiate between groups. Data are then collected, and a system for measuring differences and similarities is developed. Such measurements are based on some form of measurement of distance between two cases (x and y) in terms of each single variable scale. If the variable is a continuous one, then the simplest measure of distance between two pieces of data is the Euclidean distance, ($d[x,y]$) defined as

$$d(x,y) = (x_i - Y_i)^2$$

For categorical or discontinuous data, the simplest distance measure is the matching distance, defined as

$$d(x,y) = \text{number of times } x_i \neq Y_i$$

After a table of such distance measurements is developed for each of the local variables, some weighting factor is assigned to each variable. A weighting factor gives more importance to those variables that are believed to have more relevance or predictive value. The weighted variables are then used to assign each piece of data to a group. The actual process of developing numerically based classifications and assigning data members to them is called cluster analysis. Classification of biological data based on qualitative factors has been well discussed by Glass,[1] and Gordon[2] does an excellent job of introducing the entire field and mathematical concepts.

In developing an exposure classification for a material, the investigator must understand the process or processes involved in making, shipping, using, and disposing of the material. The Environmental Protection Agency (EPA) recently proposed guidelines for such identification and exposure quantitation.[3] The exposure groups may be very large or relatively small populations, each with a markedly different potential for exposure differences (such as the lung for inhalation exposures) and special toxicity problems (such as delayed contact dermal sensitization). Route-specific concerns for ocular and dermal exposure are the subject of the other chapters in this book.

All such estimates of exposure in humans (and of the number of humans exposed) are subject to a large degree of uncertainty.

C. Potential Hazard

Once the types of exposures have been identified and the quantities approximated, a toxicity matrix can be developed by identifying the potential hazards. Such an identification can proceed by one of three major approaches.

1. Analogy from data reported in the literature
2. Structure-activity relationships
3. Predictive testing

For the third approach, the tests and their objectives (the questions they ask) are detailed as follows.

Primary dermal irritation (PDI)* — Evaluate potential of a single dermal exposure to cause skin irritation.

Primary eye irritation* — Evaluate potential of a single ocular exposure to cause eye irritation.

Dermal corrosivity* — Determine potential of a single 4-h dermal exposure to cause skin corrosion.

Dermal sensitization* — Evaluate the potential of a material to cause delayed contact hypersensitivity.

Photosensitization* — Evaluate the potential of a material to cause a sunlight-activated delayed contact hypersensitivity.

Lethality screen** — Determine the potential of a single dose of a material, at a predetermined level, for lethality.

Carcinogenicity — Determine the potential for a material to induce and/or promote the formation of dermal neoplasms.

Mutagenicity — Determine the potential of a material to cause undesirable genetic events.

Five of these seven (*) are specific points of interest for this volume, while a sixth (**) may also be if the exposure route is dermal or ocular.

The second category contains the "shotgun" (or multi-endpoint) tests. The name "shotgun" suggests itself for these studies because it is not known in advance at which endpoints are being aimed. Rather, the purpose of the study is to identify and quantitate all potential systemic effects resulting from a single exposure to a compound. Once known, specific target organ effects can then be studied in detail if so desired. Accordingly, the generalized design of these studies is to expose groups of animals to controlled amounts or concentrations of the material of interest, then to observe for, and to measure changes in as many variables as practical over a period during or past the exposure. Further classification of tests within this category is defined either by the route by which test animals are exposed or dosed (generally oral, dermal, or inhalation are the options here) or by the length of dosing. Acute studies, for example, imply a single exposure interval (of 24 h or less) or dosing of test material. Chapter 3 of this volume addresses these tests in detail.

II. AVAILABLE TOXICOLOGY INFORMATION SOURCES AND THEIR USE

A. Sources

1. Published information sources
2. On-line literature searches
3. Colleagues
4. Building a database
5. Monitoring published literature and other research in progress

Also to be considered is maintenance of a database once it is established (in many places this is called product surveillance).

However, before review of the literature is initiated, it is important to obtain as much of the following product composition and exposure information as practical.

1. Chemical composition and major impurities
2. Chemical production and use information, i.e., manufacturing process, exposure patterns, and other commercial uses
3. Correct chemical identity including formula, Chemical Abstract Service Number, common synonyms and trade names
4. Selected physical properties, i.e., physical state, vapor pressure, chemical reactivity, and pH
5. Other chemical substances exhibiting similar toxicity and/or structure/activity relationships

Collection of the aforementioned information is not only important for the hazard assessment (high vapor pressure would indicate high inhalation potential just as high or low pH would indicate high irritation potential), but prior identification of all product uses and exposure patterns can identify alternative information sources, e.g., chemicals formerly used as anesthetics, food additives, or pesticides may have extensive toxicology data obtainable from government or private sources.

B. Published Information Sources

There are numerous published texts for use in literature reviewing. An alphabetic listing of seventeen of the more commonly used texts is provided in Table 1.

Table 1
Published Information Sources for Dermal and Ocular Toxicology Data

Title	Ref.
Burger's Medicinal Chemistry	4
Chemical Hazards of the Workplace	5
Clinical Toxicology of Commercial Products	6
Contact Dermatitis	7
Criteria Documents	8
Current Intelligence Bulletins (NIOSH)	9
Dangerous Properties of Industrial Materials	10
Documentation of the Threshold Limit Values (AIHA)	11
Handbook of Toxic and Hazardous Chemicals	12
Hygienic Guide Series (AIHA)	13
Industrial Toxicology	14
Merck Index	15
Occupational Health Guidelines for Chemical Hazards (NIOSH/OSHA)	16
Patty's Industrial Hygiene and Toxicology	17
Physician's Desk Reference	18
Registry of Toxic Effects of Chemical Substances (RETECS)	19
Toxicology of the Eye	20

Obviously, this is not a complete listing and consists of only the general multipurpose texts that have a wider range of applicability for the dermal and ocular toxicity of commercial products (current or potential). Texts dealing with specialized classes of chemicals, e.g., petroleum hydrocarbons, plastics, or those with specific target organ toxicities (neurotoxins and teratogens), are not considered to be within the scope of this section and the interested reader is referred to Parker[23] or Wexler[22] for further details.

C. On-Line Literature Searches

In the last 10 years, the use of on-line literature searches by many toxicologists has changed from an occasional, sporadic request to the semicontinuous need for computerized search capabilities. Usually nontoxicology-related search capabilities are already in place in many companies. Therefore, all that is needed is to expand the information source to include some of the databases that cover the types of toxicology information needed. A university or a private contract laboratory can often provide this service. Most of the databases of interest to toxicologists are parts of the National Library of Medicine (NLM) system.

The NLM information retrieval service[21] contains Medline, Toxline, and Cancerline databases. Databases commonly used by industrial toxicologists in the NLM service are briefly discussed below.

1. Toxline (Toxicology Information Online) is a bibliographic database covering the pharmacological, biochemical, physiological, environmental, and toxico-

logical effects of drugs and other chemicals. It contains approximately 1.7 million citations, most of which are complete with abstract index terms and Chemical Abstracts Service Registry Numbers. Toxline citations have publication dates of 198 present. Older information is on Toxback 76 (1976—1980) and Toxback 65 (pre-1965 through 1975). Information on Toxline comes from 13 secondary subfiles.

2. Medline (Medical Information Online) is a database containing approximately 800,000 references to biomedical journal articles published since 1980. These articles, usually with an English abstract, are from over 3000 journals. Coverage of previous years (back to 1966) is provided by back files searchable online, that total some 3.5 million references.

3. Toxnet (Toxicology Data Network) is a computerized network of toxicologically oriented databanks. Toxnet offers a sophisticated search and retrieval package which accesses the following subfiles:

   a. Hazardous Substances Data Bank is a scientifically reviewed and edited databank containing toxicological information strengthened with additional data related to the environment, emergency situations, and regulatory issues. Data are derived from a variety of sources including government documents and special reports. This database contains records for over 4100 chemical substances.

   b. Toxicology Data Bank is a peer-reviewed databank focusing upon toxicological and pharmacological data, environmental and occupational information, manufacturing and use data, and chemical and physical properties. References have been extracted from a selective list of standard source documents.

4. Registry of Toxic Effect of Chemical Substances (RTECS) is the NLM on-line version of the National Institute of Occupation, Safety and Health (NIOSH) annual compilation of substances with toxic activity.[19] The original collection of data was derived from the 1971 Toxic Substances Lists. The RTECS data contain threshold limit values, aquatic toxicity ratings, air standards, National Toxicology Program carcinogenesis bioassay information and toxicological/ carcinogenic review information. NIOSH is responsible for the file content in RTECS, and for providing quarterly updates to NLM. Currently, RTECS covers toxicity data on more than 61,000 substances. Greater detail on available data sources and their use in toxicology can be found in Wexler[22] or Parker.[23]

The next step, given that no data are found from any of these sources, is to perform appropriate tests. The bulk of this volume addresses specifics of performing such tests. How did we come to the current stage in employing such tests?

III. HISTORY OF EYE AND SKIN TESTING

Although the historical basis of each of the major test types is separate, they run to a common stream in the end.

A. Eye Testing

Early in the 1930s, an untested eyelash dye containing *p*-phenyl-enediamine ("Lash Lure") was brought onto the market in the U.S. This product (as well as a number of similar products) rapidly demonstrated that it could sensitize the external ocular structures, leading to corneal ulceration with loss of vision and at least one fatality.[24] This occurrence led to the revision of the Food and Drug Act, which became the Food, Drug, and Cosmetic Act of 1938. To meet the provisions of this act, a number of test methods were proposed. Latven and Molitor[25] and Mann and Pullinger[26] were among those to first report the use of rabbits as a test model to predict eye irritation in humans. No specific scoring system, however, was presented to grade or summarize the results in these tests and the use of animals with pigmented eyes (as opposed to albinos) was advocated. Early in 1944, Friedenwald et al.[27] published a method that used albino rabbits in a similar manner to that of the original Draize publication,[28] but still prescribed the description of the individual animal responses as the means of evaluating and reporting the results. Although a scoring method was provided, no overall score was generated for the test group. Draize (head of the Dermal and Ocular Toxicity Branch at the Food and Drug Administration [FDA]) modified Friedenwald's procedure and made the significant addition of a summary scoring system.

Over the 40 years since the publication of the Draize scoring system, it has become common practice to call all acute eye irritation tests performed in rabbits "the Draize eye test". However, since 1944, ocular irritation testing in rabbits has significantly changed. Clearly, there is no longer a single test design that is used, and there are different objectives that are pursued by different groups using the same test.

B. Dermal Irritation/Corrosion Testing

Evaluation of materials for their potential to cause dermal irritation and corrosion due to acute contact has been common for industrial chemicals, cosmetics, agricultural chemicals, and consumer products since at least the 1930s (generally, pharmaceuticals are only evaluated for dermal effects if they are to be administered topically, and then by repeat exposure/type tests. As with acute eye irritation tests, one of the earliest formal publications of a test method (although others were used) was that of Draize et al. in 1944.[28] The methods currently used are still basically those proposed by Draize et al.[28] and have changed little since 1944. Although (unlike their near relatives, the eye irritation tests) these methods have not caught the interest of the animal welfare movement, there are efforts underway to develop alternatives that either do not use animals, do not use rabbits, or are performed in a more humane and relevant (to human exposure) manner. These are addressed in Chapter 11.

C. Sensitization Testing

Koch's[29] initial findings of tuberculin reactivity was observed in humans and guinea pigs. Although the rabbit and guinea pig have both been considered the animals of choice for evaluating adverse skin reactions of chemicals, from the beginning guinea pigs have been the animals of choice for predictive tests. Although it is widely believed that this is due to a relatively higher degree of susceptibility to dermal sensitization, the preference was actually based on the availability, ease of handling, and the clear pale skin of an albino which is easily denuded of hair and which allows easy determination of erythema response.

Landsteiner[30] first proposed a formalized predictive test in guinea pigs. Later, he and Chase[31] used low molecular weight chemicals to sensitize guinea pigs and developed the theory of complete antigen formations being due to hapten, i.e., protein interactions.

The basis of modern predictive tests is the Draize test, as established by Draize in 1944.[28] It consists of 10 intracutaneous injections of the test compound into the skin of albino guinea pigs during the 3-week induction period and a single intracutaneous challenge application 14 days after the last induction injection. A standardized 0.1% test concentration is used for induction and challenge. This method was widely used and recommended until the end of the 1960s. Its principal disadvantage is that only strong allergens are detected, while well-known moderate allergens fail to sensitize the animals.

Starting in 1964, however, a wide variety of new test designs were proposed. Buehler[32] proposed what is now considered the first modern test (described in Chapter 5) that used an occlusive patch to increase test sensitivity. The Buehler test is the primary example of the so-called "epidermal" methods that have been criticized for giving false-negative results for moderate-to-weak sensitizers such as nickel.

A new generation of tests was established by using Freund's complete adjuvant (FCA) during the induction process to stimulate the immune system independent of the type of hapten and independent of the method or application, i.e., whether or not the substance is incorporated in the adjuvant mixture. It is claimed that this family of tests displays the same level of susceptibility to sensitization in guinea pigs as is normally observed in humans.[33] The adjuvant tests include the guinea pig maximization test,[34] optimization test,[35,36] split adjuvant test,[37] and the epicutaneous maximization test.[38]

Finally, during the last few years a test system which used albino mice instead of guinea pigs, — the mouse ear swelling test, (MEST)[39,40] has been proposed as an alternative.

This chain of development should be expected to continue and the overall quality and utility of tests should improve.

IV. OBJECTIVES BEHIND DATA GENERATION AND UTILIZATION

To understand how dermal and ocular toxicity data are used, and how the data generation process might be changed to better meet the product safety assessment need, it is essential to understand that different commercial organizations have different answers to these questions. The ultimate answer is a multidimensional matrix, with the three major dimensions of the matrix being (1) the toxicity data type (lethality, sensitization, corrosion, irritation, photosensitization, phototoxicity, etc.),

(2) exposure characteristics (extent, population size, population characteristics, etc.), and (3) type of commercial organization (community of interest). This matrix is shown in Figure 2.

Communities of interest are really defined by how the products are to be used, who regulates their use, and what benefits are expected for the consumer. There are a number of ways of classifying such communities, but for our purpose we will divide and define them as follows.

Pharmaceuticals — Materials of concern are agents intended as therapeutics (or medical devices) where patient or health care provider (doctor, nurse, or pharmacist) have a significant chance of ocular or dermal exposures — particularly if the intended primary route of administration is ocular or topical. The Food and Drug Administration (FDA) is the U.S. regulator.

Cosmetics and toiletries — The materials are cosmetics, fragrances, shampoos, hand and body soaps, hair dyes, and other materials intended to improve appearance and personal presentation. These are intended to be either applied to the skin and proximity of the eye, or to be applied or used in a manner that makes dermal or ocular exposure unavoidable. The major U.S. regulators are the FDA and CPSC.

Consumer products — Products intended to be used by the average person in and around their home can be divided into those that have a high potential for exposure (dish and laundry detergents, for example) those that have low potential for such exposure (drain cleaners, oven cleaners, etc.), or those that have a wide range in between (such as window and carpet cleaners). The primary regulators are CPSC and DOT, but the EPA also is important in terms of new chemical entities and disposal and waste management.

Agricultural products— These products are pesticides, herbicides, fertilizers (which represent a special case), and other intentional food additives (such as preservatives, sterilants, etc.). The extent of dermal and ocular exposure will vary widely in use. Note that these could be subdivided into those agents used in the field (i.e., actually used in agriculture) and those agents used in the storage and processing of foods. Ocular and dermal concerns for the former fall primarily to the EPA (under the Federal Insecticide, Fungacide, and Rodenticide Acta [FIFRA]), with secondary considerations by DOT and FDA. The second group has FDA as the primary driving force, with EPA and DOT concerns secondary.

Industrial chemicals — These are materials for which the major exposure is to workers involved in the manufacture formulation and transportation of products. In a sense, all the materials (e.g., the above categories) fall into this group at some time, plus a number of other chemicals that never appear (as such) in those categories (such as hydrofluoric acid and plasticizers). These are handled by a smaller population (relative to most other products). Eye and skin contact is never intended; in fact, active measures are taken to prevent it. The use of eye and skin data in these cases is to fulfill labeling requirements for shipping and to provide hazard assessment information for accidental exposures and their treatment. The results of such tests usually do not directly affect the economic future of a material.

Each of these communities has different needs and uses for each of the kinds of data produced and these must be examined independently. Next, we will examine each test (endpoint case), develop a decision matrix for that type of data, and then point out the short falls in existing test systems.

V. OCULAR IRRITATION

For the pharmaceutical industry (with the exception of the case of contact lenses, which will not be discussed here), eye irritation testing is done when the material is intended to be put into the eye as a means or route of application or for ocular therapy. There are a number of special tests applicable to pharmaceuticals or medical devices that are beyond the scope of this volume, as they are not intended to assess potential acute effects or irritation. In general, however, an eye irritation test that is used by this group must be both sensitive and accurate in predicting the potential to cause irritation in humans. Failing to identify human ocular irritants (lack of sensitivity) is to be avoided, but of equal concern is the occurrence of false positives. This subject is addressed in more detail in Chapter 13.

The cosmetics and toiletries industry is similar to the pharmaceutical industry in that the materials of interest are frequently intended for repeated application in the area of the eye. In such uses contact with the eye is common, although not intended or desirable. In this case the objective is also to employ a test that is sensitive, even if this results in a low incidence of false-positives. Even a moderate irritant is not desired, but might be acceptable in certain cases (such as deodorants and depilatories) where the potential for eye contact is minimal.

Consumer products that are not intended for personal care (such as soaps, detergents, and drain cleaners) are approached from a different perspective. These products are not intended to be used in a manner that either causes them to get into eyes or makes that occurrence likely, but because a large population uses them and their modes of use do not include active measures to prevent eye contact (such as goggles or face shields), severe eye irritants must be accurately identified.

For agricultural chemicals, ocular exposure is never intended. However, unless rigorous steps are taken (which is almost never the case in the field), such exposure is unavoidable to a sizeable population. The desire here is to identify severe irritants or corrosives that require use of applicator systems or other methods of use that would preclude exposure. For ocular irritation, those chemicals used in food processing are generally treated as industrial chemicals.

For industrial chemicals, eye irritation data are used primarily to fulfill labeling requirements for shipping and to provide hazard assessment information for accidental exposures and their treatment. The results of such tests do not directly affect the economic future of a material. It is desirable to identify moderate and severe irritants accurately (particularly those with irreversible effects) and to know if rinsing of the eyes after exposure will make the consequences of exposure better or worse. False-negatives for mild reversible irritation are acceptable.

The needs and uses of these different communities in terms of ocular irritation data are summarized in Table 2. Historically, the philosophy underlying the test

Table 2

Matrix of Intended Product Use vs. Required Test Features for Ocular Irritation

	Features			
Types of organization (intended product use)	Desired sensitivity (lowest level of irritation essential to detect)	Need to evaluate recovery and effects of timely irrigation	Acceptable incidence of	
			False-positives	False-negatives
Pharmaceutical	Moderate	None	None	None
Cosmetic	Moderate	Recovery-high, irrigation-none	Minimal	None
Consumer product[a] (personal use)	Moderate	Recovery-high, irrigation-low	Minimal	None
Consumer product (household use)	Severe	Medium to high	Low	Low for moderates, none for substantials
Agricultural chemical	Severe	High	Low	None for severes
Industrial chemical	Severe	High	Low	Low for substantials, none for severes

[a] Current FHSA regulations require that any consumer-used product (other than pharmaceuticals and cosmetics which are regulated by FDA), must be identified as to their potential to cause irritation as defined earlier in this section.

designs that were used to evaluate eye irritation made maximization of the biological response equivalent to being the most sensitive test. As this review of the objectives of the communities of interest has shown, the greatest sensitivity (especially at the expense of false-positive findings, which is an unavoidable consequence) is not what is universally desired. As shall be seen later, maximizing the response in rabbits does not *guarantee* sensitive prediction of the results in humans.

Method variations that are commonly used to improve the sensitivity and accuracy of describing damage in these tests are inspection of the eyes with a slit lamp and instillation of a vital dye in the eye (or, most commonly, fluorescein) to indicate increased permeability of the corneal barrier. These techniques, and an alternative scoring system which is more comprehensive than the Draize scale, have previously been presented and reviewed by Frazier et al.[41]

Another *in vivo* alternative which has been proposed (and which a survey[41] has shown to have been adopted by a number of laboratories) is using a reduced volume/weight of test materials.

In 1980 Griffith et al.[58] reported on a study in which they evaluated 21 different chemicals at volumes of 0.1, 0.03, 0.01, and 0.003 ml. These 21 chemicals were materials on which there was already human data. The volume reduction was found to not change the rank order of responses, and it was found that 0.01 ml (10 μl) gave results which best mirrored those seen in man. In 1982 Williams et al.[59] reported a comparison of 7 materials evaluated at 0.1 and 0.01 and found that the rank of results was not changed with the volume reduction while the responses were moderated.

In 1984 Freeberg et al.[60] published a study of 29 detergents (for which there was human data), each evaluated at both 0.1 and 0.01 ml test volumes in rabbits. The results of the 0.1 ml tests were reported to be more reflective of results in man. In 1986 Freeberg et al.[61] published a further evaluation of low vs. classical volume tests, in both humans and rabbits, and found that recovery times from low volume rabbit tests gave a better correlation with results seen in humans than classical volumes.

In 1985, Walker[62] reported on an evaluation of the low volume (0.01 ml) test which assessed its results for correlation with those in humans based on the number of days until clearing of injury and reported that 0.01 ml gave a better correlation than did 0.1 ml.

There are only two objections to the low volume test. These are that we would lose a screen for exquisitely toxic materials (single drops of which in the eye will kill an animal, such as was reported for an organophosphonium salt by Dunn et al. in 1982[63]) and that there may be some classes of chemicals for which low volume tests may give less representative results.

The American Society of Testing and Materials (ASTM) has published the low volume method (Method E 1055-85) as a consensus standard procedure). It seems clear that this approach should be seriously considered by those performing *in vivo* eye irritation tests.

One clear need, which should be technologically feasible, is for a simple *in vitro* screening system that will identify severe irritants or corrosives with a very low false-negative rate and an acceptable (about 5%) false-positive rate.

As the matrix in Table 2 should make clear, such a test system (although "blind" to mild and moderate irritants) would readily replace the need to use animal tests. At the same time, there are special cases where testing in the intact animal is the only means of detecting special case toxicities, such as the exquisite lethality of parathion via the ocular route.[64]

VI. INTERPRETATION AND USE OF DERMAL TOXICITY DATA

Unlike eye irritation, dermal toxicities are complex and multifaceted. In this chapter only some of the possible cases will be considered — irritation and corrosion, sensitization, light-mediated dermal toxicities (photosensitization and phototoxicity), percutaneous absorption, and systemic toxicity by the dermal route. In each of these cases, limitations of current test systems will be considered first. Then the use of the test data by each of the community of interests will be presented. Finally, needs for improvement in test systems will be proposed.

There are three major objectives behind dermal testing:

1. Labeling and Registration — Any product now in commerce must both be labeled appropriately for shipping[42] and accompanied by a material safety data sheet (MSDS) that clearly states potential hazards associated with handling. DOT regulations also prescribe different levels of packaging on materials found to constitute hazards as specified in the regulations. The EPA (under FIFRA) also has a pesticides-labeling requirement. These requirements demand absolute identification of severe irritants or corrosives and adherence to the basics of test methods promulgated by the regulations. False-positives (type I errors) are to be avoided in these usages.
2. Hazard Assessment for Accidents — For most materials, dermal exposure is not intended to occur, yet it will occur in cases of accidental spillage or mishandling. Here, the hazard associated with such exposures needs to be correctly identified giving equal concern to false-positives and false-negatives.
3. Assessment of Safety for Use — The materials at issue here are the full range of products for which dermal exposure will occur in the normal course of use. These range from cosmetics and hand soaps to bleaches, laundry detergents, and paint removers. No manufacturer wants to put a product on the market that cannot be safely used and that could lead to extensive liability. Accordingly, the need here is to accurately predict the potential hazards in humans — i.e., to have neither false-positives nor false-negatives.

A. Dermal Irritation and Corrosion

Among the most fundamental assessments of the safety of a product or, indeed, of any material that has the potential to be in contact with a significant number of people in our society, are tests in animals that seek to predict potential skin irritation or corrosion. As all the other tests in what is classically called a range-finding study, tier I or acute battery, the tests used here are both among the oldest designs and are undergoing the greatest degree of scrutiny and change. Currently, all the established test methods for these endpoints use the rabbit (almost exclusively the New Zealand White strain), although other animal models have been proposed.

Virtually all manufactured chemicals have the potential to contact the skin of people. In fact, many (cosmetics and shampoos) are intended to have skin contact. The greatest number of industrial medical problems are skin conditions, indicating the large extent of dermal exposure where none is intended.

Testing evaluates the potential occurrence of two different, yet related, endpoints. The broadest application of these is an evaluation of the potential to cause skin irritation, characterized by erythema (redness) and edema (swelling). Severity of irritation is measured in terms of both the degree and persistence of these two characteristics. There are three types of irritation tests, each designed to address a different concern.

1. Primary (or acute) irritation, which is a localized, reversible dermal response

resulting from a single application of, or exposure to, a chemical without the involvement of the immune system.

2.   Cumulative irritation, which is a reversible dermal response that results from repeated exposure to a substance (each individual exposure is not capable of causing acute primary irritation). This is beyond the scope of acute toxicology and will not be discussed in this book.

3.   Photochemically induced irritation, which is a primary irritation resulting from light-induced molecular changes in the chemical to which the skin has been exposed. This is discussed in a later section.

Although most regulations and common practices characterize an irritation that persists 14 days past the end of exposure as other than reversible, the second endpoint of concern with dermal exposure — corrosion — is assessed in separate test designs. These tests start with a shorter exposure period (4 h or less) and then simply evaluate whether tissue has been killed or not (if necrosis is present or not).

It should be clear that if a material has less than severe dermal irritation potential it will not be corrosive and, therefore, not need be tested separately for the corrosion endpoint. The reverse, of course, is not true.

B. Factors Affecting Responses and Test Outcome

The results of these tests (particularly of the irritation test, which identifies a graded response as opposed to the all-or-none response of the corrosivity test) are subject to considerable variability due to relatively small differences in test design or technique. Weil and Scala[43] arranged and reported on the best known of several intralaboratory studies which clearly established the existence of such variabilities. Although the methods most commonly used have given reproducible results in the hands of the same technicians over a period of years[44] and contain some internal controls (i.e., positive and vehicle controls in the PDI) against variabilities in results or false-positives or negatives, it is still essential to be aware of those factors which may systematically alter test results. These factors are summarized below.

In general, any factor that increases absorption through the stratum corneum will also increase the severity of any intrinsic response. Unless these factors mirror potential exposure conditions, this may, in turn, adversely affect the relevance of test results.

The physical nature of solids must be carefully considered in testing and in the interpretation of results. Shape (sharp edges), size (small particles may abrade when rubbed back and forth under the occlusive wrap), and the rigidity (stiff fibers or very hard particles are more physically irritating) of solids may all contribute to enhancing an irritation response.

As the outline for the procedures of the corrosivity test design indicates, solids frequently give different results when tested dry than when tested wet. As a general rule, solids are more irritating if moistened (this tends to enhance absorption). Care should also be taken as to what is used as a moistening agent — some few batches of USP physiological saline (used to simulate sweat) have proven to be mildly irritating to the skin on their own. Liquids other than water or saline should not be used as wetting agent.

Several older regulations require that some or all of the animals in a PDI test group have the test-site skin abraded before test material application. This is based on the assumption that abraded skin is uniformly more sensitive to irritation. Experiments have shown that this is not necessarily true; however, some materials produce more irritation on abraded skin, while others produce less.[44,45]

The degree of occlusion (tightness of the wrap over the test site) also alters percutaneous absorption and therefore irritation. One important quality control issue is achieving a reproducible degree of occlusion in dermal wrappings.

Both the age of the test animal and the location of the application site (saddle of the back vs. flank) can markedly alter test outcome. Both of these factors are also operative in humans,[46] but in these tests, the objective is to remove all such sources of variability. In general, as animal age increases, sensitivity to irritation decreases. And the skin of the middle of the back (other than directly over the spine) tends to be thicker (and therefore less sensitive to irritation) than that of the flanks.

The sex of the test animals can also alter study results, as both regional skin thickness and surface blood flow vary between males and females.[65]

Finally, the single most important (yet also most frequently overlooked) factor that may influence the results and outcome of acute studies is the training of the technical staff. In determining how test materials are prepared and applied, and in how results are "read" against a subjective scale, both accuracy and precision are extremely dependent on the technicians involved. To achieve the desired results, initial training must be careful and all inclusive. Use of a set of color photographic standards as a training and reference tool is strongly recommended; such standards should clearly demonstrate each of the grades in the Draize dermal scale. Some form of regular refresher training should also be exercised, particularly in scoring of results.

It should be recognized that the dermal irritancy test is designed with a bias to preclude false-negatives and, therefore, tends to exaggerate results in relation to what would happen in humans. Findings of negligible (or even very low range mild) irritancy should therefore not be of concern unless the product under test is to have large scale and prolonged dermal contact.

Unlike the primary dermal irritancy test, the results from the corrosivity test should be taken at face value. There is some lab-to-lab variability and the test does produce some false-positives (usually with severely irritating compounds), but does not produce false-negatives.

C. Interpretation and Use of Data and Test Results

Although in a sense, irritation and corrosion are parts of a single spectrum of results, they have very different impacts on the development and handling of products. Corrosion is generally not acceptable when there will be substantial, unavoidable, skin contact — such as with topical pharmaceuticals, cosmetics, and some consumer product (i.e., personal use) products. If a material is rapidly corrosive (such as hydrofluoric acid[47]), it would also not be acceptable for any form of consumer product or most forms of agricultural use. Dermally corrosive materials usually do not need to be evaluated for eye irritation (as it can be reasonably assumed to be at least a severe irritant in the eye).

When evaluating and making decisions based on externally derived dermal irritation data from animal tests, it is extremely useful to have some form of benchmark as part of those data. That is, if in the same laboratory, saline is a nonirritant and 0.1% sodium lauryl sulfate is a severe irritant, one can better understand the relevance of an evaluation of a *de novo* compound to the human condition. Such reference standards should be required on some regular basis from external laboratories and can frequently be incorporated into a test series by adding additional patches. Appropriate standard compounds should be selected based on the needs of the data user. Table 3 presents the summary matrix for use of dermal irritation and corrosion data by different communities of interest.

For the pharmaceutical industry the concern is for agents that will be applied topically. As these will certainly have direct (and usually repeated) application to the skin of humans, only moderate irritation (and that only if there is a marked therapeutic benefit) can be tolerated. Frequently, multiple formulations of the same basic product will be compared to select one or more with the best safety profile.

Cosmetic products call for repeated direct application to the skin or the close proximity of the skin (hair, eyebrows, nails, etc.) with no therapeutic benefit. Only truly mild irritants might possibly be tolerated here, with the preference being for nonirritants. For both the cosmetic and pharmaceutical industries, a test system with high accuracy in identifying weak irritants is called for. It should be noted that the special case of safety concerns for cosmetics has recently been the subject of an entire book.[48]

Personal use consumer products (such as hand and body soaps) are close to cosmetics in that only some irritation can be tolerated, but this tolerance is usually at the level of moderate irritation because exposure periods are generally short as opposed to being continuous. Again, it is frequently necessary to compare multiple formulations to determine relative irritation effect. Household use consumer products, meanwhile, rarely require comparison of formulations.

Household use consumer products, agricultural chemicals, and industrial chemicals generally require only that strong or severe irritants (and, of course, corrosives) be identified, unless they are to be shipped, in which case a determination for DOT labeling requirements will need to be made. This is because there is no intended human dermal contact, so that the information is only for use in setting precautions and in case of accidents. In these cases, false-positives are considerably more tolerable. And there is again an obvious role for some of the current *in vitro* screens which can seemingly identify strong irritants and corrosives effectively.

D. Problems in Testing (and Their Resolutions)

Some materials, by either their physicochemical or toxicological natures, generate difficulties in the performance and evaluation of these tests. The most commonly encountered of these problems are presented below.

Compound volatility — The evaluation of the potential irritancy of a liquid that

Table 3

Matrix of Intended Product Use vs. Required Test Features for Dermal Irrigation and Corrosion

		Features		
Community of interest (intended product use)	Desired sensitivity (lowest level of irritation/corrosion that it is essential to detect)	Need to compare multiple formulations of same basic chemical entity	Acceptable incidence of	
			False-positives	False-negatives
Pharmaceutical (topical agents)	Moderate	Essential	Low	Very low
Cosmetic	Mild	Essential	Low	Very low
Consumer product (personal use)	Moderate	Important	Low	Very low
Consumer product (household use)	Strong	Minimal	Low	Low
Agricultural chemical	Strong/severe	No	Some	Low
Industrial chemical	Strong/severe	No	Some	Low

has a boiling point between room temperature and the body temperature of the test animal is sometimes required; as a result, the liquid portion of the material will evaporate before the end of the testing period. There is no real way around the problem; one can only make clear in the report on the test that the traditional test requirements were not met, although an evaluation of potential irritant hazard was probably achieved (for the liquid phase would also have evaporated from a human that it was spilled on).

Pigmented material — Some materials are strongly colored or act to discolor the skin at the application site. This makes the traditional scoring process difficult or impossible. Removal of the pigmentation with a solvent can be attempted; if successful, the erythema can then be evaluated. If this fails or is unacceptable, the skin can be felt to determine if there is warmth, swelling, or rigidity — all secondary indicators of the irritation response.

Wrong test — Sometimes there is severe necrosis in a primary dermal irritation study or irritation (but no corrosion) in a corrosivity test. In the former, the material can be graded as a severe irritant, but cannot be evaluated as to its (as defined by regulation) corrosive potential. In the latter case, the material can be judged to not be a corrosive and to have irritating potential. The difference is, of course, in the length of the exposure period (although more recent EPA guidelines suggest making the tests both 4 h long).

Systemic toxicity — On rare occasions, animals in a dermal irritation study die very rapidly after test material application. Though it is not possible to evaluate potential irritancy in such cases, the results are even more important in the hazard-evaluation process, and such findings should be rapidly communicated to those producing and handling the subject material.

Concentration dependency — There are occasions when the responses to ques-

Table 4

Summary of Lesion Incidence in Rabbits Exposed to Dilute Hydrofluoric Acid (HF) Solutions or Distilled Water for Varying Periods of Time

Treatment	Exposure	No. of animals exposed	No. of animals w/ lesions	Mean no. of lesions per animal[a]	Size range of lesions (cm)
Distilled water	60 min	5	0	—	—
2% HF (neutralized)	60 min	5	0	—	—
2% HF	60 min	5	3	3.3(2-6)[b]	0.4 × 0.4-3.5 × 1.0
2% HF	30 min	5	4	2.8(1-6)	0.1 × 0.1-0.6 × 0.3
2% HF	15 min	5	3	2.7(2-4)	0.1 × 0.1-1.3 × 1.0
2% HF	5 min	5	2	1.5(1-2)	0.1 × 0.1-0.6 × 0.4
2% HF	1 min	5	0	—	—
0.5% HF	60 min	5	2	2.0(1-3)	0.2 × 0.2
0.5% HF	30 min	5	1	4.0	0.1 × 0.1
0.5% HF	15 min	5	1	3.0	<0.1 cm-0.2 × 0.2
0.5% HF	5 min	5	3	1.7(1-2)	0.1 × 0.1-1.0 × 0.6
0.1% HF	60 min	5	1	1.0	0.5 × 0.5
0.1% HF	30 min	5	1	1.0	0.2 × 0.1
0.1% HF	15 min	5	2	1.5(1-2)	0.1 × 0.1
0.1% HF	5 min	5	2	1.5(1-2)	<0.1 cm-0.1 × 0.1
0.01% HF	60 min	5	2	2.0	0.1 × 0.1-0.2 × 0.2
0.01% HF	30 min	5	0	—	—
0.01% HF	15 min	5	2	3.0(2-4)	0.1 × 0.1-1.0 × 0.6
0.01% HF	5 min	5	2	2.5(1-4)	0.1 × 0.1

[a] Animals with lesions.
[b] Range.

From M.J. Derelanko, S.C. Gad, F.A. Gavigan, and B.J. Dunn, *Ocul. Dermal. Toxicol.* 4(2), 74 (1985). With permission.

tions asked about the dermal irritancy or corrosivity of a material are more complicated than a yes or no. These are generally when it is known that the material, at some concentration, has the given effect and there is a need to know what a nonirritating or noncorrosive concentration might be.

For example, a polymer produced for use in cosmetics contains a certain level of unreacted monomer (such as acrylic acid) that is known to be irritating. At additional cost, the levels of residual monomer could be reduced. How much must the monomer level be reduced to make the polymer nonirritating? In this case, a dose response study could be designed and conducted to evaluate different concentrations (at different sites on the same animal). A single set of six animals could be used to evaluate as many as four concentrations, each being patched at four separate sites.

In a second example, the desire is to identify a level of exposure of a material that is not corrosive. Table 4 (from Derelanko et al.[47]) demonstrates an approach to this problem of defining exposure limits to a known hazardous case. Both concentration and time of exposure can be varied to produce two sets of dose responses.

It is equally important to evaluate and to consider the persistence and reversibility of the irritation. Obviously, the degree of concern is considerably greater for a material with a sustained irritation effect.

E. Interpretation and Use of Dermal Sensitization Data

Evaluating materials to determine if they could be delayed contact dermal sensitizers in humans is different on a number of grounds from the other predictive test problems examined so far. These differences are primarily due to the immune system (the mechanistic basis for this adverse or undesirable response) functions. Much of that functioning is discussed in detail in Chapter 5, but a brief summary is appropriate here.

Bringing about this Coombs type IV hypersensitivity response[66] (which we will call "sensitization" for short) requires more than a single exposure to the causative material, both in humans and in test animals. To a much greater degree than irritation responses, sensitization occurs in individuals in an extremely variable manner. A portion of the human population is considerably more liable to be sensitized, while others are infrequently affected. And the response, once sensitization is achieved, becomes progressively worse with each additional exposure. All three of these characteristics are due to the underlying mechanisms for the response, and influence the manner in which tests are conducted. These factors mean that *in vivo* test systems require multiple exposures of animals (unlike all of the other systems in this volume, except for photosensitization, which is related to sensitization) and tend to underpredict the potential for an adverse response in those individuals who are most susceptible to sensitization. Concern must be addressed to the area of response to repeated exposures of even minimal amounts of material in susceptible individuals producing strikingly adverse responses.

A number of factors influence the potential for a chemical to be a sensitizer in humans and, in turn, also influence the performance of test systems. These are summarized in Table 5. Various test systems manipulate these in different ways.

There are a number of references which explore and discuss the underlying immune system mechanisms involved in greater detail, such as Gibson et al.[49]

F. Objectives and General Features of Test Systems

Given the considerations of mechisms, practicality, and the degree of concern about protecting people, the desired characteristics of a sensitization test should include the following:

1. Be reproducible
2. Involve fairly low technical skills so that it may be performed as a general laboratory test
3. Involve standard animals and equipment

Table 5
Factors Influencing Delayed-Type Sensitization Responses

1.	Percutaneous absorption of agent
2.	Genetic status of host
3.	Immunological status of host
4.	Host nutrition
5.	Chemical and physical nature of potential sensitizing agent
6.	Number, frequency, and degree of exposures of immune system to potential antigen
7.	Concurrent immunological stimuli (such as adjuvants, inoculations, and infections). System can be "up modulated" by mild stimuli or overburdened by excessive stimulation
8.	Age, sex, and pregnancy (by influencing factors 1, 3, and 4 above)

4. Use relatively small amounts of test material
5. Be capable of evaluating almost any material
6. Be sensitive enough to detect weak sensitizers (those that require extensive exposure to sensitize other than the most sensitive individuals)
7. Predict the relative potency of sensitizing agents accurately

Several of these desired characteristics are mutually contradictory; as with most other test systems, the methods for detecting dermal sensitization each incorporate a set of compromises.

All the *in vivo* tests have some common features, however. The most striking is that they involve at least three (and frequently four) phases — they are multiphasic. These phases are, in order, the irritation/toxicity screen, the induction phase, the challenge phase and (often) the rechallenge phase.

Irritation/toxicity screen — All assays require knowledge of the dermal irritancy and systemic toxicity of the test material to be used in the induction, challenge, and rechallenge. These properties are defined in this pretest phase. Most tests include mild irritation in the induction phase, but no systemic toxicity. Generally, a nonirritating concentration is required for the challenge and for any rechallenge, as having irritation present either confounds the results or precludes a valid test. Even a carefully designed screen does not always proivde all of the needed guidance in selecting usable test concentrations. During this phase solvent systems are also selected.

Induction phase — This requires exposing the test animals to the test material several times over a period of days or weeks. A number of events must be accomplished during this phase if a sensitization response is to be elicited. The test material must penetrate through the epidermis and into the dermis. There, it must interact with dermis protein. The protein-test material complex must be perceived by the immune system as an allergen. Finally, the production of sensitized T cells must occur. Some assays enhance the sensitivity of the induction phase by compromising the natural ability of the epidermis to act as a barrier. These enhancement techniques include irritation of the induction site, intradermal injection, skin stripping, and occlusive dressings. In contrast, events such as the development of a scab over the induction site may reduce percutaneous absorption. The attention of the immune system can be drawn to the induction site by the intradermal injection of some form of adjuvant (which serves as a mild immunological stimulant).

Challenge phase — This consists of exposing the animals to a concentration of the test material which would not normally be expected to cause a response (usually an erythema type response). The responses in the test animals and of the control animals are then scored or measured.

Rechallenge phase — This is a repeat of the challenge phase and can be a valuable tool if used properly. Sensitized animals can be rechallenged with the same test material at the same concentration used in the challenge to assist in confirming sensitization. Sensitized animals can be rechallenged with different concentrations of the allergen to evaluate dose response relationships. Animals sensitized to an ingredient can be challenged by a formulation containing the ingredient to evaluate the potential of the product to elicit a sensitization response under adverse conditions. Conversely, animals that responded (sometimes unexpectedly) to a final formulation can be challenged with a formulation without the suspected sensitizer or to the ingredient which is suspected to be the allergen. Cross-reactivity can be evaluated. That is, the ability of one test material to elicit a sensitization response after exposure in the induction phase to a different test material. A well-designed rechallenge is important and should be considered at the same time that the sensitization evaluation is being designed since the rechallenge must be made within 1 to 2 weeks after the primary challenge. Unless plans have been made for a possible rechallenge, a test material may have to be reformulated or additional pure ingredient may have to be obtained and additional irritation/toxicity screens may need to be conducted before the rechallenge can be run. The ability of the sensitized animals to respond at a rechallenge can fade with time; this is the reason for the rechallenge being run shortly after the challenge. In addition, some tests use sham-treated controls (i.e., vehicle only) and these must be included during the induction phase. An additional piece of information which must be kept in mind when evaluating a rechallenge is that the animal does not differentiate between the induction exposure and the challenge exposure. If an assay uses three induction exposures and one challenge exposure, then at the rechallenge, the animal has received four induction exposures. This "extra" induction may serve to strengthen a sensitization response.

G. Test System Manipulation

It should be obvious at this point that there are a number of steps which can be taken to either up- or down-modulate (i.e., to either increase or decrease the sensitivity of) the *in vivo* test system. Foremost among these is increasing the rate of percutaneous absorption. Factors that will increase absorption (and techniques for achieving them) include the following:

1. Increase the surface area of solids.
2. Hydrate the region of skin exposed to the chemical. This can be done by the wetting of solids and the occlusive wrapping of application sites.

3. Irritate the application site.
4. Abrade the application site.
5. Inject the test material intradermally (if possible).
6. Properly select the solvent or suspending system. (See Christensen et al.[50] for a discussion of the effect of vehicle or even a strong sensitizer).
7. Remove part or all of the "barrier layer" (stratum corneum) by stripping the application site with tape.
8. Increase the number of induction applications.

Although it is not a factor which increases percutaneous absorption, mildly stimulating the immune system of test animals (by such means as injecting FCA alone or FCA blended with the test material) also increases responsiveness of the test system.

Also, it is generally believed that using the highest possible test material concentrations (mildly irritating for induction, just below irritating for challenge) will guarantee the greatest possible sensitization response and, therefore, also will serve to universally increase sensitivity. There are increasing numbers of reports, however,[51,52] indicating that this is not true for all compounds and that a multiple dose (i.e., two or more applications) study design would increase sensitivity. Such designs, however, would also significantly increase the time required and the cost to complete the test.

Concurrent or frequent positive and negative controls are essential to guard against test system failure. Any of these test systems should be sensitive enough to show a solution of 0.05% dinitrochlorobenzene (DNCB) in 70% ethanol to be a strong sensitizer.

H. Making Hazard Decisions for Humans: Problems and Solutions with Current Test Systems

Virtually all the general problems associated with the use of the current test systems can be thought of as being an aspect of "what do the results mean in terms of hazard to people?" These problems may arise for a number of reasons, but two major factors must always be considered. First, as a population, humans may have some individuals who are more sensitive than an animal test system. In trying to reduce this gap, we need to develop tests systems which give the best possible estimate of relative hazard (i.e., what portion of the human population will be sensitized, and how easily) in people.

Second, the material which is being evaluated in these models, if found positive, may not be the only material of concern. Rather, it may be a component of a mixture, with the mixture (such as a cosmetic) having actually been tested. Similarly, if a given chemical is found to be a sensitizer, there may be other, structurally related, compounds that may evoke a similar response in those already sensitized to the given chemical being tested (i.e., cross-sensitization).

Interpretation problems may arise when animal sensitization test data must be related to potential hazards in humans. On one end of the scale, a negative finding does not guarantee that a material will not be a sensitizer in humans, although most

Table 6
Sensitization Severity Grading Based on Incidence of Positive Responses

Sensitization rate (%)	Grade	Potency classification
>0 to 8	I	Weak
9 to 28	II	Mild
29 to 64	III	Moderate
65 to 80	IV	Strong
81 to 100	V	Extreme

From A.M. Kligman, *J. Invest. Dermatol.* 47, 393 (1966). With permission.

investigators would agree that it is unlikely that such a material would be other than a weak or mild sensitizer.

On the other hand, however, it is not always as clear what a finding of a strong or extreme sensitization means in each of these assays. For example, the guinea pig maximization test (GPMT) and split adjuvant tests, plus related maximization-style tests, tend to overpredict potency and what action a manufacturer of a material so tested may take, with respect to the test results, will usually be a reflection of both how the material is to be used, and how much of a risk it will present to humans.

Accordingly, there remain two options: (1) human patch-style tests; and (2) potency evaluations using an animal model. In patch-style tests using human subjects with large enough (100 to 200 people) test groups of a representative nature, the results will give an understanding of what to expect in humans.[7,67] This approach, however, is expensive and has both ethical and liability concerns of its own.

The second approach is to use a method that allows evaluation of potency in an animal model and extrapolate the results to humans. As both Gad et al.[51,53-55] and Thorne et al.[52] have pointed out, such potency evaluations require dose response testing and there are a number of considerations which should be taken into account. Current practice for predicting relative hazard using data obtained using an animal model is a scheme based on incidence of response in a group of animals, as shown in Table 6.

1. Potency

Starting with several assumptions, data from four animal test systems have been used to evaluate one possible procedure for ranking the potency of known sensitizers. These assumptions were

1.   As absolute (100%) responses do not give actual data points (rather, they define a portion of an unlimited response region), only partial (1 to 99%) response data from animal tests can be used to predict potency.
2.   As the probit transformation has already been shown to linearize sensitization

dose response data, probit values can be used to adjust different partial dose response values to a comparable basis. This transformation is most stable in the central region (16 to 84%) of the response range, so partial responses in this range are most desirable.

3. As individual molecules of material evoke the response being both measured and predicted, doses (or exposures) should be expressed on a molecular basis. Accordingly, data should be adjusted for molecular weight.

A method for calculating a potency index should have at least six characteristics.

1. It should be relatively easy to perform, requiring little more than the data and a calculator.
2. It should incorporate data on test concentration used, incidence of response achieved, and molecular weight of the test material.
3. The resulting potency index numbers should cover a compact scale — from 0 to 10 — and not include negative numbers nor cover more than one order of magnitude.
4. There should be a positive correlation between potency and the index number (i.e., more potent compounds should have higher index numbers).
5. The results should serve to separate materials into clear clumps or clusters that lend themselves to classification of materials into categories.
6. Data from various test systems should produce similar classification results for a compound and should predict results that are accurate in humans.

An earlier attempt at such an index calculating method[54] produced results which were promising, but clearly did not fit desired characteristics 3 and 4. This has been modified to use the formula[55]

$$\text{Potency index (PI)} = \text{Log}\left[\frac{1000\,(\text{probit of response incidence})}{(\text{test concentration})\,(\text{molecular weight})}\right]$$

where test concentration is expressed as a decimal fraction with 100% equal to 1.0. This produces results that generally fulfill the desired design characteristics. A classification scheme based on the resulting index has been devised, with the scale:

Class I	—	PI	>	4.0			Severe
Class II	—	4.0	>	PI	>	3.0	Strong
Class III	—	3.0	>	PI	>	2.0	Moderate
Class IV	—	2.0	>	PI	>	1.0	Mild
Class V	—	1.0	>	PI	>	0	Weak or Questionable

where PI = Potency Index

2. Cross-Sensitization

A frequent situation is that one member of a structural series of compounds is tested and found to be a sensitizer. If other members of the structural series will evoke a positive response in those that have been sensitized, this broader response is called "cross-sensitization". This occurs because the structures of these materials complexed with a protein are not distinguished as different by the "educated" surveillance lymphocytes.

Any of the animal tests described here can be modified to see if cross-sensitization occurs among members of a series. The test is conducted with multiple groups of animals. Those animals which are successfully sensitized are then rechallenged with other members of the class.

3. Mixtures

Mixtures become a particular problem in sensitization testing because, frequently, a complex mixture must be evaluated in an animal test system; then, if it is found to be a sensitizer, which component is the case of the positive response must be determined. If such a component can be identified, it is frequently possible to reformulate the mixture to serve the need without the problem component.

Such components can be identified by continued testing in a set of animals previously sensitized to the mixture as a whole. Groups of positively sensitized animals are rechallenged with separate samples of different suspect components to identify that which evokes a positive response. The guinea pig methods offer some advantage here, in that multiple components may be simultaneously evaluated on different sites of the same animal.

Table 7 presents the matrix summarizing the needs of the major intended product-use groups. Because once induced, subsequent sensitization responses can be evoked by small exposure and tend to become more severe, the need to accurately identify those materials with even relatively low potentials to induce sensitization is critical for those products with high consumer dermal exposures. This is reflected in the matrix, as is the fact that as such agents do not generally enter the market, it is not usually important to identify other materials for which a cross-sensitization potential exists. For industrial chemicals, however, as from time-to-time sensitizations will develop in workers, it then becomes very important to determine which materials may present a cross-sensitization hazard.

The need for both high sensitivity and high specificity in identifying sensitizers for pharmaceuticals, cosmetics, and some consumer products suggests that screening-type tests are not generally appropriate for these applications and that it is very important to have an accurate means of evaluating potency relative to known human agents.

VII. PHOTOSENSITIZATION AND PHOTOTOXICITY

Phototoxicity and photosensitization both require that a chemical or one of its metabolites absorb light energy to be activated. From that point on, however, they

Table 7
Matrix of Intended Product Use vs. Required Test Features for Dermal Sensitization

	Features			
Community of interest (intended product use)	Desired sensitivity (lowest level of irritation/corrosion essential to detect)	Need to compare multiple formulations of same basic chemical entity	Acceptable incidence of	
			False-positives	False-negatives
Pharmaceutical (topical agents)	Weak/mild	Moderate	Very low	Very low
Cosmetic	Weak	Low	Very low	Very low
Consumer product (personal use)	Mild	Low	Very low	Very low
Consumer product (household use)	Mild	Moderate	Low	Low
Agricultural chemical	Moderate	High	Low	Low
Industrial chemical	Moderate	High	Low	Low

are different, as photosensitization is immunologically mediated but phototoxicity is not. Because exposure to sunlight can (long after any exposure to the chemical agent has ceased) continue to evoke drastic adverse dermal and systemic effects once the light-mediated dermal toxicities are initiated, this class of toxicities represents a particular concern (and call for extreme hazard management) when identified.

In 1939, Epstein,[57] in studying photosensitization due to sulfonamides, clearly differentiated the two separate conditions of photosensitization and phototoxicity. This differentiation is presented in Table 8.

As with each of the other classes of toxicities evaluated here, it should be recognized that there are ways of manipulating (with and without willful intent) test results.

A. Factors Influencing Phototoxicity/Photosensitization

Many factors can influence an agent acting in either of these modes. In addition to all those factors previously reviewed in the chapter on sensitization, there are also the following

1. The quantity and location of photoactive material present in or upon the skin.
2. The capacity of the photoactive material to penetrate into normal skin by percutaneous absorption as well as into skin altered by trauma, such as maceration, irritation, or sunburn
3. The pH, enzyme presence, and solubility conditions at the site of exposure
4. The quantity of activating radiation to which the skin is exposed
5. The capacity of the spectral range to activate the materials on or within the skin
6. The ambient temperature and humidity

Table 8
Comparison of Phototoxic and Photosensitization Reactions

Reaction	Photo-toxic	Photo-sensitization
Reaction possible on first exposure	Yes	No
Delay period necessary after first exposure	No	Yes
Chemical alteration of photosensitizer	No	Yes
Covalent binding with carrier	No	Yes
Clinical changes	Usually like sunburn	Varied morphology
Flares at distant previously involved sites possible	No	Yes
Persistent light reaction can develop	No	Yes
Cross-reactions to structurally related agents	Infrequent	Frequent
Broadening of cross-reactions following repeated photopatch testing	No	Possible
Concentration of drug necessary for reaction	High	Low
Incidence	Usually relatively high (theoretically 100%)	Usually very low (but theoretically could reach 100%)
Action spectrum	Usually similar to absorption	Usually higher wavelength than absorption spectrum
Passive transfer	No	Possible
Lymphocyte stimulation test	No	Possible
Macrophage migration inhibition test	No	Possible

7.   The thickness of the horny layer
8.   The degree of melanin pigmentation of the skin

Basically, any material that has both the potential to absorb UV light and the possibility of dermal exposure or distribution into the dermal region should be subject to some degree of suspicion as to potential phototoxicity. Many of agents have been identified as phototoxic or photoallergenic agents. Of these, tetrachlorosalicylanide (TCSA) is most commonly used as a positive control in animal studies.

B. Photosensitization

Severe, disabling dermal reactions have resulted from repeated exposure to certain agents, such as halogenated salicylanilides and related compounds once used as antimicrobial agents in cosmetics and toiletries. The original observation of these effects was made on perhaps the most potent agent, 3,3′,4′,5-tetrachlorosalicylanilide. Other major contact photoimmunological agents in humans include 3,4,5-tribromosalicylanilide (TBS). 4′,5-dibromosalicylanilide (DBS), and the chemically related bithionol are agents that can cross-sensitize with themselves and other compounds. Since safer alternative antimicrobial agents are available, the halogenated salicylanilides (i.e., TCSA, TBS, DBS) and bithionol have been prohibited from use in cosmetics. Likewise, other agents identified as photosensitizers would be unacceptable in most uses where there was any degree of regular human exposure. Such effects, it should be noted, have also been seen with some orally administered agents. Such compounds capable of inducing photosensitivity are encountered relatively rarely compared with those producing ordinary contact sensitization; and, except for certain chemical types, they do not appear to pose a major practical problem insofar as the usual topical agents are concerned. A thorough safety evaluation program, even without direct tests for photosensitization, is likely to detect strong photosensitizers, since in such a program a relatively large number of people with free exposure to light would be in contact with the agent during normal use. While not giving the extra assurance that controlled specific tests would, these exposures do provide a sizable opportunity for photosensitivity to display itself, if it should occur.

The existing human and animal test systems, as presented in Chapter 6, do present some difficulties in interpretation. This is because the endpoint of effect (erythema) can be caused by photosensitization and/or by phototoxicity, delayed contact sensitization, or by primary irritation. Dissecting which of these is operative calls for careful test conduct and close consideration of the interpretation matrix presented in Table 9.

C. Strengths and Weaknesses

The most common animal test, the Armstrong assay,[68] has been found to give responses in the guinea pig that are qualitatively consistent with what has been found in humans; positive responses for 6-methyl coumarin and musk ambrette. One major disadvantage is that the procedure is time consuming, with six induction exposures; additional work might demonstrate that fewer exposures will yield the same results. The procedure is very stressful for the animals because of the injection of adjuvant and the multiple skin strippings and depilations. As with any assay involving the intradermal injection of adjuvant, there is often a problem with using the results of the irritation screen in naïve animals to accurately predict the results that will be seen in the sham controls at the challenge. If the material being tested is a nonirritant or if the concentration of an irritant is far below the irritating concentration, then the screen does an adequate job of predicting the background irritation level in the challenge controls. However, if a slightly nonirritating concentration of an irritant is used, then the screen often underpredicts the irritation response and a high background level of irritation is observed at the challenge in the sham controls. The

Table 9
Dermal Erythema Interpretation in Photosensitization

Test group		Control Group		
Irradiated	Non-irradiated	Irradiated	Non-irradiated	Interpretation
0	0	0	0	No contact photosensitization No contact sensitization No phototoxicity No primary dermal irritation
1-4	0	0	0	Contact photosensitization No contact sensitization No phototoxicity No primary dermal irritation
1-4	1-4	0	0	No contact photosensitization (unless sum of scores exceeds 2× of nonirradiated scores) Contact sensitization (nonphoto) No phototoxicity No primary dermal irritation
1-4[a]	0	1-4	0	Contact photosensitization[a] No contact sensitization Phototoxicity No primary dermal irritation
1-4	1-4	1-4	1-4	No contact photosensitization No contact sensitization No phototoxicity Primary dermal irritation

[a] Contact photosensitization only if the incidence and degree of positive responses of the test group are judged to clearly exceed those of the control group.

interpretation of the results of the challenge becomes difficult. The use of animals in the irritation screen which have had a prior injection of adjuvant might provide a viable alternative and reduce the number of times that rechallenges must be done because of high background levels of irritation. The Armstrong assay was designed to evaluate materials for their photoactivated sensitization potential and not for their potential to be nonphotoactivated dermal sensitizers. At this time, there are no background data that allow for properly positioning results of the Armstrong assay with regard to human risk if the assay indicates that a test material is a sensitizer or that a material is both a sensitizer and a photoallergen. Thus, it is highly recommended that a "standard" sensitization assay that can be related to humans be run before, or in conjunction with, the photosensitization assay. The use of a subjective grading system can be a source of significant verification.

D. Phototoxicity

Phototoxicity is a nonimmunological, light-related irritation that would occur in (almost) everyone following skin exposure to light of sufficient intensity and wavelength along with a light-activated chemical in adequate amount. It should be noted that the chemical exposure may have occurred at some earlier time, but the active chemical moiety remains available in the skin for photoactivation. Phototoxicity is neither as severe, nor as persistent, as photosensitization, so agents found to be active will not necessarily be unusable. What happens instead is that formulation and use of products which must contain phototoxic agents are controlled so that relatively few people develop dermatitis. There has been minimal impetus in the industrial sector to determine the mechanism involved in most cases.

The development of suitable animal models followed the development of a satisfactory human test model. The basic requisites are nonerythrogenic light (320 nm) and percutaneous penetration of the phototoxic agent. However, phototoxic responses, although uncommon, can result from oral as well as topical exposure.

E. Systemic Dermal Toxicity

Generalized dermal toxicity is systemic toxicity in which the skin is the portal of entry. The usual animal for testing agents for this type of toxicity is the albino rabbit, although the rat is becoming more commonly used. As with all animal tests, the basic problem is associated with the interpretation of the experimental data in terms of humans. In the widely used subchronic test involving 90 days of repeated exposure, lack of evidence of injury or toxicity affords reassurance that the material under investigation would pose no problem of percutaneous toxicity under conditions of use involving reasonable skin exposure (except for possibly carcinogenicity). If, however, evidence of systemic effects is observed, then the problem becomes essentially identical with that of assessing hazard of any other systemic toxicant. Provided the utility of the material warrants it, this would involve establishment of a "no-effect dose", a specific study to determine whether the observed effect has a counterpart in humans and development of other pertinent information, including relative ease of passage of the substance through the skin of the rabbit and of humans.

One factor that is sometimes overlooked, but which deserves consideration in acute percutaneous test procedures, is the effect of stress on the animal. Rabbits are especially susceptible to the stress of severe irritation or prolonged immobilization, and this factor can affect mortality to a marked degree (author's personal experience and unpublished data). Consequently, it is important in all (but particularly acute) percutaneous systemic toxicity work to include adequate vehicle and technique controls.

F. Percutaneous Absorption

As detailed separately in several of the chapters of this book, there are a number of specific test designs (both *in vivo* and *in vitro*) for assessing the extent to which materials are absorbed through the skin. This can be of concern in both a negative and a positive sense. In the first case, if a pharmaceutical is intended to be administered by the dermal route, but to treat a systemic disease, then a high degree of predictable absorption through the skin and into the systemic circulation is highly desirable (and in fact, to at least some degree an essential component of the agent and its formation). In the second case, which is that of major concern to toxicologists, more than minor (1 to 2%) absorption through the skin serves to markedly complicate the hazard evaluation problem.

If a material is applied to or regularly comes in contact with the skin and is not appreciably absorbed, then concerns as to hazard are limited to local effects (irritation, corrosion, sensitization, etc.). However, if there is a significant degree of systemic absorption of such agents, then a hazard assessment must also include some evaluation of systemic effects just as with other routes of administration. Such evaluations should include identification of target organs and no effect levels, plus at least some indication of the dose response curve.

Chronic dermal exposures to chemicals, it should be realized, do not fit into the above scheme, and presents the need for special evaluations, as presented in the chapter on dermal carcinogenicity.

VIII. SUMMARY

The first principle in hazard assessment is to collect test data as near to the real life situation you are concerned about as possible, i.e., the nearer the test model is to man, the better will be the quality of any prediction of any potential hazard which is derived from the data obtained from the model.

The second principle is to be able to clearly translate toxicity to hazard, and to then be able to manage such hazards. It is essential in hazard assessment to know how the chemical agent is to be used and the marketplace it is to be a part of. It is hoped that this chapter makes these relationships clear.

Finally, alternatives of both *in vitro* and *in vivo* test methods are in the process of development for almost all the different endpoints presented in this chapter. Some of these test methods have great promise, and could be used as screens for if not to replace some of the tests presented. Complete replacement is, however, not near at hand, particularly for the more complicated *in vivo* endpoints.

REFERENCES

1. L. Glass, *J. Theor. Biol.* 54, 85, (1975).
2. A.D. Gordon, *Classification.* Chapman and Hall, New York (1981).
3. EPA, Guidelines for Carcinogen Risk Assessment *Fed. Reg.* 51(185) (September 24), 33992 (1986).
4. M.E. Wolff, *Burger's Medicinal Chemistry.* John Wiley & Sons, New York, Parts II (1979) and III (1981).
5. N.H. Proctor and J.P. Hughes, *Chemical Hazards of the Workplace.* J.B. Lippincott, Philadelphia (1978).

6. R.E. Gosselin, R.P. Smith, and H.C. Hodge, *Clinical Toxicology of Commercial Products,* 5th ed. Williams and Wilkens, Baltimore (1984).
7. E. Cronin, *Contact Dermatitis.* Churchill-Livingstone, Edinburgh (1980).
8. National Institute for Occupational Safety and Health, *NIOSH Criteria for a Recommended Standard for Occupational Exposure to . . .* Cincinnati, OH, Department of Health, Education and Welfare.
9. National Institute for Occupational Safety and Health. *NIOSH Current Intelligence Bulletins.* Cincinnati, OH, Department of Health, Education and Welfare.
10. N.I. Sax, *Dangerous Properties of Industrial Materials,* 6th ed. Van Nostrand Reinhold, New York (1985).
11. American Conference of Governmental Industrial Hygienists (ACGIH), *Documentation of the Threshold Limit Values,* 5th ed. Cincinnati (1986).
12. M. Sittig, *Handbook of Toxic and Hazardous Chemicals.* Noyes, Park Ridge, NJ (1981).
13. American Industrial Hygiene Association, *Hygienic Guide Series,* Vol. I and II. Akron (1980).
14. A.J. Finkel, *Hamilton and Hardy's Industrial Toxicology,* 4th ed. John Wright PSG, Boston (1983).
15. M. Windholz, *The Merck Index,* 10th ed. Merck and Company, Rahway, NJ (1983).
16. F. Mackinson, National Institute for Occupational Health and Safety/Occupational Safety and Health Administration. *Occupational Health Guidelines for Chemical Hazards.* Department of Health and Human Services (NIOSH)/Department of Labor (OSHA) DHHS No. 81-123, Washington, D.C., Government Printing Office (1981).
17. D.G. Clayton and F.E. Clayton, *Patty's Industrial Hygiene and Toxicology,* 3rd Revised ed. Vol. 2A, 2B, and 2C. John Wiley & Sons, New York (1981).
18. E.R. Barnhart, *Physicians Desk Reference,* Medical Economics Company, Oradell, NJ (1987).
19. National Institute of Occupational Safety and Health, *Registry of Toxic Effects of Chemical Substances,* 11th ed. Vol. 1 to 3.   Washington, D.C., Department of Health and Human Services DHHS No. 83-107, 1983 and RTECS Supplement DHHS 84-101 (1984).
20. W.M. Grant, *Toxicology of the Eye,* 2nd ed. Vol. I and II. Charles C Thomas, Springfield, IL (1974).
21. National Library of Medicine, Office of Inquiries and Publications Management, 8600 Rockville Pike, Bethesda, MD, 20209.
22. P. Wexler, *Information Resources in Toxicology.* Elsevier, New York (1982).
23. C.M. Parker, *Handbook for Product Safety Evaluation.* S.C. Gad, Ed. Marcel Dekker, New York, 23 (1987).
24. A.W. McCally, A.G. Farmer, and E.C. Loomis, *JAMA* 101, 1560 (1933).
25. A.R. Latven and N. Molitor, *J. Pharm. Exp. Ther.* 65, 89 (1939).
26. I. Mann and B.D. Pullinger, *Proc. R. Soc. Med.* 35, 229 (1942).
27. J.S. Friedenwald, W.F. Hughes, and H. Herrmann, *Arch. Ophthalmol.,* 31, 279 (1944).
28. J.H. Draize, G. Woodard, and H.O. Calvery, *J. Pharmacol. Exp. Ther.* 82, 377 (1944).
29. R. Koch, *Dtsch. Med. Wochenschr.* 16, 1029, (1890).
30. K. Landsteiner and J. Jacobs, *J. Exp. Med.* 61, 643 (1935).
31. K. Landsteiner and M.W. Chase, *Proc. Soc. Exp. Biol. Med.* 49, 288 (1942).
32. E.V. Buehler, *Toxicol. Appl. Pharmacol.* 6, 341 (1964).
33. E. Cronin and G. Agrup, *Br. J. Dermatol.* 82, 428 (1970).
34. B. Magnusson and A.M. Kligman, *J. Invest. Dermatol.* 52, 268 (1969).
35. T. Maurer, P. Thomann, E.G. Weirich, and R. Hess, *Agents Actions.* 5, 174 (1975).
36. T. Maurer, E.G. Weirich, and R. Hess, *Toxicology* 15, 163 (1980).
37. H.C. Maguire and M.W. Chase *J. Invest. Dermatol.* 49, 460 (1967).
38. J.P. Guillot and J.F. Gonnet, *Curr. Probl. Dermatol.* 14, 220 (1985).
39. S.C. Gad, B.J. Dunn, and D.W. Dobbs, *In Vitro Toxicology,* Proc. 1984 Johns Hopkins Symp., A.M. Goldberg, Ed. 539 (1985).
40. S.C. Gad, B.J. Dunn, D.W. Dobbs, and R.D. Walsh, *Toxicol. Appl. Pharmacol.* 84, 93.

41. J. Fraizer, S.C. Gad, A.M. Goldberg, and J. McCaulley, *A Critical Appraisal of Alternatives to the Rabbit Eye Irritation Test.* Mary Ann Liebert, New York (1987).

42. Department of Transportation Code of Federal Regulations, Title 49, 173 (1980).

43. C.S. Weil and R.A. Scala, *Toxicol. Appl. Pharmacol.* 19, 276 (1971).

44. S.C. Gad, R.D. Walsh, and B.J. Dunn, *Ocul. Dermal Toxicol.* 5(3), 195 (1986).

45. J.P. Guillot, J.F. Gonnet, C. Clement, L. Caillard, and R. Trahaut, *Food Chem. Toxicol.* 201, 563 (1982).

46. C.G.T. Mathias, *Dermatotoxicology.* F.M. Marzulli and H.T. Maibach, Eds. Hemisphere, New York, 167 (1983).

47. M.J. Derelanko, S.C. Gad, F.A. Gavigan, and B.J. Dunn, *Ocul. Dermal Toxicol.* 4(2), 74 (1985).

48. J.H. Whittam, *Cosmetic Safety.* Marcel Dekker, New York (1987).

49. C.C. Gibson, R. Hubbard, and D.V. Parke, *Immunotoxicology.* Academic Press, New York (1983).

50. O.B. Christensen, M.B. Christensen, and H.I. Maibach, *Contact Dermatol.* 10, 166 (1984).

51. S.C. Gad, D.W. Dobbs, B.J. Dunn, C. Reilly, and R.D. Walsh, Pres. Am. Coll. Toxicol., Washington, D.C., November 1985.

52. P.S. Thorne, J.A. Hillebrand, G.R. Lewis, and M.H. Karol, *Toxicol. Appl. Pharmacol.* 87, 155 (1987).

53. S.C. Gad, R.W. Darr, D.W. Dobbs, B.J. Dunn, C. Reilly, and R.D. Walsh, *Toxicologist* (1986).

54. S.C. Gad, *Toxicologist* A343 (1987).

55. S.C. Gad, *J. Appl. Toxicol.* 8, 301 (1988).

56. A.M. Kligman, *J. Invest. Dermatol.* 47, 393 (1966).

57. S. Epstein, *J. Invest. Dermatol.* 2, 43 (1939).

58. J.F. Griffith, G.A. Nixon, R.D. Bruce, P.J. Reer, and E.A. Bannan, *Toxicol. Appl. Pharmacol.* 55, 501 (1980).

59. S.J. Williams, G.J. Graepel, and G.L. Kennedy, *Toxicol. Lett.* 12, 235 (1982).

60. F.E. Freeberg, J.F. Griffith, R.D. Bruce, and P.H.S. Bay, *J. Toxicol. Cut. Ocul. Toxicol.* 1, 53 (1984).

61. F.E. Freeberg, G.A. Nixon, D.J. Reer, J.E. Weaver, R.D. Bruce, J.F. Griffith, and L.W. Sanders, *Fund. App. Toxicol.* 7, 626 (1986).

62. A.P. Walker, *Food Chem. Toxicol.* 23, 175 (1985).

63. B.J. Dunn, C.W. Nichols, and S.C. Gad, *Toxicology* 24, 245 (1982).

64. W.J. Hayes, *Toxicology of Pesticides.* Williams and Wilkins, Baltimore, 71 (1975).

65. R.L. Bronaugh and H.I. Maibach, In Vitro Percutaneous in *Dermatotoxicology.* F.N. Marzulli and H.I. Maibach, Eds., Hemisphere, Washington, D.C., 128 (1987).

66. R.R.A. Coombs, and P.G.H. Gell, *Clinical Aspects of Immunology.* Blackwell, Oxford, 575 (1960).

67. F.N. Marzulli, and H.I. Maibach, Eds., Contact Allergy: Predictive Testing in Humans, in *Dermatotoxicology.* Hemisphere, Washington, D.C., 319 (1987).

68. H. Ichikawa, R.B. Armstrong, and L.C. Harber, *J. Invest. Dermatol.* 76, 498 (1981).

13

Statistical Considerations for Ocular and Dermal Toxicology Studies

ROBERT D. BRUCE

AND

JOHN D. TAULBEE
The Procter & Gamble Company
Miami Valley Laboratories
Cincinnati, Ohio

I. INTRODUCTION

The purpose of this chapter is to discuss statistical principles important in the design and analysis of ocular and dermal toxicology studies. In the space of a single chapter it is not possible to provide a detailed exposition of applied statistics nor even to give a "cookbook" containing all of the recipes. The reader will be referred to standard statistical references for theoretical development and computing methods for specific procedures. It is also hoped that the reader will have access to and will consult with a skilled statistician during the development of study protocols and subsequent analyses and reporting of these studies. Although many scientists have received a course in basic statistics at some point in their career, only a limited amount of material can be covered in such a course and even less is likely to be retained after many years of disuse. This problem is compounded by the widespread availability of computer programs which simplify the computational aspects of statistical analysis but do little or nothing to guide the user as to their appropriateness for the problem at hand. The following comments and references, augmented by consultation with a statistician, should do much to alleviate this problem.

II. STUDY DESIGN

A. General Considerations

The overriding principle of statistics is that analysis follows design. This is not simply a statement of a chronological truism, but of the compelling influence that the design of the experiment *must* have on the subsequent study execution and analysis. Design not only dictates the appropriate methods of analysis but also limits the conclusions that may properly be reached. While much toxicological testing is based upon routine use of standard protocols, each new protocol and each modification of an existing protocol is a candidate for careful discussion with a statistician. Where there has already been extensive experience with a particular animal model, the statistical considerations may be resolved with little difficulty. However, when working with a new animal model or new methods of measurement, it may be necessary to conduct one or more pilot experiments just to gain the experience and knowledge needed to design the first study. Such new programs proceed most smoothly if it is understood from the outset that their development will be an evolutionary process. The pressure from management or study sponsors to "do the last experiment first" may be considerable, but must be resisted unless the test system has been well studied and characterized. Such "final" experiments often raise more questions than they answer.

1. Choice of Study Groups

The choice of study groups or, more explicitly, the treatments and dose levels of each, has an important influence on study design. With the use of too many groups, the experiment may become too large to be run or there may be a temptation to reduce the number of subjects per group so far that precision is unduly compromised. Conversely, with the use of too few groups, important concurrent control groups may be missing or information about a dose response may be missing. Here are a few simple suggestions for choosing study groups.

1. Be sure that the study includes an appropriate control group (or groups), but, at the same time, do not overdo such groups. For example, in a skin painting study, it will usually be necessary to include a vehicle control group so that the effect of the test compound can be separated from that of the vehicle. However, in this same study, a naïve control group may be of little or no value. The use of two identical control groups is sometimes recommended on the grounds that, if something goes wrong with the first control group, the experimenter can then use the second control group. This practice may be discouraged on the grounds that it is usually difficult or impossible to decide after the fact which control group is the correct one to use without the appearance of equivocation. It may, however, be desirable to make the control group larger than other groups (see Section II.A.2).

2. If the study is to include several levels of the test material(s) to explore the nature of a dose-response relationship, then doses should be chosen carefully. The highest dose will usually be limited by toxicity or, in dermal and ocular studies, by irritancy. The lowest dose may be chosen as one likely to induce only a minimal biological response. Between these extremes, the experimenter may be tempted to interpolate 5 or 10 intermediate doses with the hope of sketching, in detail, the shape of the dose-response curve. This temptation

should be resisted since it will almost always be futile. With typical group sizes, the biological variation in the average response for each group will usually frustrate such efforts. Instead, one may wish to use one, two, or, at the most, three intermediate doses. If there is no biological basis for choosing the spacing of intermediate doses, then it will often be reasonable to choose doses that are equally spaced on a logarithmic scale. This is often called a geometric dose scale; each dose is a constant multiple of the dose below where the multiplier (m) may be calculated as

$$m = (d_{max}/d_{min})^{1/(n-1)}$$

where d_{max} = highest dose, d_{min} = lowest dose, and n = total number of dose levels.

3.   After the initial determination of study groups, it is wise to think through exactly which comparisons will be made at the conclusion of the study. This will often permit one to discard an unnecessary group or reveal the need for an additional study group. If in doubt, we suggest erring on the side of dropping a marginal group. In our experience, it is better to study a small number of groups well than to inadequately characterize a larger number of groups.

2. Group Size

The objective of most toxicological testing is to determine whether a certain effect results from administration of an agent of interest. In a dose-response experiment, we also want to learn at what dose level the effect first occurs. The question of "whether an effect occurs" is often formally stated as a "null hypothesis" of no effect (H_O) to be tested against the corresponding alternative hypothesis (H_A) that an effect exists. The following is an example of null and alternative hypothesis pairs:

Null H_O — The agent is associated with no increased level of eye irritation compared to its vehicle.
Alternative H_A — The agent is associated with an increase in level (e.g., mean grade) of eye irritation compared to its vehicle.

Since we are basing our decision on a sample to be tested, we can err in two ways with our hypothesis test: (1) reject a true null hypothesis (called a type I error); or (2) fail to reject a false null hypothesis (called a type II error).

The value of a statistical approach to hypothesis testing is that while we cannot eliminate the chances of error, we can determine and control the probabilities of each type of error. For convenience, we designate the probability of type I error as α and the probability of type II error as β. Directly related to this is the *power* of the test, which is the probability of correctly rejecting a false null hypothesis and is equal to $1 - \beta$.

Often, the α level (significance level, probability of type I error) will be chosen to be 0.05 because this is a "traditional" level and will not usually provoke controversy. Several other issues are related to the level of significance. These include:

Number of "sides" in the test — Two-sided tests are usually more appropriate because they allow for departures in either direction from the null hypothesis. A one-sided test may be appropriate only when one is certain that departures from the null hypothesis can occur in one direction only.

Multiple comparisons procedure — The appropriate choice here is related to the error for which α is the probability, and is usually dictated by the design of the study.

These issues do not have solutions which are universally accepted, and are best handled by consulting with a skilled statistician.

For a given α and sample size (n), the value of β (or Power = $1 - \beta$) depends upon the difference or effect of *interest*. This gives rise to a *power curve* for each combination of α and n. At values very near the null hypothesis, the power will be equal to α (and so $\beta = 1 - \alpha$). As the effect or difference increases, power increases until it is virtually equal to 1. This is reasonable, as we should have a higher probability of detecting (or, it should be "easier" to detect) a larger effect or difference. Of course, a larger sample size will result in higher power for each difference above zero. Increasing α is a way to increase power (reduce β) at a given sample size, but this is usually not a desirable alternative. Two key issues, of course, are deciding what constitutes "adequate power" and the "difference of interest". Figure 1 includes three power curves, with power increasing as the difference increases. Comparing cases 1 and 2 illustrates the effect of increased sample size. Comparing cases 2 and 3 illustrates the effect of changing α. Traditionally, the minimum power (probability) of detecting an important difference has been set at 0.80. Some applications where missing an important difference would be especially costly would require higher power.

A key element in determining group size is the *important difference*, or the difference of interest. This is the smallest difference which is of clinical or biological significance. The null hypothesis is one of no difference or effect. For practical sample sizes, power will be very low for small differences, and high for very large differences. A qualified statistician with suitable information can then compute, based on the design and the difference or effect of interest, the sample size required to detect this difference with adequate probability (power).

Of course, sometimes it is very difficult for the scientist to specify what the important difference is. If so, one possible solution is to examine power curves for various practical sample sizes with the hope that the differences which can be detected with adequate power will appear to the scientist to be small enough so that smaller differences or effects will be less important.

If this approach still does not provide a useful solution, the investigator can allocate available or reasonable resources or follow tradition if similar studies have been done before.

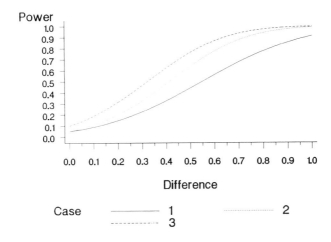

Figure 1. Power vs. difference for various values of Alpha and N. Case 1 is for Alpha = 0.05, N = 9; case 2 is for Alpha = 0.05, N = 16; case 3 is for Alpha = 0.10, N = 16.

For some applications, a sequential approach can be used. Here, experimental units are tested one (or a few) at a time. Each result is compared to a rule set that has three possible decisions: continue experimentation, stop and conclude no effect, or stop and conclude that an effect exists. These designs have the advantage of potentially using fewer experimental units, but may require a longer time to reach a conclusion than more traditional designs. An example of such methods can be found in Reference 1.

3. Additional Considerations

In studies whose intended use is support of the safety of a product or material it is especially important that any adverse effects seen be unambiguous in their interpretation. For this reason, great caution should be used in advocating safety studies where the same group of subjects is treated, successively or concurrently, with two or more test materials. This type of design should be employed only if carryover effects can be ruled out. Otherwise, there is a very real hazard that one of the test materials may be falsely implicated by the toxic effect of a material applied earlier or at another test site in the same animals.

The sensitivity of a study depends both upon group sizes, as discussed above, and upon the subject-to-subject variability in the response. Reducing this variability has the effect of increasing sensitivity or of reducing the number of subjects required to achieve a stated level of sensitivity. One method that can often be employed to achieve lower response variability is to measure the response more precisely. If the response is evaluated subjectively, consider the use of grading scales rather than simply recording presence or absence of an effect. Photographic standards may be helpful in developing and maintaining consistent use of such scales. Even more improvement may be achieved if the response can be measured quantitatively, possibly by an instrument. For dermal responses, for example, consider measuring the area involved, the diameter of a lesion, or using a colorimeter to asses the redness of skin. The gain achieved by using more precise measures of the response will depend to a large degree upon the nature of the response. If it is present to a greater or lesser degree in most or all of the subjects, then the gain may be substantial. If, however, the response is present in only a small fraction of the subjects and absent in the others, then the statistical "leverage" obtained by use of a more precise measurement may be minimal.

B. Special Considerations for Ocular Toxicity Studies

The fact that each subject used in an *in vivo* study has two eyes presents a few special considerations. One can consider tests where both eyes are used for testing. The simplest possibility is to test the same material in both eyes of any subject. If this course is chosen, care must be taken in the subsequent analysis to "count" each such subject only once in the analysis by such techniques as summing or averaging results from the two eyes. The subject, not the eye, is the proper unit of statistical analysis.

A second possibility is to test one material in the right eye and a different material in the left eye. This strategy is reasonable only if systemic effects of the test materials or cross-contamination of the eyes (by scratching, for example) can be ruled out. The statistical term for the subject, in this case, is a "block". Here, the block is said to be of size two, that is it will accommodate two different test materials. The designs for such "blocked" experiments may be categorized as

1. Complete — where the number of different test materials is equal to the block size, in this case two.
2. Incomplete — where the number of test materials exceeds the block size. In the latter case, three materials, A, B, and C, might be tested by treating one group of subjects with A in one eye and B in the other eye, a second group similarly, with the A-C pair, and a third group with the B-C pair.

These blocked designs may be further categorized as to whether they are balanced for order. In this case, balancing for order would involve testing each material equally often in left and right eyes. This type of balancing is important in blocked designs unless it is known *a priori* that the side, right or left, has no systematic effect on any aspect of the response (be it differences in the eyes, in treatment administration, or in evaluation). When these designs are not balanced for order, the treatments should be assigned to the eyes at random; the design is then called a randomized block. As a general rule it is safest to use order-balanced designs until enough experience has accrued to justify ignoring this factor.

There are two distinct advantages in using blocked designs. The first is the doubling of test sites available from a given number of test subjects. And the second is that, with proper analysis, each subject serves as its own control, thus reducing the impact of subject-to-subject variability on the test results. Both the design and the analysis of blocked studies are too specialized to provide details here. And, unless properly designed, an incomplete block study may be impossible to analyze. References giving more information on such designs include Chapters 4, 9, 11, and 13 of Reference 2 and Chapters 4, 7, and 8 of Reference 3.

Still another strategy that is often used in ocular studies is to treat only one eye of each subject (either randomly chosen or balanced, left vs. right) and to use the other eye as a control or reference point for grading. Some investigators will use subjects (animals) for two such studies, allowing ample time for the treated eyes to clear after the first study, and then treating the previously untreated eye in the second study while using the previously treated eye as the control. Of course, this latter strategy is useful only when the existence of carryover effects from the first to the second study can be ruled out. A comment: while the use of the contralateral eye as a control may be helpful to the grader as a reminder of the appearance of a normal eye, our experience is that the grades for untreated eyes are almost invariably all negative (zero). Thus, inclusion of these grades will have no effect on the subsequent statistical analysis.

C. Special Considerations for Dermal Toxicity Studies

Dermal studies present the possibility of treating two or more similar sites on the subject's body with different test materials. The number of sites depends both upon the size of a single treatment site and upon the size of the subject. With mice, it may be practical to define only a single treatment site per animal, but with rabbits it may be practical to define up to eight or ten separate treatment sites on the back of each animal and, if larger animals or humans are being used, many more sites may be defined. This opens up the possibility of using block designs in a way quite analogous to that discussed above for ocular studies. Here, though, the block size is not necessarily limited to two but may be larger, depending upon the size of the subject. Of course, the same limitations to the application of these designs apply here as already discussed for ocular studies, namely that it should be possible to rule out both systemic effects of the materials being tested and cross-contamination of the treatment sites. The latter condition may be possible to guarantee in the case of short-term studies where the sites will be occluded but, because of normal grooming behavior, may not be possible for unoccluded treatment sites and/or for longer term studies.

The advantages of blocking are, as mentioned above for ocular studies, a reduction in the number of subjects required and a potential for reducing the influence of subject-to-subject variability. Design opportunities are similar to those discussed above for ocular studies; the reader is referred to that section for more detail and references. An example of a dermal study using these techniques[4] will be discussed later in this chapter. In general, the active participation of a statistician will be most helpful to the investigator wishing to make use of these techniques.

III. STUDY CONDUCT

A few aspects of the way in which a study is executed have a direct bearing upon its subsequent statistical analysis and interpretation. Some of these aspects are discussed below.

A. Randomization

There are two reasons for employing randomization in a study. The first is to reduce biases that are likely to occur when systematic patterns of experimentation are used. The second is to validate subsequent statistical analyses. Every statistical testing procedure commonly used to analyze experimental studies rests upon the assumption that the subjects were randomly assigned to treatment groups. A statistical test of significance is designed to test whether the difference among treatments exceeds that due to chance alone; without random assignment there is no conceptual basis for such a test procedure.

Randomization is typically required at the beginning of a study in assigning treatments (test materials) to the subjects. For a simple design with one treatment per subject, this is achieved by a process that randomly selects which treatment will be received by each subject. Some common methods for doing this include drawing slips of paper from a container, using a random number table, and computer-generated randomizations. While there are many methods of achieving such a random assignment of subjects, one possible method consists of the following steps (Table 1):

1.   Make a list of the subjects to be included in the experiment.
2.   Write a random number (from a random number table or computer program which generates random numbers) next to each subject. The random number should contain enough digits to preclude the possibility of ties.
3.   Sort the list in order by the random numbers.
4.   Assign treatments sequentially to the sorted list of subjects.

A similar randomization could equally as well have been generated using one of many computer programs or even certain electronic calculators. For studies that involve blocking, randomization is still necessary, but it is rather more specialized in nature. Details may be found in the references cited in the section on study design.

Some investigators insist upon "balancing" the initial assignment of subjects to groups in addition to or instead of randomizing. For example, it is popular to balance animal studies on body weight so that all groups have nearly equal average initial weights. While this practice may appear ideal, the problem is that it overlooks other (possibly unknown or unmeasurable) factors that may be even more important in determining study outcome. If it is felt that such balancing schemes must be used, they should include an element of randomization to guard against other biases, but our preference is to avoid balancing. If the small difference in initial body weight or some other parameter that results from random assignment is felt to be influencing the results, a technique such as covariance analysis[6,7] can readily correct for this disturbance.

In animal studies, it is also important to randomize housing in studies longer in

Table 1
A Study of 10 Subjects, 4 of which are to Receive Treatment A and 3 of which are to be Assigned to Treatments B and C

Original list		Sorted list		
Subject	Random number[a]	Subject	Random number	Assigned treatment
1	4004	3	1146	A
2	9763	7	1335	A
3	1146	1	4004	A
4	9513	6	5877	A
5	8957	10	6362	B
6	5877	5	8957	B
7	1335	8	8963	B
8	8963	9	8975	C
9	8975	4	9513	C
10	6362	2	9763	C

[a] Random numbers were chosen from a table in Reference 5, page 527, 6th column reading downward from the 31st entry.

duration than a few days. Many conditions may vary systematically throughout an animal room despite our best efforts to provide a uniform environment. These include light, temperature, and noise, among others. Randomization of cage position within the room, or on the rack devoted to a single study will help to assure that these environmental factors impinge equally on each study group. If animals receiving a common treatment are all housed in adjacent positions, then their microenvironment becomes inextricably tied to their treatment. The statistician says, in this case, that the treatment and environmental factors are "confounded" and no amount of clever analysis can separate such factors. Thus, while you had hoped to compare material X with material Y, what you end up with is a test of material X and its particular environment vs. material Y and its (possibly different) environment. Randomization of housing does exact a substantial toll in handling to be sure that the correct treatments are applied; the reward is a study in which observed effects can clearly be attributed to the treatments themselves. The consequences of not randomizing have been reported in the literature,[8-10] although we suspect that these problems often remain unreported either because the study was deemed "unreportable" or because the confounding was never detected. Other, more complex schemes for assigning housing positions are sometimes advocated,[6] but these may entail special methods of study analysis, so they are best discussed with a statistician before use. Others advocate cage rotation as a remedy for this problem,[10] but this method is not foolproof. Since, with rotation, animals receiving the same treatment are usually grouped together, it does not guard against such influences as communicable disease shared among close neighbors or a localized environmental disturbance during part of an experiment.

The third place in a study where randomization should be employed is during the application of treatments and the collection of observations. Since most studies are large enough that this process, either interim treatment and/or grading or terminal studies, may take several hours or even days, care must be taken that temporal effects do not bias the results. Such effects include fatigue during subjective grading, diurnal patterns, and aging. Time of day can have substantial effects on liver weight and certain blood parameters in rodents.[11] An animal study in which terminal studies are performed group by group as the day proceeds may be severely compromised by such effects. With careful planning, it may be possible to avoid a third and separate randomization operation for the data collection step of an animal study. As long as housing order has been randomized, then the animals may be presented for treatment, grading, or terminal studies in the same order in which they were housed.

B. Blinding

Blinding refers to the practice of making observations in such a way that the observer is unaware of the identification of the test material associated with the site or subject being evaluated. Our advice is to use blind methods of evaluation for the collection of data that are, in any way, subjective in nature. Examples would include grading of skin or eyes, as well as histopathologic examination of tissue. There are two related reasons for this recommendation. The first is to eliminate bias that may occur and influence the results if the observer is aware of the treatment. To suggest that an observer may produce biased results often meets with an emotional response, since we often associate bias with unfairness or even dishonesty. This implication is certainly not intended here. Even the best-intentioned observer can be unconsciously influenced by external considerations when the treatment is identified. The intent here is to have the subjective evaluation reflect just exactly what is seen in that particular subject rather than to represent some sort of composite response possibly influenced by other subjects or by preconceived notions of what the treatment effect might or should be. The second reason for blind evaluation is to assure that the result for each subject is independent of that for each other subject. This condition, like randomization, is a requirement of every test of statistical significance. If it cannot be assured, then a subsequent statistical test of significance will be tainted. As a general rule, subjective bias will tend to work in the direction of making the results within a treatment group more uniform than they would have been otherwise. This will induce, in a statistical test, an exaggerated sensitivity and a tendency toward more "false positive" results. The need for blinding has long been recognized in the field of clinical trials.[12,13] Its need in toxicology studies, especially those designed to answer important questions of safety, is every bit as great.

The process through which blinding is usually achieved is randomization coupled with maintaining treatment records separately from evaluation records. In trials with human subjects or with small numbers of animals, it may be necessary to have different individuals administer the test materials than those who perform grading or other evaluations. With large animal studies, it should suffice to present the animals for grading in a random order identified only by their assigned numbers. If animals are re-treated immediately after interim grading, some additional precautions may be needed.

C. Data Recording

If observations will be recorded manually and later entered into a computer for storage and/or analysis, it is wise to design the data recording system with this in mind. It may be possible to save a laborious and error-prone process of transcription by use of appropriate forms for data collection.

Another precaution is that seemingly unimportant details may be useful in the later analysis of unexpected experimental results. In our laboratories we first became aware of the importance of diurnal variation in rat liver weights in the context of an experiment with an unexpected pattern of organ weights which was, at first, thought to be related to treatment. Unscrambling this puzzling result was possible because sacrifice times had been recorded for each animal. Similarly, the effects of caging position noted above[8,9] were discovered only because records of these positions had been kept.

IV. STUDY ANALYSIS

In order to discuss this topic, it is convenient to organize the subject by the type of data represented by the response. Many such categorizations can be defined. Here, we will use the following four groupings:

Attribute data — Such data are characterized by the presence or absence of a particular attribute, e.g., irritation or no irritation, skin carcinoma or no skin carcinoma.

Grading scale data — Scales such as none, minimal, marked, extreme, or 0, 1, 2, 3 are often used to describe subjectively graded responses. These are what we will refer to as grading scales. Note that the words or numbers indicate a progression in severity of the response.

Continuous data — These are responses that can take on a very large number of different values, limited only by the precision to which they are measured and recorded. Examples include body and organ weights, skin penetration rates, clinical chemistry values and tumor sizes.

"Time to event" data — These data are a combination of continuous data (e.g., length of time observed) and attribute data (whether the response under study occurred) which are used *together* in the analysis. These types of data will be discussed separately below, because very specialized techniques have been developed for their analysis.

The choice of statistical methodology for each of these types of data is a matter of finding methods whose assumptions are satisfied by the data. While this choice will be importantly conditioned by type of data, as categorized above, it will also depend upon further considerations such as the type of experimental design, group sizes, and the presence of unusual (extreme or outlying) values in the specific data at hand. Although we will suggest those methods that we have found most useful for each of these types of data, this discussion will be far from a complete inventory of the available techniques. The references will indicate more completely the methods available. It should also be apparent that a single study may present data of one, two, or even all of the types mentioned above. Thus, a variety of methods may be required for analysis of a single study.

A final introductory remark concerns the identification of the experimental unit. As a general rule, the experimental unit will be that unit to which a single treatment (or, in the case of blocked designs, a group of treatments) is randomly assigned at the outset of the experiment. For *in vivo* studies, the experimental unit will almost invariably be a subject (animal or human). Although many measurements of a particular response may be made on each subject, it is quite important that each subject appear or "count" only once in any specific statistical test procedure. This need will be further clarified by examples below. Failure to heed this advice is likely to result in completely misleading statistical analyses.

Before proceeding with a discussion of statistical methods used for evaluating specific types of data, we will discuss various ways of presenting the results.

A. Reporting Statistical Analysis Results

Regardless of the type of data represented by the response, there are at least three ways to summarize our comparison of treatments in the study. The three most frequently used are: the p value, the result of a test of hypothesis, and the confidence interval. These three are, of course, interrelated.

The p value is the probability that the difference between treatments would be as large or larger than the one actually observed, if, in fact, there were no differences between the treatments (i.e., if the null hypothesis were true). That is, the p value is the probability of a result as extreme or more extreme than that observed arising by chance alone. We would tend to believe that there is no treatment effect if the p value is "large", while we would tend to believe that a treatment effect (or group difference) exists if the p value is "small".

In constructing a test of hypothesis, we will have a "null" hypothesis (usually of no difference between treatment and control or "absence of treatment effect") and an "alternative" hypothesis (usually that there is a difference between treatment and control or "presence of treatment effect"). This concept was discussed earlier in the discussion of group size, where we introduced the concept of controlling the type I error probability to be "α". Although the choice of α is at the discretion of the investigator, it is often set at 0.05. This means that of 20 experiments when H_0 is true, we can expect, on average, to reject H_0 once ($0.05 \times 20 = 1$). The p value of a test can be used to determine whether to accept or reject the null hypothesis: If the p value is less than or equal to our prechosen α, we will reject the null hypothesis. Otherwise, we will "accept" it. Many statisticians prefer the phrase "fail to reject" rather than "accept", because of the possibility that the null hypothesis is false, but the experiment upon which we based our decision lacked the power to determine this. Hypothesis tests and p values should always include the estimates of treatment differences for the response of interest.

Another way to express the results of an experiment is to place a confidence interval on the difference between treatments. This difference is expressed in terms of the parameter of interest (e.g., percent with tumor). We can construct the confidence interval on treatment difference such that prior to running the experiment, it has a 100 $(1 - \alpha)$ probability of containing the true underlying difference between treatments $(0 < \alpha < 1)$. After the experiment has been run, we must use the term "confidence" rather than probability, because chance is no longer involved.

There is a direct correspondence between the hypothesis test and the confidence interval: any difference leading to rejection of the null hypothesis will fall outside the corresponding confidence interval. This correspondence sometimes provides the only practical way to construct a confidence interval: the 100 $(1 - \alpha)$% confidence interval consists of all hypothesized values of the estimate of interest which result in failure to reject the null hypothesis with significance level α.

As long as estimates are included with p values and hypothesis test results, the three ways of reporting results of an experiment are closely related, and there is certainly no prohibition on reporting the results in more than one form, e.g., p value and confidence interval.

If more than two treatments are involved in a study, the issue of multiple comparisons may arise. This occurs when more than one hypothesis is being tested in one study. For example, we may want to test each of several treated groups against one control. If each test of hypothesis is conducted at level of significance α, then the risk of making at least one type I error among the several tests is greater than α. The advice of a statistician on whether multiple comparison procedures are required should be obtained in these cases. It must be kept in mind that use of multiple comparison procedures reduces the power of a test of hypothesis, so the gain from reduction in type I error risk must be balanced against this loss of power.

B. Attribute Data

Data of this type, characterized by the presence or absence of a particular attribute, are technically described as binomial data and most appropriately analyzed by methods based upon the binomial distribution and the closely related hypergeometric distribution. The only test procedure of this type in common use is often referred to as Fisher's "Exact Test" and is described in Armitage's excellent textbook[14] (where it is called "the exact test for fourfold tables"). This test is used for comparing the proportion of subjects in one group having the attribute with the similar proportion in a second, independent group of subjects. The remaining methods commonly used for testing attribute data are usually described as chi-square tests; these tests are all based on approximations to the binomial distribution. Important tests include the chi-square tests for homogeneity, which may be used to compare proportions for two or more independent groups, a test for dose-related trends in proportions (the Cochran-Armitage test), and a test that can be used to compare two proportions measured on a single group of subjects or on paired sets of subjects (McNemar's test). Good references for these methods include the text by Armitage[14] and the text by Snedecor and Cochran.[7]

1. Completely Randomized Design

As an example of attribute data from this type of design let us consider microscopic evidence for skin irritation related to dermal application of alkyldimethylamine oxide (ADAO)[15] as reported in Table 4 of the cited publication. There the authors give the following results for skin irritation among male mice.

	Number (%) of animals	
Degree of irritation	Control group	High dose group
Moderate or less	63	56
Moderately severe or severe	1 (2%)	13 (19%)
Total animals	64	69

The proportions (or percentages) of animals with moderately severe or severe irritation in the two groups may be compared by Fisher's exact test which gives a one-sided p value of 0.0009. This p value may be doubled, following the advice of Armitage,[14] to give a two-sided p value of 0.0018. This very small p value indicates quite clearly that a difference in degree of irritation as large as this is not likely to have arisen by chance alone. Statistical packages such as SAS[16] and BMDP[17] contain procedures for performing this test.

As a second example, let us consider the following data for hepatocellular carcinomas in male mice exposed dermally to ADAO taken from Table 3 of Reference 15.

Group	No. of animals examined	No. (%) of animals with primary tumor
Vehicle control	64	18 (28%)
0.05% ADAO	75	15 (20%)
0.13% ADAO	75	16 (21%)
0.25% ADAO	69	23 (33%)

Applying Fisher's exact test to compare each treated group with the vehicle controls gives:

Group	Two-sided p value
0.05%	0.356 (decrease vs. control)
0.13%	0.465 (decrease vs. control)
0.25%	0.645 (increase vs. control)

Clearly, no single treated group showed evidence of a different rate of hepatocellular tumors than the vehicle control animals. A further test that is wise to apply to such data is a test for dose-related trend.[18] Such a test combines data from all groups of animals (four groups in this case) and may be somewhat more sensitive as a result. For these data, the test for trend in proportions[14] gives a chi-square statistic of 1.244 with a corresponding (two-sided) p value of 0.265. While this p value is somewhat smaller than those for any of the treated group vs. control tests reported above, it still gives no convincing evidence of any treatment-related effect.

While the occurrences of skin lesions and tumors are usually treated as attribute data, they may sometimes be more appropriately analyzed by other techniques. Often, the time at which skin lesions or tumors could first be visualized is recorded. If this is the case, then it will usually be best to compare these "times to tumor" using the time to event methods discussed below. Finally, in skin painting studies, the number of lesions (tumor burden) at the treatment site may be recorded. Such data should be analyzed by the methods used for grading scale or continuous data. Completely misleading results can be obtained if tumor burden data are analyzed as attribute data since each subject (animal or human) is no longer classified simply as a responder or a nonresponder.

2. Blocked Design

In a blocked design, two or more materials are tested in the *same* group of subjects. If we wish to compare attribute type responses for one of these materials against another, the methods discussed above cannot be used since the requirement for *independent* groups of subjects is not met. An appropriate test, in this case, is a comparison of two paired (nonindependent) proportions,[2,14] often called McNemar's test. As an example, let us consider an experiment in which isopropylmyristate (IPM) was applied to the skin of ten human volunteers.[4] Four different skin sites on each subject were exposed to either one, two, three, or four successive applications of IPM, using occluded patches left in place for 23 h. Suppose we wished to compare the number of subjects developing erythema, after three exposures with the number developing erythema after a single exposure (grading 24 h after patch removed). In order to perform this test, we must first cross-tabulate the ten original subjects (Table 2). As we can see, none of the subjects had erythema at the site receiving one application while, after three applications, seven had erythema. Applying McNemar's test to this table gives a chi-square statistic of 5.143 with a corresponding two-sided p value of 0.0233. This test provides fairly convincing evidence that the higher incidence of erythema after three applications of IPM is not due to chance alone.

As suggested by its name, comparison of paired proportions, this statistical procedure may be applied to a broader class of problems. It is applicable to paired data where each subject has been paired with a second subject, prior to the experiment, on certain criteria. Another obvious application of this method would be to the analysis of eye irritation test data sets where two materials have been tested, one in each eye, and the response is in the form of attribute data. Although beyond the scope of this chapter, an example in animal experimentation might be the use of litter-

Table 2
Ten Subjects Classified by Presence of Erythema After 1 or 3 Applications of IPM

	After 3 applications	
After 1 application	No erythema	Erythema
No erythema	3	7
Erythema	0	0

matched controls. This method is also useful in the analysis of repeated measures of attribute data, for example where a group of subjects has been evaluated for presence or absence of an attribute at two (or more) points in time, and it is desired to test whether the proportion responding has changed over time.

More general methods exist for analyzing attribute data from a blocked design that are capable of dealing with several responses at a time or with questions of the existence of dose-related or time-related trends. For these, the reader should consult References 19 and 20.

C. Grading Scale Data

Two principal methodologies may be used to analyze grading scale data, the choice between these methods being based upon considerations such as the group sizes and the nature of the grading scale. Broadly, these methodologies may be described as normal distribution methods (for example, the t test, analysis of variance, and regression analysis) and distribution-free (sometimes called nonparametric) methods such as Wilcoxon's rank sum test. While the normal distribution tests are formally applicable only to continuous measurements that follow a normal distribution, it can be shown that group *averages* tend to follow a normal distribution, even for grading scale data, providing the number of subjects per group is not too small and that the grades are not too highly concentrated at just one or two points on the scale. The distribution-free methods are, in many ways, ideally suited to grading scale data, since they make no distributional assumption. Methods that are commonly available, however, are limited to the simpler experimental designs.

As a prelude to analysis of grading scales, the responses must be expressed on a numeric scale that represents a logical progression. An example might be grades of redness of the conjunctiva, in an eye irritation study, on a scale using the values 0, 1, 2, and 3 where 0 represents no abnormality and 3 represents the maximum degree of redness. Grades on a word scale such as none, minimal, marked, severe will need to be transformed to numeric values prior to analysis by the methods described here. For the distribution-free methods, the number values used are of little consequence provided only that they are assigned in a logical fashion. For example, the verbal scale given above could be equally well transformed to the values 0, 1, 2, and 3, respectively, or to the values 0, 1, 2, and 10. This is true because the first step in the common distribution-free methods is to replace the data values by their ranks; the ranks are sensitive only to which observations are larger than others and not to the amount by which they are larger. By contrast, the values on a numeric scale are of real importance in analyses using normal distribution methods. With these methods, one of the key statistical operations is to compute the average response value for each experimental treatment (test material/dose level). Clearly, it can make a very substantial difference, with these methods, whether a "severe" response is equated to the value 3 or to the value 10. The very act of averaging grades makes the implicit assumption that a one point difference on the grading scale is of equal importance be it the difference between 0 and 1 or the difference between 2 and 3 on the scale. Said in another way, one grade of 10 would have the same impact on the average score as would ten grades of 1. Those interested in more details of scale construction can consult either Reference 21 or 22.

The arbitrary way in which numbers are assigned on many grading scales is one of the strongest reasons for us to prefer the use of distribution-free methods for grading scale data. However, for some experimental designs (the balanced incomplete block design, for example), distribution-free methods may exist, but they are not readily accessible. In the examples that follow, both methods will be illustrated.

Many computer programs exist to aid in analyses using either the normal theory or the distribution-free methods. When distribution-free methods are chosen, it is particularly important that the chosen program make adjustments for ties both when transforming the data to ranks and when computing variances. This is especially important for grading scale data because ties are likely to be very numerous in such data. Both of the computer packages referenced[16,17] include procedures for computation of Wilcoxon's rank sum test that do include these adjustments.

1. Completely Randomized Design

As an example, let us consider data drawn from a study that illustrated the effect of different dosing volumes on eye irritation grades in the albino rabbit.[23] In that study, eyes were graded at various times after a single exposure to a test material using the scale of Draize et al.[24] (Table 3).

To confirm the apparent increase in redness produced by the two higher volumes when compared with the 0.01 ml dose, Wilcoxon's rank sum test[14] was computed using the SAS program NPAR1WAY.[16] Comparing the 0.01 and 0.03 ml doses gave a two-sided p value of 0.0034 and comparing the 0.01 and 0.1 ml doses gave a two-sided p value of 0.0007. The tests clearly confirm that these differences are unlikely to have arisen by chance alone.

This method might equally as well have been applied to other components of the Draize scoring system or to the total scores reported in the published study. Although distribution-free methods that test for dose response do exist,[25] we have not found them to be particularly useful for two reasons. First, the nonparametric tests for dose response (also called trend or regression) are not widely discussed in basic statistics books nor are they readily available in computer packages. Second, these methods usually test for a very general pattern among the successive dose groups called an ordered alternative. This means, roughly, that at some point in the progression of

Table 3
For 3% Acetic Acid, Scores for Individual Animals 1 Day After Instillation for Three Different
Dosing Volumes

	Conjunctival redness score[a]		
Dosing volume	(0.01 ml)	(0.03 ml)	(0.1 ml)
	2	3	3
	1	3	3
	1	3	3
	1	3	3
	1	3	3
	1	1	2
	1		
	1		
	1		

[a] Draize scoring system — scale of 0 to 4, ranging from normal (0) to a maximum degree of redness
(4).

dose groups, there is evidence for a change in the level of the response. We have not
found this concept of dose response to be particularly helpful to biological scientists
in their interpretation of experimental data.

2. Blocked Design

In the experiment with IPM referenced earlier,[4] the authors present mean
erythema scores for rabbits exposed a single time to various doses of the test
material. The experiment used four different doses of IPM and also included vehicle
control treatment for a total of five test materials. Four test sites were defined on the
back of each of ten animals. Four of the five test materials were applied to each
animal following an incomplete Latin square design (Table 4).[2]

The analysis provides adjusted means for each test material which allow for the
fact that each animal received only four of the five test materials. Because of this
adjustment process, the means can sometimes lie slightly outside the range of the
original grading scale, but this need not interfere with their use and interpretation.

Clearly both the 23 and 85 mg/cm^2 doses produce higher erythema grades than
does the vehicle alone. The lower two doses differ from the vehicle by no more than
might easily be due to chance variation alone (Table 5).

D. Continuous Data

As with grading scale data, both normal distribution methods and distribution-free
methods may be useful. Although with continuous data, the usual choice would be
the normal distribution methods, there are frequently instances where outlying data
values or skewed distributions (a preponderance of values far from the mean in one

Table 4

Design and Erythema Grades 48 h After Patch Removal

| | | Test material (grade) | | |
	Animal	Site A	Site B	Site C	Site D
	1	2(0)	3(0)	4(1)	5(2)
	2	1(0)	2(0)	3(1)	4(2)
	3	4(1)	3(0)	2(0)	1(0)
	4	1(0)	5(2)	4(1)	3(0)
	5	3(0)	4(1)	5(2)	1(0)
	6	4(1)	5(2)	1(0)	2(0)
	7	5(3)	1(0)	2(0)	3(0)
	8	3(0)	2(0)	1(0)	5(2)
	9	2(0)	1(0)	5(2)	4(1)
	10	5(3)	4(1)	3(0)	2(0)

Note: 1 = Vehicle control; 2 = 1.7 mg/cm^2; 3 = 6.3 mg/cm^2; 4 = 23 mg/cm^2; 5 = 85 mg/cm^2. Erythema was scored on a scale of 0 — 4. These data have been analyzed by normal distribution methods.[2]

Table 5

Results of Using General Linear Models (GLM) Procedure of the SAS Package[16]

IPM dose mg/cm^2	Adjusted mean grade	*p* value vs. vehicle control
0 (vehicle control)	0.00	—
1.7	–0.03	0.8187
6.3	0.10	0.4937
23	1.13	0.0001
85	2.30	0.0001

Note: To obtain these results from SAS, the LSMEANS option must be used. The MEANS option alone will not make the adjustments necessary in an incomplete block design such as this.

direction) may make the adoption of normal distribution methods a bit questionable. When the experimenter has doubts about this matter, our recommendation would be either to transform the data, as illustrated below, or to use the distribution-free methods. While some statisticians complain that there is a loss in statistical efficiency entailed by using distribution-free methods when they are not needed, this loss is only about 5% (for the Wilcoxon rank sum test vs. the *t* test). In our experience, this slight loss in efficiency is a small price to pay since in many cases, where normality is not present, the distribution-free methods will actually be more sensitive than normal distribution methods. An exception to this would be with extremely small group sizes. For example, with two groups of size 3 or less, the smallest p value that can be achieved using the rank sum test is 0.100. In this situation, there is little point in performing a statistical test with this method; indeed, any statistical analysis may be futile in this situation.

1. Completely Randomized Design

This design type will be illustrated with data from a series of experiments that measured the flux rates of 36 compounds through human skin using *in vitro* methods.[26] Five replicate determinations were made, using separate skin samples, for each compound. Here, we will analyze the data for just four of the compounds that were studied in a single experiment. The flux rate, J_m, is the steady-state flux as measured during the final 24 h of a 3-day experimental run. In their publication, the original authors reported that the data exhibited a distribution that was markedly skewed toward large values. Furthermore, they reported that taking logarithms resulted in normally distributed data. We will follow their advice and use logarithms (to the base 10) in the present analysis (Table 6).

The data were analyzed using the GLM procedure of the SAS package[16] to perform a one-way analysis of variance on the logarithms of these values. Geometric means were calculated by taking anti-logarithms of the means of the log values (Table 7).

The overall F test of the analysis of variance confirms the existence of significant differences among some of the compound mean flux rates (p value <0.0001). Clearly, the flux rate for Griseofulvin is lower than for the other three compounds. This is confirmed by p values $= 0.0001$ obtained when it is compared with each of the other three compounds. What of the other three compounds? Pairwise p values for these comparisons range from 0.46 to 0.72, thus providing no basis to refute the hypothesis that the other three are equal in mean flux rate.

Had the materials tested represented different concentrations or dosages of the same material, the methods used above would still have been useful. However, a useful ancillary procedure would have been a test for dose-related trends. As mentioned earlier, this test may be a bit more sensitive than pairwise comparisons among treatment groups. This test may be performed by the method of simple linear regression analysis[7,14] or by the method of linear contrasts (Snedecor and Cochran[7] pp. 268 to 271 or Armitage[14] pp. 202 to 205). Because of the substantial amount of subject-to-subject variability present in most biological data, it will generally be difficult to either justify or fit more complex, curved relationships unless these are clearly supported by underlying theory, though it may be useful to consider a simple transformation of the "dose" variable. A good discussion of this problem is given by Tukey et al.[27] This reference also presents a technique for identifying the highest dose level which may be administered without producing a statistically significant effect (often called a "no effect level" by toxicologists). A final method that is recommended after a regression has been performed is a test for lack of fit. This method checks the adequacy of the regression fit to the data and can be performed whenever there are two or more data values for at least one of the "doses". Since this is typically the case in biological testing, this test can usually be applied. Unfortunately, it is not

Table 6
Geometric Means for Four Selected Compounds Using Logarithms (to the Base 10)

Compound	Observed flux rates ($\mu g/cm^2/h$)				
Dextromethorphan	16.02	26.09	5.62	6.43	6.58
Dextromethorphan HBr	11.41	5.79	9.66	9.17	9.66
Griseofulvin	0.16	0.44	0.35	0.20	0.16
Indolyl-3-acetic acid	8.33	7.76	10.66	10.68	24.66

Table 7
Geometric Means Calculated by Taking Anti-Logarithms of the Means of the Log Values

Compound	Geometric mean (flux rate, $\mu g/cm^2/h$)
Dextromethorphan	10.0
Dextromethorphan HBr	8.9
Griseofulvin	0.24
Indolyl-3-acetic acid	11.3

a standard part of many statistical programs nor is it always discussed in basic statistics books. The test for lack of fit is presented by Armitage[14] under the heading "test for linearity", as well as in the more specialized texts dealing with regression and linear models.[28,29]

2. Blocked Design

Blocked designs with continuous response variables are not very common in the area of toxicology because of concerns, when two or more different treatments are applied to the same subject, about systemic effects (if the treatments are applied simultaneously) or carryover from one treatment to the next (if the treatments are applied sequentially). However, one area where this design is fairly common is in human trials to ascertain bioavailability of drugs or other materials. Subjects are typically monitored, by periodic sampling of blood and/or urine, for remaining levels of the test material. These measurements themselves permit one to determine when re-dosing with a different dosage of the same material or with a different material can be undertaken. The general methods of analysis used are similar to those already illustrated in the section on blocked designs for grading scale data. A completely worked out example is given by Wagner.[30] Although the example is for oral dosing, the method is immediately applicable to transcutaneous dosing as well.

Another closely related design is that in which repeated measurements of the same response are made on subjects over the course of time. Simple examples include observation of body weights or food consumption periodically throughout the course of a study. While different groups of subjects may be compared with each other separately at each time point by the methods of the previous section (completely

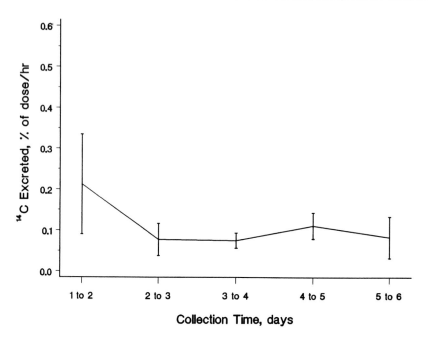

Figure 2.   Urinary excretion of testosterone. Bars are ±1 STD error.

randomized design), special considerations come into play if we wish to study the time course of the response within a single group of subjects. This is illustrated by an experiment reported by Hunziker et al.[31] in which they followed urinary excretion of ^{14}C after topical application of a single dose of ^{14}C-labeled testosterone to the neck skin of four hairless Mexican dogs. A plot of the mean daily excretions and their standard errors is shown in Figure 2.

On inspection, it appears that the excretion rate is nearly constant across days 2 to 6 of the experiment (the experiment was terminated at day 6). Although there is no doubt that excretion would eventually diminish, one might wish to test the hypothesis that, for days 2 to 6, the excretion rate was constant. While one might be tempted to apply the regression methods discussed in the previous section, this may be done only in a very special manner which will be described below. A straightforward regression analysis would be invalid, since the measurements on successive days are made upon the same animals. This is quite different than the situation of the previous section, where testing for dose response involved different subjects at each dose. The correct method here is to fit a linear regression equation to each separate animal. Then the slopes of these lines tell us, for each animal, whether excretion went up (positive slope), held steady (zero slope), or decreased (negative slope) over time (Table 8).

Now, the hypothesis of constant excretion rate is tested by performing a *paired t* test[7,14] on the column of slopes to see whether it differs systematically from zero.

Table 8
Data and Linear Regression Slopes

			^{14}C Excretion		
Dog	Days 2–3	Days 3–4	Days 4–5	Days 5–6	Slope
1	0.027	0.026	0.037	0.019	–0.001
2	0.003	0.078	0.187	0.234	0.080
3	0.174	0.082	0.089	0.030	–0.042
4	0.106	0.115	0.132	0.050	–0.051
Mean	**0.078**	**0.075**	**0.111**	**0.083**	**0.005**

The result of this test is a p value of 0.853, and so these data support the hypothesis of constant excretion rate across this limited period of time.

E. Time to Event Data

The sole factor separating this type of data from continuous data is that we may not know the time to event for a particular experimental unit simply because the event has not occurred prior to the termination of the experiment or some other occurrence which would make observing the event impossible. Data from such an experimental unit are said to be censored. If this censoring (inability to wait long enough to observe the event) did not occur, then the handling of time to event data would be exactly as for continuous data. However, censoring often cannot be avoided. For time until death data, it may not be possible or practical to wait until all experimental units have died to obtain results. For time until tumor data, death or some other event may preclude the observation of tumor formation.

So, for experimental units *with* the event, we record and utilize both the continuous response (time of event), and the attribute response (event actually occurred). For experimental units *without* the event, we record and utilize both the continuous response (length of time observed) and the attribute response (event did *not* occur).

An example where such data were analyzed is found in Cardin et al.,[15] where commercial ADAO was dermally applied in 0.1 ml of aqueous solution once daily three times per week for 104 weeks at concentrations of 0.0 (control), 0.05, 0.13, and 0.269. Mortality checks were made twice daily and times of death recorded. A contrived example of such data is shown in Table 9.

There are numerous methods available which take into account the censoring. One parametric method is the likelihood ratio method assuming exponentially distributed survival times. One nonparametric method is an extension of the Wilcoxon rank sum test (also called the Mann-Whitney test) for uncensored continuous data. This extension to accommodate censoring is called the Gehan-Wilcoxon test. Another nonparametric method is the log rank test.

A graphical representation of the survival data given above is shown in Figure 3. For a test of the null hypothesis of equal times until event for the two groups, the

Table 9
Contrived Example of Time to Event Data

Treated		Control	
Time (weeks)	Censoring[a]	Time (weeks)	Censoring[a]
1	1	2	1
2	1	3	1
3	1	5	1
3	1	6	1
5	1	10	0
6	1	10	0
7	1	10	0
8	1	10	0
9	1	10	0
10	1	10	0

[a] If censoring = 1, time = time of occurrence of event. If censoring = 0, event was not observed during length of observation and Time = time observed.

Gehan-Wilcoxon test gives a p value of 0.06, the log rank test gives a p value of 0.02, and the likelihood ratio method gives a p value of 0.02. The usual t test applied to the survival times results in the higher p value of 0.14. Further information on these methods can be found in Reference 32.

V. CONCLUSIONS

This chapter has provided an overview of statistical concepts and methods that the toxicologist should find useful when conducting oral or dermal toxicity studies. We conclude by emphasizing the following points:

1. The methods of statistical analysis appropriate for a study are dictated by the design of the study and the nature of the data observed. Thus, the methods of analysis and the goals of the study must be carefully considered when planning the study.
2. The statistical literature is rich and varied. Although the t test is an excellent procedure, its use is not always appropriate or optimal.
3. The field of statistics is broad and constantly expanding. While the statistical procedures you learned in school may be perfectly adequate, it would be best to consult regularly with a statistician who is well versed in this field. This consulting process should begin when the research is being planned, rather than at the conclusion of the study.

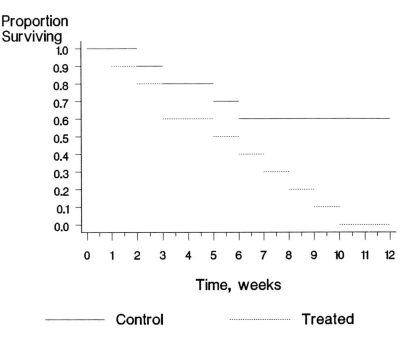

Figure 3.   Survival curves.

REFERENCES

1. R. D. Bruce, *Fund. Appl. Toxicol.* 5, 151 (1985).
2. W. G. Cochran and G. M. Cox, *Experimental Designs,* 2nd ed., John Wiley & Sons, New York (1957).
3. G. E. P. Box, W. G. Hunter, and J. S. Hunter, *Statistics for Experimenters,* John Wiley & Sons, New York (1978).
4. R. L. Campbell and R. D. Bruce, *Toxicol. Appl. Pharmacol.* 59, 555 (1981).
5. D. B. Owen, *Handbook of Statistical Tables,* Addison-Wesley, Reading, MA (1962).
6. D. S. Salsburg, *Statistics for Toxicologists,* Marcel Dekker, New York (1986).
7. G. W. Snedecor and W. G. Cochran, Statistical Methods, 7th ed., Iowa State University Press, Ames (1980).
8. S. Lagakos and F. Mosteller, *J. Natl. Cancer Inst.* 66, 197 (1981).
9. S. S. Young, *Fund. Appl. Toxicol.* 8, 1 (1987).
10. J. K. Haseman, *Environ. Health Perspect.* 58, 385 (1984).
11. D. L. Rothacker, R. L. Kanerva, W. E. Wyder, C. L. Alden, and J. K. Maurer, *Toxicologic Pathology,* 16, 22 (1988).
12. M. C. Miller, III, Ed., *Mainland's Elementary Medical Statistics,* Biometry Imprint Series Press, Ann Arbor, MI (1978).
13. S. H. Shipper and T. A. Lovis, Eds., *Clinical Trials,* Marcel Dekker, New York (1983).
14. P. Armitage, *Statistical Methods in Medical Research,* John Wiley & Sons, New York (1971).
15. C. W. Cardin, B. E. Domeyer, and L. Bjorkquist, *Fund. Appl. Toxicol.* 5, 869 (1985).

16. SAS Institute, *SAS User's Guide: Statistics,* SAS Institute, Cary, NC (1985).

17. W. J. Dixon, Ed., *BMDP Statistical Software,* University of California Press, Los Angeles (1985).

18. N. Mantel, *Biometrics* 36, 381 (1980).

19. G. G. Koch, J. R. Landis, J. L. Freeman, D. H. Freeman, and R. G. Lehnen, *Biometrics* 33, 133 (1977).

20. Y. M. M. Bishop, S. E. Fienberg, and P. W. Holland, *Discrete Multivariate Analysis: Theory and Practice,* MIT Press, Cambridge, MA (1975).

21. A. L. Edwards, *Techniques of Attitude Scale Construction,* Appleton-Century-Crofts, New York (1957).

22. W. S. Torgerson, *Theory and Methods of Scaling,* John Wiley & Sons, New York (1958).

23. J. F. Griffith, G. A. Nixon, R. D. Bruce, P. J. Reer, and E. A. Bannan, *Toxicol. Appl. Pharmacol.* 55, 501 (1980).

24. J. H. Draize, G. Woodard, and H. O. Calvery, *J. Pharmacol. Exp. Ther.* 82, 377 (1944).

25. M. Hollander and D. A. Wolfe, *Nonparametric Statistical Methods,* John Wiley & Sons, New York (1973).

26. G. B. Kasting, R. L. Smith, and E. R. Cooper, *Pharmacol. Skin* 1, 138 (1987).

27. J. W. Tukey, J. L. Ciminera, and J. F. Heyse, *Biometrics* 41, 295 (1985).

28. J. Neter, W. Wasserman, and M. H. Kutner, *Applied Linear Statistical Models,* 2nd ed. Irwin, Homewood, IL (1985).

29. N. R. Draper and H. Smith, *Applied Regression Analysis,* 2nd ed. John Wiley & Sons, New York (1981).

30. J. G. Wagner, *Fundamentals of Clinical Pharmacokinetics,* Drug Intelligence Publications, Hamilton, IL (1975).

31. N. Hunziker, R.J. Feldman, and H. Maibach, *Dermatologica* 156, 79 (1978).

32. E. T. Lee, *Statistical Methods for Survival Data Analysis,* Lifetime Learning Publications, Belmont, CA (1980).

Ocular Toxicology

14

Comparative Anatomy and Physiology of the Mammalian Eye

DAVID A. WILKIE
Assistant Professor
College of Veterinary Medicine
The Ohio State University
Columbus, Ohio

AND

MILTON WYMAN
Professor
College of Veterinary Medicine
The Ohio State University
Columbus, Ohio

I. INTRODUCTION

The eye and adnexa are very similar in all mammals; however, some ocular anatomical and physiological differences are important to recognize when selecting animal models to test drugs, cosmetics, dentifrices, shampoos, and other compounds intended for human use. Because certain animals are also predisposed to ocular diseases more so than others, this should be taken into consideration when designing a biologic model for the eye. Anatomical and physiological differences should also be weighed heavily against the cost associated with using one species or another because a low cost model with little or no validity is worth very little. Unfortunately, cost has long been the basic prerequisite for *in vivo* ocular model selection and selection on the basis of cost does not necessarily result in the most effective or applicable ocular model for each situation. The gross similarities of the animal eye to that of humans also may be misleading. For example, in a survey of over 2000 Rhesus monkeys, none of them had any overt, provocative, or occult signs of glaucoma, even though this animal has an eye very similar to that of humans.[1]

In order to provide the most information possible, this chapter will discuss the general characteristics and identify some important differences among common

433

laboratory animals. It is intended as an introduction for the toxicologist or pharmacologist to the ocular anatomy and physiology of a variety of species. It will present and contrast species differences to better enable the reader to determine if a species might be an appropriate model for use in the type of study they plan to perform. In addition, the reader will become familiar with correct ocular anatomical terminology to use when describing aspects of an ocular toxicology study and also with the various responses of the eye to toxic insult.

This chapter is by no means intended to stand alone as the reader's sole source of ophthalmic information. It must be used in conjunction with the additional chapters in this text concerned with ocular toxicology and also with the references and suggested reading given at the conclusion of this chapter. Our knowledge of ocular physiology, immunology, and toxicology is increasing each day and it is essential that scientists concerned with this aspect of ophthalmology remain current.

II. THE GLOBE

The eye itself can be divided into three concentric tunics plus the internal components (Figure 1). The three tunics from the outside surface of the eye inward are (1) the fibrous tunic (cornea and sclera), (2) the vascular tunic (iris, ciliary body, and choroid), and (3) the neuroectodermal (nervous) tunic (retina). These tunics inwardly decrease in size as they become more specialized. This chapter will first discuss these three tunics followed by the lens, orbit, and adnexal tissues.

A. Fibrous Tunic

The fibrous tunic is comprised of two components, the clear cornea and the opaque sclera. They differ primarily in the arrangement or organization of their collagen and water content with this accounting for the difference in their transparency. Each will be discussed separately.

1. Cornea

a. Anatomy

The mammalian cornea is a transparent, avascular structure that functions to transmit and refract light, and as a protective barrier for the internal ocular contents. In general, the cornea of various species is remarkably similar, differing in thickness, curvature, and shape only. The cornea is thickest peripherally and its thickness increases during sleep.[2] The histologic layers of the cornea, from the outside in, are the epithelium, stroma, Descemet's membrane, and endothelium (Figure 2). In addition in humans, primates, and cattle a layer beneath the epithelium, Bowman's layer, is present under light microscopy, but disappears under electron microscopy.[2,3]

The outermost layer of the cornea is a stratified squamous, nonkeratinized cell layer, the epithelium. Embryologically, the cells of the epithelium originate from surface ectoderm. The number of epithelial cell layers varies in thickness with the species and increases in number toward the limbus.[3] Epithelial cells are anchored to each other by desmosomes and to the basement membrane by hemidesmosomes and

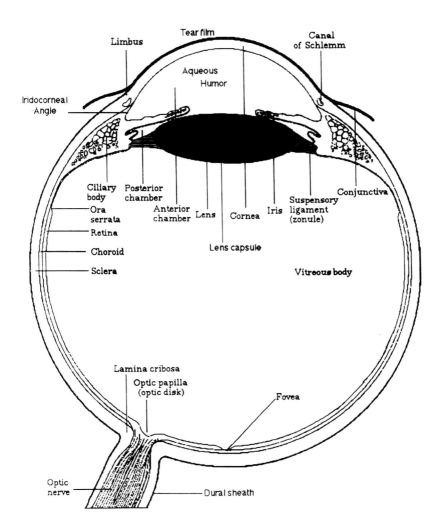

Figure 1.   Representative cross-section of the globe of a primate. Note the presence of the fovea which is absent in most lower species. (Modified from TNA Graphics, Fullpaint for the Macintosh.)

anchoring fibrils (Figure 3).[3-5] The desmosomes are most numerous in the outermost cells of the epithelium.[3,6] External to the epithelium is the tear film. The tear film supplies oxygen to the cornea and approximately 10% of the glucose used by the cornea.[6] The epithelial cells are divided into basal, wing, and superficial cells. The superficial cells have microvilli, microplicae, and a glycocalyx on their surface to stabilize the tear film.[3,7] These microplicae are reported to be very sensitive to topical chemical agents.[6] Beneath the superficial cells are the wing cells, which represent a transition zone between the basal and superficial cells. A single layer of tall, colum-

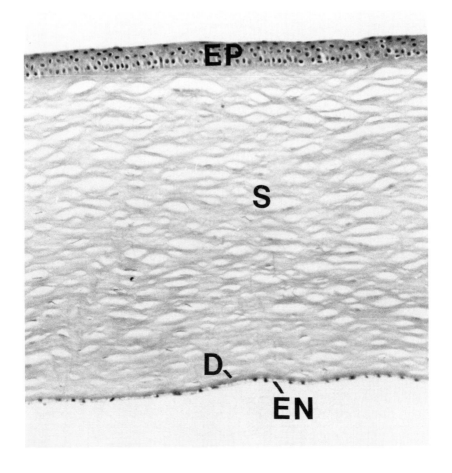

Figure 2.   Full-thickness, histologic view of the cornea of a primate. Epithelium (EP); stroma (S); Descemet's membrane (D); endothelium (EN). The stromal clefts are a sectioning artifact and should not be confused with corneal edema. (H&E; magnification × 39.)

nar basal cells is the deepest layer of the epithelium. These cells contain abundant glycogen stores and are the source of cells for the renewal of the outer layers.[6] The turnover rate of these cells is 7 days.[3,6] In addition to the outer layers of the epithelium, the basal cell layer is also responsible for the secretion of the basement membrane. This membrane is composed of type IV collagen and glycosaminoglycans (GAGs).[6]

The corneal stroma comprises 90% of the corneal thickness. It is composed of collagen fibers, keratocytes, and GAGs which account for 22% of the stroma.[3,6] The remaining 78% of the stroma is water.[6] The corneal fibers measure 24 to 30 nm in diameter and have a 64 nm banding periodicity.[6] The collagen fibers are regularly

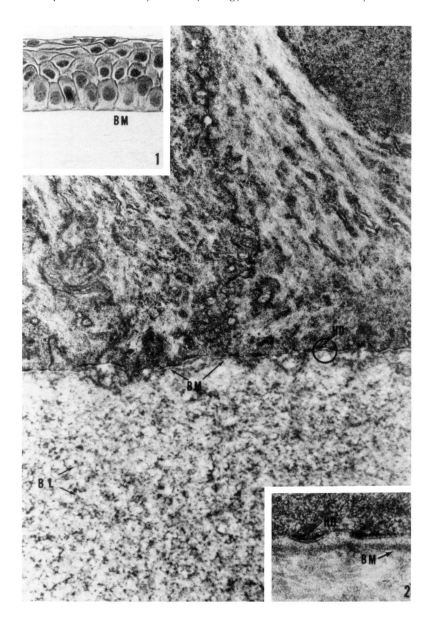

Figure 3.   Basal layer of the corneal epithelium. The hemidesmosomes (HD) and basement membrane (BM) are evident along with the underlying Bowman's layer (BL). (Magnification × 16,000.) Inset 1: Light micrograph of the corneal epithelium and the adjacent Bowmans' membrane (BM). (Magnification × 485.) Inset 2: Magnification of the hemidesmosome circled in the main photograph. The basement membrane (BM) and its electron dense and electron lucent regions is evident. (Magnification × 48,000.) (From B.S. Fine and M. Yanoff, *Ocular Histology,* 2nd ed. Harper & Row, Hagerstown, MD, 165 (1979). With permission.)

Figure 4. The posterior corneal stroma of a cat. The collagen fibrils within a lamellae (L) are parallel, but vary between the lamellae. A single keratocyte (K) is present. (Magnification × 9800.) (From K.N. Gelatt, *Veterinary Ophthalmology*, Lea & Febiger, Philadelphia, 34 (1981). With permission.)

arranged in lamellar sheets and span the entire diameter of the cornea.[3,6] These lamellae lie parallel to the corneal surface, but obliquely to adjacent lamellae (Figure 4).[3,6] The posterior lamellae are more regularly arranged than those in the anterior stroma. The transparency of the cornea depends on the arrangement of these lamellae and the distance between them. It is hypothesized that an interfiber distance of less than a wavelength of light allows transmission with minimal interference.[6] The normal collagen in the corneal stroma is type I and is more heavily glycosylated than in skin.[6] When injured, this is replaced by type III collagen.[6] Between the corneal collagen fibers are flat cells, the keratocytes, and GAGs. The keratocytes are capable of fibroblast transformation and phagocytosis, but normally are a quiescent population of cells.[6] The GAGs of the stroma affect hydration, thickness, and transparency.[2,6] The GAGs act as anions and bind cations and water.[2] The most common GAGs are keratan sulfate, chondroitin, and chondroitin sulfate A.[2] Of these, chon-

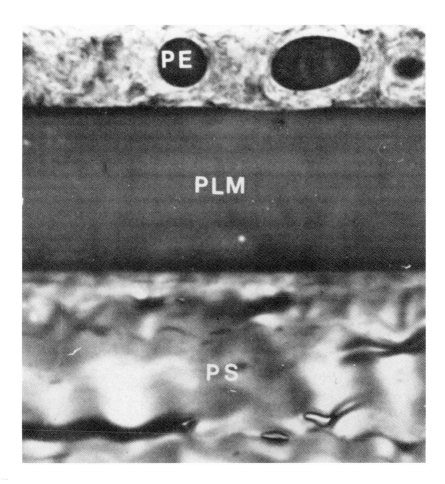

Figure 5.   The posterior portion of the cornea of a horse demonstrating the posterior corneal endothelium (PE), the posterior limiting membrane (PLM), or Descemet's membrane, and the posterior corneal stroma (PS). (Magnification × 1200.) (From K.N. Gelatt, *Veterinary Ophthalmology,* Lea & Febiger, Philadelphia, 37 (1981). With permission.)

droitin is found exclusively in the cornea.[2] Lymphocytes and Langerhans' cells are also reported in the cornea, especially near the limbus.[8]

The posterior limiting membrane or Descemet's membrane is a basal lamina secreted by the endothelium (Figure 5). It is comprised of type IV collagen and is produced throughout life, continuing to increase in thickness with age.[3,6] It is insoluble except in strong acid or alkali and is more resistant to collagenase than is the corneal stroma.[2]

The innermost layer of the cornea is the endothelium. The origin of this layer is thought to be the neural crest cells.[6,9,10] The endothelium is a single cell layer of

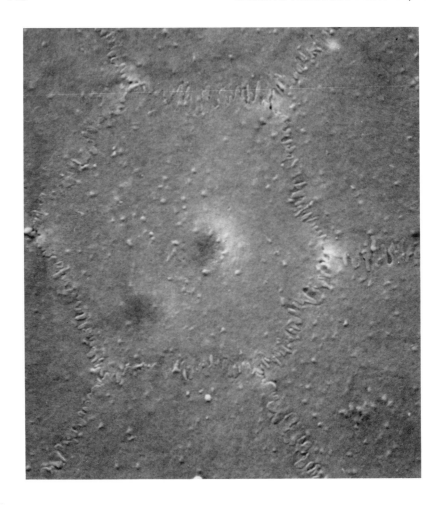

Figure 6.   The monolayer of corneal endothelial cells in the dog are hexagonal in shape with numerous intercellular interdigitations. (SEM; magnification × 2760.)

hexagonal cells, 15 to 20 μm in diameter, attached to each other by terminal bars (Figure 6).[5,6] In most species these cells are capable of mitosis only very early in life. The exception to this is the rabbit, in which the endothelium is capable of undergoing mitosis.[10] The greatest cell density of the endothelial cells is in the neonate.[10] Overall, their number per mm^2 continues to decrease with age, first as they spread to cover the enlarging cornea and later as they decline with aging.[6,10,11] Surprisingly, it would appear that the adult cornea of most species has a central endothelial cell density of approximately 2500 cells/mm^2.[10,11] Once damaged, endothelial cells repair defects through enlargement and spreading rather than by replacement. A minimum cell density of 1000 cells/mm^2 appears to be required to maintain corneal deturgesence. In addition to the number of endothelial cells, it appears that the morphology correlates with function. With age comes an increase in variability in cell size and shape which correlate with a decrease in function.[11,12]

Although the cornea is avascular, it does contain nerve fibers. These are branches of the ophthalmic branch of cranial nerve 5 and are primarily concentrated in the anterior third of the cornea. Upon entering the cornea these nerves lose their myelin sheath, but retain the sheath of Schwann.[3] These nerve endings terminate between the wing cells of the corneal epithelium and in the stroma.[3,13] Each corneal nerve has a receptive field of 5 to 20% of the cornea, with extensive overlapping between receptive fields.[13] It has been demonstrated that there is a species variation in the presence of cholinergic, adrenergic, and substance P nerve fibers in the cornea.[13] It would appear that the cholinergic fibers are the principal sensory fibers.[3] When stimulated, these nerve fibers can result in an axonal reflex that results in miosis, hyperemia, and an increase in the aqueous humor protein content.

b. Physiology

The cornea is responsible for approximately 70% of the refractive power of the eye. It is able to transmit light because of the regular arrangement of its stromal collagen, its avascular nature, and because of its state of relative deturgesence. These properties result in some unique requirements. Since it is avascular, the cornea obtains its oxygen primarily from the tear film and its glucose from the aqueous humor.[2,6] Some oxygenation also occurs from the aqueous humor.[2] The majority of the glucose is consumed by the epithelium and endothelium and is metabolized predominantly by the hexose monophosphate shunt pathway.[2] In addition to metabolizing glucose, the corneal epithelium also stores glucose in the form of glycogen.[2]

In order to maintain its transparency the cornea must actively maintain its stromal water content at 75 to 80%.[2] The stroma has a natural tendency to absorb water and swell because of the osmotic force of the GAGs.[2] This swelling is seen clinically as an increase in corneal thickness which can be measured by pachymetry. It is a function of the epithelium and endothelium to prevent this imbibition of water by the stroma. The epithelium is primarily a mechanical barrier to fluid passage because of the desmosomes between cells,[6] but also has a Cl-dependent pump which has a minor role in fluid removal.[14] The endothelium has a dual role, that of a mechanical barrier and of a metabolic pump.[2,6] The mechanical barrier is dependent on Ca^{2+} and this should be included in any intraocular irrigating solution. There is debate as to whether the pump is dependent on Cl^-, HCO_3, carbonic anhydrase, or is a Na-K ATPase system.[6,15,16] Because of the limited regenerative capabilities of the endothelium any insult — inflammatory, chemical, or mechanical — to the endothelium can result in permanent stromal swelling. Loss of the corneal epithelium results in a 200% increase in corneal thickness, while loss of the endothelium will result in a 500% increase in thickness over a 24 h period.[2] Both the barrier and pump functions of the corneal endothelium and epithelium can be disrupted. The pump is affected by ouabain, carbonic anhydrase inhibition, hypoxia, cytochalasin, and hypothermia, while the barrier can be disrupted by elevations in intraocular pressure, perfusion of the anterior chamber with a Ca^{2+}-free media, and by hypothermia.[6] In addition to the epithelium and endothelium, evaporation of fluid from the tear film, resulting in an increase in its osmolality, aids in the dehydration of the cornea.[6] It is for this reason that the corneal thickness is increased following a period of eyelid closure such as sleep.[17] Any compound which interferes with the tear film or its osmolality, oxygenation of the cornea, the integrity of the epithelium or endothelium, or the enzymatic function of either of these cell layers will result in an increase in corneal thickness. In addition to stromal dehydration the regular arrangement of the stromal collagen lamellae is vital to the maintenance of corneal transparency. Infiltration by cells or fluid, or replacement by type III collagen as is seen in repair, will all result in a loss of transparency.

The barrier of the corneal epithelium and endothelium and the hydrophilic nature of the stroma are also significant in the selection of topical ocular medication. The optimum drug for intraocular penetration is one with biphasic solubility. In addition, using surface-tension-reducing agents and altering the concentration, osmolality, or pH of the drug will also influence intraocular penetration.

When damaged, the cornea may respond with stromal edema, epithelial cell migration and mitosis, migration of polymorphonuclear and monocyte cells to the injured site, secretion of collagenase, activation of stromal keratocytes, vascularization, secretion of new collagen, and a shift in GAGs.[2,6] The initial response to surface injury is the exfoliation of damaged epithelial cells, followed within 24 h by the infiltration of polymorphonuclear cells (PMN) from the tear film, and possibly by migration from the limbus. This is followed by the infiltration of the stroma by mononuclear cells and the activation of stromal keratocytes to become phagocytic. Collagenase is secreted by the invading cells, epithelial cells, as well as the keratocytes and serves to remove the damaged stroma. This is replaced by type III collagen that is produced from the keratocytes. The epithelial cells will form a monolayer of cells at the margin of the wound and migrate to cover the wound surface. Once covered, they will undergo mitosis to return to the normal stratified, squamous, nonkeratinized epithelium and will secrete a new basement membrane. In those species which possess a Bowman's layer, this is not replaced when lost to injury. Many of these processes are controlled or modulated by fibronectin, extracellular growth factor, prostaglandins, and leukotrienes.[18-20] As noted previously, intraocular injury to the endothelium differs in that the endothelium does not possess the mitotic capabilities of the epithelium, with the possible exception of the rabbit.

Arachidonic acid (AA) metabolites have demonstrated a role in the PMN response to corneal injury.[21] Topical AA, prostaglandin E_1 (PGE$_1$), prostaglandin E_2 (PGE$_2$), and prostaglandin I_2 (PGI$_2$) all resulted in an increase in the PMN response by enhancing cell migration from the vascular space into the tear film.[21] Also, following corneal injury tear samples contain increased levels of PGE$_2$ activity.[21] Indomethacin and ketoprofen, both cyclooxygenase inhibitors, inhibit the PMN response to corneal injury in a dose-dependent manner when given topically or intraperitoneally.[21-23]

In addition to their role in corneal inflammation, prostaglandins may assist in the maintenance of normal corneal deturgesence. It has been suggested that the normal endothelial polygonal shape is maintained by PGE_2.[24] Endothelial cells synthesize significant amounts of PGE_2 *in vitro*.[24] In the presence of indomethacin or flurbiprofen, these cells are also seen to lose their normal polygonal shape and become elongated. Although both flurbiprofen and indomethacin are specific cyclooxygenase inhibitors, flurbiprofen has less effect on endothelial cell shape than does indomethacin.[24] The significance of this effect of PGE_2 on the corneal endothelium is not known at this time, but it is thought it may have a role in the healing of endothelial damage.[24]

When selecting animals for corneal investigation, it is important to be aware of the prevalence of corneal abnormalities that are related to the environment and to the breed and species. Corneal dystrophy/degeneration have been reported to occur in the dog, rabbit, mouse, and rat.[25-28] In addition, a lipid keratopathy is reported to occur secondary to diet in the rabbit[29] and in rabbits with inherited hyperlipidemia.[30] A primary endothelial dystrophy has also been reported in several breeds of dogs.[31,32]

Toxic injury to the cornea can result from caustic agents, solvents, surfactants, and detergents. In general, these agents may produce a sensation of itching or burning, pain, conjunctival and episcleral hyperemia, corneal ulceration, deposits in the corneal epithelium and/or stroma, endothelial damage, and possibly inflammation of the anterior portion of the eye, termed anterior uveitis. These are discussed in detail in Chapter 15.

2. Sclera

The sclera constitutes the major portion of the fibrous tunic. It is covered by conjunctiva beneath which is the episclera, the scleral stroma, and the lamina fusca.[3,4] The episclera is composed of loose bundles of collagen and elastic fibers, blood vessels, nerves, fibroblasts, and, depending on the species and color pattern, melanocytes. Like the corneal stroma, that of the sclera consists of collagen lamellae but, unlike the cornea, they are obliquely arranged and lack uniformity.[4] This lack of uniformity, along with the variability in collagen fiber diameter and the relative deficiency in water-binding GAGs, accounts for the lack of transparency to the sclera.[4] The thickness of the sclera varies with the species and is usually thinnest at the equator and thickest at the posterior pole in ungulates and man, but at the area of the scleral venous plexus in animals such as the dog and cat.[3,4] The lamina fusca is the innermost region of the sclera that joins with the vascular tunic below. The sclera possesses several channels or emissaria to allow the passage of nerves and blood vessels. Examples of these are the optic nerve, the vortex veins, long posterior ciliary arteries, and the long and short ciliary nerves.

B. Vascular Tunic

The vascular tunic is comprised of three distinct regions (1) the iris, (2) the ciliary body, and (3) the choroid. In addition, the vascular tunic is responsible for the production of both the aqueous humor and the vitreous and as such they will be discussed in this section. The vascular tunic is mesodermal in origin and is situated between the outer fibrous tunic and the inner nervous tunic. The vascular tunic is also

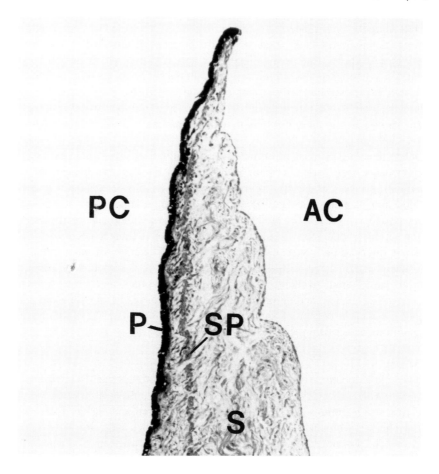

Figure 7. Histologic view of a normal feline iris. Anterior chamber (AC); iris stroma (S); sphincter muscle (SP); posterior pigmented and nonpigmented epithelium and dilator muscle (P); posterior chamber (PC). (H&E; magnification × 15.)

referred to as the uvea. The anteriormost portion of the uvea is the iris, with the ciliary body immediately posterior to the iris, and the choroid situated beneath the retina.

1. Iris

a. Anatomy

The iris is the anteriormost portion of the vascular tunic and functions as a moveable diaphragm between the anterior and posterior chambers (Figure 7). Embryologically, the iris is comprised of components that originate from neural ectoderm, the sphincter and dilator muscles and the two-layered posterior epithelium, and mesodermal components that form the iris stroma.[3,33,34] The iris is divided into the pupil margin, the collarette zone, and the iris root. The pupil margin is that portion of the iris adjacent to the aperture of the pupil. This aperture varies in shape between species and can be circular (dog, rabbit), oval (horse, cow), or vertical (cat).[3] This central portion of the iris normally rest against the anterior lens surface and in the absence of this support, the iris will tremble. This is termed iridodenesis. The collarette zone is that portion of the iris that demarcates the junction of the thinner pupillary and thicker peripheral zones of the iris, and is the site of origin of the fetal pupillary membrane. The color of the iris can vary with species, age, and sex, and can even differ between eyes or within eyes in the same animal. The color of the iris depends on the pigmentation of both the stromal melanocytes and the posterior epithelium. In animals with blue eyes the pigment is absent in the stroma, while in the albino it is lacking in both the stroma and the posterior epithelium.[33,35]

The anterior border of the iris is often discussed as possessing an epi- or endothelial cell layer.[36] This is incorrect.[3,33] The anterior border is formed by fibroblasts and melanocytes.[3,33] They form an almost continuous border, but in those regions where they are absent, a crypt will be seen clinically.[33] Peripherally, this layer ends abruptly in man with iris processes extending forward to Schwalbe's line.[33] In lower animals this peripheral portion of the iris inserts as the pectinate ligaments in the area where Descemet's membrane ends. This angle formed by the cornea and the root of the iris is termed the iridocorneal angle and is the entrance to the trabecular meshwork, the site of outflow of the aqueous humor. This region is discussed further in the section on the aqueous humor.

The iris stroma is a loose tissue comprised of collagen bundles, blood vessels, nerves, melanocytes, fibroblasts, and the sphincter muscle of the iris.[3,33] The melanocyte is often the most prominent cell type and the pigment granules in these cells vary in shape between species.[3] The blood vessels of the iris usually enter at the 3 and 9 o'clock positions and form a major arterial circle at the base of the iris. From this circle vessels radiate toward the pupillary margin, where they may or may not form a minor arterial circle. The vessels themselves are endothelial-lined structures with a basement membrane, smooth muscle, and pericytes.[33] The capillary endothelium is not fenestrated and the type of intercellular junctions varies with the species. The mouse, primate, and man possess terminal bars while the rat, cat, and pig have a 40 Å gap junction.[37] This is typical of a nonfenestrated capillary and is part of the blood-eye barrier. Both myelinated and nonmyelinated nerves are present in the iris stroma.[3] They follow a similar pattern to the blood vessels.

The sphincter muscle is found within the iris stroma. It is smooth muscle in mammals and striated in lower vertebrates. As stated previously, it has its origin from neural ectoderm.[3,33] It is situated in the pupillary zone and varies in shape depending on the shape of the pupil. Contraction of the sphincter muscle results in a decrease in the pupil diameter, termed miosis. The sphincter muscle is innervated by parasympathetic fibers from the Edinger-Westphal nucleus, the fibers of which travel with the third cranial nerve (oculomotor). In addition to this, there is evidence of adrenergic innervation in the form of α-adrenergic stimulation and β-adrenergic inhibition to the iris sphincter in the cat, dog, rabbit, monkey, and man.[38-41] The distribution of these adrenergic fibers differs however between species with either α- or β-adrener-

Table 1
The Distribution of the Anterior Uveal Adrenergic Receptors in Various Species

	Dilator	Sphincter	Ciliary muscle
Cat	Mainly alpha, some beta	Mainly beta, some alpha	Mainly beta, some alpha
Rabbit	Mainly alpha, few beta	Mainly beta, few alpha	Mainly alpha, few beta
Monkey	Mainly alpha, very few beta	Mainly alpha, perhaps beta	Exclusively beta, no alpha
Man	Mainly alpha, very few beta	Alpha and beta in equal amounts	Mainly beta, very few or no alpha

Modified from G.W. Van Alpen, *Invest. Ophthamol. Vis. Sci.* 15, 502 (1976).

gic fibers predominating. With the clinical interest in both α- and β-adrenergic pharmacology for the treatment of glaucoma, it is important to attempt to select the species of interest to investigate these drugs because of this wide range in species differences (Table 1).

The posterior epithelium of the iris is actually two cell layers, both of neuroectodermal origin. Because of their origin from the invaginating optic cup, these cells are situated apex to apex and remnants of the optic vesicle may occasionally persist between these cell layers.[33] The posteriormost cell layer is pigmented and is continuous with the nonpigmented layer of the ciliary body and ultimately with the neural retina (Figure 8).[33] It is separated from the posterior chamber by a basal lamina. Adjacent, posterior epithelial cells are attached by desmosomes and terminal bars to each other.[33] The anterior epithelial layer consists of a pigmented apical portion and a myoepithelial basal portion which forms the dilator muscle of the pupil. The dilator muscle is a radially arranged, smooth muscle and is innervated primarily by sympathetic fibers. Like the sphincter muscle, there is evidence that it too may have a dual innervation by both adrenergic and cholinergic fibers.[42] A complete discussion of the pharmacology of the sphincter and dilator musculature is beyond the scope of this chapter and the reader is directed to alternative sources for this information.[34,43]

b. Physiology

When light is directed into the eye, the normal iris will constrict, resulting in miosis. In addition, the contralateral pupil will also constrict and this is termed a consensual response. In all animals, except for primates and man, the consensual response will be less than the direct because of the disproportionate crossover of fibers of the optic nerve in the chiasm and the midbrain.[43]

In addition to its role in the control of light entering the posterior portions of the eye, the iris also actively participates in inflammation. Inflammation of the iris is termed iritis, but if the inflammation involves both the iris and the ciliary body it is termed iridocyclitis, or more commonly anterior uveitis. Because of its vascular

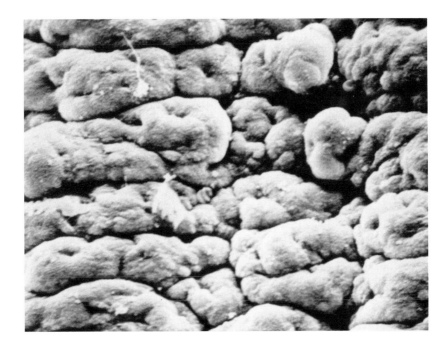

Figure 8. The posterior epithelium of the iris. (SEM; magnification × 500.) These cells originate from neural ectoderm and are continuous posteriorly with the epithelium of the ciliary body and ultimately the neural retina.

nature, the iris is very sensitive to systemic toxins and infectious agents. In addition, a reflex pathway is present that results in inflammation of the anterior uveal tissue when the sensory innervation of the cornea, via the fifth cranial nerve, is stimulated. It is for this reason that animals with severe corneal disease often have a concomitant anterior uveitis. Also, when ulcerated, the cornea may allow passage of exotoxins into the anterior chamber, or alter the aqueous humor pH or oxygenation and result in inflammation. The clinical signs of inflammation are miosis, ocular hyperemia, increased protein, and cells in the aqueous humor, photophobia, and either an increase and/or a decrease in the intraocular pressure, depending on the species involved and the duration of the inflammation.

Prostaglandins (PGs) are considered to be primary mediators of ocular inflammation.[34] The major role played by PGs in ocular inflammation is evidenced by the increased levels of PGs present in inflamed animal and human eyes, the increased amount of PGs synthesized by inflamed uveal tissue, the ability of PGs to reproduce the signs of ocular inflammation, and the fact that inhibitors of PGs reduce the signs of ocular inflammation.[44-46]

Both the cyclooxygenase and lipoxygenase pathways have been shown to be present in the eyes of various species.[45-54] The anterior uvea is capable of synthesizing cyclooxygenase products from AA [45,47,48,55] with PGE_2 and $PGF_{2\alpha}$ being the main products.[47,55] The predominate lipoxygenase products, 12-HETE, 5-HETE, and 5,12-HETE, are also produced in lesser amounts.[54] The eye has very little PG 15-dehydrogenase, the enzyme required to inactivate PG.[34,46] Once PGs are produced, the eye relies on active transport by the ciliary body to remove PG.[34,46] This is reported to be a saturable, energy- and Na^+-dependent pathway and to be impaired in the presence of inflammation.[34]

Although many species have both cyclooxygenase and lipoxygenase activity in their uvea, conjunctiva, and cornea, there are species differences in the rate and amount of endproducts produced and the effect of these on the eye.[34] In the rabbit, for example, the iris synthesizes PGE_2 at a rate many times that of other species.[34,56] Also, ocular prostaglandins result in a biphasic effect on intraocular pressure (IOP) in the rabbit, but not in the cat.[57]

PGs result in miosis, ocular hyperemia, and an increase in ocular vascular permeability.[44,46,57] They act directly on the iris musculature of the dog and cat, and not through the release of adrenergic or cholinergic transmitters.[44,58] Their main effect is to stimulate contraction of the sphincter muscle.[58] In the rabbit, topical application of PGE_1, PGE_2, and $PGF_{2\alpha}$ results in uveal vasodilation and increased uveal vascular permeability.[59] PGE_2 affects the blood-eye barrier at the tight junctions of the ciliary body nonpigmented epithelium (NPE), resulting in the disruption of cell junctions.[44,60] The effect of this disruption of the blood-eye barrier and the uveal vasodilation is an increase in the protein content of the anterior chamber, seen clinically as flare.[60,61]

When injected intracamerally in the cat, leukotriene C_4 (LTC_4) and leukotriene D_4 (LTD_4) cause miosis which is dose dependent.[62] This action is not blocked by indomethacin, a specific cyclooxygenase inhibitor, or by atropine, a muscarinic-blocking agent.[62] In addition to miosis, decreased ocular blood flow and slightly decreased IOP occurred.[62] No effect on the blood-aqueous barrier of the cat was observed, as judged by aqueous protein values.[62] Also, LTD_4 had no effect on the guinea pig uveal vascular permeability,[63] although it did increase the permeability of the conjunctival vasculature.[63] In addition to these effects, LTB_4 and other lipoxygenase products are important in cell chemotaxis.[53,54,64,65]

2. Ciliary Body

a. Anatomy

The ciliary body, like the iris, contains both neurectodermal and mesodermal tissue. It is divided into two parts, the anterior pars plicata and the posterior pars plana.[33,66] In sagittal section the ciliary body is triangular in shape with its base at the iris and the apex posteriorly at the ora ciliaris retinae. The pars plicata possesses 70 to 100 major ciliary processes and between these ciliary processes are valleys in which can be found smaller minor processes. The posterior two-layered epithelium of the iris continues over the ciliary body as the outer pigmented and inner NPE (Figure 9). As is the case in the iris, these epithelial cells are arranged apex to apex.[33] At the junction of the ciliary body and the retina, the single layered NPE continues

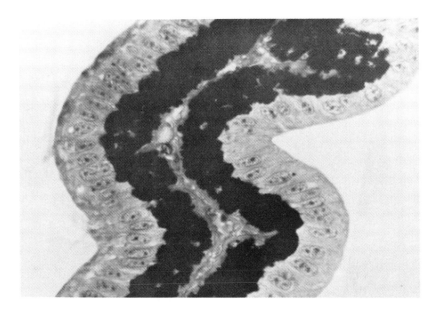

Figure 9.   A longitudinal cross-section of an equine ciliary process. The nonpigmented epithelium is outermost and adjacent to the aqueous humor of the posterior chamber. The inner pigmented epithelium is situated between the outer nonpigmented epithelium and an inner core of capillaries and collagen. (Magnification × 650.) (From K.N. Gelatt, *Veterinary Ophthalmology,* Lea & Febiger, Philadelphia, 54 (1981). With permission.)

as the multilayered retina and the pigmented epithelium of the ciliary body continues as the retinal pigment epithelium. Unlike in the retina and the iris, the two cell layers of the ciliary epithelium are strongly attached to each other by terminal bars.[33] In addition, there are zonulae adherentes and zonulae occludens between the apices of the nonpigmented cells.[33] This is the site of the ciliary body blood-aqueous barrier.[3,33] The basal aspect of the nonpigmented cells is covered by the basal lamina, which intermingles with the adjacent vitreous to form the vitreous base and gives support to the vitreous.[33] It is, in fact, the nonpigmented cells that are responsible for the production of the acid mucopolysaccharide component of the vitreous.[33] In addition, it is the basal lamina that gives anchorage to the zonules that support the lens. These zonules originate in the pars plana and travel forward in the valleys between the ciliary processes (Figures 10 and 11). The zonules ensheath the ciliary processes as they insert on the lens capsule both anterior and posterior to the lens equator. The zonules appear to be elastic microfibrils.[67] The pigmented cells are joined to each other by desmosomes and the intercellular space of these cells is therefore permeable.[3,33,66] The basal portion of the cell faces the ciliary body stroma and is covered by a basal lamina which is a continuation of Bruch's membrane. Beneath the pigmented cell layer is the stroma of the ciliary body. This is mesodermal in origin and contains vessels, nerves, collagen bundles, smooth muscle, melanocytes, and

Figure 10.   Numerous interweaving of lenticular zonules is seen as they travel forward from the pars plana to the valleys between the ciliary processes. The zonules are actually seen to ensheath the ciliary processes. (SEM; magnification × 180.)

fibroblasts. The stroma of the ciliary processes is a highly vascularized connective tissue with capillaries containing large fenestrations, 300 to 1000 Å.[3,33,66] The ciliary muscle is smooth and in all primates is comprised of three parts: the longitudinal, circular, and radial portions.[33] This muscle is poorly developed in most other mammals, with the longitudinal portion predominating.[3] This muscle is primarily innervated by parasympathetic fibers, but also has some adrenergic innervation (Table 1).[3,33] Between the ciliary body and the sclera lies the supraciliary space.

b. Physiology

The ciliary body has three basic functions. The tight junctions of the NPE are the site of the blood-aqueous barrier, the ciliary muscle serves to alter the dioptric power of the lens, termed accommodation, and the ciliary processes are the site of production of the aqueous humor.

The tight junctions of the NPE cells prevent protein and other larger molecules from gaining access to the anterior chamber. In man, for example, the normal plasma protein is 6 g/100 ml, while that of the aqueous is less than 20 mg/100 ml.[66] Experimentally, it has been demonstrated that the tight junctions of the NPE cells and the iris vessels are impermeable to horseradish peroxidase.[66] This barrier can be damaged during inflammation, specifically by prostaglandins, as was discussed in

Figure 11.   Higher magnification of the association of the lenticular zonules and the underlying ciliary epithelium of the pars plana. (SEM; magnification × 1840.)

the section on iris physiology. Also, paracentesis, the removal of aqueous from the anterior chamber, will result in a breakdown of the blood-eye barrier by fragmentation of tight junctions.[68,69] In rabbits, paracentesis will also result in an elevation in the IOP, a phenomenon that appears unique to the species.[66,69]

Accommodation is a process where the eye is adjusted to bring near objects into focus.[67] This occurs through the constriction of the pupil, the anterior displacement of the lens, an increase in the convexity of the anterior lens, and an increase in the axial thickness of the lens, resulting in an increase in its dioptric power.[67] These changes in the lens are brought about through the relaxation of the ciliary body musculature, which releases the tension on the ciliary zonules and allows the axial length of the lens to increase passively.[67] The process of accommodation occurs in man, primates, and some lower animals such as birds, but is poor at best in most other mammals due to their relatively poorly developed ciliary musculature.

3. Aqueous Humor

The aqueous humor is the sole source of nutrition for the lens, a major portion of the cornea, and the trabecular meshwork. The aqueous humor is formed by diffusion, ultrafiltration, and active secretion. Of these, active secretion is the most significant.[66] The active transport mechanism depends on the Na-K-ATPase pump located along the lateral interdigitations of the NPE cells.[66] Interference with this enzyme by ouabain will result in a significant decrease in the production of aqueous humor. It is suggested that it is the Na^+ ion that is actively transported, and depending on the species, either Cl^- or HCO_3^- follow to maintain neutrality.[66] In the rabbit and guinea pig, the aqueous humor concentration of Cl^- is low and that of HCO_3^- is high, while in man, goats, and horses it is the opposite.[66,70,71] In the cat there is an excess of HCO_3^-, but no deficiency of Cl^-, and in the dog and monkey both anions are slightly above that of plasma.[71] In addition to Na-K-ATPase, carbonic anhydrase, adenylate cyclase, and the nucleotide phophatases are also present.[66] Interference with these enzyme systems will, likewise, decrease the active production of aqueous humor. It would appear that the production of aqueous humor is not maintained at a constant level throughout the day. Most species appear to exhibit a circadian rhythm, with the IOP highest in the morning in diurnal animals and in the evening in nocturnal animals.[72-74] It is suggested that this may relate to fluctuations in the levels of endogenous cortisol or epinephrine.[74-77]

The aqueous humor contains electrolytes, glucose, oxygen, amino acids, protein, immunoglobulins, ascorbate, urea, and lactate (Table 2).[66,70] The major cations in the aqueous humor are sodium, potassium, calcium, and magnesium, while the major anions are chloride, bicarbonate, ascorbate, phosphate, and lactate.[70] The concentration of glucose is usually 80% of the plasma value, while the concentration of lactate and ascorbate exceed that of plasma.[66,70] Lactate arises from the metabolism of glucose by the ocular tissues, while ascorbate is actively secreted by the ciliary epithelium.[66,70] The oxygen tension of the aqueous humor is approximately 30 to 40 mmHg.[66] Because of the blood-aqueous barrier, the large molecular weight proteins are excluded and the low molecular weight albumins and β-globulins predominate.[66] IgG is the predominant immunoglobulin in the aqueous while IgD, IgA, and IgM are not detectable.[66] In addition to these, there are also trace amounts of lipids, complement components, hyaluronic acid, and α and γ lens crystallins.[66]

The pressure in the posterior chamber slightly exceeds that of the anterior chamber and this, combined with thermal convection currents and head and eye movements, is responsible for the flow of aqueous from the posterior to the anterior chamber through the pupil. Once in the anterior chamber, the aqueous drains via the iridocorneal angle (Figures 12 and 13). First, the aqueous passes the pectinate ligaments and enters the uveoscleral trabecular meshwork. From here, it enters the corneoscleral trabecular meshwork, the primary site of the resistance to outflow. In man and primates, the aqueous then enters the canal of Schlemm and thereby gains access to the vascular compartment. In most other mammals, however, a distinct canal of Schlemm is absent and, instead, is replaced by a scleral venous plexus.[78] The development of the iridocorneal angle in man and primates is complete by the third trimester, but in cats and dogs is not developed until several weeks or months of age.[78,79] It appears to develop by growth, rearrangement, and a resorption of cells and extracellular matrix.[78] The trabecular meshwork is lined by cells which have a common origin with the corneal endothelial cells. In addition, the trabecular mesh-

Table 2
Concentrations of the Principal Components of the Aqueous Humor as Compared to Plasma of Various Species

Substance (units)		Aqueous	Plasma	Species	Ref.
Ascorbate	(μmol/ml)	1.18	0.02	Monkey	100
		0.96	0.02	Rabbit	101
		1.06	0.04	Human	102
	(mg/dl)	20.0	—	Horse	103
		5.5	—	Dog	103
		1.0	—	Cat	103
		21.0	—	Cow	103
Bicarbonate	(μmol/ml)	22.5	18.8	Monkey	100
		27.7	24.0	Rabbit	101
		20.2	27.5	Human	102
	(mm/g H$_2$O)	Ratio of aqueous/ plasma	0.82	Horse	103
		Ratio of aqueous/ plasma	1.13	Dog	103
		30.4	25.3	Cat	104
		36.0	—	Cow	104
Calcium	(μmol/ml)	2.5	4.9	Monkey	106
		1.7	2.6	Rabbit	104
		—	—	Human	
	(mEq/l)	3.0	5.5	Horse	103
		2.9	5.24	Dog	104
		2.7	4.8	Cat	104
		—	—	Cow	103
Chloride	(μmol/ml)	—	—	Monkey	
		105.1	111.8	Rabbit	105
		131.0	107.0	Human	102
	(mEq/L)	12.1	10.1	Horse	103
		Ratio of aqueous/ plasma	1.07	Dog	103
		—	—	Cat	
		Ratio of aqueous/ plasma	1.15	Cow	104
Glucose	(μmol/ml)	3.0	4.1	Monkey	101
		4.9	5.3	Rabbit	105
		2.8	5.9	Human	102
	(mg/dl)	98	91	Horse	103
		51	70	Dog	103
		45	56	Cat	103
		33	57	Cow	103
Hyaluronate	(μmol/ml)	—	—	Monkey	
		—	—	Rabbit	
		1.1	—	Human	107
		—	—	Horse	

Table 2 (continued)
Concentrations of the Principal Components of the Aqueous Humor as Compared to Plasma of Various Species

Substance (units)		Aqueous	Plasma	Species	Ref.
		—	—	Dog	
		—	—	Cat	
		4.4	—	Cow	107
Lactate	(μmol/ml)	4.3	3.0	Monkey	100
		9.3	10.3	Rabbit	105
		4.5	1.9	Human	102
		—	—	Horse	
		—	—	Dog	
		—	—	Cat	
		—	—	Cow	
Oxygen	(mmHg)	—	—	Monkey	
		30	77	Rabbit	108
		53	—	Human	109
		—	—	Horse	
		45	—	Dog	104
		—	—	Cat	
		—	—	Cow	
Phosphate	(μmol/ml)	0.14	0.68	Monkey	100
		0.89	1.49	Rabbit	105
		0.62	1.11	Human	110
		0.33	0.31	Horse	111
		0.53	1.26	Dog	103
		0.48	1.87	Cat	103
		—	—	Cow	
Potassium	(μmol/ml)	3.9	4.0	Monkey	100
		5.1	5.6	Rabbit	105
		—	—	Human	
	(mEq/l)	5.1	5.5	Horse	111
		5.0	4.4	Dog	104
		4.4	4.0	Cat	104
		7.1	4.7	Cow	104
Protein	(mg/100 ml)	33.3	—	Monkey	100
		25.9	—	Rabbit	112
		23.7	—	Human	112
		20.0	730	Horse	111
		38.0	650	Dog	113
		15—55	780	Cat	113
		17.0	750	Cow	111
Sodium	(μmol/ml)	152	148	Monkey	100
		143	146	Rabbit	105
		—	—	Human	
		117.4	143.5	Horse	111
		149.4	154	Dog	104
		158.5	163.6	Cat	104
		149.5	143	Cow	103

Table 2 (continued)
Concentrations of the Principal Components of the Aqueous Humor as Compared to Plasma of Various Species

Substance (units)		Aqueous	Plasma	Species	Ref.
Urea	(μmol/ml)	6.1	7.3	Monkey	100
		7.0	9.1	Rabbit	104
		—	—	Human	
	(mg/dl)	28	27	Horse	111
		Ratio of aqueous/ plasma	0.70	Dog	103
		Ratio of aqueous/ plasma	0.73	Cat	103
		—	12	Cow	103
Creatinine	(μmol/ml)	0.04	0.03	Monkey	100
		0.11	—	Rabbit	114
		—	—	Human	
		0.18	0.18	Horse	111
		—	—	Dog	
		—	—	Cat	
		—	—	Cow	

Modified from G.M. Schmidt and D.B. Coulter, *Veterinary Ophthalmology*, K.N. Gelatt, Ed., Lea & Febiger, Philadelphia, 129 (1981).

work contains GAGs which are postulated to play a role in the regulation of IOP through the regulation of the flow of water.[80-82] This is suggested by the fact that the enzymatic degradation of these by hyaluronidase decreases the resistance to outflow.[81,82]

The majority of aqueous humor in most species drains by the iridocorneal angle, the so-called conventional pathway. An additional path for outflow is the uveoscleral pathway which accounts for 4 to 14% of the drainage in man, 30 to 65% in primates, 3% in cats, 13% in rabbits, and 15% in dogs.[83-88] In this pathway, the aqueous leaves by passing posteriorly through the interstitial spaces of the ciliary body musculature and the suprachoroidal space. This pathway is dependent on both pressure and particle size. It is also decreased by parasympathomimetic agents that act to contract the ciliary body musculature.

The control of the production of aqueous humor appears to be complex and not completely understood. It is suggested that the adenylate cyclase receptor complex is responsible for the formation of cyclic AMP, an intracellular messenger that results in a decrease in aqueous humor production.[66] This receptor complex apparently is affected by a number of humoral, neurohumoral, and pharmacologic processes and is the final common pathway in the control of aqueous humor inflow.[66]

It has been shown that cells of the trabecular meshwork produce PGs, especially

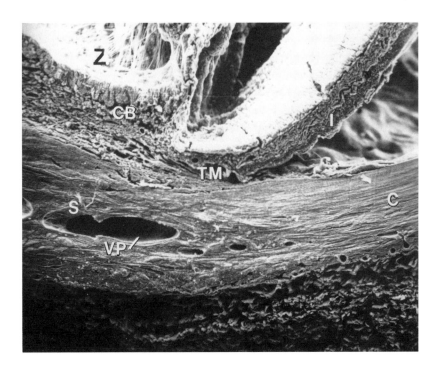

Figure 12.   Low power magnification of the region of a normal iridocorneal angle of a dog. The cornea (C), iris (I), ciliary body (CB), lenticular zonules (Z), trabecular meshwork (TM), sclera (S), and scleral venous plexus (VP) are seen. (SEM; magnification × 90.)

PGE_2 and $PGF_{2\alpha}$,[89,90] and lipoxygenase products, principally 12- and 15-HETE and LTB_4.[90] These may also have a role in the local regulation of IOP and may account for the effect of exogenous ocular prostaglandins in lowering IOP.[57,91-94]

The effects of topical PGs on IOP are variable depending on the species of animal involved. Topical $PGF_{2\alpha}$ is a potent hypotensive agent in the rabbit, cat, monkey, and human.[57,91-94] Of these species, cats appear most sensitive to the hypotensive effects of PGs.[57] In rabbits and monkeys, there is an initial hypertensive phase prior to the hypotensive effect.[57,59,61] Tonography shows that the hypotensive effect is the result of an increase in outflow.[57] Topical PGs do not alter the aqueous flow rate, episcleral venous pressure, or outflow facility.[95] Topical PG does, however, lower the resting tension in the ciliary muscle.[95,96] This effect is blocked by pretreatment with pilocarpine.[95,96] Therefore, it is assumed that the increase in outflow is the result of an increase in uveoscleral outflow.[95,96]

In addition to the reduction in IOP, $PGF_{2\alpha}$ results in mild miosis in rabbits and monkeys, and in cats, a marked miosis.[57] Flare is evident in most rabbits, some cats, and no monkeys.[57] Unlike $PGF_{2\alpha}$, PGE_2 in cats will lower IOP without resulting in

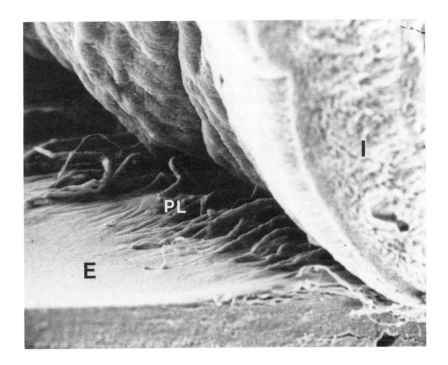

Figure 13.   A view of a normal iridocorneal angle from the anterior chamber demonstrates the iris (I), corneal endothelium (E), and the pectinate ligaments (PL) present at the base of the iris. (SEM; magnification × 200.)

miosis and flare.[93,94] In humans, $PGF_{2\alpha}$ lowers IOP without affecting pupil diameter, aqueous cells, or protein. Prostaglandin E_2 will also lower IOP in monkeys[93,94] and rabbits, but with an initial hypertensive phase in the rabbit.[59,61] Prostaglandin E_3 and PGD_3 in rabbits lower IOP without resulting in miosis, flare, or an initial rise in IOP.[97] A stable PGI_2 analog has been shown to reduce IOP in rabbits and in glaucomatous beagles (37% reduction).[98,99] In rabbits it resulted in a biphasic response with an initial rise in IOP followed by a reduction.[98,99] Also in rabbits, miosis and an increase in the protein content of the aqueous humor was seen.[98,99] In beagles there was no initial rise in IOP and pupil size and aqueous protein were unchanged.[98,99]

The derived PGs of the A and B types are more potent hypotensives than those of the E, F, or D types.[95,96] Also PGs of the A and B types lower IOP in cats without flare, cells, or miosis, and have efficacy in lowering IOP even after 4 months of application.[96] The conclusion from this is that various species demonstrate different responses to ocular PGs in terms of IOP, miosis, and flare, the only consistent finding being a reduction in IOP.[93]

4. Choroid

a. Anatomy

The choroid is the third component of the uvea and, in combination with the iris and ciliary body, comprises the vascular tunic. It extends from the edge of the optic nerve to the pars plana. The choroid is loosely attached to sclera and this zone of transition is termed the suprachoroidal space, a site of uveoscleral outflow. Within this space are found the long posterior ciliary arteries and their corresponding long ciliary nerves which supply the choroid and travel forward to the ciliary body and iris.[33]

The choroidal stroma varies from being darkly pigmented to a complete absence of pigmentation depending on the species, breed, and individual. In those animals lacking pigmentation, the fundic reflex appears red and on ophthalmoscopic examination the choroidal vessels are visible. The stroma itself is composed of collagen fibrils, melanocytes, fibroblasts, nerves, and larger blood vessels.[3,33] The majority of the vessels are veins with arteries situated among them. The arteries have a capillary-free zone surrounding them. Associated with the arteries are nonmyelinated nerve fibers, most of which are motor fibers to the smooth muscle of the arteries.

The capillary layer of the choroid, the choriocapillaris, is found in the inner portion just below the retinal pigment epithelium (RPE). It is responsible for the nutrition of the RPE and the outer retina and, in some species, for the entire retinal nutrition. The capillaries form a lobular network and are comprised of typical fenestrated endothelial cells surrounded by a basal lamina that surrounds and is shared by pericytes and smooth muscle cells.[3,33,115] These lobules are supplied by a central arteriole and are surrounded by a ring of postcapillary venules.[33] These vessels are permeable to fluorescein and horseradish peroxidase. Between the choriocapillaris and the RPE is Bruch's membrane which, when fully developed, is comprised of five layers: (1) the basement membrane of the RPE, (2) an inner collagenous zone, (3) elastic layer, (4) outer collagenous zone, and (5) the basement membrane of the capillary endothelial cells.[3,33] In animals with a cellular tapetum the basal lamina of the RPE and the choriocapillaris fuse, obliterating the other three layers in the region overlying the tapetum.[3] Venous drainage from the choroid occurs in four quadrants where blood collects in an ampulla and then drains via one of four vortex veins which penetrate the sclera.

In addition to above structures, some species also contain a layer in the choroid, the tapetum lucidum, situated between the larger choroidal vessels and the choriocapillaris (Figure 14). Because of its location, the vascular communications between the larger vessels and the choriocapillaris must traverse the tapetum. The tapetum is responsible for the so-called "eye-shine" or bright, colored reflection seen in certain species. It is situated in the superior one half to one third of the choroid. The color of the tapetum varies with and within species and can be green, blue, yellow, orange, or a variation of these. The tapetum is found in dogs, cats, ferrets, horses, ruminants, and nocturnal animals. It is absent in humans, other primates, pigs, rabbits, rats, guinea pigs, and mice. The tapetum can be cellular (dog, cat, ferret, and other carnivores) or fibrous (horse, ruminants, and other ungulates).[3] The tapetum varies

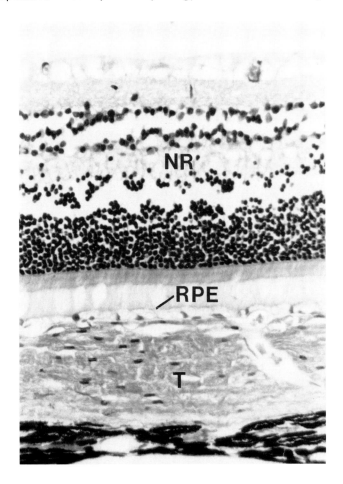

Figure 14.   Histologic section of the neural retina (NR), retinal pigment epithelium (RPE), and the inner choroid containing the tapetal cells (T) in a cat. The pigment of the underlying vascular choroid is seen external to the tapetum. The retinal pigment epithelium is nonpigmented in this region of the retina overlying the tapetum. The neural retina has undergone a degree of postmortem autolysis. (H&E; magnification × 170.)

in morphology, numbers of layers, and in its biochemical makeup.[3,116] The central tapetum is the thickest with the periphery reduced to a single cell (or lamella).[3] The tapetum of the dog and ferret contains large quantities of zinc and cysteine, while that of the cat contains riboflavin, has no detectable cysteine, and has a zinc concentration that is 15 to 18 times lower than that of the dog or ferret (Table 3).[3,116] In the cellular tapetum, the cells contain numerous, slender electron-dense rods in the cytoplasm (Figure 15).[3] These rods are oriented with their long axis parallel to the retinal

Table 3

Comparison of the Tapetum in the Cat, Dog, and Ferret

Parameter	Dog	Cat	Ferret
Number of central cell layers	9	16—20	7—10
Thickness of central tapetum	26—33 μm	61—67 μm	23—24 μm
Presence of microtubule-like structure in tapetal rod	Present	Absent	Present
Presence of electron-dense core in tapetal rod	Absent	Present	Absent
Presence of electron-dense cores in tapetal rods after prolonged glutaraldehyde fixation	Absent	Present	Absent
Retention of tapetal color after prolonged glutaraldehyde fixation	Lost	Retained	Lost
Tapetal zinc concentration	26,000 ppm	1497 ppm	22,500 ppm
Tapetal cysteine concentration	241 μmol/g	0	216 μmol/g

Modified from G.Y. Wen, J.A. Sturman, and J.W. Shek, *Lab. Anim. Sci.* 35, 200 (1985).

surface. In ungulates, the fibrous tapetum is acellular and appears as a dense collagenous layer with its long axis parallel to the retinal surface.[3]

b. Physiology

The main purpose of the choroid is to serve the retina. In some species, such as the rabbit, horse, guinea pig, and chinchilla, the retina is almost totally dependent on the choroid for nutrition since the retinal vessels are distributed to only a small area or are totally absent. In higher animals with holangiotic retinal vasculature, the choroid supplies nutrition to the level of the middle limiting membrane. These animals include humans, primates, dogs, cats, rats, and mice. The choroid is also responsible for the majority of the nutrition supplied to the fovea.[117]

The choroidal blood flow is extremely high and the rate of oxygen extraction very low. The rate of choroidal blood flow is the fastest in the body; 809 ± 73 mg/min in rabbits, 734 ± 94 mg/min in cats, and 677 ± 67 mg/min in monkeys.[118] The venous oxygen content is 95% that of the arterial blood.[70] This compares to an oxygen extraction rate of 38% from the retinal circulation.[117] Even with this low oxygen extraction rate, the choroid still supplies 80% of the retinal oxygen, even to holangiotic retinas.[117] This high flow rate ensures retinal oxygen supply and also serves to protect the retina from thermal damage.[70,119] Unlike the retina and the optic nerve, the choroidal vessels have no autoregulatory mechanism and no precapillary sphincters. The rate of blood flow through the choroid is steady and not alternating.[117] Carbon dioxide is a potent vasodilator of choroidal vessels, but because of the high

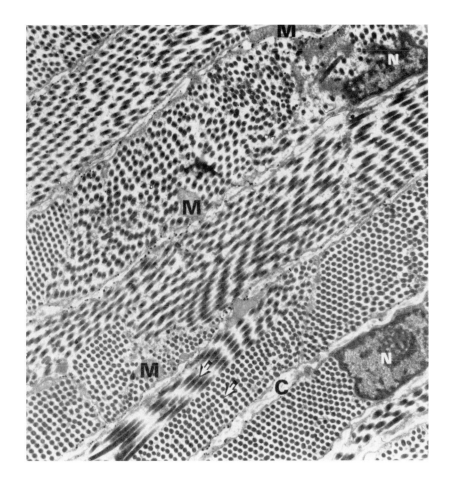

Figure 15.   The tapetal cells of the cat choroid contain numerous rods. These are uniformly arranged within a group, but more than one group can be present in a cell with the arrangement of the rods varying between them (arrows). Infrequent mitochondria (M) are present and cell nuclei (N) can also be seen. The intercellular spaces contain scattered collagen fibrils (C). (Magnification × 7750.) (From K.N. Gelatt, *Veterinary Ophthalmology*, Lea & Febiger, Philadelphia, 89 (1981). With permission.)

rate of flow and low oxygen extraction, the level of carbon dioxide does not increase substantially even with a marked reduction in the choroidal blood flow.[117]

Both cholinergic and adrenergic nerve fibers have been found in the choroid associated with the vasculature.[117] Sympathetic stimulation will result in a decrease in blood flow in all parts of the uvea in rabbits, cats, and monkeys, but has no effect on the retinal circulation.[117] The adrenergic receptors appear to be α in the species studied.[70,117]

5. Vitreous

The vitreous humor is a hydrogel of 98% water that occupies the major portion of the globe and serves as a transparent media and aids in the support of the retina.[70,120,121] The framework of the vitreous is collagen with hydrated hyaluronic acid filling the spaces. In addition, there are a few cells of mesodermal origin, hyalocytes, present predominantly in the peripheral vitreous.[121] These have a short half-life and are thought to be migrating hematogenous monocytes.[120]

Embryologically, the vitreous is developed in three stages. The primary vitreous fills the space between the invaginating optic cup and the lens placode. It is, in turn, filled by vascularized mesoderm which extends from the optic nerve to the posterior lens and connects with the tunica vasculosa lentis (Figure 16). The secondary vitreous forms around the primary and is secreted by the neuroectoderm. The primary vitreous will eventually degenerate leaving its remnant, Cloquet's canal, and the secondary vitreous to fill the posterior cavity of the eye. Failure of this atrophy results in a persistent hyaloid artery and possibly an associated cataract. Although the primary vitreous atrophies in all mammals, it does not do so at the same time. In humans, the primary vitreous does not carry blood beyond the seventh month,[120] while in the dog and rat, these vessels can remain patent for up to 17 days postpartum.[122,123] In addition, it has been estimated that 60% of Sprague-Dawley and 90% of Ola:Wistar rats at 6 weeks of age have a persistence of the hyaloid vessels which may be associated with vitreal hemorrhage.[123] The tertiary vitreous is the lenticular zonules.

The collagen of the vitreous measures 80 to 160 Å and has lacks obvious banding periodicity.[121] This is unlike any other collagen and supports its neuroectodermal origin. The vitreous collagen is firmly anchored at the optic nerve, the posterior lens capsule, and the pars plana of the ciliary body. Hyaluronic acid, a glycosaminoglycan with a negative charge and a high affinity for water, is in highest concentration, like collagen and hyalocytes, in the peripheral vitreous.[120] The concentrations of collagen and mucopolysaccharide vary between species, evident by the fluid vitreous of the owl monkey and the gel vitreous of humans (Table 4).[121]

In addition to collagen, hyaluronic acid, and cells, the vitreous also contains electrolytes, glucose, and other chemical components (Table 5).[120,124] The concentrations of these substances has been shown to be relatively stable postmortem and may be a better indicator of antemortem values than postmortem serum.[125,126] It has been calculated that half the water content of the vitreous is replaced every 10 to 15 min and that the flow of water is from the ciliary body to the posterior pole.[120]

C. Nervous Tunic

1. Retina

a. Anatomy

The retina is the most complex of all the ocular tissues. The classic description of the retina includes 10 layers which, from the outside in, are as follows (see Figures 17, 18, and 19):

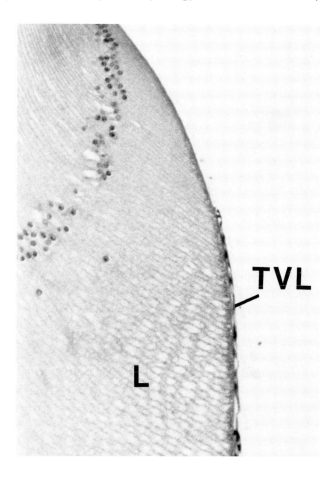

Figure 16.   The posterior lens (L), lens capsule, and tunica vasculosa lentis (TVL) of a 3-day-old dog. The nuclei of the lens epithelial cells are seen at the top of the photograph in the region of the so-called lens bow. The TVL and hyaloid artery system has already begun to atrophy at this time. The remnants seen in this photograph will atrophy in all normal dogs by 3 weeks old. (H&E; magnification × 39.)

1.   Retinal pigment epithelium
2.   Photoreceptor outer segments
3.   Outer limiting membrane
4.   Outer nuclear layer
5.   Outer plexiform layer
6.   Inner nuclear layer
7.   Inner plexiform layer
8.   Ganglion cell layer
9.   Nerve fiber layer
10.   Inner limiting membrane

Table 4

The Concentration of Mucopolysaccharide and Collagen of the Vitreous in Various Species

Species	Mucopolysaccharide (µg/ml)	Collagen (µg/ml)
Rabbit	31	104
Guinea pig	37	134
Human	240	286
Owl monkey	423	25
Steer	710	57

Modified from B.P. Gloor, *Adler's: Physiology of the Eye,* 8th ed. R.A. Moses and W.H. Hart, Eds., C.V. Mosby, St. Louis, 246 (1987).

Table 5

Concentration of the Various Components of the Vitreous[103,124]

Constituent	Cattle[177]	Rabbit[177]	Pig[124]	Human[177]	Horse[103]
Inorganic constituents (mmol/kg H_2O)					
Sodium	130.5	133.9—152.2	142	137	118—153
Potassium	7.7	5.1—10.2	5.0	3.8	4.9—7.3
Calcium	3.9	1.5	5.7	—	4.9—7.3
Magnesium	0.8	—	2.6	—	—
Chloride	115.6	104.3	118	112.8	112—120
Water and organic constituents (mg/100 ml H_2O)					
Creatinine	1.0	—	0.5	—	—
Water	99%	99%	99%	99%	99%
Glucose	55—62	55—80	—	30—70	57—100
Lactic acid	14.8	65	—	70	17.5

Modified from J. Nordman, *Biologie et Chirurgie du Corps Vitre,* A. Brini, Ed. Masson et Cie, Paris (1968).

Embryologically, the retina is neuroectodermal in origin and forms first as an optic vesicle and later invaginates to form the optic cup. The posterior wall of the optic cup gives rise to the RPE, while the anterior wall gives rise to the remaining nine layers of the retina. The space between these layers that was present embryologically remains as a potential space and it is here, between the RPE and the sensory retina, that retinal detachments usually occur. In humans, the retina is mature at birth but, in many laboratory animals, the development of the retina continues postnatally. In the dog and cat, the retina is not mature until approximately 3 to 5 weeks centrally and 8 to 9 weeks peripherally.[127,128] Once formed, the retina undergoes maturation from central to peripheral and, in most laboratory animals, this process is not complete until several days to weeks postpartum.

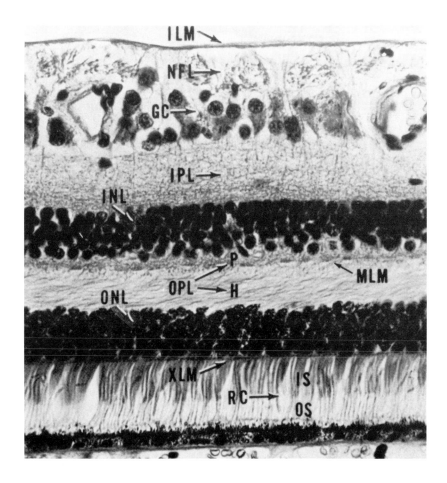

Figure 17. Cross-section of a normal retina, from the area centralis, demonstrating the layered arrangement. Internal limiting membrane (ILM); nerve fiber layer (NFL); ganglion cell (GC) layer containing the larger blood vessels; inner plexiform layer (IPL); inner nuclear layer (INL); outer plexiform layer (OPL); plexiform (P); Henle fibers (H); middle limiting membrane, (MLM); outer nuclear layer (ONL); external limiting membrane (XLM); layer of rods and cones, the photoreceptors; inner segments (IS); outer segments (OS). (Magnification × 390.) (From B.S. Fine and M. Yanoff, *Ocular Histology,* 2nd ed., Harper & Row, Hagerstown, MD, 72 (1979). With permission.)

The RPE is the homolog of the epithelium of the choroid plexus of the brain.[129] It is a monolayer of hexagonal cells which is continuous anteriorly with the pigmented epithelial layer of the ciliary body and the anterior epithelium of the iris which, in turn, gives rise to the dilator muscle. External to the RPE is Bruch's membrane and the choriocapillaris. The basal layer of the RPE is in contact with Bruch's membrane and has marked basal infoldings. The inner apical surface of the

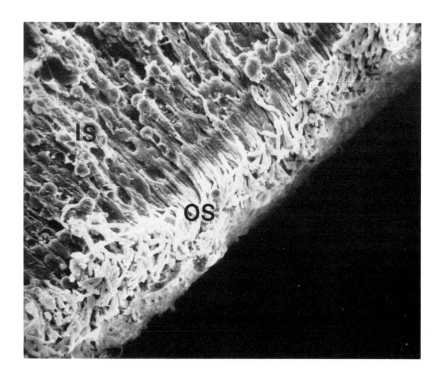

Figure 18.   SEM of retinal inner (IS) and outer segments (OS) of canine retina. The retina has been detached from the retinal pigment epithelium which is not seen in this photograph. (Magnification × 160.)

RPE has numerous long villous processes which surround the outer segment of the photoreceptors. There is no attachment between these cells, but an acid mucopoly-saccharide ground substance exists between them.[3] The lateral RPE cell margins, near the apical surface, have terminal bar attachments (zonula occludens, zonula adherens) which are one portion of the blood-retinal barrier.[4] As the name suggests, the RPE contains melanin granules. These are found predominantly in the inner third of the cytoplasm.[4] An important exception to this is seen in animals possessing a tapetum where the RPE overlying the tapetal portion of the choroid is nonpig-mented.[3]

The sensory retina is thickest near the optic nerve and thins toward the periphery. External to the sensory retina is the RPE and internally is the vitreous humor. The photoreceptors are in the outer portion of the sensory retina and can be divided into the cell body, inner segment, and outer segment. In addition, photoreceptors are divided into rods and cones based on their morphology, physiology, and sensitivity. As suggested by their name, the rods are long and slender, while the cones tend to have a shorter, wider appearance. The cones are further divided according to their

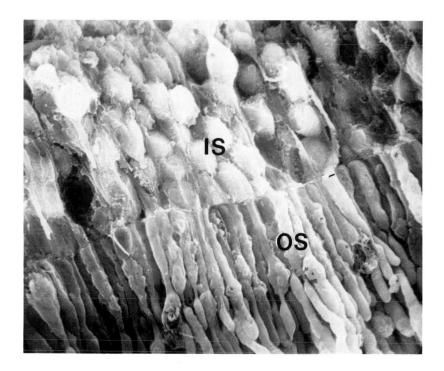

Figure 19.   Higher magnification of the inner (IS) and outer segments (OS) of the photo-receptors of a dog. (SEM; magnification × 3000.)

spectral absorption characteristics into the red-sensitive (570 nm), green-sensitive (540 nm), and the blue-sensitive (440 nm) cones.[130]

The cell body of the photoreceptors is found in the outer nuclear layer which lies just internal to the outer limiting membrane, formed by the terminal bar attachments between the photoreceptors and the Muller cells. The cell body of cones is larger than that of rods and is found in the outer portion of the outer nuclear layer.[4] The apical extension of the cell body, the inner segment, lies just external to the outer limiting membrane. It is here that the Golgi complex, mitochondria, glycogen, smooth endoplasmic reticulum, and free ribosomes are found.[4] Projecting sclerad from the inner segment is a highly modified cilium which consists of a 9 + 0 microtubule arrangement, not the typical 9 + 2 microtubule arrangement seen elsewhere.[4] The cilium is surrounded by the plasma membrane of the apex of the inner segment which projects as a cup or calyx.[4] Beyond the cilium the cell plasmalemma balloons to form a cylindrical sac, the outer segment. In the rod the outer segment is filled with a stack of superimposed discs with double lamellae.[4] These are free of attachment to the plasma membrane or adjacent discs. In the cone these lamellae are less easily separated as they possess attachments to each other and to the cell plasma membrane.[4] The rod outer segment is longer than that of the cone, extending to the apex of the RPE, while that of the cone lies short of the RPE and is enveloped by the long villous processes of the RPE.[4]

The inner extension of the photoreceptor is the outer plexiform layer and is comprised of the axons of the photoreceptor cells, the dendrites of the bipolar cells, and Muller cells.[4] The synapse between the photoreceptors and the bipolar cells differs between rods and cones. The cone synapse is termed a pedicle, while that of the rod is a spherule. Both contain synaptic vesicles which are believed to contain acetylcholine.[4] By light microscopy, these synapses appear as a series of dashes in the innermost part of the outer plexiform layer and this region is often termed the middle limiting membrane.

The inner nuclear layer contains the cell bodies of the bipolar, Muller, amacrine, and horizontal cells.[4] The Muller cells extend through the entire thickness of the sensory retina as a supporting framework and are, therefore, termed ependymoglia as they serve as both glial and support cells.

The inner plexiform layer is formed by the axons of the bipolar and amacrine cells and the dendrites of the ganglion cells.[4] The ganglion cells form a single layer over most of the retina, the exception being adjacent to the fovea in those species in which this is present. The axons of the ganglion cells are aggregated into nerve fiber bundles which make up the nerve fiber layer. These bundles travel parallel to the retinal surface through arcades formed by the foot processes of the Muller cells.[4] These are nonmyelinated nerve fibers until they reach the optic nerve where they acquire myelination. Depending on the species and where the myelination begins, it may be evident clinically. The innermost layer of the retina is the internal limiting membrane. This is a thick basement membrane that is smooth on its internal surface, but conforms to the uneven Muller cell basal plasma membrane externally.[4]

In addition, there are accessory glia present in the retina. These consist of astrocytes and oligodendrocyte-like cells and are usually only found in the nerve fiber layer.[4]

The vascular system of the retina varies greatly between species (Table 6). In those animals possessing retinal blood vessels, the arterial supply enters as either a central retinal vessel (human, primate) or via the short posterior ciliary arteries which give rise to the cilioretinal arteries.[3] The number of larger arterioles and venules varies with the species, but all large vessels are contained in the nerve fiber layer. The arterioles are, in general, internal to the venules and are surrounded by a very large capillary-free zone.[4] The smaller vessels extend deeper into the retina and the capillaries can extend as far as the middle limiting membrane. All structures external to this, the photoreceptor cell bodies, inner and outer segments, and the RPE are supplied by the choroid via the choriocapillaris.[4] The capillaries have a single layer of endothelial cells surrounded by a basement membrane and an interrupted layer of pericytes which are, in turn, surrounded by their own basement membrane.[4] The capillary endothelial cells are attached to each other by terminal bars and are, like the RPE, part of the blood-retinal barrier.

It has been demonstrated that the maximum distance for oxygen diffusion in the

Table 6
Comparison of the Type of Retinal Vasculature of Various Species

Species	Holangiotic	Merangiotic	Paurangiotic	Anangiotic
Dog	X			
Cat	X			
Human	X			
Primate	X			
Rabbit		X		
Rat	X			
Mouse	X			
Gerbil	X			
Cattle	X			
Horse			X	
Guinea pig				X
Chinchilla				X
Degu				X
Bird				X

Note: Holangiotic — The retinal blood supply is from a central retinal or cilioretinal arteries and extends over the entire retina. Merangiotic — A portion of the retina is supplied by retinal vessels. Paurangiotic — Retinal vessels are small and extend only a very short distance from the optic nerve. Anangiotic — The retina is without vessels.

retina is 143 μm.[131] Animals with retinal blood vessels have retinae that are up to twice this distance in thickness, while avascular portions of the retina, be they part or all of the retina, do not exceed this calculated maximum diffusion distance.[131] In addition, avascular retinae show increased glycogen deposition in the Muller cells, have shortened photoreceptors, and lack a tapetum.[131] All of these appear to be adaptations to an avascular environment.

Certain species possess a highly specialized area of the retina, the macula, which contains the fovea (Figure 20). This is found in humans, primates, birds, and some fish and reptiles. In addition, an area centralis is described in several other laboratory animals such as the cat, but these animals do not possess a true fovea. In the center of the macula is an area, the fovea, which is the area of highest visual acuity. It is completely rod free, containing only cone photoreceptors.[4] The fovea is an avascular area and lacks all retinal layers except the photoreceptors and their nuclei.[4]

b. Physiology

It is beyond the scope of this chapter to give a thorough discussion of the physiology of the retina. If the reader requires further information they are directed to seek additional sources.[70,129,132]

The RPE has several responsibilities in the maintenance of the health of the retina.[133] It along with the retinal vasculature forms the blood-retinal barrier. It aids in the attachment of the retina by regulating the volume of fluid that enters the ocular ventricle and by a barb action of the elongated melanosomes in the villous proc-

Figure 20.   Section through the center of the fovea of a Rhesus monkey. At the floor of the fovea all retinal layers except the internal limiting membrane, the photoreceptors, and the retinal pigment epithelium are absent. The underlying choroid and sclera are also present. (Magnification × 150.) (From B.S. Fine and M. Yanoff, *Ocular Histology,* 2nd ed. Harper & Row, Hagerstown, MD, 115 (1979). With permission.)

esses.[129,133] The RPE has phagocytic properties, removing the shed outer segments of the photoreceptors. The bleached visual pigment is regenerated in the RPE and also the RPE is responsible for the transport and storage of vitamin A, an important component of rhodopsin, the rod visual pigment.[129,133] In addition to vitamin A, the RPE is also responsible for the transport of other metabolites to the retina. The importance of the RPE is evident in the Royal College of Surgeons (RCS) rat in which the phagocytosis of shed outer segments is defective and results in the accumulation of the shed outer segments and subsequent photoreceptor degeneration.[132]

The discs of the photoreceptor outer segments contain the visual pigments which capture the photons to initiate the visual process. The visual pigments are embedded in a lipid bilayer of the photoreceptor membrane.[130] They are regenerated and shed throughout life. Rod outer segments are usually shed at the onset of a period of light, while the shedding of cone outer segments is variable, but usually occurs following the onset of a period of darkness. The rate of renewal of rod outer segments has been calculated to be 9 to 13 days, which in the dog corresponds to 2.3 µm/24 h.[130,134]

There is a steady dark current that flows along the length of the photoreceptors. This current is maintained by Na^+ which is extruded from the inner segment by a Na^+-K^+-ATPase pump and then enters passively through Na^+ channels in the outer segment.[129,135] Thus, during darkness, the photoreceptors are continuously discharging. Exposure to light results in the closure of the Na^+ channels in the outer segment and a resultant hyperpolarization and a decrease in the release of neurotransmitter. This process is termed transduction and involves the absorption of a photon by rhodopsin. This, in turn, communicates via a molecule termed transducin to activate the enzyme phosphodiesterase which will hydrolyze many cyclic GMP molecules which are responsible for maintaining the Na^+ channels.[129] It is this phosphodiesterase enzyme that has been found to be abnormal in animals, such as the Irish Setter and Collie dogs and the Rd mouse, that undergo hereditary retinal degeneration. The signal from the photoreceptor cell is transmitted inwards to the ganglion cells via the bipolar cell with input from the horizontal and amacrine cell. The neurotransmitters involved in this process may include GABA, glycine, acetylcholine, monoamines, peptides, substance-P, aspartate, and glutamate.[136,137]

Several dietary deficiencies have been described which result in retinal damage. The photoreceptors contain large quantities of taurine and it has been shown in the cat and primate that deprived of taurine the photoreceptors undergo degeneration, with the cones and the area centralis affected first.[129,138-141] This damage occurs independent of light and appears to result in shortened outer segments, swollen inner segments, and a loss of synaptic terminals in the outer nuclear layer.[138,139] In addition to taurine, the retina has a requirement for two lipid-soluble vitamins. Vitamin E is a powerful, naturally occuring antioxidant and it serves to protect the polyunsaturated fatty acids from oxidation.[130,142,143] Animals suffering from vitamin E deficiency demonstrate an accumulation of lipofuscin in the RPE as a result of the degeneration of the receptor outer segments.[142,144] Vitamin A (retinol) is required for the visual pigment of the photoreceptor outer segments.[130] The vitamin A in the retina, in the form of 11-cis-retinaldehyde, is attached to an apoprotein, opsin, to form the molecule rhodopsin, the rod visual pigment.[129] In addition, it is incorporated into the three cone visual pigments in the human retina.[130] Deprived of vitamin A, the outer segments will undergo degeneration in addition to the corneal and conjunctival changes associated with vitamin A deficiency. Although not a significant problem in the U.S., vitamin A deficiency is the primary cause of blindness in both the human and livestock of Third World countries.[145]

In addition to nutritional deficiencies, the retina is also sensitive to toxic injury. Zinc pryidinethione, an antifungal and antibacterial agent used as an antidandruff agent in shampoo, has been demonstrated to result in injury to the tapetum lucidum with a resultant retinitis, retinal hemorrhage, retinal detachment, and blindness in cats and dogs.[146] The lesions were not present in the nontapetal region of affected eyes. 6-Aminonicotinamide, a potent nicotinamide antagonist, results in vacuolation of the RPE and outer plexiform layer in rabbits.[147,148] The anticancer agent, adriamycin, has been shown to be both cardio- and retinotoxic, likely through the formation of lipid peroxides.[142] Metal ions, especially iron, are toxic to the retina.[142] The phenothiazine and quinolines compounds are also toxic to the retina, possibly by disrupting the rod outer segments and resulting in the accumulation of visual pigment in the RPE.[149,150] Urethane results in a toxic retinopathy in pigmented rats.[141] In addition, compounds such as ethambutol, diphenylthiocarbazone, hydroxypyridinethione, naphthalene, chloroquine, and numerous others, have been demonstrated to result in degeneration of the photoreceptors or ganglion cells.[149] Low illumination for a prolonged period (photic injury) or intense illumination for a short period (mechanical injury) are also damaging to the retina.[151] Phototoxicity is influenced by the spectral wavelength of the light, the intensity of the light, and the photoperiodicity.[152] It also appears to be influenced by the light intensity under which the animal was raised.[153] This has importance in the laboratory animal environment, as inappropriate levels of illumination can result in retinal damage that could be confused with a toxic insult. It would appear that albino rats are more susceptible to phototoxic injury than other laboratory animals.[152] One study indicates that the threshold to phototoxic damage appears to be 1.3 log units greater than the light intensity under which the animal was raised.[153] Attention must also be paid to the illumination at various areas of the room, for example, the difference in light intensity between the top and bottom row of cages.

While nutritional and toxic injury can result in damage to the retina, many species also exhibit hereditary retinal degeneration which must be differentiated from the acquired retinal degeneration. Hereditary degeneration of the retina has been described in dogs,[134,154,155] cats,[156,157] horses,[158] rats,[159,160] mice,[161,162] and chickens.[163,164]

D. Optic Nerve

The optic nerve is comprised of the axons of the ganglion cells of the retina, myelin, blood vessels, glial cells, and the three meningeal sheaths of the central nervous system (dura, arachnoid, and pia). The optic nerve fibers originate from the ganglion cells of the retina which give rise to the nerve fiber layer, the innermost layer of the retina. These fibers take an arcuate course, merge, and exit the eye at the posterior pole at an area that can be seen clinically as the optic disc. Depending on the species, this intraocular portion of the optic nerve may or may not be myelinated and varies in size and shape. The nerve then courses posteriorly past the choroid, where it is surrounded by astrocytes, and the sclera where both astrocytes and scleral collagen surround the nerve, comprising the lamina cribrosa. As the nerve exits the globe it acquires its meningeal covering, travels craniad, and leaves the orbit via the bony optic foramen. The right and left optic nerve meet at the optic chiasm and, depending on the species, a percent of the fibers will decussate (Table 7).[3]

The vascular supply to the optic nerve arises from the pial vessels and the intraocular portion receives blood from the pia, short posterior ciliary arteries, and the retinal capillaries in the species studied to date.[165] In man and other primates there is, in addition to this, a central retinal artery which arises from the internal ophthalmic, a branch of the internal carotid, but this is not thought to be responsible for a significant portion of the blood supply to the optic nerve. The vessels of the optic nerve, like those of the retina, is thought to possess autoregulatory capacity.[166]

The predominant glial cell in the optic nerve is the astrocyte. Astrocytes weave together to form tunnels through which the axon bundles must pass. They serve, therefore, in a support role.[166] In addition, they also interpose themselves between the optic nerve and all mesodermal elements and, in doing so, presumably serve to

Table 7
Percentage of Optic Nerve Fibers Decussating at the Optic Chiasm

Species	Decussation (%)
Human	50
Primate	50
Dog	75
Cat	65-70
Horse	81
Cow	83
Pig	88
Bird	100

moderate the environment.[166] Oligodendroglia are also present and serve to form and maintain the myelin sheaths.[166]

Intracytoplasmic transport of molecules and organelles occurs within the optic nerve by bidirectional rapid axoplasmic flow, and also slow axonal flow.[166] Impulses in the optic nerve are carried by saltatory conduction, as is typical for white matter tracts and myelinated peripheral nerves.[166]

E. Lens

1. Anatomy

The lens is a biconvex transparent structure and is the second refracting unit of the eye. It lies posterior to the iris and is suspended from the ciliary body by the zonular fibers (Figure 21). It also has a posterior attachment to the anterior vitreous face where it lies in a depression of the vitreous, the patellar fossa. It is a unique tissue in that it is avascular, transparent, lacks nerve supply, and has the highest concentration of protein and glutathione in the body.[3] Embryologically, the lens originates from the surface ectoderm which is induced to form the lens placode and invaginate by the advancing optic vesicle and cup.[167]

The lens is surrounded by a basement membrane, the lens capsule, which is secreted by the lens epithelial cells anteriorly and the cortical fibers posteriorly (Figure 22). The capsule is predominantly type IV collagen and its hydrolysines are almost completely glycosylated.[168] It is thickest anteriorly and thinnest at the posterior pole and continues to grow throughout life. Beneath the lens capsule, anteriorly, is the lens epithelium. These form a monolayer of cuboidal cells whose apices face the cortical fibers, with the basal portion of the cell adjacent to the lens capsule. The apical portion of the epithelial cells have terminal bars.[168] Adjacent to the lens equator is the preequatorial or germinative zone where the lens epithelial cells replicate. It is these cells that are susceptible to radiation and result in the cataract seen with radiation exposure. The newly formed cells migrate equatorially where they elongate, differentiate into cortical fibers, are displaced inward, compressed, and lose their nuclei.[3,167,168] This replicative process occurs throughout life. As these

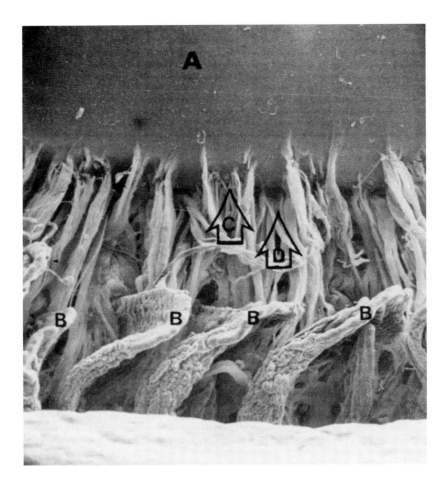

Figure 21.   Caudal view of the ciliary processes, lenticular zonules, and peripheral lens of the cat. The lenticular zonules are seen to travel forward in the valleys between the ciliary processes of the pars plicata, ensheathing the ciliary processes, and attaching to the anterior and posterior lens capsule at its equator. Posterior lens (A); ciliary processes (B); posterior lenticular zonules (C); anterior lenticular zonules (D). (SEM; magnification × 40.) (From K.N. Gelatt, *Veterinary Ophthalmology*, Lea & Febiger, Philadelphia, 79 (1981). With permission.)

cortical fibers elongate anteriorly and posteriorly, they attach at a line, the lens suture. In human, dogs, and cats these suture lines are in a Y shape, upright anteriorly and inverted posteriorly, but this varies with other species. This results in a layered arrangement of the lens fibers with the oldest fibers in the center and the newest fibers surrounding them. These layers are divided into regions that can be seen clinically using the biomicroscope. These regions are, from anterior to posterior, the anterior lens capsule, lens epithelium, anterior adult cortex, anterior adult nucleus, anterior fetal nucleus, embryonal nucleus, posterior fetal nucleus, posterior adult

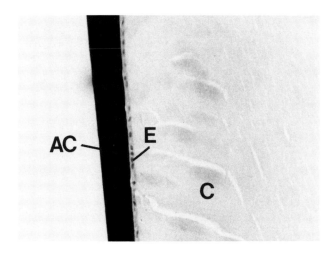

Figure 22.   The anterior lens capsule (AC), lens epithelium (E), and anterior cortex (C) of a mature dog. (PAS; magnification × 39.)

nucleus, posterior adult cortex, and the posterior lens capsule. On cross-section, the lens fibers are hexagonal and adjacent cells interdigitate with each other in the form of microplicae and by a ball-and-socket arrangement (Figure 23).[167] The space between these fibers is extremely small, accounting for only 5% of the lens volume.[169]

2. Physiology

Since the lens is avascular, it must obtain its nutrition and oxygen from the surrounding aqueous and vitreous humors. These materials and waste products traverse the lens capsule to the lens epithelium which is the area of the lens with the highest metabolic rate.[169] Here, glucose and oxygen are used by the lens to transport carbohydrates, electrolytes, and amino acids into the lens, and to synthesize new lens protein.[169]

The lens is 33% protein and 66% water and is, therefore, a dehydrated organ.[169] This state of hydration is maintained by an active Na^+ ion-water pump in both the epithelium and the lens fibers.[169] Also, the lens is electronegative with a –64 to –78 mV potential difference across the lens capsule and a –23 mV difference between the anterior and posterior lens surface.[170]

The lens proteins are divided into water-soluble and insoluble.[168] Immunologically, these proteins are organ, but not species specific.[70,169] The water soluble proteins are mostly crystallins (α, β, and γ) and comprise 85% of the lens protein in young animals, decreasing with age (Table 8).[70,168] The γ crystallin constitutes 1.5% of the lens proteins in the adult, but can constitute as much as 60% in weanling animals.[168] In addition in the avian species, there is an additional crystallin, δ.[168] The precise role of the crystallins is not known, but they are likely to be a major

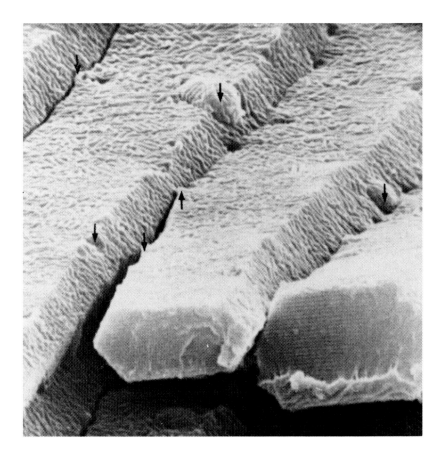

Figure 23.   Sagittal section of the lens fibers of a dog. The individual fibers are hexagonal in shape with surface irregularities. Regions of indentations and protruberances are seen (arrows) and represent the ball and socket interdigitations. (SEM; magnification × 4212.) (From K.N. Gelatt, *Veterinary Ophthalmology,* Lea & Febiger, Philadelphia, 77 (1981). With permission.)

contributor to the cytoarchitecture. The insoluble proteins are associated with the membranes of the lens fibers, are found primarily in the nucleus, and increase with age and in a cataractous lens.

The concentrations of K^+, amino acids, glutathione, and inositol in the lens exceeds that of the aqueous, while the concentration of Na^+, Cl^-, and water is lower (Figure 24).[70,169] To maintain this balance, the lens uses energy to extrude Na^+ through a Na^+-K^+-ATPase pump.[70,169] This pump mechanism is inhibited by the cardiac glycoside, ouabain, and by anticholinesterase agents such as phospholine iodide.[169] These will result in opacification of the lens by interfering with Na^+-K^+-ATPase and, subsequently, lens protein synthesis. In addition to the above, the concentration of ascorbate in the lens exceeds that of the aqueous humor, except in

Table 8

Characteristics of Major Lens Proteins from Adult Mammals

Protein	Molecular weight (Da)	Lens proteins (%)	Site of first appearance
α-Crystallin	80×10^4	32	Epithelium
β-Crystallin		55	As cortical fibers
$β_H$	$10—55 \times 10^4$		begin to differentiate
$β_L$	$5—8 \times 10^4$		
γ-Crystallin	20×10^4	1.5	As cortical fibers begin to differentiate

Modified from J. G. Jose, *Biochemistry of the Eye*, R.E. Anderson, Ed., Am. Acad. Ophthalmol. Manual, 111 (1983).

the dog, rabbit, and guinea pig.[70] The lens contains a high concentration of glutathione which is synthesized in the lens. Most of the glutathione in the lens exists in the reduced state (GSH) with only 6.8% in the oxidized state (GSSG).[169] The level of glutathione in the lens cortex is higher than that of the nucleus. Glutathione is maintained in the GSH state by NADPH which is generated from the pentose shunt metabolism of glucose.[169] The role of GSH is to preserve the lens proteins by maintaining high levels of reduced sulfhydrl groups and to maintain the transport pumps.[169] The level of GSH is found to be reduced in the cataractic lens.

The limiting enzyme in the metabolism of glucose in the lens is hexokinase which forms glucose-6-phosphate. Glucose-6-phosphate is metabolized in the lens primarily (80%) by anaerobic glycolysis and therefore does not require oxygen. In addition, 15% of the glucose-6-phosphate is metabolized by the pentose shunt pathway. The Krebs cycle utilizes only 5% of the lens glucose.[169] The pentose pathway serves to generate pentoses needed for RNA synthesis and to maintain glutathione in a reduced state.[169] In addition to these a fourth pathway, the sorbitol pathway, is present in the lens. This is an alternative pathway that is utilized when the concentration of glucose exceeds the capacity of the rate-limiting enzyme, hexokinase, such as is seen in diabetes mellitus. Debate exists as to whether the sorbitol pathway is nonenzymatic or dependent on the enzyme aldose reductase.[169] Regardless, the endproduct of this pathway is sorbitol. Because of the slow breakdown of sorbitol by the lens an osmotic shift occurs, resulting in the accumulation of water and a decrease in protein synthesis in the lens and resulting in cataract formation. Deprived of glucose, the lens in culture will gain water and lose transparency, but given glucose and deprived of oxygen the lens can maintain transparency.[169]

In general, the formation of cataracts is associated with[169]

1. Increased efflux of K^+
2. Gain in Na^+ in the lens
3. Loss of glutathione, amino acids, inositol, and K^+ from the lens

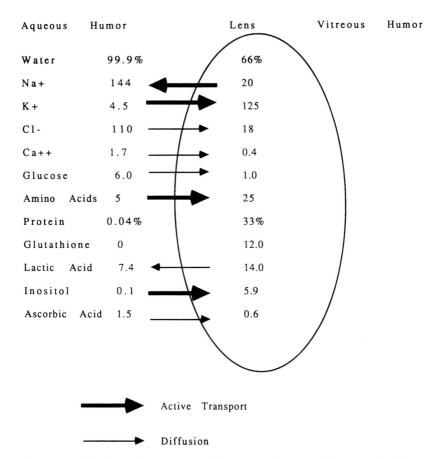

Figure 24.   The chemical composition of the aqueous humor and lens. (Modified from R.A. Moses and W.H. Hart, Eds., *Adler's: Physiology of the Eye*, 8th ed., C.V. Mosby, St. Louis, 268, 1987.)

4. Compensatory increase in NADPH synthesis
5. Decreased protein synthesis
6. Decrease in soluble and an increase in insoluble protein
7. Increase in protein S–S groups and Ca^{2+}
8. Decreased ATP
9. Decreased enzymatic activity[169]

The majority of cataractogenic agents damage the ability of the lens to maintain GSH synthesis or alter membrane permeability allowing a shift in Na^+ and K^+. This can occur by one of eight mechanisms:

1.   **The production of harmful metabolites** — Examples of this include disophe-

nol and quinone as toxic agents. Also, inflammatory cataracts are considered in this category.

2. **Accumulation of metabolites such as is seen in diabetes mellitus or galactosemia** — This is seen in humans, dogs, rats, marsupials, and other species, but is rare or not seen in cats and mice.

3. **Loss of a metabolite** — Examples of this include chlorpromazine which inhibits Ca^{2+} ATPase, triparinol which interferes with cholesterol synthesis, and ouabain which inhibits Na^+-K^+-ATPase.

4. **Dietary deficiency** — Several reports of dietary deficiency resulting in cataracts in laboratory animals and fish exist. These usually involve amino acids or vitamin deficiencies (or excess). For example, a deficiency of tryptophan has been shown to result in cataract formation in guinea pigs and rats.

5. **Radiation** — All electromagnetic wavelength outside the visible spectrum can result in cataracts.

6. **Hereditary** — These are reported to occur in humans, dogs, cats, birds, mice, rats, guinea pigs, horses, cattle, and many other species. Some, such as the Nakano mouse cataract which has a low molecular weight peptide interfering with Na^+-K^+-ATPase, are well understood, while others have simply been described.

7. **Senile** — This is associated with the shift in soluble to insoluble protein, but likely has additional factors as not all senile lenses are cataractic. Regardless of the cause, the end result of all of these processes is opacification, partial or total, of the lens.

8. **Secondary to retinal degeneration** — This is seen in the RCS rat, where posterior subcapsular cataracts are seen following the degeneration of the retina secondary to defective retinal pigment epithelial cell phagocytosis.[132] It is suggested that the cataracts are the result of toxic products released from the retina. The incidence of these cataracts can be decreased in the RCS rat by dietary manipulation.[171]

Not all areas of the lens experience the same result when undergoing cataractogenesis. For example, the equator of the lens is the site of mitotic activity, protein synthesis, and the transformation of epithelial cells into cortical fibers. It is here that the vacuolation of carbohydrate cataracts first occurs and this is also the site of damage by ionizing radiation.[169] The posterior lens is susceptible to metabolic changes occurring between the vitreous cavity and the posterior lens. This is often the site of toxic cataracts, either from intravitreal injection of the compound or because of the accumulation of toxic products in the vitreous because of its slow metabolic rate.[169]

III. ORBIT

A. Anatomy
The question has often been forwarded that asks, does the orbit function to protect

the eye from external injury or the brain from the eye? There are arguments for both. For example, rabbits and rodents have very prominent eyes which project anterior to the orbital rim while, in contrast, most primates present with globes deeply set within the orbit and the orbital rim extends well anterior to the globe.

It is conjectured that the activity and habits of an animal may have had an influence on the anatomy of their orbit. Activity can be divided into diurnal, active during daylight; nocturnal, active during night, and arrhythmic; active day and night. Habits can be broadly subdivided into animals which hunt (predators) and those which are hunted (prey). Ultimately, these activities and habits may have contributed significantly to the orbital and ocular arrangements required for the perpetuation and survival of various species. It is interesting to consider the influence evolution has had in changes of function for survival irrespective of protection of the eye or brain.

The orbits are "sockets" that vary in depth between and within species. They also vary in the completeness of the orbital rim and floor. Carnivores have an incomplete bony orbital rim which is completed by dense connective tissue referred to as the orbital ligament. This is a condensation of periosteum which extends from the frontal process of the zygomatic bone to the zygomatic process of the frontal bone. The floor of the bony orbit is also incomplete in the carnivore. Dense fibrous connective tissue completes this bony defect. This tissue is a part of the periorbita which varies in thickness, being thickest where there is no bone.

The bones of the orbit may include the frontal, lacrimal, zygomatic, palatine, ethmoid, sphenoid, and maxillary, but differences between species is marked. The size and positions of these bones are variable between species as are the foramina through which nerves, arteries, and veins enter and exit the orbit. Cranial nerves 2, 3, 4, 5, and 6 enter the orbit through these emissaries. The ophthalmic arteries and orbital veins also pass through their respective foramina.

The type of orbit, complete or incomplete, does not necessarily indicate the placement of the eyes within the orbit or the optic axis. For example, the angle between the two eyes in humans and cats is 0 to 20°, humans having a complete and cats an incomplete orbit. In contradistinction, horses have a complete orbit and an optic axis of 70 to 80° and rats have an incomplete orbit and an optic axis of 120 to 130°. The angle of the two orbits differ within species as well. Skull types result in variation of optic axis, as exemplified in dogs with dolicocephalic (elongated noses) contrasted with brachyocephalic (short noses) skull types. The dolicocephalic skull has a more lateral orbital placement while the brachyocephalic animal has an anterior placement. The theory that the hunting animal (dog) has a more anterior ocular placement than the hunted (i.e., rabbit) does not seem completely appropriate when looking at the greyhound, which is reportedly a "sight hound", with marked lateral placement of its eyes. Lateral placement provides a greater panoramic view for the animal, while anterior placement provides more binocular (stereoscopic) vision. The primate and feline are the most appropriate examples of anterior placement.

The globe is suspended in the orbit by muscles, ligaments, conjunctiva, and other orbital contents. In order to simplify a very complex system, it may be best to start with the optic nerve (second cranial nerve) as it exits the orbit toward the brain. The dural sheath that surround the optic nerve continue forward as portions of the orbital fascia.[3,172] The dura bifurcates at the optic foramina, forming the periorbita (periosteum) covering the orbital bones which is continuous with the periosteum of the facial bones at the orbital rim. At the orbital rim is the orbital septum, which is considered an extension of the periorbita as it continues into the lids. The orbital septum prevents the escape of orbital fat, etc., from the orbit as well as preventing orbital infection from entering externally.

The inner dura continues along the optic nerve and becomes Tenon's capsule.[3,172] This is a very complex structure, but it can be simplified by explaining it as a fibrous, elastic tissue sheath that envelopes the extrinsic muscles, forms the check ligaments between muscles, and covers and blends into the sclera posteriorly as well as extending anteriorly to the limbus. At the orbital apex, this dural sheath is concentrated into a dense collagenous ring, the Annulus of Zinn. The dural sheath (Tenon's capsule) that covers the extrinsic muscles also has thickened condensations that are especially well developed where one muscle passes over another. They function as check ligaments which serve as a fluid brake during and after muscle contraction. These are especially evident in the area of the trochlea of the superior oblique as well as in the Annulus of Zinn.

Tenon's capsule is firmly attached to the conjunctiva and sclera 1 to 2 mm posterior to the limbus. It is thickest at the muscle insertion and thins posteriorly. Orbital fat and glands (orbital gland in carnivores) are also found within the orbit.

Species variability is also manifest with respect to the extrinsic musculature of the eye. The major difference is in the presence or absence of the retractor oculi muscle. Primates, including humans, either do not have this muscle or it is vestigial; however, all other common laboratory animals do. Although present in these animals, the extent of development varies, for example, dogs and cats have well-developed retractor muscles while rats and mice do not. Those animals that do have well-developed retractors can withdraw their globes deep within the orbit. The rodent cannot and it is easy to prolapse the globe by minor digital pressure in these animals. This phenomenon is due to a combination of orbital depth and poor development of retractor muscles. Ocular movement is also less in those animals whose extrinsic muscular development is lacking. Although not directly related to our interest in this chapter, the bird has little ability to move the globe and the extrinsic muscles are very poorly developed.

The extrinsic muscles are the rectus, oblique, and retractor muscles. There are four rectus muscles; dorsal, medial, ventral, and lateral. Contraction of each moves the eye in the direction of its name. The four recti originate adjacent to the optic foramen, from the Annulus of Zinn, and insert on the globe at variable distances from the limbus in the general region of the globe from which the name of the muscle is derived. There are two oblique muscles, the dorsal and ventral. The dorsal oblique muscle also originates from the annulus and courses forward through a condensation of fibroelastic connective tissue, the trochlea, in the medial aspect of the orbit. The dorsal oblique then extends laterally to insert both below and above the insertion of the lateral rectus. Although complex, the primary action of the superior oblique is intorsion. The ventral oblique originates from the medial orbital wall and inserts close to the insertion of the lateral rectus from a ventral position. As in the case of the superior oblique, its action is complex but its primary action is extorsion.

The retractor oculi muscles are the largest of the extrinsic muscles in those animals in which they are present and well developed. They consist of bundles of muscles, usually four, which originate from the annulus and insert posterior to the equator, forming a smaller cone than that of the recti. Its primary function is to withdraw the globe within the orbit.

Another muscle which is not directly attached to the globe but integral to ocular mobility is the levator palpebral superioris. This muscle originates from the annulus, courses superiorly, fanning out to interdigitate with the orbicularis oculi muscles and insert in the upper lid. Its primary function is to elevate the upper lid.

The innervation of these muscles are provided by cranial nerves 3, 4, and 6. A useful mnemonic has the formula $(LR_6 \, SO_4)_3$. L = lateral rectus and R = retractor oculi, innervated by the sixth cranial nerve (note: in those animals in which a retractor is absent, LR = lateral rectus). SO = superior oblique, innervated by the fourth cranial nerve. All the rest $[(\quad)_3]$ = medial rectus, ventral rectus, dorsal rectus, inferior oblique, and the levator palpebral superioris are innervated by the third cranial nerve. As stated earlier, the degree of muscular development determines mobility. Evolution certainly contributes to the way these modifications develop. Those characteristics which provide the best opportunity for survival are those that persist; hence, the variations between species.

IV. ADNEXA

A. Eyelid

The eyelids are modified integumental folds which protect the eye from injury and dehydration, spread the precorneal tear film, assist in propelling lacrimal fluid toward the lacrimal drainage apparatus and out of the interpalpebral space, remove foreign bodies from the surface of the eye, exclude light during sleep and provide some of the constituents of the precorneal film. The lids are divided into three structural planes; the external (skin and subcutaneous fascia) plane, the muscular plane, and the fibrous plane.

The upper eyelid is more mobile than the lower, and very delicate, in most animals. Pliability and thickness varies with sex and age, with older animals having less elasticity and greater thickness. This is especially true in male cats. The lids are relatively fat free and they are provided with many hairs of variable size. The carnivores have a general almond-shaped palpebral fissure, most apparent in the dog. Rodents and laprines have more round palpebral margins, the rabbit being an example. There is a zone relatively free from surface hairs near the free lid margin. This zone varies in width from 1 to 3 mm or more. On the eyelid margin at the junction of the skin and conjunctiva is a row of ducts, most numerous in the upper lid, which are the openings of the tarsal (Meibomian) glands. These glands provide the lipid layer of the precorneal tear film which serve to prevent tear overflow and evaporation. The tarsal glands are contained in an area referred to as the tarsal plate, best developed in the primate. Other laboratory animals have a poorly developed tarsal plate. In addition to the tarsal glands, the lids contain the apocrine glands of Moll and the holocrine sebaceous glands of Zeis which empty their secretions in the follicles of the cilia.[3] The sensory innervation of the eyelids is provided by the fifth cranial nerve, while the motor function comes from the seventh cranial nerve.

The cilia (eyelashes) are found external to the ducts of the tarsal glands and are usually more plentiful and longest in the upper lid. Cilia are absent on the lower eyelid of the dog and horse and are absent altogether in the cat.[3] They are extremely sensitive and along with the specialized facial hairs, vibrissae, are the first line of protection for the eye. They are named according to location: superciliary, mystacial, genal, submental, and interramal. These hairs are innervated by the fifth cranial nerve and touch or air current will elicit a blink. The cilia of most animals are usually more heavily pigmented and are more coarse than other facial hair except in white animals or albino animals. They are replaced at 4 or 5 month intervals, but this also varies with the species.

The musculature of the eyelids includes the muscles which close the eye, the orbicularis oculi, and those that open the eye, the levator palpebrae being the major muscle. A smooth muscle (Muller's muscle) is found deep, extending along the course of the levator. It is under adrenergic innervation and results in a widened palpebral margin when stimulated. The orbicularis muscle fibers are arranged in bundles parallel to the free lid margin. The muscle is restricted in its circular sphincter action by restrictions in the nasal and temporal extremities. A firm attachment is accomplished by the medial palpebral ligament in all animals except the primate, which has a common tendon of the superficial heads of the pretarsal orbicularis oculi that inserts in the medial orbital wall. On the temporal side, the common tendon of the temporal pretarsal orbital axis assumes the function of the lateral ligament in primates, while in the other animals it is the retractor anguli muscle which serves this function. These structures allow the lid to close from laterally to medially, propelling tears to the medial aspect where the lacrimal puncta are found. The innervation to the eyelids includes sensory, motor, and autonomic portions. The sensory to the upper lid is the frontal branch of the ophthalmic division of the fifth cranial nerve, while the lower is the maxillary division of the fifth cranial nerve. Motor innervation is provided by the third (levator) and seventh (orbicularis) cranial nerves. The autonomic innervation is sympathetic to Muller's and the smooth muscle of the third eyelid.

The upper and lower lacrimal puncta are the openings of the lacrimal drainage apparatus. They are found at the medial canthus adjacent to the junction of the skin and conjunctiva. From the puncta, the upper and lower canaliculi course in the lids beneath the medial palpebral ligament where they are joined to form the lacrimal sac. The primate has a well-developed lacrimal sac; however, other laboratory animals only have a junction between the two canaliculi. The sac drains into the nasolacrimal duct which continues forward to end near the external nares. The mechanical closure of the lids propels the tear film medially and the pressure on the lacrimal sac results in an abrupt negative pressure within the sac and canaliculi when the lid relaxes results in a sucking effect at the puncta, drawing tears into the lumen. Further blinking then forces the fluid through the nasolacrimal duct and into the nose.

B. Conjunctiva

The conjunctiva is a modified integumental mucous membrane that aids in the anterior suspension of the eye. It is vascular, possesses lymphatics, glands, and nerves. The conjunctiva can be divided into two layers, the epithelium and the substantia propria. The epithelium originates from surface ectoderm and is composed of nonkeratinizing, cuboidal, epithelial cells. Between these cells are also found focal areas of goblet cell differentiation. The goblet cells are mucous secretors, supplying the layer of the precorneal tear film adjacent to the cornea, which covers the glyco-calyces of the corneal epithelial cells. They are most plentiful in the inferior, nasal fornix.[173] Goblet cells appear to be absent from the bulbar conjunctiva.[173] The substantia propria is found immediately beneath the basement membrane of the epithelium and can be subdivided into the superficial adenoid layer and the deeper fibrous layer. The adenoid layer contains lymphocytes, a few mast cells, and plasma cells, lymphatics, and blood vessels. The deep layer is a thick meshwork of col-lagenous and elastic fibers in the bulbar conjunctiva and fornices. The majority of vessels and nerves supplying the conjunctiva are found in this layer.

The conjunctiva is divided into the following three regions: (1) that which lines the inner surface of the eyelids (palpebral or tarsal); (2) that which covers the globe (bulbar); and (3) the junction (fornix) between the two. It is in the fornix that the ducts of the lacrimal gland open, depositing the aqueous portion of the precorneal tear film on the surface of the eye.

The palpebral conjunctiva begins at the mucocutaneous junction of the lid margin. Here, the substantia propria is firmly attached to the underlying tarsus and is not freely movable. The substantia propria of the bulbar conjunctiva is very delicate and freely movable and only firmly attached to the globe and Tenon's capsule at the limbus. The substantia propria is redundant in the fornices. There are large aggre-gates of lymphocytes and plasma cells in the fornices. IgA is produced by these lymphocytes and by those in the lacrimal gland. IgA is present in tears in a concen-tration equal to that of IgG and as it crosses the conjunctival epithelium IgA acquires a secretory piece.[174] Inflammatory conditions cause lymphoid hyperplasia which results in clinical folliculosis or follicular conjunctivitis. Infections, toxic, or allergic reactions are potential causes of this phenomenon. Chronic inflammation also causes increases in the goblet cell population, producing a thickened conjunctiva with increased mucous secretion.

The deeper layers of the bulbar conjunctival epithelium often have pigment granules in pigmented animals. They are most plentiful in the conjunctiva overlying the caruncle and edge of the third eyelid.

The blood supply to the conjunctiva is complex and varies among species.

Generally the blood supply is from two sources: (1) the palpebral branches of the nasal and lacrimal arteries and (2) the anterior ciliary arteries. The palpebral arteries send out multiple branches from the tarsus and terminate as glomerular-like capillary tufts which results in the redder appearance of the palpebral conjunctiva as compared to the bulbar conjunctiva. This phenomenon is important when evaluating the conjunctiva for erythema or inflammation. The venous drainage of lids and conjunctiva is via the palpebral and conjunctival veins.

The response of the conjunctiva to topical irritants includes hyperemia, hemorrhage, edema (chemosis), depigmentation, necrosis, decrease in the number of goblet cells, increased secretion of mucus, lymphoid hyperplasia, and hypertrophy and hyperplasia. The agents which result in this are similar to those which are toxic to the cornea and include caustic agents, solvents, surfactants, and detergents. These are discussed in detail in Chapter 15. In addition, certain systemic agents can result in conjunctival damage, an example of which would be the photosensitization seen with systemic phenothiazine.

C. Nictitating Membrane (Third Eyelid)

All of the common laboratory animals possess a nictitating membrane (nictitans, third eyelid). Humans possess a plica semilunaris which is a vestigial nictitating membrane. The nictitating membrane is located in the inferonasal aspect of the eye, except in birds where it is found in the superior nasal fornix. It moves passively superior and temporal across the eye as the eye is withdrawn into the orbit, displacing the bulk of the base of the third eyelid resulting in this sweeping action. Most animals possess some smooth muscle which is under autonomic innervation (sympathetic) and which may contribute to the movement of the membrane or at least its position.

The third eyelid is is formed by a cartilaginous skeleton. This is in the form of a "T" with the leading edge fine and delicate and tightly adherent to the overlying conjunctiva. The cartilage is hyaline in the dog and cow and elastic in the horse and cat.[3] It is concave in shape, conforming to the cornea, and is covered by conjunctiva.[3] The base of the third eyelid is surrounded by glandular tissue which is serous in the horse and cat and seromucoid in the dog and cow.[3] This gland contributes to the aqueous portion of the precorneal tear film and empties via ductules onto the bulbar surface of the nictitans. Deep to this gland, in the rat, mouse, and rabbit, can be found an additional gland, the Harders gland. The deep gland has been identified with four glandular types. Lipid secretor with large amounts of porphyria, mucous secretor, seromucous secretor, and mixed. Other animals only possess a superficial gland. The bulbar surface of the third eyelid also contains lymphoid follicles which, in chronic irritation, will hypertrophy, resulting in follicular conjunctivitis. In the rabbit, there are also a series of lymphoid follicles on the palpebral free lid margin. The nictitating membrane is functional in cats, dogs, and rabbits. Its function may be vestigial in rats, mice, and primates.

The nictitans is considered part of the conjunctiva and its reactions to toxicity are similar.

D. The Lacrimal Apparatus and Tear Film
1. Anatomy and Physiology

The lacrimal apparatus includes the lacrimal gland, the gland of the third eyelid, Harders gland when present, accessory lacrimal glands, goblet cells, meibomian glands, and the drainage apparatus. These glands secrete the precorneal tear film. The precorneal tear film serves to maintain an optically uniform ocular surface, is a source of corneal oxygen and a small percent of corneal glucose, provides lubrication, maintains corneal hydration, contains antibacterial lysozyme, and aids in the mechanical removal of debris. Once formed, tears must be removed and this is achieved by the movement of the eyelids in the act of blinking, which close from lateral to medial, displacing the tears toward the lacrimal drainage apparatus. The drainage system consists of the upper and lower puncta, at the medial canthus, the upper and lower canaliculi into which they drain, the nasolacrimal sac where the canaliculi meet, and the nasolacrimal duct proper which drains into the nasal passage in most animals. Exceptions to this include the rabbit and the pig, which possess a single punctum in the lower and upper eyelids. Although the majority of tears leave via the nasolacrimal system, 10 to 25% are lost through evaporation.[174]

The products of the lacrimal glands form the three layers of the precorneal film. The corneal epithelium is hydrophobic and water is not retained on its surface without a wetting agent. This may be due to the lipid concentration of the epithelium. The mucin produced by the goblet cells provide the wetting agent which is intertwined with the microvilli and glycocalyces of the surface epithelial cells and is the first layer of the precorneal film. This layer serves to provide an optically smooth surface, traps debris and bacteria, and contributes to the local immunity by providing a medium for the adherence of IgA and lysozymes. The second layer is a serous layer and is provided by the primary lacrimal glands (lacrimal, nictitans, Harders) and by the glands of Krause and Wolfring.[174] This is the thickest portion of the tear film measuring 6.5 to 7.5 μm thick.[174] The inorganic salts, glucose, urea, proteins, and glycoproteins are found in this layer.[174] The third and most superficial layer of the precorneal tear film is the lipid layer provided by the tarsal and possibly the Zeis glands. Simplistically, this layer prevents overflow of tears by increasing surface tension and reduces evaporation.[174]

The osmolality of the tear film is equal to 0.9% sodium chloride if evaporation is prevented.[174] Changes in tonicity are seen in the syndrome of keratitis sicca (dry eye) and also with inappropriately formulated topical medications. Keratitis sicca results in a hypertonic tear film and secondary corneal dehydration. Also, changes in tonicity can result in irritation, blepharospasm, and changes in the corneal epithelium.

Several systemic medications have been demonstrated to result in a decrease in the production of the precorneal tear film and keratitis sicca. These include phenazopyridine,[175] sulfonamides,[176] beta-blocking agents, diazepam, phenothiazines, and others.[174] Topical medications, if irritating, will result in increased tear production. This is mediated through the stimulation of the ophthalmic branch of the trigeminal nerve as the afferent branch and the parasympathetic fibers of cranial nerve seven as the efferent branch. To prevent irritation topical medication must be similar in pH and osmolality to tears. The normal pH of human tears is 7.3 to 7.2 and a pH outside the range of 6.6 to 7.8 will result in irritation.[174]

The quantity of tears for a given stimulus differs among species. For example, humans produce approximately 15 µl of tear absorption in 5 min, whereas a dog will produce approximately 15 µl of tear absorption in 1 min for the same stimulus; a horse produces even greater quantities. Clinically, the quantity of tear production is evaluated using the Schirmer tear test which serves to measure the serous portion of the tears. This test uses a 5 × 30 mm strip of #41 Whatman® filter paper inserted into the conjunctival sac.

An additional method for evaluating the tear film is the tear break-up time. This is performed by instilling fluorescein solution in the conjunctival sac and scanning the cornea for the appearance of dry (fluorescein-free) spots. These will appear in approximately 20 s. If reduced, this is an indication of a mucin deficiency and suggests dysfunction of the goblet cells.[173]

V. SUMMARY

At this point the reader may have a feeling of being overwhelmed by the differences in the ocular anatomy and physiology of the various species. While this is true in many regards, there are also many similarities. It is up to the investigator to be aware of the differences and to account for them when selecting an animal model or describing or interpreting an ocular response. It is also important to realize that while the mechanism of a response to an ocular toxin may be similar between species, the degree of the response may not. This is especially true of the rabbit, which unfortunately has been the accepted ophthalmic model for years, and yet reacts in a much more severe manner to a toxic insult than do many other species. The optimum investigational model is always, therefore, the species that the results are to be applied toward. If this is not possible, then the use of cell cultures, from the eye of the species of choice, is an alternative that in many instances may be better than selecting a different species. The last choice, and unfortunately the most common, is to perform the investigation on one species and apply the results to a different species.

For additional information the reader is directed to the remaining chapters in this text. If you are involved in the area of ocular toxicology, pharmacology, or physiology and are using animal models in your studies, then additional resources that may prove helpful include: (1) W. M. Grant, *Toxicology of the Eye,* 2nd ed., Charles C Thomas, Springfield, IL, 1974; (2) K. F. Tabbara and R. M. Cello, *Animal Models of Ocular Diseases,* Charles C Thomas, Springfield, IL, 1984.

REFERENCES

1. R.E. Schmidt, *Vet. Pathol.* 8, 28 (1971).
2. S.R. Waltman and W.M. Hart, *Adler's: Physiology of the Eye,* 8th ed. R.A. Moses and W.H. Hart, Eds. C.V. Mosby, St. Louis, 36 (1987).
3. C.L. Martin and B.G. Anderson, *Veterinary Ophthalmology,* K.N. Gelatt, Ed., Lea & Febiger, Philadelphia, 12 (1981).
4. B.S. Fine and M. Yanoff, *Ocular Histology,* 2nd ed. Harper & Row, Hagerstown, MD 163 (1979).
5. I.K. Gipson, S.J. Spurr-Michaud, and A.S. Tisdale, *Invest. Ophthalmol. Vis. Sci.* 28, 212 (1987).
6. R.M. Stanifer, R.K. Synder, and F.L. Kretzer, *Biochemistry of the Eye,* R.E. Anderson, Ed. American Academy of Ophthalmology Manual, 23 (1983).
7. B. Nichols, C.R. Dawson, and B. Togni, *Invest. Ophthalmol. Vis. Sci.* 24, 570 (1983).
8. L. Vantrappen et al., *Invest. Ophthalmol. Vis. Sci.* 26, 220 (1985).
9. C.F. Bahn et al., *Ophthalmology* 91, 558 (1984).
10. C.F. Bahn et al., *Invest. Ophthalmol. Vis. Sci.* 27, 44 (1986).
11. R.M. Gwin et al., *Invest. Ophthalmol. Vis. Sci.* 22, 267 (1982).
12. M.R. O'Neal and K.A. Polse, *Invest. Ophthalmol. Vis. Sci.* 27, 457 (1986).
13. H. Burton, *Adler's: Physiology of the Eye,* 8th ed. R.A. Moses and W.H. Hart, Eds. C.V. Mosby, St. Louis, 60 (1987).
14. W.H. Beekhuis and B.E. McCarey, *Exp. Eye Res.* 43, 707 (1986).
15. J. Fischbarg, *Exp. Eye Res.* 15, 615 (1973).
16. S. Hodson and F. Miller, *J. Physiol.* 263, 563 (1977).
17. T. Chan-Ling, N. Efron, and B.A. Holden, *Invest. Ophthalmol. Vis. Sci.* 26, 102 (1985).
18. H. Mishima et al., *Invest. Ophthalmol. Vis. Sci.* 28, 1521 (1987).
19. K. Watanabe, S. Nakagawa, and T. Nishida, *Invest. Ophthalmol. Vis. Sci.* 28, 205 (1987).
20. M.M. Jumblatt and C.A. Paterson, *Invest. Ophthalmol. Vis. Sci.* 32, 360 (1991).
21. B.D. Srinivasan and P.S. Kulkarni, *Invest. Ophthalmol. Vis. Sci.* 19, 1087 (1980).
22. P.S. Kulkarni and B.D. Srinivasan, *Exp. Eye Res.* 41, 267 (1985).
23. P.S. Kulkarni, P. Bhattacherjee, K.E. Eakins, and B.D. Srinivasan, *Curr. Eye Res.* 1, 43 (1981).
24. R.N. Weinreb, M.D. Mitchell, and J.R. Polansky, *Invest. Ophthamol. Vis. Sci.* 24, 1541 (1983).
25. C.P. Moore, R. Dubielzig, and S.M. Glaza, *Vet. Pathol.* 24, 28 (1987).
26. A.D. MacMillan et al., *J. Am. Vet. Med. Assoc.* 175, 829 (1979).
27. M.B. Ekins et al., *Exp. Eye Res.* 36, 279 (1983).
28. R.W. Bellhorn, G.E. Korte, and D. Abrutyn, *Lab. Anim. Sci.* 38, 46 (1988).
29. A. Sebesteny et al., *Lab. Anim.* 19, 180 (1985).
30. B.A. Garibaldi and M.E. Pecquet Goad, *Vet. Pathol.* 25, 173 (1988).
31. R.M. Gwin et al., *J. Am. Anim. Hosp. Assoc.* 18, 471 (1982).
32. C.L. Martin and P.F. Dice, *J. Am. Anim. Hosp. Assoc.* 18, 327 (1982).
33. B.S. Fine and M. Yanoff, *Ocular Histology,* 2nd ed. Harper & Row, Hagerstown, MD, 197 (1979).
34. A.A. Abdel-Latif, *Biochemistry of the Eye,* R.E. Anderson, Ed. American Academy of Ophthalmology Manual, 48 (1983).
35. L.N. Thibos, W.R. Levick, and R. Morstyn, *Invest. Ophthalmol. Vis. Sci.* 19, 475 (1980).
36. J.N. Shively and G.P. Epling, *Am. J. Vet. Res.* 30, 13 (1969).
37. J. Szalay, B. Nunziata, and P. Henkind, *Exp. Eye Res.* 21, 531 (1975).
38. U. Schaeppi and W.P. Koella, *Am. J. Physiol.* 207, 273 (1968).
39. G.W. van Alphen, R. Kern, and S.L. Robinette, *Arch. Ophthalmol.* 74, 253 (1965).
40. B. Ehinger, B. Flack, and H. Persson, *Acta Physiol. Scand.* 72, 139 (1968).
41. T. Yoshitomi and Y. Ito, *Invest. Ophthalmol. Vis. Sci.* 27, 83 (1968).
42. B. Ehinger, *Arch. Ophthalmol.* 77, 541 (1967).
43. H.S. Thompson, *Adler's: Physiology of the Eye,* 8th ed. R.A. Moses and W.H. Hart, Eds. C.V. Mosby, St. Louis, 311 (1987).
44. W.H. Havener, Ed., *Ocular Pharmacology,* C.V. Mosby, St. Louis, 1983.

45. M.A. Kass, N.J. Holmberg, and M.E. Smith, *Invest. Ophthalmol. Vis. Sci.* 20, 442 (1981).
46. S. Mishima and K. Masuda, *Metab. Pediatr. Ophthalmol.* 3, 179 (1979).
47. P. Bhattacherjee, P.S. Kulkarni, and K.E. Eakins, *Invest. Ophthalmol. Vis. Sci.* 18, 172 (1979).
48. M.A. Kass and N.J. Holmberg, *Invest. Ophthalmol. Vis. Sci.* 18, 166 (1979).
49. P. Engstrom and E.W.Dunham, *Invest. Ophthalmol. Vis. Sci.* 22, 757 (1982).
50. R.N. Williams, N.A. Delamere, and C.A. Paterson, *Exp. Eye Res.* 41, 733 (1985).
51. P.L. Cooley, R. Milvae, R.C. Riis, and L.J. Larratta, *Am. J. Vet. Res.* 45, 1383 (1984).
52. A. Regnier et al., *J. Ocul. Pharm.* 2, 165 (1986).
53. P.S. Kulkarni, A.V. Rodriguez, and B.D. Srinivasan, *Invest. Ophthalmol. Vis. Sci.* 25, 221 (1984).
54. R.N. Williams, P. Bhattacharjee, and K.E. Eakins, *Exp. Eye Res.* 36, 397 (1983).
55. T.M. Richardson et al., *Exp. Eye Res.* 41, 31 (1985).
56. S.Y.K. Yousufzai and A.A. Abdel-Latif, *Exp. Eye. Res.* 37, 279 (1983).
57. P. Lee, S.M. Podos, and C. Severin, *Invest. Ophthalmol. Vis. Sci.* 25, 1087 (1984).
58. T. Yoshitomi and Y. Ito, *Invest. Ophthalmol. Vis. Sci.* 29, 127 (1988).
59. R.A.F. Whitelocke and K.E. Eakins, *Arch. Ophthalmol.* 89, 495 (1973).
60. A.H. Neufeld and M.L. Sears, *Exp. Eye Res.* 17, 445 (1973).
61. P.S. Kulkarni and B.D. Srinivasan, *Invest. Ophthalmol. Vis. Sci.* 23, 383 (1982).
62. J. Stjernschantz, T. Sherk, and M. Sears, *Prostaglandins* 27, 5 (1984).
63. D.F. Woodward and S.E. Ledgard, *Invest. Ophthalmol. Vis. Sci.* 26, 481 (1985).
64. A.J. Higgins, *J. Vet. Pharmacol. Ther.* 8, 1 (1985).
65. P.S. Kulkarni and B.D. Srinivasan, *Invest. Ophthalmol. Vis Sci.* 24, 1079 (1983).
66. J. Caprioli, *Adler's: Physiology of the Eye,* 8th ed. R.A. Moses and W.H. Hart, Eds. C.V. Mosby, St. Louis, 204 (1987).
67. R.A. Moses, *Adler's: Physiology of the Eye,* 8th ed. R.A. Moses and W.H. Hart, Eds. C.V. Mosby, St. Louis, 291 (1987).
68. W.G. Unger, D.F. Cole, and B. Hammond, *Exp. Eye Res.* 20, 255 (1975).
69. A. Al-Ghadyan, A. Mead, and M. Sears, *Invest. Ophthalmol. Vis. Sci.* 18, 361 (1979).
70. G.M. Schmidt, and D.B. Coulter, *Veterinary Ophthalmology,* K.N. Gelatt, Ed. Lea & Febiger, Philadelphia, 129 (1981).
71. M.V. Riley, *Biochemistry of the Eye*, R.E. Anderson, Ed. American Academy of Ophthalmology Manual, 79 (1983).
72. P. Henkind, M. Leitman, and E. Weitzman, *Invest. Ophthalmol. Vis. Sci.* 12, 705 (1973).
73. C.D. Phelps et al., *Am. J. Ophthalmol.* 77, 367 (1974).
74. J.M. Rowland, D.E. Potter, and R.J. Reiter, *Curr. Eye Res.* 1, 169 (1981).
75. G.R. Reiss et al., *Invest. Ophthalmol. Vis. Sci.* 25, 776 (1984).
76. J.E. Topper and R.F Brubaker, *Invest. Ophthalmol. Vis. Sci.* 26, 1315 (1985).
77. E.D. Weitzman et al., *Br. J. Ophthalmol.* 59, 566 (1975).
78. E.M. Van Buskirk, *Invest. Ophthalmol. Vis. Sci.* 18, 223 (1979).
79. D.A. Samuelson, G.G. Gum, and K.N. Gelatt, *Invest. Ophthalmol. Vis. Sci.* 30, 550 (1989).
80. T.M. Richardson, *Invest. Ophthamol. Vis. Sci.* 22, 319 (1982).
81. Y. Ohnishi and Y. Taniguchi, *Invest. Ophthamol. Vis. Sci.* 24, 697 (1983).
82. P.A. Knepper, A.I. Farbman, and A.G. Tesler, *Invest. Ophthamol. Vis. Sci.* 25, 286 (1984).
83. P.R.B. McMaster and F.J. Macri, *Arch. Ophthamol.* 79, 297 (1968).
84. A. Bill and C.I. Phillips, *Exp. Eye Res.* 12, 275 (1971).
85. A. Bill, *Exp. Eye Res.* 5, 185 (1966).
86. W.L. Fowlks and V.R. Havener, *Invest. Ophthalmol. Vis. Sci.* 3, 374 (1964).
87. H. Inomata, A. Bill, and G.K. Smelser, *Am. J. Ophthalmol.* 73, 1893 (1972).
88. D.A. Samuelson, G.G. Gum, and K.N. Gelatt, *Am. J. Vet. Res.* 46, 242 (1985).
89. R.N. Weinreb, M.D. Mitchell, and J.R. Polansky, *Invest. Ophthalmol. Vis. Sci.* 24, 1541 (1983).
90. R.N. Weinreb, J.R. Polansky, J.A. Alvarado, and M.D. Mitchell, *Invest. Ophthamol. Vis. Sci.* 29, 1708 (1988).
91. C.B. Camras et al., *Invest. Ophthalmol. Vis. Sci.* 28, 921 (1987).

92. C.B. Camaras et al., *Invest. Ophthalmol. Vis. Sci.* 28, 463 (1987).
93. F.A. Stern and L.Z. Bito, *Invest. Ophthalmol. Vis. Sci.* 22, 588 (1982).
94. L.Z. Bito, A. Draga, J. Blanco, and C.B. Camaras, *Invest. Ophthalmol. Vis. Sci.* 24, 312 (1983).
95. M. Hayashi, M.E. Yablonski, and L.Z. Bito, *Invest. Ophthalmol. Vis. Sci.* 28, 1639 (1987).
96. L.Z. Bito, R.A. Baroody, and O.C. Miranda, *Exp. Eye Res.* 44, 825 (1987).
97. P.S. Kulkarni and B.D. Srinivasan, *Invest. Ophthalmol. Vis. Sci.* 26, 1178 (1985).
98. P.F.J. Hoyng and N. de Jong, *Invest. Ophthalmol. Vis. Sci.* 28, 470 (1987).
99. P.F.J. Hoyng, *Exp. Eye Res.* 40, 161 (1985).
100. D.F. Gaasterland, *Invest. Ophthalmol. Vis. Sci.* 24, 153 (1983).
101. V.E. Kinsey, *Arch. Ophthalmol.* 40, 401 (1953).
102. E. DeBarnadinis, *Exp. Eye Res.* 4, 179 (1965).
103. C.N. Graymore, *Biochemistry of the Eye,* Academic Press, New York (1970).
104. H. Davson and L.T. Graham, Comparative Aspects of the Intraocular Fluids, in *The Eye,* Vol. 5, H. Davson, Ed. Academic Press, New York (1974).
105. V.E. Kinsey and D.V.N. Reddy, *The Rabbit in Eye Research,* J.H. Prince, Ed. Charles C Thomas, Springfield, IL (1964).
106. L. Bito, *Exp. Eye Res.* 10, 102 (1970).
107. V.B.G. Laurent, *Exp. Eye Res.* 33, 147 (1981).
108. J.K. Wegener and P.M. Moller, *Acta Ophthalmol. (Copenh)* 49, 577 (1971).
109. O. Kleinfeld and H.C. Neumann, *Klin. Monatsbl. Augenheilkd.* 135, 224 (1959).
110. A.M. Walker, *J. Biol. Chem.* 101, 269 (1933).
111. Sir S. Duke-Elder, *Physiology of the Eye,* Vol. 4 of System of Ophthalmology, C.V. Mosby, St. Louis, (1968).
112. J.P. Dernouchamps, *Doc. Ophthalmol.* 53, 193 (1982).
113. J.R. Blogg and E.H. Coles, *Vet. Bull.* 40, 347 (1970).
114. C. Furuichi, *Acta Soc. Ophthalmol.* 65, 561 (1961).
115. J.M. Risco and W. Nopanitaya, *Invest. Ophthalmol. Vis. Sci.* 19, 5 (1980).
116. G.Y. Wen, J.A. Sturman, and J.W. Shek, *Lab. Anim. Sci.* 35, 200 (1985).
117. A. Alm and A. Bill, *Adler's: Physiology of the Eye,* 8th ed. R.A. Moses and W.H. Hart, Eds. C.V. Mosby, St. Louis, 183 (1987).
118. A. Bil, and G. Sperber, *K. Ujiie, Int. Ophthalmol.* 6, 101 (1983).
119. S. Yoneya and M.O.M. Tso, *Arch. Ophthalmol.* 105, 681 (1987).
120. B.P. Gloor, *Adler's: Physiology of the Eye,* 8th ed. R.A. Moses and W.H. Hart, Eds. C.V. Mosby, St. Louis, 246 (1987).
121. L.A.C. Moorehead, *Biochemistry of the Eye,* R.E. Anderson, Ed. American Academy of Ophthalmology Manual, 145 (1983).
122. J.A. Duddy, W.C. Powzaniuk, and L.F. Rubin, *Am. J. Vet. Res.* 44, 2344 (1983).
123. R. Poulsom and J. Marshall, *Exp. Eye Res.* 40, 155 (1985).
124. P.S. McLaughlin and B.G. McLaughlin, *Am. J. Vet. Res.* 48, 467 (1987).
125. V.M. Lane and S.D. Lincoln, *Am. J. Vet. Res.* 46, 1550 (1985).
126. S.D. Lincoln and V.M. Lane, *Am. J. Vet. Res.* 46, 160 (1985).
127. A. Donovan, *Exp. Eye Res.* 5, 249 (1966).
128. G.G. Gum, K.N. Gelatt, and D.A. Samuelson, *Am. J. Vet. Res.* 45, 1166 (1984).
129. A.I. Cohen, *Adler's: Physiology of the Eye,* 8th ed. R.A. Moses and W.H. Hart, Eds. C.V. Mosby, St. Louis, 458 (1987).
130. R.E. Anderson, Ed., *Biochemistry of the Eye,* American Academy of Ophthalmology Manual, 164 (1983).
131. J. Chase, *Ophthalmology* 89, 1518 (1982).
132. J.S. Zigler and H.H. Hess, *Exp. Eye Res.* 41, 67 (1985).
133. S.F. Basinger and R.T. Hoffman, *Biochemistry of the Eye,* R.E. Anderson, Ed. American Academy of Ophthalmology Manual, 256 (1983).
134. G. Aquirre et al., *Invest. Ophthalmol. Vis. Sci.* 23, 610 (1982).

135. L. Stryer, *Sci. Am.* July, 42 (1987).
136. N.W. Daw, M. Ariel, and J.H. Caldwell, *Retina* 2, 322 (1982).
137. B. Ehinger, *Retina* 2, 305 (1982).
138. H. Pasantes-Morales et al., *Exp. Eye Res.* 43, 55 (1986).
139. N. Lake and N. Malik, *Exp. Eye Res.* 44, 331 (1987).
140. K.C. Hayes, A.R. Rabin, and E.L. Berson, *Am. J. Pathol.* 78, 505 (1975).
141. R.W. Bellhorn et al., *Invest. Ophthalmol.* 12, 65 (1973).
142. R.E. Anderson and J.G. Hollyfield, *Biochemistry of the Eye,* R.E. Anderson, Ed. American Academy of Ophthalmology Manual, 243 (1983).
143. W.G. Robinson, T. Kuwabara, and J.G. Bieri, *Retina* 2, 263 (1982).
144. R.C. Riis et al., *Am. J. Vet. Res.* 42, 74 (1981).
145. E.L. Berson, *Retina* 2, 236 (1982).
146. G.G. Cloyd et al., *Toxicol. Appl. Pharmacol.* 45, 771 (1978).
147. J.A. Render, J.J. Turek, E.J. Hinsman, and W.W. Carlton, *Vet. Pathol.* 22, 475 (1985).
148. J.A. Render and W.W. Carlton, *Vet. Pathol.* 22, 72 (1985).
149. L.F. Rubin, *Atlas of Veterinary Ophthalmoscopy,* Lea & Febiger, Philadelphia, 425 (1974).
150. P.B. Koneru et al., *Drug Design and Toxicology,* D. Hadzi and B. Jerman-Blazic, Eds. Elsevier, New York, 313 (1987).
151. M.O.M. Tso and B.J. Woodford, *Ophthalmology* 90, 952, (1983).
152. R.W. Bellhorn, *Lab. Anim. Sci.* 30, 440 (1980).
153. S.L. Semple-Rowland and W.W. Dawson, *Lab. Anim. Sci.* 37, 289 (1987).
154. N.J. Millichamp, R. Curtis, and K.C. Barnett, *J. Am. Vet. Med. Assoc.* 192, 769 (1988).
155. G.M. Acland and G.D. Aquirre, *Exp. Eye Res.* 44, 491 (1987).
156. K. Narfstrom and S.V. Nilsson, *Invest. Ophthalmol. Vis. Sci.* 27, 1569 (1986).
157. L. West-Hyde and N. Buyukmihci, *J. Am. Vet. Med. Assoc.* 181, 243 (1982).
158. D.A. Witzel et al., *Invest. Ophthalmol. Vis. Sci.* 17, 788 (1978).
159. J.R. Cotter and W.K. Noell, *Invest. Ophthalmol. Vis. Sci.* 25, 1366 (1984).
160. D.W. Batey, J.F. Mead, and C.D. Eckhert, *Exp. Eye Res.* 43, 751 (1986).
161. M. Doshi, M.J. Voaden, and G.B. Arden, *Exp. Eye Res.* 41, 61 (1985).
162. R.K. Hawkins, H.G. Jansen, and S. Sanyal, *Exp. Eye Res.* 41, 701 (1985).
163. C.D.B. Bridges et al., *Invest. Ophthalmol. Vis. Sci.* 28, 613 (1987).
164. R.J. Ulshafer et al., *Exp. Eye Res.* 39, 125 (1984).
165. D.E. Brooks et al., *Am. J. Vet. Res.* 50, 908 (1989).
166. D.R. Anderson, *Adler's: Physiology of the Eye,* 8th ed. R.A. Moses and W.H. Hart, Eds. C.V. Mosby, St. Louis, 491 (1987).
167. B.S. Fine and M. Yanoff, *Ocular Histology,* 2nd ed. Harper & Row, Hagerstown, MD, 149 (1979).
168. J.G. Jose, *Biochemistry of the Eye,* R.E. Anderson, Ed. American Academy of Ophthalmology Manual, 111 (1983).
169. E. Cotlier, The lens, in *Adler's: Physiology of the Eye,* 8th ed. R.A. Moses and W.H. Hart, Eds. C.V. Mosby, St. Louis, MO, 268 (1987).
170. G. Duncan, *Ciba Found. Symp.* 19, 99 (1973).
171. H.H. Hess et al., *Lab. Anim. Sci.* 35, 47 (1985).
172. J.H. Prince, C.D. Diesem, I. Eglitis, and G. Ruskell, *Anatomy and Histology of the Eye and Orbit in Domestic Animals,* Charles C Thomas, Springfield, IL (1960).
173. C.P. Moore et al., *Invest. Ophthalmol. Vis. Sci.* 28, 1925 (1987).
174. B. Milder, The lacrimal apparatus, in *Adler's: Physiology of the Eye,* 8th ed. R.A. Moses and W.H. Hart, Eds. C.V. Mosby, St. Louis, MO, 15 (1987).
175. D.H. Slatter, *J. Small Anim. Pract.* 14, 749 (1973).
176. R.V. Morgan and A. Bacharach, *J. Am. Vet. Med. Assoc.* 180, 432 (1982).
177. J. Nordmann, *Biologie et Chirurgie du Corps Vitre,* A. Brini, Ed. Masson & Cie, Paris (1968).

15

Pathological Processes of the Eye Related to Chemical Exposure

H. HUGH HARROFF, JR.
Administrative Director
The Children's Hospital Research Foundation
Columbus, Ohio

I. INTRODUCTION

The eye is not a common target organ for pathological effects produced by chemicals to which exposure has occurred systemically; however, a great many chemicals produce ophthalmic pathology when the exposure to the eye is topical. It is clearly indicated that the eyes be carefully examined, and all pathological lesions fully evaluated when there has been a direct chemical exposure to them. Despite the fact that the eyes are rarely a target organ for pathology caused by systemic chemical exposure, it is scientifically prudent and has become more commonplace to include an ophthalmic examination as part of the overall toxicity evaluation done on test chemicals. It is the purpose of this chapter to describe methodology for conducting evaluations for ophthalmic pathology, and to detail some specific chemicals and groups of chemicals and the ophthalmic lesions they produce.

II. ASSESSMENT OF OPHTHALMIC PATHOLOGY

Proper assessment of ophthalmic pathology from chemically produced injury requires a combination of thorough clinical examination and, where possible, gross necropsy and histopathological evaluation. Gross and microscopic postmortem evaluation of ocular lesions should be accomplished by a veterinary pathologist or someone trained and experienced in conducting such examinations on eyes from the species being studied. Likewise, clinical ophthalmic examinations should be done by a veterinary ophthalmologist or someone trained and experienced in doing clinical eye evaluations on the species of interest. The completeness and quality of the examinations performed and the information generated will be enhanced if the observer has had appropriate training and sufficient experience.

A. Clinical Ophthalmic Examination

In properly examining the eyes of any animal species clinically, the observer must separately evaluate the ocular fundus and the anterior segment and adnexa. The ocular fundus refers to that portion of the eye extending posteriorly from the anterior edge of the retina; including the retinal vasculature, the choroid, the optic nerve, and in some instances, the sclera.[1] The anterior segment includes the iris, the lens, the anterior chamber, and the cornea. The adnexa refers to the eyelids and lacrimal apparatus and the conjunctivae (palpebral and bulbar).[2] Examination of these structures requires the use of appropriate instrumentation and, when indicated, mydriatics or staining agents.

1. Examination of the Ocular Fundus

For adequate broad visualization of the fundus, particularly in the case of the small rodent species, mydriatics are necessary. A variety of mydriatics are available for use, but the one most commonly used is a parasympatholytic agent, tropicamide (0.1 ml of a 0.5% solution), because of its rapid onset and rapid recovery. In the small rodents, tropicamide produces maximum effect in approximately 5 min following instillation, and recovery occurs in about 1 h. In the dog, cat, and monkey, maximum effect occurs 20 to 30 min after instillation, and recovery does not occur until 2 h in the monkey and 12 to 18 h in the dog and cat.[3]

When the animal's pupil has completely dilated, fundoscopic examination can be accomplished with the aid of any of several instruments. With the direct ophthalmoscope (Figure 1), the fundus is viewed with the instrument held in very close proximity to the cornea (2 cm or closer). The closer the observer is to the animal's eye, the larger will be the field of view. First, the optic disc is located, and then the retina is systematically examined in quadrants by slight head movements on the part of the observer. In animals that have much eye movement such as the dog, cat, and monkey, the observer may want to keep his head stationary and allow the animal to sweep its fundus over the illuminated area.[1]

The most common method of indirect ophthalmoscopy requires the use of a light source mounted on a headband (Figures 2 and 3), a binocular indirect ophthalmoscope and a hand-held condensing lens. The examiner stands about 2 to 3 ft from the animal and the light is directed through the dilated pupil. The hand-held condensing lens is then placed between the observer's light source and the animal's eye and is slowly moved toward the observer until the retinal image is clear and seems to fill the entire lens. Then, by gently moving the lens and the light source (keeping them in the same plane), all quadrants of the retina may be visualized. A small pupil indirect ophthalmoscope is also available for those relatively rare instances in which a fundus examination must be performed and the pupil cannot be dilated.[1]

The ocular fundus can also be examined using the slit-lamp biomicroscope (Figure 4), although this instrument is ordinarily used for examination of the anterior segment. One way of using the slit-lamp biomicroscope is to superimpose the concave –55 Diopter Hruby lens which converts it to a telescope and allows a focus to be obtained on the fundus.

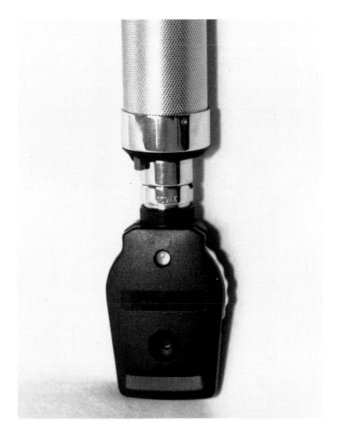

Figure 1. Direct ophthalmoscope. This instrument is hand held. The examination focal point can be adjusted by turning the white dial. This allows the operator to compensate for varied eye shapes and different areas of interest within the eye.

2. Examination of the Anterior Segment and Adnexa

Much of the examination of the anterior segment and adnexa can be accomplished with the naked eye and a hand-held light source. The observer methodically visualizes the eyelids, nicitating membrane (if present), and the conjunctivae by gently everting the eyelids. The cornea may also be examined grossly in this manner.

For examination of the lens and iris and for closer examination of the cornea and the adnexa, the use of a slit-lamp biomicroscope is indicated. Slit-lamp examination can be performed with a mydriatic if good visualization of the lens is desired, or without a mydriatic if the iris is of concern. The slit-lamp biomicroscope is a necessity in performing eye irritation evaluations (e.g., the Draize test).

In performing slit-lamp biomicroscopic examination of the eye, the observer is well advised to develop a standard procedure and follow it every time in order to avoid omitting some important part of the evaluation (see Figure 5 for a sample

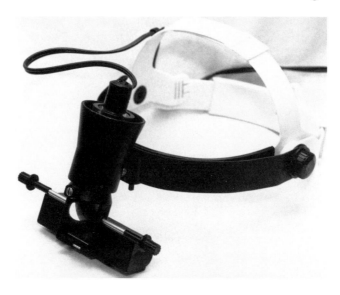

Figure 2. Indirect ophthalmoscope with associated headpiece. Operator looks through the two oculars in making the examination. The light source is built in. A hand-held lens used in conjunction with this instrument is advisable to help with magnification and facilitate the examination.

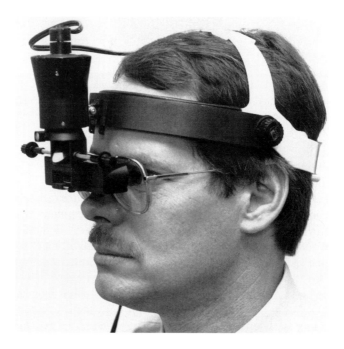

Figure 3. Operator wearing an indirect ophthalmoscope. A hand-held lens is advisable to help with magnification and facilitation of the examination.

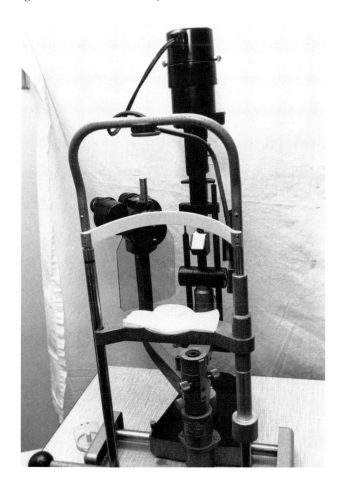

Figure 4. Front view of a slit-lamp biomicroscope mounted on an examination table (hand-held models are available). In using this model the technician holds the restrained animal at the area of the chinpiece (white piece), and the operator manipulates the oculars and light source so as to complete the examination.

checklist for slit-lamp biomicroscopic examination). The author prefers to begin with the light on full open and, looking through the scope, get an appreciation for the presence or absence of discharge, redness of the adnexa, or small lesions on the cornea. The eyelids are then everted to allow for thorough examination of the palpebral and bulbar conjunctivae for such conditions as injected vasculature or petechial or ecchymotic hemorrhages. With the globe proptosed as much as possible, the light is reduced to the thinnest slit and passed slowly over the corneal surface. The width of the slit as it passes over the cornea gives a good indication of the thickness of the cornea and, thereby, will demonstrate corneal edema if present (see

Figure 6). The slit-lamp is then moved toward the animal's eye and refocused on the lens surface. As the slit of light is passed slowly over the lens surface, any lesion such as cataracts will be visualized.

Simultaneous with lens examination, the iris can also be evaluated. The final step in evaluating the anterior segment is to reduce the light beam to pinpoint size and focus on the corneal surface. There should be no light visible from the cornea to the lens. At the lens the light beam will radiate posteriorly. If there is light present between the cornea and the lens, it is indicative of something (e.g., pus, cellular debris) floating in the anterior chamber (see Figure 7).

If during examination of the cornea, the observer sees what appears to be an ulceration, the depth of this lesion, its size, and its stage of healing can be further evaluated by staining. A sterile, cotton-tipped applicator soaked with sterile fluorescein dye is gently touched to the bulbar conjunctiva. The eyelids are held in approximation so that the dye covers the corneal surface. The cornea is examined with the light beam of the slit-lamp biomicroscope filtered through a cobalt blue filter. Retention of the dye, as evidenced by fluorescence, indicates absence of the corneal epithelium (ulceration).

3. Other Clinical Ophthalmic Evaluations

There are many other evaluative procedures that can be used on eyes clinically such as tonometry, gonioscopy, the schirmer test for tearing, and exophthalmometry.[3] Special tests may be indicated in specific instances, but the examination procedures described above for the ocular fundus, the anterior segment, and the adnexa will be sufficient to clinically evaluate most lesions produced by chemical exposures to the eye.

B. Postmortem Ocular Evaluation

Postmortem ocular evaluation consists of gross examination of the globe and adnexa and histopathological examination of prepared sections. During gross examination, care should be taken to observe any visible lesions or abnormalities associated with the eyes, and include such lesions in the sections collected for microscopic examination. Routine sectioning of eyes should be done in such a manner as to include all anatomical structures and areas for histopathological evaluation. Extra sections should be prepared from areas where lesions have been seen clinically or during gross postmortem examination. Special preservatives or staining techniques may be used when indicated because of a particular type of lesion or a specific causative chemical.

III. OPHTHALMIC PATHOLOGY FROM CHEMICAL EXPOSURE

In general, two types of chemical exposure can occur to the eyes and result in some pathological condition. Topical exposure can occur as a result of a direct splash into the eyes, exposure of the eyes to an aerosol, or some similar circumstance that results in the chemical coming into contact with the corneal surface or the adnexa.

**CHECKLIST FOR SLIT-LAMP BIOMICROSCOPIC
EXAMINATION OF THE EYES
(check appropriate response)**

I.   Light Beam on Full Open
   A.   Adnexa
      1.   Conjunctivae
          a.   Redness
              1)   Vessels Normal ----------------------------- __
              2)   Vessels Slightly Injected ------------------ __
              3)   Diffuse Redness --------------------------- __
              4)   Diffuse Deep Redness (Beefy) --------------- __
          b.   Chemosis (Swelling)
              1)   No Swelling ------------------------------- __
              2)   Slight Swelling --------------------------- __
              3)   Obvious Swelling - Some Eversion of Lids --- __
              4)   Swelling with Lids Half Closed ------------ __
              5)   Swelling with Lids Closed ----------------- __
          c.   Discharge
              1)   No Discharge ------------------------------ __
              2)   Slight Discharge -------------------------- __
              3)   Discharge with Moistening of Hair ---------- __
              4)   Discharge with Much Moistening of Hair ----- __
   B.   Cornea
              1)   No Opacity -------------------------------- __
              2)   Scattered Opacity - Iris Visible ----------- __
              3)   Slight Opacity - Iris Slightly Obscured ---- __
              4)   Opalescent Area - No Details of Iris ------- __
              5)   Complete Opacity - Iris Invisible --------- __
   C.   Iris
              1)   Normal ------------------------------------ __
              2)   Marked Folds, Congestion, Swelling --------- __
              3)   No Reaction to Light ---------------------- __
II.   With Light Beam On Slit
   A.   Cornea
              1)   Normal Thickness ------------------------- __
              2)   Thickened -------------------------------- __
   B.   Lens
              1)   Normal Thickness ------------------------- __
              2)   Thickened -------------------------------- __
III.   With Light Beam On Pinpoint
   A.   Anterior Chamber
              1)   No Evidence of Debris --------------------- __
              2)   Evidence of Debris ------------------------ __
IV.   Using Cobalt Blue Filter and Fluorescein Dye
   A.   Cornea
              1)   No Evidence of Ulceration (No Stain)-------- __
              2)   Evidence of Ulceration (Staining)----------- __

Figure 5.   This checklist was designed by the author as an aid in performing slit-lamp bi-omicroscopy in a manner similar to his own procedure. It is intended to give the novice a method of assuring a thorough examination, and as a reminder of some of the changes that may be encountered.

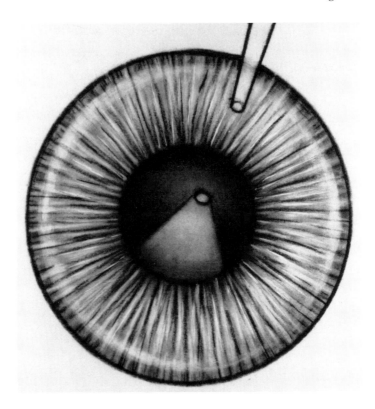

Figure 6.   Artist's conception of the image produced when the light from the biomicroscope in a "slit" mode impinges on the cornea. The inner dark line represents the corneal endothelium and the outer dark line represents the corneal epithelium. Between the lines is the stroma. The distance between lines indicates the corneal thickness. This distance is greater in corneal edema.

Pathological lesions produced by topical exposure vary in nature and severity according to the properties of the insulting chemical, but are generally those associated with local irritation. Characteristic of this type of chemical injury are swelling and edema, hyperemia, and lacrimal discharge. More severe sequellae to specific topical chemical exposures are discussed later in this chapter.

The second type of chemical exposure to the eyes is systemic. Such exposure results from a chemical or its metabolites being transported to the eye via the bloodstream after percutaneous absorption, ingestion, injection, or inhalation. The types of pathological lesions produced by systemic exposure also vary with the nature and amount of the chemical involved, but can be degenerative, reactive, mechanical, hyperplastic, or neoplastic. Pathological processes produced by specific chemicals are also discussed later in this chapter.

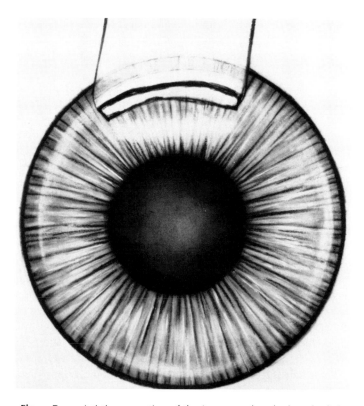

Figure 7. Artist's conception of the image produced when the light from the biomicroscope in a "dot" mode impinges on the cornea. The flare in the middle represents the dot hitting the lens and being refracted. There should be no light visible between the dot on the cornea and the dot on the lens. If there is light visible between the two dots, it is indicative of foreign matter floating in the anterior chamber.

A. Pathological Effects Produced by Topical Chemical Exposure to the Cornea and Adnexa

Some chemicals produce serious injury or pain almost instantly after exposure. Others may seem rather mild at first, but then become progressively more severe after a latent period. Some chemicals produce reversible damage, and others produce damage of a permanent nature. Some of these differences will be further highlighted in the following discussions.

1. Chemicals Producing Immediate Effects

The first group of chemicals that can produce an immediate effect upon the cornea and adnexa by topical exposure includes caustics such as acids and alkalies. These chemicals principally produce their effects by the very rapid action on the tissues resulting from an extreme change of pH within the tissues. Physical changes such as dissolution of the epithelium of the cornea, and clouding of the corneal stroma occur with alkalies. Coagulation of the corneal epithelium may occur a short time following exposure to acids.

In general, acids produce hyperemic, hemorrhagic, and chemotic conjunctivae, whereas alkalies usually produce thrombus formation in conjunctival blood vessels with ischemic necrosis and edema.

Some examples of caustic chemicals in the acids group which may produce ocular damage are acids — sulfuric, hydrochloric, nitric, phosphoric, chromic, liquid sulfur dioxide, and trichloroacetate. Some examples of caustic chemicals in the alkalies group which may produce ocular damage are calcium hydroxide, tetraethylammonium hydroxide, barium hydroxide, strontium hydroxide, lithium hydroxide, potassium hydroxide, sodium hydroxide, and ammonium hydroxide.

A second group of chemicals that produce local effects are solvents. Most solvents are neither strongly acidic nor strongly alkaline in nature and do not react chemically with the ocular tissues. Topical exposures to these chemicals may elicit some pain reaction and may produce a transient loss of corneal epithelium, swelling of the corneal stroma, and a wrinkled appearance to the posterior corneal surface. These effects are particularly pronounced if the chemical is a fat solvent. Generally, the eye can recover well from a solvent exposure. Example chemicals in this group are toluene, xylene, methyl acetate, ethyl acetate, butyl acetate, and n-hexane.

A third group of chemicals that are immediately irritating are surfactants and detergents. These substances may be either cationic, anionic, or nonionic. The type and severity of surfactant-produced lesions vary according to the chemical characteristics of each surfactant. In general, cationic surfactants are most damaging, anionic are less damaging, and nonionic are the least damaging. Damage caused by these chemicals ranges from evidence of slight discomfort (e.g., stinging) with little or no injury, to corneal edema and loss of corneal epithelium with conjunctival swelling, petechiation, and discharge. Examples of surfactant chemicals are cationic surfactants — benzalkonium chloride, cetylpyridinium chloride, and decyltrimethylammonium bromide; anionic surfactants — sodium lauryl sulfate, Ivory® soap, Duponol, Entsufon sodium, and Triton® W 30; nonionic surfactants — Tween, laurithyl, Triton® X 155, Span, Renex, and tridecylhexaethoxylate.[4]

A fourth group of chemicals producing immediate irritation are lacrimators. These chemicals cause a stinging and smarting sensation in the eyes, with profuse tearing. The action is apparently the result of stimulation of sensory nerve endings in the cornea, and the most chemically reactive lacrimators produce the most intense tearing.[5] In most instances, there is no injury to the cornea at low concentrations of these chemicals, but at high concentrations they can cause edema and petechiation of the conjunctiva, and corneal epithelial erosion and opacity.[5] Chemicals in this group include: chloroacetone, acrolein, chloroacetophenone, lewisite, ethyl benzene, methyl vinyl ketone, O-chlorobenzidene malonitrile, nitroethylene, and onion vapor.

2. Chemicals Producing Delayed Effects

There are some chemicals that do not seem to have an immediate effect on the eye, but with either continuous exposure to vapors for several hours, or in some cases, several hours after a discrete exposure, they will produce damage to the cornea and adnexa. This group of chemicals are usually cytotoxic to a particular type of cell in the affected tissue as a result of a biochemical reaction. These chemicals produce edema, hyperemia, petechiation, and varicose distortions of the blood vessels in the conjunctiva, as well as epithelial erosion; swelling of the epithelium, stroma, and endothelium of the cornea; and wrinkling of the posterior surface of the cornea. The severe lesions produced by these chemicals often progress to infiltration of inflammatory cells, invasion by interstitial vessels, fibrosis, and finally, permanent scarring, vascularization, and opacity of the cornea. Such extreme reactions are usually rare, but they may be produced by chemicals such as dimethyl sulfate and mustard gas.[6]

Corneal epithelial vacuoles, which are seen with the slit-lamp biomicroscope as shiny, colorless spheres that look like gas bubbles within the epithelium, have been produced in cats with n-butanol, ethyl acetate, methyl acetate, toluene, and xylene.[4] These same chemicals did not produce corneal vacuoles in rabbits or guinea pigs. Examples of other chemicals producing delayed effects are allyl alcohol, diethylamine, dimethylamine, ethylene oxide, digitalis glycosides, hydrogen sulfide, methyl bromide, and trimethyl siloxane.

B. Pathological Effects Produced by Systemic Chemical Exposure to the Cornea and Adnexa

Although most chemical injuries to the cornea and adnexa occur as a result of topical exposure, some chemicals produce lesions in the cornea and conjunctivae through systemic exposure. Such chemicals cause pathological conditions ranging from lacrimation to corneal deposits and injuries. Some examples of these chemicals and the injuries they produce follow.

1. Lacrimation, burning, or itching — bismuth subnitrate, cyclohexanol, dichlorophenoxyacetic acid, dimidium bromide, hexachloronaphthalene, phthalofyne, and pyrithione
2. Corneal and conjunctival inflammation — allyl cyanide in rats, *Lantana* spp. in sheep and cattle, and phenazopyridine in dogs
3. Photosensitized keratitis — hypericum, methoxsalen, phenothiazine, and *Lantana* spp.
4. Corneal epithelial deposits — amiodarone, chloroquine, chlorpromazine, hydroxychloroquine, and monobenzone
5. Corneal stromal deposits — copper, gold, mercury, silver, chlorpromazine, clofazimine, and *Indomethecine* spp.
6. Corneal endothelial injury — 1,2-dibromoethane, 1,2-dichloroethane, 1,2-dichloroethylene, and methylhydrazine; dogs are particularly susceptible to these injuries[4]

C. Pathological Effects Produced on the Lens, Iris, and Anterior Chamber by Chemical Exposure

Cataract is partial or total opacity of the lens and/or its capsule. Cataracts may be clinically diagnosed using a slit-lamp biomicroscope and scanning the lens for areas of cloudiness or opacity. Histopathologically, the capsule may be thickened, thinner, or ruptured; there may be degeneration, necrosis, and proliferation of the lens epithelium, and there may be degeneration and sclerosis of the lens nucleus.[7-9] The specific pathological changes observed vary according to the cause.

Since the lens has no blood supply, the source of its nutrients is the aqueous humor and to a lesser extent, the vitreous. To produce cataracts, chemicals must be capable of traversing the blood-aqueous barrier.[10] This results in a limitation on the number and nature of cataractogenic chemicals, and a necessity for a high daily dosage, slow excretion, or long-term administration before toxic levels can be reached.[11]

More than 50 chemical substances have been shown to produce cataracts in animals when there is a systemic exposure. More chemicals have been shown to cause cataracts in animals than have been incriminated in man for several reasons: (1) drugs and chemicals are tested in animals; (2) animals may be experimentally exposed to unusually high concentrations of chemicals; and (3) animals are often exposed at a younger age when they are more susceptible to cataractogenic effects. Chemicals producing cataracts and similar effects on the lens may be categorized as follows.

1. Hyperglycemia-Producing Chemicals

Sugars such as galactose, xylose, and glucose, and chemicals such as Alloxan and Streptozotocin which are diabetogenic, all result in hyperglycemia which, in turn, can produce cataracts. Mechanistically, sugar enters the lens where it is converted to a sugar alcohol. This alcohol osmotically attracts water causing excessive hydration of the lens, with cellular swelling and intercellular fluid buildup. Secondarily, some ions leak out of the lens, causing sodium chloride and more water to enter the lens. Ultimately, the swelling results in disruption of nerve fibers, changes in protein structure, and irreversible cataracts.[12] In the case of Alloxan, there is some debate about whether it is the hyperglycemia or a direct toxic effect that produces cataracts.[13]

Medications used to treat diabetes — Some chemical medications used to treat diabetes melitus in humans have produced keratitis and cataracts in rats. These drugs are carbutamide, chloropropamide, and tolbutamide.[14]

Corticosteroids — Some glucocorticoids administered for long periods produce posterior subcapsular cataracts. Dichlorisone and methyldichlorisone are two such chemicals.[15]

Antimitotic antineoplastic chemicals — These chemicals affect the most actively mitotic cells at the equator of the lens, causing them to become gradually more and more opaque. Eventually a cataract is formed. Examples of the chemicals are busulfan, dibromomannitol, dimethylaminostyryloquinoline, iodoacetate, nitrogen mustard, tretamine, and triaziquone.[16]

Antimetabolites — This is a group of chemicals that in some way interfere with enzyme action or formation of lens proteins. The result is cataract formation. Examples of these chemicals are chlorophenylalanine, phenylhydrazopropionitrile, mimosine, dichloronitroaniline, and naphthalene and related compounds.

Photosensitizers — Hematoporphyrin and methoxsalen are two chemicals that have been shown to produce cataracts in animals in the presence of bright light. The light may have some effect in causing the chemical to bind to the proteins of the lens, thus producing a cataract.[17]

Metal chelators — These chemicals cause damage elsewhere in the eye and, therefore, may have only an indirect effect in cataract production. Nevertheless, cataracts result from toxic doses of such chemicals as deferoxamine, dithizone, and pyrithione.

Dimethyl sulfoxide (DMSO) — DMSO is a fairly commonly used chemical (often as a pharmaceutical vehicle), and it has been shown to produce a peculiar change in the refractive power of the nucleus of the lens, without causing opacity, in dogs, rabbits, guinea pigs, and rats. With the slit-lamp biomicroscope, this lesion has the appearance of a lens within a lens.[18]

In addition to the chemicals producing cataracts, there are more than 20 chemicals that have been found to produce a transient myopia in humans. This problem is idiosyncratic and does not occur in association with pupillary constriction. The exact mechanism of this phenomenon is not known, but in some instances it has been associated with a shallowing of the depth of the anterior chamber, and in even rarer instances, there has been cellular debris floating in the aqueous humor. As a rule, these lesions have been completely reversible. Examples of chemicals producing these lesions are acetazolamide, sulfanilamide, dichlorphenamide, phenformin, chlorthalidone, and tetracycline.[4]

D. Pathological Effects Produced on the Retina, Choroid, and Optic Nerve by Chemical Exposure

Pathological damage done to the structures of the ocular fundus is invariably caused by a systemic exposure to the offending chemical. The retina is highly vascular and is, therefore, susceptible to circulating toxicants. Lesions of the ocular fundus vary from mild to severe, and affect different structures, depending on the nature and concentration of the chemical causing the damage. Below are listed some of the chemicals that have been incriminated in producing lesions in the structures of the ocular fundus.

Retinal edema —There are more than 20 chemicals currently known to produce retinal edema in animals. The mechanism by which they produce this lesion is not currently understood. Examples of some chemicals in this group are acridine, cobalt, cyanide, dithizone, fluoride, glutamate, *Helichrysum* spp., iminodipropionitrile, iodate, naphthalene, phosphorus, pyrithione, streptomycin, and triaziquone.[4]

Retinal hemorrhages — A variety of chemicals have been incriminated in causing retinal hemorrhages in animals. Some of them are known to cause blood dyscrasias or clotting abnormalities, but for the most part, the mechanism by which they produce retinal hemorrhage is not understood. Examples of chemicals that produce this lesion are alloxan, desoxycortone acetate, dextran, diquat, dithizone, epinephrine, naphthalene, 2-(2-naphthyloxy)ethanol, phosphorus, and pyrithione.[4]

Retinal vessel narrowing — Once again, the mechanism that results in this lesion is not clearly understood, but it is believed that it may be associated with a reduced demand for nutrients by the tissues, especially in those instances where optic nerve atrophy is an accompanying defect. Examples of chemicals that cause this lesion are aspidium, diaminodiphenoxypentane, ethylenimine, lead, hyperbaric oxygen, and P-1727.[19]

Retinal ganglion cell pathology — In order to accurately diagnose damage to the ganglion cells of the retina, it is essential to do histopathology. This lesion is most commonly seen in conjunction with optic nerve atrophy. Retinal ganglion cell pathology may be caused by direct chemical intoxication, but it is often associated with the same conditions that cause optic nerve atrophy, such as excessive elevation of intracranial pressure, elevation of intraocular pressure, or optic neuritis. Examples of chemicals that can produce this lesion are arsanilic acid, aspidium, carbon disulfide, cysteine, ergot, glutamate, *Astragalu* spp., quinine, quinoline, and tellurium.[4]

Retinal photoreceptor pathology — This is another lesion that must be diagnosed histopathologically. Some chemicals affect the retinal pigmented epithelium, and others destroy the rods and/or cones. Examples of chemicals damaging the photoreceptors are urethane, aramite, benzoic acid, bromoacetate, colchicine, ethylenimine, fluoride, fluothane, hexachlorophene, iodoacetate, and sodium azide.[4]

Retinal pigmented epithelial pathology — Damage to the retinal pigmented epithelium may be diagnosed either clinically or histopathologically. These lesions can be visualized with either a direct or an indirect ophthalmoscope. They appear as a loss of color to the retinal epithelium. These lesions may be associated with concurrent damage to the photoreceptors. Examples of chemicals that have been shown to produce damage to the pigmented epithelium in animals are alloxan, aspartate, bilirubin, bromoacetate, colchicine, dibutyl oxalate, epinephrine, fluoride, glutamate, lead, urethane, vincristine, and vitamin A.[4]

Choroidal edema, exudate, and detachment — This type of lesion has been described in animals, but has not been seen in human beings. Chemicals that produce this lesion are diethyldithiocarbamate, diquat, dithizone, epinephrine, iodate, naphthalene, hyperbaric oxygen, pyrithione, cobalt salts, mercuric chloride, P-1727, and thioridazine.[4]

Tapetal pathology — Lesions have been described, particularly in dogs, in which there is depigmentation of the tapetum lucidum. Most of the chemicals incriminated in producing these lesions are metal chelators. It is widely believed that these chemicals produce their effect by interacting with zinc, which is an important element in the chemical makeup of the tapetum. Examples of chemicals producing this lesion are diethyldithiocarbamate, dithizone, edetate, ethambutol, ethylenediamine derivatives, hydroxychloroquine, imidazoquinazoline, oxypertine, and pyrithione.[4]

Papilledema — Three chemicals have been reported to cause papilledema in the dog. They are *p*-dichlorobenzene, helichrysum, and hexachlorophene. In addition, there are many chemicals that have been reported to produce this lesion in human beings.[20] Papilledema can accompany optic neuritis, or can occur as a sequellae to an elevation of intracranial pressure. Diagnosis can be made by using a direct or an indirect ophthalmoscope and visualizing swelling and congestion of the optic nerve.[20]

Optic neuropathy or neuritis — This lesion is simply a degenerative or inflammatory lesion along the optic nerve. An interesting phenomenon is that the site of the lesion along the optic nerve varies according to the chemical causing the lesion. In animals, the chemicals ethambutol, helichrysum, *Stypandra imbricata*, and triethyl tin affect the optic chiasma, whereas hexanedione and tellurium have their sites of action at the lateral geniculate bodies and optic tracts. Some other chemicals that produce optic neuropathy are acrylamide, arsanilic acid, clioquinol, filcin, glutamate, indarsol, lead, and sodium azide.[21,22]

Optic nerve atrophy — This lesion can be a sequellae to optic neuritis, papilledema, glaucoma, or brain tumors. Optic nerve atrophy can be diagnosed with a direct or indirect ophthalmoscope by visualizing an abnormally white or pale appearance, and a decreased number of blood vessels on the optic disc. Histopathologically, there will be a lack of nerve fibers and ganglion cells in the retina. Examples of chemicals that have been shown to cause optic nerve atrophy in animals are arsanilic acid, aspidium, cyanide, formic acid, hexachlorophene, and methanol.[20]

IV. ADDITIONAL INFORMATION

Throughout this chapter, examples of chemicals that produce specific types of lesions have been cited. It is the objective of the chapter to describe the myriad lesions that can occur to the eyes of animals exposed, either experimentally or accidentally, to chemical insult, and to suggest methods of evaluation of these lesions. It is not within the scope of the chapter to provide a comprehensive reference of all ocular chemical toxicants. If the reader desires more information about any particular chemical and its affect upon the eye, a very complete listing of such chemicals can be found in the book, *Toxicology of the Eye,* 3rd ed., by W. Morton Grant.

REFERENCES

1. L. F. Rubin, *Atlas of Veterinary Ophthalmoscopy,* Lea & Febiger, Philadelphia (1974).
2. D. Vaughan and T. Asbury, *General Ophthalmology,* 9th ed. (1980).
3. R. A. Moses, Ed., *Adler's Physiology of the Eye — Clinical Application,* 7th ed. C.V. Mosby, St. Louis (1981).
4. W. M. Grant, *Toxicology of the Eye,* 3rd ed. (1986).
5. D. G. Cogan, *JAMA* 122, 435 (1943).
6. W. J. Geeraets, S. Abedi, and R. V. Blanke, *South. Med. J.* 70, 348 (1977).

7. J. G. Bellows, *Cataract and Abnormalities of the Lens,* Grune & Stratton, New York (1975).

8. Sir S. Duke-Elder, *System of Ophthalmology,* Vol. 3, Part 2, *Normal and Abnormal Development: Congenital Deformities,* C.V. Mosby, St. Louis (1963).

9. M. J. Hogan and L. E. Zimmerman, *Ophthalmic Pathology,* 2nd ed. (1962).

10. R. Heywood, *Br. Vet. J.* 127, 301 (1971).

11. R. L. Peiffer, Jr., Ed., *Comparative Ophthalmic Pathology,* Charles C Thomas, Springfield, IL (1983).

12. L. T. Chylack and H. M. Cheng, *Survey Ophthalmol.* 23, 26 (1978).

13. R. Bernat and K. Bombicki, *Acta Physiol. Pol.* 19, 205 (1968); (Chem. Abstr. 69: 9445, 1968).

14. H. N. Wright, *Diabetes* 12, 550 (1963).

15. S. I. Griboff and W. Futterweit, *J. Mt. Sinai Hosp. NY* 32, 121 (1967).

16. K. W. Christenberry et al, *Arch. Ophthalmol.* 70, 250 (1963).

17. D. G. Cogan, *Arch. Ophthalmol.* 66, 612 (1961).

18. D. C. Wood and N. V. Wirth, *Ophthalmol. Addit. Ad.* 158, 488 (1969).

19. F. A. Patty, *Interscience,* Vol. 2. 2172 (1962).

20. G. Martin-Bouyer et al., *Lancet* 1, 91 (1982).

21. T. R. Hedges and H. A. Zaren, *Neurology* 19, 359 (1969).

22. *The Merck Index Of Chemicals And Drugs,* 8th ed. Merck Sharpe & Dohme, Rahway, NJ (1968).

16

Ocular Irritation Testing

GEORGE P. DASTON
The Procter & Gamble Company
Miami Valley Laboratories
Cincinnati, Ohio

AND

F. E. FREEBERG
The Procter & Gamble Company
Ivorydale Technical Center
Cincinnati, Ohio

I. INTRODUCTION

Acute accidental exposure of the body's external tissues is a possible (and perhaps inevitable) event associated with the use of any chemical or chemical formulation in widespread commercial and industrial use. Severe irritation or corrosion of the anterior structures of the eye can result in persistent or permanent loss of vision, even after healing has taken place, if their optical properties are disturbed. For this reason, virtually all safety assessments of new commercial products include an evaluation of ocular irritating potential.

Although clinical observations of chemically induced injuries to the eye have been made for centuries[1] the first attempts to quantitate the severity of ocular irritation in controlled experimental settings were carried out in the 1940s. These included studies on the effects of strong acids and bases[2-4] and various organic chemicals[5] on the cornea of the rabbit.

These studies were significant in that they introduced the albino rabbit as a model for ocular irritancy testing, and provided a methodological framework for establishing procedures for ocular irritancy testing. Also of significance was the introduction of a standardized scoring system for grading ocular reactions to irritants. Friedenwald et al.[2] reported their results using a numerical scoring system for the intensity and duration of corneal opacity, edema, ulceration, and the development of pannus; and for conjunctival and iridial irritation. Although this scoring system is subjective, it provided a standard for comparing the irritation responses of different animals, and the irritant potential of different chemical substances.

Draize et al.[6] modified this scoring system, and described a method for exposure of the rabbit eye and subsequent observation of irritant responses. This methodology, known as the Draize test for eye irritation, was the accepted standard for eye irritation for decades. Although it has been modified several times in government regulations, the procedure described in the original manuscript is still recognizable as the basis for the protocols used today.

II. METHODS FOR ASSESSING OCULAR IRRITANCY POTENTIAL

The European Economic Community (EEC) regulatory agencies' guidelines for assessing ocular irritancy were established internationally by the Organization for Economic Cooperation and Development (OECD) and the EEC. U.S. regulatory agencies use two different sets of guidelines: one for substances regulated by the Toxic Substances Control Act (TSCA) and the Federal Insecticide, Fungicide, and Rodenticide Act (FIFRA); and the other for materials regulated by the Federal Hazardous Substances Act (FHSA). Most consumer products are regulated under FHSA, but any with antiseptic properties may be regulated under FIFRA.

An attempt was made to devise a common set of guidelines for all U.S. regulatory agencies by the Interagency Regulatory Liaison Group (IRLG). IRLG published guidelines in 1981 which are virtually identical to the OECD's; however, the IRLG has since disbanded, and its guidelines carry no regulatory force. Still, since the U.S. has accepted the OECD guidelines and is a member of OECD, ocular irritancy data using either the OECD or IRLG guidelines should be acceptable to U.S. regulatory agencies.

These methods will be described in detail below. First, however, it is important to understand the original Draize method and its interim updates, as many researchers and organizations continue to rely on these earlier protocols or grading schemes in conducting screening or in comparing data on new materials to a large historical database.

A. The Draize Method

The procedure published by Draize et al.[6] for testing materials for ocular irritancy involves the introduction of 0.1 ml of liquids, solutions, or ointments into the lower conjunctival cul-de-sac of the albino rabbit eye. Observations of injury to the cornea, conjunctiva, and iris are made 1, 24, and 48 h after exposure to the test material. If injury persists after 48 h, a fourth observation is made 96 h after instillation of the test material. (Subsequent amendments to the protocol extended this to 7 days.) Although the number of animals to be used for each material was not stated in the original manuscript, it was later accepted that nine rabbits should be treated in one eye, with the other eye serving as a control. However, some of these rabbits were used to determine the effect of irrigation on the irritant response.

A scoring system was devised which assigns numerical grades for various aspects of irritation. This system is reproduced in Table 1. Ranks are assigned for the degree

Table 1

Scale of Weighted Scores for Grading the Severity of Ocular Lesions[6]

I. Cornea
A. Opacity-Degree of Density (area which is most dense is taken for reading)
    Scattered or diffuse area — details of iris clearly visible 1
    Easily discernible translucent areas, details of iris clearly visible 2
    Opalescent areas, no details of iris visible, size of pupil barely discernible 3
    Opaque, iris invisible 4
B. Area of Cornea Involved
    One quarter (or less) but not zero 1
    Greater than one quarter — less than one half 2
    Greater than one half — less than three quarters 3
    Greater than three quarters — up to whole area 4

    Score equals A × B × 5   Total maximum = 80

II. Iris
A. Values
    Folds above normal, congestion, swelling, circumcorneal injection (any one or all of these or
    combination of any thereof), iris still reacting to light (sluggish reaction is positive) 1
    No reaction to light, hemorrhage; gross destruction (any one or all of these) 2

    Score equals A × 5   Total possible maximum = 10

III. Conjunctivae
A. Redness (refers to palpebral conjunctivae only)
    Vessels definitely injected above normal 1
    More diffuse, deeper crimson red, individual vessels not easily discernible 2
    Diffuse beefy red 3
B. Chemosis
    Any swelling above normal (includes nictitating membrane) 1
    Obvious swelling with partial eversion of the lids 2
    Swelling with lids about half closed 3
    Swelling with lids about half closed to completely closed 4
C. Discharge
    Any amount different from normal (does not include small amounts observed in inner canthus
    of normal animals) 1
    Discharge with moistening of the lids and hairs just adjacent to the lids 2
    Discharge with moistening of the lids and considerable area around the eye 3

Score (A + B + C) × 2   Total maximum = 20

Note: The maximum total score is the sum of all scores obtained for the cornea, iris, and conjunctivae.

of corneal opacity and the approximate area of corneal involvement; irritation of the
iris; and redness, swelling (chemosis), and discharge from the conjunctiva. The score
for each structure is weighted by multiplying it by a scaling factor: five for the cornea
and iris, two for the conjunctiva. The weighted scores can be considered separately,
or added together as a composite score for ocular irritation. The maximum composite

score is 110, of which 80 points (73%) are for corneal injury, 10 (9%) are for iridial inflammation, and 20 (18%) are for conjunctival damage. Thus, corneal injury is by far the most significant aspect of ocular irritation by the Draize scoring system.

There is merit to providing such gravity for corneal involvement, since injury to the cornea may directly lead to impairment or total loss of vision. However, it is at least theoretically possible that relying only on composite Draize score could result in oversight of significant conjunctival or iridial damage.[7]

B. The FHSA Method

A modified Draize protocol was described in the FHSA which was passed into law in 1964,[8] to be used as the officially prescribed eye irritancy assessment of the Bureau of Product Safety, Food and Drug Administration (FDA). This was updated in 1979.[9] The FHSA protocol is very similar to the Draize, with a few changes.

First, the number of rabbits to be used was decreased from nine to six. Second, the stipulated intervals for observing irritation were limited to 24, 48, and 72 h after administration of the test material. Third, a procedure was described for rinsing the insulted eye, although this was not required. Rinsing involves gently irrigating the eye with an isotonic sodium chloride solution, to be done 24 h after the administration of the test substance. Fourth, and perhaps most significantly, the numerical scoring system was changed to a set of descriptions of lesions that would be considered to be evidence that the substance that produced them was an "irritant".

The FHSA mandated the labeling of irritating materials with warnings to the consumer; thus, eye irritation results needed only to indicate whether a material is a corrosive, irritant, or nonirritant. Although there are still graded scales for scoring lesions, a numerical score for irritation is not designated under this statute. In practice, most industrial toxicologists find it necessary and useful to discriminate degrees of irritation; thus, use of the Draize scoring system has persisted.

Under the FHSA scoring scheme, conjunctival, corneal, and iridial damage are each considered separately. No weighting of scores or composite scores are used. A positive ocular reaction to a test material is considered to be (1) any corneal ulceration or opacity greater than a fine stippling, (2) any corneal opacification greater than a loss of normal luster, (3) any iridial inflammation greater than a slight deepening of the iridial rugae or slight injection of the circumcorneal blood vessels, (4) conjunctival swelling sufficient to cause partial eversion of the eyelids, or (5) a diffuse redness of the conjunctiva which obscures the details of individual blood vessels.

The test material is considered to be an irritant if four or more of the six rabbits have a positive response, and a nonirritant if no or one rabbit has a positive reaction. If two or three rabbits show ocular irritation, the test should be repeated in a second group of six animals. If three or more animals in this second block demonstrate irritation, then the material is classified as an irritant. If only one or two rabbits are irritated in the second test, a third block of six animals should be tested. If any of these six have a positive reaction, the test material is considered to be an irritant.

Responsibility for administering the FHSA was transferred to the Consumer

Product Safety Commission (CPSC) when it was formed in 1972. The CPSC published an *Illustrated Guide for Grading Eye Irritation Caused by Hazardous Substances* which contained a grading system based on the FHSA scoring method, and a series of photographs which illustrate each of the grades. This guide is out of print, but we have reproduced the grading system in Table 2, and the photographs in Plates 1 to 6.*

In addition to modifying the Draize protocol, the FHSA procedure also made some noteworthy additions. These included a provision for testing solid materials, and suggested the use of a slit lamp, binocular loupe, and fluorescein dye to facilitate the detection of corneal injury.

FHSA specifies that 100 mg of solid materials should be tested, except when the solids are in flake, granular, or powdered form such that a 0.1 ml volume weighs considerably less than 100 mg. In that case, the material is compacted as much as possible without crushing individual particles into a 0.1 ml volume, and this is used. In either case, the material is instilled into the lower conjunctival sac, as with liquids.

The use of fluorescein to detect small corneal lesions is recommended but not required by FHSA. One drop of sodium fluorescein (fluorescein sodium ophthalmic solution USP or equivalent) is dropped directly on the cornea, which is then rinsed with isotonic saline. Injured areas of the cornea retain more fluorescein, and appear yellow under blue or long wavelength UV illumination. Plate 3 depicts examples of corneal lesions stained with fluorescein.

FHSA also recommends the use of a magnifying device to aid in detecting ocular injury. Use of the slit-lamp biomicroscope by a trained expert is the most exact means of examining the cornea and anterior chamber *in situ* (see Chapter 15). Fine structures such as the corneal endothelium and the small, tertiary vessels of the iris can be visualized. It is also possible to make objective measurements of anterior chamber damage, such as corneal thickening and light flare in the aqueous humor. These techniques and their significance will be discussed later.

C. The OECD Method

The OECD established guidelines for ocular irritation testing in 1981[10] and updated these in 1987.[11] The OECD guidelines specify the use of young adult rabbits of either sex. These will probably weigh 2 to 3 kg. A minimum of three rabbits should be tested initially, with an additional group of three added if the test material is irritating. The eyes of the second group must be rinsed 30 s after exposure. (In the original 1981 guidelines a third group of three was specified, with rinsing 4 s after exposure.)

The rabbits must be observed for irritation 1, 24, 48, and 72 h after exposure to the test material. Observations may also be made up to 21 days after exposure, at the discretion of the investigator, to determine the reversibility of persistent effects.

The scoring system in the OECD guidelines is the same as that developed by CPSC, listed in Table 2. The criteria for classifying a test material as an ocular irritant are the same as in the FHSA guidelines.

---

* Plates 1—6 appear after page 520.

Table 2
Grades for Ocular Lesions (CPSC Scale)

Cornea

Opacity: degree of density (area most dense taken for reading)

No ulceration or opacity 0

Scattered or diffuse areas of opacity (other than slight dulling of normal luster, details of iris clearly
visible) .... 1[a]

Easily discernible translucent areas, details of iris slightly obscured 2

Nacreous areas, no details of iris visible, size of pupil barely discernible 3

Opaque cornea, iris not discernible through the opacity 4

Iris

Normal 0

Markedly deepened rugae, congestion, swelling, moderate circumcorneal hyperemia, or injection, any
of these or any combination thereof, iris still reacting to light (sluggish reaction is positive) .... 1[a]

No reaction to light, hemorrhage, gross destruction (any or all of these) 2

Conjunctivae

Redness (refers to palpebral and bulbar conjunctivae excluding cornea and iris)

Blood vessels normal 0

Some blood vessels definitely hyperemic (injected) 1

Diffuse, crimson color, individual vessels not easily discernible .... 2[a]

Diffuse beefy red 3

Chemosis: lids and/or nictitating membranes

No swelling 0

Any swelling above normal (includes nictitating membranes) 1

Obvious swelling with partial eversion of lids .... 2[a]

Swelling with lids about half closed 3

Swelling with lids more than half closed 4

[a]   Readings at these numerical values or greater indicate positive responses.

The OECD guidelines include two innovations: (1) a procedure for testing aerosols; and (2) recommendations on the use of topical anesthetics. For testing aerosols, the cornea is sprayed for 1 s from a distance of approximately 10 cm. The velocity of the aerosol should be insufficient to damage the eye. The amount of material used should be approximated by weighing the aerosol container before and after treating each rabbit.

The guidelines also permit the use of topical anesthetics in the eye if there is the likelihood that the test will be extremely painful. The anesthetic may be applied only once, just before administration of the test material, and must also be applied to the animal's unexposed (control) eye. The guidelines do not specify the type or dosage of anesthetic to be used.

In the 1987 update of the OECD guidelines, provisions were made for permitting limited or no testing of materials likely to be severely irritating or corrosive. Specifically, materials with high (≥ 11.5) or low (≤ 2) pH will be considered to be irritating and need not be tested. Severe skin irritants or corrosives need not be tested for ocular irritancy. The OECD will also consider results from a single rabbit if marked effects are observed. Lastly, the OECD will consider results from well-validated alternative tests as a substitute for more traditional data if severe irritant or corrosive potential is established.

D. The IRLG Method

In the late 1970s the Environmental Protection Agency (EPA) developed its own testing guidelines for ocular irritancy of pesticides under FIFRA, which requires labeling irritant products with warnings to the consumer, and for other chemical substances and complex mixtures under TSCA. In order to resolve any potential inconsistencies with eye irritation methods prescribed by different legislation and required by different regulatory agencies, the IRLG, a consortium of U.S. regulatory agencies, attempted to establish a consensus method. The IRLG, composed of members from EPA, CPSC, FDA, the Occupational Safety and Health Administration, and the Department of Agriculture's Food Safety and Quality Service published a test guideline.[12]

This guideline was an attempt at harmonizing various procedures then in use by the separate regulatory agencies. The IRLG was discontinued in 1981, before any of the participating agencies had formally adopted its recommendations. However, some of the ideas contained in them were incorporated in the OECD Guidelines. There are now different guidelines published for FIFRA- and TSCA-regulated materials.

E. The FIFRA/TSCA Method

The U.S. EPA has devised its own set of guidelines for testing the ocular irritancy of materials regulated under FIFRA[13,14] and TSCA.[13,15] These guidelines are virtually identical to each other, but diverge considerably from previous standards in some aspects. However, the fundamental procedure — observation and scoring of ocular damage after deposition of 0.1 ml of test material into the lower conjunctival sac — is unchanged. The procedure uses the FHSA grading system to assess the severity of ocular lesions. Eyes are observed 1, 24, 48, and 72 h after instillation of the test material, or longer (up to 21 days) if irritation persists.

A major difference from previous guidelines is that the test species is not strictly prescribed. It is recommended that adult albino rabbits be used, but other species may be substituted. If another species is used, its selection must be justified in the study report. Regardless of species selected, at least six animals must be used unless a lesser number can be justified.

The FIFRA guidelines specify that solid materials should be ground to a fine dust and instilled at a volume of 0.1 ml, rather than keeping the solid particles in their native form.

As with the OECD guidelines, any materials with a pH ≥ 11.5 or ≤ 2 will be considered *a priori* to be a severe ocular irritant, as will anything which is severely irritating or corrosive to the skin. All other specifications in the protocol are comparable to the OECD methods.

F. The NAS Method

In 1977 the CPSC requested that the National Research Council of the National Academy of Sciences (NAS) review the FHSA method of ocular irritation testing. The NAS Subcommittee on Dermal and Eye Toxicity, composed of academic and industrial scientists, proposed several changes in the way in which eye irritation could be assessed.[16] This protocol has never been widely used; however, it contains several noteworthy alternatives to traditional methods. Of particular interest is the recommendation that materials be tested using at least two different volumes, in groups of four rabbits (minimum) per dose. The test volumes should be selected based on the anticipated typical exposure, but the NAS suggests volumes of 0.1 and 0.05 ml as benchmarks.

The NAS document also comments that exposure should be directly to the surface of the cornea, with the lids retracted, as this more closely models accidental exposure in humans than does deposition into the lower conjunctival sac.

The NAS recommended a new hazard assessment system which evaluates the length of time required for the eye to return to normal, the severity of lesions, and the probability that these lesions will be irreversible. Since the time required to heal is of such importance in this scheme, it was recommended that animals routinely be examined until the eyes were normal, up to 21 days after exposure.

G. The Low Volume Method

The Draize procedure and all of its subsequent modifications have been criticized for overestimating human ocular irritation response. Although there are a number of factors which may contribute to this, including the sensitivity of the rabbit eye and the unrealistic method of instillation, the use of the standard 100 μl volume of test material greatly increases the irritancy score relative to what is obtained from identical exposures of human eyes.[17]

Griffith et al.[18] determined in rabbits the dose-response characteristics of 21 chemical solutions and mixtures with varying irritant potentials (Table 3). These substances were tested at volumes of 0.003, 0.01, 0.03, and 0.1 ml, and were placed directly on the corneal surface. Results were compared to published results[19] on the human irritancy of these substances. They found that a dose volume of 0.01 ml produced irritation which was most consistent with that observed in humans. In addition, the best separation of irritancy scores was achieved at the lower dosing volumes, especially at 0.01 ml. These data indicate that the 0.1 ml volume of the Draize test may be off the linear portion of the dose-response curve; hence, it is not as good a differentiator of irritant potentials. The data in Table 3 may also be useful as a guideline for the degree of irritation (from none to severe) produced by various chemicals at 0.01 and 0.1 ml volumes.

Table 3
Rabbit Eye Irritation Scores at Various Times After Instillation of Materials (Draize Score $\bar{X} \pm SE$)[a]

Material	Dose volume[b]	N[c]	Day 1	3	7	14	21	Maximum	Median day to clear
Hexane, 100%	0.01	9	0 ± 0	0 ± 0	0 ± 0	0 ± 0	0 ± 0	0 ± 0	1
	0.03	6	0 ± 0	0 ± 0	0 ± 0	0 ± 0	0 ± 0	0 ± 0	1
	0.10	6	0 ± 0	0 ± 0	0 ± 0	0 ± 0	0 ± 0	0 ± 0	1
Benzalkonium chloride, 0.1%	0.01	6	0 ± 0	0 ± 0	0 ± 0	0 ± 0	0 ± 0	0 ± 0	1
	0.03	6	0 ± 0	0 ± 0	0 ± 0	0 ± 0	0 ± 0	0 ± 0	1
	0.10	6	0 ± 0	0 ± 0	0 ± 0	0 ± 0	0 ± 0	0 ± 0	1
Triethanolamine, 98%	0.01	6	0 ± 0	0 ± 0	0 ± 0	0 ± 0	0 ± 0	0 ± 0	1
	0.03	6	1 ± 1	0 ± 0	0 ± 0	0 ± 0	0 ± 0	1 ± 1	1
	0.10	6	4 ± 1	2 ± 2	2 ± 2	0 ± 0	0 ± 0	6 ± 1	3
Silver nitrate, 1%	0.01	6	2 ± 1	0 ± 0	0 ± 0	0 ± 0	0 ± 0	2 ± 1	1
	0.03	6	3 ± 1	1 ± 1	0 ± 0	0 ± 0	0 ± 0	3 ± 1	3
	0.10	6	12 ± 3	4 ± 4	1 ± 1	0 ± 0	0 ± 0	13 ± 4	3
Acetic acid, 3%	0.01	9	4 ± 0	0 ± 0	0 ± 0	0 ± 0	0 ± 0	4 ± 0	3
	0.03	6	23 ± 4	8 ± 3	1 ± 1	0 ± 0	0 ± 0	23 ± 4	7
	0.10	6	36 ± 2	20 ± 6	6 ± 2	2 ± 1	2 ± 1	37 ± 3	14
Isopropyl alcohol, 70%	0.01	9	21 ± 3	4 ± 1	0 ± 0	0 ± 0	0 ± 0	21 ± 3	7
	0.03	6	36 ± 4	19 ± 4	4 ± 1	2 ± 2	2 ± 2	36 ± 4	14
	0.10	6	37 ± 1	18 ± 3	4 ± 2	1 ± 1	1 ± 1	37 ± 1	14
Sodium hypochlorite, 5%	0.01	6	11 ± 3	4 ± 1	1 ± 1	0 ± 0	0 ± 0	11 ± 3	7
	0.03	6	28 ± 3	26 ± 12	17 ± 9	15 ± 11	12 ± 10	39 ± 9	18
	0.10	6	31 ± 2	25 ± 6	20 ± 10	23 ± 13(5)	15 ± 9(5)	40 ± 6(5)	>21

Table 3 (continued)
Rabbit Eye Irritation Scores at Various Times After Instillation of Materials (Draize Score $\overline{X} \pm$ SE)[a]

Material	Dose volume[b]	N[c]	Day					Maximum	Median day to clear
			1	3	7	14	21		
Sodium lauryl sulfate, 10%	0.01	9	25 ± 2	6 ± 2	1 ± 0	0 ± 0	0 ± 0	25 ± 2	7
	0.03	6	36 ± 1	17 ± 4	2 ± 2	1 ± 1	1 ± 1	36 ± 1	7
	0.10	6	40 ± 2	23 ± 4	3 ± 2	2 ± 2	1 ± 1	40 ± 2	7
Sodium lauryl sulfate, 29%	0.01	9	35 ± 2	26 ± 4	4 ± 2	1 ± 1	1 ± 1	36 ± 2	14
	0.03	6	38 ± 2	35 ± 1	8 ± 2(5)	2 ± 1(5)	4 ± 2(5)	39 ± 1(5)	>21
	0.10	6	40 ± 2	36 ± 5	11 ± 6	3 ± 2	2 ± 2	42 ± 2	>21
Acetic acid, 10%	0.01	9	38 ± 2	36 ± 5	22 ± 5	13 ± 6	6 ± 2	42 ± 4	>21
	0.03	6	43 ± 2	52 ± 3	46 ± 3	24 ± 6	13 ± 2	53 ± 3	>21
	0.10	6	54 ± 5	51 ± 5	50 ± 9	42 ± 13	16 ± 4(3)	64 ± 11(3)	>21
Calcium hydroxide, 100%	0.01	9	45 ± 6	40 ± 6	38 ± 6	36 ± 6	21 ± 6(6)	38 ± 6	>21
	0.03	6	72 ± 8	70 ± 8	66 ± 9	57 ± 7(4)			[d]
	0.10	6	90(1)	90 ± 0(2)	88(1)				[d]
Formaldehyde, 38%	0.01	6	57 ± 4	49 ± 2	48 ± 4	61 ± 7(5)			[d]
	0.03	6	56 ± 4	49 ± 2	52 ± 3	53 ± 8(3)			[d]
	0.10	6	56 ± 5	66 ± 3	68 ± 5	74 ± 7(3)			[d]

a Each value is the average ± SE for the indicated number of animals. Parenthesized values give the number of animals when the original group size was reduced due to death, test termination, or eye too severely afflicted to grade.
b ml of liquid, or dry weight equivalent.
c Number of animals tested.
d Terminated before 21 days, but it was judged that the injury would not have cleared.

From J.E. Griffith, G.A. Nixon, R.D. Bruce, P.J. Reer, and E.A. Bannan, *Toxicol. Appl. Pharmacol.* 55, 501 (1980). With permission.

Table 4
Comparison of Statistical Parameters for Mean Time-to-Clear: Best Fit

Comparison	p value[a]	Correlation coefficient
0.01 ml Human vs. low-volume animal	<0.037	0.66
0.1 ml Human vs. Draize animal	>0.316	0.35
0.01 ml Human vs. Draize animal	>0.256	0.40
0.1 ml Human vs. low-volume animal	<0.018	0.72

[a] Probability that the relationship is due to chance.

From Freeberg et al., *Fund. Appl. Toxicol.* 7, 626 (1986). With permission.

The low volume test is essentially similar to other modified Draize protocols except in volume of material used and placement of the material. It can be used with the Draize or CPSC scoring system. In addition, data on the length of time required for recovery of the lesions in the irritated eye is more meaningful with this procedure since these lesions are typically less severe and there is a greater probability that recovery will occur within 21 days.

Using the time for the eye to return to normal facilitates comparisons between animal experiments and human data, since for humans this is the only quantitative value that can be obtained from most accidental eye exposures. Considerable data have been collected for accidental human exposures to commercial products, and these have been compared to low volume data in rabbits.[20,21] These data indicate that low volume test results correlate well with human exposure data. Freeberg et al.[17] empirically determined in human volunteers that the irritant response in the human eye most favorably compares to rabbit eye irritation after exposure to 0.01 ml of material. In fact, correlations for mean time-to-clear were only statistically significant between human and rabbit low volume exposures (Table 4).

A number of researchers have reported on the use of a low volume protocol, and corroborate that it is a reliable test for ocular irritation of commercial and industrial materials.[22-25] However, as of this date, neither U.S. nor any international regulatory agencies have accepted the low volume test as an alternative to higher volume Draize procedures.

The procedures described above are for the assessment of hazard associated with acute accidental exposure to commercial or industrial products. There are other test methods prescribed for assessing the toxicity of materials intended for use in the eye. These are described in Chapter 19 of this volume.

III. METHODS FOR OBSERVING OCULAR IRRITATION *IN VIVO*

Testing for ocular irritation involves close examination of the anterior structures of the eye: the conjunctiva, cornea, and iris. The following is a brief description of the anatomical structures which are pertinent to the observation of ocular irritation. A more complete description of ocular anatomy is provided in Chapter 14 of this volume.

A. Cornea

The cornea is composed of five distinct layers. From the outside in these are corneal epithelium, its basement membrane (Bowman's membrane), a connective tissue stroma, the basement membrane for the corneal endothelium, and the corneal endothelium which borders on the aqueous humor.

The corneal epithelium is a stratified columnar epithelium, typically 5 to 6 cells thick in the rabbit, 7 to 10 in the human. Bowman's membrane is thick in the human (8 to 12 μm), but much thinner (approximately 2 μm) in rabbits.[5] The endothelium is a single layer of low cuboidal cells. Both epithelia form a coherent barrier to diffusion. The corneal stroma, which is by far the thickest part of the cornea, is composed of collagen fibrils which are arranged in a highly regular pattern in order to achieve transparency. The fluid content of the stroma is higher than that of the adjacent sclera,[26] and this probably also contributes to its transparency.

Irritating materials applied to the eye are in direct contact with the corneal epithelium, and this is the layer most likely to be damaged by direct action of the chemical. If the material is sufficiently corrosive, then the stroma may also be eroded. Erosion of the corneal epithelium can be observed without the aid of optical instruments and/or fluorescein as long as the lesion is extensive. Otherwise, observations of corneal damage are facilitated by their use.

As noted above, fluorescein staining is accomplished by placing one or two drops of an ophthalmic solution of sodium fluorescein directly on the cornea, then washing off the excess after a few seconds with isotonic saline. Damaged areas retain more of the dye than the intact epithelium, and appear brighter yellow. Fluorescein is excited maximally by blue light, but fluorescein-stained corneas have been traditionally observed using a long wavelength UV light source, in a darkened room. Since observations are strictly qualitative, controlling the actual wavelength of the light source is not critical. Fluorescein emits light in the yellow-green portion of the spectrum. Damaged areas of the cornea appear bright yellow, and intact portions are a duller green (Plate 3).

Damage to the cornea may also be manifested as opacity. Opacity of the cornea is a result of stromal edema. The stroma is almost always edematous when the corneal epithelium has been eroded, since this allows the leakage of fluid into the stroma. Leakage of fluid may also occur in the absence of obvious damage to the epithelium if the barrier function of the epithelium has been compromised by

Plates 1 to 6 were reproduced directly from the *Consumer Product Safety Commission's Illustrated Guide for Grading Eye Irritation Caused by Hazardous Substances*. The plates are intended to illustrate the subjective grades for corneal, conjunctival, and iridial manifestations of ocular irritation. The descriptions of these grades are in Table 2.

Plate 1 (page one of color pages): A series of four photographs indicating grades of conjunctival redness. Conjunctival redness is typically not homogeneous, as can be seen in these photographs; therefore, only the most severely affected area of the conjunctiva should be graded. The graded areas are denoted by arrows in the photographs.

Plate 2 (page two of color pages): A series of photographs depicting the four grades of corneal opacity. Since the lesion is not homogeneously distributed, the most severely affected part of the cornea, denoted by arrows in the photographs, is graded.

Plate 3 (page three of color pages): The eyes in Plate 2, stained with fluorescein and photographed under UV light. Fluorescein stains areas of corneal epithelial erosion, not opacity. Thus, the areas stained with fluorescein do not necessarily correspond to the grades for opacity.

Plate 4 (page four of color pages): Grades of iritis. Grade 1 is a deepening of iridial rugae, or injection (hyperemia) of iridial vessels. The upper left photograph clearly shows injection of the secondary vessels of the iris. The upper right photograph also demonstrates this, but is more difficult to perceive due to loss of corneal clarity. Grade 2 iritis involves hemorrhage or destruction of the iris. Since this is almost invariably accompanied by significant corneal opacity, it may be difficult to see. Hemorrhage is best seen in the lower right photograph.

Plate 5 (page five of color pages): Grades of conjunctival chemosis. Note the caveat placed on the plate by the CPSC, that these photographs may not be absolutely accurate. However, the photographs do indicate the degree of difference between each chemosis grade.

Plate 6 (page six of color pages): This is a time sequence of the same eye, from before administration of an irritant to seven days after exposure. The grades for all four lesions are printed under each photograph.

REPRESENTATIVE ILLUSTRATIONS OF
DRAIZE EYE IRRITATION SCORES

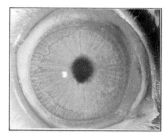

Normal Eye

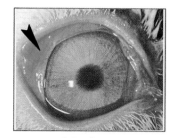

1 Redness

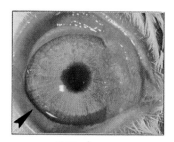

2 Redness

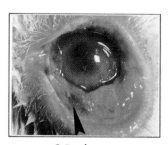

3 Redness

om CPSC, 1972

REPRESENTATIVE ILLUSTRATIONS OF DRAIZE EYE IRRITATION SCORES

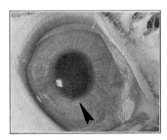

1 Opacity

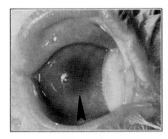

2 Opacity

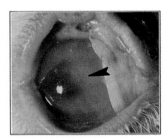

3 Opacity

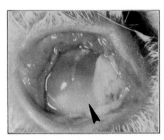

4 Opacity

om CPSC, 1972

REPRESENTATIVE ILLUSTRATIONS OF
DRAIZE EYE IRRITATION SCORES

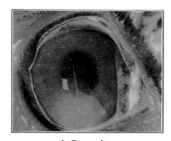

1 Opacity

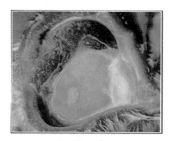

2 Opacity

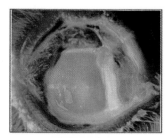

3 Opacity

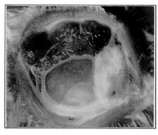

4 Opacity

REPRESENTATIVE ILLUSTRATIONS OF
DRAIZE EYE IRRITATION SCORES

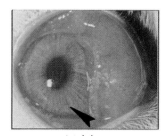

1 Iritis

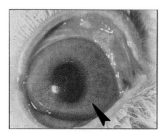

1 Iritis

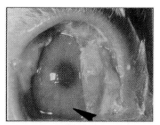

2 Iritis

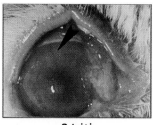

2 Iritis

REPRESENTATIVE ILLUSTRATIONS OF
DRAIZE EYE IRRITATION SCORES

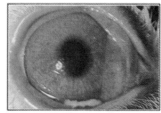

1 Chemosis

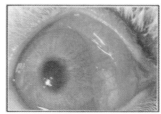

2 Chemosis

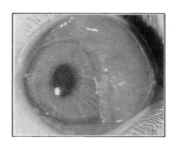

3 Chemosis

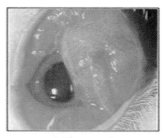

4 Chemosis

REPRESENTATIVE ILLUSTRATIONS OF DRAIZE EYE IRRITATION SCORES

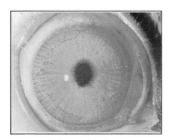

Normal Eye

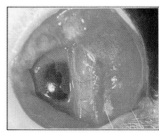

1 Hour
2-3 Redness   >2 Opacity
1 Iritis   4 Chemosis

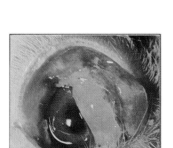

24 Hours
3 Redness   1 Opacity
2 Iritis   >3 Chemosis

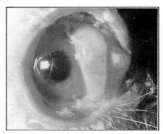

48 Hours
3 Redness   >1 Opacity
2 Iritis   3 Chemosis

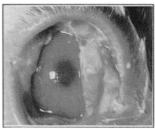

72 Hours
3 Redness   >1 Opacity
2 Iritis   >2 Chemosis

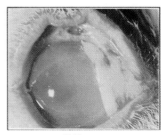

7 Days
3 Redness   4 Opacity
2 Iritis   2 Chemosis

From CPSC, 1972

exposure to a toxicant. Other possible sources of fluid seepage are from the anterior chamber, through a damaged corneal endothelium, or from the circumcorneal capillary plexus. The anterior portion of the stroma typically swells first after chemical insult,[27] indicating that leakage through the corneal epithelium is the principal source of excess fluid, at least initially.

Opacity can be observed with or without optical aids. Both the severity and extent of opacity should be noted, although the CPSC modification of the Draize scoring system only grades severity (Table 2). Opacity can range from a slight dulling of the normal reflectance and luster of the cornea, to translucency, to complete opacity (Plate 2). If there are varying degrees of opacity, the most severe area of opacity should be reported. The Draize scoring system (Table 1) also ranks the area of opacity, from 1 to 4, depending on whether the area of opacity is less than 25%, less than 50%, less than 75%, or greater than 75% of the total area of the cornea.

The slit-lamp biomicroscope is an excellent tool for detecting subtle toxicity in the cornea such as mild edema and minor damage to the corneal epithelium, and for more accurately determining the extent of more severe damage (see Chapter 15). Scoring systems have been suggested which take advantage of these properties of slit-lamp microscopy,[16,27] but none are recognized in official test guidelines.

There are several other more severe manifestations of chemical injury to the cornea, including pannus, a vascularization of the stroma. Under normal conditions, the cornea is avascular, with the exception of a few capillary loops at the edges of the stroma; however, after extensive corneal damage the stroma may become highly vascularized. The development of pannus is significant in corneal healing, and may promote re-epithelialization of the cornea by conjunctival cells. These conjunctival cells typically undergo metaplasia to corneal epithelium; however, if blood vessels are present in the cornea, the new epithelium retains its conjunctival character.[28] Thus, pannus may lead to persistent or permanent visual impairment, and its presence after chemical exposure should be noted.

There are a number of objective measurements of corneal damage which have been developed. These include the measurement of corneal thickness to assess stromal edema, corneal permeability, intraocular pressure, and leakage of protein into the anterior chamber (alterations in the latter two parameters may also result from iridial irritation). These measurements will be described later in the section dealing with objective measurements of ocular toxicity.

B. Iris

Irritation of the iris is assessed in standard ocular irritation screening by observing hyperemia (injection) of the iridial blood vessels, and edema which appears as deepening of the iridial rugae (folds). Iridial irritation may result from direct interaction of an irritating substance that has entered the anterior chamber, or indirectly, through the release of inflammatory mediators by the damaged cornea.

The permeability of the cornea to a number of chemical irritants has been reported,[29-32] and it appears that a number of these, especially surfactants, pass rapidly into the anterior chamber. Thus, it is probable that the iris is exposed to measurable concentrations of many topically applied irritants.

The cornea and iris also have closely associated blood supplies and innervation. Thus, it is also possible that inflammatory responses mediated by the vasculature and nervous system in the cornea may also be produced in the iris.

The blood supply to the iris consists of four primary vessels which are visible through the cornea, one for each quadrant of the iris. These can be seen in the photograph of the normal eye in Plate 1 or 6. They are the vessels which form a diamond-shaped pattern at the periphery of the iris. The most obvious primary vessel in these photographs runs from the left (9:00 position) to the top (12:00 position) of the iris.

Secondary vessels emanate perpendicularly from the primary vessels, and run parallel to each other from the primary vessels to the border of the pupil. These are also visible in Plates 1 and 6. Tertiary vessels emanate from the secondary vessels, but these are difficult or impossible to see without magnification.

The surface of the iris also has folds, or rugae, which run radially from the pupil to the periphery (limbus).

Irritation of the iris may include deepening of the folds of the iris, hyperemia of the iridial blood vessels, and edema of the iridial stroma. These are all relatively easy to observe, and typical photographs are provided on Plate 4. Also, pigmented tissue may slough from the posterior surface. Severely irritating materials may damage the iris enough to cause hemorrhaging into the anterior chamber.

Irritation of the iris is frequently accompanied by exudation of plasma proteins from the iridial vessels into the aqueous humor. The relative amount of protein present in the aqueous can be assessed using the slit-lamp. The presence of protein changes the refractive index of the aqueous humor, causing aqueous flare, also known as the Tyndall phenomenon, which results in the scatter of the light by insoluble protein macromolecules. Under normal conditions, light from the slit-lamp is not visible as it passes through the anterior chamber. The presence of insoluble protein makes the light beam visible. The intensity of light refracted is directly proportional to the amount of protein present. The aqueous becomes noticeably turbid when a large quantity of protein is present. Subjective systems for grading aqueous flare and anterior chamber turbidity have been published.[33] It should be noted that quantitative measurements of plasma protein exudation into the aqueous do not correlate well with severity of iridial irritation;[34] thus, the significance of aqueous flare measurement as an accessory to iridial assessment is dubious.

C. Conjunctiva

The conjunctiva is a squamous, nonkeratinized epithelium which lines the inner surface of the eyelid (palpebral conjunctiva) and the surface of the nonvisual parts of the eye (bulbar conjunctiva). It contains numerous mucous-secreting goblet cells, especially in the palpebral conjunctiva. The conjunctiva has a rich supply of blood vessels which are visible through the conjunctival epithelium.

Like the cornea, the conjunctiva is in direct contact with the test material, and the epithelium is prone to damage. However, in standard practice lesions of the conjunctival epithelium are not evaluated. Instead, observations are made on the inflammatory response of the conjunctiva. Inflammation of the conjunctiva is typical of that in any vascularized tissue. The vessels become congested and ischemic. This causes the conjunctiva to become red. Redness of the conjunctiva is described and graded in the Draize and CPSC scoring schemes (Tables 1 and 2). As the redness becomes more pronounced, it becomes increasingly difficult to discern individual blood vessels. Examples of conjunctival redness grades are depicted in Plate 1.

Ischemia in the conjunctiva promotes edema (chemosis), causing the conjunctiva to bulge. Chemosis is graded on the severity of swelling, and the degree to which it distorts the eyelids (Plate 5).

Damage to the conjunctiva may also alter its secretions. Abnormal secretions of fluid or mucus should be noted, although there is no scoring scheme or reporting requirement for this under current regulations (the Draize scoring system does grade discharge). The discharge from the irritated eye may be purulent, indicating that there is significant activity of neutrophils in the damaged tissue.

Severe and persistent conjunctival effects that can result from chemical insult include a chronic condition, the formation of scar tissue, and a more acute effect, phlyctenae (blisters).

IV. CRITICISMS OF CURRENTLY MANDATED OCULAR IRRITATION SCREENS

The Draize procedure and its updated versions in federal regulations have been critized for a number of reasons which can be grouped into three broad categories. First, the rabbit is an inappropriate animal model, with sufficient anatomical and physiological differences from the human that it routinely overestimates ocular irritancy potential. Second, the exposure conditions, particularly the volume of test material used and the means of instillation are inconsistent with typical human accidental exposures, also leading to overestimation of irritancy potential. Third, the methods rely on descriptive assessments based on subjective scoring schemes, leading to an unacceptable degree of variability in the interpretation of test results. These three criticisms will be discussed in detail.

A. Choice of Species

The albino rabbit was chosen as an animal model for ocular irritation testing because its eye is relatively large and lacks pigmentation, thus facilitating the observation of damage, especially of the iris. However, in many ways the rabbit eye differs substantially from the human eye, which decreases its applicability as a model for predicting human ocular toxicity.

The rabbit eye contains a third eyelid or nictitating membrane, a membrane internal to the eyelids. The nictitating membrane probably serves to protect the eye from physical injury. However, after instillation of a chemical irritant directly onto the cornea or into the lower conjunctival sac the nictitating membrane may retard the removal of the substance from the eye, prolonging contact with the cornea and conjunctiva. Buehler and Newmann[35] tested this belief by surgically removing the nictitating membrane, but were unable to demonstrate that this had any effect on the healing rate of a damaged cornea.

The rabbit possesses secretory glands around the eye which are not present in humans, including the Harderian gland, and nictitan gland. These may alter the composition of the liquid film covering the surface of the eye, although the extent of such differences, or what effect they would have on chemical irritancy, are not known. Buehler and Newmann[35] communicated that tear production in the rabbit is restricted relative to human responses, which could also retard the clearance of a toxicant from the eye, potentiating the irritant response. There is no empirical evidence in the literature to support or refute this observation, however.

The thickness of the rabbit cornea is approximately 400 μm, roughly equivalent to that of the human (530 to 540 μm). However, the area of the eye covered by the cornea is 25% in the rabbit vs. 7% in the human.[36] Thus, the rabbit would have an increased likelihood of corneal damage following chemical exposure than would result from accidents involving humans where the chemical splashes in the eye.

The histological structure of the cornea is also different in rabbit and human, and probably predisposes the rabbit to greater sensitivity to ocular irritants. Specifically, the rabbit's corneal epithelium is one to three cell layers thinner than the human's, and the underlying Bowman's membrane is almost an order of magnitude thinner in the rabbit.[5] This would tend to facilitate the penetration of irritants through the corneal epithelium, exacerbating surface lesions and promoting stromal edema, and possibly increasing the exposure of the iris to the irritant.

The rabbit eye may be more easily anesthetized by irritants than is the human eye. Harris et al.[37] determined that commercially available soaps and shampoos produced corneal anesthesia in the rabbit which was comparable to that produced by an 0.5% solution of tetracaine. The anesthetic effect persisted for several hours. None of these substances produced any corneal anesthesia in human subjects. This effect may exacerbate chemically induced irritation, since the rabbits would not blink excessively in response to the irritation, thus inhibiting clearance of the irritant from the eye.

B. Comparative Studies

A number of comparative studies have been carried out to determine which species is the most predictive of human eye irritation. Comparisons were made to published human clinical data, or in some cases to empirically derived human data. These studies demonstrate that the rabbit is very sensitive to ocular irritants, apparently much more sensitive than humans.

Beckley[38] compared the ocular response of rabbits, dogs, rhesus monkeys, and humans to a detergent liquid. The response of the monkey's eye was most similar to humans. Rabbits were least similar, and exhibited the highest degree of irritation.

Beckley et al.[39] compared the ocular responses of the rabbit, rhesus monkey, and human to two surfactant-containing solutions, using a 0.1 ml volume for each species. Again, they concluded that the response of the monkey was much more similar to the human response than was the rabbit.

Buehler and Newmann[35] compared the response of the rabbit and rhesus monkey eye to a number of surfactant-based solutions, and to a 1% sodium hydroxide solution. They reported that the rabbit is much more sensitive to these irritants than the monkey, and that the responses elicited in the monkey eye are similar to those reported for accidental human exposure to the tested materials.

On the other hand, Elliot and Schut[40] discerned no differences in the corneal changes induced by cytarabine HCl (cytosine arabinoside) in rabbit, stump-tailed macaque, or human eyes.

Gershbein and McDonald[41] compared the ocular irritancy of eight surfactant formulations of varying potencies in rabbits, guinea pigs, rats, mice, hamsters, rhesus monkeys, dogs, cats, and chickens. All were instilled with 0.1 ml of the test substance except the small species, which received 0.05 ml. Corneal and conjunctival responses were graded. It was concluded that the rabbit, hamster, and mouse are the most sensitive species, the rat and guinea pig are less sensitive, and the dog, cat, rhesus monkey, and chicken are the least sensitive. Table 5 is a summary of these data.

Since previous studies[35,38,39] had concluded that the rhesus monkey is most like the human, Gershbein and McDonald's study corroborated that the rabbit was overly sensitive as a model to predict human irritation, and suggests that the dog, cat, and chicken could be explored further to determine as a potential animal model to better predict human ocular irritation.

C. Administered Volume

The Draize procedure has also been criticized for its use of exposure volumes which are probably far in excess of the typical human accidental exposure volume, and that this is a source of overestimation of risk.

As with any toxic response, ocular toxicity is a dose-response phenomenon. Carpenter and Smyth[5] recommended testing of graded volumes and concentrations to arrive at a dose of chemical that would give a specific response. Griffith et al.[38] conducted a dose-response study for a number of different irritants in the rabbit eye, and compared the responses to published information on the human ocular response to these agents.[19] They concluded that the rabbit response was most like the human response when a 0.01 ml dose volume was used, and indicated that ranking the irritancy of materials was easier when this volume was used.

Freeberg et al.[20] compared the recovery time of human eyes accidentally exposed to consumer products to recovery time in rabbits treated with 0.1 or 0.01 ml, and cynomolgous monkeys exposed to 0.1 ml of the same products. They reported that all three animal protocols overestimated the human response, and that only the rabbit 0.01 ml volume data significantly correlated with the human data.

Johnson[22] also advocated the use of a 0.01 ml volume in eye irritation tests. In the event that this dosage produces no irritation, the material can be retested at 0.1 ml.

Table 5
Corneal Findings in Various Animal Species Following Instillation of the Eyes with Test Shampoos and Detergents

Product	Rabbits		Guinea pigs		Rats		Mice		Hamsters		Dogs		Cats		Monkeys		Chickens	
	No. of lesions[a]	Mean score (%)[b]	No. of lesions[a]	Mean score (%)[b]	No. of lesions[a]	Mean score (%)[b]	No. of lesions[a]	Mean score (%)[b]	No. of lesions[a]	Mean score (%)[b]	No. of lesions[a]	Mean score (%)[b]	No. of lesions[a]	Mean score (%)[b]	No. of lesions[a]	Mean score (%)[b]	No. of lesions[a]	Mean score (%)[b]
SP-1	11(11)	37.5 ± 6.9	0(16)	—	5(15)	11.4 ± 3.9	11(20)	53.7 ± 12.3	9(12)	25.9 ± 10.3	1(10)	17	1[e](10)	13[e]	0(3)	—	0(4)	—
SP-2	9(12)	35.7 ± 13.3	0(15)	—	0(17)	—	1(20)	8	5(12)	26.6 ± 8.0	0(10)	—	0(10)	—	0(3)	—	0(4)	—
SP-3	11(10)	44.4 ± 7.7	2(15)	3:3	1(18)	8	5(20)	46.0 ± 16.2	7(11)	59.0 ± 37.4	2(10)	17:17	0(10)	—	0(3)	—	0(4)	—
SP-4	11(11)	37.7 ± 4.5	0(17)	—	2(17)	13; 34[c]	12(20)	74.4 ± 9.9	7(11)	84.0 ± 17.4	1(10)	11[d]	0(10)	—	0(6)	—	—	—
Roccal (2%)	9(10)	62.1 ± 7.4	6(19)	15.3 ± 4.4	10(14)	23.5 ± 3.5	12(12)	45.7 ± 10.1	12(13)	54.6 ± 10.6	4(13)	6.7 ± 3.5	1[e](10)	4[e]	—	—	—	—
Isothan Q-15 (2%)	7(11)	59.2 ± 6.3	6(18)	19.0 ± 5.0	6(15)	8.7 ± 2.2	12(12)	86.2 ± 3.9	12(13)	57.1 ± 9.1	2(11)	8; 17	0(10)	—	1(3)	25	—	—
Neutronyx 600 (20%)	10(10)	34.4 ± 5.9	14(15)	41.4 ± 7.1	8(15)	30.8 ± 8.3	11(12)	70.7 ± 8.8										
Isothan Neutronyx[f]	10(10)	74.8 ± 6.4	17(18)	63.4 ± 5.5	16(16)	29.7 ± 6.3	12(12)	87.1 ± 6.8									1[e](4)	4[e]

[a] With total number of eyes examined given in parenthesis.

[b] The mean score (±SEM) applies to those eyes displaying corneal lesions and is calculated on the basis of the day 7 readings or the most extreme ratings. In rabbits, lesions of greater severity generally persisted for 30 to 90 days, but recovery was more rapid with many of the other species. The single lesion noted in the monkey 48 h after treatment was absent on day 4.

[c] Findings on day 3; the respective scores were 0 and 8% on day 13.

[d] Scoring on day 3; the value was 6% on day 6 and the four lesions elicited by 2% Roccal had disappeared by day 30.

[e] Score 24 h after treatment; lesions were absent 15 h later.

[f] Mixture of 20% Neutronyx 600-2% Isothan Q-15, 1:1 (v/v).

From L.L. Gershbein and E.McDonald, *Food Cosmet. Toxicol.* 15, 131 (1977). With permission.

Table 6
Mean Time-to-Clear (Days) in Animals and Humans Exposed Accidentally to Surfactant-Containing Products

Product[a]	Current human data	Low volume rabbit	FHSA rabbit
Liq. L.P. #1	1.92	26.6	35.0
Liq. D.P. #1	0.77	8.2	25.7
Dry D.P. #1	0.59	4.6	18.3
Liq. D.P. #2	0.43	7.7	11.7
Liq. H.C.P. #1	0.38	—	11.1
Liq. D.P. #3	0.30	3.9	22.2
Liq. H.C.P. #2	0.23	4.0	15.2
Dry H.C.P. #1	0.19	1.3	29.2
Shampoo #1	0.11	11.6	30.3
Shampoo #2	0.10	3.7	7.0
Dry D.P. #1	0.08	2.1	13.8
Dry D.P. #2	0.06	2.9	15.1

[a] L.P., laundry products: additives, main wash detergents, and fabric softeners; D.P., dishwashing products: automatic and hand detergents; H.C.P., household cleaning products: hard surface cleaners and nonabrasive cleaners.

From F.E. Freeberg, D.T. Hooker, and J.F. Griffith, *J. Toxicol. Cut. Ocul. Toxicol.* 5(2), 115 (1986). With permission.

Like Griffith et al.,[18] he points out that the 0.01 ml volume often discriminates differences between irritants which are not apparent at 0.1 ml, where they may have reached the upper asymptote of the dose-response curve.

Williams[25] compared several materials using the 0.1 and 0.01 ml dosing regimens, and concluded that although the low volume procedure decreases absolute irritation scores, it does not alter the rank ordering of the irritancies of the materials.

MacRae et al.,[42] in studying the ocular toxicity of diethyltoluamide, found that 10 µl appeared to be the best volume to predict ocular toxicity.

Freeberg et al.[20,21] correlated data on human responses to accidental exposure to commercial products, and compared these with animal data. The human database was extensive, involving several hundred exposed to over 23 different formulations. They concluded that the low volume procedure correlated much better with the human data than did protocols using a 0.1 ml dosing volume (Table 6).

Human eye irritation data were collected empirically in human volunteers exposed to four commercial products, and these data were compared to rabbit data after exposure to 0.1 and 0.01 ml dosing volumes. The 0.01 ml volume correlated much better with the human data.[17]

The results of all of these studies strongly argue that the low volume procedure is much more predictive of human ocular exposure than are protocols using a 0.1 ml dosing volume. As Johnson[22] points out, nothing is lost even if the test material is not irritating at the low volume, as it can be retested at a higher volume.

D. Lack of Reproducibility of the Draize Procedure

It must be recognized that the Draize and CPSC scoring systems are subjective, and may be prone to differences in interpretation among laboratories and individuals. Therefore, it has been criticized as a measurement with a low level of repeatability. A number of studies have been conducted to determine the magnitude of variability in these subjective assessments.

Weil and Scala[43] coordinated a study in 25 industrial, contract, and government laboratories to characterize the intra- and interlaboratory variability in Draize results. They tested 12 to 16 materials in each laboratory, using both a standardized protocol and whatever protocol the lab typically used. The standardized protocol was the FHSA method, but used the Draize scoring system. Both intra- and interlaboratory variability were high — unacceptably so in the opinion of the authors. Unanimous agreement over whether a substance was an irritant or nonirritant was rare. The authors concluded that most of the variability was attributable to the reading of eye irritation, but that some must also be due to interpretation and execution of the test guidelines.

Marzulli and Ruggles[44] conducted their own collaborative study in response to the Weil and Scala report. They coordinated the testing of seven materials in ten government, industrial, and contract laboratories. Their criterion for comparison was the ability to correctly identify a material as an irritant or nonirritant using the FHSA guidelines and grading system. These authors concluded that there was reasonably good correlation of results between laboratories when all four components of the grading system (corneal opacity, iridial effects, conjunctival redness, and conjunctival chemosis) were used in the irritancy determination. However, interlaboratory variation for each individual endpoint was substantial.

McDonald and Shadduck[27] reported the results of a study designed to assess the consistency between investigators, and between readings by the same individual, in grading ocular irritation. The investigators graded ocular irritation in rabbits which had been exposed to chemical irritants (the investigators were unaware of the type of treatment). Each investigator was given two trials with the same group of rabbits to determine the consistency in his grading. It was observed that interindividual variability was low for most parameters, with the exception of conjunctival congestion and discharge. Similarly, the ability of each investigator to correctly reproduce an observation was good for all parameters with the possible exception of discharge. The authors offer this type of comparative study as a training exercise within laboratories as a means of increasing the reliability and consistency of results. Weil and Scala[43] had also suggested that some sort of training, especially centralized training, may be beneficial for beginning toxicologists and technicians.

All in all, it appears that the standard protocols for ocular irritation assessment are probably sufficient to classify materials as irritants or nonirritants according to an arbitrary scheme such as that of the FHSA, but there is little indication that they can reliably distinguish between irritant potencies. Such severe limitations may be acceptable for regulatory purposes, as indicated by Marzulli and Ruggles.[44] However, a high degree of inconsistency is unacceptable in product development, where decisions on marketing a product or making a product improvement may hinge on comparisons to irritancies of old formulations or other standards. There is an indication from these studies that intralaboratory variability may be controlled by rigorous and consistent training of all personnel in a standardized method of ocular irritation assessment.

It may also be true that certain procedures and measurements may be inherently variable. We have noted above that the large volume of test material used in standard protocols is far in excess of the typical human exposure, and is also in excess of the capacity of the rabbit eye. This may lead to variable exposures of the eye due to a lack of control of the actual exposure to the eye (how much stays in contact with the eye). Although there is much less historical data generated using the 0.01 ml volume procedure, it appears that it produces consistent responses and interpretable results.

It is also disturbing that there is little agreement between laboratories in either of the collaborative studies[43,44] as to the severity of effects on individual ocular tissues. This indicates that either (1) the procedure is not the best model for ocular irritation, (2) it does not adequately control the known sources of variation, or (3) that the endpoints being measured are inappropriate. As for the former thought, it has already been noted that several aspects of traditional procedures are inconsistent with human injury, including dose volume and placement of material in the eye (conjunctival sac vs. direct corneal application), and that the rabbit eye is substantially more sensitive to ocular irritants.

As for the idea that the endpoints of the traditional protocols are inappropriate, it may be beneficial to rely on other measurements of the irritant response, such as time for recovery from toxic effects, or to include more objective and quantitative endpoints. Time-to-recovery is an attractive endpoint because its use requires no modifications to existing methodology, it is probably easier for most observers to discern than subjective grades of irritation, and it allows direct comparison of animal data to human accidental exposure data. For human exposures, time-to-recovery is generally the only reliable index of the severity of irritation. It is of interest that time-to-clear measurements from the low volume procedure correlate well with a human database on accidental exposure to commercial products.[20,21]

The lack of consistency in quantitative Draize results may have important consequences in the development and validation of *in vitro* methods for assessing ocular irritation. Most of these methods have focused on modeling a particular aspect of ocular irritation, especially primary damage to the corneal epithelium. The high variability of the Draize procedure, especially the complete lack of consistency between the scores for individual components of ocular toxicity, may make it difficult or impossible to use the historical database for *in vivo* rabbit ocular irritation as a yardstick for assessing the performance of *in vitro* assays. The use of the 0.01

ml dose volume may help, but perhaps the *in vitro* tests would better be compared with objective measurements of ocular irritation *in vivo*.

V. OBJECTIVE MEASUREMENTS OF OCULAR IRRITATION

It can be concluded from Weil and Scala's[43] collaborative study that most of the variability in the Draize procedure is attributable to varying levels of skill in observing, interpreting, and categorizing ocular lesions in a subjective scoring system. The use of objective measurements of ocular irritation may increase the reliability and reproducibility of ocular irritation assessments by removing the subjective components of the procedure. A number of objective assessments of ocular structure and function have been applied to irritant screening. These include measurements of corneal edema (thickness and water content), conjunctival edema, intraocular pressure, conjunctival capillary permeability, corneal permeability, protein leakage into the anterior chamber, area of corneal damage by fluorescein staining, rate of re-epithelialization of induced corneal wounds, and cytological analysis of exfoliated corneal and conjunctival cells.

A. Intraocular Pressure

Intraocular pressure tends to increase in a dose-related manner after the eye is exposed to an irritant.[45] Walton and Heywood[46] measured intraocular pressure in rabbit eyes using an applanation tonometer, a devise which determines the force required to flatten the eye to a known extent. Treatment with sodium dodecyl sulfate caused a rapid increase (a 70% increase in 30 min), then a rapid decline over the next 24 h, in intraocular pressure. Grossly observable irritation persisted after intraocular pressure had returned to normal.

Measurement of intraocular pressure, at least by applanation tonometry, may be difficult to carry out. The tonometer must be adjusted for changes in corneal thickness which are likely to occur during corneal injury; there appears to be measurable interoperator variability in reading the tonometer; the cornea must be anesthetized; and a contrasting medium must be introduced on the surface of the eye in order to visualize the applanation rings of the tonometer (Walton and Heywood used evaporated milk).

B. Corneal Thickness

Swelling of the corneal stroma is a typical early response to irritant exposure. Burton[47] described a method to measure corneal thickness (pachymetry) using a depth-measuring attachment on a slit-lamp microscope. He used the ratio of corneal thickness at the corneal apex after the instillation of an irritant to the preexposure thickness as an index of irritation. Burton found that there is a strong positive correlation between corneal thickness and the Draize scores for corneal opacity, overall corneal irritation, and overall conjunctival irritation (although corneal swelling had apparently reached a maximum when conjunctival irritation was only moderate). Furthermore, subtle differences in corneal thickness were measured in eyes which would have been given the same subjective Draize score. Thus, corneal thickness measurement may be a more powerful method for ranking the relative irritancies of similar materials.

There has been little follow-up to this work, although Burton et al.[48] have reported that corneal thickness measurements are possible in enucleated rabbit eyes, which they propose as an *in vitro* method for assessing ocular irritation.

Heywood and James[49] criticized corneal thickness measurements on the basis that the cornea is not uniformly thick, toxic substances are not uniformly distributed, and exfoliation of the corneal epithelium would invalidate thickness measurements. There appears to be little merit to these criticisms. Burton[47] stated that measurements were always made at the corneal apex and that preexposure measurements were made so that each animal serves as its own control; therefore, the criticism of irregularity of corneal thickness is moot. It is possible that an interlaboratory comparison study may show this method more consistent than the Draize scoring procedure.

As for uneven distribution of toxic effects, Burton reported that substances which were at all irritating caused at least a 30% increase in corneal thickness. While this does not guarantee that toxicity was evenly distributed, it does indicate that the effect is of sufficient magnitude that it will not be hidden by uneven distribution of toxic effects on the cornea. It would also be possible to place test materials directly on the cornea instead of into the conjunctival sac to ensure that the toxicant was always placed in direct contact with the area of the cornea where thickness would be measured.

Loss of corneal epithelium does not invalidate the measurement. Most of the cornea (about 90%) is stroma; thus, even a substantial loss of epithelium is negligible in the face of a 30% or greater increase in stromal thickness. Furthermore, extensive or complete exfoliation of the corneal epithelium would only be likely to occur after exposure to the most irritating or corrosive substances. Such extreme irritation would be recognizable grossly and the materials could be classified on this basis alone. There should be no argument over the classification of extreme irritants, and there is certainly no need to make any sensitive measurements of structure or function.

Conquet et al.[34] also measured corneal thickness in rabbits exposed to organic solvents, and found that this parameter was significantly correlated with the Draize corneal score, at least for the first day after exposure.

These results indicate that corneal thickening is a common event in ocular irritation, and that assessment of this endpoint may be a sensitive and predictive index of chemically induced corneal injury. Additional studies are needed to further define this response and to validate the method as a screen.

C. Edema and Capillary Permeability

Conquet et al.[34] and Laillier et al.[50] examined a number of different endpoints associated with edema of ocular tissues, including corneal and conjunctival water content as an index of swelling, and conjunctival and aqueous humor content of Evans blue dye to detect increased capillary permeability in the conjunctiva and iris and ciliary body. (Evans blue dye injected intravenously associates with plasma proteins. Its presence in extravascular spaces indicates extravasation of plasma proteins.) There was a strong correlation between conjunctival water content and Draize conjunctival score, and between corneal water content and Draize corneal score, at least for the first day after exposure to irritating solvents. There was also a good correlation between dye content of the conjunctiva and Draize conjunctival score. On the other hand, there was little relationship between dye concentration in the aqueous and visible iridial irritation.

Measurements of edema may be valuable as objective measurements of ocular irritation; however, they are impractical for routine screening purposes because the measurements require the sacrifice of animals at each observation point in order to measure wet and dry weights of the tissues of interest, and Evans blue dye concentrations. At least for the cornea, thickness measurements appear to be as indicative of edema and irritation as water content, and thickness measurement is not invasive.

D. Corneal Permeability

It is likely that corneal fluid accumulation resulting from irritant exposure is caused by a loss of barrier function, and therefore increased permeability, of the epithelium. Corneal epithelium normally has a low permeability to water[51] and actively transports ions, presumably maintaining ionic gradients.[32] Quantitative measurements of corneal permeability have been made by measuring the passage of fluorescent molecules through the cornea, or by measuring electrical resistance across the cornea. These measurements have been applied to ocular irritation assessment.

Green and Tonjum[52] described a method for determining the flux of sodium fluorescein across the excised rabbit cornea. Passage of fluorescein was low in the undisturbed cornea, but increased sharply when the corneas were exposed to solutions of surfactants, an effect which was concentration related.

Corneal permeability to sodium fluorescein has also been measured in the intact mouse.[53,54] This required the use of an epifluorescence attachment (as a monochromatic light source to excite the fluorescein) and a photomultiplier attachment (to quantitate fluorescence) to a microscope. They reported that entry of fluorescein into the cornea increased after exposure to irritating surfactants such as benzalkonium chloride, but not to nonirritating materials such as Tween 80 (a polysorbate). Although the experiment was conducted in living mice, the procedure was somewhat invasive in that the mice were under general anesthesia.

Maurice and Singh[55] examined the permeability of the cornea to two fluorescent ions, fluorescein and sulforhodamine B. Sulforhodamine B was preferred over fluorescein because it has a much lower lipid solubility at physiological pH (thus decreasing its permeability into the undamaged cornea), and because the spontaneous background fluorescence of the eye is much higher at the green wavelength emitted by fluorescein than at the red wavelength emitted by rhodamine. Exposure to sodium hydroxide or benzalkonium chloride caused concentration-related increases in corneal permeability to sulforhodamine B. No permeability increase was observed after instillation of water, isotonic saline, or Tween 80.

Corneal permeability changes appear to be indicative of corneal injury, and this method of assessment is sensitive. On the negative side, it requires expensive equipment (although Maurice and Singh[55] indicate that a simpler and less expensive apparatus is being developed) and it is invasive to the animal. One technique required excision of the cornea,[52] another involved the sacrifice of the animals before making fluorescence measurements,[55] and the third procedure called for general anesthesia prior to examination.[53,54]

Increased permeability of the cornea to ions can also be measured electrophysiologically. A decrease in the potential difference across the rabbit cornea has been measured after exposure to benzalkonium chloride[31] and cetylpyridinium chloride.[32] Electrical resistance, a more directly interpretable measure of decreased barrier function, was measured in explanted bovine corneas and found to be decreased in the presence of surfactants.[30] As with other measures of corneal permeability, electrophysiological techniques are sensitive to injury of the corneal epithelium and may be reliable indices of ocular toxicity. On the other hand, they require specialized equipment and are invasive.

E. Exfoliative Cytology

In addition to functional assessments, it is also possible to examine histological aspects of the eye in order to assess irritation. Obviously, traditional methods of histological preparation require sacrifice of the animal. Walberg[56] has described a method to collect and examine the cells which are sloughed from the cornea and conjunctiva after chemical injury. The total number of cells collected correlated well with Draize score. Cells were stained and classified in an attempt to determine which tissues were being exfoliated. Walberg was able to distinguish neutrophils (present as a consequence of irritation), corneal epithelium, and conjunctival epithelium, but differential counts were only reliable when a relatively large number of cells was collected. Even when they were available, these counts did not increase the overall sensitivity of the method for predicting severity of irritation.

VI. ALTERNATIVES TO TRADITIONAL OCULAR IRRITATION TESTING

A great deal of effort has gone into identifying alternatives to traditional determinations of ocular irritancy potential. One of the driving forces in developing alternatives is the inherent unreliability of the traditional methods, and their lack of relationship to human experience after accidental exposure or direct clinical testing. Ethical questions have also been raised as to whether traditional methods induce unnecessary discomfort in the animals used. Both concerns have impelled the search for alternatives.

A principal focus of alternative methods development has been the development and validation of *in vitro* models. Progress in this area will be reviewed in a subsequent chapter in this volume (Chapter 18). It should be noted that ocular toxicity is a phenomenon which lends itself to *in vitro* modeling. Specific target tissues are known (and limited), the exact dosage to the targets is known, the effects observed are acute and can be fully assessed in a short period of time. On the other hand, it should also be realized that even with these advantages, modeling any biological phenomenon (especially toxicological phenomena) is complicated. A large number of *in vitro* screens are now under development and validation in industrial and academic laboratories; however, we still do not have sufficient experience with or confidence in any of these methods to supplant current *in vivo* methods.

A. The Low Volume Procedure

There has been significant effort in adapting existing *in vivo* procedures to make them more predictive of human ocular injury, and less traumatic to laboratory animals. We have already discussed the low volume procedure,[18] which appears to more closely model human accidental exposures to commercial products, correlates much more closely to injury reports in humans[20,21] and to human eye irritation data generated under controlled laboratory conditions[22] than do traditional procedures using a 0.1 ml dose volume. Since the low volume test produces less ocular irritation, it is also less traumatic.

B. Use of Local Anesthetics

Local anesthesia of the cornea may alleviate pain associated with ocular irritation testing. This may not be a truly alternative method, since the 1987 OECD guidelines sanction the use of local anesthetics. However, theirs may be an offhand suggestion since the guideline does not suggest the type or dosage of anesthetic to be used, and their use is restricted to a single dosage prior to the administration of the test agent.

The use of topical anesthetics has been criticized on the basis that it exacerbates ocular irritancy. There is limited evidence to support this. Gunderson and Liebman[57] report that cocaine, larocaine, butacaine, tetracaine, and phenacaine all delay the healing of the abraded cornea, but phenacaine and tetracaine were the least toxic in this respect. Bykov and Semenova[58] indicate that regeneration of the corneal epithelium after physical damage is slowed by cocaine or dicaine treatment, but not by lidocaine. Hood et al.[59] reported that cocaine actually loosens the corneal epithelium and causes stromal damage.

Leuenberger[60] studied the ultrastructure of the surface of the corneal epithelium after treatment with benoximate, cocaine, lidocaine, and tetracaine. All produced changes in the plasma membrane, and appeared to alter membrane-associated structures such as microvilli and desmosomes, which may have led to the loss of surface epithelial cells. He reported no difference in the severity of effects of any of the four. In contrast to this, Pfister and Burstein[61] found no effects on cell surface ultrastructure of corneal epithelium after proparacaine or tetracaine treatment.

Grant[19] stated that topical anesthetics increased the permeability of the corneal epithelium. However, actual measurements of mouse corneal permeability indicate some increase with tetracaine, but none with proparacaine.[55]

Thus, it appears that most or all topical anesthetics are somewhat damaging to the cornea, and may, therefore, potentiate the irritancy produced by the test substance. It is difficult at this time to recommend that a topical anesthetic be used in conjunction with eye irritation tests because there are no data in the peer-reviewed literature which have evaluated the relative contribution of anesthesia to the irritancy of commercial or industrial products. However, the small amount of data which exist suggests that some topical anesthetics, particularly proparacaine, may have minimal or no adverse effects. The pursuit of research to determine the possible potentiative effects of this anesthetic on the ocular response to known irritants should be encouraged.

As for the present, no scientifically based recommendation can be made on the use of a topical anesthetic. To follow published guidelines and use an anesthetic when severe irritation is anticipated is to follow circular reasoning. If an investigator has sufficient information to convince himself that a material will be severely irritating, then there is little need to test the material. On the other hand, anesthetics should probably not be used in testing materials with low irritancy until we know that the anesthetic will not exacerbate a mild response and result in the misclassification of the material.

C. Reliance on Historical Data

Another reasonable alternative to testing is to rely on historical data regarding the irritancy of test materials. The OECD guidelines specify that a material with a pH greater than 11.5 or less than 2 need not be tested because they have a high probability of causing severe irritation and should be labeled as such. This stipulation is based on a strong historical database on the ocular effects of acids and bases.[2-4] It may be possible to make similar judgments on other new products which contain substantial quantities of substances previously tested and known to cause substantial irritation. There is little sense in carrying out an ocular irritation screen for risk assessment when it is known *a priori* that the tested material is an irritant.

D. Reliance on Dermal Irritation Data

The use of dermal irritation results has also been advanced as a substitute for ocular irritation testing, the idea being that any substance which is irritating to the skin, with its thick layer of keratinized epithelium will be at least as irritating to ocular tissues.

There have been a few published studies which have directly compared the skin and eye irritating potentials of a representative sampling of materials. For example, Dutertre-Catella et al.[62] compared these two parameters for a series of ethanolamines. They reported good correlation between the dermal and ocular irritancies of these compounds. Ciuchta and Dodd[63] compared the ocular and dermal irritancy of seven surfactants and reported good agreement between the tests in how they rank ordered the seven.

Williams[64] surveyed the historical files of duPont's Haskell Laboratories to determine the degree of fidelity between ocular and dermal irritation results. Of 60 substances which were severely irritating or corrosive to rabbit skin, 39 were also severe ocular irritants, 6 were moderate, and the remaining were either mild or nonirritants. Gad et al.[65] compared the dermal and ocular irritancy of 72 materials. They reported reasonably good correlation between the two tests, with some exceptions. In six cases, skin results failed to identify severe ocular irritants. In one case, only mild irritation was observed for a substance which was a significant dermal irritant.

These results indicate that there is no simple relationship between dermal and ocular irritancy, and that one cannot be used to predict the other. A decision to consider all dermal irritants *a priori* to be ocular irritants may be a conservative decision, in that it will result in the mislabeling of nonirritants as irritants, but not of irritants as nonirritants. However, in light of the above comparisons this decision will result in a lot of mislabeling. It appears that dermal test results should not be used to predict ocular test results until there is some clearer understanding of the relationship between the two.

E. Using Fewer Animals

Six rabbits have traditionally been used in ocular irritation testing,[8,9] decreased from the nine prescribed by Draize et al.[6] It may be possible to use even fewer rabbits in each test. The 1981 OECD guidelines[10] indicate that preliminary testing may be conducted in only three rabbits. The update of the OECD guidelines in 1987[11] calls for only three rabbits for a test, with a single rabbit screening test. If the material is found to be severely irritating or corrosive in those rabbits, no further testing is necessary. A 1977 NAS report,[16] which advocated the use of multiple dose levels, stipulated that a minimum of four rabbits be used at each dose level unless unequivocal evidence of severe irritancy could be obtained with fewer animals.

There are a number of considerations which must be made in determining whether animal number can be reduced in traditional ocular irritation tests. As noted earlier in this chapter, the results of these procedures may be quite variable,[43] and it has been argued that reducing the number of animals per test would increase the incidence of classification errors.[66] However, Heywood and James[49] have made the observation that increasing the size of the test group does not increase the precision of the test. If, as Marzulli and Ruggles[44] indicate, the purpose of the test is simply to indicate irritancy or nonirritancy, as in the case with the U.S. FHSA, then the ability to distinguish severity of response is not necessary in the regulatory process, and the number of animals needed to assess irritancy need not be high. If, on the other hand, it is of value to differentiate between grades of irritancy, as is often the case during new product development, then the number of animals used can not be reduced, and there may in fact be valid arguments for increasing animal number.

DeSousa et al.[67] used data from eye irritation tests, using six rabbits per group, on 67 materials, to determine the loss of predictiveness of the test when animal number was reduced. They report that subsamples of 2, 3, 4, or 5 rabbits were 88, 93, 95, and

96% accurate, respectively, when compared to the six rabbit tests, in ranking materials into five categories of irritation from minimal to extreme. In all cases except the subsample of two, errors in classification involved a difference of only one category. These data are encouraging in that they support the OECD guidelines in limiting tests to three rabbits when a material is severely irritating, and indicate that this limit could also be placed on materials which are nonirritating in three rabbits. On the other hand, results which are not definitively at either end of the scale raise problems, as decisions on whether to add warning labels, or even to market certain products may hinge on as little as one irritancy grade (as defined by DeSousa et al.). In these cases there is little choice but to use a larger sample size.

VII. CONCLUSIONS

This chapter has concentrated on the methodology used in assessing ocular irritancy potential of commercial and industrial materials. At the present time, it is mandated that ocular irritation be assessed using methodology developed over 4 decades ago, with some modifications. Although widespread usage and regulatory proliferation suggests that this methodology has withstood the test of time, a number of valid criticisms have been made concerning its shortcomings in modeling human accidental exposures, the extreme sensitivity of the test species used, the high degree of inter- and intralaboratory variability in its results, its lack of objectivity, and most recently its tendency to induce undue discomfort in the test animals. These criticisms were discussed, and potential alternatives have been presented.

The most attractive alternative available now is the low volume procedure. This correlates much better with human data, and appears to be more reliable than traditional methods. Since the low volume method is similar in execution to traditional methods, it can be easily performed by any investigator or technician familiar with standard ocular toxicity testing. Since it relies on traditional grading systems, its results can be directly compared to the historical database for the Draize test.

Several objective measurements of ocular toxicity were presented. Although none has been extensively validated, a few show promise as being predictive indicators of ocular toxicity. Further developments of these tests should be encouraged, since they have the potential of being the best discriminators of degrees of irritation.

Possible means were discussed for reducing the number of ocular irritation tests performed, including reliance on historical data, or the use of data from dermal toxicological assessments. In specific cases such data may be acceptable, but at present there are no generic substitutes for ocular irritation testing. Reduction of the number of animals used in each test may be considered if the test is only used to discriminate between irritants and nonirritants; however, it is usually desirable to have greater resolving power which can only be achieved by using an appropriate sample size.

ACKNOWLEDGMENTS

The authors wish to thank J. F. Griffith, G. A. Nixon, P. J. Reer, J. F. Powers, and L. H. Bruner for their thoughtful criticisms of this manuscript.

REFERENCES

1. G.J. Beer (1813) Lehre von den Augenkrankheiten, als Leitfaden zu seinen offentlichen Vorlesungen entworfen, Vienna, Camesina, Heubner and Volke, Vol. 1, as cited by W.F. Hughes, *Arch. Ophthalmol.* 35, 423 (1946).
2. J.S. Friedenwald, W.F. Hughes, and H. Herrmann, *Arch. Ophthalmol.* 31, 279 (1944).
3. J.S. Friedenwald, W.F. Hughes, and H. Herrmann, *Arch. Ophthalmol.* 35, 98 (1946).
4. W.F. Hughes, Jr., *Arch. Ophthalmol.* 35, 423 (1946).
5. C.P. Carpenter and H.F. Smyth, *Am. J. Ophthalmol.* 29, 1363 (1946).
6. J.H. Draize, G. Woodard, and H.O. Calvery, *J. Pharm. Exp. Ther.* 82, 377 (1944).
7. J.H. Kay and J.C. Calandra, *J. Soc. Cosmet. Chem.* 13, 281 (1962).
8. Federal Register, Sept. 17, Federal Hazardous Substances Act, Title 21, CFR 13009 — 191.12 Test for eye irritants (1964).
9. Federal Register, Consumer Product Safety Commission, Code of Federal Regulations Title 16, part 1500.42, U.S. Govt. Printing Office, Washington, D.C., (1979).
10. Organization for Economic Cooperation and Development, Acute eye irritation and corrosion, OECD Publications and Information Center, Suite 1207, 1750 Pennsylvania Ave. NW, Washington, D.C., 20006 (1981).
11. Organization for Economic Cooperation and Development, Acute eye irritation and corrosion, OECD Publications and Information Center, Suite 1207, 1750 Pennsylvania Ave. NW, Washington, D.C. 20006 (1987).
12. Federal Register, Interagency Regulatory Liaison Group Recommended Guideline for Acute Eye Irritation Testing, National Tech. Info. Service, PB82-117557 (1981).
13. U.S. EPA, Primary eye irritation study in "Pesticide assessment guidelines, subdivision F hazard evaluation: human and domestic animals", EPA guideline 81-4, 51 (1982).
14. U.S. EPA, Data requirements for pesticide registration, final rule, EPA 40 CFR 158 (1984).
15. U.S. EPA, Toxic Substances Control Act test guidelines, part 798: health effects testing guidelines: subpart B — general toxicity testing, *Fed. Reg.* 50(188): 39398 (1985).
16. National Academy of Sciences, Principles and procedures for evaluating the toxicity of household substances, National Academy of Sciences, Washington, D.C., 4 (1977).
17. F.E. Freeberg, G.A. Nixon, P.J. Reer, J.E. Weaver, R.D. Bruce, J.F. Griffith, and L.W. Sanders, III, *Fund. Appl. Toxicol.* 7, 626 (1986).
18. J.F. Griffith, G.A. Nixon, R.D. Bruce, P.J. Reer, and E.A. Bannan, *Toxicol. Appl. Pharmacol.* 55, 501 (1980).
19. W.M. Grant, *Toxicology of the Eye,* 2nd ed. Charles C Thomas, Springfield, IL, (1974).
20. F.E. Freeberg, J.F. Griffith, R.D. Bruce, and P.H.S. Bay, *J. Toxicol. Cut. Ocul. Toxicol.* 3(1), 53, (1984).
21. F.E. Freeberg, D.T. Hooker, and J.F. Griffith, *J. Toxicol. Cut. Ocul. Toxicol.* 5(2), 115 (1986).
22. A.W. Johnson, CTFA ocular safety testing workshop: in vivo and in vitro approaches, October 6 and 7, 1986, Washington, D.C., Cosmetic, Toiletry and Fragrance Association (1981).
23. Ch. Gloxhuber, *Food Chem. Toxicol.* 23, 187 (1985).
24. A.P. Walker, *Food Chem. Toxicol.* 23, 175 (1985).

25. S.J. Williams, *Food Chem. Toxicol.* 23, 189 (1985).
26. D.M. Maurice, *The Cornea and Sclera in the Eye*, Vol. 1. H. Davson, Ed. Academic Press, New York, 489 (1969).
27. T.O. McDonald and J.A. Shadduck, *Eye Irritation, Advances in Modern Toxicology*, Vol. 4, *Dermatotoxicology and Pharmacology*, 139 (1977).
28. A.E. Maumenee, *Trans. Am. Ophthalmol. Soc.* 77, 133 (1979).
29. D.G. Cogan and E.O. Hirsch, *Arch. Ophthalmol.* 32, 276 (1944).
30. L.M. Carter, G. Duncan, and G.K. Rennie, *Exp. Eye Res.* 17, 409 (1973).
31. K. Green and A.M. Tonjum, *Acta Ophthalmol.* 53, 348 (1975).
32. K. Green, *Acta Ophthalmol.* 54, 145 (1976).
33. H.A. Baldwin, T.O. McDonald, and C.H. Beasley, *J. Soc. Cosmet. Chem.* 24, 181 (1973).
34. Ph. Conquet, G. Durand, J. Laillier, and B. Plazonnet, *Toxicol. Appl. Pharmacol.*, 39, 129 (1977).
35. E.V. Buehler and E.A. Newmann, *Toxicol. Appl. Pharmacol.* 6, 701 (1964).
36. H. Davson, *Vegetative Physiology and Biochemistry*, in *The Eye*, Vol. 1. Academic Press, New York (1962).
37. L.S. Harris, Y. Kahanowicz, and M. Shimmyo, *Ophthalmologica* 170, 320 (1975).
38. J.H. Beckley, *Toxicol. Appl. Pharmacol.* 7, 93 (1965).
39. J.H. Beckley, T.J. Russell, and L.F. Rubin, *Toxicol. Appl. Pharmacol.* 15, 1 (1969).
40. G.A. Elliott and A.L. Schut, *Am. J. Ophthalmol.* 60, 1074 (1965).
41. L.L. Gershbein and J.E. McDonald, *Food Cosmet. Toxicol.* 15, 131 (1977).
42. S.M. MacRae, B.A. Brown, J.L. Ubels, H.F. Edelhauser, and C.L. Dickerson, *J. Toxicol. Cut. Ocul. Toxicol.* 3, 17 (1984).
43. C.S. Weil and R.A. Scala, *Toxicol. Appl. Pharmacol.* 19, 276 (1971).
44. F.N. Marzulli and D.I. Ruggles, *J. Assoc. Off. Anal. Chem.* 56, 905 (1973).
45. B. Ballantyne, M.F. Gazzard, and D.W. Swanston, *J. Physiol.* 226, 12P (1972).
46. R.M. Walton and R. Heywood, *J. Soc. Cosmet. Chem.* 29, 365 (1978).
47. A.B.G. Burton, *Food Cosmet. Toxicol.* 10, 209 (1972).
48. A.B.G. Burton, M. York, and R.S. Lawrence, *Food Cosmet. Toxicol.* 19, 471 (1981).
49. R. Heywood and R.W. James, *J. Soc. Coismet. Chem.* 29, 25 (1978).
50. J. Laillier, B. Plazonnet, and J.C. LeDouarec, *Proc. Eur. Soc. Toxicol.* 17, 336 (1975).
51. K. Green and M.A. Green, *Am. J. Physiol.* 217 (1969).
52. K. Green and A. Tonjum, *Am. J. Ophthalmol.* 72, 897 (1971).
53. J-Cl. Etter and A. Wildhaber, *Pharm. Acta Helv.* 59, 8 (1984).
54. J-Cl. Etter and A. Wildhaber, *Food Chem. Toxicol.* 23, 321 (1985).
55. D. Maurice and T. Singh, *Toxicol. Lett.* 31, 125 (1986).
56. J. Walberg, *Toxicol. Lett.* 18, 49 (1983).
57. T. Gundersen and S.D. Liebman, *Arch. Ophthalmol.* 31, 29 (1944).
58. N.F. Bykov and D.I. Semenova, *Oftalmol. Zh.* 27, 173 (1972).
59. C.A. Hood, A.R. Gasset, E.D. Ellison, and H.E. Kaufman, *Am. J. Ophthalmol.* 71, 1009 (1971).
60. P.M. Leuenberger, *Albrecht Graefes Arch. Klin. Exp. Ophthal.* 186, 73 (1973).
61. R.R. Pfister and N.Burstein, *Invest. Ophthalmol.* 15, 246 (1976).
62. H. Dutertre-Catella, N.P. Lich, V.N. Huyen, and R. Truhaut, *Arch. Mal. Prof.* 43, 455 (1982).
63. H.P. Ciuchta and K.T. Dodd, *Drug Chem. Toxicol.* 1(3), 305 (1978).
64. S.J. Williams, *Food Chem. Toxicol.* 22, 157 (1984).
65. S.C. Gad, R.D. Walsh, and B.J. Dunn, *J. Toxicol. Cut. Ocul. Toxicol.* 5(3), 195 (1986).
66. A.S. Weltman, S.B. Sharber, and T. Jurtshuk, *Toxicol. Appl. Pharmacol.* 7, 308 (1965).
67. D.J. DeSousa, A.A. Rouse, and W.J. Smolon, *Toxicol. Appl. Pharmacol.* 76, 234 (1984).

17

Principles and Methods of Ocular Pharmacokinetic Evaluation

KEITH GREEN
Department of Ophthalmology
Medical College of Georgia
Augusta, Georgia

I. BACKGROUND

The determination of the kinetics of chemicals reaching the eye by a variety of routes is fundamental to the development of ocular pharmaceuticals, and to the assessment of the ocular toxicity of any agent that either directly or indirectly reaches the eye. Without this knowledge, drug-dosing regimens in humans cannot be predicted with any degree of faith that pharmacologically effective concentrations will be achieved at the site of action, that the desired responses will occur, and that any toxicological effects will be minimized.

Pharmacokinetics is the study of the time course of absorption, distribution, metabolism, and elimination of any agent.[1] Toxic ocular responses may occur either through the actions of agents deliberately administered to influence the eye tissues, or to treat a disease or infection of the eye, or through the inadvertent exposure of the eye to agents not designed or anticipated to reach the eye. Such effects may be realized after either topical or systemic exposure to the agent. Under any circumstance the assessment of the pharmacokinetics is important to establish not only tissue concentrations, but also the rate of clearance of agents from these tissues. It is the latter, with potentially long periods of exposure of ocular tissues to sufficiently deleterious concentrations of agents, that leads to expressions of overt toxicity.

While this chapter will deal with the methodology associated with the determination of ocular pharmacokinetics, an associated topic that will be combined with the primary topic will be the pharmacological or toxicological responses of the ocular tissues. The latter response can affect the pharmacokinetics of compounds having similar or dissimilar physicochemical characteristics; this may have marked effects on the distribution of those compounds, and may also affect the ocular behavior of drugs used simultaneously. Thus, while the pharmacokinetics of a particular compound may be well described in the literature, one must remember that other compounds, used either simultaneously or concurrently with the compound under test, can have a marked effect on both the pharmacokinetics and the pharmacological or toxicological action of that compound. The expression of overt toxicity by any ocular tissue will undoubtedly alter the pharmacokinetics of any substance that is tested simultaneously. These simultaneous physical effects can be either direct, i.e., physical interaction or disruption of the tissue, or indirect, i.e., alteration in membrane characteristics or metabolism of the tissue.

It is the intent of this chapter to provide a functional description of various methodologies, both *in vitro* and *in vivo*, for the determination of ocular pharmacokinetics of a variety of compounds. We will illustrate the various approaches with procedures currently employed. Where several procedures are available, they will be contrasted on the basis of their relative merits. For the most part, the techniques are the same whether considering the approaches taken for the evaluation of therapeutic agents, or toxicants, including delivery of pharmaceuticals beyond their therapeutic range. Complete references will be provided so that the reader may refer directly to the source of the methodological description when space does not allow the full description to be made in this chapter. The choice of references is somewhat selective, rather than being fully comprehensive, to make it easier for the reader to find those references that specifically address each point, or provide a further detailed reference source.

Access of any compound to the eye represents a rather unique situation because different barriers exist depending upon the chosen route of application. Consider the illustration shown in Figure 1, where topically applied drugs that are required to reach the posterior pole of the eye initially face the large washout with tears, the rapid loss of material from the ocular surface and loss across the much greater area of vascularized conjunctiva compared to the cornea. During an initial washout phase, which severely restricts the time available for ocular absorption, the drug also faces the problem of crossing a highly lipid, five-cell-layered, tight epithelium, with the outermost layer of cells being tightly linked to form a sheet of lipid plasmalemma, before it encounters the corneal stroma, which is 75% water. The added compound may penetrate the epithelium via cellular or paracellular routes, although the molecular size would need to be small to enter the paracellular pathway at the apices of the corneal epithelial cells. Past these superficial barriers, the endothelial permeability is high to most compounds up to the size of hemoglobin (equivalent radius, 35 to 40 Å), and passage into the anterior chamber is easy.

Once in the anterior chamber, the drug must traverse this 3 mm deep, liquid-filled space through which the fluid is passing at 1% of the total volume per minute. Since the volume is about 250 μl, the flow rate is roughly 2 to 2.5 μl/min. This fluid is formed at the ciliary processes and passes into the anterior chamber through the pupil. If the pupil is relatively small, i.e., 3 to 4 mm, then the velocity of the fluid passing between the anterior lens surface and the posterior face of the iris is fairly high. If it has overcome these impediments to diffusion, then the compound must pass either through or around the lens, and into the gel-like vitreous in order to reach the posterior pole of the eye. The characteristics of the vitreous humor, however,

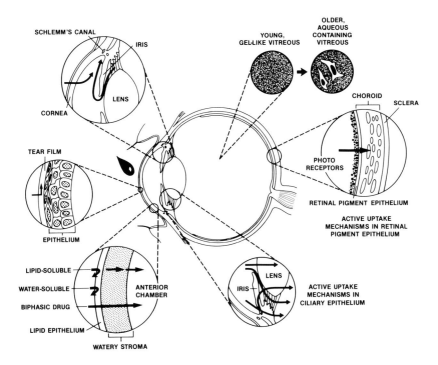

Figure 1. Illustration of eye showing the serial impediments in the passage of a topically applied agent to the posterior uvea. The agent must face washout by tears, penetration across the lipid corneal epithelium, removal by drainage of aqueous humor through outflow pathways during diffusion across the anterior chamber, diffusion through the iris and/or lens, uptake mechanisms in the ciliary body, diffusion through the vitreous, and outwardly directed transport systems in the retinal pigment epithelium. Each aspect is discussed in the text.

change with age, causing the vitreous to become more liquid.[2] Thus, the penetration characteristics of this bulk compartment also may change with aging of the eye.

Diffusion is the primary means of movement through any of these tissues or fluids, but removal from the vitreous humor may occur via transport systems[3] that can actively remove certain chemicals. The latter process of transport out of the eye can also occur in the ciliary processes,[4] thereby presenting another means of reducing the posteriorly directed movement of a compound within the eye. Mechanisms exist, therefore, for the removal of certain substances from the eye across the blood-retinal and blood-aqueous barriers. These transport systems can have a pronounced effect on the composition of the aqueous and vitreous humor through the removal of selected chemicals. It is evident, therefore, that the effective delivery of any agent to the posterior pole of the eye by a topical drop delivery system requires facing a variety of different challenges from the tear film to the posterior uveal tissues.

A compound given systemically must pass across the blood vessel walls and across other barriers into the eye. In the anterior segment, for example, a compound must pass across either the iris vessels that are relatively impermeant[5] (at least in the outward direction), or the capillaries of the ciliary body that are fenestrated.[6] In the ciliary body, a drug must then pass across the bilayered ciliary epithelium that requires passage across lipid membranes before entering the posterior chamber.[7] In the posterior pole of the eye, passage is required across Bruch's membrane and the retinal pigment epithelium for penetration of material from the choroid into the vitreous humor. Diffusion of material into the eye assumes that the compound is not bound to some degree to plasma proteins, or subjected to metabolism, through a first pass effect (the latter after oral intake of the compound), after systemic absorption.

To achieve good drug penetration into the eye, therefore, requires a specific set of characteristics to be achieved, depending upon the chosen route of administration. Conversely, one needs to minimize uptake of potentially toxic compounds, or to maintain levels of therapeutic agents below those capable of inducing a toxic reaction. With respect to the eye, pharmacokinetics includes the quantitation of such elements of drug bioavailability as tear dilution after topical application, drug binding after both topical and systemic administration, and vehicle effects, as well as any factors affecting corneal absorption (e.g., polymers, surfactants, etc.) and subsequent tissue and fluid distribution.[1]

Various methodologies have been developed to investigate the pharmacokinetics of drugs in the eye whether after topical, iontophoretic, or systemic application in order to delineate the various factors that modify drug or chemical behavior. Such studies are vital in the development of drugs and in designing structural modifications to improve ocular penetration and therapeutic efficacy.

Pharmacokinetic evaluations are also vital in toxicology, where the kinetics need to be known in order to evaluate any potential effects on intraocular tissues and to minimize toxic effects of both therapeutic agents and compounds of inherent toxic capacity. Much of the toxicological data generated from such studies is used in predictive models for human exposure to the materials under study.[8,9] We are exposed to a multitude of potentially toxic agents, whether by design or accident, and an understanding of how these agents can interact with ocular tissues is essential. Only through an understanding of ocular pharmacokinetics can an appreciation be obtained of the potential toxicity of agents to the eye.

II. *IN VITRO*

A. Partition Coefficients

One of the primary determinants of the penetration of any chemical into ocular tissues and compartments is the partition coefficient of the compound under investigation. The partition coefficient is a measure of the relative solubility of a compound in a lipid vs. an aqueous phase. These values are often given as the ratio of solubility of a substance in an organic solvent (e.g., ether, isobutanol, octanol) or oil (e.g., olive oil) to that in water. High values (e.g., 10 to 10^{-2}) indicate that a compound is very lipid soluble, while low values (e.g., 10^{-4} to 10^{-6}) conversely indicate a high aqueous solubility, or lipophobicity. It is important to have some knowledge of this determinant, especially in relation to the route of access that the compound will have relative to the eye. For example, a compound that is lipid soluble will have less difficulty passing through the cornea if applied as a topical drop or if it has access to the ocular surface, than will a more water-soluble compound. This is due to the composition of the cornea, with its highly lipid-containing, cellular epithelium covering the anterior surface. An excellent illustration of this might be the transcorneal penetration of lipid- and water-soluble steroids. With the epithelium present, only the lipid-soluble drugs penetrated the entire cornea, while in the absence of epithelium, no marked differences occurred in the penetration of lipid- and water-soluble steroids.[10]

The inherent chemical nature of a compound also influences its penetration into ocular tissues. An alkaloid such as pilocarpine (but certainly not all alkaloids), for example, penetrates the cornea with relative ease due to its ability to exist in both an ionic and a nonionic form. For passage through the lipid epithelium, the nonionic form assumes a greater role since in this lipophilic form the drug can enter the epithelium with relative ease. The ionized, lipophobic form, as with most polar compounds, is far less lipid soluble, and will not enter the corneal epithelium. When crossing the stroma, which consists of at least 75% water, the ionic (hydrophilic) form is predominant and passage from the stroma to the aqueous humor can occur with relative ease. Access to sites on the surface of iris or ciliary muscle cells can be achieved by a combination of both the free base (nonionic) form or the ionic form, with predominance occurring by the dictates of the local environment and concentration gradients that occur between compartments.

The advantages of the use of compounds with similar overall physicochemical characteristics to pilocarpine, i.e., an ability to exist in ionic and nonionic forms, for drug penetration into the eye are obvious. Conversely, if the disease entity for which the compound is to be used is confined to the ocular surface, or a toxic compound is to be excluded from penetration into the eye, then the choice of drug would be made to fulfill requirements of the lack of ocular penetration.

For any agent reaching the eye, knowledge of its physicochemical characteristics can provide some initial estimate of the potential for entry into the eye tissues and fluids. As in many other situations, however, there are variations to such predictions which are characteristic of the particular material and the particular membrane involved.

B. Isolated Tissue Preparations

In order to determine the penetration of a compound across individual tissue layers or into different ocular tissue compartments, there are several approaches. The most obvious method for intraocular tissues, such as the iris and lens, is immersion of the isolated tissues in a suitable incubation medium containing the compound under test. Determination of the amount of the compound in these tissues, at different times after incubation is initiated, will provide data on the rate of uptake and the maximal uptake from the concentration of compound included in the medium.

Care must be exercised with all tissues isolated from the eye since damage from excess handling can ensue. This damage can be either physically or chemically induced. The iris from certain species, if handled, will produce various inflammatory agents, such as prostaglandins, and these may be produced in varying quantities by different species. These agents may alter membrane permeability or change the metabolism of the compound under test. In either case the kinetics of the test compound would be altered, and interpretation of the comparative *in vitro* and *in vivo* data can be compromised by these induced changes.

Any physical disruption of the lens capsule or epithelium caused by undue manipulation can result in swelling and disruption of its internal milieu. The lens zonule fibers that connect the lens capsule to the inner limiting membrane or basement membrane of the nonpigmented ciliary epithelium, should be cut with care to ensure that the bounding membranes remain intact. The lens can be handled gently and transported using a glass attachment that will fit snugly onto the lens surface; a gentle suction can be applied to transport the lens from the eye to the incubation solution. In this way the bounding membranes of the lens are not physically disrupted.

While the cornea, sclera, and conjunctiva can be handled with reasonable assurance of the absence of damage, albeit with great care, other intraocular tissues such as the retina require even greater care. The cornea requires considerable caution in its handling, since wrinkling of the tissue can markedly influence the permeability of either bounding membrane.

The accepted methodology for *in vitro* preparation of the cornea is well described by Dikstein and Maurice.[11] The whole eye is excised, with the epithelium moistened with isotonic saline or Ringer's solution, and is then placed with the cornea down on a lucite column through which a hole has been drilled. The column must be about 12 to 13 mm in diameter to be large enough to accept most corneas, and the central hole is about 4 to 5 mm diameter to provide a suitable corneal area to hold with a vacuum. The head of the column is concave to hold the curved corneal surface, and Ringer's solution is placed both on the top of the column, as well as into the tube, to provide a moist environment for the epithelium. The conjunctiva is drawn over the top of the column, to which is attached a corneal mounting ring. The conjunctiva is tied in a groove in the mounting ring with cotton thread, and a small vacuum placed in the tube in order to hold the cornea in a position where it cannot become wrinkled (Figure 2). The eye is then cut at the equator using a razor blade and the remaining ocular contents gently removed using forceps. The posterior, endothelial surface of the cornea is then rinsed gently with 37°C Ringer's solution and an appropriate chamber placed around the cornea. (Chamber design is discussed in Section II.C.) A viable corneal preparation that maintains a constant thickness and permeability characteristics of the bounding membranes equal to those *in vivo* can be sustained *in vitro* for up to 8 h.

The choice of bathing solutions for the isolated cornea, and other ocular tissues that contain cellular elements, can be critical to the *in vitro* longevity of the preparation. For the cornea, detailed examinations have shown that the inclusion of reduced glutathione (0.3 mM) and adenosine (0.5 mM) in the bicarbonate-rich (25 to

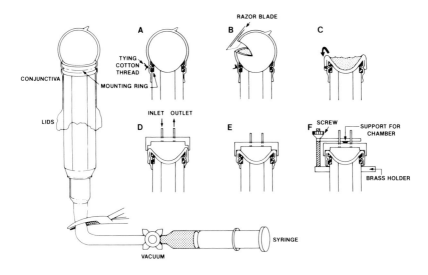

Figure 2.   Preparation of isolated cornea for mounting in chambers. Mounting rod with mounting ring in place; the central hole in the mounting rod has been filled with Ringer's solution (loss is prevented by clamping the rubber tube attached to the opposite end of the rod) and the mounting ring is filled with Ringer's solution. The enucleated eye is placed over the rod and the conjunctiva and lids pulled into position. A small vacuum is then placed on the system via the rubber tube at the bottom of the rod. In our laboratory we pull back a syringe plunger 1 ml to give sufficient vacuum. The rubber tube is then reclamped to maintain the vacuum. (A) The conjunctiva is tied to the mounting ring using cotton thread, and the remaining conjunctiva and lids are removed. (B) The eye is cut at the equator and the ocular contents removed. (C) The endothelial surface of the cornea is rinsed with Ringer and is now ready for mounting in a suitable chamber. (D) The top portion of the chamber holder is placed over the mounting rod before preparation of the eye while the posterior lucite chamber is to be placed on the tissue. (E) The chamber holder is brought up to the mounting ring and the endothelial-facing lucite chamber put into place. (F) The lucite chamber is held in place with the clamp attached to the brass holder. The chamber is removed from the mounting rod after release of the vacuum, turned to provide the correct orientation, solution or oil, placed on the outward epithelial-facing surface and placed in a water-jacketed sleeve.

40 mM), glucose-containing, bathing medium prolongs the ability of the cornea to maintain a constant thickness, enhances the ability of a preswollen cornea to deturgesce to normal thickness, and allows the revelation of the full complement of ion transport systems.[11-17] For other tissues, the inclusion of glutathione and adenosine does not appear to be necessary for the maintenance of full physiological function, although the presence of essential ions and metabolic substrates (Na^+, K^+, Ca^{2+}, Mg^{2+}, Cl^-, $SO_4^=$, HCO_3^- and glucose) are minimal requirements for the ability of the tissue to function normally in an *in vitro* setting. For tissues of the anterior segment the inclusion of ascorbic acid (at 1 mM) could be beneficial; this is a normal ingredient of aqueous humor and can act as an antioxidant. For epithelial permeabil-

ity determinations it is not necessary to remove the endothelium, since the endothelium is not rate limiting, but for endothelial permeability measurements the epithelium must be removed prior to placement of the eye on the mounting rod. Epithelium is conveniently removed by scraping the anterior corneal surface with a Gill corneal knife, or razor blade with removal of debris using a cotton-tipped applicator.

Conjunctiva can be removed simply from the eye and placed between two half chambers. Also sclera can be handled in a similar manner once the choroid has been removed from the internal surface. Caution must prevail when working with the sclera as it is perforated by blood vessels; thus, an area must be chosen that does not contain any trans-scleral perforations. Inclusion of these perforations in any preparation used for trans-scleral permeability determinations can lead to erroneous data.

The permeability of different compounds across all the isolated tissues can then be evaluated using any convenient means. It is imperative that the choice of the chemical concentrations be realistic in terms of the amounts likely to be available at the site of penetration. This may require some chemical estimate of the concentration of the test material after exposure of the eye under simulated conditions in which the test material could reach the eye. Many times, very large concentrations of material are used *in vitro* which bear no relationship to the concentration that is likely to be present in the fluid bathing the tissue under study.

The form of the drug preparation should also be considered since, as in the case of radioactively labeled pilocarpine, it is supplied from the manufacturers as a labeled compound in an ethanolic vehicle. Direct use of this material will give data that more readily reflect the penetration of the labeled vehicle because if the drug or chemical label is ^{3}H it exchanges for the H of the vehicle. Thus, as soon as the ^{3}H is in contact with any H-containing material (such as water or the ethanol vehicle used by the manufacturers), reverse exchange will result. Suitable evaporation, washing, and reevaporation procedures can assure that only ^{3}H-labeled drug is used. This is not true of compounds in which the ^{3}H is chemically incorporated into the integral molecular structure of the drug, but the investigator should always check as to how the material was synthesized before its use.[18,19]

C. Chamber Design

The chamber design chosen for permeation determinations will be dictated to some degree by the area of tissue available (especially for retina and sclera, where the exposed area may be small). If the area is small, then the chamber volumes will also need to be relatively small, i.e., <0.5 ml, in order to be able to detect the compound regardless of the chosen method of analysis (i.e., spectrophotometric, HPLC, radioactive tracers, chemical). Another consideration is the method of sampling; e.g., the chambers must allow access to each bathing solution.

Several chamber designs exist ranging from classic Ussing-type chambers (Figure 3), specular microscope chambers (especially for corneas where corneal thickness can be monitored, Figure 4), or modified Ussing chambers (Figure 5). Each approach has advantages and disadvantages compared to another. The Ussing-type design in Figure 3 has the advantage of ready access to the bathing solutions, yet has the

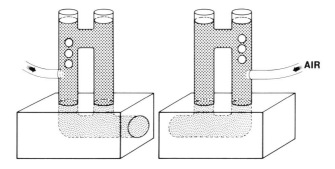

Figure 3. Ussing-type chambers used for permeability measurements. This diagram shows an overall design: custom modifications to this basic structure can be made to hold a variety of tissues, including the cornea. Appropriate changes are required in one of the hemi-chambers to allow a mounting ring to be placed between the chambers (see Figure 5 for an example of such modifications). The air input provides both aeration of the solution with the appropriate gas mixture, as well as circulation of the fluid through the chambers.

disadvantages of requiring large volumes on either side of the tissues, as well as not providing a sealed environment, particularly on the endothelial side of the cornea. The latter is very important for pH control, especially with bicarbonate-buffered bathing solutions.

The specular microscope chambers shown in Figure 4 have the advantage that any concurrent deleterious effect of the test compound on either the epithelium or endothelium can be detected by measurements of corneal thickness. The disadvantage is that in order to sustain a viable endothelium, there must be a minimal flow rate through the small volume ($\cong 200$ µl) chamber on that surface. Fast flow rates can cause excessive turbulence and resultant physical damage to the endothelium, with localized swelling adjacent to the inlet needle to the chamber. This swelling could be misinterpreted as a chemically induced phenomenon if adequate precautions are not taken against excessive flow rates through the chamber.

The advantage of the sealed chamber design shown in Figure 5 is that the chambers are sealed from the environment and thereby the pH can be carefully controlled; this is especially true if the bathing solution is made in bulk and taken up in sealed syringes that are attached to the chambers. The disadvantages are that one must sample by total replacement of the chamber volumes which includes a dead space of the needle volumes, and one can only judge the viability of the preparation from the resultant data. Practice is also needed to achieve the correct balance of applied pressure and flow rate of the replacement solution to ensure the best replacement solution conditions. In our laboratories, we normally use a 5 ml flow-through volume delivered over about 2 min.

Similar procedural steps can be applied to the isolated iris-ciliary body preparation in which the uptake of drugs or agents can be measured. Additionally, iris-ciliary

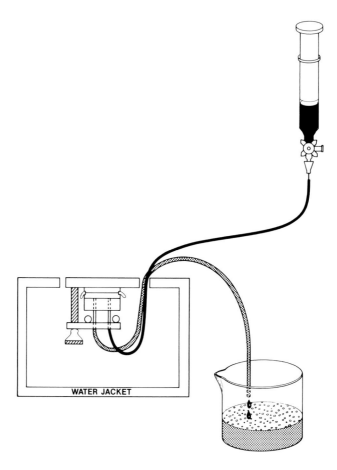

Figure 4.   Specular microscope chambers. The endothelium is perfused either by a gravity system (shown) or by a perfusion pump. The rate of fluid flow in the gravity-fed system is regulated by the height differential of the inflow and outflow tube above the cornea. The entire corneal chamber is placed within a water-jacketed sleeve.

body preparations can also be clamped between two Ussing-type chambers so that access of the material under test can be restricted to one side of the tissue. Normally the major source of any topically applied drug to the iris would be via the aqueous humor, hence the anterior-facing aspect of the iris would face the highest concentration. The methodology, including modifications to reduce edge damage in this delicate tissue preparation, has been well described in the literature.[20-23] The disadvantage to this type of experiment, of course, is that there is no blood flow occurring that would markedly alter the kinetics. Baseline tissue uptake data can be obtained, however, that provide comparisons to *in vivo* studies.

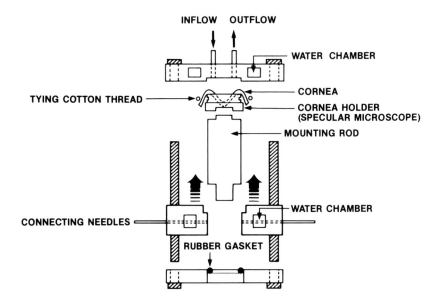

Figure 5.   Sealed chambers for permeability measurements. The cornea is prepared in a manner identical to that shown in Figure 2. Caution: if endothelial permeability is to be determined the corneal epithelium must be removed *before* any dissection occurs. The bottom chamber is placed over the mounting rod before the eye is placed on the mounting ring. The bottom of the mounting rod is connected to a rubber tube as shown in Figure 2, and the appropriate Ringer's addition and vacuum application steps are followed. After preparation of the cornea on the mounting ring, the two water-jacketed half chambers are brought together (arrows) with the cornea between them. The details of the mounting ring/cornea/chamber interface are as shown in Figure 2. After release of the vacuum, the cornea and chamber are turned over and the bottom plate holding the rubber gasket put into position. The assembled chamber is then held in position for sampling. In this system the endothelium remains facing upward. A small Teflon®-coated magnetic stirring bar can be placed on the anterior-facing surface of the cornea, and driven by an external horseshoe magnet, to provide adequate mixing of this chamber.

D. Permeability Measurements

In order to assess the ocular kinetics of any compound, knowledge of the rate of exchange between compartments is essential. While the *in vivo* approach is obviously the most realistic, it is difficult to assess several characteristics *in vivo* because of the periodic sampling requirement and access to the two compartments. Thus, *in vitro* determinations of permeability are often needed to completely evaluate the movement of compounds across different ocular barriers. As indicated above, straightforward measurements of tissue uptake on some tissues provide all that is required to assess permeability, particularly in terms of pharmacokinetics, e.g., for the lens.

For water-soluble compounds equilibration with the lens probably occurs relatively quickly, even with the diffusion distances involved, although the cell membranes offer substantial resistance. For lipid-soluble compounds, because the cell membranes of the lens fibers are distributed widely and are small in total mass, the equilibration would be likely to occur more rapidly.

For permeability measurements across any tissues, the experimentalist must decide whether to measure uptake or steady-state kinetics. For uptake determinations, the points of interest lie between the time of immediate access to the compound and equilibrium, whereas for steady-state kinetics the measurements will be taken only after equilibration of the membrane or tissue has occurred with the compound in the bathing solution.

1. Measurement Techniques

The choice of measurement technique for determining the compound passing through the membrane is dependent on the concentration of the material, whether it can be labeled, and its chemical form. If the compound is to be used at a low concentration, or a low concentration is anticipated at the membrane across which the movement is to be detected, then the most convenient means of measurement is by using a labeled version of the material. The label can be radioactive (^{3}H, ^{14}C, ^{35}S, ^{32}P, ^{125}I), the compound may be fluorescent labeled with a dye such as fluorescein, or with an enzymatic label such as horseradish peroxidase.

Ideally, the attachment of a label to the compound of interest should not result in a change in either the physical (i.e., charge, molecular size) characteristics of the compound, nor the chemical (i.e., the label be attached to a part of the compound that may be cleaved from the rest of the "active" moiety) nature of the compound. Such behavior should be checked using high performance liquid chromatography (HPLC), thin-layer chromatography (TLC), gas chromatography (GC), or gas chromatography-mass spectrometry (GC-MS) before permeability measurements occur.

Care should also be taken to ensure that the process used for tissue digestion, whether a quaternary ammonium compound (such as Protosol), or an alkali, such as sodium hydroxide, does not cause a loss of the drug under test. For this reason, it is always profitable to subject nonexperimental tissues, harvested using the same techniques as for experimental tissues, to the same digestion process immediately after adding known quantities of labeled material. This provides values for the recovery of the material after tissues are subjected to harsh extraction techniques. More gentle approaches such as the disruption of tissues in distilled water, which causes osmotic breakdown of the cells, can be used, but are generally only successful with small tissue masses such as corneal endothelium, where all cells can be assuredly exposed to the bathing solution.

2. Viability Assessment

Some means should be employed to assess the viability status of the excised tissue after or during its experimental use, whether this be morphological, physiological, or biochemical, in order to ensure that the tissue is functionally as close as possible to that found *in vivo*. An example is the maintenance of corneal thickness *in vitro*, a characteristic that can be readily measured (e.g., with specular microscopy during perfusion of the posterior corneal surface) and provides an excellent index of functional viability of the cornea and the integrity of its membranes. Similarly, the maintenance of lens clarity *in vitro* is indicative of the integrity of the ion pump systems and membrane integrity. This ensures that the *in vitro* tissue has the same characteristics as when *in situ* and that its isolation has been achieved with minimal trauma.

3. Example Procedures

An example of this type of *in vitro* to *in vivo* transference of permeability values has been used in the evaluation of topically administered carbonic anhydrase inhibitors (CAIs). *In vitro* studies on isolated corneas were first used to evaluate the permeation rates of structurally modified CAIs. The data from these studies were used to identify those compounds that had higher rates of entry into the aqueous humor and to eliminate those with poor permeability characteristics.[24,25] *In vivo* studies were then undertaken to provide data on the pharmacological efficacy of the selected compounds on intraocular pressure.[24,26,27] Another example is the development of an erodable gel that provided a longer contact time for pilocarpine from *in vitro* studies,[28] a finding which was subsequently confirmed by *in vivo* studies.[29]

A similar approach has been used to determine the various corneal compartments that determine the rate of penetration of pilocarpine from different solutions varying in pH, tonicity, concentration, etc.[30] These studies revealed that the epithelium played a dual role in the transcorneal penetration of pilocarpine. First, it acted as a barrier to drug penetration, and second, it acted as a reservoir for drug in the intact cornea. No rate-determining role was found for the endothelium.[31]

The measurement of the movement of compounds across the crystalline lens can be achieved using a chamber such as that shown in Figure 6. Here, the anteroposterior measurement, or vice versa, of compounds can be determined with ease.

For tissue uptake measurements the tissues must be weighed and then the concentration of the compound determined within the tissue. This can be achieved by quantitative measurements of the level of the compound after tissue digestion or disruption to release the compound. It can then be measured by one of several techniques, including HPLC, spectrophotometry, GC, GC-MS, or by radioactive counting procedures.

For tissues that are subjected to radioactive counting, one must prepare quench curves from tissues digested in the same manner as for those used for the compound, but without exposure to any of the material. The handling of these tissues follows the same procedures as for the uptake measurements, except that a known quantity of the tracer material (^3H, ^{14}C, etc.) is added to the digests or aliquots thereof. These samples provide an estimate of the degree of quenching of the tracer count rate under the experimental conditions. They also give a correction factor that must be applied to the tissue in question (the quench factor will usually differ from one tissue to another) in order to estimate the true count rate from that actually measured. From the amount taken up by the tissue and its weight, the quantity of material per unit weight can be

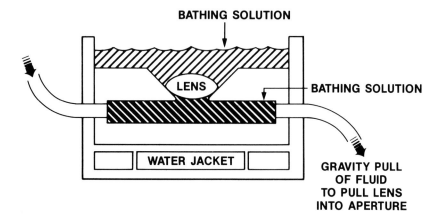

Figure 6. Drawing of a chamber for use in determination of translens permeability. The lens is held in position by application of a gentle suction applied to the lower chamber which serves to keep the lens in contact with the chamber and enables the lens to act as the membrane. A light coating of silicone grease aids in forming a complete seal between the lens and the chamber.

expressed. For some tissues, such as the retina and the iris, uptake measured using an isolated preparation cannot reflect that in the living eye, due to the absence of blood flow through the tissue *in vitro*. This factor plays an important role, especially in the choroid, for instance, which is perfused by a large percentage of ocular blood flow.[32,33]

For permeability measurements the solutions used on both sides of the membrane are usually identical and thus the quenching will be the same for all bathing solutions. The samples are weighed, or determined volumetrically, and subjected to the appropriate counting procedures. Inclusion of another compound, which will not interfere with the behavior of the agent under test but for which the permeability is known, provides a cross-check that the tissue has not been damaged. For example, in the corneal endothelium the permeability to nonelectrolytes such as sucrose, mannitol, inulin, and dextran is well characterized and may be used as a standard to verify the absence of tissue damage.[14,17]

E. Calculations

For tissue uptake, the determination of the tissue to medium ratio is usually expressed in terms of per milligram of water, rather than total tissue weight, in order to provide a more accurate value for the distribution. Thus, both wet weight and dry weight measurements of the tissues are needed. For dry weight, drying to constant weight is usually employed where tissues are dried at about 100°C for at least 12 h, cooled over a dessicant (e.g., anhydrous calcium sulfate or silica gel), and weighed to constant weight. A further drying period at 100°C is followed again by cooling and reweighing. If the dry weights are the same then the value is taken but if the second weight is significantly lower than the first (e.g., 5%), the process is repeated.

For membrane permeabilities, the values are usually calculated from equations such as those given below:

$$J = \frac{dC_b^* \times V_b \times C_a}{C_a^* \times t \times A}$$

where J is the flux of the compound in μmol cm^{-2} min^{-1} (or μeq cm^{-2} h^{-1}) from compartment a to compartment b; dC_b^* is the change in concentration of the labeled compound (per unit time, t) *into* which tracer movement occurs (counting rate cm^{-3}); V_b is the volume of the chamber (cm^3) *into* which the diffusion of the compound occurs (the volume of the other chamber is considered infinite as regards tracer concentration, such that the counting rate cm^{-3} does not decrease by >5% during the experiment); C_a is the total concentration (moles, equivalents) of the compound in the chamber *from* which diffusion occurs (μ equivalents cm^{-3}); C_a^*, concentration of labeled compound in compartment *from* which tracer movement occurs (counting rate cm^{-3}); and A is the exposed area of tissue (cm^2). Care should be exercised with expression of the area, A, since tissue convolutions such as in the ciliary epithelium can increase an exposed tissue area by a factor of three to four times.[34]

Maffly et al.[35] also provided an equation for the permeability (P) of a membrane which has been used frequently by many authors:

$$P = \frac{\text{increase in counts on the unlabeled side} \times \text{volume of unlabeled side}}{\text{area of tissue exposed} \times \text{time} \times \text{concentration of counts on labeled side}}$$

where P has the dimensions of centimeters per unit time (s, min, or h).

From these two equations given above it can be seen that the relationship between permeability and flux is

$$P = \frac{\text{flux}}{\text{concentration}}$$

III. *IN VIVO*

A. Dosing Frequency

In the determination of ocular pharmacokinetics *in vivo* it is imperative to select a dosing or exposure regimen which reflects that which would occur in the real life situation. Usually at least a two-step procedure is required for therapeutic agents, with both a single- and multiple-drop application regimen. In addition, one must select an animal with an appropriate age for the compound. For example, if the compound is designed for use in young animals or humans, or there is a greater possibility of accidental exposure to the test compound by young eyes, then a young animal (of the same species for which the compound is designed for veterinary application or to which accidental exposure may occur) should be employed.

Toxicological responses may be seen after single exposures, but more frequently do not occur until after multiple applications or exposures. External ocular toxicity may be seen more frequently after single exposures, while cataracts or retinal changes occur after multiple applications. These responses are the consequence of the pharmacokinetics of the material in the eye, since the accumulation of any material is governed by the chemistry of the material, the concentration available to the particular tissue, and the nature of the tissue (e.g., lipid-rich tissues will tend to act as a long-standing reservoir for lipid-soluble compounds).

For a topically applied pharmacological agent that would be potentially used twice a day for glaucoma treatment, then determination of the pharmacokinetics on a six times a day application regimen would provide less useful data. Similarly, if the compound under test would likely be an additive therapeutic agent (e.g., topical pilocarpine and topical epinephine) then a dosing regimen should be chosen that takes into account the inherent effect of one topical drop when given immediately after another. Chrai et al.[36,37] have investigated this phenomenon in detail and shown that when two drops are given sequentially the time between drop applications profoundly affects the kinetics of both drugs.

Chrai et al.[36] showed that when a second drop was applied within 10 min of the first drop, the kinetics of each drug were markedly affected. When the second drop was administered within 1 min of the first drop, the drug entering the eye from the first drop was reduced by 40%; when the time between drop applications was 2 min, the amount of drug reduction was 25%, etc. These authors also showed that the loss in fractional concentration at zero time and at 5 min was due principally to dilution, whereas at times between 0 and 5 min the loss in fractional concentration was due to drainage and dilution. Thus, if there is potential for the drug to be used under conditions of multiple application in combination with other topical drugs, the pharmacokinetics can be affected and this should be allowed for in the experimental design. An obvious practical utilization of the washout of material from the eye by excess dilution is the use of an eye wash when chemicals accidentally splash onto the eye (in the absence of either a screen or safety glasses). Here, dilution occurs into a large excess volume held in an eye cup.

Such information is highly pertinent to any consideration of chemicals reaching the eye by a variety of routes. It has been only though a study of the pharmacokinetics that these characteristics have been observed and their role in modulation of drug effects on the eye brought into everyday use. Consideration of drop size, sites of absorption, drug interactions, regional blood flow, tear flow kinetics, viscosity enhancers as solution additives, to name a few, have lead to substantial changes in the comprehension and modeling of ocular pharmacokinetics. These physiologically based models of kinetics have a great impact on the utilization of drugs in ophthalmology and understanding the accumulation and effects of toxic agents. Appreciation of the time course of tissue uptake and loss has led also to an understanding of how toxic reactions can occur to compounds used over long time periods. Protein binding of drugs or chemicals can lead to substantial accumulation of those materials, a fact that has been realized relatively recently.

B. Vehicle

The choice of vehicle for the application of a compound to the eye can severely affect the distribution, especially for topically applied drugs. Various chemicals have been used to enhance the retention of a topically applied drop on the ocular surface. Such compounds range from hydroxymethylcellulose, polyvinyl alcohol, polyvinylchloride, polyvinylpyrrolidone, and lanolin to the use of micronized materials, contact lenses soaked in the drug, and dissolving vehicles.[19,38-43] Some materials have been used to increase the viscosity of the solutions and aid in the penetration of drugs into the eye through increasing the exposure time of the cornea to a higher concentration. Additionally, compounds such as EDTA are often used in ophthalmic preparations. The use of such calcium-chelating agents can enhance epithelial permeability. Calcium is a necessary ion for the maintenance of cell membrane integrity and cellular metabolism and its depletion could enhance epithelial permeability.

The desire behind all of these choices of pharmaceutical delivery systems is to attempt to reach zero order kinetics, i.e., a constant delivery rate of drug to the eye.[1] The closest approach to zero order kinetics in drug delivery is achieved using an Ocusert® or related delivery system wherein the compound is held between a sandwich of two membranes across which the compound diffuses at a fixed rate for one or more days. Newer approaches to this problem of topical drug delivery include the incorporation of drugs into semisolid vehicles that dissolve or erode in the tear film at a constant rate, thereby delivering a more-or-less constant amount of drug per unit time. Only though the use of systems in which particles or delivery modes are retained on the ocular surface can any significant enhancement (i.e., >50 to 100% of that available through current approaches) be made in drug delivery. Almost all delivery systems that include vehicle manipulations such as viscosity enhancer or micronized particles are lost quickly from the tear film by the rapid drainage from the ocular surface.

In addition to the viscosity enhancers, other additives such as benzalkonium chloride (BAK) can have a pronounced effect on the pharmacokinetics of drug in the eye upon topical delivery. BAK is known to increase the permeability of the corneal epithelium to many drugs.[19,39,44,45] Carbachol is one of the earliest examples of the role of BAK in enabling a drug known to have a potential pharmacological effect on the pupil, but otherwise unable to penetrate the cornea, to exert its effects within the eye. The addition of BAK to the solution allowed the intraocular penetration of carbachol and revelation of its pupillary-constrictive effect after topical application.[45] BAK increases corneal epithelial permeability by lysing some of the surface cells and increasing the paracellular permeability.[46-49] BAK does not penetrate into the adult eye beyond the cornea,[50-53] and even in the eye of the neonate, only minute quantities enter the aqueous humor.[52] Thus, while BAK has great potential for inducing damage to intraocular membranes[54] when applied to a normal cornea, BAK does not enter ocular tissues except the cornea, and even here is mainly restricted to the corneal epithelium. The situation may be different where there is a lesion in the cornea or the epithelium is abraded. This is an example of a potentially toxic agent being inherently restricted in ocular penetration and thus reducing the possibility of intraocular toxicological effects. This is of particular importance given the marked sensitivity of the corneal endothelium to BAK.

Liposomes have been suggested to provide a specific targetry system for drugs applied to the eye,[55-57] and have been used both as external and internal ocular delivery systems. These approaches suggest that topically applied liposomes disappear from the tear film both by drainage and by absorption to the exposed surfaces of the ocular tissues, i.e., corneal epithelium and conjunctival mucosa. The absorption of inulin into the cornea from liposomes seems to be a function of the number of liposomes absorbed at the corneal epithelial surface rather than of the number of liposomes administered.[58] The difficulty with liposome systems lies more with the inability to form a stable association with the cell membrane surfaces and thereby be unable to release entrapped drug at a rate sufficient to sustain high intraocular drug concentration.[59] That liposomes can deliver agents to the cornea is undisputed,[60,61] and liposome binding can be enhanced by incorporating ligands for receptors in the corneal epithelial surface.[62] The latter approach may offer some advantages in a topical delivery system, since even with liposomes the same washout and limited time availability for absorption phenomena exist that face any liquid application to the ocular surface.

C. Route of Administration

1. Topical

This is the most common route of application for ophthalmic drugs due to the practicality of application by patients. It is also the primary route of exposure to toxic agents that are used daily and have the potential for accidental exposure to the eye. To make pharmaceutical use of this route, one must overcome not only the effect of other topical drugs instilled concomitantly on the kinetics (see Section III.A), but also tear film kinetics and blinking.

The initial reaction to the application of a topical drug is to blink. Given the normal ophthalmic drop size of about 35 μl, this means that at the moment that the drop touches the eye surface there are about 45 μl (35 μl added plus 10 μl of tears) on the eye. One blink and the volume is reduced to 10 μl, which means that one has 10/45 or about 20% of the original dose remaining on the eye. The dramatic loss of material from the tear film in such a short time after application is at once both a process that must be overcome for improved drug delivery and a protective mechanism against foreign substances entering the tear film and ocular tissues. Such a calculation of loss of applied material does not take into account any adsorption of drug on the cilia or lid margins that can alter initial assumptions regarding loss via the tear film.

Depending upon the degree of irritation caused by the combination of drug and vehicle, the next blink (and in man, the blink rate is about 10 to 12 blinks per minute) will dilute the drug even more. Given a basal tear flow rate of about 1 μl/min[63] (about 14 to 17%/min) the drug will be diluted by approximately 10% at each blink thereafter. Obviously, the smaller the initial drop size applied to the eye, the better the pharmacokinetics in terms of drug retention on the ocular surface.[40,64]

Both the degree of irritation caused by the test material and its pH when in the tear film will alter the tear kinetics. Induction of lacrimation will dilute the material at an even faster rate and even the application of a "nonirritating" drop can increase basal tear flow, thereby enhancing the dilution of any compound in the tears. This also places constraints on the concentrations of drugs that may be applied to the eye due to the tolerance of the tissues. Based upon this knowledge, it is easy to perceive that only small amounts of topically applied material reach the anterior segment tissues (and even less reach the posterior uveal tissues), and the residence time of any material in the tear film is <2 min.

If a solid is delivered to the eye in a normal ocular irritancy test, the usual quantity employed is 100 mg (0.1 g). This total quantity is not retained on the ocular surface, however, even if delivered into the lower cul-de-sac. After one blink, as little as 5 to 10 mg of material remains on the external ocular surface.[65] The characteristics of the compound or powder will dictate the precise amount remaining, but almost any solid material in powder form will be lost and only variable amounts remain.

The maximum delivery of fluid to the rabbit eye is 100 μl, although to achieve retention of this volume requires considerable care in delivery.[66] Differences in ocular toxicity have been noted, depending upon the volume of material instilled into the cul-de-sac,[67-70] with lesser responses occurring with decreasing volumes. A major difference occurs between the instillation of 100 and 50 μl, with the greater response at 100 μl being attributed to a longer time lag in the initiation of disappearance of this fluid volume.[66] Use of volumes below 50 μl tends to result in the same degree of ocular irritancy. A volume of 10 μl in the rabbit eye has been shown to provide a better predictive model of the human response, based on correlations between rabbit irritation studies and the human response to various products that accidentally reach the ocular surface.[8,9,67]

The route of penetration of a topically applied compound has usually been considered to be solely though the corneal epithelium. Some evidence exists, however, which suggests that at least some drugs can penetrate through the conjunctiva and then the sclera to reach the base of the iris and ciliary processes.[71] The quantity of drug passing this route is estimated at about 1 to 2% of the total drug entering the eye. In consideration of a full pharmacokinetic description of drug behavior in the eye, this possible route must be considered. It is also pertinent to consider the ratio of conjunctival to corneal surface area in any chosen model. While this ratio may be high in man (approximately 17:1), the ratio is much less in rabbit (approximately 8:1).[72] The loss of material from the ocular surface via absorption though blood vessels is thus much lower in the rabbit.

Since at best the aqueous humor concentration (per milliliter) after topical application will be $1/_{500}$ to $1/_{1000}$ of the mass of drug applied to the ocular surface, then due consideration must be taken regarding the concentration of the test drug that is applied. For tests of ocular pharmacokinetics, a radioactively labeled compound is usually employed as a tracer. For a drug such as pilocarpine, for instance, where a test concentration could be as high as 2%, the labeled material is supplied at a high enough specific activity (Bq/mg) that the use of only a trace amount of label will provide an adequate count rate in eye tissues.[18,19] In contrast, for a compound such as benzalkonium chloride, where the highest concentration used pharmaceutically is 0.02% and the specific activity is not as high, the entire quantity of BAK needs to be labeled in order to provide adequate counts in tissues that would not require count times in excess of 20 min per sample.

All of the above considerations, such as tear flow rate, blink rate, conjunctival area available, conjunctival blood flow, drug binding by either tissue or tear proteins, and the rate of corneal epithelial absorption and release of compounds, serve to minimize the uptake of agents into the eye. These systems serve to restrict the bioavailability of drugs within the eye while also serving as a means of preventing toxic compounds from entering the eye tissues. The net result of these systems is that only a small percentage of any compound reaching the ocular surface actually enters the eye tissues and fluids. From a pharmacological perspective as many of the protective mechanisms as possible must be overcome for the effective delivery of drugs.

Many agents have been studied for their ocular pharmacokinetics ranging from drugs used therapeutically including pilocarpine,[18,19,37] epinephrine and dipivalylepinephrine,[73] bunolol,[74] timolol,[75] prednisolone phosphate,[39] prednisolone acetate,[76] to other agents such as benzalkonium chloride,[50-53] and sodium lauryl sulfate.[51,52,77] All of these studies have added to our overall understanding of how drugs interact with ocular tissues, and how each has a different behavior in the tissues despite the similarities in kinetics in the tears and eye fluids.

All of the above considerations assume that the test animal is conscious, yet the penetration of materials into the eye is influenced by the status of the animal since either local or general anesthesia can influence drug penetration into the eye. It should also be noted that local anesthetics are an option in certain guidelines for ocular irritation tests. The use of such anesthetics, while reducing the possibility of any anticipated unneccessary pain or discomfort to the experimental animal, can serve to alter the response to the test substance.[78] In addition, animal position can play a role (i.e., whether upright or lying on its side).

While the distribution of agents after true systemic absorption is addressed below, it is worthy of note that many topically applied drugs are absorbed into the systemic circulation via the nasal mucosa.[75,79-86] The loss of compounds into the nasal canaliculi from the ocular surface represents a significant systemic absorption pathway. This route avoids the first-pass effect normally observed after oral presentation of a compound, and the pharmacological sequelae resemble those seen after an intravenous administration. It is important to minimize absorption though this route. Indeed, recommendations for overcoming this pharmacokinetic phenomenon have been given[87-89] in clinical situations, since the systemic effects of drugs such as timolol can be quite pronounced.[81,90] Occlusion of the nasal punctum for 3 to 5 min is an effective means of both increasing drug uptake by the cornea and reducing systemic side effects. Punctal occlusion obstructs the canaliculi, thereby allowing the possibility of greater ocular uptake of drug by modifying the drainage of fluid from the ocular surface.

2. Systemic

This route of application for ocular therapeutic reasons is used for a few drugs, e.g., carbonic anhydrase inhibitors. For this route, aspects that must be considered include protein binding, the dose of the agent given systemically relative to that needed for biological activity in the eye, and the vehicle. Obviously, a drug that has great binding capacity for protein and a requirement for high dose levels to achieve biological activity requires special consideration.

The pharmacokinetics of systemically applied ocular medications or toxicants are likely to be far different from those after topical application. Systemic ocular administration of a drug will enter the eye via the blood supply, and since the latter proceeds in the order choroid > iris > ciliary process >> retina, the greatest source of potential drug mass would be delivered via the choroid. The interface between the choroid and retina, however, is the retinal pigment epithelium that presents a marked barrier. If a drug did penetrate the retinal pigment epithelium, it would have to have characteristics that would allow passage across the highly lipid retina (with as much as 50% of the lipid being polyunsaturated fatty acids) before gaining entry into the vitreous humor. Such movement might have to occur against transport systems that are directed outward into the choroid.

The most likely source of ocular entry of a systemically administered drug or a toxicant into the vascular system would be via the circulation of the iris/ciliary processes. Here, especially in the ciliary processes, the capillaries are fenestrated, allowing penetration to the outer limiting membrane, the basement membrane of the pigment epithelial cell layer that faces the ciliary body stroma.

The kinetics of drug entry into and loss from the eye are quite different when this route of administration is considered. First, the aqueous humor of the posterior chamber receives the highest initial concentration, followed by the anterior chamber fluid as the aqueous flows from the posterior chamber though the pupil.[7] The ciliary process, of course, is the actual site where the highest initial concentration occurs. The iris then takes up significant quantities, while loss to the cornea and vitreous humor from the aqueous leads to lower concentrations at those sites. The further distribution of drug depends on its physicochemical characteristics and its rate of diffusion into the tissues, whether it is absorbed or transported by the tissues, and its metabolism.

3. Local Injections

This includes both injection of drug into areas outside of the eye itself, i.e., subconjunctival or sub-Tenon's, as well as direct injection into the eye.

Injection into the subconjunctival space or sub-Tenon's capsule is usually made in four injection sites, one in each quadrant of the eye. The sites are chosen to be

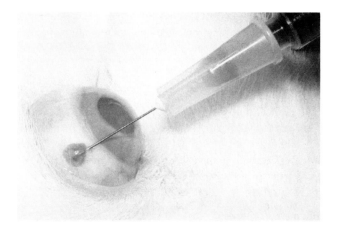

A

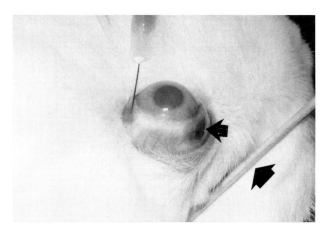

B

Figure 7. Technique for subconjunctival injection. A fluorescein solution was used to aid in visualization. The conjunctiva is lifted with a pair of forceps and the 30-gauge needle inserted between the extraocular muscles. (A) Site of one injection between extraocular muscles. (B) For two or more injections the needle is inserted subconjunctivally at other locations. Large arrow indicates wooden applicator used to hold tissue; small arrow indicates site of first injection.

between the most anterior portions of the extraocular muscles, thus avoiding the thinnest parts of the sclera (under the extraocular muscle insertions onto the sclera) to reduce the potential for direct injection into the eye (Figure 7). The injection of 100 to 200 µl of drug-containing solution subconjunctivally leaves a substantial bleb.

While the intent of subconjunctival injection is to provide drug in closer apposition to the sclera, across which drug penetration would occur more readily (since the permeability is much higher than that of the cornea due to its structural configuration and absence of limiting membranes), this notion has been modified. Current thinking suggests that the injection perforation per se leaves a route for drug to enter the tear film and thus enter the eye via the corneal route.

A similar logic does not apply to sub-Tenon's injections, since the depot source is further back in relation to the corneoscleral junction, and drug emerging from the Tenon's capsule could reenter the subconjunctival space. Thus, while subconjunctival injections usually result in pharmacokinetics akin to that seen after topical administration, sub-Tenon's injections provide drug to the posterior area at higher concentrations than can be achieved though topical or subconjunctival injections.

Direct injections into the eye are usually reserved for severe cases of endophthalmitis and these injections are made into the vitreous cavity. The dangers of this route of injection are the following: physical damage to the retinal tissues, direct-needle induced damage to the lens, and loss of fluid though the injection site. If a site is chosen that is 4 to 5 mm posterior to the corneoscleral junction, the possibility of creating a ciliary body detachment is far reduced. Injections should also be made though one of the extraocular muscles which acts as a tamponade on the ocular surface after needle withdrawal. Use of this route minimizes fluid loss through the scleral perforation. In addition, a few seconds should be allowed after the injection has been performed (using as small a volume as is possible and practical), to allow the intraocular pressure to fall. The injection of even microliter volumes of fluid into the eye causes a temporary rise in intraocular pressure,[91] but this falls quickly to normal levels (Figure 8).

The pharmacokinetics of intravitreally injected materials is most important in discerning the concentration that is therapeutically useful without inducing a toxic response. The kinetics will depend on the degree of binding by protein evoked by the inflammatory reaction in the vitreous, the degree of inhibition of active transport systems out of the eye both by the agent and by the inflammatory response, any vascular effects (mainly vasodilation, as a result of the inflammation), and metabolism of the materials. The toxicity of a compound on retinal function, and its clearance will, therefore, be different in an inflamed vs. a noninflamed eye. This is a point of practical importance in evaluating drugs for retinal toxicity after intravitreal injection. Retinal toxicity may be evaluated by electroretinograms or by microscopy.

4. Iontophoresis

This mode of application has limited value since it must be performed with suitable supervision. Nevertheless, iontophoresis offers a means of enhancing the penetration of charged drugs into the eye that could be helpful in the management of certain diseases such as endophthalmitis and herpes infections.[92]

The apparatus (Figure 9) consists of a controlling device that supplies current to the electrodes. Either an anionic or cationic electrode is connected to the drug source

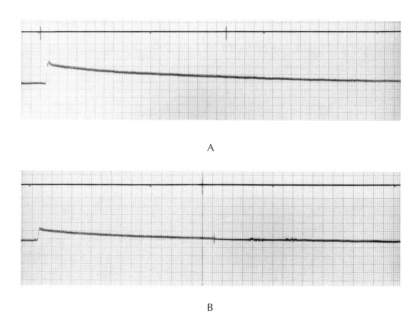

A

B

Figure 8.   Transient intraocular pressure increases following intravitreal injection. (A) Trace obtained after a volume of 10 µl was injected intravitreally. Note the rapid increase in intraocular pressure followed by an initial rapid fall and an exponential fall over the next minute or so. (B) Trace obtained after the injection of 5 µl into the vitreous. The intraocular pressure change is smaller than that in (A) and the intraocular pressure falls to normal values more quickly. In both traces the time marker (small marks on the top trace) indicates 1 min. The time course of the intraocular pressure illustrates the need to allow some time to pass between volume injection and withdrawal of the needle.

that can be placed on the eye and the other electrode in contact with the skin. The choice of which electrode contacts the drug solution is dictated by the charge on the drug. If it is anionic, then the cathode of the iontophoresis device is connected to the drug solution, and vice versa.

Drug can be either in solution or in an agar gel. In either case the solution must be distilled water, or at least contain no anions or cations since they, being more mobile, will carry the current in preference to the drug and enhanced drug levels will not ensue. Similarly, the compound to be iontophoresed must be charged. Solutions can be placed in a cup shaped to fit the eye at one end as shown in Figure 10. For an agar solution of nonthermolabile drugs, the procedure is that distilled water is heated to boiling and, with the heat turned off, agar, to give a final concentration of 4% w/v, is added to the drug solution. As the solution cools, the agar can be either injected or poured into appropriate tubings. The agar/drug solution is then placed in contact with the appropriate electrode, and a small amount of agar is extruded from the end of the tubing (Figure 11). The latter procedure ensures that good contact is

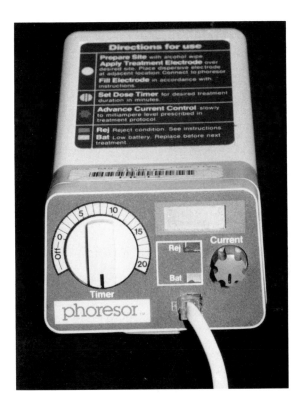

Figure 9.   Iontophoresis apparatus. This apparatus (Phore-sor™) has a battery check, a timer, and a current control button (with a digital display).

made between the ocular tissue and the drug-containing agar. This procedure is used for iontophoresis of fluorescein for aqueous humor turnover studies in man[93-96] and for the enhancement of drug penetration in experimental animals.[97,98] Only small currents (about 200 μA) for short time periods (a few seconds to a minute) can result in greatly enhanced tissue penetration. Kinetics depend on the final drug concentration reached in the various compartments and their subsequent clearance rates.

The procedure of iontophoresis is currently being tested for direct drug delivery into the vitreous in cases of endophthalmitis where delivery of drug to this ocular compartment, at concentrations sufficient to depress bacterial growth, is difficult.[99,100] Direct injection of antibiotics is used but carries the risk of the injection per se, and cannot be repeated frequently.

D. Model Systems

The choice of the model system with which to study the pharmacokinetics of any compound is dependent upon the use of an eye that has similar characteristics to that

A

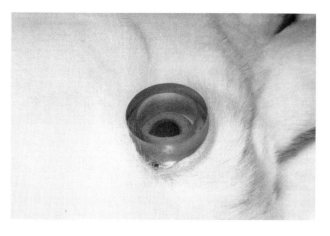

B

Figure 10.   Eye cup for iontophoresis of fluid. (A) Eye cup showing
flanged bottom for contact with curvature of ocular surface. (B) Eye cup
in place on rabbit eye; held in place by lids. A light pressure applied
to the cup easily prevents loss of fluid placed into the cup.

of the human. The most widely used model for size of the eye, ready availability,
cost, the presence of a large database in the literature, and, in many regards (for
pharmacokinetic purposes), similarity to the human eye is that of the rabbit. Most of
the physical characteristics of the tear film of rabbit and man are similar, e.g.,
thickness, viscosity, etc., while the turnover rate of rabbit tears is about one half that
of man. Few of these characteristics are known for other species. Unfortunately, the
rabbit blink rate is far less than that of man, thus the tear film kinetics also differ

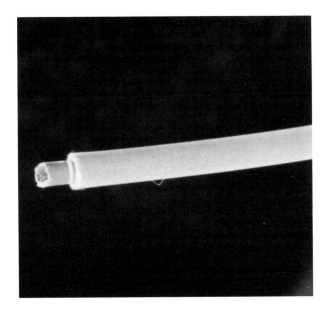

Figure 11.   Agar bridges showing extruded agar (containing fluorescein for better visualization).

markedly. This large difference in blink rate strongly suggests that the tear film composition (mucin, lipids) is vastly different between the two species, and may also reveal differences in behavior toward drugs and their vehicles that may or may not be reflected in different pharmacokinetics.

The retinal circulation of the rabbit eye is also different from the human eye.[101] Other factors include differences in corneal permeability (lower in humans to those agents studied to date) and a different aqueous outflow system (a complex system of interconnecting channels, forming a plexus, in the rabbit compared to the single Schlemm's canal in the human). The cost also plays an important role since the establishment of appropriate data points requires large numbers of samples. The samples must be taken at intervals sufficient to allow the definition of the absorption, distribution, and loss of material from the ocular tissues. The necessity for such large numbers is the most frequent determinant in deciding the species of choice.

Pasteurellosis is frequently endemic in colonies of New Zealand albino rabbits and can cause intermittent blockade of the outflow pathway for tears. For this reason the patency of the canaliculi should be checked in this species using topical fluorescein. Open canaliculi are indicated by appearance of the dye in the nostrils.

The use of pigmented rabbits is absolutely advisable for any pharmacokinetic studies where extrapolation to humans is desired, since the presence of pigment can markedly influence the kinetics of drug uptake and/or washout. Binding of drugs and chemicals to melanin is a known phenomenon and can extend the half-life of drug release by several orders of magnitude.

The factors discussed above render the rabbit less than ideal for interspecies extrapolation, yet many other factors, including cost, must be considered. The precise relationship of the ocular physiology of the rabbit to that of man, especially in terms of pharmacokinetics, has not been delineated. This is especially true for topically applied materials where the much smaller ratio of body weight to eye size makes systemic absorption and distribution to the untreated contralateral eye a distinct possibility in the rabbit.

Use of other species is restricted primarily by cost and availability. For example, the cat offers an eye that is more closely akin to that of man, yet this animal is more costly than the rabbit and is under greater restrictions regarding its availability. The most obvious model to use is that of the subhuman primate, yet these animals are very expensive (about $400 for a squirrel monkey and approximately $1000 to $1500 for a rhesus monkey) with many species (i.e., owl and rhesus monkeys) being virtually unavailable due to import restrictions, although some rhesus monkeys are bred for research purposes from already-existing stock in this country. Their cost, however, is almost prohibitive in terms of experiments where a large number of animals is required to produce unequivocal data. Not only is the initial purchase outlay expensive, but primate care is also more costly as is the need for more elaborate facilities and the possibility of infecting primates, or vice versa, with human pathogens.

The use of expensive animals may be justified under some circumstances. The distribution of a drug in ocular fluids can be characterized with only a few replicate eyes, to obtain relevant data from aqueous or vitreous humor samples of these fluids. In this way, the eyes can be reused if aqueous and vitreous humor taps (samples obtained by paracentesis) are taken only at intervals of at least 2 weeks. If the relative distribution of a drug can be evaluated from knowledge of the aqueous or vitreous to tissue ratios, then analysis of these fluids can be used to substantiate pharmacokinetic data in other species such as the rabbit. The use of primates is usually prohibited by cost if a detailed analysis that requires corneal epithelial and ocular tissue analysis is required for the complete description of the distribution kinetics or histotoxicity of any compound.

An obvious difference between the eyes of rabbits, cats, and dogs, and those of primates (including man) is the presence of the nictitating membrane in the nonprimate species. The nictitating membrane can act as a reservoir for topically applied drugs and provide an overestimate of the pharmacokinetics of drugs in the eye. This membrane reservoir can provide a source of drug that may return to the tear film over a considerable time period. Surgical excision of the nictitating membrane in nonprimate species can markedly reduce the impact of this tissue on pharmacokinetic parameters used for extrapolation to man after topical drug administration.

Depending upon availability, the pig offers another viable model system. The pig has an eye that is in many respects physiologically similar to that of the human. Unfortunately, there is a paucity of data in the literature from which to draw useful comparisons.

Another important consideration for pharmacokinetic analysis is the status of the eye during drug administration. For example, the determination of steroid pharmacokinetics in a normal eye may lead to different conclusions regarding the drug behavior from data obtained in an inflamed eye. Steroids are often used clinically in an eye that is inflamed, either on or in the cornea or in the anterior chamber. Antibiotics are used in cases where there is either an endophthalmitis where intraocular penetration is a prerequisite for therapeutic efficacy, or in cases of external ocular infection where the drug of choice would be one showing an absolute minimum of intraocular penetration.

Under conditions of inflammation, normal ocular physiology does not exist, and breakdown of the blood-aqueous barrier occurs with increased cellular and protein infiltration into various ocular compartments. With a corneal inflammation, epithelial or endothelial permeability may be compromised and the responses to drugs will be different from that of a normal cornea. It is quite conceivable that a drug used to treat an external ocular infection would show no intraocular penetration in a normal eye, yet have significant penetration though the compromised epithelium of an infected cornea. Under these circumstances a compound that would not normally initiate a toxic intraocular response could well induce such a reaction with access though a permeabilized corneal epithelium. It is imperative, therefore, to choose a model that represents the clinical condition under which the drug will be used.[102] Such considerations also extend to the potential use of drugs in aphakic eyes, where the absence of the lens allows access of materials to the posterior portion of the eye to a greater extent than if the lens is present.

Furthermore, intraocular metabolism of a compound may occur in one species relative to another. This is well illustrated by chloramphenicol, where about 45% of the delivered drug is recovered in the form of metabolites in the rabbit aqueous humor, whereas little or no metabolism occurs in man.[38] Awareness of the possibility of metabolism either under normal or pathological conditions must be uppermost in the mind of the experimentalist. For this reason HPLC analyses of fluid and tissue samples are almost mandatory for a full and correct interpretation of the data.

Information gathered from experiments as described above is extremely relevant to toxicological studies, since such chemicals might be expected to induce morphological or pathophysiological changes in tissues which would in turn further enhance their penetration rate. In this way the agent itself could modify the pharmacokinetic data. This represents a potential artifact that must be determined in any model system. Verification of any effect on the permeabilities of membranes can be made using nonmetabolizable, nonelectrolytes such as sucrose, inulin, or dextran. Other factors such as disturbances of the ocular surface (corneal and/or conjunctival) that may cause changes in the tear film composition (e.g., protein increase) could also significantly alter drug pharmacokinetics in the tear film. The type of injury to the eye surface can also be an important factor that must be appropriately considered in any testing procedure.

The choice of *in vitro* or *in vivo* model is dependent upon the nature of the toxic reaction either observed and requiring further detailed analysis, or anticipated by the use or exposure to a chemical. *In vitro* models provide the opportunity to study a response in a well-controlled environment but lacks the interactive influences that occur *in vivo*. While *in vivo* models have the complexity of being controlled by multifactorial forces, exposure or use of a chemical will occur *in vivo* and thus cannot be replaced as the ultimate model.

E. Analytical Methods

Various methods exist for determining the presence of any drug in the eye. These include the use of radiolabeled materials (^{3}H, ^{14}C, ^{35}S, ^{32}P, ^{131}I, ^{36}Cl, etc.). It is necessary that the label be in a stable position in the molecule, and if it is metabolized (as with a prodrug type of material[73]), that the label be on the portion of interest. This can be confirmed using either HPLC, TLC, or column chromatography with subsequent counting of the spot or elution volume to determine the radioactivity.

With the use of any tracer material the radiochemical purity should be checked before use to avoid the measurement of compounds other than those of interest in the study. This can be performed with either HPLC or chromatography using a sample of the test material. As indicated above for *in vitro* tests, it is necessary to ensure that the material is not lost during any tissue extraction whether this be mild (i.e., osmotic shock using distilled water to disrupt cells) or more severe (i.e., the use of alkali or quaternary ammonium compounds to solubilize tissues or tissue digestion with enzymes).

Performance of this test requires harvesting the tissues in the same manner as for an experimental procedure, adding a known quantity of the labeled material and subjecting it to the same digestion process normally used. Subsequent analysis will reveal whether any degradation or loss of the material has occurred. In addition, each tissue analyzed should also be prepared according to the experimental procedure, and digested first with subsequent sampling, using the same procedures as for experimental tissues. A known quantity of the labeled drug is then added to each of the tissues under examination and this is compared to the same amount of labeled drug added to the scintillation fluid alone. Depression of the count rate caused by digested or solubilized tissue can be determined and a quench correction curve constructed. Different tissues may yield varying degrees of quenching dependent upon pigmentation, native protein content, and differences in other composition (e.g., lipid content). Chemiluminescence of samples (particularly those that are from heavily pigmented tissues) can be reduced by the addition of hydrogen peroxide and/or glacial acetic acid, or even small volumes of water with certain scintillation fluors. These steps are given in Table 1. Direct comparison can then be made between all experimental tissues when all count rates are converted to a common, unified basis and expressed in the same units.

When harvesting tissues for analysis of any tracer, it is essential to avoid any source of cross-contamination. This is true for control vs. experimental eyes in the same animal or from animal to animal. Some means should be available to ensure that clean instruments are used for each eye, whether this be in the form of cleaned instruments or fresh instruments for each eye. These considerations apply whether or not the eye is dissected in a fresh or frozen state. Freezing of the whole eye

Table 1
Determination of Quench Curve

1.	Harvest tissues from animals not receiving test drug.
2.	Digest or disrupt tissues as for experimental tissues (i.e., distilled water, alkali, or quaternary ammonium compounds).
3.	Add a known quantity of labeled test drug to each vial in exactly the same way as for experimental tissue (i.e., same dilution of sample, etc.).
4.	Count all samples, including test drug added to scintillation fluid alone.
5.	Convert all counts from cpm to dpm.
6.	The relative depression of the count rate of each test tissue compared to drug in fluid alone will give the quench caused by that tissue.
7.	After performing experiments, convert all dpm using the quench values.

Example:
cpm to dpm
Counts per minute of test drug in fluid alone for known amount of material of known specific activity, i.e., cpm/mass of drug.

$$\text{Known efficiency of scintillation counter, cpm} \times \frac{1}{\text{efficiency}} = \text{dpm}$$

1 Ci = 3.2×10^{12} dpm.
Test drug alone in fluid = 10,000 dpm
Test drug added to digested cornea = 6000 dpm
Test drug added to digested retina = 3000 dpm
Therefore, all cornea counts of experimental tissues must be multiplied by 10,000/6000 to correct for corneal quenching, and all retina counts of experimental tissues must be multiplied by 10,000/3000 to correct for retinal quenching. This allows all count rates to be directly compared.

provides a means of removing tissues while markedly retarding the possibility of diffusional loss of material from any tissue. Whatever approach is taken, all possibilities for cross-contamination should be examined and prevented.

Labeling of drugs with radioisotopes is not the only means of detection available. Chemical methods can be used if a suitably accurate method is available. Other labels include fluorescein or horseradish peroxidase, although their molecular size may preclude their use unless the test material is large enough that the attachment of the 320 Da fluorescein or the 50,000 Da horseradish peroxidase would not significantly alter the molecular weight of the labeled compound.

The distribution in the transparent tissues of the eye can be achieved noninvasively with fluorescein, or a fluorescein-labeled compound, using fluorophotometry if the concentrations are high enough for detection. The minimum detectable level is about 10^{-9} g/ml. Since the levels of drug in the anterior chamber have been found to be roughly $1/_{1000}$ or less of that added to the external ocular surface, it is obvious that large quantities would need to be added to the cornea. Detecting fluorescein in the transparent ocular media (cornea, aqueous humor, lens, vitreous humor) may be accomplished with a fluorophotometer. These exist in several forms: fully assembled, primary functional components, and self-made. The Fluorotron Master (Coherent, Inc.),[103-105] is a fully assembled, dedicated, commercially available unit. The primary functional components category includes Gamma Scientific™ eyepieces, fiberoptic probes, filters, and radiometer that can be attached to a slit-lamp.[106] The final category represents a self-made unit based on the publication of Brubaker and Coakes.[107]

Qualitative, not quantitative, determination of a fluorescent label could be made using fluoresence microscopy, although the preparation procedures for microscopy may dilute the drug concentration as the tissue is processed though different solutes before embedding. For water-soluble drugs, the best approach is lyophilization of the tissue prior to examination[108] or precipitation of the molecule to which the label is attached, which minimizes loss or intratissue diffusion of the labeled compound.

With horseradish peroxidase, only large test compounds could be used due to the high molecular weight of the peroxidase. The detection of this material is qualitative and indicates the presence of the label in some tissue locations. Sections of tissues are prepared as usual either for microscopy or in frozen sections with subsequent exposure of horseradish peroxidase by incubation in diaminobenzidine (DAB) which reacts with the peroxidase to provide a dark-colored reaction product. The reaction will continue as long as the substrate (DAB) is present; thus, the optimum conditions for adequate detection without overproduction of product must be sought for each experimental condition.

For the counting of radioactively labeled samples, it is preferable to perform the tissue digestion and counting in the same vial. There is a tendency for some test compounds to adsorb to glass, thus it is possible to have a perfectly good extraction of tissues, but lose some material to the glass walls of the container. Since digestion or solubilization of tissues using quaternary ammonium or alkali compounds cannot be performed in anything but a glass vial, whenever possible avoid transference of samples. A step-by-step procedure is given in Table 2.

While it should be fairly self-evident, when performing pharmacokinetic analysis it is highly desirable, if not required, to separate the corneal epithelium from the rest of the cornea. The epithelium is a repository for many topical agents and needs to be considered as a distinctly separate tissue. A number of early experiments on drug pharmacokinetics analyzed only the whole cornea, and provided data clouded by the lack of distinction between epithelium and the rest of the cornea.

Another analytical method is bioassay. This approach is useful for testing antibiotics. Tissue samples can be extracted and the material placed on agar plates of colonies of bacteria.[109,110] Concentrations are obtained from the distribution of the colonies and the areas in which bacteria are prevented from growing.

F. Calculations

While an outline to equate all count rates to the same relative dpm is given in Table 2, the methods of calculation to provide a basis for pharmacokinetic analysis are given here. To determine the kinetics after either single or multiple drops, the ocular fluids and tissues must be analyzed at different times.

Table 2

General Method for the Determination of the Distribution of a Radiolabeled Compound within Ocular Tissues

---

1. Estimate the relative proportion of labeled material to total mass of test drug needed to give adequate concentrations in aqueous humor and tissues using the rule-of-thumb that only 1/1000th of the total mass of drug will be present in the aqueous humor. Thus, if the concentration is 1% (1 g/100 ml) and a 20 μl drop is applied to the eye, only 0.2 mg will be present in the eye. Peak aqueous concentrations of about 50 μg/ml can be anticipated.

2. Make up test solutions as fresh as possible before use, especially with ^{3}H-labeled compounds that were labeled by tritium exchange.

3. Check that label is retained on the compound and the degree of metabolism if it occurs using HPLC or other analytic chromatography methods.

4. Perform quench curve determinations (see Table 1).

5. Perform experiments, preferably without the use of anesthetics or sedatives that can alter tear flow, blink rate, aqueous humor turnover rate, blood flow, etc., and compromise the data.

6. Remove tissues. The test animal is killed quickly and using a method that does not provoke a response (e.g., intravenous overdose of sodium pentobarbital). The eye is proptosed (rabbit) or enucleated rapidly and rinsed quickly in normal, 0.9% saline solution. An anterior chamber paracentesis is made using either a glass capillary with a 30-gauge needle glued to the end or a 1 ml tuberculin syringe and a 30-gauge needle. The cornea is then penetrated near the limbus and a circumferential cut made to excise this tissue. The epithelium is harvested using a Gill corneal knife or razor blade either before or after removal of the cornea; collection is easier with the cornea *in situ*. The lens can be easily removed by splitting the capsule and squeezing the lens out of the remaining capsule. By grasping the iris and rotating it gently to loosen its adherence at the iris root, the entire iris-ciliary body and vitreous are usually removed. The vitreous can be separated from the iris-ciliary body with scissors or by sliding a pair of forceps over the iris-ciliary body to dislodge the vitreous gel. Ciliary body can be collected by scraping of the processes or by separation from the iris using scissors. The remainder of the eye (sclera, retina, and choroid) is cut into quarters beginning at the limbus and ending at the optic nerve. After the tissue is laid flat the retina can be collected; and the choroid scraped from the scleral shell. With practice, this entire procedure can be achieved in about $1\frac{1}{2}$ to 2 min per eye.

7. Digest tissues. If using alkalis to digest tissues, do preliminary experiments to determine the amount of tissue needed to give adequate count rate, and digest that amount of tissue in the same vial as will be used for counting. Use glass vial with a polyethylene insert in the cap.

8. Cool tissue samples, add hydrogen peroxide to decolor samples. Usually 100 μl of 30% H_2O_2 will suffice for a digested sample volume of up to 3 ml. Glacial acetic acid may also be added to decolor samples.

9. Allow sample chemiluminescence to decrease by storing in a dark area.

10. If scintillation counter is refrigerated, store samples in a refrigerator.

11. Count samples.

The conversion of all cpm to dpm negates the quenching effect of the different tissues and allows direct comparisons to be made. Calculations for a quench curve are given in Table 1. All dpm of the tissue samples from eyes receiving the test compound must be corrected to either a total tissue weight basis or on a unit weight basis. For this reason all tissue samples must be weighed as wet weight; this approximates a more realistic measure of tissue or fluid distribution. Expression of the data as per milliliter suffices for aqueous and vitreous humor. Aqueous humor has a specific gravity of near 1 as does the vitreous humor.

G. Pharmacological Modulation of Kinetics

While determination of the ocular pharmacokinetics of a drug in a model system may provide information pertinent to its distribution, note that the drug under test must act in the chosen model in a manner similar to that in man. For example, while some authors have indicated that timolol maleate reduces the intraocular pressure of rabbits with artificially elevated intraocular pressure, induced by the introduction of water into the stomach,[111-113] there are many reports that this does not occur in normotensive rabbits.[111,114,115] Timolol is active in reducing intraocular pressure in the cat; this animal represents a suitable model for pharmacokinetic studies and provides an accurate representation of the drug distribution under conditions where aqueous humor inflow is reduced in a manner similar to that in man.

As mentioned elsewhere, one should be cognizant also of the potential intervention of drugs in the evaluation of kinetics. Thus, not only should the model be evaluated for the appropriateness of its pharmacological actions toward the test drug, but also with other drugs with which it is likely to be used. In ophthalmology, as well as internal medicine, it is not uncommon for several drugs to be used simultaneously. This is particularly true in ophthalmology with the treatment of glaucoma, where several pharmacologically active drugs may be employed to reduce intraocular pressure. The potential effects of the application of two or more topical agents on the penetration of drugs across the cornea have been addressed elsewhere (Section III.C), but here we consider the relative pharmacological effects of drug combinations.

One such combination used to reduce intraocular pressure comprises topical pilocarpine, topical epinephrine, and systemic diamox (acetazolamide). Pilocarpine reduces intraocular pressure by increasing the outflow of fluid drainage from the eye.[116] Epinephrine has mixed actions both on aqueous humor inflow and outflow,[117-120] and diamox decreases aqueous humor inflow.[121,122] While diamox is not known to have other pharmacological effects in the eye, pilocarpine causes pupillary constriction[121] and epinephrine causes vasoconstriction of the anterior segment blood vessels of primates[123] and rabbits.[33,123]

Inference of the relative kinetics in this combination will be examined using pilocarpine as the labeled test drug for which determination of the pharmacokinetics is required. If pilocarpine kinetics are determined alone, then a baseline is established from which to relate other findings. Addition of epinephrine to this baseline could change the pharmacokinetics by its actions in decreasing aqueous humor inflow into the eye. First, this is achieved by increasing pilocarpine availability for tissue uptake since aqueous humor flow though the anterior chamber would be reduced. Second, the uptake of pilocarpine predictably would be further increased due to anterior segment ischemia, since a much reduced iris blood flow could not remove as much pilocarpine from the aqueous humor as would occur when pilocarpine alone was used. The superimposition of aqueous humor inflow reduction by about 50% upon this already complex system with the addition of diamox would result in additional pilocarpine retention in the aqueous humor. Thus, not only could the kinetics of pilocarpine be extended, but the pharmacological sequelae would be anticipated to be accentuated due to higher pharmacologically active concentrations in the tissues and fluids of the eye for longer periods of time. This is but one illustration of the potential pharmacological interaction of drugs that can markedly alter the kinetics of an applied drug. A potential for such interactions must be investigated as it is possible to shift a drug that has desirable therapeutic effects into the range of inducing toxicological effects.

H. Prodrugs

There is increased emphasis on the use of prodrugs that are metabolized in eye tissues by endogenous enzymes to provide a pharmacologically active drug in the eye. This approach reduces the possibility of toxicological effects on the ocular surface, as well as reducing the systemic absorption of any drug. This is of importance since beta-blockers, for example, can induce marked cardiovascular effects (bradycardia) even when applied topically, and adrenergic agents (such as epinephine) can induce a tachycardia, reactive conjunctival hyperemia, and deposits of metabolites in eye tissues. These problems can be overcome though the use of prodrugs as illustrated by dipivalyl epinephrine[73,125,126] which is converted to epinephrine in the cornea by esterases, thereby substantially reducing ocular surface and systemic toxicity. Similar approaches are being used for beta-blockers[127,128] with compounds that are converted by ocular esterases into pharmacologically active metabolites, thus avoiding the systemic bradycardia often caused by the systemic absorption of beta-blockers. Thus, not only ocular, but also systemic, toxicity is reduced though the use of prodrugs. The esterases are not specific to ocular tissues, but are present at high concentrations.

Increasingly, prodrugs are being used experimentally, as well as clinically, to reduce side effects. For example, one of the difficulties with the potential topical use of delta-9- tetrahydrocannabinol is the systemic absorption that occurs. This could lead to undesirable effects. With the use of a water-soluble prodrug, a maleate salt of (+)-delta-6-tetrahydrocannabinol dimethylheptyl, Mechoulam et al.[129] overcame this problem by using the native corneal enzymes to convert this water-soluble form into a lipid-soluble drug. Once the drug enters the epithelium and becomes hydrophobic, there is no loss into the watery tear film. Thus, the drug is trapped and remains exclusively in ocular tissues. Use has been made of this approach by using the plentiful esterases in the cornea to transform prodrugs into active compounds that can then enter both anterior segment tissues and the more posterior ocular tissues.[130]

I. Compartmental Analysis

Several models for compartmental analysis have been presented over the last few years that reflect the incorporation of both extra- and intraocular compartments, as we have become more aware of the role of different cell layers in tissues in the regulation of drug distribution. Maurice and Mishima[131] have presented a comprehensive model system that incorporates many needed features for modeling of drug kinetics. These authors have shown that many investigations of drug penetration into the eye have only addressed the problem of finding whether a therapeutic concentration can be achieved rather than specifically addressing questions of kinetics. They examined the literature for information regarding drug passage across and into ocular tissues to determine the applicability of kinetic models to the data.

They contended that the distribution of material that penetrates the corneal epithelium and its changes with time are dominated by the corneal and anterior chamber compartments and are mirrored by the transfer coefficients between these compartments and between the aqueous humor and blood. Based upon the accumulated data, a two-compartment model was presented as a reasonable approximation. The cornea and its components remain the primary determinants of how much material enters the eye after topical administration. The permeability of the epithelium and endothelium to any material, and its absorption by the epithelium and stroma are properties that need determination. These factors are illustrated by studies on the pharmacokinetics of cationic and anionic surfactants[132] represented by benzalkonium chloride and sodium lauryl sulfate. Both agents enter the corneal epithelium, which acts as a reservoir, but benzalkonium chloride is lost only to the tear film while sodium lauryl sulfate penetrates into the remainder of the ocular tissues with apparent ease. A comparison of the ocular penetration of these compounds with several others (for which adequate data exists) has been made.[132] This work illustrates the extreme length of time that both of these surfactants are retained by the corneal epithelium relative to any other tested compound to date.

Robinson and his co-workers[31,133] describe the interplay of various corneal compartments together with extraocular characteristics in regulating drug penetration. They have investigated the disposition of drugs when placed onto the corneal surface and into the tear film. Their results showed that the corneal epithelium played a dual role, first as a barrier to drug penetration and second as a reservoir for drug storage in the intact cornea.

The importance of drug elimination via lacrimal drainage was highlighted; also that uptake into the cornea ceases once the concentration of material in the tear film falls below that in the surface epithelium. The role of precorneal tear film kinetics in determining the absorption of material into the cornea was well established with these studies.[31] Further examination of the pharmacokinetics of drug absorption not only confirmed that this route of material loss was important, but also illustrated the involvement of conjunctival uptake, alterations in conjunctival blood flow, and changes in tear production on the disposition of material reaching the ocular surface.[133] These factors were evaluated using mathematical models formulated from experimental findings in the rabbit. The relative effectiveness of precorneal factors affecting the disposition of material was found to be the following: drainage $\cong$ vasodilation > nonconjunctival (nictitating membrane) loss > lacrimation induced by the material $\cong$ conjunctival absorption > normal tear turnover.

Any kinetic analysis of drug penetration into the eye must include the variety of factors that influence distribution. Loss from the precorneal area, the tear flow rate, absorption by conjunctiva, nictitating membrane, solution drainage, and lacrimation are all considered. Absorption into the corneal epithelium and loss from this site to the stroma and subsequently into the aqueous humor and tissue reservoirs are considered in order (Figure 12). Differential equations have been written to incorporate these processes.[59] An example of an equation that takes into account the change in the mass of drug in various compartments (precorneal, corneal epithelium, corneal stroma-endothelium-aqueous chamber, and tissue reservoirs) is given below

$$\frac{dV_A(t)[A]}{dt} = -p_p([A] - [B]) - k_d[V_A(t) - V_o][A] - Q_T(t)[A] - p_n[A]$$

where $V_A(t)$ is the volume of fluid in the tears at a given time, V_o is the normal resident tear volume, $[A]$ is the drug concentration in the precorneal area, $[B]$ is the drug concentration in the corneal epithelium, P_p is the transfer coefficient between the precorneal area and corneal epithelium, k_d is drainage coefficient from the tear film, $Q_T(t)$ is the coefficient associated with tear flow including both normal tear turnover and induced lacrimation, p_n is the coefficient associated with nonproductive loss of drug (conjunctival absorption, nonconjunctival loss, and any drug effects on the vascular status). Drug transfer is thus considered as a series of simple diffusional steps, each occurring simultaneously.[133] Thus,

$$\frac{d[A]}{dt} = \frac{(p_p + p_n}{V_A(t)} + \frac{Q_T(t))}{V_A(t)} [A] + \frac{p_p}{V_A(t)} [B]$$

Miller and Patton[134,135] have investigated drug disposition in young vs. adult eyes. They reported that age-related differences exist that can be related to size differences, but which also must include functional and developmental differences. These findings have been substantiated by findings with surfactants in different age groups (Table 3).[50-52,77,132]

Miller and Patton[134] indicated that a reasonable approximation of aqueous humor drug concentrations could be predicted based upon the aqueous humor volume ratio. For each age group (selected to be 20- and 60-day-old rabbits) drug uptake into various tissue or fluid compartments of the eye was compared after the administration of the same quantity of drug to the ocular surface. Consideration of size differences alone between these two groups of rabbits was not sufficient to explain all of the age-related differences. A greater fraction of applied drug was absorbed in younger rabbits, and other factors influencing drug disposition in the eye were suggested as differences in aqueous humor dynamics, such as drug metabolism in fluids and tissues. Corneal permeability was another factor. For nonionized pilocarpine, corneal permeability was greater in young relative to adult animals. For ionized pilocarpine, however, the corneal permeability was identical at either age.[136]

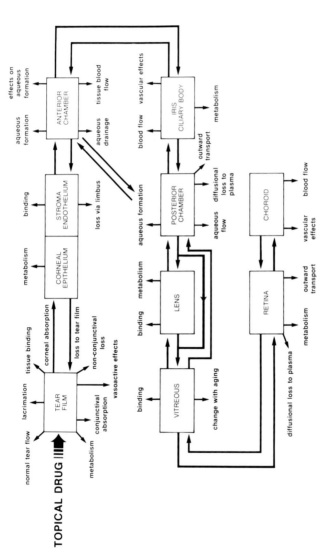

Figure 12. Illustration of many factors to be taken into account when considering drug disposition following topical ocular application. The many subcompartments are shown with the variety of potential influences on drug penetration into ocular tissues.

Table 3
Comparison of Ocular Tissue Content of Sodium Lauryl Sulfate from Different Aged Rabbits 6 h after a Single Drop Application

Tissue	Adult	Juvenile	Neonate
Cornea	1.372 ± 0.292	1.022 ± 0.120	12.368 ± 1.47
Epithelium			571.7 ± 101.3
Aqueous	0.015 ± 0.002	0.039 ± 0.004	0.159 ± 0.014
Iris	0.037 ± 0.003	0.049 ± 0.015	0.488 ± 0.285
Lens	0.004 ± 0.0004	0.009 ± 0.001	0.144 ± 0.046
Vitreous	0.001 ± 0.0001	0.003 ± 0.0001	0.026 ± 0.040
Retina	0.013 ± 0.002	0.201 ± 0.168	0.748 ± 0.138
Choroid	0.060 ± 0.030	0.545 ± 0.331	1.917 ± 0.457
Conjunctiva	—	—	36.40 ± 5.57

Note: Values are the mean ±SE of at least four tissues in ng/mg tissue wet weight or ng/mg fluid. Data adapted from Clayton et al.[77] and Green et al.[51,52] Sodium lauryl sulfate concentration was 1.3%; drop size was 50 μl in adults and juveniles and 10 μl in neonates; thus the mass of drug applied to adult and juvenile eyes was five times more than that applied to the eyes of neonates.

IV. CONCLUSION

It is evident from the immediately foregoing that pharmacokinetic models have been and continue to be useful in their application to several problems related to ocular pharmacokinetics. While primarily addressing the delivery of drugs to the eye, the same considerations apply equally to the study of the pharmacokinetics of toxic agents and to the comprehension of the development of toxicity after long-term treatment with therapeutic agents. The influence of each factor on the distribution and kinetics of any agent must be assessed individually, particularly compounds that cause alterations in conjunctival blood flow, permeability of the cornea or conjunctiva, or tear secretion. These factors play a predominant role in the pharmacokinetics of agents on the ocular surface.

As with considerations of drug pharmacokinetics, the purpose of the application of these models is for extrapolation to the human case. Direct applicability has been demonstrated in terms of extension of contact time for drugs by punctal occlusion that leads to less systemic drug absorption and an extension of uptake into ocular tissues (not necessarily an increase in absorption per unit time, but providing a longer period for absorption to occur).

Ocular drug-induced side effects are numerous and occur frequently,[137] and occur after materials reach the eye by a variety of routes. Both intra- and extraocular structures can be affected, although the most severe lesions tend to appear in the cornea, lens, or retina. The effects are related to the concentration and the duration of exposure, and can only be assessed by kinetic studies that involve examination of metabolism of the agent, tissue retention, and considerations of permeation of the agent to ocular tissues and fluids after reaching the eye by different routes.[138] These processes are the fundamentals of ocular pharmacokinetics, and form the basis for the comprehension of the physiological, pharmacological, and toxicological effects of any agent which may come into contact with ocular tissues.

Only though the evaluation of pharmacokinetic parameters can the potential of any chemical for inducing toxicological responses in ocular tissues be evaluated. The application of pharmacokinetic principles and analysis has allowed enormous progress in our understanding of those characteristics that must be taken into account in ocular drug delivery. This has been and continues to be of importance as drugs used in the treatment of ocular disease have become more complex and often have characteristics that do not allow good penetration into the eye.[59] In addition, the application of pharmacokinetics to ocular drug delivery has provided a rational basis for the design of new approaches to overcome the many characteristics that normally impair drug penetration. Considerable progress has also been made in evaluating the underlying causes of toxicological reactions within the eye. Determinations of ocular pharmacokinetics play an important role in both the development of new therapeutic agents, as well as in the assessment of the potential toxicity of either therapeutic compounds or those agents reaching the eye by accident and should always be considered for incorporation into evaluatory procedures where such concerns are evident.

ACKNOWLEDGMENTS

Some of the work included here from the author's laboratory was supported by PHS Grants EYO4558 and EYO4559 from the National Eye Institute. I thank Lisa Cheeks for her invaluable help with many of the technical and editorial aspects of this chapter, Brenda Sheppard for her enduring patience and skill in translating the information into a readable form, Dr. Donald L. MacKeen for his valuable critical reading of the chapter and for his many suggestions, Heather Unwin for her editorial assistance, Laura McKie for her excellent illustrations, and Frank Lazenby and Mike Stanley for their photographic expertise.

REFERENCES

1. J.W. Shell, *Surv. Ophthalmol.* 26, 207 (1982).
2. E.A. Balazs, *Int. Ophthalmol. Clin.* 13, 169 (1973).
3. D.M. Maurice, *Invest. Ophthalmol.* 6, 464 (1967).
4. E.H. Barany, *Acta Physiol. Scand.* 93, 250 (1975).
5. G. Raviola and J.M. Butler, *Invest. Ophthalmol. Vis. Sci.* 25, 827 (1984).
6. A. Bill, *Acta Physiol. Scand.* 73, 204 (1968).
7. V.E. Kinsey and D.V.N. Reddy, *The Rabbit in Eye Research,* J.H. Prince, Ed. Charles C Thomas, Springfield, IL, 218 (1964).

8. F.E. Freeberg, J.F. Griffith, R.D. Bruce, and P.H.S. Bay, *J. Toxicol. Cut. Ocul. Toxicol.* 3, 53 (1984).

9. F.E. Freeberg, D.T. Hooker, and J.F. Griffith, *J. Toxicol. Cut. Ocul. Toxicol.* 5, 115 (1986).

10. D.S. Hull, J.E. Hine, H.F. Edelhauser, and R.A. Hyndiuk, *Invest. Ophthalmol.* 13, 457 (1974).

11. S. Dikstein and D.M. Maurice, *J. Physiol. (London)* 221, 29 (1972).

12. E.I. Anderson, J. Fischbarg, and A. Spector, *Biochim. Biophys. Acta* 307, 557 (1973).

13. J. Fischbarg, J.J. Lim, and J. Bourquet, *J. Membr. Biol.* 35, 95 (1977).

14. K. Green, L. Laughter, and D.S. Hull, *Curr. Eye Res.* 2, 797 (1983).

15. H.F. Edelhauser, D.L. Van Horn, R.A. Hyndiuk, and R.O. Schultz, *Arch. Ophthalmol.* 93, 648 (1975).

16. H.F. Edelhauser, R. Gonnering, and D.L. Van Horn, *Arch. Ophthalmol.* 96, 516 (1978).

17. D.S. Hull, K. Green, and R. Berdecia, *Curr. Eye Res.* 5, 321 (1986).

18. S.S. Chrai and J.R. Robinson, *Am. J. Ophthalmol.* 77, 735 (1974).

19. K. Green and S.J. Downs, *Arch. Ophthalmol.* 93, 1165 (1975).

20. K. Green and J.E. Pederson, *Am. J. Physiol.* 222, 1218 (1972).

21. J.E. Pederson and K. Green, *Exp. Eye Res.* 15, 277 (1973).

22. J. Brodwall and J. Fischbarg, *Exp. Eye Res.* 34, 121 (1982).

23. T. Krupin, P.S. Reinach, O.A. Candia, and S.M. Podos, *Exp. Eye Res.* 38, 115 (1984).

24. T.H. Maren, L. Jankowska, G. Sanyal, and H.F. Edelhauser, *Exp. Eye Res.* 36, 457 (1983).

25. M.G. Eller, R.D. Schoenwald, J.A. Dixson, T. Segarra, and C.F. Barfknecht, *J. Pharm. Sci.* 74, 155 (1985).

26. A. Stein, R. Pinke, T. Krupin, E. Glabb, S. Podos, J. Serle, and T. Maren, *Am. J. Ophthalmol.* 95, 222 (1983).

27. M.W. Duffel, I.S. Ing, T.M. Segarra, J.A. Dixson, C.F. Barfknecht, and R.D. Schoenwald, *J. Med. Chem.* 29, 1488 (1986).

28. G.M. Grass, J. Cobby, and M.C. Makoid, *J. Pharm. Sci.* 73, 618 (1984).

29. P.P. Ellis and M.A. Riegel, *J. Ocul. Pharmacol.* 3, 121 (1987).

30. J.W. Sieg and J.R. Robinson, *J. Pharm. Sci.* 66, 1222 (1977).

31. J.W. Sieg and J.R. Robinson, *J. Pharm. Sci.* 65, 1816 (1976).

32. A. Bill, *Physiol. Rev.* 55, 383 (1975).

33. T.R. Morgan, K. Green, and K. Bowman, *Exp. Eye Res.* 32, 691 (1981).

34. D.F. Cole, *Doc. Ophthalmol.* 21, 116 (1966).

35. R.H. Maffly, R.M. Hays, E. Landin, and A. Leaf, *J. Clin. Invest.* 39, 630 (1960).

36. S.S. Chrai, M.C. Makoid, S.P. Eriksen, and J.R. Robinson, *J. Pharm. Sci.* 63, 333 (1974).

37. S.S. Chrai, T.F. Patton, A. Mehta, and J.R. Robinson, *J. Pharm. Sci.* 62, 1112 (1973).

38. K. Green and D.L. MacKeen, *Invest. Ophthalmol.* 15, 220 (1976).

39. K. Green and S. Downs, *Invest. Ophthalmol.* 13, 316 (1974).

40. S.S. Chrai and J.R. Robinson, *J. Pharm. Sci.* 63, 1218 (1974).

41. T.F. Patton and J.R. Robinson, *J. Pharm. Sci.* 64, 1312 (1975).

42. T.F. Patton and J.R. Robinson, *J. Pharm. Sci.* 65, 1295 (1976).

43. J.W. Sieg and J.R. Robinson, *J. Pharm. Sci.* 64, 931 (1975).

44. K. Green and A.M. Tonjum, *Am. J. Ophthalmol.* 72, 897 (1971).

45. C.S. O'Brien and K.C. Swan, *Arch. Ophthalmol.* 27, 253 (1942).

46. K. Green and A.M. Tonjum, *Acta Ophthalmol.* 53, 348 (1975).

47. A.M. Tonjum, *Acta Ophthalmol.* 53, 335 (1975).

48. R.R. Pfister and N. Burstein, *Invest. Ophthalmol.* 15, 246 (1976).

49. N.L. Burstein and S.D. Klyce, *Invest. Ophthalmol. Vis. Sci.* 16, 899 (1977).

50. K. Green and J.M. Chapman, *J. Toxicol. Cut. Ocul. Toxicol.* 5, 132 (1986).

51. K. Green, J. Chapman, L. Cheeks, and R.M. Clayton, *Drug-Induced Ocular Side Effects and Ocular Toxicology,* O. Hockwin, Ed. S. Karger, Basel, 126 (1987).

52. K. Green, L. Cheeks, and J.M. Chapman, *Ocular Drug Delivery: Biopharmaceutical, Technological and Clinical Aspects,* M.F. Saettone, M. Bucci, and P. Speiser, Eds. Liviana Press, Padova, 177 (1987).

53. E.J. Champeau and H.F. Edelhauser, *The Preocular Tear Film in Health, Disease, and Contact Lens Wear*, F.J. Holly, D.W. Lamberts, and D.L. MacKeen, Eds. Dry Eye Institute, Lubbock, TX, 292 (1986).

54. K. Green, D.S. Hull, E. Vaughn, A. Malizia, and K. Bowman, *Arch. Ophthalmol.* 95, 2218 (1977).

55. J.M. Megaw and S. Lerman, *Invest. Ophthalmol. Vis. Sci.* 28, 1429 (1987).

56. D.L. Krohn and J.M. Breitfeller, *Invest. Ophthalmol.* 13, 312 (1974).

57. H.E. Schaeffer and D.L. Krohn, *Invest. Ophthalmol. Vis. Sci.* 22, 220 (1982).

58. V.H.L. Lee, K.A. Takemoto, and D.S. Imoto, *Curr. Eye Res.* 3, 585 (1984).

59. V.H.L. Lee and J.R. Robinson, *J. Ocul. Pharmacol.* 2, 67 (1986).

60. R.E. Stratford, D.C. Yang, M.A. Redell, and V.H.L. Lee, *Curr. Eye Res.* 2, 377 (1983).

61. R.E. Stratford, D.C. Yang, M.A. Redell, and V.H.L. Lee, *Int. J. Pharm.* 13, 263 (1983).

62. H.E. Schaeffer, J.M. Breitfeller, and D.L. Krohn, *Invest. Ophthalmol. Vis. Sci.* 23, 530 (1982).

63. A. Jordan and J. Baum, *Ophthalmology* 87, 920 (1980).

64. R.M. Rossomondo, W.H. Carlton, J.H. Trueblood, and R.P. Thomas, *Arch. Ophthalmol.* 88, 523 (1972).

65. K. Green, K.A. Bowman, R.D. Elijah, R. Mermelstein, and R.W. Kilpper, *J. Toxicol. Cut. Ocul. Toxicol.* 4, 13 (1985).

66. G. Jacobs, M. Martens, and J. De Beer, *J. Toxicol. Cut. Ocul. Toxicol.* 6, 109 (1987).

67. J.F. Griffith, G.A. Nixon, R.D. Bruce, P.J. Reer, and E.A. Bannan, *Toxicol. Appl. Pharmacol.* 55, 501 (1980).

68. S.J. Williams, G.J. Graepel, and G.L. Kennedy, *Toxicol. Lett.* 12, 235 (1982).

69. S.J. Williams, *Food Chem. Toxicol.* 23, 189 (1985).

70. H.A. Bradford, G.L. Kennedy, S.C. Pennis, H. North-Root, L.C. Dispasquale, D.A. Penney, T. Re, H.J. Scherke, and J. Dinardo, *J. Toxicol. Cut. Ocul. Toxicol.* 5, 215 (1986).

71. M.G. Doane, A.D. Jensen, and C.H. Dohlman, *Am. J. Ophthalmol.* 85, 383 (1978).

72. M.A. Watsky, M.M. Jablonski, and H.F. Edelhauser, *Curr. Eye Res.* 7, 483 (1988).

73. C. Wei, J.A. Anderson, and I. Leopold, *Invest. Ophthalmol. Vis. Sci.* 17, 315 (1978).

74. C.C. Chen, J. Anderson, M. Shackleton, and J. Attard, *J. Ocul. Pharmacol.* 3, 149 (1987).

75. C.J. Schmitt, V.J. Lotti, and J.C. Le Douarec, *Arch. Ophthalmol.* 98, 547 (1980).

76. A. Kupferman and H.M. Leibowitz, *Am. J. Ophthalmol.* 82, 109 (1976).

77. R.M. Clayton, K. Green, M. Wilson, A. Zehir, J. Jack, and L. Searle, *Food Chem. Toxicol.* 23, 239 (1985).

78. H.E. Seifried, *J. Toxicol. Cut. Ocul. Toxicol.* 5, 89 (1986).

79. T. Kaila, L. Salminen, and R. Huuponen, *J. Ocul. Pharmacol.* 1, 79 (1985).

80. J.A. Anderson, *Arch. Ophthalmol.* 98, 350 (1980).

81. V. Kumar, R.D. Schoenwald, W.A. Barcellos, D.S. Chien, J.C. Folk, and T.A. Weingeist, *Arch. Ophthalmol.* 104, 1189 (1986).

82. E. Duzman, J. Anderson, J.B. Vita, J.C. Lue, C.C. Chen, and I.H. Leopold, *Arch. Ophthalmol.* 101, 1122 (1983).

83. C.H. Chiang and R.D. Schoenwald, *J. Pharmacokinet. Biopharm.* 14, 175 (1986).

84. R.G. Janes and J.F. Stiles, *Am. J. Ophthalmol.* 56, 84 (1963).

85. C.W. Chiang, G. Barnett, and D. Brine, *J. Pharm. Sci.* 72, 136 (1983).

86. S.C. Chang and V.H.L. Lee, *J. Ocul. Pharmacol.* 3, 159 (1987).

87. T.J. Zimmerman, K.S. Kooner, A.S. Kandarakis, and L.P. Ziegler, *Arch. Ophthalmol.* 102, 551 (1984).

88. F.T. Fraunfelder, *Trans. Am. Ophthalmol. Sci.* 74, 457 (1976).

89. F.T. Fraunfelder and S.M. Meyer, *J. Ocul. Pharmacol.* 3, 177 (1987).

90. M.S. Passo, E.A. Palmer, and E.M. van Buskirk, *Ophthalmology* 91, 1361 (1984).

91. M.E. Langham, *Br. J. Ophthalmol.* 43, 705 (1959).

92. B.S. Kwon, L.P. Gangarosa, K. Green, and J.M. Hill, *Invest. Ophthalmol. Vis. Sci.* 22, 818 (1982).

93. R.F. Brubaker and J.W. McLaren, *Ophthalmology* 92, 884 (1985).

94. R.F. Brubaker, S. Nagataki, D.J. Townsend, R.R. Burns, R.G. Higgins, and W. Wentworth,

Ophthalmology 88, 283 (1981).

95. R.F. Jones and D.M. Maurice, *Exp. Eye Res.* 5, 208 (1966).

96. J.E. Pederson, D.E. Gaasterland, and H.M. McLellan, *Invest. Ophthalmol. Vis. Sci.* 17, 190 (1978).

97. J.M. Hill, Y. Haruta, and D.S. Rootman, *Curr. Eye Res.* 6, 1065 (1987).

98. L. Hughes and D.M. Maurice, *Arch. Ophthalmol.* 102, 1825 (1984).

99. M. Barza, C. Peckman, and J. Baum, *Ophthalmology* 93, 133 (1986).

100. M. Barza, C. Peckman, and J. Baum, *Invest. Ophthalmol. Vis. Sci.* 28, 1033 (1987).

101. G.L. Ruskell, *The Rabbit in Eye Research*, J.H. Prince, Ed. Charles C Thomas, Springfield, IL, 514 (1964).

102. A. Kupferman and H.M. Leibowitz, *Arch. Ophthalmol.* 92, 329 (1974).

103. W.E. Plehwe, P.S. Chahal, T.J. Fallon, J.R. Cunningham, M.J. Neal, and E.M. Kohner, *Exp. Eye Res.* 44, 209 (1987).

104. C. Ohrloff and M. Spitznas, *Graefes Arch. Klin. Exp. Ophthalmol.* 225, 84 (1987).

105. J.R. Gray, M.A. Mosier, and B.M. Ishimoto, *Graefes Arch. Klin. Exp. Ophthalmol.* 222, 225 (1985).

106. S.R. Waltman and H.E. Kaufman, *Invest. Ophthalmol.* 9, 247 (1970).

107. R.F. Brubaker and R.L. Coakes, *Am. J. Ophthalmol.* 86, 474 (1978).

108. M.C. Grayson and A.M. Laties, *Arch. Ophthalmol.* 85, 600 (1971).

109. S.C. Edberg and L.D. Sabath, *Antibiotics in Laboratory Medicine*, V. Lorian, Ed. Williams & Wilkins, Baltimore, 201 (1980).

110. W.M. Jay, P. Fishman, M. Aziz, and R.K. Shockley, *J. Ocul. Pharmacol.* 3, 257 (1987).

111. P. Vareilles, D. Silverstone, B. Plazonnet, J.C. Le Douarec, M.L. Sears, and C.A. Stone, *Invest. Ophthalmol. Vis. Sci.* 16, 987 (1977).

112. R.L. Radius, G.R. Diamond, I.P. Pollack, and M.E. Langham, *Arch. Ophthalmol.* 96, 1003 (1978).

113. L. Bonomi, S. Perfetti, E. Noya, R. Bellucci, and F. Massa, *Graefes Arch. Klin. Exp. Ophthalmol.* 210, 1 (1979).

114. S.P. Bartels, H.O. Roth, M.M. Jumblatt, and A.H. Neufeld, *Invest. Ophthalmol. Vis. Sci.* 19, 1189 (1980).

115. A. Bar-Ilan, *Curr. Eye Res.* 3, 1305 (1984).

116. P.L. Kaufman and E.H. Barany, *Invest. Ophthalmol.* 15, 793 (1976).

117. J.V. Thomas and D.L. Epstein, *Br. J. Ophthalmol.* 65, 596 (1981).

118. M.N. Cyrlin, J.V. Thomas, and D.L. Epstein, *Arch. Ophthalmol.* 100, 414 (1982).

119. R.G. Higgins and R.F. Brubaker, *Invest. Ophthalmol. Vis. Sci.* 19, 420 (1980).

120. D.J. Townsend and R.F. Brubaker, *Invest. Ophthalmol. Vis. Sci.* 19, 256 (1980).

121. W.H. Havener, *Ocular Pharmacology*, C.V. Mosby, St. Louis, (1974).

122. B. Becker, *Am. J. Ophthalmol.* 37, 13 (1954).

123. J. Caprioli, M. Sears, and A. Mead, *Exp. Eye Res.* 39, 1 (1984).

124. T.R. Morgan, D. Mirate, K. Bowman, and K. Green, *Arch. Ophthalmol.* 101, 112 (1983).

125. J.A. Anderson, W.L. Davis, and C.P. Wei, *Invest. Ophthalmol. Vis. Sci.* 19, 817 (1980).

126. R.D. Tamaru, W.L. Davis, and J.A. Anderson, *Arch. Ophthalmol.* 101, 1127 (1983).

127. E.M. van Buskirk and J.R. Samples, *Ophthalmology* 92, 811 (1985).

128. E.M. van Buskirk, R.N. Weinreb, D.P. Berry, J.S. Lustgarten, S.M. Podos, and M.M. Drake, *Am. J. Ophthalmol.* 101, 531 (1986).

129. R. Mechoulam, N. Lander, M. Srebnik, I. Zamir, A. Breuer, B. Shalita, S. Dikstein, E.A. Carlini, J.R. Leite, H. Edery, and G. Porath, *The Cannabinoids: Chemical, Pharmacologic and Therapeutic Aspects*, S. Agurell, W.L. Dewey, and R.E. Willette, Eds. Academic Press, New York, 777 (1984).

130. V.H.L. Lee, K.W. Morimoto, and R.E. Stratford, *Biopharm. Drug Disp.* 3, 291 (1982).

131. D.M. Maurice and S. Mishima, *Pharmacology of the Eye*, M.L. Sears, Ed. Springer-Verlag, New York, 19 (1984).

132. K. Green, J.M. Chapman, L. Cheeks, R.M. Clayton, A. Zehir, and M. Wilson, *J. Toxicol. Cut. Ocul. Toxicol.* 6, 89 (1987).

133. V.H.L. Lee and J.R. Robinson, *J. Pharm. Sci.* 68, 673 (1979).
134. S.C. Miller and T.F. Patton, *Biopharm. Drug Disp.* 2, 215 (1981).
135. S.C. Miller and T.F. Patton, *Biopharm. Drug Disp.* 3, 115 (1982).
136. M. Francoeur, I. Ahmed, S. Sitek, and T.F. Patton, *Int. J. Pharm.* 16, 203 (1983).
137. P. Lapalus, D. Fredj-Reygrobellet, and F. Bourett, *Drug-Induced Ocular Side Effects and Ocular Toxicology,* O. Hockwin, Ed. S. Karger, Basel, 215 (1987).
138. J.W. Shell, *J. Toxicol. Cut. Ocul. Toxicol.* 1, 49 (1982).

18

Alternative Methods for Assessing the Effects of Chemicals in the Eye

LEON H. BRUNER
The Procter & Gamble Company
Miami Valley Laboratories
Cincinnati, Ohio

JOHN SHADDUCK
Office of the Dean
College of Veterinary Medicine
Texas A & M University
College Station, Texas

AND

DIANE ESSEX-SORLIE
College of Veterinary Medicine
University of Illinois
Urbana, Illinois

I. THE NEED FOR ALTERNATIVES

The pursuit of alternative methods for the evaluation of ocular irritation responses has become an area of increasing interest during the past decade. This is because the standard *in vivo* tests have been criticized for their subjectivity, for their over-prediction of human response, and their use of animals. Furthermore, recent advances in biotechnology have provided a new range of biological endpoints that has made the development of viable alternatives feasible. This research has resulted in development of a variety of techniques that may provide useful information in the ocular safety assessment process.

Alternative methods in ocular toxicology, in the broadest interpretation, include any test that refines, reduces, or replaces a standard *in vivo* test. Reduction involves any change in method or procedure that decreases the number of animals used per test. Refinements are changes in procedure that decrease any discomfort of the test animals. Replacements are alterations in procedure that eliminate the need for

585

animals in testing. This chapter will present a discussion of proposed replacement tests that use living cells, tissues, or recently enucleated eyes evaluated *in vitro*. Recent reviews cover in considerable detail many of the points to be touched upon in this chapter and the interested reader is referred to these sources for additional information.[1,2]

Although several individual tests will be described, it should be noted at the outset that it is unlikely that a single test can provide a complete replacement to rabbit eye irritation tests. The eye is complex, both structurally and functionally, and is capable of a wide variety of responses to injurious insults, several of which are evaluated in ocular irritation tests.[3-9] Thus, it is most likely that batteries of tests will be needed to evaluate the many different classes of materials that are assessed for ocular safety.[1,2,10,11] For example, it is quite likely that an *in vitro* acute cytotoxicity test will be an accurate single measure of the capability of a material to injure the eye via direct cytolytic damage of corneal epithelial cells. Such a test, however, probably would be inadequate if the injury produced by the test compound is via noncytotoxic release of chemical mediators of inflammation. Since very few classes of chemicals act in simplistic ways on complex tissues, it is unlikely that a single test will suffice if development of complete replacement tests is to be achieved.

II. CHARACTERISTICS OF *IN VITRO* ALTERNATIVES

A. Advantages and Disadvantages of *In Vitro* Testing

Not unexpectedly, there are both advantages and disadvantages associated with the use of *in vitro* tests for ocular irritation. The primary advantage of developing viable *in vitro* alternatives for ocular irritation testing is that there is a good possibility they may ultimately eliminate the need for animals.[4] In addition to this important value, there are other attractive features that make developing *in vitro* alternatives desirable. For example, *in vitro* tests often are more controllable and reproducible than the *in vivo* ocular irritation tests. Furthermore, the mechanisms underlying the tests may be better understood. Additionally, *in vitro* tests often lend themselves to the collection of large numbers of observations which can be obtained relatively rapidly, offering a particularly satisfactory system to study dose-response relationships. Replacing the animal test would also eliminate the necessity of having trained observers who are competent to evaluate and record animal responses and ensuring that their evaluations are consistent from animal to animal and test to test.[4,12]

Disadvantages of *in vitro* tests also are distinct and significant. It is not yet possible to reproduce either the complex ocular tissues or all their characteristic responses *in vitro*. Several responses of the eye to injury are very difficult to measure in today's *in vitro* assays.[13] Examples include repair of injury caused by exposure to various test substances, and inflammation (Chapters 15 and 16) produced as a result of ocular immune responses and other effects. Further, it seems unlikely that some responses of particular importance in man, such as tearing and pain,[14] will be easily reproduced *in vitro*. In addition, the task of developing *in vitro* tests that will accurately predict human ocular irritation is complicated by the fact that very little is known about the cellular or molecular mechanisms by which ocular irritants act in either man or rabbits.[5] Some have argued that knowledge of mechanism of action is not required for an *in vitro* test to be useful, and this may be true when the test is limited to a group of materials with similar modes of action. If tests are to be used broadly however, increased knowledge of the mechanistic basis for the tests will be required.[1,15-17]

Other disadvantages include the significantly increased cost that will result from adoption of *in vitro* techniques for ocular irritation testing. This increased cost results from the additional time that is required to conduct an *in vitro* test battery, the need for highly skilled technicians to perform the assays, and for instrumentation required for conducting the testing. Another important impediment to the use of *in vitro* tests today is that large-scale validation studies have not been completed. Such validation studies will be required before alternative tests will be accepted by regulatory agencies.[1,15,16]

The absence of a large human database from which to extrapolate and the difficulty of relating *in vitro* test results with human experiences are additional problems associated with alternative tests. The purpose and goal of conducting alternative ocular irritation tests is to predict accurately and reliably the human risk potential posed by the test material. Of necessity, the data against which all *in vitro* alternative tests are now compared are derived from rabbit ocular irritation tests and it is difficult to extrapolate the animal test results to man (Chapter 16). *In vitro* tests are even further removed from direct human responses. Thus, the extension of *in vitro* data to predict human response will require considerable systematic investigation and validation before investigators are confident that one or several *in vitro* tests are satisfactory alternatives.[15-18]

B. Differences Between *In Vivo* and *In Vitro* Tests

There are several differences between the *in vivo* ocular irritation test and *in vitro* tests that may have significant effects on the development of viable alternatives for ocular irritation testing. Usually only one concentration of test material is applied directly to the cornea in the *in vivo* test. The effect of the test material is then evaluated by observing multiple responses of the eye and associated vessels, secretory tissues, and accessory glands. The irritancy potential of a test substance is recorded as the sum of clinically detectable changes in integrity of the corneal epithelium, corneal edema, episcleral and iridal vascular engorgement and edema, and periocular discharges[9,19] (Chapter 16). Very frequently, the mechanisms by which these responses are initiated are unknown and it is almost never clear which are primary responses induced by the test material and which are secondary.

In vitro, this situation is typically quite different. Generally, *in vitro* systems are biologically simpler than the eye and therefore produce a more limited range of responses. *In vitro*, the test material is prepared in a series of diluted samples, and a concentration-dependent endpoint, such as an ED_{50}, is determined. Sometimes other endpoint approximations are obtained, such as the rate of occurrence of a particular biological event. The incorporation of a labeled precursor into a macro-molecule, or measuring the rate of an enzyme-mediated reaction are typical examples.[1] Whether these dose-dependent endpoints are related to events occurring *in vivo* is still unclear and a critical issue that requires additional research.

C. Attributes of the Ideal *In Vitro* Alternative Test

There is currently no ideal *in vitro* alternative test. It is useful to identify at least some of the attributes of an ideal test. Such a definition provides a frame of reference against which systems can be developed. Frazier and colleagues[1] have suggested that *in vitro* alternatives should be evaluated on the basis of (a) logistics, (b) economy, and (c) scientific merits.

Logistics include considerations related to the standardization of the test, whether the test is transportable across laboratories, and the commercial availability of test components. An ideal test will produce results that are reproducible consistently across time, both within and between laboratories. It also must be possible to conduct the test with reasonable speed and it must not be so complex that an operator cannot conduct multiple tests simultaneously in a relatively short period of time. The ideal test also should allow the evaluation of test substances in a broad range of physical forms. The tests should be designed to permit examination of responses relative to internal standards. Furthermore, the observed responses are best when they can be interpreted relative to background, such as might be associated with cell damage or lysis due to normal handling and washing of cells in culture. Also, observed responses must be viewed in relation to the maximum potential response, such as total lysis or destruction.

Economy is reflected in the cost of the test system. Factors that must be considered include operational, instrument, reagent, test subject, operator, and data analysis costs. The instrumentation required should not be overly complex, and the reagents, cell lines, and other materials required should be readily available, safe to use, and easily destroyed at the end of the study. An ideal *in vitro* test should be efficient in design, meaning that a balance should exist between the precision of estimation, quality of validity tests, number of dilutions, and number of observations at each dilution.

Evaluation of the scientific merits of a test includes consideration of whether the test is mechanistically based (evaluates a critical step in the irritation/injury response) or correlation based (measures a parameter which is associated with the response). Mechanistically based tests are more likely to be readily interpretable and accepted by the scientific and regulatory communities.[1,15,16] However, because so little is known about the mechanisms that lead to chemically induced ocular injury, development of correlation-based tests has been a necessary prelude to the development and refinement of the mechanistically based alternatives that are ultimately needed. As additional research is conducted, greater understanding of the important mechanisms will be gained. From that understanding, better, more highly predictive *in vitro* tests will be developed.

In addition to the factors already described, there are several important characteristics relating specifically to the data available from the *in vitro* tests that should be considered. First, *in vitro* tests should have an objectively defined and quantitatively estimated endpoint. These endpoints should be parametric, rather than nonparametric, and be measured continuously, rather than categorically. Second, the estimated endpoint should be adequately precise. Third, a numerical assessment of the precision of the estimated endpoint should be obtained,[20] and fourth, the tests of the fundamental validity of the assay and the statistical analyses should be included.[20]

The need for the first characteristic should be obvious. Without an objectively defined endpoint, the probability is minimal that different operators, across time, within the same laboratory or between laboratories, will describe the same outcome. Further, if one is to assess the extent to which *in vitro* results are reliable predictors of either animal or human responses, then *in vitro* responses must be quantitative; appropriate statistical analyses must be utilized systematically to establish beyond chance levels that such predictions are indeed justifiable.

The second and third characteristics relate to the precision of the estimated endpoint. Precision describes the ability of a test to properly predict the endpoint of interest. The *in vitro* assays must accurately identify genuinely unsafe (injurious, toxic) materials or chemicals as unsafe. Also, the tests must have the capacity to accurately identify genuinely safe (noninjurious, nontoxic) materials or chemicals as safe. Tests meeting these criteria will provide product developers and regulators assurance that consumers will not be exposed to dangerous materials. Furthermore, losses caused by elimination of safe and useful products by falsely classifying them as dangerous materials would be minimized.

When estimating an endpoint obtained from an *in vitro* test, it is important to generate and use all of the data available. For example, it is possible to generate an assay endpoint (usually an estimated ED_{50} value) based on a very limited number of test material concentrations.[21] This type of evaluation has limits because it only provides a small part of all the information available from a full dose-response curve. If a well-defined dose-response curve is generated during the assay procedure, the ED_{50} concentration can be determined with precision, and other useful data such as the shape and slope of the dose-response curve can be obtained. These data, although not widely used by developers of *in vitro* alternatives, may provide a wealth of additional information on test substance characteristics that may be useful for making a safety assessment with the *in vitro* test. For example, dose-response curve slope data may be useful in grouping materials according to common modes of action for further evaluation. Also, the additional data may be useful for pointing the way to further experiments that will help elucidate the mechanism(s) by which the test materials irritate and injure target tissues.

The fourth characteristic, a numerical assessment of the precision of the estimated endpoint in the form of confidence intervals or other error term, permits one to determine whether two materials with similar effects can be differentiated reliably. This final characteristic, that of validity of both the assay and statistical analysis, is important. If the assay itself fails to meet the conditions essential to its conduct, then no statistical analysis, regardless of complexity, can render the results meaningful.

Further, if the analysis is predicated on conditions that cannot be met reasonably by the data (such as linearity for a linear regression analysis of a dose-response relationship), then the estimated endpoint and the characterization of the dose-response curve may not truthfully represent the observed responses.

III. APPROACHES TO DEVELOPING ALTERNATIVES TO OCULAR IRRITATION TESTS

Frazier et al.[1] have conveniently organized the current *in vitro* alternative methods into categories based on the endpoints measured (Table 1). This overview is quite useful and offers an easy appreciation of the various approaches currently under evaluation. Since the monograph was published, more data have accumulated on the *in vitro* assays and several new tests have been developed and evaluated. The following provides an update on the currently available tests using this framework as a guide for discussion.

The *in vitro* assays proposed as alternatives for ocular irritation testing range from simple to relatively complex. For example, one complex test uses as its endpoint the population of cells that survive the treatment with test substance, remain adhered to an underlying substrate (a plastic dish), and proliferate *in situ* to sufficient numbers to be scored as a colony.[22] Conversely, a more simple test measures membrane integrity and requires only that injured cells release an intracytoplasmic marker.[23] Although one might predict that complex tests rank materials more accurately; this is not necessarily the case. Indeed, several of the simple cytotoxicity assays seem to compare well with the results of *in vivo* ocular irritation tests. This may be expected, in part, by the fact that some of these tests have been evaluated with classes of compounds that are likely to be acutely cytotoxic *in vivo*.[23] The complexity of the ocular irritation response has already been noted; it is again important to appreciate that a battery of tests that evaluate different endpoints may be required to properly predict the ocular irritancy potential of a wide variety of test materials.

The problems associated with the development of *in vitro* alternatives for the *in vivo* ocular irritation test were introduced earlier. Much research effort will be required to deal with these problems before the *in vivo* test can be completely replaced. Some of the problems that make development of *in vitro* alternatives for ocular irritation particularly difficult include the following: (1) *in vivo* data are not easy to generate consistently, making comparisons with *in vitro* results very difficult; (2) the *in vivo* data are subjective assessments of injury even though each observation is converted into a numerical score;[4] (3) mechanisms of injury are poorly or not at all understood in either the animal[13] or *in vitro* systems; (4) rabbit eye irritation tests were not designed to discriminate among materials of similar irritancy potential;[19] (5) data generated from some *in vivo* procedures do not properly predict the human experience;[24,25] and (6) the rabbit eye irritation tests do not lend themselves to statistical analyses much beyond counts of the occurrence of different scores (Chapter 13).

With all of these problems, how will it be possible to evaluate and validate *in vitro*

Table 1
List of Test Protocols

Morphology

Enucleated Superfused Rabbit Eye
Balb/c 3T3 Cells/Morphological Assays (HTD)

Cell Toxicity Assays

Adhesion/Cell Proliferation
    BHK Cells/Growth Inhibition
    BHK Cells/Colony Formation Efficiency
    BHK Cells/Cell Detachment
    SIRC Cells/Colony Forming Assay
    Balb/c 3T3 Cells/Total Protein
    BCL D1 Cells/Total Protein
Membrane Integrity
    LS Cells/Dual Dye Staining
    Thymocytes/Dual Fluorescent Dye Staining
    LS Cells/Dual Dye Staining
    RCE SIRC P815 YAC-1/Cr Release
    L929 Cells/Cell Viability
    Bovine Red Blood Cell/Hemolysis
    MDCK Cells/Transepithelial Barrier Integrity
Cell Metabolism
    Rabbit Corneal Cell Cultures/Plasminogen Activator
    LS Cells/ATP Assay
    Balb/c 3T3 Cells/Uridine Uptake Inhibition Assay
    Balb/c 3T3 Cells/Neutral Red Uptake
    Balb/c 3T3 Cells/Neutral Red Release
    HeLa Cells/Metabolic Inhibition Test (MIT-24)
    NEHK/Silicon Microphysiometer
    Photobacterium phosphoreum/Luminescent Bacterial Test (LBT)

Cell and Tissue Physiology

Epidermal Slice/Electrical Conductivity
Rabbit Ileum/Contraction Inhibition
Bovine Cornea/Corneal Opacity
Proposed Mouse Eye/Permeability Test

Inflammation/Immunity

Chorioallantoic Membrane (CAM)
    CAM
    CAMVA
    HET-CAM
Bovine Corneal Cup Model/Leukocyte Chemotactic Factors
Rat Peritoneal Mast Cells/Histamine Release
Rat Peritoneal Mast Cells/Serotonin Release
Rat Vaginal Explant/Prostaglandin Release
Bovine Eye Cup/Histamine (Hm) and Leukotriene C4 (LT-C4) Release

Table 1 (Continued)
List of Test Protocols

<hr>

Recovery/Repair

Rabbit Corneal Epithelial Cells/Wound Healing

Other

Protein Denaturation Procedures
Computer Based/Structure Activity Relationship (SAR)
Tetrahymena thermophila Motility

<hr>

From Frazier et al., *Alternative Methods in Toxicology*, Vol. 4, Mary Ann Liebert, New York (1987). With permission.

<hr>

methods for use in the ocular safety assessment process? Significant effort has been directed toward developing validation schemes that will provide data necessary for introduction of alternative assays including those for ocular irritation assessment.[2,15-17,26] A practical approach recently described by Wilcox and Bruner[2] has proven useful for evaluating currently available *in vitro* procedures, providing enough information to allow incorporation of *in vitro* testing into an ocular safety assessment program.[27] Because of its practical application for currently available technology, the process will be described below. The three steps followed are preliminary evaluation, parallel testing and expanded test set evaluation and tier testing.

Preliminary evaluation — This first step is designed to identify a limited number of *in vitro* tests that may be useful in the ocular safety assessment of relevant test substances. This evaluation is accomplished by testing a small set of materials in several promising *in vitro* assays. The tests considered may be those already developed by others, or may represent new technology. After the results from each assay are obtained, the results can be interpreted independently and in comparison with the other tests to determine whether they are internally consistent, i.e., whether the same endpoint is generated repeatedly for the same material,[28] if any of the tests predict ocular irritancy correctly, and if any tests appear to be more predictive than others. There are now several examples of studies where these types of evaluations have been completed.[29-31]

Parallel testing and expanded test set evaluation — Since the preliminary evaluation is generally undertaken with a limited number of materials, additional testing must be done to prove the utility of the identified assays. If the developer of *in vitro* alternatives is part of an organization that is conducting *in vivo* ocular irritation tests, any material assayed *in vivo* should be evaluated in parallel in the selected *in vitro* assays. This should be done for a time long enough to provide necessary additional data supporting the utility of the assays. If the developer has no access to *in vivo* testing, then another set of well-characterized materials must be assembled and evaluated *in vitro*. This additional testing increases the numbers of materials examined in both *in vivo* and *in vitro* tests allowing assessment of the correctness of the conclusions reached in the preliminary evaluation. It also allows toxicologists an opportunity to learn how to use *in vitro* data in the ocular safety assessment process, where the tests are useful, and when the tests fail to be predictive. When enough data and experience are gained, it allows implementation of the next step in the process called tier testing.

Tier testing — Tier testing (Figure 1) is the process where *in vitro* alternatives are actually incorporated into ocular safety assessment.[2,10,29] This process begins with an evaluation of historical eye irritation data and composition of the product. If it is possible to make a safety assessment based on that information alone, then no additional testing is needed. If additional data are necessary, then the materials are assayed in the *in vitro* test battery. The results from the *in vitro* assays are then evaluated relative to all the other available data, and if possible, a safety assessment is made. If the *in vitro* data are insufficient to make an assessment, then *in vivo* testing with limited numbers of animals becomes necessary.

Under such a scenario, the *in vitro* results would provide just one part of all the information needed to make a safety assessment. This process will allow use of currently available *in vitro* alternatives to refine *in vivo* testing procedures and reduce the numbers of animals required to make a safety decision. As work is done with the *in vitro* systems, it will provide much information on when the assays predict ocular irritation correctly and when they do not. As more predictive, mechanism-based tests are developed and validated, it may be possible to place more confidence in the *in vitro* test results, further decreasing the need for animals. Ultimately it may be possible to completely eliminate the need for animals in ocular safety assessment.

A. Functional Descriptions

Having made these general remarks regarding development of alternative tests, it will now be useful for us to consider some of the specific *in vitro* assays that are being evaluated as *in vitro* alternatives for ocular irritancy testing. The following sections provide brief descriptions of each of the general approaches listed in Table 1, followed by more detailed information about selected techniques in each test category.

1. Morphologic Endpoints

Morphologic methods use grossly or microscopically visible alterations in tissue or cell structure as endpoints. Two major systems have been evaluated. In one, changes in the thickness of rabbit corneas are evaluated using enucleated, perfused rabbit eyes[32-35] or isolated bovine corneas.[36] In the second, morphologic changes in living cells in culture are scored.[37] The enucleated, perfused rabbit eye and bovine cornea system offer the advantage of being able to observe significant functional changes in the thickness of the cornea which may reflect important effects of ocular irritants *in vivo*. The difficulties with these systems are the technical skills required to conduct the assays, the relative complexity of the isolated eye procedures, and

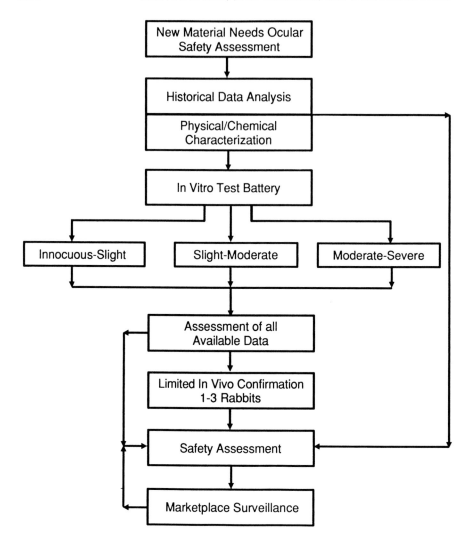

Figure 1. Tier Testing Process. This diagram indicates how *in vitro* alternatives may be incorporated into the ocular safety assessment process. An evaluation of all previous testing data and physical and chemical characteristics is completed first. If necessary, the materials are then evaluated in a battery of *in vitro* assays. After the *in vitro* testing, all the data would be assessed again. Then either a safety assessment would be made or an *in vivo* test in a limited number of animals would be performed prior to making the final safety assessment.

occasional lack of agreement in rank order between *in vivo* and *in vitro* tests.[38] Further, the test requires the sacrifice of animals, unless eyes can be obtained from animals killed for other purposes.

The tissue culture morphology system is useful because of its technical simplicity and the rapidity with which results can be obtained. It may serve as an excellent screening tool to set doses for other more definitive *in vitro* procedures. Morphologic observations are difficult to quantify consistently making this test most accurately performed by persons who have experience interpreting changes in cells in culture.

Selected technique — Highest tolerated dose.[37,39-41] BALB/c 3T3 cells are seeded into 96-well culture plates at densities sufficient to obtain semiconfluent cultures 24 h after seeding. Growth medium is removed and replaced with medium containing the test chemicals; then, the cells are incubated in the presence of the chemical for 24 h. Cells are observed microscopically and evaluated for morphologic alterations using phase contrast microscopy. Various concentrations of the test chemical are used, and the highest concentration of the chemical that does not produce morphological alterations is designated the highest tolerated dose. Unexposed BALB/c 3T3 cells are used for comparison. The indices evaluated include decreased cell density, increased cell roundedness, and cellular cytoplasmic granularity. In terms of ranking a chemical relative to others in a set, agreement with other *in vitro* tests and rabbit eye irritation test data has been good in a limited number of evaluations.

2. Cell Toxicity Assays

Cell toxicity assays can be categorized into three groups: (1) adhesion/cell proliferation; (2) membrane integrity; and (3) cell metabolism. Several assays have been developed in each of these groups. The advantages of cell toxicity assays include their technical simplicity and the fact that endpoints are usually quantitative and objective. Most cytotoxicity tests can be performed relatively rapidly. Drawbacks to individual tests include the use of radioisotopes and the requirement for complex, expensive equipment. A significant amount of work has been done with these assay systems. The results indicate many are promising procedures.

a. Cell Adhesion and Cell Proliferation Assays

Several assays have been developed which have some common features.[11,22,42-45] Usually, a well-characterized, anchorage-dependent cell line is used in a typical tissue culture system. Cells are exposed to the test material, the test substance is washed away, and after a recovery period, an endpoint is measured. Parameters evaluated include growth inhibition, colony formation efficiency, cell detachment, or total cellular protein remaining after treatment.

Selected technique — SIRC Colony Forming Assay.[22,46] The SIRC colony-forming assay is an example of a test that requires both adhesion and proliferation of cells to obtain the endpoint. SIRC cells are an established line derived from rabbit corneas available from the American Type Culture Collection. Four hundred SIRC cells are plated on 60 mm culture dishes in Ham's F12S medium. After 18 h of incubation at 37°C, cells are washed twice with medium, and fresh medium containing the test chemical is added to 3 plates for each concentration of chemical tested. After 1 h of additional incubation at 37°C, the cells are washed, provided fresh culture medium, and incubated for 7 to 8 days to allow surviving cells to form visible colonies. Cells are fixed, stained, and the colonies (10 or more cells) are counted and compared with the number of colonies in untreated control plates. The endpoint is expressed as the concentration of the test chemical giving 50% survival relative to the control.

In order to be scored as a colony, cells must survive exposure to the chemical (e.g., not undergo lysis), must remain adherent, and be capable of multiplying (or migrating) since a colony is scored only if 10 cells or more are grouped together. There is some subjectivity since standard criteria must be developed by the operator who counts the colonies. The colonies formed by SIRC cells tend to be somewhat loose and diffuse, so deciding precisely what constitutes a colony is a source of potential variation between both operators and laboratories. Nevertheless, two independent laboratories have ranked the same compounds similarly,[28] and ranks obtained with this test agree with ranks of the same materials tested in rabbit eyes.

b. Membrane Integrity Assays

There are several tests that use one or more markers of membrane integrity as a toxicity endpoint.[23,28,47-52] Some tests use cells freshly collected from animal tissues (e.g., rat thymus or bovine red blood cells), but most use established cell cultures. Systems for assaying membrane integrity include hemoglobin release from red cells,[52] release of chromium-tagged macromolecules following chromium labeling of the target cells,[23] retention of fluorescein diacetate in viable cells,[47,48] and uptake of ethidium bromide or propidium iodide by dead cells.[47-49]

Each of these tests has an objective endpoint that can be quantified, and each lends itself to multiple repeat tests of many concentrations of each material. These assays tend to be technically simple, reproducible, and relatively inexpensive. Several studies have shown good agreement when materials were ranked for ocular irritancy in rabbit eyes and cytotoxicity *in vitro*. Although it is often assumed that the endpoints of these *in vitro* tests are the result of events that also occur *in vivo*, this is not proven in most cases.

Selected technique — P815-^{51}Cr release assay.[23,28] The P815 cytotoxicity test uses a continuous line of murine mastocytocytoma cells which grow in suspension that also are available from the American Type Culture Collection. The cells are labeled with radioactive chromium (^{51}Cr) in the form of sodium chromate (1 μCi ^{51}Cr/10^7 cells). Cells are labeled for 4 h and dispensed into 96-well microtiter plates at 50,000 cells per well. Cells are exposed for 2 or 4 h, in the absence of serum, to various concentrations of the test chemical. At the end of the exposure period, microtiter plates are centrifuged and supernatants are collected and counted in a gamma counter. Background is determined by measuring the amount of ^{51}Cr released from cells cultured in medium free of the test chemical. A maximum (100% release) is determined by completely lysing the cells with a detergent. The endpoint is calculated by determining the radioactive counts in the supernatant of each test well minus the number of counts released by normal cells. This value is expressed as a percent of maximum release minus background. The release caused by multiple doses of test materials is measured, and the concentration of test substance that causes 50% release of the ^{51}Cr (CD_{50}) is the endpoint of the test.

This assay provides a quantitative and objective endpoint. It is also rapid, technically simple, and uses a readily available cell. Its major disadvantage is that it uses a radioactive isotope. Endpoints obtained in this test are similar to those obtained in the SIRC colony-forming assay, bovine red cell hemolysis test, and the dual fluorescence test (unpublished results, Shadduck et al.). The P815 cytotoxicity test correctly ranked most materials when *in vitro* CD_{50} values and rabbit eye irritation scores were compared (unpublished results, Shadduck et al.).

c. Epithelial Barrier Integrity Assays

Integrity of the corneal epithelium is important to prevent water movement into the cornea. The barrier is maintained by tight junctions and desmosomes that form between the epithelial cells. When the cells of the corneal epithelium are damaged, this barrier can be lost resulting in corneal edema. An *in vitro* assay system has been developed that attempts to model this function.[53] The assay uses a renal tubular epithelial cell called Madin-Darby canine kidney cells (MDCK). When MDCK are grown on a permeable support, tight junctions form between the cells that prevent movement of large molecules across the monolayer. If this monolayer is damaged by a test substance, the permeability barrier can be disrupted. The results of Tchao[53] suggest this test may be useful for broadly classifying irritancy potential of test materials. Shaw et al.[54] have also evaluated a modification of the Tchao procedure that involves exposing MDCK cells to high concentrations of test substance for short periods of time. Their results also support the potential utility of this approach.

d. Cell Metabolism Assays

Both primary cell cultures and continuous cell cultures have been used for these assays.[29,37,41,55-62] Endpoints include evaluation of plasminogen activator activity,[56] ATP production,[57] uridine uptake inhibition,[58,59] neutral red uptake,[37,41] and pH change.[60] Recently, two other promising assays in this category have been evaluated. These are the silicon microphysiometer[29,61-63] and the luminescent bacteria test.[29,63,64] Assays that measure metabolic endpoints offer a number of advantages, most notably the potential of detecting more subtle cellular alterations than techniques that measure cell death.

Selected technique — The neutral red uptake assay is a very versatile procedure that has been evaluated using several different cell lines and protocols.[29,37,39,41,43,65-72] This is a simple, rapid test that depends upon the accumulation of the dye, neutral red, in the lysosomes of living cells. The endpoint is quantitative, and data from several evaluation studies indicate it may be a useful procedure. Only simple equipment is required, and the assay does not use radioisotopes. The protocols that have been developed include: (1) neutral red uptake assay; (2) the neutral red release assay; and (3) the agarose diffusion assay.

1. **Neutral Red Uptake Assay** — This procedure has been evaluated most extensively as an *in vitro* alternative for ocular irritation testing. Several different cell lines, different tissue culture media and exposure regimes have been evaluated with this test (for review see Babich and Borenfreund[72]). In one

of the more common protocols, BALB/c 3T3 mouse fibroblasts are seeded at 9×10^3 cells per well in a 96-well tissue culture microtiter test plate and incubated overnight at 37°C. Cells are washed, and fresh medium containing various concentrations of test material is added to each of four replicate wells per concentration. Incubation continues for another 24 h. Control cells are provided untreated medium only. At the end of the incubation period, medium is removed, cells are washed, neutral red is added to each well. Incubation is continued for 3 h at 37°C. Then the medium with neutral red is removed and the cells are washed in formaldehyde-based fixative followed by a mixture of acetic acid and ethanol to extract the neutral red which is measured in a microtiter plate reader. The concentration of test substance that inhibits neutral red uptake by 50% is the endpoint of the assay.

2. **Neutral Red Release Assay** — This modification of the neutral red uptake assay was developed to establish a procedure that allows assessment of cell damage after treatment with high concentrations of test materials for short periods of time. Confluent monolayers of mouse fibroblasts pre-loaded with neutral red are exposed to neat or minimally diluted test substance for 1 min. The test substance is aspirated from the cell surface, the cells are washed, and then neutral red de-stain solution is added to the cells. The amount of neutral red remaining within the cells after the short exposure is measured spectrophotometrically. The remaining dye gives an indication of the cells susceptibility membrane damage due to the treatment. Developers of this assay report mixed results on the ability of this assay to predict the ocular irritancy potential of test substances,[73-75] and additional research will be required to ultimately determine its utility.

3. **Agarose Diffusion Method** — Another modification of the neutral red assay has been proposed which may be useful for cytotoxicity testing of solids.[76-78] In this assay, mouse fibroblasts (L929 cells) are grown as monolayers in 25 cm² tissue culture flasks. Confluent monolayers are overlaid with agarose, and sterile membrane filter disks impregnated with the test material are laid on the agar. The assay depends upon diffusion of the test material from the impregnated membrane through the agar and into the cells. The cells which have been previously labeled with neutral red and damaged by the diffusing test material release the neutral red which is seen as a zone of relative clarity adjacent to the test disk. The endpoint is objective (presence or absence of a zone of lysis), but not directly quantitative (the size of the clear zone probably is not directly related to the toxicity of the material), although quantitation might be achieved by using different concentrations of test material. The modified test is advantageous because it allows testing of solids, but the toxic components must still diffuse through the aqueous culture medium and agarose matrix to exert their effects.

Silicon Microphysiometer — The Silicon Microphysiometer is a newly developed system that indirectly assesses cellular metabolism by measuring the production of acid metabolic byproducts.[29,61-63,79] Cells attached to glass cover slips are placed inside a flow chamber that contains a light addressable potentiometric sensor. This sensor is capable of measuring milli pH units in 100 nl volumes inside the chamber. A low-buffered cell culture medium is pumped through the chamber. When the flow is stopped, the metabolic byproducts released from the cells accumulate in the stagnant medium and reduce the intra-chamber pH. This change is easily detected and recorded.

The first step in an assay is to establish the baseline metabolic rate of the cells. The cells are then exposed to increasing concentrations of test materials for a fixed period of time. After each exposure, the test substance is flushed from the chamber and the metabolic rate is measured. The concentration of test material that decreases the metabolic rate by 50% is the endpoint. Preliminary data suggest that this is a promising alternative system.[29,61-63]

Luminescent Bacteria Test (LBT) — The LBT is an assay that has been used primarily for evaluating the toxicity of environmental and biomedical samples.[80,81] Recently, this system has been evaluated as an alternative for ocular irritancy testing.[29,63,64] The LBT measures light produced by a luminescent strain of bacteria called *Photobacterium phosphoreum*. The light is produced as a byproduct of the bacterial energy metabolic pathways. A decrease in light output is an indication that these pathways have been altered by the toxic effects of a test substance. When assessing test materials, increasing concentrations of sample are added to a suspension of the bacteria and incubated for a short period. The light output is measured. The concentration of test material that decreases the light output by 50% is the endpoint. Preliminary data suggests this system may provide useful information in the ocular safety assessment process.

3. Cell and Tissue Physiology

These tests generally measure one or more functions of tissues grown in culture. Some of these tests measure endpoints that have little direct relationship to ocular toxicity. Tests in this group that have been proposed as viable alternatives for ocular irritation testing include evaluation of electrical conductivity in rat skin slices,[82] measuring alterations in smooth muscle contractility of isolated rabbit ileum,[83] and changes in corneal opacity or permeability following exposure to the test material.[36,84-86] In each instance, animals must be sacrificed to obtain the target tissue, or the tissue must be obtained from food animals at slaughter. These tests offer the advantage of more direct evaluation of tissue rather than isolated cells, and in some instances, evaluation of tissue from the eye itself. None of these tests has been evaluated extensively.

4. Inflammation/Immunity

There are several tests in this category, most of which use freshly prepared cells or tissues. Included are tests using the chorioallantoic membrane (CAM) of the developing chick embryo,[87-91] excised bovine corneas,[92] rat peritoneal exudate cells,[93,94] and explants of rat vaginal tissue.[95] These tests differ from tissue culture systems because they offer the opportunity to evaluate a number of features of a more complex tissues. Endpoints range from morphologic observations on the treated CAM, to quantifiable measurements of released chemical mediators of inflammation such as histamine, serotonin, prostaglandins, and other cytokines. The tests have the disadvantage of requiring freshly collected cells or tissue from donor animals.

Selected technique — Chorioallantoic Membrane (CAM).[87,96] The CAM assays have generated considerable interest as alternatives for ocular irritation testing. The CAM is the vascularized respiratory membrane found immediately beneath the shell of a developing bird inside an incubated egg. Initially it was thought the CAM was analogous to the conjunctiva and therefore a good conjunctival inflammation model. However, additional studies have shown that classical inflammatory responses are not observed after treatment of the CAM[91,97] and the primary lesions assessed after treatment with test substance are observations of necrosis or vascular changes. Thus, these assays should more properly be considered morphology endpoint procedures.

Fertile hen's eggs are incubated at 37°C. On day 3 of incubation, the shell is penetrated and approximately $1^1/_2$ to 2 ml of albumin is removed from the end of the egg opposite the air cell. This creates an airspace on the side of the egg over which a window is cut through the shell. The window is closed with tape and the eggs replaced in the incubator. On day 14, the window is reopened and a 10 mm Teflon® ring is placed on the CAM. Aliquots of liquid or solid test samples are placed in the ring, the window is closed, and the egg is returned to the incubator for three additional days. On day 17 of incubation, the CAM is evaluated using a standard set of criteria. The test can be semi-quantified by measuring the size of the lesions and by scoring the various components of the lesion.

There are two other variations of the CAM procedure that have also been evaluated. First is the CAM vascular assay (CAMVA).[98-100] In the CAMVA, eggs are prepared as in the CAM procedure, but the lesions on the membrane are scored at 30 min after treatment instead of at 3 days. The most significant lesions scored are changes in the CAM vasculature. If hemorrhage occurs or if blood vessels are obstructed after treatment, the test is considered positive.

Another variation of the CAM assay is the Hen's Egg Test-CAM (HET-CAM) developed by Luepke et al.[101,102] The CAM, under the air cell of 10-day-old chicken eggs, is exposed and test materials are applied for 20 s. The reactions are observed, scored, and classified over a 5 min period according to a standard scoring scheme. This procedure has also been conducted in combination with an assay that evaluates the effects of test materials on isolated bovine corneas. This test is the Bovine Eye-CAM (BE-CAM) assay.[29,103,104] In addition to evaluation in the HET-CAM, test materials are placed onto the corneal surface of freshly excised, undamaged bovine eyes for 30 s. Corneal lesions are graded for the amount of damage observed. An irritancy prediction is made based on the results from either of the two tests that provide the most severe score.

Recently, Rosenbruch[105,106] suggested it may be possible to apply test materials to the yolk sac of 2-day-old chicken eggs and evaluate morphologic changes after treatment. Work with this system is in the early stages, and more must be done in order to assess its ultimate utility.

The CAM assays make use of living tissue in which there is an active blood supply. It has been suggested, however, that the CAM has many limitations when compared with mammalian tissue.[91,97] Also, the normal cornea is not perfused by blood vessels. Nevertheless, a variety of reactions such as hemorrhage, and necrosis can be detected, although histologic examination of the CAM lesions reveals minimal inflammation.[91,97]

Reports on the utility of the CAM assays are mixed. Several studies have indicated the test provides information useful for categorizing the ocular irritancy potential of test materials.[96,100,107,108] Others have concluded the CAM has no utility for assessing the ocular safety because of high false positive rates, embryotoxicity, and inaccurate categorization of irritancy potential.[89,97,109] Also, the British Animals (Scientific Procedures) Act of 1986 defines developing birds as protected animals after half the incubation has elapsed. An assay using a chicken egg greater than 10.5 days therefore is not considered an *in vitro* procedure.

5. Recovery and Repair

Recovery and repair are important issues in evaluating the rabbit ocular irritation response even though such observations are not a standard part of acute rabbit ocular irritation tests. Many investigators, however, extend the rabbit eye irritation tests up to 21 days or even longer to determine whether corneal lesions or other evidence of ocular injury are repaired. The entire issue of recovery from acute topical ocular injury is, of course, of paramount importance in man and is critical in assessing the long-term effects of ocular irritants. Relatively little progress has been made in the development of *in vitro* alternatives for assaying wound healing.

Selected technique — Wound healing in corneal cells in culture.[110-112] Primary rabbit corneal epithelial cells are cultured in medium supplemented with cholera toxin, epidermal growth factor, insulin, and dimethylsulfoxide. After 7 to 10 days of growth in 35 mm diameter culture dishes, the cells form confluent multilayers which are then subcultured and plated into 24-well multiplates. Disks 16 mm in diameter cut from Millipore filters are placed in the center of each culture, and a 6 mm stainless steel probe chilled in liquid nitrogen is placed against the plastic surface opposite the disk for 5 s. After the probe is removed, medium is added and the disk is carefully lifted out, which removes the frozen cells and leaves a discrete circular defect in the cell layer. Test chemicals are added and the culture is incubated for 42 h. Cultures are then drained of medium, washed, and stained. The size of the remaining defect, revealed as an unstained, cell-free area, is determined by projecting the plates onto a screen with an overhead projector at a fixed distance and tracing the unstained area onto paper which is cut out and weighed. Alternatively, any of several standard morphometric systems can be used to determine the total area of the defect. The endpoint is determined either by comparing the size of the remaining cell-free area to that of the initial defect or determining the area in square millimeters for comparison of wound sizes.

The test system objectively measures cell migration and perhaps some local cell proliferation. Cell migration is an important component of the early closure of corneal wounds and, thus, is an important parameter to evaluate. Despite the fact that the cells are growing in multiple layers, this system, of course, is not the same as the intact corneal epithelium. In particular, there is no basement membrane. The technique is relatively laborious and requires tissue culture skills sufficient to consistently produce multilayers of corneal epithelium. Animals must be sacrificed as donors. Nevertheless, this assay does allow some approximation of corneal healing, although testing to date is insufficient to determine whether there is agreement between this *in vitro* assay and corneal healing in rabbits or man.

IV. NEXT STEPS

As indicated throughout this chapter, currently available tests will not completely replace the need for animals in ocular safety assessment. It will only be possible to eliminate animal use when we have a battery of assays that reliably assess the key components of ocular injury known to occur *in vivo*. Today, we do not yet know what these key components are. Much basic research must be done in order to identify them. With greater understanding of the mechanisms that underlie chemically induced ocular injury, it will be possible to develop *in vitro* tests that directly assess relevant processes that occur in eyes exposed to test substances. If it is possible to develop such assays, toxicologists and regulators using the tests will have much greater confidence that the *in vitro* tests are providing predictive information on irritancy potential.

There are several areas where additional basic research is needed. Assessing the direct effects of test materials on cellular function will be useful in determining the ability of test materials to disrupt cellular physiologic processes. Modern biotechnology is making new methods available for culturing cells and measuring endpoints that will allow investigators to probe the major cellular functions that may be altered by test substances. Procedures such as flow cytometry and image analysis will provide better information than is available with current assays.

Test substances can induce ocular inflammation through a number of mechanisms. Direct damage to ocular cells can illicit these responses. In addition to gross damage, an inflammatory reaction may be initiated by far more subtle effects not associated with overt tissue destruction. Since inflammation plays a key role in ocular injury, and repair, it will be necessary to develop better procedures to determine whether test substances induce inflammatory processes. Developing such tests will be a challenge. Many of the mediators associated with inflammation are continuously produced, even under normal circumstances. It will therefore be necessary to develop precise measurement techniques that distinguish between normal and inflammatory situations, or to identify mediators that are expressed only under postexposure conditions. Our understanding of inflammation is growing rapidly, and as more information becomes available, it may become possible to add assessment of inflammatory properties to the *in vitro* test battery.

Models that estimate how long it will take to repair damage and to determine the consequences of the tissue repair process would also be very useful. Initial work has been done to assess the effects of test substances on wound healing.[110-112] These approaches are reasonable starts, but much more development is needed in this area. An approach that appears particularly promising is development of the so-called tissue equivalent systems.[113-115] These are usually composed of a collagen or nylon matrix that contains fibroblasts from the tissue of interest. Once this matrix is developed *in vitro*, it is seeded with epithelial cells. It may be possible to reconstruct a three-dimensional matrix composed of corneal epithelium, stromal fibroblasts, and endothelial cells that provides a good model for *in vitro* ocular safety assessment.

V. SUMMARY

In vitro alternative systems offer a number of potential advantages ranging from rapidity and simplicity of the test system through quantitation of data and automation of test endpoints. No single system has yet proved satisfactory as a complete replacement for the rabbit eye irritation test. It is likely that a battery of tests will be required. Although *in vitro* and *in vivo* tests often produce similar ranks for the ocular irritation, little is known about mechanisms responsible for producing injury in man, animals, or alternative systems. Nevertheless, the research conducted to date suggests many approaches are promising and that it is possible to begin using *in vitro* methods to reduce animal use in ocular safety assessment. Additional research and application of new biotechnology will lead us closer to the ultimate goal of eliminating the need for animals in ocular safety testing.

REFERENCES

1. J.M. Frazier, S.C. Gad, A.M. Goldberg, and J.P. McCulley, A Critical Evaluation of Alternatives to Acute Ocular Irritation Testing, in *Alternative Methods in Toxicology,* Vol. 4, Mary Ann Liebert, New York (1987).
2. D.K. Wilcox, and L.H. Bruner, *ATLA,* 18, 117 (1990).
3. W.M. Grant, Effects on the Eyes and Visual System from Chemicals, Drugs, Metals and Minerals, Plants, Toxins and Venoms; also Systemic Side Effects from Eye Medications, in *Toxicology of the Eye,* Charles C Thomas, Springfield, IL (1986).
4. T.O. McDonald, V. Seabaugh, J.A. Shadduck, and H.F. Edelhauser, *Dermatotoxicology,* F.N. Marzulli and H.I. Maibach, Eds. Hemisphere Publishing, New York, 555 (1983).
5. B. Ballantyne and D.W. Swanston, *Current Approaches in Toxicology,* B. Ballantyne, Ed. John Wright & Sons, Bristol, 139 (1977).
6. P. Duprat and P. Conquet, *Adv. Vet. Sci. Comp. Med.,* 31, 173 (1987).
7. L.Z. Bito, *Exp. Eye Res.,* 39, 807 (1984).
8. E. Bosshard, *Food.Chem. Toxicol.,* 23, 149 (1985).
9. P.K. Chan and A.W Hayes, in *Principles and Methods of Toxicology,* A.W. Hayes, Ed., Raven Press, New York, 169 (1989).

10. J. Jackson and D.A. Rutty, *Food Chem. Toxicol.*, 23, 293 (1985).
11. C.A. Reinhardt, D.A. Pelli, and G. Zbinden, *Food Chem. Toxicol.*, 23, 247 (1985).
12. C.S. Weil and R.A. Scala, *Toxicol. Appl. Pharmacol.*, 19, 276 (1971).
13. S.J. Williams, *Food Chem. Toxicol.*, 23, 189 (1985).
14. D.M. Maurice, In Vitro Toxicology, in *Alternative Methods in Toxicology*, Vol. 3, A.M. Goldberg, Ed., Mary Ann Liebert, New York, 343 (1991).
15. M. Balls, B. Blaauboer, D. Brusick, J.M. Frazier, D. Lamb, M. Pemberton, C. Reinhardt, M. Roberfroid, H. Rosenkranz, B. Schmid, H. Spielmann, A.L. Stammati, and E. Walum, *ATLA*, 12, 313 (1990).
16. C. Atterwill, P. Bach, M. Balls, D. Bawden, P. Botham, J. Bridges, D. Clark, P. Duffy, D. Esdaile, C. Garner, I. Kimber, R. Morrod, W. Parish, J. Southee, D. Swanston, M. Tute, G. Volans, C. Walker, and S. Walker, *ATLA*, 19, 116 (1991).
17. O.P. Flint, *ATLA*, 18, 11 (1990).
18. R.A. Scala, In Vitro Toxicology, Approaches to Validation, in *Alternative Methods in Toxicology*, Vol. 5, A.M. Goldberg, Ed., Mary Ann Liebert, New York, 1 (1987).
19. J.H. Draize, G. Woodard, and H.O. Calvery, *J. Pharmacol. Exp. Ther.*, 82, 377 (1944).
20. D.J. Finney, *Statistical Methods in Biological Assay*, Charles Griffin, London, 133 (1978).
21. D.S. Salsburg, *Statistics for Toxicologists*, Marcel Dekker, New York (1986).
22. H. North-Root, F. Yackovich, J. Demetrulias, M. Gacula, Jr., and J.E. Heinze, *Toxicol. Lett.*, 14, 207 (1982).
23. J.A. Shadduck, J. Everitt, and P.H.S. Bay, *In Vitro Toxicology. A Progress Report from The Johns Hopkins Center for Alternatives to Animal Testing*, Vol. 3, A.M. Goldberg, Ed., Mary Ann Liebert, New York, 641 (1985).
24. J.F. Griffith, G.A. Nixon, R.D. Bruce, P.J. Reer, and E.A. Bannan, *Toxicol. Appl. Pharmacol.*, 55, 501 (1980).
25. F.E. Freeberg, G.A. Nixon, P.J. Reer, J.E. Weaver, R.D. Bruce, J.F. Griffith, and L.W. Sanders, III, *Fund. Appl. Toxicol.*, 7, 626 (1986).
26. N.P. Luepke, *Br. J. Dermatol. Suppl.*, 115, 133 (1986).
27. L.H. Bruner, C.L. Alden, and J. Yam, *FASEB J.*, 5, A486 (1991).
28. J.A. Shadduck, J. Render, J. Everitt, R.A. Meccoli, and D. Essex-Sorlie, In Vitro Toxicology. Approaches to Validation, in *Alternative Methods in Toxicology*, Vol. 5, A.M. Goldberg, Ed., Mary Ann Liebert, New York, 75 (1987).
29. L.H. Bruner, D.J. Kain, D.A. Roberts, and R.D. Parker, *Fundam. Appl.Toxicol.*, in press (1991).
30. K.A. Booman, J. De Prospo, J. Demetrulias, A. Driedger, J.F. Griffith, G. Grochoski, B.M. Kong, W.C. McCormick, H. North-Root, M.G. Rozen, and R.I. Sedlak, *J.Toxicol. Cut. Ocular Toxicol.*, 8, 35 (1989).
31. S.D. Gettings, L.C. Dipasquale, D.M Bagley, M. Chudkowski, J. Demetrulias, P.I. Feder, K.L. Hintze, K.D. Marenus, W. Pape, M., Roddy, R. Schnetzinger, P. Silber, J.J. Teal, and S.L. Weise, *In Vitro Toxicol.*, 3, 293 (1990).
32. A.B. Burton, M. York, and R.S. Lawrence, *Food Cosmet. Toxicol.*, 19, 471 (1981).
33. H.B.W.M. Koeter and M.K. Prinsen, CIVO Institutes TNO Report No. V5.188/140322 (1985).
34. J.B. Price and I.J. Andrews, *Food Chem. Toxicol.*, 23, 313 (1985).
35. M. York, R.S. Lawrence, and G.B. Gibson, *Int. J. Cosmet. Sci.*, 4, 223 (1982).
36. H. Igarashi and A.M. Northover, *Toxicol. Lett.*, 39, 249 (1987).
37. E. Borenfreund and J.A. Puerner, *Toxicol. Lett.*, 24, 119 (1985).
38. H. Igarashi, *ATLA*, 15, 8 (1987).
39. E. Borenfreund and C. Shopsis, *Xenobiotica*, 15, 705 (1985).
40. C. Shopsis, E. Borenfreund, J. Walberg, and D.M. Stark, *Food Chem. Toxicol.*, 23, 259 (1985).
41. E. Borenfreund and J.A. Puerner, *J. Tissue Culture Meth.*, 9, 7 (1984).
42. C. Shopsis and B. Eng, *Toxicol. Lett.*, 26, 1 (1985).
43. D.M. Stark, E. Borenfreund, J. Walberg and C. Shopsis, In Vitro Toxicology. A Progress Report from The Johns Hopkins Center for Alternatives to Animal Testing. *Alternative Methods in Toxicology*, Vol. 3, A.M. Goldberg, Ed., Mary Ann Liebert, New York, 373 (1985).

44. M. Balls and S.A. Horner, *Food Chem. Toxicol.*, 23, 209 (1985).
45. M. Watanabe, K. Watanabe, K. Suzuki, O. Nikaido, I. Ishii, H. Konishi, N. Tanaka, and T. Sugahara, *Toxicol. In Vitro*, 3, 329 (1989).
46. H. North-Root, F. Yackovich, J. Demetrulias, M. Gacula, Jr., and J.E. Heinze, *Food.Chem. Toxicol.*, 23, 271 (1985).
47. M.C. Scaife, *Int. J. Cosmet. Sci.*, 4, 179 (1982).
48. M.C. Scaife, *Food Chem. Toxicol.*, 23, 253 (1985).
49. M. Aeschbacher, C.A. Reinhardt, and G. Zbinden, *Cell Biol. Toxicol.*, 2, 247 (1986).
50. R.B. Kemp, R.W. Meredith, S. Gamble, and M. Frost, *Cytobios* 36, 153 (1983).
51. H.J. Douglas and S.D. Spillman, Product Safety Testing, in *Alternative Methods in Toxicology*, Vol. 1, A.M. Goldberg, Ed., Mary Ann Liebert, New York, 205 (1983).
52. C.K. Muir, C. Flower, and N.J. Van Abbe, *Toxicol. Lett.*, 18, 1 (1983).
53. R. Tchao, Progress in *In Vitro* Toxicology, in *Alternative Methods in Toxicology*, Vol. 6, A.M. Goldberg, Ed., Mary Ann Liebert, New York, 271 (1988).
54. A.J. Shaw, R.H. Clothier, and M. Balls, *ATLA*, 18, 145 (1990).
55. K.Y. Chan, In Vitro Toxicology. A Progress Report from The Johns Hopkins Center for Alternatives to Animal Testing, in *Alternative Methods in Toxicology*, Vol. 3, A.M. Goldberg, Ed., Mary Ann Liebert, New York, 407 (1985).
56. K.Y. Chan, *Curr. Eye Res.*, 5, 357 (1986).
57. R.B. Kemp, R.W. Meredith, and S.H. Gamble, *Food Chem. Toxicol.*, 23, 267 (1985).
58. C. Shopsis and S. Sathe, *Toxicology* 29, 195 (1984).
59. C. Shopsis, *J. Tissue Culture Meth.*, 9, 19 (1984).
60. J. Selling and B. Ekwall, *Xenobiotica* 15, 713 (1985).
61. J.W. Parce, J.C. Owicki, K.M. Kercso, G.B. Sigal, H.G. Wada, V.C. Muir, L.J. Bousse, K.L. Ross, B.I. Sikic, and H.M. McConnell, *Science* 246, 243 (1989).
62. L.H. Bruner, K.M. Kercso, J.C. Owicki, J.W. Parce, and V.C. Muir, *Toxicol. In Vitro*, in press (1991).
63. L.H. Bruner, *Proceedings of the 2nd CTFA Ocular Safety Testing Workshop: Evaluation of In Vitro Alternatives*, S.D. Gettings and G.N. McEwen, Eds., The Cosmetic, Toiletry, and Fragrance Association, Washington, D.C., 94 (1991).
64. A.A. Bulich, K.K. Tung, and G. Scheibner, *J. Biolumin. Chemilumin.*, 5, 71 (1990).
65. E. Borenfreund and O. Borrero, *Cell Biol. Toxicol.*, 1, 55 (1984).
66. K. Hockley and D. Baxter, *Food Chem. Toxicol.*, 24, 473 (1986).
67. R.J. Riddell, R.H. Clothier, and M. Balls, *Food Chem. Toxicol.*, 24, 469 (1986).
68. E. Borenfreund, H. Babich, and N. Martin-Alguacil, *Toxicol. In Vitro* 2, 1 (1988).
69. C. Shopsis, *Alternative Methods in Toxicology*, Vol. 7, in press (1989).
70. M. Bracher, C. Faller, J. Spengler, and C.A. Reinhardt, *Mol. Toxicol.*, 1, 561 (1987).
71. M.J. Arranz and M.F.W. Festing, *Toxicol. In Vitro* 4, 211 (1990).
72. H. Babich and E. Borenfreund, *ATLA*, 18, 129 (1990).
73. S.J. Reader, V. Blackwell, R. O'Hara, R. Clothier, G. Griffin, and M. Balls, *Toxicol. In Vitro*, 4, 264 (1990).
74. S.J. Reader, V. Blackwell, R. O'Hara, R.H. Clothier, G. Griffin, and M. Balls, *ATLA*, 17, 28 (1989).
75. M.S. Dickens, *Proceedings of the 2nd CTFA Ocular Safety Testing Workshop: Evaluation of In Vitro Alternatives*, S.D. Gettings and G.N. McEwen, Eds., The Cosmetic, Toiletry, and Fragrance Association, Washington, D.C., 62 (1991).
76. R.F. Wallin, R.D. Hume, and E.M. Jackson, *J. Toxicol. Cut. Ocular Toxicol.*, 6, 239 (1987).
77. E.M. Jackson, R.D. Hume, and R.F. Wallin, *J. Toxicol. Cut. Ocular Toxicol.*, 7, 187 (1988).
78. K.A.F. O'Brien, P.A. Jones, and J. Rockley, *Toxicol. In Vitro*, 4, 311 (1990).
79. D.G. Hafeman, W.J. Parce, and H.M. McConnell, *Science*, 240, 1182 (1988).
80. A.A. Bulich, Aquatic Toxicology, ASTM 667, L.L. Markings and R.A. Kimerle, Eds. American Society for Testing and Materials, Philadelphia, 98 (1979).
81. A.A. Bulich, Toxicity Testing Using Micro Organisms, Vol. 1, G. Bitton and B.J. Dutka, Eds., CRC Press, Boca Raton, FL, 57 (1986).

82. G.J. Oliver and M.A. Pemberton, *Food Chem. Toxicol.*, 23, 229 (1985).

83. C.K. Muir, *Toxicol. Lett.*, 19, 309 (1983).

84. C.K. Muir, *Toxicol. Lett.*, 22, 199 (1984).

85. C.K. Muir, *Toxicol. Lett.*, 24, 157 (1985).

86. D. Maurice and T. Singh, *Toxicol. Lett.*, 31, 125 (1986).

87. J. Leighton, J. Nassauer, and R. Tchao, *Food Chem. Toxicol.*, 23, 293 (1985).

88. B.M. Kong, C.J. Viau, P.Y. Rizvi, and S.J. DeSalva, In Vitro Toxicology, Approaches to Validation, in *Alternative Methods in Toxicology*, Vol. 5, A.M. Goldberg, Ed., Mary Ann Liebert, New York, 59 (1989).

89. J.B. Price, M.P. Barry, and I.J. Andrews, *Food Chem. Toxicol.*, 24, 503 (1986).

90. N.P. Luepke, *Food Chem. Toxicol.*, 23, 287 (1985).

91. R.S. Lawrence, M.H. Groom, D.M. Ackroyd, and W.E. Parish, *Food Chem. Toxicol.*, 24, 497 (1986).

92. S.A. Elgebaly, K. Nabawi, N. Kerkbert, J. O'Rourke, and D.L. Kruetzer, *Invest. Ophthalmol. Vis. Sci.*, 26, 320 (1985).

93. R.B. Jacaruso, M.A. Bartlett, S. Carson, and L.D. Trombetta, *J. Toxicol. Cut. Ocular Toxicol.*, 4, 39 (1985).

94. M. Chasin, C. Scott, C. Shaw, and F. Persico, *Int. Arch. Allergy Appl. Immunol.*, 58, 1 (1979).

95. N.H. Dubin, M.C. Wolff, C.L. Thomas, and M.C. DiBlasi, *Toxicol. Appl. Pharmacol.*, 78, 458 (1985).

96. J. Leighton, J. Nassauer, R. Tchao, and J. Verdone, *Alternative Methods in Toxicology*, 1, 165 (1983).

97. W.E. Parish, *Food Chem. Toxicol.*, 23, 215 (1985).

98. D.M. Bagley, B.M. Rizvi, B.M. Kong, and S.J. DeSalva, Progress in In Vitro Toxicology, in *Alternative Methods in Toxicology*, Vol. 6, A.M. Goldberg, Ed., Mary Ann Liebert, New York, 131 (1988).

99. D.M Bagley, B.M. Kong and S.J. DeSalva, In Vitro Toxicology: New Directions, in *Alternative Methods in Toxicology*, Vol. 7, A.M. Goldberg, Ed., Mary Ann Liebert, New York, 265 (1989).

100. M.J. Muscatiello and D.M. Bagley, *Proceedings of the 2nd CTFA Ocular Safety Testing Workshop: Evaluation of In Vitro Aliteratives*, S.D. Gettings and G.N. McEwen, Eds. The Cosmetic, Toiletry, and Fragrance Association, Washington, D.C., 71 (1991).

101. N.P. Luepke and F.H. Kemper, *Food Chem. Toxicol.*, 24, 495 (1986).

102. N.P. Luepke, unpublished (1987).

103. P.J. Weterings and Y.H.M. Van Erp, In Vitro Toxicology. Approaches to Validation, in *Alternative Methods in Toxicology*, Vol. 5, A.M. Goldberg, Ed., Mary Ann Liebert, New York, 515 (1987).

104. Y.H.M. Van Erp and P.J. Weterings, *Toxicol. In Vitro*, 4, 267 (1990).

105. M. Rosenbruch, *J. Comp. Pathol.*, 101, 363 (1989).

106. M. Rosenbruch and A. Holst, *Toxicol. In Vitro*, 4, 327 (1990).

107. B.M. Kong, C.J. Viau, P.Y Rizvi, and S.J. DeSalva, In Vitro Toxicology. Approaches to Validation. *Alternative Methods in Toxicology*, Vol. 5, A.M. Goldberg, Ed., Mary Ann Liebert, New York, 59 (1987).

108. B.M. Kong, Soap/Cosmetics/Chemical Specialties, July, 40 (1987).

109. R.S. Lawrence, D.M. Ackroyd, and D.L. Williams, *Toxicol. In Vitro*, 4, 321 (1990).

110. M.M. Jumblatt and A.H. Neufeld, *Invest. Ophthalmol. Vis. Sci.*, 27, 8 (1986).

111. M.M. Jumblatt and A.H. Neufeld, In Vitro Toxicology. A Progress Report from The Johns Hopkins Center for Alternatives to Animal Testing, in *Alternative Methods in Toxicology*, Vol. 3, A.M. Goldberg, Ed., Mary Ann Liebert, New York, 393 (1985).

112. S.J. Simmons, M.M. Jumblatt, and A.H. Neufeld, *Toxicol. Appl. Pharmacol.*, 88, 13 (1987).

113. E. Bell, B. Ivarsson, and C. Merrill, *Proc. Natl. Acad. Sci. U.S.A.*, 76, 1274 (1979).

114. E. Bell, H.P. Ehrlich, D.J. Buttle, and T. Nakatsuji, *Science* 211, 1052 (1981).

115. C.B. Weinberg and E. Bell, *Science* 231, 397 (1986).

19

Preclinical Toxicology/Safety Considerations in the Development of Ophthalmic Drugs and Devices

ROBERT B. HACKETT
Alcon Laboratories, Inc.
Fort Worth, Texas

AND

MICHAEL E. STERN
Allergan Pharmaceuticals, Inc.
Irvine, California

I. INTRODUCTION

Toxic responses in the eye may result from topical ocular administration of drugs, intraocular administration, or implantation of drugs or devices, or may be the result of target organ toxicity following systemic administration of a drug. The primary responsibility of the toxicologist is to establish the safety profile for a drug or device under development and thus provide an adequate risk analysis of the drug/device for human use. For the ophthalmic toxicologist this safety profile must include the appropriate toxicological evaluations to place in perspective the intended use of the drug or device, its effect on the relevant ocular tissues, its potential for adverse systemic effects if warranted, and the potential risk to the patient in the clinical setting. The intent of this chapter is to provide the standard methods for determining the ophthalmic toxicological potential of a drug or device intended for an ophthalmic indication. It is assumed that the reader is familiar with the necessary parameters evaluated for assessing the systemic toxicity potential of a drug/device.

II. REGULATORY REQUIREMENTS

Regulatory requirements and nonclinical safety study guidelines are issued by the various governmental health offices throughout the world to provide industry with the appropriate steps necessary to develop a safety profile for drugs and devices. These guidelines often include the types of nonclinical safety studies required by the health agency for registration of a drug candidate in that country. In general, testing requirements for ophthalmic drugs are more specific in the European Economic Community (EEC) with regard to the number of studies, study design, and length of studies necessary for market approval than are the guidelines provided by the U.S. The only published guidelines specific for ophthalmic drugs offered by the U.S. Food and Drug Administration (FDA) for an ophthalmic drug date back to 1968.[5b] Additional insight into the contents of the nonclinical drug safety profile can be obtained from Weissinger.[20] The types and length of studies presented below for ophthalmic drugs are, in the opinion of the authors, appropriate for the listed therapeutic classes but should not be construed as the requirements promulgated by any governmental regulatory agency. The design, appropriateness, and the adequacy of the nonclinical safety program for ophthalmic drugs rest with the toxicologist.

Testing requirements for market approval of devices (i.e., contact lenses and companion solutions, intraocular lenses and ophthalmic prosthetic devices) are available from the health offices of the EEC as well as the U.S. FDA. The guidelines for market approval in the U.S. are presented below under the device category.

III. DESIGN OF PRECLINICAL SAFETY PROFILE

A. Ophthalmic Drugs

The extent of the preclinical safety tests necessary to provide an adequate safety profile and risk assessment for a topical ophthalmic drug is contingent upon several considerations:

- Intended use of the drug
- Probable systemic exposure
- Anticipated length of clinical exposure
- Severity of the condition requiring treatment relative to visual loss and patient well-being

The intended use of the drug may involve single administration such as is the case with a diagnostic drug used for producing mydriasis, a topical anesthetic for producing corneal anesthesia or ophthalmic diagnostic dyes such as fluorescein or rose bengal. The extent of testing necessary for single-use diagnostics can be limited to evaluation for acute topical ocular irritation potential, short-term repeated topical dose studies, and an assessment of the acute and subacute oral toxicity potential. A typical drug safety profile for such ophthalmic drugs may include the following tests:

- Evaluation for mutagenic/genotoxicity potential to include *in vitro* and *in vivo* assays
- Acute topical ocular irritation evaluation — rabbit
- 7-day topical ocular toxicity evaluation — rabbit

- Acute oral toxicity limit test — rat
- 14-day repeated oral toxicity test — rat

Most drugs, however, are intended for multiple-dose therapy and hence the testing requirements are more extensive in order to develop an adequate drug safety profile. We may break the intended use of a drug into two arbitrary therapy categories of intermediate use and chronic use.

Intermediate-use drugs are typically those drugs employed in external eye disease and are represented by the anti-inflammatories and antimicrobials. These drugs are generally administered up to several times per day for treatment periods of 2 weeks to 3 months. A typical drug safety profile may include the following tests:

- Evaluation for mutagenic/genotoxicity potential to include *in vitro* and *in vivo* assays
- Acute topical ocular irritation evaluation — rabbit
- 1-month ocular/systemic toxicity evaluation — rabbit
- 3- to 6-month ocular/systemic toxicity evaluation — rabbit and/or dog/monkey, if warranted
- Appropriate systemic toxicity evaluations following parenteral administration, if warranted

Drugs intended for chronic administration, such as for the treatment of glaucoma, are generally administered daily for years. These classes of therapeutics require a more comprehensive safety evaluation since the length of exposure is extensive or may represent a remaining lifetime exposure. A typical drug safety profile may include the following tests:

- Evaluation for mutagenic/genotoxicity potential to include *in vitro* and *in vivo* assays
- Acute topical ocular irritation evaluation — rabbit
- 3-month ocular/systemic toxicity evaluation — rabbit and/or dog/monkey if warranted
- 12-month ocular/systemic toxicity evaluation — rabbit and/or dog/monkey if warranted
- Subchronic systemic toxicity evaluation — rodent and nonrodent
- Chronic systemic toxicity evaluation — rodent and nonrodent
- Evaluation of effect on reproduction, fetal development, and fertility (Segments I, II, and III)
- Lifetime carcinogenicity bioassays

It is recommended that topical ocular drugs be evaluated for systemic absorption. If systemic absorption is negligible, the nature and degree of systemic toxicity evaluations utilizing a parenteral route of administration for drugs intended for intermediate or chronic topical ophthalmic administration may be omitted or highly modified. In addition, conducting lifetime carcinogenicity bioassays may be unnecessary if the carcinogenic potential of compounds of similar structures is known, the compound is not mutagenic or genotoxic, and the metabolic profile of the drug in man has been established and can be correlated to chemical classes of compounds known to be noncarcinogenic. It is prudent to discuss regulatory requirements with the appropriate governmental agency prior to submission of the nonclinical safety profile.

B. Contact Lens Solutions

Presented herein are the 1985 draft guidelines for the testing criteria and methodologies used to determine the safety of solutions for use with Class III contact lenses. These guidelines were issued by the Division of Ophthalmic Devices, Center for Devices and Radiological Health of the U.S. FDA, and are readily obtainable through request to the Division.

1. Solution Testing

This section of the guideline discusses the acute toxicology considerations that must be addressed before an original Investigational Device Exemption/Premarket Approval (IDE/PMA) application for a Class III contact lens solutions (i.e., daily cleaners, rinsing solutions, chemical or heat disinfection solutions, rinsing and heat disinfection solutions, chemical disinfection/rinsing/storage solutions, lubricating and rewetting solutions, etc.) is approved by the FDA and/or an IRB and before initiating a clinical investigation of contact lens solutions. In addition, this section of the guideline includes a discussion of the suggested acute toxicology tests that can be performed to provide the necessary preclinical toxicology and biocompatibility information to allow the FDA and/or an IRB to predict with reasonable assurance that patients participating in a clinical investigation of a contact lens solution will not be placed at undue risk. A general description of the suggested acute toxicology test procedures are presented. The suggested tests discussed are intended for use as a general guideline for testing that the FDA believes to be adequate and appropriate for the preclinical acute toxicological evaluation of a Class III contact lens solution.

For purposes of an original IDE/PMA application, a sponsor should assess the potential of the contact lens solution to produce:

1. An acute oral toxicity
2. An acute systemic toxicity
3. An acute ocular irritation and cytotoxicity
4. An allergic response due to sensitization

The toxicology data submitted in support of an original IDE application should also be submitted in support of a PMA. Generally, these data are sufficient to establish an acute toxicity profile for the solution. While such data may support the safety of the solution for its intended use, additional *in vitro* or *in vivo* toxicity or biocompatibility testing may be required by the FDA if it is deemed appropriate and necessary to assess product safety and make risk/benefit decisions prior to premarket approval. The Division of Ophthalmic Devices may be consulted prior to the initiation of any tests if, after reading the guideline, questions remain concerning a specific test requirement for a contact lens product.

The Agency is aware of new and developing *in vitro* and *in vivo* methods to evaluate products for ocular toxicity. Cell culture methods using corneal epithelial, stromal, and endothelial cell lines are currently being researched as *in vitro* alternatives to the Draize test for identification of ocular irritants and for in cytotoxicity evaluations. In addition, similar *in vitro* cell culture research is being investigated using a protozoan species as a model for identifying potential ocular irritants. Corneal perfusion studies using excised corneas from animal or eyebank eyes, performed according to an *in vitro* procedure described by McCarey et al. [12] or an *in vivo* procedure described by Marzulli et al. [11] can provide valuable data on the toxic potential of drugs, preservatives, or other chemicals to the corneal structures (e.g., epithelium, stroma, and endothelium). Quantitative measures of corneal permeability, uptake, and release of a test substance can be demonstrated using these techniques. Scanning (SEM) and transmission electron microscopy (TEM) following corneal perfusion studies can provide graphic details on the toxic effects of a test substance to the corneal structures. A new *in vivo* model for ophthalmic toxicity studies measures the regeneration rate of rabbit corneal epithelial cells in the presence of a test substance following the removal of a known (e.g., measurable) area of epithelial cells from the cornea. Because the regeneration rate of corneal epithelial cells is predictable, the time for re-epithelialization of the area in the presence of a test substance is being studied as a method to predict potential ocular toxicity. The Agency encourages the development and use of such new test procedures that would be applicable to the assessment of your product for potential ocular toxicity.

Acute Oral Toxicology Assessment — The purpose of this study is to assess the potential of a contact lens solution to produce acute oral toxicity in rodents. While the Division of Ophthalmic Devices (DOD) believes that acute oral toxicity studies should be performed, the division recognizes that alternative testing to the classic LD_{50} assay may be appropriate under certain circumstances for assessing risks to humans from the proper or improper use (i.e., accidental misuse) of a contact lens solution. DOD will accept for review oral toxicity test data for a contact lens solution as formulated that does not show signs of toxicity, i.e., mortality, morbidity, and/or pathogenesis, if the solution has been given orally to a group of test animals in a single large dose, generally referred to as the maximum tolerable dose. For rodent testing, the maximum volume of an aqueous solution generally should not exceed 2 ml/100 g body weight. These data will be considered sufficient for predicting with reasonable assurance that a contact lens solution does not present an acute oral concern. This interpretation is based on the division's historical experience with toxicology test data from OTC solutions for contact lens use and is consistent with current trends in safety evaluation testing. However, should signs of toxicity be demonstrated at the maximum tolerable dose, further testing consistent with accepted toxicological practices is recommended in order for the Division of Ophthalmic Devices to complete its risk/benefit assessment of the contact lens solution.

Acute Systemic Toxicity Assessment — The purpose of this study is to assess the potential of the solutions as formulated to produce an acute systemic toxicity in mice. The individual solutions are injected into mice, and the mice observed for acute systemic toxicity.

Acute Ocular Irritation and Cytotoxicity Assessment — The purpose of this assessment is to evaluate the potential for ocular irritation or a cytotoxicity resulting from the solutions as formulated. The effects are assessed *in vitro* using cytotoxicity studies (i.e., tissue culture-agar overlay method of a suitable alternative) and *in vivo* using the 3-week ocular safety study in rabbits.

Sensitization/Allergic Response Assessment — The purpose of this assessment is to evaluate the *in vitro* and/or *in vivo* potential for the induction of a sensitivity/ allergic response resulting from an uptake and release of preservatives used in lens care solutions when in conjunction with representative lenses from each polymer group.

Preservative Uptake and Release Test — All contact lens polymers will either absorb or adsorb preservatives or other chemical components used in lens care solutions. Manufacturers should provide FDA with the amount of preservative uptake per lens, the amount released, and the time course of release. The suggested test procedure for this study is presented below. This study is a quantitative analysis that should be conducted with the lens care solutions. At the FDA, the toxicologist will evaluate the quantitative findings of the study once the chemist has determined that the methodology used to perform the study is appropriate and the data are accurate for toxicology review. The results of these test data will be used to predict the potential for a preservative-related toxicity, as well as the potential for inducing a sensitivity/allergic response associated with the lens group. The suggested test procedures have been established for the quantitative analysis of the uptake of the preservatives *thimerosal* and *chlorhexidine* in contact lenses. If preservatives other than thimerosal or chlorhexidine are used in the recommended lens care solutions (sorbic acid, quaternary ammonium compounds, etc.), it is the responsibility of the sponsor to select an appropriate chemical method for the quantitative analyses of the uptake and release of the preservative from the lens material for each group.

Guinea Pig Maximization Test — The purpose of this test is to grade or rank the active ingredients of the lens solution on a scale of I through V as to their potential for inducing sensitivity response in the guinea pig model. The grades or rankings are based on the number of animals sensitized and the results are classified on an ascending scale from a weak sensitizing agent (grade I) to an extreme sensitizing agent (grade V). It is the FDAs intent to use this test primarily to assess the sensitization potential of *new* preservatives for contact lens solutions.

Container/Accessory Testing Requirements — The purpose of these testing requirements is to assess the potential toxicity of any constituent(s) that may be extractable from the solution container and accessories when the solution, as formulated, will come in contact with the container or accessory for a prolonged period of time. The following *in vitro* and *in vivo* test procedures are recommended by the FDA and are consistent with USP XXII Tests for Plastic Ophthalmic Containers: (1)

in vitro cytotoxicity testing — tissue culture-again overlay (cytotoxicity) test — or an alternate test agreed to by the FDA and manufacturer; (2) systemic toxicity testing — systemic toxicity studies using extractives obtained from container/accessory components, utilizing USP XXII Tests for Plastic Ophthalmic Containers. The following methodologies should be observed.

The acute systemic toxicity study in mice is performed using extracts of the container material obtained according to procedures outlined in the USP XXII Tests for Plastic Ophthalmic Containers.

1. Ten healthy, not previously used, albino mice weighing approximately 20 g each should be divided into two groups of five mice each.

2. Group I should receive an intraperitoneal injection of cottonseed oil which contains the extracted components(s). Each mouse receives a dose of 50 ml/kg at an injection rate of 0.1 ml/s.

3. Group II should receive an intravenous injection of sodium chloride which contains the extracted component(s). Each mouse receives a dose of 50 ml/kg at an injection rate of 0.1 ml/s.

4. The mice are observed immediately after injection at 6, 24, 48, 72, and 96 h for clinical signs of acute systemic toxicity.

5. Clinical Observations — All toxicological and pharmacological signs should be recorded including time of onset, intensity, and duration. The time of death, if applicable, should also be noted. Individual records should be maintained for each animal.

6. Weight Change — Animals must be weighed individually on the day the test substance is administered and prior to sacrifice.

7. Necropsy — A complete gross necropsy should be performed on all animals that die during the course of the test. When appropriate, complete gross necropsy should be performed on all animals at termination of the test. Microscopic examination (i.e., histopathology) of gross lesions should also be performed.

Primary Ocular Irritation Testing — Primary ocular irritation studies using extractives obtained from container/accessory components utilizing USP XXII Tests for Plastic Ophthalmic Containers. The following methodologies should be observed:

1. Physiologic saline accessory material extracts should be tested with three healthy albino rabbits for each test extract. Animals determined to be free of eye lesions by slit-lamp examination with fluorescein staining should be used.

2. A volume of 0.2 ml saline accessory material extract will be instilled into the cup formed by gently pulling the lower eye lid away from the eye. The lids will then be held together for 30 s. Similarly, 0.2 ml of blank saline will be instilled into the other eye which will serve as a control. Both eyes should be observed at 24 and 48 h with a hand ophthalmoscope and fluorescein staining. At 72 h,

both eyes in each animal should be examined with a slit-lamp and fluorescein stain. In addition, both eyes should be scored for gross signs of eye irritation at the same intervals using the Draize or McDonald-Shadduck method.[13]

C. Contact Lens Material

In order to establish the safety of new polymers for use as contact lenses, the Office of Device Evaluation, Center for Devices and Radiological Health, of the U.S. FDA has issued draft guidelines. The most recent guidelines from 1988 are presented herein.

The toxicology data submitted in support of an IDE application will also be submitted in support of a PMA. However, the manufacturing and chemistry procedures used to fabricate a contact lens as well as the material itself should dictate, in general, the extent of the toxicology testing necessary to provide reasonable assurance of the safety of a contact lens. Therefore, the toxicologist must have any *in vitro* or *in vivo* toxicology or biocompatibility test available that will provide the necessary data to assess product safety and make risk/benefit decisions prior to premarket approval. Center for Devices and Radiological Health (CDRH) would like to remind all applicants that it is their responsibility to develop an appropriate toxicology and biocompatibility profile for the specific lens material of their device.

For purposes of submission of a PMA, CDRH reminds all applicants that as required by 21 CFR 814.20(b)(6)(i), all nonclinical laboratory studies shall include a statement that each study was conducted in compliance with Good Laboratory Practice (GLP) Regulation for Nonclinical Laboratory Studies, (21 CFR Part 58) or, if the study was not conducted in compliance with GLP regulations, a justification of the noncompliance must be submitted.

CDRH is aware of the ongoing research efforts to achieve the goal of eventual substitution of *in vitro* tests for certain biological tests utilizing animals. Cell culture methods using corneal epithelial, stromal, and endothelial cell lines are currently being researched as *in vitro* alternatives to the Draize test for identification of ocular irritants and for use in cytotoxicity evaluation. Another method being considered as an alternative is the chorioallantoic membrane (CAM) assay. The CAM assay is based on the fact that the CAM has anatomical components that are similar to the structure of the eye and reacts to insults with inflammatory responses. However, at present, *in vitro* alternatives to animal testing have not been sufficiently developed or validated for use. Therefore, CDRH regrets that toxicology tests involving animals must continue to be used at this time in order to adequately assess risks and evaluate safety of ocular products prior to approval for general marketing. CDRH will continue to monitor the developments of alternatives to animal testing and will encourage their use once such studies have been validated.

1. Recommended Toxicology Test Procedures for Contact Lens Materials

Systemic Injection Test (USP)* — The purpose of this study is to assess the

---

*   United States Pharmacopeia XXI/National Formulary XVI (or current update) — Containers for Ophthalmics, Plastics, and Biological Test Procedures.

potential of leachable chemical constituents from a contact lens material to produce an acute systemic toxicity in mice. Extracts of the lens material are prepared in two types of solvents (polar and nonpolar), injected into mice, and the mice observed for acute systemic toxicity.

Eye Irritation Test (USP)* — The purpose of this study is to evaluate the potential for ocular irritation resulting from residual chemical leachables in contact lens materials. The effects are assessed *in vivo* using rabbits.

Cytotoxicity Test — The purpose of this study is to evaluate the potential for cytotoxicity resulting from residual chemical leachables in contact lens materials. The effects are assessed *in vitro* using cytotoxicity studies; e.g., tissue culture-agar overlay method[4] or a suitable alternative.

Sensitization Tests — These tests are required, especially for uniquely new monomers, or for new additives used in contact lens polymers, or to assess the sensitization from interaction between new lens materials and preservatives in contact lens solutions. The purposes of these tests are to evaluate the *in vitro* and/or *in vivo* potential for the induction of a sensitivity/allergic response resulting from an uptake and release of preservatives used in lens care solutions or from residual chemical leachables; e.g., residual unreacted monomers, additives or other impurities, in contact lens materials.

Three Week Ocular Irritation Test in Rabbits — The *in vivo* test of the contact lenses in rabbits may be used as a biocompatibility test as well as a toxicity test of the lens material to assess the *in vivo* effects of the ocular environment on the lens material as well as the *in vivo* effects of the lens material on the ocular tissues.

Presented here is an example of a test design that can be used as general guidance in developing an appropriate *in vivo* ocular irritation test of a contact lens made of a material other than plastic. It is the responsibility of the sponsor to design an appropriate *in vivo* test using a sufficient number of animals to assess the safety and biocompatibility when using the recommended lens care regimen proposed for use in the clinical investigation.

CDRH suggests that this *in vivo* ocular irritation test be performed in the rabbit model. Test lenses for the test should be the thickest lenses in the product line or lenses of the greatest mass (with and without the use of the recommended lens care regimen) as they are to be used in a clinical investigation. Appropriate controls (e.g., eyes receiving no lens or eyes receiving a control lens) should be included in the test design. A *minimum* of 12 rabbits, determined to be free of corneal defects by initial slit-lamp examination with fluorescein staining, should be distributed into groups similar to the proposed groupings outlined in the example that follows:

12 Rabbits — 6 male/6 female (24 total eyes).
Group 1 (males) — 3 normal eyes fitted with the test lenses that have been treated with disinfection procedures proposed for patient use.
Group 2 (females) — 3 normal eyes fitted with the test lenses that have been treated with disinfection procedures proposed for patient use.

Group 3 (males) — 3 normal eyes fitted with the test lenses that have *not* been treated with disinfection procedures proposed for patient use.

Group 4 (females) — 3 normal eyes fitted with the test lenses that have *not* been treated with disinfection procedures proposed for patient use.

When appropriate, all lenses used in the test should be retrieved for purposes of assessing the *in vivo* effects of the ocular environment on the lens material when using the recommended lens care regimen. The information submitted in the report should include, but not be limited to, data that compare the physical and optical parameters of the lens, such as physical appearance, e.g., lens discolorization, protein deposits, chipped or pitted lenses, center thickness, and lens powers as measured before starting the test and after termination of the test.

D. Ocular Devices (Prosthetics and Intraocular Lenses)

Prosthetic devices or intraocular lenses represent a potential toxicity due to the elution of monomers, catalysts, etc., from the polymer matrix. In 1987, the Office of Device Evaluation, Center for Devices and Radiological Health distributed the Tripartite Biocompatibility Guidelines for establishing the safety of devices. Prosthetic devices and intraocular lenses may be classified as an internal device with exposure to tissues and tissue fluids. The Tripartite Biocompatibility Guidelines recommend the following biological tests.

Sensitization assay — Estimates the potential for sensitization of a test material and/or the extracts of a material using an animal model and/or human.

Irritation tests — Estimate the irritation and sensitization potential of test materials and their extracts, using appropriate site or implant tissue such as skin and mucous membrane in an animal model and/or human.

Cytotoxicity — With the use of cell culture techniques, determines the lysis of cells (cell death), the inhibition of cell growth, and other toxic effects on cells caused by test materials and/or extracts from the materials.

Acute systemic toxicity — Estimates the harmful effects of either single or multiple exposures to test materials and/or extracts, in an animal model, during a period of <24 h.

Hemocompatiblity — Evaluates any effects of blood contacting materials on hemolysis, thrombosis, plasma proteins, enzymes, and the formed elements using an animal model.

Pyrogenicity (material mediated) — Evaluates the material-mediated pyrogenicity of test materials and/or extracts.

Hemolysis — Determines the degree of red blood cell lysis and the separation of hemoglobin caused by test materials and/or extracts from the materials *in vitro*.

Implantation tests — Evaluate the local toxic effects on living tissue, at both the gross and microscopic levels, to a sample material surgically implanted into appropriate animal implant animal implant site or tissues (e.g., muscle, bone) for 7 to 90 days.

Mutagenicity (genotoxicity) — The application of mammalian or nonmammalian cell culture techniques for the determination of gene mutations, changes in chromosome structure and number, and other DNA or gene toxicities caused by test materials and/or extracts from materials.

Subchronic toxicity — The determination of harmful effects from multiple exposures to test materials and/or extracts during a period of 1 day to <10% of the total life of the test animal (e.g., up to 90 days in rats).

Chronic toxicity — The determination of harmful effects from multiple exposures to test materials and/or extracts during a period of 10% to the total life of the test animal (e.g., over 90 days in rats).

Carcinogenesis bioassay — The determination of the tumorogenic potential of test materials and/or extracts from either a single or multiple exposures, over a period of the total life (e.g., 2 years for rat, 18 months for mouse or 7 years for dog).

Pharmacokinetics — To determine the metabolic processes of absorption, distribution, biotransformation, and elimination of toxic leachables and degradation products of test materials and/or extracts.

Reproductive and developmental toxicity — The evaluation of the potential effects of test materials and/or extracts on fertility, reproductive function, and prenatal and early postnatal development.

The tests for leachables, such as contaminants, additives, monomers, and degradation products, must be conducted by choosing appropriate solvent systems that will yield a maximal extraction of leachable materials to conduct biocompatibility testing.

The effects of sterilization on device materials and potential leachables, as well as, toxic byproducts as a consequence of sterilization should be considered. Therefore, testing should be performed on the final sterilized product or representative samples of the final sterilized product.

IV. CHOICE OF SPECIES

In general practice, the rabbit, generally albino, remains the standard test system utilized for evaluating the safety of an ophthalmic drug or establishing the *in vivo* safety of an ophthalmic device and contact lens solutions. The popularity of the albino rabbit as a test system for ophthalmic drugs and devices is due to the relatively large size of the animal's eye and the easily observable bulbar conjunctival and corneal surface, the ability to evaluate the iridal vasculature, and, of course, the large historical database with this animal.

Special anatomical and physiological aspects of the rabbit eye must, however, be kept in mind and may make this species inappropriate as a predictor of clinical safety under special circumstances. The rabbit cornea is thinner than the human cornea due in part to a thinner epithelial layer and a virtually absent Bowman's membrane.[13] The rabbit possesses an arterial connection between the two internal ophthalmic arteries which has been postulated to be responsible for the consensual irritative response of the rabbit eye.[2] Unlike the primate eye, the rabbit possesses a prominent nictitating membrane in which test formulations, especially viscous formulations, may become entrapped and thus yield false positive results relative to actual clinical experience. The entrapment of the test article may result in adverse corneal or conjunctival changes in the inferonasal quadrant which would not be seen using the same treatment regimen in the dog,[1a] the nonhuman primate or in the clinical setting.[5b] Tear flow in the rabbit is quantitatively and qualitatively different than the human and may affect a drug's response to ocular tissue. The outer layer of the tear film in rabbits is oily from Meibomian gland secretion.[18] The pH of the lacrimal fluid of rabbits is 8.2 compared to a pH of human tears of 7.1 to 7.3.[16] The lacrimal fluid volume of the rabbit is 7.5 µl, which is similar to the 7.0 µl lacrimal fluid volume of humans. The turnover rate of lacrimal fluid in rabbits is only 7.1%/min compared to a tear turnover rate of 16% in humans.[1] Similarly, the blink rate in the rabbit is significantly slower than the human[9a] and thus ocular retention of a drug is typically longer in the rabbit than in the human. In contrast to many other species, the oily nature of the rabbit tear film in conjunction with the slow blink rate results in a lack of mixing of the precorneal tear film with the reservoir tear fluid.[18] As drugs pass through the cornea, toxicity may be expressed at the endothelial monolayer. In the rabbit the endothelial cells readily divide as part of the healing process, whereas in the primate these cells are incapable of dividing but must enlarge and migrate to cover an endothelial defect. Green et al.[5] report on an extensive study comparing the ocular responses of the albino rabbit and the rhesus monkey to chemical substances and present a comparison of the anatomical and physiological differences of the ocular systems of rabbits and humans. These differences generally render the rabbit a more sensitive species to chemically induced ocular changes than the nonhuman primate and, presumably, man. Nonetheless, adverse findings of a mild degree involving the cornea or structures unique to the rabbit do not necessarily indicate that similar toxicity can be expected in man. Generally, the presence of such adverse findings in the rabbit indicate the need to evaluate the drug in a test system more anatomically and physiologically similar to man; i.e., either the dog or the nonhuman primate such as the cynomolgus or rhesus monkey.

Regardless of the choice of species for use in evaluating the ocular safety/toxicity of a drug under development, the means for evaluating ocular changes are generally uniform throughout the industry. We shall not detail the use of the various equipment/procedures, since individual investigators show preference as to the type and manufacturer. Ocular toxicity may involve external structures such as the lacrimal puncta and nasolacrimal duct, extraocular muscles or eyelids, or may involve the ocular tissues, thereby affecting the conjunctiva, cornea, iris, lens, retina, or optic nerve, either directly as a result of cellular or biochemical toxicity or indirectly as a result of alteration of blood flow or release of cellular mediators of inflammation. The latter cascade may result in autoimmune destruction. It is important, therefore, to perform thorough ocular evaluations in order to obtain the maximum amount of information and thereby assess the presence of any ocular lesions. In general, routine evaluations should include a biomicroscopic and ophthalmoscopic evaluation of the ocular tissues. More specialized techniques may be employed should the need arise.

V. TECHNIQUES

A. Slit-Lamp Biomicroscopy

The slit-lamp is the standard ophthalmic instrument used in routine evaluation of the external features of the eye and the anterior portion of the globe (conjunctiva, cornea, iris, lens, anterior portion of the vitreous). The inter- and intralaboratory variations observed among different investigators using subjective scoring methods has been well documented. Thus, subjective descriptors should be avoided. The ocular grading scale of McDonald et al.[13] for the assignment of numerical scores to ocular changes is preferred since this scoring method assesses 12 parameters in a semiquantitative manner. The scoring system is the most comprehensive system in use and the data can readily be presented in tabular format by treatment group vs. parameter, thereby facilitating data analysis. Generally, biomicroscopic evaluations should be performed prior to the initiation of treatment with the test article, once or twice during the first week of the study, every week during the first month, and either once or twice a month thereafter for the duration of the study. Through the use of a photobiomicroscopic slit-lamp, a photographic record may be maintained.

B. Ophthalmoscopy

Although binocular viewing of the optic nerve head and retina may be performed with a biomicroscope and the use of a specialized contact lens, ophthalmoscopy is the technique routinely used for evaluation of the fundus. In the albino rabbit, the retinal and choroidal vessels are readily observable and changes to the optic nerve head may be assessed by determination of the cup:disc ratio at appropriate intervals. In pigmented animals, the choroidal vessels are obscured. Changes to the macula and fovea in primates as well as changes to the optic nerve head and retinal vasculature are critical in assessing a drug's potential for ocular toxicity. Generally, ophthalmoscopic evaluations should be performed prior to the initiation of treatment, following 1 month of treatment and every 3 months thereafter for the duration of the study. There are two types of ophthalmoscopes: direct and indirect. Direct ophthalmoscopy provides high magnification of the fundus and generally may be performed without the use of mydriatics. The main disadvantage of direct ophthalmoscopy is the limited viewing area. Indirect ophthalmoscopy, by contrast, is stereoscopic providing a wide area of view, thereby allowing better evaluation of the peripheral retina, and greater illumination. Unlike direct ophthalmoscopy however, indirect ophthalmoscopy requires the use of mydriatics to retard the constriction of the pupil due to the bright illumination producing an inverted image of the fundus. A fundus camera may be used to maintain a photographic record.

C. Pachymetry

This technique utilizes either an optical or ultrasonic pachymeter to determine the degree of corneal swelling following a toxic ocular insult. Morgan et al.[14] reported that corneal pachymetry performed 3 days after application of a variety of test materials to the rabbit eye was as predictive as the eye irritation classification determined by observing the ocular response for 21 days. Similarily, Kennah et al.[8] utilized changes in corneal thickness as a means of providing an objective quantitation of ocular irritation. The authors state that the use of changes in corneal thickness possesses a greater sensitivity or detection of corneal healing than subjective scoring using the Draize scale.

D. Electroretinography

This technique takes advantage of the electrical changes which occur in the retina in response to exposure to light. The electroretinogram is a record of the difference in electrical potential between an electrode placed on the cornea and an electrode placed superior to the orbit. The electroretinogram consists of four wave components:

1. **a wave** — This wave represents the response of the photoreceptor cell layer and consists of an initial negative response following the light stimulus.
2. **b wave** — This wave represents the response of the bipolar cell layer and produces a positive deflection on the electroretinogram.
3. **c wave** — This wave is a slight positive deflection.
4. **d wave** — When the light is turned off, a positive potential called the d wave is produced.

This technique is useful in the determination of diffuse retinal damage from compounds. The sensitivity of this technique allows for early diagnosis of retinal damage prior to ophthalmoscopic findings or is useful in the determination of the functional integrity of the retina when fundoscopic viewing is impaired due to lenticular opacities.

E. Specular Microscopy

Corneal specular microscopy utilizes the light reflected from an optical interface of the tissue for image formation and is extremely useful in evaluation of the cornea endothelium.[10] As previously stated, the corneal endothelium of rabbits is capable of regeneration, whereas in primates the endothelium does not regenerate. Morphometric analysis of *in vivo* endothelial photomicrographs to include mean endothelial cell size, total endothelial cell number, percent hexagonality, and coefficient of variation (SD mean cell area/mean cell area) will provide an excellent measurement of the health of the endothelium. An excellent review of the principles and use of specular microscopy is presented by Leibowitz and Laing.[10]

F. Scheimpflug Photography

Although the use of biomicroscopy allows for adequate viewing of the crystalline lens, the lack of a sufficient depth of field with photobiomicroscopy has imposed serious limitations in the documentation of developing cataracts. This limitation has recently been overcome through the use of Scheimpflug photography with subsequent densitometric image analysis of the film.[6] This principle was first described by Scheimpflug in 1906 and basically involves intersecting the object plane from the slit beam and the image plane or film plane at one point with resulting identical angles. This configuration produces a nondistorted photograph of sufficient depth of field to allow for analysis of lens transparency. The use of this technique is invaluable in the documentation of compound-related toxicity to the crystalline lens.

VI. SPECIALIZED TECHNIQUES FOR THE CORNEA

Apart from ocular changes limited primarily to conjunctival irritation, the cornea typically represents the ocular tissue most often acutely affected by topical ophthalmic drug candidates during the drug development process. There are several specialized methods available through which the cornea can be evaluated. These methods can be divided as to the primary tissue of the cornea which they evaluate; however, it is necessary to note that the cornea is very syncytial in nature and not totally separated by its laminar anatomy. A brief description of these techniques is presented below, starting with those techniques that primarily examine effects on the corneal epithelium and then proceeding to those techniques evaluating stromal effects, and finally the endothelium. The techniques to be described will be, in most cases, those most useful for screening compounds. This may eliminate description of some techniques which are too complex to be used as routine screens.

A. Epithelium
1. Corneal Re-epithelialization

This model serves to evaluate the ability of corneal epithelial cells to migrate over a denuded surface with the basement membrane intact. This model could be used to simulate clinical findings due to contact lens abrasion trauma or a corneal ulcer.

Although this procedure could be performed in several different species, a primary screen in rabbits is usually adequate and provides a good indicator of whether a drug retards, has no effect, or enhances corneal re-epithelialization. The rabbit is anesthetized systemically with Ketamine (30 mg/kg) and xylazine (6 mg/kg), and topically with one drop of a topical ophthalmic anesthetic. The rabbit is then laid on its side and the eye is proptosed. The epithelium is then removed with a corneal gill knife, taking care not to compromise the basement membrane. The entire corneal epithelium can be removed, in which case the corneal surface will be repopulated with conjunctival epithelial cells, including the mucin-producing goblet cells. If corneal epithelial cells are specifically desired then a specific area of cells can be removed by lightly marking the epithelial surface with a guarded trephine, so as not to penetrate the basement membrane, and then removing the cells within the circle. After the cells are removed, the defect area is stained with 20 µl of liquid fluorescein and photographed using a photo slit-lamp equipped with a cobalt blue exciter filter.

The eyes are photographed at predetermined time points during the healing process and the slides subjected to computer image analysis to quantitate the size of the remaining defect and generate a healing curve. When testing a specific formulation, a dosing regimen can be adjusted to expected clinical use.

2. Anterior Keratectomy

This model will evaluate the ability of corneal epithelial cells to migrate and adhere to a denuded stromal surface without the presence of an intact basement membrane. This model may be clinically indicative of a deep stromal ulcer or healing in the diabetic after epithelial removal.

This procedure is performed by using a guarded trephine measured to extend to the mid-stroma. A pair of forceps is used to grab the edge of the cut circle and remove the epithelium and anterior stroma along a lamellae leaving a smooth stromal bed. From this point on the procedure is much like the re-epithelialization. The denuded area is stained with 20 µl of liquid fluorescein and photographed at predetermined time points with a photo slit-lamp equipped with a cobalt blue exciter filter. A healing curve is generated using the same procedure as in the corneal re-epithelialization model.

3. Conjunctival Cup

This model is used to show the effects of an exaggerated exposure of a formulation on the intact corneal epithelium. A New Zealand albino rabbit is anesthetized systemically by subcutaneous injection of Ketamine (30 mg/kg) and xylazine (6 mg/kg). The rabbit is then laid on its side and the eyelid is held open using a pediatric speculum. The conjunctiva thus forms a "cup" around the ocular surface. This cup is then filled immediately, to avoid drying of the ocular surface, with a physiological formulation. It is necessary to adjust formulations to physiological pH and osmolarity levels in order to avoid inducing changes in the corneal epithelium which are not related to the formulation test component. The solution is changed every 5 min to avoid buildup of cellular wastes. After 15 min, the formulation is removed and the rabbit is euthanized and the cornea isolated and fixed for SEM. A series of SEMs at 300×, 500×, and 1000× are taken and are scored using the following scoring system.

0 — Totally normal corneal epithelium. Superficial cell layer is intact with exfoliation holes, tight junctions, and normal microvilli.

1 — Minor damage to the superficial cell layer. Some cell sloughing and a decrease in the density of microvilli is observed. Some "pitting" of superficial cells is also seen.

2 — The entire superficial cell layer is gone, leaving an intact secondary cell layer. No exfoliation holes are evident.

3 — Damage to the secondary cell layer is evident. Cell sloughing is seen exposing the thin tertiary cell layer. This cell layer can be identified by the well-defined nuclear bulges. Loss of secondary cell junctions and microvilli is also evident.

4 — Severe damage. Exposure of the rounded deep epithelial cells. Loss of superficial cell layers. Much cell debris is usually evident.

Usually four to six corneas per group are evaluated in a masked fashion by two separate individuals who are familiar with the scoring system. Depending on the type of formulations used it may be necessary to adjust the scoring system to suit individual needs.

4. Corneal Nerve Conduction Model

This model is included in this chapter because it is the only technique currently available to preclinically evaluate topical ocular comfort. It is, however, very complex and not suitable for routine comfort screening. A rabbit or other species is placed under deep urethane anesthesia. A pharyngeal ring is affixed to the eye using the conjunctiva. A 2 ml chamber is attached to the pharyngeal ring and a moist chamber is formed over the cornea to protect the corneal epithelium. The orbital tissue is then excavated to expose the optic nerve. The long ciliary nerve, sensory to the cornea, runs just superficially to the optic nerve. A hook electrode is placed under the long ciliary nerve where it monitors nerve traffic moving centrally from the ocular surface. Once baseline nerve traffic is determined, formulations can be placed into the chamber and the reactive nerve traffic is monitored.

B. Stroma
1. Wound Bursting Technique

Evaluation of the strength and ability of the corneal stroma to heal and withstand trauma is extremely important in evaluating the effects of topical and intraocular formulations. This impacts on postoperative healing of corneal transplants, cataracts, and refractive surgeries such as radial keratotomy and epikeratophakia. In this technique a 9 mm corneal incision is placed in the central cornea and closed with four interrupted 10.0 nylon sutures. The cornea is allowed to heal for a predetermined period of time (usually 6 to 9 days). At this time the rabbit is humanely euthanized and a needle is placed in the anterior chamber which is attached to an infusion pump. Another needle is placed in the anterior chamber which is attached to a pressure transducer and physiograph. As saline is pumped into the eye, the pressure increases and is recorded by the transducer and physiograph. The pressure increases until the incision fails, which is marked by a sudden decrease in monitored pressure. The peak of the pressure curve is called the bursting pressure.[9] This pressure can be compared to others in corneas treated with test formulations, either topically or intraocularly. Steroids, for example, are known to decrease the bursting pressure of a corneal wound at various time points.

2. Wound Tensile Strength Technique

The initial surgery for this technique is the same as for the wound bursting technique. At the end of the healing period, the rabbit is sacrificed and the cornea is isolated. A corneal strip is cut perpendicular to the original wound, with the incision traversing the short axis of the strip, and the tissue is clamped into a materials testing device. This device pulls at an increasing strain until the incision is ruptured. This force is directly correlated to the amount of corneal stromal healing.[17]

C. Endothelium

1. In Vitro Specular Microscopy

The corneal endothelium is a monolayer of cells on the internal aspect of the cornea (in the anterior chamber) which is bathed by the aqueous humor. These cells are primarily responsible for maintaining the proper level of corneal hydration. The corneal water content is approximately 72% in the normal, nonswollen, state. If this level is allowed to increase, the resulting corneal edema will cause back-scattering of light and eventual corneal opacity. This technique takes advantage of the direct correlation between corneal thickness and endothelial compromise.

Isolated corneas used in this technique can be obtained from rabbits, human eye bank tissue, and, with some equipment modification, cats. The cornea is isolated and mounted in the perfusion chamber.[12] The chamber temperature is kept at 35°C and 15 mmHg pressure to simulate the normal physiological state. The corneal endothelium is perfused using syringes mounted in a dual infusion pump at a rate of 0.096 cm^3/min. One cornea serves as the control, being perfused with a vehicle solution, while the other cornea is perfused with the vehicle containing the test compound. Thickness measurements are taken every 15 min for each cornea and corneal swelling curves are generated. Following the perfusion, the corneas can be fixed with glutaraldehyde for electron microscopic evaluation. It is also possible to collect the effluent perfusate and analyze it for corneal glycosaminoglycans.[7]

2. In Vivo Endothelial Perfusion

It is possible to evaluate the reaction of the corneal endothelium to surgical stress in the live animal. Glasser et al.[3] compared the response to two different intraocular irrigating solutions after perfusion of the anterior chamber for 15, 30, or 60 min. The study was performed in cats because the cat corneal endothelium is similar to humans in that it does not readily divide during wound healing. In this study, the animals were anesthetized and the anterior chamber was perfused for the designated period of time by placing inflow and outflow cannulas opposite each other at the limbus into the anterior chamber. Following the perfusion, the punctures were closed with single suture placement and the animal was allowed to recover. During the first 30 days of the postsurgical course, the corneal endothelium was examined by wide-field endothelial microscopy and the photographs were evaluated by computer morphometric analysis. Alterations in cell shape (pleomorphism), cell size (polymegathism), and coefficient of variation were found as a result of the individual irrigating solution used.[3] In addition, actual corneal endothelial wounds can be induced using a monofilament nylon or an olive-tipped cannula, after which endothelial migration and wound closure can be evaluated. Topical or intraocular wound healing agents can be evaluated in this manner.

VII. SUMMARY

In this chapter we have provided an approach toward the development of the preclinical safety profile for an ophthalmic drug: types of studies and choice of the appropriate test system were reviewed, with emphasis on anatomical and physiological differences between the rabbit and human. Despite these differences, the rabbit remains the most useful test system in the development of ophthalmic drugs. The various techniques used on a routine basis by the ophthalmic toxicologist have been presented and references given to allow the interested reader the springboard to expand their depth of knowledge in the area of ophthalmic toxicology.

ACKNOWLEDGMENT

The authors express their sincere appreciation to Mrs. Dorothy L. Watson for her expert assistance in the preparation of this manuscript.

REFERENCES

1. S.S. Charai, T. F. Patton, A. Mehta, and J. R. Robinson, Lacrimal and Instilled Fluid Dynamic in Rabbit Eyes, *J. Pharm. Sci.* 62(7), 1112 (1973).

1a. G. Durand-Cavagna, P. Delfort, P. Duprat, Y. Bailly, B. Plazonnet, and L.R. Gordon, Corneal Toxicity Studies in Rabbits and Dogs with Hydroxyethyl Cellulose and Benzolkonium Chloride, *Fund. Appl. Toxicol.* 13, 500 (1989).

2. S. Forster, A. Mead, and M. Sears, An Interophthalmic Communicating Artery as Explanation for the Consensual Irritative Response of the Rabbit Eye, *Invest. Ophthalmol. Vis. Sci.* 18(2), 161 (1979).

3. D.B. Glasser, M. Matsuda, J.G. Ellis, and H.F. Edelhauser, Effects of intraocular irrigating solutions on the corneal endothelium after in vivo anterior chamber irrigation, *Am. J. Ophthalmol.* 99, 321 (1985).

4. W.L. Guess, S.A. Rosenbluth, B. Schmidt, and J. Autian, Agar Diffusion Method for Toxicity Screening of Plastics on Cultured Cell Monolayers, *J. Pharm. Sci.* 54, 1545 (1965).

5. W.R. Green, J.B. Sullivan, R.M. Hehir, L.G. Scharpl, and A.W. Dickinson, A Systemic Comparison of Chemically Induced Eye Injury in the Albino Rabbit and Rhesus Monkey, Soap and Detergent Association, New York (1978).

5a. R.B. Hackett, Species Selection in the Preclinical Development of Ophthalmic Drugs: The Rabbit versus the Nonhuman Primate, DruSafe East Meeting, Cherry Hill, NJ, November 30, 1989.

5b. E.I. Goldenthal, Current Views on Safety Evaluation of Drugs, FDA Papers 2, 13 (1968).

6. O. Hockwin, V. Dragomirescu, H. Laser, A. Wegener, and U. Eckerskorn, Measuring Lens Transparency by Scheimpflug Photography of the Anterior Eye Segment: Principle, Instrumentation and Application to Clinical and Experimental Ophthalmology, *J. Toxicol. Cut. Ocul. Toxicol.* 6(4), 251 (1987).

7. G.I. Kaye, H.F. Edelhauser, M.E. Stern, N.D. Cassai, P. Weber, and Z. Dische, *Structure of the Eye IV*, J.G. Hollyfield, Ed. Elsevier, New York, New York, 271 (1982).

8. H.E. Kennah, S. Hignet, P.E. Laux, J. D. Dorko, and C. S. Barrow, An Objective Procedure for Quantitating Eye Irritation Based Upon Changes of Corneal Thickness, *Fund. Appl. Toxicol.* 12, 258 (1989).

9. B.P. Lee, A. Kupferman, and H.M. Leibowitz, Effect of Suprofen on corneal wound healing, *Arch. Ophthalmol.* 103, 95 (1985).

9a. V.H.L. Lee and J.R. Robinson, Review: Topical Ocular Drug Delivery. Recent Developments and Future Challenges, *J. Ocular Pharmacol.* 2, 67 (1986).

10. H.M. Leibowitz and R.A. Laing, Specular Microscopy, in *Corneal Disorders: Clinical Diagnosis and Management,* H.M. Leibowitz, Ed. W.B. Saunders, Philadelphia (1984).

11. F.N. Marzulli, D.W.C. Brown, and M.E. Simon, A Continuous Sampling Technique for Quantitatively Measuring Transcorneal Penetration *In Vivo, Invest. Ophthalmol.* 6, 93 (1967).

12. B.E. McCarey, H.F. Edelhauser, and D.L. Van Horn, Functional and Structural Changes in the Corneal Endothelium During In Vitro Perfusion, *Invest. Ophthalmol.* 12, 410 (1973).

13. T.O. McDonald, V. Seabaugh, J. A. Shadduck, and H. F. Edelhauser, Eye Irritation, in *Dermatotoxicology,* 3rd ed. F. N. Marzulli and H. I. Maibach, Eds. Hemisphere, Washington, D.C. (1987).

14. R.L. Morgan, S.S. Sorenson, and T.R. Castles, Prediction of Ocular Irritation by Corneal Pachymetry, *Food Chem. Toxicol.* 25, 609 (1987).

15. T. Scheimpflug, Der Photoperspektograph und seine Anweendung, *Photor. Korr.* 43, 516 (1906).

16. H.E. Seifried, Eye Irritation Testing: Historical Perspectives and Future Directions, *Cut. Ocul. Toxicol.* 5(2), 89 (1986).

17. A.H. Simonsen, T.T. Andreassen, and K. Bendix, Healing strength of corneal wounds in the human eye, *Exp. Eye Res.* 35, 287 (1982).

18. D.W. Swanston, Assessment of the Validity of Animal Techniques in Eye-Irritation Testing, *Food Chem. Toxicol.* 23(2), 169 (1985).

19. H.W. Thompson, B. Dupui, R.W. Beuerman, J.W. Hill, D. Rootman, and Y. Haruta, Y., Neural activity from the rabbit cornea following HSV-1 infection, *Curr. Eye Res.* 7, 147 (1988).

20. J. Weissinger, Nonclincal Pharmacologic and Toxicologic Considerations for Evaluating Biologic Products, *Reg. Toxicol. Pharmacol.* 10, 255 (1989).

Index

INDEX

A

Q

R

S